Cambridge Studies in Morphology and Molecules: New Paradigms in Evolutionary Biology

This new Cambridge series addresses the interface between morphological and molecular studies in living and extinct organisms. Areas of coverage include evolutionary development, systematic biology, evolutionary patterns and diversity, molecular systematics, evolutionary genetics, rates of evolution, new approaches in vertebrate palaeontology, invertebrate palaeontology, palaeobotany, and studies of evolutionary functional morphology. The series invites proposals demonstrating innovative evolutionary approaches to the study of extant and extinct organisms that include some aspect of both morphological and molecular information. In recent years the conflict between "molecules vs. morphology" has given way to more open consideration of both sources of data from each side making this series especially timely.

Carnivoran Evolution: New Views on Phylogeny, Form and Function
Edited by Anjali Goswami and Anthony Friscia

Evolutionary History of Bats: Fossils, Molecules and Morphology
Edited by Gregg F. Gunnell and Nancy B. Simmons

Evolution of the House Mouse

EDITED BY

Miloš Macholán
Academy of Sciences of the
Czech Republic and
Masaryk University, Brno

Stuart J. E. Baird
University of Porto, Portugal
and INRA, France

Pavel Munclinger
Charles University in Prague,
Czech Republic

Jaroslav Piálek
Academy of Sciences of the
Czech Republic, Brno

CAMBRIDGE
UNIVERSITY PRESS

CAMBRIDGE UNIVERSITY PRESS
Cambridge, New York, Melbourne, Madrid, Cape Town,
Singapore, São Paulo, Delhi, Mexico City

Cambridge University Press
The Edinburgh Building, Cambridge CB2 8RU, UK

Published in the United States of America by Cambridge University Press, New York

www.cambridge.org
Information on this title: www.cambridge.org/9780521760669

First published 2012

Printed in the United Kingdom at the University Press, Cambridge

A catalogue record for this publication is available from the British Library

Library of Congress Cataloguing in Publication data
Evolution of the house mouse / edited by Miloš Macholán ... [et al.].
p. cm. – (Cambridge studies in morphology and molecules)
ISBN 978-0-521-76066-9
1. Mice – Evolution. I. Macholán, Miloš
QL737.R6E93 2012
599.35–dc23
2012015491

ISBN 978-0-521-76066-9 Hardback

Contents

The colour plates are situated between pages 268 and 269.

Contributors

Ken P. Aplin, Australian National Wildlife Collection, CSIRO Division of Sustainable Ecosystems, Canberra, Australia

Jean-Christophe Auffray, Institut des Sciences de l'Evolution de Montpellier, CNRS, Université Montpellier 2, Montpellier, France

Stuart J. E. Baird, Centro de Investigação em Biodiversidade de Recursos Genéticos, Campus Agrário de Vairão, Vairão, Portugal

Hermann Bauer, Department of Developmental Genetics, Max Planck Institute for Molecular Genetics, Berlin, Germany

R. J. Berry, Department of Genetics, Evolution & Environment, University College London, United Kingdom

François Bonhomme, Institut des Sciences de l'Evolution de Montpellier, CNRS, Université Montpellier 2, Montpellier, France

Janice Britton-Davidian, Institut des Sciences de l'Evolution de Montpellier, CNRS, Université Montpellier 2, Montpellier, France

Anna Büntge, Max Planck Institute for Evolutionary Biology, Plön, Germany

Thomas Cucchi, Département Ecologie et Gestion de la Biodiversité, Muséum national d'Histoire naturelle, Paris, France; and Department of Archaeology, University of Aberdeen, Aberdeen, United Kingdom

Jiří Forejt, Laboratory of Mouse Molecular Genetics, Institute of Molecular Genetics, Academy of Sciences of the Czech Republic, Prague, Czech Republic

Guila Ganem, Institut des Sciences de l'Evolution de Montpellier, CNRS, Université Montpellier 2, Montpellier, France

Mabel D. Giménez, Facultad de Ciencias Exactas, Químicas y Naturales, Universidad Nacional de Misiones, Félix de Azara 1552, N3300LQH Posadas, Misiones, Argentina

Joëlle Goüy de Bellocq, Evolutionary Biology Group, University of Antwerp, Antwerp, Belgium

Heidi C. Hauffe, Department of Biodiversity and Molecular Ecology, Research and Innovation Centre, Fondazione E. Mach, S. Michele all'Adige (TN), Italy

Bernhard G. Herrmann, Department of Developmental Genetics, Max Planck Institute for Molecular Genetics, Berlin, Germany

Kateřina Janotová, Biodiversity Research Group, Faculty of Science, Charles University in Prague, Prague, Czech Republic

Robert C. Karn, Department of Medicine, College of Medicine, University of Arizona, Tucson, AZ, USA

Christian Peter Klingenberg, Faculty of Life Sciences, University of Manchester, Manchester, United Kingdom

Barbara König, Institute of Evolutionary Biology and Environmental Studies, University of Zurich, Zurich, Switzerland

Christina M. Laukaitis, Department of Medicine, College of Medicine, University of Arizona, Tucson, AZ, USA

Anna K. Lindholm, Institute of Evolutionary Biology and Environmental Studies, University of Zurich, Zurich, Switzerland

Ken Luzynski, Department of Ecology and Evolutionary Biology and Museum of Zoology, University of Michigan, Ann Arbor, MI, USA

Miloš Macholán, Laboratory of Mammalian Evolutionary Genetics, Institute of Animal Physiology and Genetics, Academy of Sciences of the Czech Republic, Brno; and Department of Botany and Zoology, Faculty of Science, Masaryk University, Brno, Czech Republic

Kazuo Moriwaki, BioResource Center, RIKEN Tsukuba Institute, Tsukuba, Japan

Kerstin Musolf, Department of Integrative Biology and Evolution, Konrad Lorenz Institute of Ethology, Veterinary University of Vienna, Vienna, Austria; and Institute of Evolutionary Biology and Environmental Studies, University of Zurich, Zurich, Switzerland

Nicolas Navarro, Ecole Pratique des Hautes Etudes (EPHE) – Paléobiodiversité et Evolution – 6 Bld Gabriel, 21000, Dijon, France

Ayako Oka, Research Organization of Information and Systems, Transdisciplinary Research Integration Center, Tokyo, Japan; and Mammalian Genetics Laboratory, National Institute of Genetics, Mishima, Japan

Dustin J. Penn, Department of Integrative Biology and Evolution, Konrad Lorenz Institute of Ethology, Veterinary University of Vienna, Vienna, Austria

Jaroslav Piálek, Department of Population Biology, Studenec, Institute of Vertebrate Biology, Academy of Sciences of the Czech Republic, Brno, Czech Republic

Alexis Ribas, Laboratory of Parasitology, Faculty of Pharmacy, University of Barcelona; and Biodiversity Research Institute (IRBIO), Faculty of Biology, University of Barcelona, Barcelona, Spain

Jun J. Sato, Laboratory of Animal Cell Technology, Faculty of Life Science and Technology, Fukuyama University, Fukuyama, Japan

Jeremy B. Searle, Department of Ecology and Evolutionary Biology, Cornell University, Ithaca, NY, USA

Toshihiko Shiroishi, Mammalian Genetics Laboratory, National Institute of Genetics, Mishima, Japan

Pavel Stopka, Biodiversity Research Group, Faculty of Science, Charles University in Prague, Prague, Czech Republic

Romana Stopková, Biodiversity Research Group, Faculty of Science, Charles University in Prague, Prague, Czech Republic

Hitoshi Suzuki, Laboratory of Ecology and Genetics, Graduate School of Environmental Earth Science, Hokkaido University, Sapporo, Japan

Diethard Tautz, Max Planck Institute for Evolutionary Biology, Plön, Germany

Katherine Teeter, Department of Ecology and Evolutionary Biology and Museum of Zoology, University of Michigan, Ann Arbor, MI, USA; and Department of Biology, Northern Michigan University, Marquette, MI, USA

Meike Teschke, Max Planck Institute for Evolutionary Biology, Plön, Germany

Zdeněk Trachtulec, Laboratory of Mouse Molecular Genetics, Institute of Molecular Genetics, Academy of Sciences of the Czech Republic, Prague, Czech Republic

Priscilla K. Tucker, Department of Ecology and Evolutionary Biology and Museum of Zoology, University of Michigan, Ann Arbor, MI, USA

Jean-Denis Vigne, Département Ecologie et Gestion de la Biodiversité, Muséum national d'Histoire naturelle, Paris, France

Liuyang Wang, Department of Ecology and Evolutionary Biology and Museum of Zoology, University of Michigan, Ann Arbor, MI, USA

Hiromichi Yonekawa, Department of Laboratory Animal Science, Tokyo Metropolitan Institute of Medical Science (Igakuken), Tokyo, Japan

Preface

As we write this preface in May 2011, it is 70 years since publication of the first book on the biology and genetics of the house mouse, authored by the staff of the Jackson Laboratory (*Biology of the Laboratory Mouse* [1941]), and 30 years since publication of R. J. Berry's seminal edited monograph, *Biology of the House Mouse* (1981). In this vein it is perhaps interesting to further explore the historical context of the house mouse in evolutionary biology: this year we celebrate the 145th anniversary of Gregor Mendel's seminal paper 'Versuche über Pflanzen-Hybriden' (*Verhandlungen des naturforschenden Vereins Brünn*, 4, 3–47 [1866]). Though his laws of inheritance are described using experiments on peas, it has been speculated that Mendel initially bred and crossed mice in his rooms of the Augustinian Abbey of St Thomas in Brno, not far from the laboratory of one of the present editors. In 2012, when *Evolution of the House Mouse* is to be published, it will be 110 years since L. Cuénot in France and A. D. Darbishire in the United States used house mice to provide the first independent confirmation of Mendel's laws; 80 and 70 years since J. B. S. Haldane's *The Causes of Evolution* and J. Huxley's *Evolution: The Modern Synthesis*, respectively; 60 years since Ursin's paper on the Danish house mouse hybrid zone; 40 years since K. Theiler's *The House Mouse*; and ten years since the first draft of the mouse genome (Mouse Genome Sequencing Consortium) was made available online.

We are living in an ever-accelerating world of information. Scientific data are gathered with increasing pace, and mouse-related research is at the forefront of this flood. Our intention is to reflect this rapid development and bring together in one easy reference a snapshot of current knowledge regarding the evolution of the house mouse. In this we want to follow in the footsteps of excellent and influential books such as *Origins of Inbred Mice* (Morse, 1978), *Biology of the House Mouse* (Berry, 1981), *The Mouse in Biomedical Research* (Foster *et al.*, 1981), *Genetics in Wild Mice* (Moriwaki *et al.*, 1994), and special issues of *Current Topics in Microbiology and Immunology* (1978) and the *Biological Journal of the Linnean Society* (1990, 2005). Since the latter publication, which was based on the proceedings of the symposium on *The Genus Mus as a Model for Evolutionary Studies* (edited by J. Britton-Davidian and J. B. Searle, 2005), significant progress in almost every field of mouse research has been published, while areas not covered in the symposium have seen a growth of interest (e.g. olfactory and acoustic communication, morphology and development, MHC, the *t*-haplotype, phylogeography, parasites, hybrid sterility).

Although this book is targeted primarily at professionals working in mouse-related research, we hope it will also be of interest to scholars of evolutionary biology, and perhaps also to scholars of human history, as it appears the genetic record of association between mouse and man can tell us of events for which there is no written record. We hope this book will serve not only as a resource for researchers entering these fields, but also to highlight the cooperative nature of research across labs and generations. In this we are inspired by Barbara Dod, who throughout her career at University of Montpellier, working on the first studied region of the European house mouse hybrid zone, the Jutland peninsula, sought to break down barriers between house mouse research groups, and in doing so started many lasting friendships.

We thank all who helped with preparation of the book, especially all the contributors and those colleagues who took time out from busy schedules to review particular chapters. We are also grateful to Ivan Horáček, Gregg F. Gunnell, and Russell L. Ciochan for encouraging us to edit this monograph and the staff at Cambridge University Press for their help and support throughout its preparation.

Miloš Macholán
Stuart J. E. Baird
Pavel Munclinger
Jaroslav Piálek

Foreword: Mice and (wo)men – an evolving relationship

House mice and humans have shared their environment over many millennia, presumably ever since mice spread from their origin in northern India and found a welcoming habitat in the cereal stores of early farmers. Mice have made much use of their human hosts, who have fought an often losing battle against their unwanted guests (Meehan, 1984; Young, 1986; Drummond, 2005). In AD 930, a Welsh king, Hywel Dda ('Hywel the Good'), published a standard price list for cats: one penny for a newborn kitten, tuppence [two pennies] for an inexperienced youngster, but fourpence for a cat once it had caught a mouse. Keeler (1931) points out that the word for mouse in Sanskrit (*mūṣaka*) is derived from a verb *mūṣ* [*mush*] meaning 'to steal' and suggests that this implies the species' pilfering habits were well known at least as early as the second millennium BC. Keeler writes of a 'mouse cult' in Asia Minor and records a number of examples of domestication and even worship by the ancient Greeks and Romans. Such familiarity with mice meant that variant animals would be noted. White mice are mentioned by many classical authors. It is said that the first written record of a variant was in a Chinese dictionary around 1100 BC, describing dominant spotting.

The Latin word for a mouse is *mus*. Pliny the Elder (AD 23–79) named the house mouse *musculus*, the little mouse, to distinguish it from the rat, its larger relative. Formally, of course, the binomial *Mus musculus* dates from Linnaeus, who presumably based his knowledge on the light-bellied mice around his home in Uppsala, Sweden, although no specimen exists in his collections. He wrote that it is 'an animal that needs no description: when found white it is very beautiful, the full bright eye appearing to great advantage amidst the snowy fur. It follows mankind, and inhabits all parts of the world except the Arctic.' Because of the priority of naming in taxonomy, the light-bellied species or subspecies is properly labelled *Mus musculus* (or *M. m. musculus*), meaning that another name is needed for the dark-bellied form. Albert Magnus, writing in Paris in the first half of the thirteenth century, referred to *Mus domesticus*, as did Conrad Gesner in Zurich in the mid-sixteenth century, and in Britain, Robert Sibbald in 1684 and John Ray in 1693. The first post-Linnean reference to *M. domesticus* was by John Rutty, who published *An Essay towards a Natural History of the County of Dublin* in 1772. In it, he lists '*Mus domefticus*. The Houfe-Moufe', with no embellishment or description. The International Commission on Zoological Nomenclature (1990) have now agreed that the name *domesticus* is legitimate, and is available for use either as *Mus domesticus* (Schwarz & Schwarz, 1943), or as *Mus musculus domesticus* (Schwarz & Schwarz, 1943).

In the early days of modern science, Robert Hooke (1635–1703) used a mouse to study the effects of increased air pressure. He does not state where the mouse came from. William Harvey (1578–1657) used mice in his anatomical studies, and Joseph Priestley (1733–1804) gives a delightful account in his *Experiments and Observations on Different Kinds of Air* (1775) of his work with mice, including how he trapped and maintained them. Half a century later, a Genevan pharmacist named Coladon bred large numbers of white and 'grey' mice and obtained segregations in agreement with Mendelian expectation 36 years before Mendel published his work on peas (Grüneberg, 1957). Indeed Sturtevant (1965) has even suggested that Mendel originally worked out his 'laws' in mice, but suppressed his results for fear of antagonizing his ecclesiastical superiors – medieval churchmen frequently commented on the 'voluptuous and libidinous habits' of mice (and, it is said, bred them so as surreptitiously to observe their wicked behaviour).

Sturtevant's idea of Mendel as a mouse geneticist is not too far-fetched. R. A. Fisher showed that Mendel's published work on peas was much closer to statistical expectation than expected (Fisher, 1936); he suggested that perhaps the pea work was really a demonstration of a factorial scheme which Mendel had previously discovered. Support for this comes from Mendel's biographer, who notes 'Mendel used to breed mice in his rooms, grey mice as well as white mice, crossing these varieties ... Mendel tells us nothing about this matter ... The silence is readily comprehensible ... [He] had to walk warily, for his bishop had a prejudice against him' (Iltis, 1932: 105).

Keeler (1931: 16) records that '[d]uring the nineteenth century, a number of European zoologists bred fancy mice [mainly coat colour variants] ... They accumulated valuable information, but the meaning of these data remained unknown until the rediscovery of Mendel's work in 1900 (Davenport, 1900).' New breeding experiments were quickly set up to test the truth and extent of Mendel's work, and within months Cuénot (1902) in France, Darbishire (1902) in the United States, and Bateson (1903) in Britain showed unequivocally that Mendelian segregations operated in mice. Haldane *et al.* (1915) used Darbishire's data to demonstrate that linkage, discovered in *Drosophila* by Morgan (1910), operated also in mammals, thus laying the basis for genome mapping. Searle (1981) has described the exponential increase in knowledge of inherited variants and linkage groups. The mouse genome was published in 2002 (Mouse Genome Sequencing Consortium, 2002).

Meanwhile, Robert Yerkes (1876–1956) at Harvard began a series of investigations in 1903 on behaviour in 'waltzing' or 'dancing' mice, a variant known in China from at least AD 80, although Yerkes obtained his animals from animal dealers in the eastern United States (Yerkes, 1907). His technique of comparing genetically related variant and non-variant animals to explore the effects of genetic change was very powerful and has burgeoned ever since.

Early mouse studies were carried out on mice obtained from a variety of sources. Laboratory mice in the modern sense can be said to date from 1907, when a Harvard undergraduate, C. C. Little (1888–1971), began to study the inheritance of coat colour under the supervision of W. E. Castle (1867–1962) at the Bussey Institute of Harvard University. Two years later, Little obtained, probably from Miss Lathrop (see below), a pair of 'fancy mice' carrying alleles for the recessively inherited traits dilute (*d*), brown (*b*) and non-agouti (*a*). He inbred their descendants by brother–sister mating, and produced thereby the first inbred mouse strain (DBA).

Castle did not believe in the value of so-called 'pure strains', and after army service in the First World War, Little moved to the Carnegie Institute at Cold Spring Harbor, where he was joined by L. C. Strong (1894–1982) and began the development of a range of inbred strains, largely descended from mice provided by Abbie Lathrop. Strong joined at a time when a paratyphoid epidemic had killed most of Little's animals. Strong (1978: 49) recorded,

> I was obliged to capture wild mice and start sorting out their hereditary traits through the tedious processes of mate, wait, select, and mate again . . . We kept the wild mice under the bed in our honeymoon tent . . . Meanwhile it became obvious that susceptibility and resistance to transplanted tumours were indeed genetically controlled [which] bent my mind to the task of remodelling *Mus musculus*.

Abbie Lathrop (1868–1918) was a failed poultry farmer, who turned to raising small animals for sale as pets. She started with a single pair of waltzing mice she obtained near her home in Granby, Massachusetts, and advertised for more animals as orders came in. The *Springfield Sunday Republican* for 5 October 1913 (cited by Morse, 1978: 11) wrote of her: 'After she had sold 200 or 300 mice, Miss Lathrop thought that the resource of mouse farming as a business must be very nearly at an end, since the offspring from that number would be enough to supply pets for the entire younger generation, but the orders continued to come in.' More and more of these orders came from research laboratories, including the Bussey Institute where Castle and Little were working. From around 1910 until her death in 1918, Miss Lathrop's sheds contained more than 11 000 mice, several hundred guinea pigs, rabbits, and rats, and occasional ferrets and canaries.

Little left Cold Spring Harbor in 1922 to become President of the University of Maine, and then in 1925 President of the University of Michigan at Ann Arbor, taking his mice with him on both occasions. Disagreements over administration issues forced him to leave Ann Arbor, and he moved with his mice to a site on

Mount Desert Island in Maine, to land donated by a family friend, George Dorr. There he established the Roscoe B. Jackson Memorial Laboratory, largely financed by Roscoe Jackson, head of the Hudson Motor Company, and by Edsel Ford, son of Henry Ford, founder of the Ford Motor Company. The Jackson Laboratory is now a world centre of mouse research (Lauber, 1971; Holstein, 1979; Rader, 2004). Festing and Fisher (2000) list 17 Nobel Prizes awarded for research associated with or stemming from the Jackson Laboratory; five more prizes have been given since their 2000 list. Little's main emphasis was on the determinants of cancer; he saw his and the laboratory's task as 'building a better mouse' for research (Rader, 2004: 11). The staff of the Laboratory collaborated in a major volume of information about laboratory mice, which summarized most of the information available at the time (Snell, 1941; Green, 1966).

For many years, research at the Jackson Laboratory and other centres used mice as little more than experimental tools, with strictly enforced breeding protocols and defined husbandry techniques, developing animal strains which performed uniformly in controlled experiments (Fox and Witham, 1997; Taft *et al.*, 2006). However, increasingly, signs appeared that standardized laboratory mice might give an incomplete or even a biased picture of mouse biology. Interest in transplantation led to mice being brought into laboratories from many places (Klein, 1975, 1986). Inherited variation was the basis of the differences between inbred strains, but the commonly used strains only carried a small part of the variation found in wild mice. House mice are found almost throughout the world – in deserts and tropical islands, from sea level to 3000 metres or so above, in mountains, usually but certainly not always in contact with humans. Genes are certainly not distributed uniformly throughout the animals' range. Taxonomists have long recognized local forms of the species (or species group). At least some of this diversity is related to maintaining body temperature. Many authors have shown that energy-dependent traits are variable and adaptive (Berry and Bronson, 1992). Lynch (1994) found that the architectural complexity of mouse nests (which is an inherited trait) is greater in mice from cold latitudes than from warm ones. Furthermore, in regions where they have no enemies, mouse populations may reach plague proportions, with the amplitude of different outbreaks dependent on environmental factors (Singleton *et al.*, 2003).

All this means that genetic variation and its distribution cannot properly be neglected in mouse studies. House mice are very variable: Ellerman (1941) listed 189 house mouse taxa in 43 species. Schwarz and Schwarz (1943) sought to bring some order to this. They lumped 133 named forms into 15 subspecies of a single polytypic species, *Mus musculus Linnaeus*, which they bunched into four groups (*wagneri*, *spicilegus*, *manchu*, and *spretus*). They believed *M. musculus wagneri* had given rise to nine commensal forms, including the common western European form, *domesticus*, while *spicilegus* and *manchu* gave rise to one each; in their scheme, Linnaeus' *musculus* was derived from *spicilegus*. This somewhat crude taxonomy has been progressively refined, with the genetic relationships and origins of the different groups codified and clarified. The results of many of these studies are described in this volume.

Two positive outcomes of the classical period of taxonomy are worth noting. Gropp *et al.* (1970) found that a population of mice from a Swiss valley originally described as a distinct species (*Mus poschiavinus*) on the basis of its colour and size (Fatio, 1869) had seven pairs of Robertsonian translocations in its chromosome set, giving a diploid count of 26 instead of the normal 40. Such Robertsonian races have been found to be comparatively widespread, particularly in *domesticus*. Second, Selander *et al.* (1969) used inherited biochemical variants to investigate a hybrid zone between *domesticus* and *musculus* in Denmark, originally described by Ursin (1952). Both of these studies have given rise to enormously profitable research, described herein.

Laboratory mice have contributed enormously to our knowledge of the genetics and reaction systems of mammals under prescribed and repeatable conditions, but they can only be indicators of what happens in the real world (Berry, 1980; Miller *et al.*, 2000, 2002). The findings of multiple Robertsonian races and the intricacies of the hybrid zone leads to the crucial consideration that house mice have an ecology as well as a limited existence as constrained confinement as laboratory animals. Although humans have long fought with mice as pests and indulged in them as show animals, it was only when the necessity of reducing the damage to stored food during the Second World War was faced that the lack of knowledge of mouse ecology was realized. The Oxford University ecologist Charles Elton was commissioned by the British government to study and advise on the control of rodent pests. He assigned H. N. Southern to work on mice. Southern's work (Southern, 1954) was continued by Crowcroft (1966) and led the way to a host of ecological and behavioural investigations.

Geneticists (or evolutionists) and ecologists routinely proclaim the need to integrate their findings, but too often ignore results outside their own discipline. This lack of cooperation began early: with the hiatus about evolutionary mechanisms which lasted for decades following the findings of the geneticists in the early 1900s; it was only finally repaired by the publication of Julian Huxley's *Evolution: The Modern Synthesis* in 1942 (Mayr and Provine, 1980). Then another period of mutual misunderstanding arose during the 1970s and 1980s, precipitated by the discovery of the large amounts of inherited variation shown by applying electrophoresis to population samples, destroying the wisdom of the time which stated that evolutionary change was constrained by limits imposed by genetic loads and the cost of natural selection. The resulting debates again drove the recognition that geneticists and ecologists need each other to fully understand evolutionary situations (Berry *et al.*, 1992). This book is a mature fruit of that recognition. It is a fitting successor to a number of publications which have integrated mice into evolutionary disciplines with varying but growing success (Lindzey and Thiessen, 1970; Berry, 1981; Foster *et al.*, 1981; Potter *et al.*, 1986; Brain *et al.*, 1989; Berry and Corti, 1990; Moriwaki *et al.*, 1994; Britton-Davidian and Searle, 2005). It will not be the last word on the subject, but it is an indispensable milestone for those taking the mouse route to understand evolutionary processes.

The house mouse has been a model and a tool for biology ever since its first recorded use in comparative anatomy by William Harvey in 1616. It has illuminated evolutionary thought since the earliest days of genetics (and, as noted above, perhaps even earlier). It took its place in the neo-Darwinian synthesis alongside *Oenothera*, *Gammarus*, *Drosophila*, *Equus*, and other classical organisms. It has become a model in evo-developmental comparisons and as such gives increasing grist to biomedical pathology. And as our knowledge of its genetics and ecology grows, the species must have considerable potential to act as a motor to drive evolutionary understanding to new heights (Berry and Scriven, 2005).

Professor R. J. Berry

REFERENCES

Bateson, W. (1903). The present knowledge of colour heredity in mice and rats. *Proceedings of the Zoological Society of London*, **2**, 71–99.

Berry, R. J. (1980). The great mouse festival. *Nature*, **283**, 15.

Berry, R. J. (ed.) (1981). *Biology of the House Mouse*. London: Academic Press.

Berry, R. J., and Bronson, F. H. (1992). Life history and bioeconomy of the house mouse. *Biological Reviews*, **67**, 519–50.

Berry, R. J. and Corti, M. (1990). Proceedings of a symposium held at the 5th International Theriological Congress, Rome. *Biological Journal of the Linnean Society*, **41**, 1–300.

Berry, R. J. and Scriven, P. N. (2005). The house mouse: a model and motor for evolutionary studies. *Biological Journal of the Linnean Society*, **84**, 335–47.

Berry, R. J., Crawford, T. J. S., and Hewitt, G. M. (eds) (1992). *Genes in Ecology*. Oxford: Blackwell Scientific.

Brain, P. F., Mainardi, D., and Parmigiani, S. (eds) (1989). *House Mouse Aggression*. Chur: Harwood Academic.

Britton-Davidian, J. and Searle, J. B. (eds) (2005). The genus *Mus* as a model for evolutionary studies. *Biological Journal of the Linnean Society*, **84**, 321–674.

Crowcroft, W. P. (1966). *Mice All Over*. London: Foulis.

Cuénot, L. (1902). La loi de Mendel et l'héréditié de la pigmentation de la souris. *Archives de Zoologie Expérimentale et Générale*, **10**, 27–30.

Darbishire, A. D. (1902). Note on the results of crossing Japanese waltzing mice with European albino races. *Biometrika*, **2**, 101–4.

Davenport, C. B. (1900). Review of von Guita's experiments in breeding mice. *Biological Bulletin*, **2**, 121–8.

Drummond, D. (2005). *Mouse Traps*. Galloway, OH: North American Trap Collectors Association.

Ellerman, J. R. (1941). *Families and Genera of Living Rodents, vol. 2*. London: British Museum (Natural History).

Fatio, V. (1869). *Faune des Vertétébrates de la Suisse*. Geneva and Basel: H. George.

Festing, M. F. and Fisher, E. M. F. (2000). Mighty mice. *Nature*, **404**, 815.

Fisher, R. A. (1936). Has Mendel's work been rediscovered? *Annals of Science*, **1**, 115–37.

Foster, H. L., Small, J. D., and Fox, J. G. (eds) (1981). *The Mouse in Biomedical Research, vol. 1: History, Genetics and Wild Mice*. New York: Academic Press.

Fox, R. R. and Witham, B. (eds) (1997). *Handbook on Genetically Standardized JAX Mice*. Bar Harbor, ME: Jackson Laboratory.

Green, E. L. (ed.) (1966). *Biology of the Laboratory Mouse*, 2nd edn. New York: McGraw-Hill.

Gropp, A., Tettenborn, U., and von Lehmann, E. (1970). Chromosome variation vom Robertson' schen typus bei der tabakmaus, *M. poschiavinus* und hybriden mit der Laboratoriumsmaus. *Cytogenetics*, **9**, 9–23.

Grüneberg, H. (1957). *Genes in Mammalian Development*. London: H.K. Lewis.

Haldane, J. B. S., Sprunt, A. D., and Haldane, N. M. (1915). Reduplication in mice. *Journal of Genetics*, **5**, 133–5.

Holstein, J. (1979). *The First Fifty Years of the Jackson Laboratory*. Bar Harbor, ME: Jackson Laboratory.

Huxley, J. S. (1942). *Evolution: The Modern Synthesis*. London: George Allen & Unwin.

Iltis, H. (1932). *Life of Mendel*. Translated by Paul, E. and Paul, C. London: George Allen & Unwin.

International Commission on Zoological Nomenclature (1990). Opinion 1607. *Mus musculus domesticus* Schwarz & Schwarz, 1943 (Mammalia, Rodentia): specific name conserved. *Bulletin of Zoological Nomenclature*, **47**, 171–2.

Keeler, C. E. (1931). *The Laboratory Mouse: Its Origin, Heredity, and Culture*. Cambridge, MA: Harvard University Press.

Klein, J. (1975). *Biology of the Mouse Histocompatibility-2 Complex*. Berlin and New York: Springer-Verlag.

Klein, J. (1986). *Natural History of the Major Histocompatibility Complex*. New York and Chichester: John Wiley.

Lauber, P. (1971). *Of Man and Mouse*. New York: Viking.

Lindzey, G. and Thiessen, D. D. (eds) (1970). *Contributions to Behavior-Genetic Analysis: The Mouse as a Prototype*. New York: Appleton-Century-Crofts.

Lynch, C. B. (1994). Evolutionary inferences from genetic analyses of cold adaptation in laboratory and wild populations of the house mouse. In *Quantitative Genetic Studies of*

Behavioral Evolution, ed. C. R. B. Boake. Chicago, IL: Chicago University Press, pp. 278–301.
Mayr, E. and Provine, W. B. (eds) (1980). *The Evolutionary Synthesis*. Cambridge, MA: Harvard University Press.
Meehan, A. P. (1984). *Rats and Mice and Their Control*. East Grinstead: Rentokil.
Miller, R. A., Dykso, R., Crisp, C. *et al.* (2000). Mouse stocks derived from tropical islands: new models for genetic analysis of life history traits. *Journal of Zoology*, **250**, 95–104.
Miller, R. A., Harper, J. M., Dykso, R. C., Durkee, S. J., and Austad, S. (2002). Longer life spans and delayed maturity in wild-derived mice. *Experimental Biology and Medicine*, **227**, 500–8.
Morgan, T. H. (1910). Sex-limited inheritance in *Drosophila*. *Science*, **32**: 120–2.
Moriwaki, K., Shiroishi, T., and Yonekawa, H. (eds) (1994). *Genetics in Wild Mice*. Tokyo: Japan Science Society.
Morse, C. C. (1978). Historical perspective on the development of inbred mice. In *Origins of Inbred Mice*, ed. C. C. Morse. New York and London: Academic Press, pp. 3–21.
Mouse Genome Sequencing Consortium (2002). Initial sequencing and comparative analysis of the mouse genome. *Nature*, **420**, 520–62.
Potter, M., Nadeau, J. H., and Cancro, M. P. (eds) (1986). The wild mouse in immunology. *Current Topics in Microbiology and Immunology*, **127**, 1–395.
Priestley, J. (1775). *Experiments and Observations on Different Kinds of Air*. London: J. Johnson.
Rader, K. (2004). *Making Mice*. Princeton, NJ: Princeton University Press.
Schwarz, E. and Schwarz, H. K. (1943). The wild and commensal forms of the house mouse *Mus musculus* Linnaeus. *Journal of Mammalogy*, **24**, 59–72.
Searle, A. G. (1981). Comparative and historical aspects of the mouse genome. *Symposia of the Zoological Society of London*, **47**, 63–84.
Selander, R. K., Hunt, W. G., and Kaufman, D. W. (1969). Protein polymorphism and genic heterozygosity in two European subspecies of the house mouse. *Evolution*, **23**, 379–90.
Singleton, G. R., Hinds, L. A., Krebs, C. J., and Sprat, D. M. (eds) (2003). *Rats, Mice and People: Rodent Biology and Management*. Canberra: Australian Centre for International Agricultural Research.
Snell, G. D. (ed.) (1941). *Biology of the Laboratory Mouse*. Philadelphia, PA: Blakiston.
Southern, H. N. (ed.) (1954). *Control of Rats and Mice, vol. 3: House Mice*. Oxford: Clarendon Press.
Strong, L. C. (1978). Inbred mice in science. In *Origins of Inbred Mice*, ed. C. C. Morse. New York and London: Academic Press, pp. 45–67.
Sturtevant, A. H. (1965). *A History of Genetics*. New York: Harper and Row.
Taft, R. A., Davisson, M., and Wiles, M. V. (2006). Know thy mouse. *Trends in Genetics*, **22**, 649–53.
Ursin, E. (1952). Occurrence of voles, mice and rats (Muridae) in Denmark, with a special note on a zone of intergradation between two subspecies of the house mouse (*Mus musculus* L.). *Videnskabelige Meddelelser Dansk Naturhistorisk Forening*, **114**, 217–44.
Yerkes, R. M. (1907). *The Dancing Mouse*. New York: Macmillan.
Young, L. (1986). *Eeeek! The Amazing Mouse Book*. London: Hodder & Stoughton.

Scientists stumped by worldwide scale of prehistoric horror death ceremony.
Picture by Jan Hošek.

1

The house mouse and its relatives: systematics and taxonomy

JEAN-CHRISTOPHE AUFFRAY AND
JANICE BRITTON-DAVIDIAN

Introduction

A quarter of a century before the advent of biochemical tools for taxonomic approaches, Schwarz and Schwarz (1943) published a pioneering survey of the systematics of the house mouse. All taxa were then considered as subspecies of the house mouse (*Mus musculus*). In other words, there was only one species within the subgenus *Mus*. It was a relatively large-scale study, but the taxonomic criteria were obviously based solely on external morphology and distribution of species. It is nowadays well established that the subgenus *Mus* exhibits high inter-specific morphological conservatism and that all species share an important part of their intra-specific variation. Most, if not all, classical external criteria such as coat colour, length of head, body, tail, ear, or foot, are poorly discriminating between species, and, except for very few (e.g. the zygomatic index for *M. musculus*; Orsini *et al.*, 1983), none has a general and specific diagnostic value. When only one or two species occur sympatrically, some characters may be diagnostic, such as the length of the tail, which is useful to distinguish *Mus musculus domesticus*, one of the longest-tailed mice, from all sympatric species. Thus, there is only a poor correspondence between the 15 subspecies of *Mus musculus* described by Schwarz and Schwarz (1943) and the species that are now diagnosed by taxonomists.

Biochemical and molecular tools have considerably clarified our knowledge of the taxonomy within the subgenus *Mus*. The revision began with the deciphering of the *Mus musculus* complex of species in Europe (Selander *et al.*, 1969), though the two subspecies *M. m. musculus* and *M. m. domesticus* had already been recognized by their respective tail lengths. This was followed by the establishment of *Mus spretus* as a valid species by Britton *et al.* (1976) and the recognition of the other European species, *M. macedonicus* and *M. spicilegus*. By thc late 1990s, 9 of the 14 taxa known today had been clearly identified, with criteria that left no doubt as to their species status: *musculus*, *spretus*, *spicilegus*, *macedonicus*, *cervicolor*, *caroli*, *cookii*,

Evolution of the House Mouse, ed. Miloš Macholán, Stuart J. E. Baird, Pavel Munclinger, and Jaroslav Piálek. Published by Cambridge University Press.

booduga, and *terricolor*. It was only in 2003 that a second wave of amazing reviews or descriptions of new species began, which led to an increase by more than half of the number of species diagnosed by taxonomists in the subgenus *Mus*: *M. famulus* (Chevret *et al.*, 2003), *M. fragilicauda* (Auffray *et al.*, 2003), *M. cypriacus* (Cucchi *et al.*, 2006), *M. nitidulus* (Shimada *et al.*, 2007), and *M. lepidoides* (Shimada *et al.*, 2010). Several subspecies were also recognized, not only in the *Mus musculus* complex but also within *M. macedonicus* (Orth *et al.*, 2002a) and *M. spicilegus* (Kryštufek and Macholán, 1998). Our knowledge of *Mus* taxonomy has thus increased much faster than for mammals in general, for which the number of new species described between 1993 and 2008 is estimated at ≈10% (Ceballos and Ehrlich, 2009). Most of these new *Mus* species were cryptic, which means that the distribution areas of the formerly described species were actually occupied by more than one species. When the number of *Mus* species that have been correctly assigned to a specific name is cumulated since Schwarz and Schwarz's 1943 seminal paper, a clear exponential curve is produced (not shown here). Considering the high number of taxa (species and subspecies) that have recently been described, it is likely that the total diversity of the genus *Mus* has not yet been fully uncovered. In particular, Southeast Asia, which has provided four of the five new species, appears as the key geographic area where more new species may be found.

This review of *Mus* taxonomy follows those of Schwarz and Schwarz (1943), Marshall (1977), Marshall and Sage (1981), Corbet (1990), Musser and Carleton (1993, 2005) and Tucker (2007). Since the latter two were published, three new species have been described, justifying a taxonomic update. After a few considerations on the four subgenera composing the genus, we have chosen to focus on the species of the subgenus *Mus* and to summarize the knowledge as it stands today on the taxonomy and distribution of the 14 recognized species that are classified following the species groups (SG) proposed by Suzuki *et al.* (2004) and Shimada *et al.* (2010) (see also Suzuki and Aplin, Chapter 2 in this volume).

Taxonomic position of the subgenus *Mus*

The genus *Mus* is a monophyletic clade comprising four subgenera – *Mus*, *Coelomys*, *Pyromys*, and *Nannomys* – totalling 41 extant species as of today. Members of this genus can be distinguished from other genera belonging to the same murine subfamily by a suite of morphological characters (Chevret *et al.*, 2005; Tucker, 2007). The geographical distribution of the genus encompasses all of Eurasia and sub-Saharan Africa, and its presence elsewhere results from human-mediated introductions during the last millennia. The highest taxonomic diversity occurs in Asia (19 species/3 subgenera), where this genus likely originated (see Suzuki *et al.*, 2004; Chevret *et al.*, 2005; Suzuki and Aplin, Chapter 2 in this volume). Whereas the monophyly of the four subgenera is supported by a series of molecular and morphological analyses (Chevret *et al.*, 2003, 2005), the relationship between them remained elusive until the advent of comparative cytogenomics. Using a cross-species chromosome-painting approach, Veyrunes *et al.* (2006) were able to resolve unambiguously the sister-clade relationships within the genus. The subgenus *Coelomys* was the first to diverge, followed by *Nannomys* as a sister group to the *Mus Pyromys* clade. In addition, this study confirmed that diversification within the genus, which occurred 6.7–8 million years ago (Mya) (Chevret *et al.*, 2005), took place very rapidly (within 1 Myr) and was accompanied by a burst of chromosomal change. Although there has been some debate on the taxonomic rank of the four subgenera (Bonhomme *et al.*, 1984; She *et al.*, 1990), a recent reappraisal has argued in favour of maintaining the present taxonomic designation, as the extent of their divergence is no larger than that within other genera such as *Apodemus* (Chevret *et al.*, 2005).

The four subgenera differ in species richness, range, and biological features (Tucker, 2007). The two Asian subgenera, *Pyromys* and *Coelomys*, comprise five and four species, respectively. From an ancestral Eurasian distribution, the subgenus *Mus* is now present worldwide owing to the commensal association of *M. musculus* with humans. Compared to the other subgenera, the subgenus *Mus* is remarkable for its chromosomal uniformity, with all members exhibiting a $2n = 40$ acrocentric chromosome karyotype. Only two taxa deviate from this pattern: the western European house mouse, *M. m. domesticus*, some populations of which have accumulated centric fusions (Hauffe *et al.*, Chapter 16 in this volume), and the Indian pygmy mouse, *M. terricolor*, in which several centric fusions and pericentric inversions have been observed (see below). The subgenus *Nannomys*, the African pygmy mouse, is the most speciose with 18 recorded species (Musser and Carleton, 2005). This subgenus is noteworthy for its highly conserved morphology, which contrasts with an extensive karyotypic diversity (Matthey, 1966; Jotterand, 1975). This chromosomal change comprises several unique innovations in mammals: a high diversity of X-autosome centric fusions and the existence of sex-reversed (XY) females (Veyrunes *et al.*, 2004, 2010a, b). The morphological, ecological, chromosomal, and genetic diversity of this genus provides a valuable comparative tool for testing hypotheses on rates and patterns of evolutionary change (Britton-Davidian *et al.*, 2012).

Whereas many species of the Asian and African subgenera are often absent from comparative approaches (but see Macholán, 2006), the subgenus *Mus* has represented an ideal taxonomic framework with which to investigate biological and molecular processes of evolution, and in doing so has provided important milestones for biomedical research (Peters *et al.*, 2007). The subgenus *Mus* harbours two highly prolific models for evolutionary studies on speciation modes at different timescales: the *M. musculus* lineage and *M. spretus*. The availability of both natural hybrid zones and laboratory resources (inbred and wild strains, backcross and recombinant panels) coupled with the house mouse genome sequence assembly opens enriching research avenues in a variety of fields: genetics of speciation (Oka *et al.*, 2007; Payseur and Place, 2007; Macholán *et al.*, 2007a, 2008a, 2011; Geraldes *et al.*, 2008; Piálek *et al.*, 2008; Dean and Nachman, 2009; see also in this volume Laukaitis and Karn, Chapter 7; Baird and Macholán, Chapter 14; Ganem, Chapter 15; Oka and Shiroishi, Chapter 17; Forejt *et al.*, Chapter 19; Goüy de Bellocq *et al.*, Chapter 18), tempo and modes of molecular evolution, chromosomal evolution (Hauffe *et al.*, Chapter 16), mating systems and social behaviour (Frynta *et al.*, 2005; see also König and Lindholm, Chapter 5; Ganem, Chapter 15), genetics of complex traits (Salcedo *et al.*, 2007; Laurie *et al.*, 2007; Takada *et al.*, 2008; see also Hermann and Bauer, Chapter 12), determinism of morphological variation (Kawakami and Yamamura, 2008; Burgio *et al.*, 2009; see also Klingenberg and Navarro, Chapter 6), chemical and acoustic communication (Stopka *et al.*, Chapter 8; Penn and Musolf, Chapter 9; Musolf and Penn, Chapter 10), reproductive traits (Fennessy, 1997; Elliott *et al.*, 2001; L'Hote *et al.*, 2007; Dean and Nachman, 2009; Laissue *et al.*, 2009; see also Forejt *et al.*, Chapter 19), adaptation (Ilmonen *et al.*, 2007; Kurzweil *et al.*, 2009; see also Teschke *et al.*, Chapter 13). In addition, the discovery of aberrant sex chromosome determinism systems (XY females, XO/XO males and females) within the African pygmy mouse suggests that the subgenus *Nannomys* has the potential to become a reference model for the identification of genes involved in the mammalian male-determining gene pathway (Jotterand-Bellomo, 1988; Veyrunes *et al.*, 2010a).

Taxonomic diversity in the subgenus *Mus*

The *M. musculus* species group

M. musculus Linnaeus, 1758, including subspecies *musculus*, *domesticus*, *castaneus*, *gentilulus*, *molossinus*

Type locality
Uppsala County, Uppsala, Sweden.

Taxonomic issues
From the 1970s to the end of the 1990s, controversy prevailed as to the taxonomic status that should be given to the subunits of commensal mice. From their very early studies characterizing allozyme variation among *musculus*, *domesticus*, and *castaneus*, the group of L. Thaler in Montpellier kept the subspecific denomination of these units. As stated in Auffray *et al.* (1990a), their conception of

M. musculus as a polytypic species composed of several subspecies able to exchange genes was in accordance with the recommendation of taxonomists (Corbet, 1990). Subspecific status was also used by the group of K. Moriwaki in their early studies on *M. m. molossinus* (Moriwaki *et al.*, 1981). The alternative approach was to give species status to the major genetic lineages of commensal mice (Sage *et al.*, 1993). In doing so, the subspecific denomination was available for populations differing by their coat colour and geographic range, such as the well-known *poschiavinus* or *praetextus*, which are now, however, listed among the synonyms of *domesticus* (Musser and Carleton, 2005). Another rationale for maintaining the species status of the commensal mice was its use by the companies providing the mouse strains, such as the Jackson Laboratory (JAX®; see table 2 in Sage *et al.*, 1993). Here, again, these companies have now adopted the subspecific nomenclature (Eppig, 2010). Among evolutionary biologists, the concept of the house mouse as a species complex is being abandoned with time. In the 1990 special issue of the *Biological Journal of the Linnean Society* on the house mouse, one-third of the papers considered *Mus musculus* as a species complex, while there were only 13% in the second special issue in 2005. In the present volume, none of the chapters dealing with the polytypic aspect of the house mouse consider it as a species complex.

Distribution

Five subspecies are currently recognized on the basis of distinct molecular-based lineages (Tucker, 2007; see also Suzuki and Aplin, Chapter 2; Yonekawa *et al.*, Chapter 4; Bonhomme and Searle, Chapter 11). *M. m. musculus* is found from central and northern Europe to northeast Asia. *M. m. domesticus* occurs in western Europe (including Norway; Jones *et al.*, 2010), Africa (coastal areas and Maghreb), and the Middle East to southwest Iran; it subsequently expanded its range to the Americas and Oceanian islands through passive transport with humans. The distribution area of *M. m. castaneus* extends from central to Southeast Asia. The two other subspecies have a much more restricted range: the recently rehabilitated *M. m. gentilulus* (Harrison, 1970; Duplantier *et al.*, 2002) occurs along the south-eastern coast of Arabia to Oman, as well as Madagascar, and *M. m. molossinus*, which is a hybrid subspecies between *M. m. musculus and M. m. castaneus*, is found in Japan (Yonekawa *et al.*, Chapter 4). Numerous studies indicate that these subspecies are not completely reproductively isolated, and in regions of secondary contact there is evidence of genetic exchanges ranging from limited introgression (i.e. *M. m. domesticus* and *M. m. musculus* in Europe; see also Cucchi *et al.*, Chapter 3; Bonhomme and Searle, Chapter 11; Baird and Macholán, Chapter 14) to broad overlap (i.e. *M. m. musculus* and *M. m. castaneus* in China; Guénet and

Bonhomme, 2003), and complete admixture (i.e. *M. m. molossinus*; see Yonekawa *et al.*, Chapter 4).

Phylogenetic relationships between subspecies

Progress has been made in unravelling the history of the radiation of *M. musculus* from its palaeontological Indo-Pakistani cradle. Two alternative hypotheses have been put forward (see Tucker, 2007: figure 2.2). The first considers that the three most widespread subspecies diverged following expansion from this cradle (Boursot *et al.*, 1996; Din *et al.*, 1996), whereas the second proposes a more western area of origin located within the current range of *M. m. domesticus*, with a subsequent expansion giving rise to the subspecies (Prager *et al.*, 1998). The answer may well lie in between these two areas (Duvaux *et al.*, 2011). Only the analysis of additional specimens from west-central Asia may provide the key to the geographic origin of the radiation as well as to its mode and tempo.

M. spretus Lataste, 1883

Type locality

Oued Magra, north of Hodna, Algeria.

Distribution

Endemic to the western Mediterranean climatic zone of western Europe (southern France, Spain, Portugal, Balearic Islands) and north Africa (Morocco, Algeria, Tunisia, Libya) (1.1a).

Description and diagnosis

M. spretus is a short-tailed wild mouse (head and body length / tail length (HB / T): *ca.* 1.51; Cucchi *et al.*, 2006). Morphometric, allozymic, and molecular criteria provide species-specific reliable diagnostic characters (Britton-Davidian and Thaler, 1978; Darviche and Orsini, 1982; Boursot *et al.*, 1985; Macholán, 1996a, b; Khammes and Aulagnier, 2006; but see below). This mouse possesses the standard G-banded karyotype with two distinctive features: a small Y chromosome and a set of subtelomeric rDNA clusters, which is a unique trait in the genus (Brannan *et al.*, 2001; Cazaux *et al.*, 2011). In addition, small differences in chromosomal architecture compared to the house mouse genome have been documented (Palmer *et al.*, 1995; Perry and Ashworth, 1999).

Phylogenetic relationships

Molecular markers place *M. spretus* within the Palaearctic European *M. musculus* SG along with *M. musculus*, *M. spicilegus*, *M. macedonicus*, and *M. cypriacus*

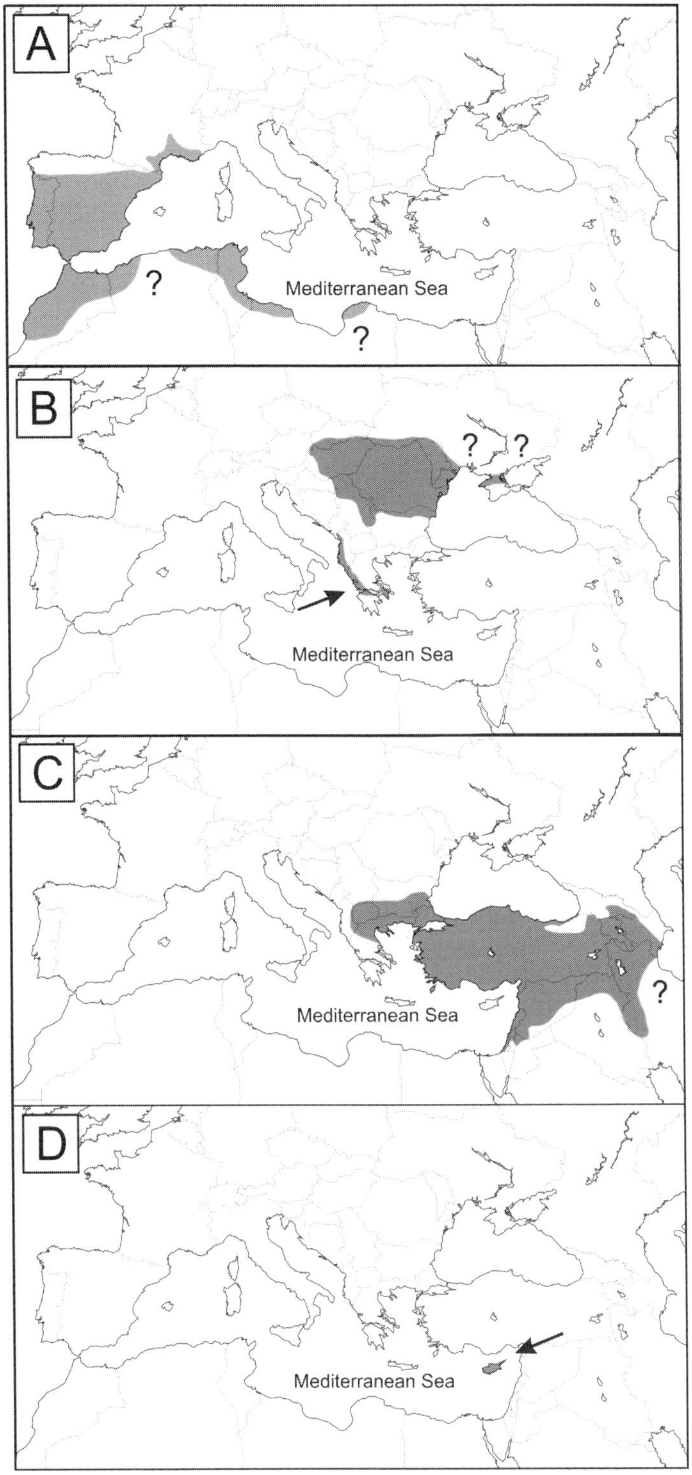

Figure 1.1 Distribution of the species of the *Mus musculus* subgroup. (a) *Mus spretus* from Amori *et al.* (2008). (b) *Mus spicilegus* from Coroiu *et al.* (2008) and Mitsainas *et al.* (2009). (c) *Mus macedonicus* from Kryštufek and Vohralík (2008). (d) *Mus cypriacus* from Amori and Hadjisterkotis (2008).

(Suzuki *et al.*, 2004; see also the '*M. musculus*' SG in Suzuki and Aplin, Chapter 2). Sister-species relationships within this group are, however, less clear, as nuclear and mtDNA sequences often lead to conflicting tree topologies (Tucker *et al.*, 2005; Tucker, 2007). *M. spretus* evolved in north Africa, where the oldest fossil was discovered at a date which is in agreement with molecular-based estimates (1.5 Mya in Dejager *et al.*, 2009). Phylogeographical analyses have highlighted a genetic diversity gradient from highly variable populations in north Africa to almost monomorphic ones in southern France, with intermediate values in the Iberian Peninsula (Boursot *et al.*, 1985). This diversity pattern supports the north African origin of this species, with a colonization of Europe via Spain involving two founder and/or bottleneck events through Gibraltar and the Pyrenees.

Ecology/behaviour

This species is commonly found in grasslands, dry shrublands, and agricultural fields. In southern France, its eastern distribution limit coincides with that of the Mediterranean oak (*Quercus coccifera*) (Orsini *et al.*, 1982). Over its whole distribution area, this species is sympatric with *M. m. domesticus*, but rarely syntopic, as *M. spretus* has limited water requirements and typically inhabits drier areas than the house mouse (Sicard *et al.*, 1985; Pouliquen-Young, 1994). Traces of horizontal transfer through introgression have been documented between *M. spretus* and *M. m. domesticus*, indicating that the two species do meet on rare occasions in the wild and are capable of producing offspring (Greene-Till *et al.*, 2000; Hardies *et al.*, 2000; Orth *et al.*, 2002b). The population and social structure of this species have been extensively studied by Hurst and co-workers, providing information on micro-habitat use and dispersion (Gray *et al.*, 1998; Gray and Hurst, 1997), social and territorial organization (Hurst *et al.*, 1994, 1996, 1997), as well as an original hygienic behaviour (Hurst and Smith, 1995). As is the case for *M. spicilegus*, studies on the reproductive system of *M. spretus* have revealed a suite of behavioural traits all of which, except for testis size (Frynta *et al.*, 2009), are indicative of a monogamous mating system (Cassaing and Isaac, 2007; Cassaing *et al.*, 2010).

Points of interest

The discovery that crossbreeding *M. spretus* and laboratory mouse strains produced viable and fertile progeny (Bonhomme *et al.*, 1978), provided an invaluable tool for the construction of high-density mouse genetic maps (Dejager *et al.*, 2009). This unique contribution stemmed from the high degree of genetic divergence between *M. spretus* and the laboratory strains. Various backcross panels were established and made accessible to the scientific community through the European Collaborative Interspecific Backcross network. Since then, the production of interspecific recombinant congenic strains and the ongoing genome sequencing project of an *M. spretus* strain now constitute a set of genetic resources that are fundamental for studying the genetics of complex traits (see Dejager *et al.*, 2009). As *M. spretus* and *M. musculus* are nevertheless valid species, the ability to produce interspecific progeny has paved the way for the analysis of genomic interactions during speciation (Hale *et al.*, 1993; Biddle *et al.*, 1994; Pilder *et al.*, 1997; Hemberger *et al.*, 2001; Zechner *et al.*, 2002; Shi *et al.*, 2004).

Unresolved issues

What is the short-tailed mouse reported in the Cyrenaic, Libya (Orsini *et al.*, 1982)? The sister-species relationships within the European SG need to be clarified.

M. spicilegus Petényi, 1882, including subspecies *spicilegus*, *adriaticus*

Type locality

Rakos Plains, Budapest, Hungary.

Distribution

Southeast Austria, south Slovakia, Hungary, Romania, Bulgaria, Moldavia, Serbia, Montenegro, and south Ukraine (Fig. 1.1b). Recent studies show an extension into Albania, northwest, west, southwest, and southeast Greece (Macholán and Vohralík, 1997; Macholán *et al.*, 2007b; Bauer *et al.*, 1998; Mitsainas *et al.*, 2009). It occurs sympatrically with *M. musculus* over its whole range.

Description and diagnosis

M. spicilegus is a short-tailed field mouse (HB / T: 1.35–1.39; Orsini *et al.*, 1983) characterized by a unique behaviour: mound-building (Holz *et al.*, 2009; see below for details). Morphometric and morphological characters discriminate this wild mouse from its European congeners (Orsini *et al.*, 1983; Macholán, 1996a, b; Cserkesz *et al.*, 2008). Chromosomal studies document the presence of the standard karyotype with $2n = 40$ acrocentric chromosomes, the only exception being a minute Y chromosome as in the other wild-living European species (Bulatova and Kotenkova, 1990).

Phylogenetic relationships/subspecies

M. spicilegus clearly belongs to the Palaearctic *M. musculus* SG within the subgenus *Mus*, where it shares a sister-taxon relationship with *M. macedonicus*

(Suzuki *et al.*, 2004; but see Tucker *et al.*, 2005). *M. spicilegus* is sympatric with both *M. m. musculus* and *M. m. domesticus* in the northern and southern part of its range, respectively. Recent analyses have supported the existence of a distinct subspecies in Montenegro, based on morphological and molecular evidence: *M. s. adriaticus* (Kryštufek and Macholán, 1998). In addition, the study of new samples throughout Greece has uncovered a highly distinct mtDNA lineage in Sterea Ellada (southeast Greece; Mitsainas *et al.*, 2009).

Ecology/behaviour

This species is the only one in the genus to display mound-building activity. During autumn, individuals cooperate in collecting various seeds and plant stems that are then covered with earth, forming a tumulus (up to almost 2 m in diameter and 40 cm high; Holz *et al.*, 2009). This communal nesting in tumuli occurs as an over-wintering strategy to limit food shortage, predation, and thermoregulatory losses (Holz *et al.*, 2009). Microsatellite analyses have identified occupants as male and female parents and their juvenile progeny, ranging between 1 and 21 mice (Garza *et al.*, 1997; Patris and Baudoin, 2000; Dobson and Baudoin, 2002). *M. spicilegus* was the first mouse species to be described with a monogamous mating system. Indeed, although polygyny may occur, social monogamy is supported by several associated behavioural traits: paternal care of pups, social and pair bonding, agonistic interactions, socio-spatial associations (Gouat and Feron, 2005; Feron and Gouat, 2007; Poteaux *et al.*, 2008).

Points of interest

The unique behavioural particularities of *M. spicilegus* (mound-building, monogamy) undoubtedly represent the main attractions of this species. Given the available phylogenetic context, comparative studies between closely related species using various modern approaches will enlighten the evolutionary processes involved in their emergence.

Unresolved issues

The geographical distribution and taxonomic status of the newly identified lineages within *M. spicilegus* warrant further investigation.

M. macedonicus Petrov and Ruzič, 1983, including subspecies *macedonicus* and *spretoides*

Type locality

Macedonia.

Distribution

From the Balkans to Georgia and northwest Iran to Syria, Lebanon and Israel (Fig. 1.1c). *M. macedonicus* occurs strictly in the Mediterranean climatic zone, in the eastern part of the Mediterranean region (Auffray *et al.*, 1990b). It is the ecological vicariant of *M. spretus*, which occupies the western Mediterranean zone (see above). *M. macedonicus* was formerly described in Cyprus, where it has recently been shown that this species is absent due to confusion with the cryptic *M. cypriacus* (Cucchi *et al.*, 2006). Over its whole range, *M. macedonicus* occurs sympatrically with *M. musculus* and is found with *M. spicilegus* in only one locality (Boursot *et al.*, 1993).

Description and diagnosis

M. macedonicus is a short-tailed mouse (HB / T: *M. m. macedonicus*: *ca.* 1.16; *M. m. spretoides*: *ca.* 1.28; Cucchi *et al.*, 2006).

Phylogenetic relationships/subspecies

M. macedonicus belongs to the *M. musculus* SG (Suzuki *et al.*, 2004), and is closely related to *M. cypriacus* (Bonhomme *et al.*, 2004; Cucchi *et al.*, 2006; Macholán *et al.*, 2007b). Two deeply divergent mitochondrial clades along with nuclear genes differentiation led to the consideration of two subspecies *M. macedonicus macedonicus* and *M. m. spretoides*. The former is present in the major part of the species range, while the latter is restricted to the very southern part of the range, since *spretoides* has so far been described only in Israel and Lebanon. The morphological differentiation between the two groups is more subtle and essentially corresponds to a divergence in size, even though a few diagnostic parameters have been described (Macholán *et al.*, 2008b). The origin of these two subspecies may be due to two different Pleistocene refuges, which is in agreement with the divergence time, estimated between 200 000 and 300 000 years.

Ecology/behaviour

M. macedonicus is wild-living, and occurs both in agricultural fields and natural habitats (grasses and shrubs). In Israel, *M. macedonicus* and *M. m. domesticus* show a strong habitat partition: *M. macedonicus* is excluded from the anthropogenic environment where only the house mouse is found. In contrast, *M. m. domesticus* may be observed outdoors outside the range of *macedonicus*. *M. macedonicus* is the first invader species in Mediterranean post-fire habitats. This ability is related with physiological adaptations such as low metabolic rate and high lower critical point of thermoneutral zone, similar to desert rodent species (Izhaki *et al.*, 1993; Shabtay *et al.*, 2000).

Unresolved issues

The transition between the two subspecies most likely occurs in north Lebanon, and has yet to be precisely localized (Macholán *et al.*, 2007b); the nature of this transition is unknown. What is the short-tailed mouse reported in the Cyrenaic, Libya (Orsini *et al.*, 1982)?

M. cypriacus Cucchi, Orth, Auffray, Renaud, Fabre, Catalan, Hadjisterkotis, Bonhomme and Vigne, 2006

Type locality

Alassa, Troodos Region, Cyprus.

Distribution

This species is only known from Cyprus, where it was sampled in the southern part of the island in the original description (Cucchi *et al.*, 2006) (Fig. 1.1d). Macholán *et al.* (2007b; 2008b) reported its occurrence along the northern coast, suggesting that *M. cypriacus* likely covers the whole island, where it occurs sympatrically with *M. m. domesticus*.

Description and diagnosis

This medium- to long-tailed mouse (HB / T: *ca.* 1.16; Cucchi *et al.*, 2006) has long been confused with *M. macedonicus*, on whole-body as well as on skull and molar criteria (Cucchi *et al.*, 2002). Molecular approaches (Bonhomme *et al.*, 2004) first identified this new taxon, which was later formerly described and named by Cucchi *et al.* (2006). On the basis of the molecular identification, specific morphological traits were retrieved, in particular a clearly larger skull, which might be ascribed to insular gigantism. Additionally, geometric morphometrics provided unambiguous discrimination from all *M. macedonicus* subspecies, *M. m. domesticus* and *M. spretus* on skulls, mandibles, and molars (Cucchi *et al.*, 2002, 2006; Cucchi, 2008; Macholán *et al.*, 2008b).

Phylogenetic relationships

This species belongs to the *M. musculus* SG; it is related to *M. macedonicus*, but is almost equidistant with *M. spicilegus*.

Ecology/behaviour

M. cypriacus inhabits abandoned cultivation terraces, vineyards, grassy fields, and bushes up to 900 m above sea level. At lower altitudes, it may be syntopic with the house mouse (*M. m. domesticus*) in forested riverine areas. It seems to be absent from highly anthropogenic areas including orchards, where the house mouse is dominant (Cucchi *et al.*, 2005). Habitat partition between these two species is very similar to that observed between *M. macedonicus* and *M. m. domesticus* in Israel (Auffray *et al.*, 1990b; Auffray and Britton-Davidian, 1992).

Points of interest

M. cypriacus is an endemic species of Cyprus which has been isolated from the continent since the Messinian Crisis (6–5.3 Myr). Based on mitochondrial control region sequences, the divergence time with *M. macedonicus* is estimated at *ca.* 0.5 Mya (Cucchi *et al.*, 2006; Macholán *et al.*, 2007b). This date corresponds to the Mindel Glaciation, during which the over-water distance between Cyprus and the continent was reduced. This may have favoured the colonization of Cyprus by the ancestor of *M. cypriacus* and *M. macedonicus*, likely by natural rafts. *M. cypriacus* is, with *Crocidura zimmermanni* in Crete, one of the only two small mammals endemic to a Mediterranean island which survived after the colonization by humans (Vigne and Cucchi, personal communication).

The M. booduga species group

M. booduga Gray, 1837

Type locality

South Mahratta, India.

Distribution

Sri Lanka, Peninsular India, Bangladesh, south Nepal, central Myanmar, Pakistan (Fig. 1.2a).

Description and diagnosis

M. booduga is a short-tailed pygmy field mouse (HB / T: *ca.* 1.19; Marshall, 1977), morphologically very similar to *M. terricolor*, which is present in the same habitats. These two species can be discriminated by their distinctive molar patterns (Matthey and Petter, 1968) and their karyotype, which in *M. booduga* is identical to the standard 2n = 40, including the Y chromosome (G- and C-banding; Sharma, 1996), while *M. terricolor* has several submetacentric autosomes and a metacentric X chromosome (see below). The ranges of both pygmy mouse species largely overlap.

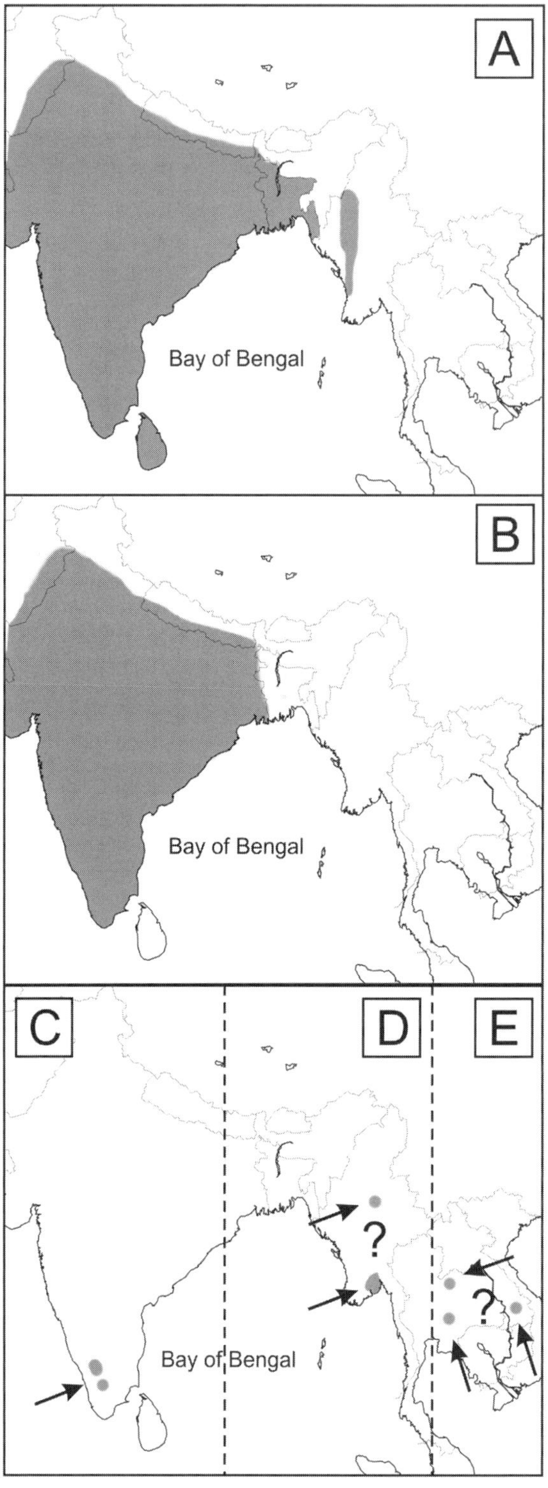

Figure 1.2 Distribution of the species of the *Mus booduga* subgroup. (a) *Mus booduga* from Aplin *et al.* (2008c). (b) *Mus terricolor* from Aplin (2008). (c) *Mus famulus* from Pradhan *et al.* (2008). (d) *Mus nitidulus* from Shimada *et al.* (2007). (e) *Mus fragilicauda* from Aplin and Lunde (2008b) and S. Morand, Y. Chaval, V. Herbreteau, and A. Orth, personal communication.

Phylogenetic relationships
Within the subgenus *Mus*, this species belongs to the *M. booduga* SG of which it is the representative species (Suzuki *et al.*, 2004). Allozyme and molecular markers clearly distinguish it from the other species in this subgroup (Awasthi *et al.*, 1999; Suzuki *et al.*, 2004; Shimada *et al.*, 2010; see also Suzuki and Aplin, Chapter 2).

Ecology/behaviour
M. booduga is only found in anthropogenic habitats such as rice and wheat fields.

Points of interest
Genetic analyses indicate that this species shows much higher levels of diversity than European samples of *Mus* (Singh and Sharma, 1997; Awasthi *et al.*, 1999). In addition, the pericentromeric constitutive heterochromatin typical of *M. musculus* is reduced or lacking in five chromosome pairs of the complement, suggesting that qualitative evolution of the mouse-like AT-rich satellite DNA sequences has occurred in the subgenus (Balajee and Sharma, 1994).

Unresolved issues
The distribution is poorly known due to confusion with *M. terricolor*; the sister-species relationships are unresolved.

M. terricolor Blyth, 1851

Type locality
Bengal, south India.

Distribution
Peninsular India, Nepal, Bangladesh, Pakistan (Fig. 1.2b). It also occurs in northern Sumatra, where it may have been introduced.

Description and diagnosis
Formerly referred to as *Mus dunni*, *M. terricolor* is a pygmy field mouse (HB / T: *ca.* 1.05; Marshall, 1977) very similar in morphology to *M. booduga*. The most distinctive feature of this species is its karyotype, which consists of 2n = 40 chromosomes as the standard, but includes 1–3 pairs of submetacentric autosomes and a metacentric X chromosome due to the presence of heterochromatic arms. The Y chromosome is also larger than the one in the standard *Mus*

karyotype (Sharma, 1996). *M. terricolor* and *M. booduga* are sympatric and even syntopic since they can occur in the same agricultural fields.

Phylogenetic relationships

Molecular analyses place *M. terricolor* clearly within the *M. booduga* SG (Suzuki *et al.*, 2004). As with the other Indian species studied, *M. terricolor* is characterized by a high level of genetic variability (Singh and Sharma, 1997; Awasthi *et al.*, 1999). Three chromosomal forms are described within *M. terricolor* based on karyotypes; they are named *terricolor I*, *terricolor II*, and *terricolor III*. Whereas *terricolor I* has minute heterochromatic short arms, *terricolor II* and *III* possess two and three pairs of submetacentric chromosomes, respectively, due to large heterochromatin additions. Genetic analyses indicate that *terricolor I* and *II* are more closely related to each other than to *terricolor III*. The more basal position of the latter suggests that a reversal of chromosomal variation may have occurred, the karyotype reverting to the ancestral one as in *terricolor I*. The distribution of these three forms is apparently parapatric: *terricolor I* is present in northern India, *terricolor II* in the interior of southern India, and *terricolor III* in the coastal region of southern India.

Ecology/behaviour

Few taxonomically verified ecological data are available for this pygmy mouse. Reciprocal crosses performed between the three chromosomal forms suggest that a postzygotic reproductive barrier is present between them, resulting from genetic incompatibilities and meiotic perturbations (Sharma, 1996; Sharma *et al.*, 2003). In addition, the three forms express a mating preference for their own form in behavioural tests, suggestive of pre-mating isolation (Sharma *et al.*, 1986; Sharma, 1996).

Points of interest

This species is the only one of this SG in which evolution of centromeric repetitive sequences and modifications in their organization have been described (rDNA sites, satellite DNA sequences, telomeric repeats; Balajee and Sharma, 1994; Sharma and Sharma, 1998; Bardhan and Sharma, 2000a; Sharma *et al.*, 2002). In fact, the X chromosome is one of only two chromosomes that have retained the mouse-like AT-rich satellite DNA. Additional chromosomal variability consists of several centric fusions and two pericentric inversions, one of which may be frequent (Sen and Sharma, 1983; Bardhan and Sharma, 2000b). However, misidentification of several chromosomal pairs (3, 4, 5, 7, 12, 15) is apparent in the original G-banded karyotypes published in 1983, which require a revised update. It is interesting to note that *M. terricolor* and *M. m. domesticus* are the only *Mus* taxa to exhibit extensive chromosomal diversity; in addition, both carry centric fusions involving centromeric regions that apparently differ in sequence.

Unresolved issues

The distribution limits of the three chromosomal forms need to be precisely determined as well as the nature of their contact; sister-species relationships within the group are unresolved.

M. fragilicauda Auffray, Orth, Catalan, Desmarais, Gonzales and Bonhomme, 2003

Type locality

Ban Nong Sanga (14°32′33″N, 101°57′44″E), Nahkon Ratchasima (Khorat) Province, Thailand.

Distribution

In the original description, *M. fragilicauda* was known only from the type locality and another close site, Tumbon (Fig. 1.2e). Since then, four individuals were trapped in two neighbouring localities of Loei Province, Thailand (17°23′36″N, 101°48′24″E), approximately 300 km north from the type locality (Morand, Chaval, Herbreteau, and Orth, personal communication). It has also been reported from Lamam, Sekong Province, and Laos PDR (Aplin and Lunde, 2008b). These new observations extend the distribution area of *M. fragilicauda* from central Thailand to central Laos. It occurs in sympatry with *M. cervicolor*, *M. caroli*, and *M. cookii*. There is as yet no evidence of an overlap in range with *M. m. castaneus*.

Description and diagnosis

M. fragilicauda is a short-tailed mouse (HB / T: *ca.* 1.21; Auffray *et al.*, 2003). Considered as a sibling species of *M. cervicolor*, it is extremely difficult to diagnose this species in the field without molecular markers. A typical trait of *M. fragilicauda* is its ability to lose its tail integument when handled; this feature is not observed in any other species of the subgenus *Mus*. Due to the fragility of the tail integument, the frequent presence of one or several ring scar marks along the tail may be one of the rare morphological traits that can be used in the field. The tail is not bicoloured.

Phylogenetic relationships

From a phylogenetic point of view, *M. fragilicauda* was first included in the European taxa on the basis of 16S rRNA sequences. Using additional markers, Suzuki *et al.* (2004) assigned *M. fragilicauda* to the *M. booduga* SG (including *M. nitidulus* and *M. terricolor*). These results provided geographic coherence to the structure of the subgenus *Mus* with four clades: an Indian, an Asian, a European, and a fourth one for *M. lepidoides* (see Suzuki and Aplin, Chapter 2).

Ecology/behaviour

The distribution of this species is still poorly known. It was trapped in dry grasses bordering rice fields or at the immediate periphery of dwellings. It was also found in pygmy bamboo. In these different types of habitat, it is often trapped together with one of the following species: *M. cervicolor*, *M. caroli* (Auffray *et al.*, 2003), and *M. cookii* (S. Morand, Y. Chaval, and V. Herbreteau, personal communication). According to recent trapping efforts (a French, Thai, and Lao scientific programme in Thailand and Laos on rodents and their pathogens – Ceropath), *M. fragilicauda* is quite rarely captured, its populations apparently remaining at low densities relative to those of *M. cervicolor* and *M. caroli.*

Unresolved issues

The ecology and particularly the niche partition among species from Southeast Asia should be investigated.

M. nitidulus Blyth, 1859

Type locality

Yangon Division, Myanmar.

Distribution

Lower central basin of Myanmar, where this species is endemic (Shimada *et al.*, 2007) (Fig. 1.2d). Its distribution may overlap with that of *M. cervicolor*, *M. m. castaneus*, *M. caroli*, and *M. lepidoides*.

Description and diagnosis

M. nitidulus is a medium-sized mouse (HB / T: *ca.* 1.28; from adult specimen measurements provided in Shimada *et al.*, 2007) with no particular morphological features. It is very similar in appearance to *M. fragilicauda* and particularly to *M. cervicolor*, with which it has long been confused. Several cranial features may, however, discriminate *M. nitidulus* specimens from *M. cervicolor* (Shimada *et al.*, 2007).

Phylogenetic relationships

The phylogenetic analyses based on mitochondrial and nuclear genes provide strong support for the species status of *M. nitidulus* and its inclusion within the *M. booduga* SG (Suzuki *et al.*, 2004) along with *M. booduga*, *M. terricolor*,

M. fragilicauda, and *M. famulus* (see also Suzuki and Aplin, Chapter 2). However, the relationships within the *M. booduga* SG are poorly resolved. The *M. nitidulus* lineage is thought to have originated during the Early Pleistocene (1.3 Mya).

Ecology

This species is present in anthropogenic habitats (lowland rice-fields). The original vegetation in this area is reconstructed as a mosaic of evergreen and deciduous monsoon forest.

Points of interest

M. nitidulus is one of the three mammal species endemic to Myanmar. The geographically isolated topology of the central basin of Myanmar may constitute a centre of mammalian diversification during the Tertiary and Quaternary (Shimada *et al.*, 2007; see also Suzuki and Aplin, Chapter 2).

Unresolved issues

The northern geographic limit of the range is not known; the karyotype needs to be determined; the sister-species relationships are unresolved.

M. famulus Bonhote, 1898

Type locality

Nilgiri Hills, Tamil Nadu, south India.

Distribution

This species is endemic to India, and has only been recorded in the Nilgiri Hills in southern India (Fig. 1.2c). Given its small distribution area and the ongoing modifications of its habitat, this species has recently been tagged as endangered by the IUCN (Pradhan *et al.*, 2008).

Description and diagnosis

M. famulus is a long-tailed, long-nosed, small-eyed, large mouse (≤30 g), shrew-like in appearance (HB / T: *ca.* 1.04; Chevret *et al.*, 2003). The G-banded karyotype is identical to that of the house mouse, including the Y chromosome (seven mice studied; J. Catalan, unpublished observations).

Phylogenetic relationships

Formerly assigned to the subgenus *Coelomys*, it is now included in the subgenus *Mus* on molecular and morphological criteria (Chevret *et al.*, 2003). Recent molecular analyses support its assignment to the *M. booduga* SG (Suzuki *et al.*, 2004; Shimada *et al.*, 2007).

Ecology/behaviour

This species is present in the montane tropical evergreen rainforest covering the Nilgiri Hills (1500 m) in areas with high ground cover and low canopy height (Shanker, 2001).

Points of interest

Morphological analyses (cranial, dental, and pelage characters) support a close relationship with the Asian species of the subgenus *Mus*, particularly *M. cookii*, in conflict with the molecular approach (Chevret *et al.*, 2003). However, genetic comparison with members of the *M. booduga* SG was lacking in that study.

Unresolved issues

The distribution limit of this species needs to be localized; the sister-species relationships within the Indian SG are unresolved.

The *M. cervicolor* species group

M. cervicolor Hodgson, 1845

Type locality

Nepal.

Distribution

From northern India to southern China through Nepal, to continental Southeast Asia; its distribution has recently been extended to southern Thailand (Fig. 1.3a). Musser and Carleton (2005) reported it from Sumatra and Java, and the Andaman Islands.

Description and diagnosis

This short-tailed mouse (HB / T: *ca.* 1.37; Marshall, 1977) is morphologically similar to *M. fragilicauda*, from which it is very difficult to distinguish. The tail is slightly bicoloured.

Phylogenetic relationships/subspecies

M. cervicolor belongs to the *M. cervicolor* SG along with *M. caroli* and *M. cookii* (see Suzuki and Aplin, Chapter 2). Marshall (1977) considered that there were two subspecies, *M. c. cervicolor* and *M. c. popaeus*, the second being larger,

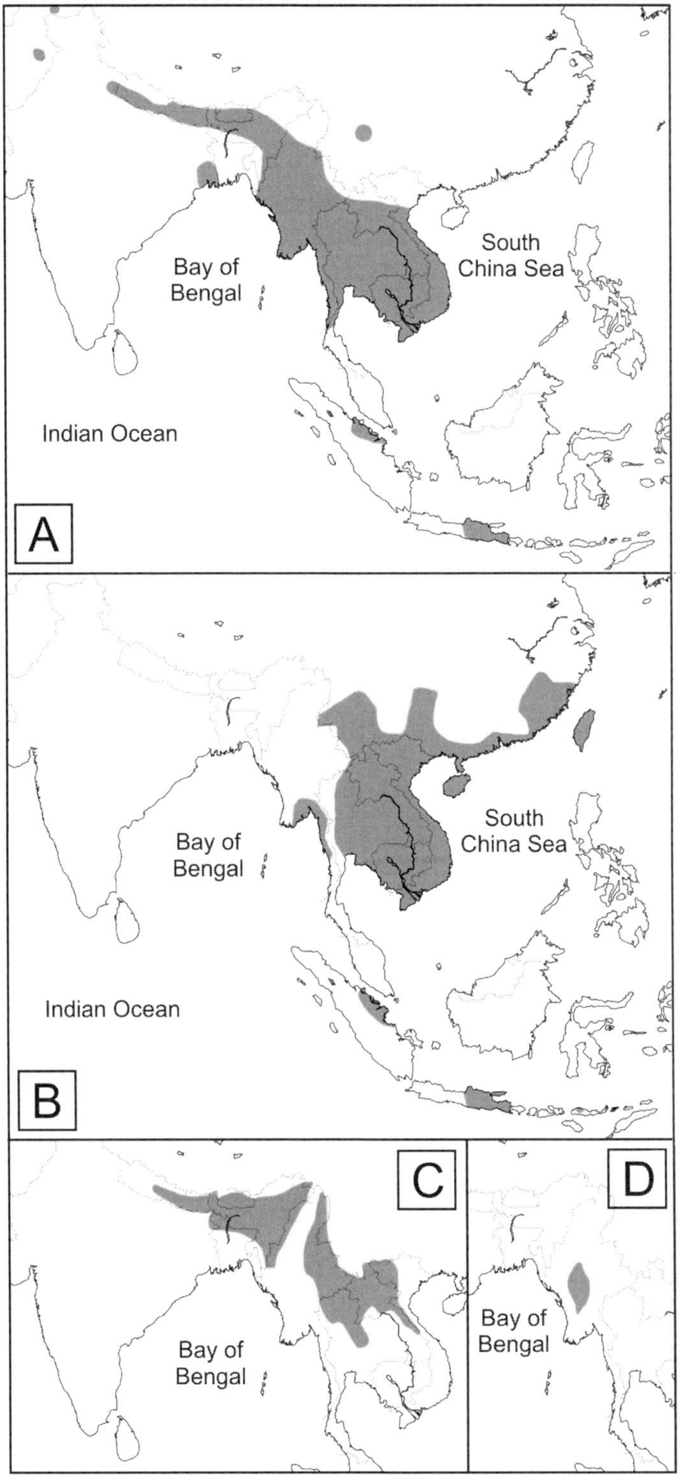

Figure 1.3 Distribution of the species of the *cervicolor* and *lepidoides* subgroups. (a) *Mus cervicolor* from Aplin *et al.* (2008a) and Aplin *et al.* (2003). (b) *Mus caroli* from Aplin and Lunde (2008a). (c) *Mus cookii* from Aplin *et al.* (2008b). (d) *Mus lepidoides* from Shimada *et al.* (2010).

inhabiting respectively rice and grass fields and forests. When mice from the supposed geographic area of both subspecies were investigated, morphological and molecular analyses failed to confirm the existence of these two subspecies (Macholán, 2001; Auffray *et al.*, 2003). It occurs sympatrically with *M. caroli*, *M. fragilicauda*, *M. cookii*, and *M. m. castaneus*.

Ecology/behaviour

This species may be found in syntopy with *M. caroli* and *M. fragilicauda*. It lives in grasslands and fields.

Points of interest

M. cervicolor is the host from which the M813 murine leukaemia virus was isolated. Strains of *M. cervicolor* are regularly used for studying the expression and regulation of endogenous retroviruses in rodents (e.g. Tarlington *et al.*, 2008; Prassolov *et al.*, 2001).

Unresolved issues

The niche partition among the species from Southeast Asia needs to be assessed.

M. caroli Bonhote, 1902

Type locality

Okinawa, Ryukyu Island, Japan.

Distribution

Southern China and continental Southeast Asia (Myanmar, Thailand, Laos Cambodia, Vietnam), Hainan, Taiwan and Ryukyu Islands, as well as Sumatra and Java (Fig. 1.3b). Musser and Carleton (2005) also reported it from Malay and Madura.

Description and diagnosis

M. caroli is a small, long-tailed mouse (HB / T: *ca.* 1.01; Marshall, 1977); its tail is sharply bicoloured and it has a short snout. In Taiwan, the tail is even longer than the head and body (Motokawa *et al.*, 2003). Two studies have succeeded in identifying a morphological differentiation between distant populations, which was interpreted as resulting from geographic distance or insularity (Macholán, 2001; Motokawa *et al.*, 2003). It occurs sympatrically with *M. cervicolor*, *M. cookii*, *M. fragilicauda*, and *M. m. castaneus*.

Phylogenetic relationships
M. caroli belongs to the *M. cervicolor* SG (Suzuki *et al.*, 2004).

Ecology/behaviour
M. caroli occurs sympatrically with *M. cervicolor* and *M. fragilicauda* in grasslands and rice fields. It is also found in sugar cane fields (Motokawa *et al.*, 2003).

Points of interest
As with *M. cervicolor*, *M. caroli* is extensively used as a model in biomedical research (e.g. Hollyoake *et al.*, 2005). It is interesting to note that several parts of the species range are disconnected from the main distribution, e.g. Myanmar, Malay, Ryukyu Islands. This suggests that *M. caroli* has the ability to expand its range with human activities. Colonization of islands suggests that this species may benefit from passive transport.

Unresolved issues
The niche partition among the species from Southeast Asia needs to be investigated.

M. cookii Ryley, 1914

Type locality
Gokteik, northern Myanmar.

Distribution
The IUCN Red List map for *M. cookii* (Aplin *et al.*, 2008b) presents two important disconnected ranges, separated by the Ayeyarwady River valley in Myanmar. The first involves eastern Nepal, Nepal, Bhutan, northern Bangladesh, Assam (northeast India) and eastern Myanmar (Fig. 1.3c). The second includes western Myanmar, southwest Yunnan (China), central Thailand, Laos, and central Vietnam. Marshall (1977), Corbet and Hill (1991), and Aplin *et al.* (2003) did not indicate this gap in the distribution of *M. cookii*. It occurs sympatrically with *M. cervicolor*, *M. caroli*, *M. cookii*, and possibly with *M. m. castaneus* on some parts of their ranges.

Description and diagnosis
In this medium- to long-tailed mouse (HB / T: *ca.* 1.11; Marshall, 1977), the tail is reported to be longer than the head and body in Laos by Aplin *et al.* (2003). *M. cookii* is one of the largest species of the subgenus *Mus*.

Phylogenetic relationships
M. cookii belongs to the *M. cervicolor* SG (Suzuki *et al.*, 2004).

Ecology/behaviour
While *M. cervicolor*, *M. caroli*, and *M. fragilicauda* are difficult to distinguish from an ecological point of view, *M. cookii* presents more characteristic features. It is essentially found in upland forests, but is not restricted to pine forests as suggested by Marshall (1977); it may also be trapped in grasslands and rice fields (Morand, Chaval and Herbreteau, personal communication).

Points of interest
M. cookii is the only species in the genus *Mus* in which B chromosomes have been found (Badenhorst *et al.*, 2009). B chromosomes are supernumerary and unpaired chromosomes that modify the chromosomal number. The origin and the impact of these chromosomes on the genomes are still debated.

The *M. lepidoides* species group

M. lepidoides Fry, 1931

Type locality
Mount Popa, Myanmar.

Distribution
Central dry zone (CDZ) of Myanmar and along the Irrawaddy River to the south (Fig. 1.3d). Its distribution may overlap with that of *M. nitidulus*, *M. cervicolor*, *M. m. castaneus*, and *M. caroli*.

Description and diagnosis
This species is an extremely small (<8 g), short-tailed mouse (HB / T: *ca.* 1.31 from adult specimen measurements provided in Shimada *et al.*, 2010). *M. lepidoides* is characterized by distinctive external and cranial morphological characteristics, some of which may serve to distinguish it from other members of the genus, although it shares many features with *M. booduga* and *M. terricolor* (Shimada *et al.*, 2010).

Phylogenetic relationships
Formerly considered as a subspecies of *M. booduga*, *M. lepidoides* has recently been reinstated as a valid species due to its high degree of molecular and

morphological distinctiveness (Shimada *et al.*, 2010). Although the phylogenetic analyses did not provide a clear-cut resolution of the relationships within the subgenus, these authors suggest its inclusion as a fourth SG on the grounds of a moderately supported topology (see also Suzuki and Aplin, Chapter 2). This topology is in agreement with a divergence of the SG lineages during the Late Pliocene (2.4–2.8 Myr).

Ecology/behaviour

All studied specimens were captured in agricultural fields, where this species is considered a pest. The CDZ of Myanmar consists of lowland, poor, sandy soils, where rainfall is lower than in surrounding areas. Current vegetation is highly degraded, but formerly included patches of dry deciduous forest and thorny scrub. It does not seem to occur on the floodplain of the Irrawaddy River where *M. cervicolor* and *M. nitidulus* are present in agricultural fields.

Points of interest

M. lepidoides may represent the sister lineage to the *M. musculus* SG, and if confirmed may represent the link between the radiation of the southeast and South Asian SGs and the western Eurasian SG (Shimada *et al.*, 2010). This scenario is supported by the fact that *M. lepidoides* is the only other member of the subgenus in addition to those of the *musculus* SG to be found in xeric habitats. Shimada *et al.* (2010) suggest an evolutionary scenario involving *M. lepidoides* as a possible *Mus* relict of the niche transition from forest to drier habitats that characterized a tract connecting eastern India to central Myanmar throughout the Pliocene (see Suzuki and Aplin, Chapter 2).

Unresolved issues

The limits of the distribution area need to be localized; the karyotype is unknown; the phylogenetic relationships with other SGs should be investigated.

REFERENCES

Amori, G., Aulagnier, S., Hutterer, R., *et al.* (2008). *Mus spretus*. In *IUCN 2010: IUCN Red List of Threatened Species. Version 2010.4*. www.iucnredlist.org. Downloaded on 4 December 2010.

Amori, G. and Hadjisterkotis, E. (2008). *Mus cypriacus*. In *IUCN 2010: IUCN Red List of Threatened Species. Version 2010.4*. www.iucnredlist.org. Downloaded on 4 December 2010.

Aplin, K. (2008). *Mus terricolor*. In *IUCN 2010: IUCN Red List of Threatened Species. Version 2010.4*. www.iucnredlist.org. Downloaded on 4 December 2010.

Aplin, K., Brown, P. R., Jacob, J., Krebs, C., and Singleton, G. R. (2003). *Field Methods for Rodent Studies in Asia and the Indo-Pacific*. Canberra: Australian Center for International Agricultural Research.

Aplin, K. and Lunde, D. (2008a). *Mus caroli*. In *IUCN 2010: IUCN Red List of Threatened Species. Version 2010.4*. www.iucnredlist.org. Downloaded on 4 December 2010.

Aplin, K. and Lunde, D. (2008b). *Mus fragilicauda*. In *IUCN 2010: IUCN Red List of Threatened Species. Version 2010.4*. www.iucnredlist.org. Downloaded on 4 December 2010.

Aplin, K., Lunde, D., Frost, A., and Molur, S. (2008a). *Mus cervicolor*. In *IUCN 2010: IUCN Red List of Threatened Species. Version 2010.4*. www.iucnredlist.org. Downloaded on 4 December 2010.

Aplin, K., Lunde, D., and Molur, S. (2008b). *Mus cookii*. In *IUCN 2010: IUCN Red List of Threatened Species. Version 2010.4*. www.iucnredlist.org. Downloaded on 4 December 2010.

Aplin, K., Molur, S., and Nameer, P. (2008c). *Mus booduga*. In *IUCN 2010: IUCN Red List of Threatened Species. Version 2010.4*. www.iucnredlist.org. Downloaded on 4 December 2010.

Auffray, J.-C. and Britton-Davidian, J. (1992). When did the house mouse colonize Europe? *Biological Journal of the Linnean Society*, **45**, 187–90.

Auffray, J.-C., Marshall, J. T., Thaler, L., and Bonhomme, F. (1990a). Focus on the nomenclature of European species of *Mus*. *Mouse Genome*, **88**, 7–8.

Auffray, J.-C., Orth, A., Catalan, J., *et al.* (2003). Phylogenetic position and description of a new species of mouse of subgenus *Mus* (Rodentia, Mammalia). *Zoologica Scripta*, **32**, 119–27.

Auffray, J.-C., Tchernov, E., Bonhomme, F., *et al.* (1990b). Presence and ecological distribution of *Mus 'spretoides'* and *Mus musculus domesticus* in Israel Circum-Mediterranean vicariance in the genus *Mus*. *Zeitschrift für Säugetierkunde*, **55**, 1–10.

Awasthi, M., Bhat, K. V., and Anand, R. K. (1999). Allozymic variation in four Indian species of *Mus*: a comparative analysis. *Journal of Genetics*, **78**, 73–80.

Badenhorst, D., Herbreteau, V., Chaval, Y., *et al.* (2009). New karyotypic data for Asian rodents (Rodentia, Muridae) with the first report of B-chromosomes in the genus *Mus*. *Journal of Zoology*, **279**, 44–56.

Balajee, A. S. and Sharma, T. (1994). The chromosomal distribution of *Mus musculus*-like AT-rich heterochromatin in the *M. dunni* complex as revealed by *Alu*I digestion on metaphase chromosomes. *Cytogenetics and Cell Genetics*, **66**, 89–92.

Bardhan, A. and Sharma, T. (2000a). Dynamic of nucleolus organizer regions and karyotype evolution in Indian pygmy field mice. *Cytogenetics and Cell Genetics*, **91**, 47–51.

Bardhan, A. and Sharma, T. (2000b). Meiosis and speciation: a study in a speciating *Mus terricolor* complex. *Journal of Genetics*, **79**, 105–11.

Bauer, K., Spitzenberger, F., and Unterholzner, K. (1998). The mound-building mouse (*Mus spicilegus*) is part of the Slovakian fauna. *Folia Zoologica*, **47**, 158–60.

Biddle, F. G., Eales, B. A., and Dean, W. L. (1994). Haldane's rule and heterogametic female and male sterility in the mouse. *Genome*, **37**, 198–202.

Bonhomme, F., Catalan, J., Britton-Davidian, J., *et al.* (1984). Biochemical diversity and evolution in the genus *Mus*. *Biochemical Genetics*, **22**, 275–303.

Bonhomme, F., Martin, S., and Thaler, L. (1978). Hybridation en laboratoire de *Mus musculus* L. et *Mus spretus* Lataste. *Experientia*, **34**, 1140–1.

Bonhomme, F., Orth, A., Cucchi, T., *et al.* (2004). Découverte d'une nouvelle espèce de souris sur l'île de Chypre. *Comptes Rendus Biologies*, **347**, 501–7.

Boursot, P., Auffray, J.-C., Britton-Davidian, J., and Bonhomme, F. (1993). The evolution of house mice. *Annual Review of Ecology and Systematics*, **24**, 119–52.

Boursot, P., Din, W., Anand, R., *et al.* (1996). Origin and radiation of the house mouse: mitochondrial DNA phylogeny. *Journal of Evolutionary Biology*, **9**, 391–415.

Boursot, P., Jacquart, T., Bonhomme, F., Britton-Davidian, J., and Thaler, L. (1985). Différenciation géographique du génome mitochondrial chez *Mus spretus* Lataste. *Comptes Rendus de l'Académie des Sciences – Series III – Sciences de la Vie*, **301**, 161–5.

Brannan, C. I., Disteche, C. M., Park, L. S., Copeland, N. G., and Jenkins, N. A. (2001). Autosomal telomere exchange results in the rapid amplification and dispersion and *Csf2ra* genes in wild-derived mice. *Mammalian Genome*, **12**, 882–6.

Britton, J., Pasteur, N., and Thaler, L. (1976). Les souris du Midi de la France: Caractérisation génétique de deux groupes de populations sympatriques. *Comptes Rendus de l'Académie des Sciences – Series III – Sciences de la Vie*, **283**, 515–18.

Britton-Davidian, J., Robinson, T. J., and Veyrunes, F. (2012). Systematics and evolution of the African pygmy mice, subgenus Nannomys: a review. *Acta Oecologica*. doi: 10.1016/j.actao.2012.01.001.

Britton-Davidian, J. and Thaler, L. (1978). Evidence for the presence of two sympatric species of mice (genus *Mus* L.) in southern France based on biochemical genetics. *Biochemical Genetics*, **16**, 13–25.

Bulatova, N. and Kotenkova E. (1990). Variants of the Y-chromosome in sympatric taxa of *Mus* in southern USSR. *Bolletino di Zoologia*, **57**, 357–60.

Burgio, G., Baylac, M., Heyer, E., and Montagutelli, X. (2009). Genetic analysis of skull shape variation and morphological integration in the mouse using interspecies recombinant congenic strains between C57BL/6 and mice of the *spretus* species. *Evolution*, **63**, 2668–86.

Cassaing, J., Cervera, S., and Isaac, F. (2010). Laboratory and field evidence of paternal care in the Algerian mouse (*Mus spretus*). *Journal of Ethology*, **28**, 7–13.

Cassaing, J. and Isaac, F. (2007). Pair bonding in the wild mouse *Mus spretus*: inference on the mating system. *Comptes Rendus de l'Académie des Sciences – Series III – Sciences de la Vie*, **330**, 828–36.

Cazaux, B., Catalan, J., Veyrunes, F., Douzery, E. J. P., and Britton-Davidian, J. (2011). Are ribosomal DNA clusters rearrangement hotspots? A case study in the genus *Mus* (Rodentia, Muridae). *BMC Evolutionary Biology*, **11**, 124.

Ceballos, G. and Ehrlich, P. R. (2009). Discoveries of new mammal species and their implications for conservation and ecosystem services. *Proceedings of the National Academy of Sciences of the United States of America*, **106**, 3841–6.

Chevret, P., Jenkins, P., and Catzeflis, F. (2003). Evolutionary systematics of the Indian mouse *Mus famulus* Bonhote, 1898: molecular (DNA/DNA hybridization and 12S rRNA sequence) and morphological evidence. *Zoological Journal of the Linnean Society*, **137**, 385–401.

Chevret, P., Veyrunes, F., and Britton-Davidian, J. (2005). Molecular phylogeny of the genus *Mus* (Rodentia: Murinae) based on mitochondrial and nuclear data. *Biological Journal of the Linnean Society*, **84**, 417–27.

Corbet, G. B. (1990). The relevance of metrical chromosomal and allozyme variation to the systematics of the genus *Mus*. *Biological Journal of the Linnean Society*, **41**, 5–12.

Corbet, G. B. and Hill, J. E. (1991). *A World List of Mammalian Species*. London and Oxford: Natural History Museum Publications and Oxford University Press.

Coroiu, I., Kryštufek, B., and Vohralík, V. (2008). *Mus spicilegus*. In *IUCN 2010: IUCN Red List of Threatened Species. Version 2010.4*. www.iucnredlist.org. Downloaded on 4 December 2010.

Cserkesz, T., Gubantyl, A., and Farkas, J. (2008). Distinguishing *Mus spicilegus* from *Mus musculus* (Rodentia, Muridae) by using cranial measurements. *Acta Zoologica Academiae Scientiarum Hungaricae*, **54**, 305–18.

Cucchi, T. (2008). Uluburun shipwreck stowaway house mouse: molar shape analysis and indirect clues about the vessel's last journey. *Journal of Archaeological Science*, **35**, 2953–9.

Cucchi, T., Orth, A., Auffray, J.-C., *et al.* (2006). A new endemic species of the subgenus *Mus* (Rodentia, Mammalia) on the Island of Cyprus. *Zootaxa*, **1241**, 1–36.

Cucchi, T., Vigne, J.-D., and Auffray, J.-C. (2005). First occurrence of the house mouse (*Mus musculus domesticus* Schwarz and Schwarz, 1943) in Western Mediterranean: a zooarchaeological revision of sub-fossil house mouse occurrences. *Biological Journal of the Linnean Society*, **84**, 429–45.

Cucchi, T., Vigne, J.-D., Auffray, J.-C., Croft, P., and Peltenburg, E. (2002). Introduction of the house mouse (*Mus musculus domesticus*) to Cyprus at the Early Preceramic Neolithic (late 9th and 8th millennia cal. BC). *Comptes Rendus Palevol*, **1**, 235–41.

Darviche, D. and Orsini, P. (1982). Morphological and biometrical characters of two sympatric species of mice – *Mus spretus* and *Mus musculus domesticus*. *Mammalia*, **46**, 205–17.

Dean, M. D. and Nachman, M. W. (2009). Faster fertilization rate in conspecific versus heterospecific matings in house mice. *Evolution*, **63**, 20–8.

Dejager, L., Libert, C. and Montagutelli, X. (2009). Thirty years of *Mus spretus*: a promising future. *Trends in Genetics*, **25**, 234–41.

Din, W., Anand, R., Boursot, P., *et al.* (1996). Origin and radiation of the house mouse: clues from nuclear genes. *Journal of Evolutionary Biology*, **9**, 519–39.

Dobson, F. S. and Baudoin, C. (2002). Experimental tests of spatial association and kinship in monogamous mice (*Mus spicilegus*) and polygynous mice (*Mus musculus domesticus*). *Canadian Journal of Zoology*, **80**, 980–6.

Duplantier, J.-M., Orth, A., Catalan, J., and Bonhomme, F. (2002). Evidence for mitochondrial lineage originating from the Arabian peninsula in the Madagascar house mouse (*Mus musculus*). *Heredity*, **89**, 154–8.

Duvaux, L., Belkhir, K., Boulesteix, M., and Boursot, P. (2011). Isolation and gene flow: inferring the speciation history of European house mice. *Molecular Ecology*, **20**, 5248–64.

Elliott, R. W., Miller, D. R., Pearsall, R. S., *et al.* (2001). Genetic analysis of testis weight and fertility in an interspecies hybrid congenic strain for chromosome X. *Mammalian Genome*, **12**, 45–51.

Eppig, J. T. (2010). Guidelines for nomenclature of mouse and rat strains (revised September 2010). International Committee on Standardized Genetic Nomenclature for Mice. http://www.informatics.jax.org/mgihome/nomen/strains.shtml

Fennessy, P. (1997). Interspecies hybrids: some thoughts. *Journal of Heredity*, **88**, 353–4.

Feron, C. and Gouat, P. (2007). Paternal care in the mound-building mouse reduces inter-litter intervals. *Reproduction Fertility and Development*, **19**, 425–9.

Frynta, D., Slábová, M., Váchová, H., Volfová, R., and Munclinger, P. (2005). Aggression and commensalism in house mouse: a comparative study across Europe and the Near East. *Aggressive Behavior*, **31**, 283–93.

Frynta, D., Slábová, M., and Vohralík, V. (2009). Why do male house mice have such small testes? *Zoological Science*, **26**, 17–23.

Garza, J. C., Dallas, J., Duryadi, D., *et al.* (1997). Social structure of the mound-building mouse *Mus spicilegus* revealed by genetic analysis with microsatellites. *Molecular Ecology*, **6**, 1009–17.

Geraldes, A., Basset, P., Gibson, B., *et al.* (2008). Inferring the history of speciation in house mice from autosomal, X-linked, Y-linked and mitochondrial genes. *Molecular Ecology*, **17**, 5349–63.

Gouat, P. and Feron, C. (2005). Deficit in reproduction in polygynously mated females of the monogamous mound-building mouse *Mus spicilegus*. *Reproduction Fertility and Development*, **17**, 617–23.

Gray, S. J. and Hurst, J. L. (1997). Behavioural mechanisms underlying the social dispersion of commensal *Mus domesticus* and grassland *Mus spretus*. *Animal Behaviour*, **53**, 511–24.

Gray, S. J., Hurst, J. L., Stidworthy, R., *et al.* (1998). Microhabitat and spatial dispersion of the grassland mouse (*Mus spretus* Lataste). *Journal of Zoology*, **246**, 299–308.

Greene-Till, R., Zhao, Y. and Hardies, S. C. (2000). Gene flow of unique sequences between *Mus musculus domesticus* and *Mus spretus*. *Mammalian Genome*, **11**, 225–30.

Guénet, J.-L. and Bonhomme, F. (2003). Wild mice: an ever-increasing contribution to a popular mammalian model. *Trends in Genetics*, **19**, 24–31.

Hale, D. W., Washburn, L. L., and Eicher, E. M. (1993). Meiotic abnormalities in hybrid mice of the C57BL/6J × *Mus spretus* cross suggest a cytogenetic basis for Haldane's rule of hybrid sterility. *Cytogenetics and Cell Genetics*, **63**, 221–34.

Hardies, S. C., Wang, L., Zhou, L., *et al.* (2000). LINE-1 (L1) lineages in the mouse. *Molecular Biology and Evolution*, **17**, 616–28.

Harrison, D. L. (1970). Some observations on the status of the South Arabian house mouse *Mus musculus gentilulus* Thomas, 1919 (Rodentia: Muridae). *Mammalia*, **34**, 244–8.

Hemberger, M., Kurz, H., Orth, A., *et al.* (2001). Genetic and developmental analysis of X-inactivation in interspecific hybrid mice suggests a role for the Y chromosome in placental dysplasia. *Genetics*, **157**, 341–8.

Hollyoake, M., Campbell, R. D., and Aguado, B. (2005). *NKp30* (*NCR3*) is a pseudogene in 12 inbred and wild mouse strains, but an expressed gene in *Mus caroli*. *Molecular Biology and Evolution*, **22**, 1661–72.

Holz, M., Hoi, H., Darolova, A., Kristokif, J., and Penn, D. J. (2009). Why do the mounds of *Mus spicilegus* vary so much in size and composition? *Mammalian Biology*, **74**, 308–14.
Hurst, J. L., Gray, S. J., Davey, P., *et al.* (1997). Social interaction alters attraction to competitor's odour in the mouse *Mus spretus* Lataste. *Animal Behaviour*, **54**, 941–53.
Hurst, J. L., Hall, S., Roberts, R. C., and Christian, C. (1996). Social organization in the aboriginal house mouse, *Mus spretus* Lataste: behavioural mechanisms underlying the spatial dispersion of competitors. *Animal Behaviour*, **51**, 327–44.
Hurst, J. L., Hayden, L., Kingston, M., Luck, R., and Sorensen, K. (1994). Response of the aboriginal house mouse *Mus spretus* Lataste to tunnels bearing the odors of conspecifics. *Animal Behaviour*, **48**, 1219–99.
Hurst, J. L. and Smith, J. (1995). *Mus spretus* Lataste: a hygienic mouse? *Animal Behaviour*, **49**, 827–34.
Ilmonen, P., Penn, D. J., Damjanovich, K., *et al.* (2007). Major histocompatibility complex heterozygosity reduces fitness in experimentally infected mice. *Genetics*, **176**, 2501–8.
Izhaki, I., Haim, A., and Zohar, O. (1993). Rodent populations recovering from fire in an East Mediterranean woodland. *Water Science and Technology*, **27**, 539–45.
Jones, E. P., van der Kooij, J., Solheims, R., and Searle, J. B. (2010). Norwegian house mice (*Mus musculus musculus/domesticus*): distributions, routes of colonization and patterns of hybridization. *Molecular Ecology*, **19**, 5252–64.
Jotterand, M. (1975). The African *Mus* (pygmy-mice): the role of chromosomal polymorphism in speciation. *Caryologia*, **28**, 335–44.
Jotterand-Bellomo, M. (1988). Chromosome analysis of five specimens of *Mus bufo-triton* (Muridae) from Burundi (Africa): three cytogenetic entities, a special type of chromosomal sex determination, taxonomy, and phylogeny. *Cytogenetics and Cell Genetics*, **48**, 88–91.
Kawakami, M. and Yamamura, K. I. (2008). Cranial bone morphometric study among mouse strains. *BMC Evolutionary Biology*, **8**, 73.
Khammes, N., Lek, S., and Aulagnier, S. (2006). Identification biométrique de deux espèces sympatriques de souris *Mus musculus domesticus* et *Mus spretus* en Kabylie du Djurdjura (Algérie). *Revue Suisse de Zoologie*, **113**, 411–19.
Kryštufek, B. and Macholán, M. (1998). Morphological differentiation in *Mus spicilegus* and the taxonomic status of mound-building mice from the Adriatic coast of Yugoslavia. *Journal of Zoology*, **245**, 185–96.
Kryštufek, B. and Vohralík, V. (2008). *Mus macedonicus*. In *IUCN 2010: IUCN Red List of Threatened Species. Version 2010.4*. www.iucnredlist.org. Downloaded on 4 December 2010.
Kurzweil, V. C., Getman, M., Green, E. D., and Lane, R. P. (2009). Dynamic evolution of VIR putative pheromone receptors between *Mus musculus* and *Mus spretus*. *BMC Genomics*, **10**, 74.
L'Hote, D., Serres, C., Laissue, P., *et al.* (2007). Centimorgan-range one-step mapping of fertility traits using interspecific recombinant congenic mice. *Genetics*, **176**, 1907–21.
Laissue, P., L'Hote, D., Serres, C., and Vaiman, D. (2009). Mouse models for identifying genes modulating fertility parameters. *Animal*, **3**, 55–71.
Laurie, C. C., Nickerson, D. A., Anderson, A. D., *et al.* (2007). Linkage disequilibrium in wild mice. *PloS Genetics*, **3**, e144.
Macholán, M. (1996a). Key to European house mice (*Mus*). *Folia Zoologica*, **45**, 209–17.

Macholán, M. (1996b). Multivariate morphometric analysis of European species of the genus *Mus* (Mammalia, Muridae). *Zeitschrift für Säugetierkunde*, **61**, 304–19.
Macholán, M. (2001). Multivariate analysis of morphometric variation in Asian *Mus* and sub-Saharan *Nannomys* (Rodentia: Muridae). *Zoologischer Anzeiger*, **240**, 7–14.
Macholán, M. (2006). A geometric morphometric analysis of the shape of the first upper molar in mice of the genus *Mus* (Muridae, Rodentia). *Journal of Zoology*, **270**, 672–81.
Macholán, M., Baird, S. J. E., Dufková, P., *et al.* (2011). Assessing multilocus introgression patterns: a case study on the mouse X chromosome in Central Europe. *Evolution*, **65**, 1428–46.
Macholán, M., Baird, S. J. E., Munclinger, P., *et al.* (2008a). Genetic conflict outweighs heterogametic incompatibility in the mouse hybrid zone? *BMC Evolutionary Biology*, **8**: 271.
Macholán, M., Mikula, O., and Vohralík, V. (2008b). Geographic phenetic variation of two eastern-Mediterranean non-commensal mouse species, *Mus macedonicus* and *M. cypriacus* (Rodentia: Muridae) based on traditional and geometric approaches to morphometrics. *Zoologischer Anzeiger*, **247**, 67–80.
Macholán, M., Munclinger, P., Šugerková, M., *et al.* (2007a). Genetic analysis of autosomal and X-linked markers across a mouse hybrid zone. *Evolution*, **61**, 746–71.
Macholán, M. and Vohralík, V. (1997). Note on the distribution of *Mus spicilegus* (Mammalia: Rodentia) in the south-western Balkans. *Acta Societatis Zoologicae Bohemicae*, **61**, 219–26.
Macholán, M., Vyskočilová, M., Bonhomme, F., *et al.* (2007b). Genetic variation and phylogeography of free-living mouse species (genus *Mus*) in the Balkans and the Middle East. *Molecular Ecology*, **16**, 4774–88.
Marshall, J. T. (1977). A synopsis of Asian species of *Mus* (Rodentia, Muridae). *Bulletin of the American Museum of Natural History*, **158**, 173–220.
Marshall, J. T. and Sage, R. D. (1981). The taxonomy of the house mouse. In *The Biology of the House Mouse*, ed. R. J. Berry. London: Academic Press, pp. 15–26.
Matthey, R. (1966). Le polymorphisme chromosomique des *Mus* africains du sous-genre *Leggada*: révision générale portant sur l'analyse de 213 individus. *Revue Suisse de Zoologie*, **73**, 585–607.
Matthey, R. and Petter, F. (1968). Existence de deux espèces distinctes, l'une chromosomiquement polymorphe, chez des *Mus* indiens du groupe *booduga*: etude cytogénétique et taxonomique. *Revue Suisse de Zoologie*, **75**, 461–98.
Mitsainas, G. P., Tryfonopoulos, G. A., Thanou, E. G., *et al.* (2009). New data on the distribution of *Mus spicilegus* Petényi, 1882 (Rodentia, Muridae) and a distinct mtDNA lineage in the southern Balkans. *Mammalian Biology*, **74**, 351–60.
Moriwaki, K., Yonekawa, W. A., Minezawa, M., *et al.* (1981). Genetic status of Japanese wild mouse, *Mus musculus molossinus*. *Hereditas*, **94**, 15.
Motokawa, M., Lin, L.-K., and Motokawa, J. (2003). Morphological comparison of Ryukyu mouse *Mus caroli* (Rodentia: Muridae) populations from Okinawajima and Taiwan. *Zoological Studies*, **42**, 258–67.
Musser, G. G. and Carleton, M. D. (1993). Family Muridae. In *Mammal Species of the World: A Taxonomic and Geographic Reference*, 2nd edn, ed. D. E. Wilson and D. M. Reeder. Washington, DC: Smithsonian Institution Press, pp. 501–755.

Musser, G. G. and Carleton, M. D. (2005). Family Muridae. In *Mammal Species of the World: A Taxonomic and Geographic Reference*, 3rd edn, ed. D. E. Wilson and D. M. Reeder. Washington, DC: Smithsonian Institution Press, pp. 894–1531.

Oka, A., Aoto, T., Totsuka, Y., *et al.* (2007). Disruption of genetic interaction between two autosomal regions and the X chromosome causes reproductive isolation between mouse strains derived from different subspecies. *Genetics*, **175**, 185–97.

Orsini, P., Bonhomme, F., Britton-Davidian, J., *et al.* (1983). Le complexe d'espèces du genre *Mus* en Europe centrale et orientale: II. Critères d'identification, répartition et caractéristiques écologiques. *Zeitschrift für Säugetierkunde*, **48**, 86–95.

Orsini, P., Cassaing, J., Duplantier, J.-M., and Croset, H. (1982). Preliminary observations on the natural populations of *Mus spretus* and *Mus musculus domesticus* in the South of France. *Revue d'Ecologie: La Terre et la Vie*, **36**, 321–36.

Orth, A., Auffray, J.-C., and Bonhomme, F. (2002a). Two deeply divergent mitochondrial clades in the wild mouse *Mus macedonicus* reveal multiple glacial refuges south of Caucasus. *Heredity*, **89**, 353–7.

Orth, A., Belkhir, K., Britton-Davidian, J., *et al.* (2002b). Natural hybridization between two sympatric species of mice *Mus musculus domesticus* L. and *Mus spretus* Lataste. *Comptes Rendus de l'Académie des Sciences – Series III – Sciences de la Vie*, **325**: 89–97.

Palmer, S., Perry, J., and Ashworth, A. (1995). A contravention of Ohno's law in mice. *Nature Genetics*, **10**, 472–6.

Patris, B. and Baudoin, C. (2000). A comparative study of parental care between two rodent species: implications for the mating system of the mound-building mouse *Mus spicilegus*. *Behavioral Processes*, **51**, 35–43.

Payseur, B. A. and Place, M. (2007). Searching the genomes of inbred mouse strains for incompatibilites that reproductively isolate their wild relatives. *Journal of Heredity*, **98**, 115–22.

Perry, J. and Ashworth, A. (1999). Evolutionary rate of a gene affected by chromosomal position. *Current Biology*, **9**, 987–9.

Peters, L. L., Robledo, R. F., Bult, C. J., *et al.* (2007). The mouse as a model for human biology: a resource guide for complex trait analysis. *Genetics*, **8**, 58–69.

Piálek, J., Vyskočilová, M., Bímová, B., *et al.* (2008). Development of unique house mouse resources suitable for evolutionary studies of speciation. *Journal of Heredity*, **99**, 34–44.

Pilder, S., Olds-Clarke, H. P., Orth, J. M., Jester, W. E., and Dugan, L. C. (1997). A male sterility mutation perturbing sperm motility, flagellar assembly, and mitochondrial sheath differentiation. *Journal of Andrology*, **18**, 663–71.

Poteaux, C., Busquet, N., Gouat, P., Katona, K., and Baudoin, C. (2008). Socio-genetic structure of mound-building mice, *Mus spicilegus*, in autumn and early spring. *Biological Journal of the Linnean Society*, **93**, 689–99.

Pouliquen-Young, O. (1994). Thermoregulatory behavior of outdoor and commensal populations of mice (*Mus musculus domesticus* and *Mus spretus*) in response to cold. *Journal of Thermal Biology*, **19**, 213–17.

Pradhan, M. S., Molur, S. and Nameer, P. O. (2008). *Mus famulus*. In *IUCN 2010: IUCN Red List of Threatened Species. Version 2010.4*. www.iucnredlist.org. Downloaded on 4 December 2010.

Prager, E. M., Orrego, C., and Sage, R. D. (1998). Genetic variation and phylogeography of central Asian and other house mice, including a major new mitochondrial lineage in Yemen. *Genetics*, **150**, 835–61.

Prassolov, V., Hein, S., Ziegler, M., *et al.* (2001). *Mus cervicolor* murine leukemia virus isolate M813 belongs to a unique receptor interference group. *Journal of Virology*, **75**, 4490–8.

Sage, R. D., Atchley, W. R., and Capanna, E. (1993). House mice as models in systematic biology. *Systematic Biology*, **42**, 523–61.

Salcedo, T., Geraldes, A., and Nachman, M. W. (2007). Nucleotide variation in wild and inbred mice. *Genetics*, **177**, 2277–91.

Schwarz, E. and Schwarz, H. (1943). The wild and commensal stocks of the house mouse *Mus musculus* Linnaeus. *Journal of Mammalogy*, **24**, 59–72.

Selander, R. K., Hunt, W. G., and Yang, S. H. (1969). Protein polymorphism and genic heterozygosity in two European subspecies of the house mouse. *Evolution*, **23**, 379–90.

Sen, S. and Sharma, T. (1983). Role of constitutive heterochromatin in evolutionary divergence: results of chromosome banding and condensation inhibition studies in *Mus musculus*, *Mus booduga* and *Mus dunni*. *Evolution*, **37**, 628–36.

Shabtay, A., Haim, A., and Arad, Z. (2000). Metabolic rate and thermoregulation in the Macedonian mouse *Mus macedonicus*. *Israel Journal of Zoology*, **46**, 305–12.

Shanker, K. (2001). The role of competition and habitat in structuring small mammal communities in a tropical montane ecosystem in southern India. *Journal of Zoology*, **253**, 15–24.

Sharma, G. G. and Sharma, T. (1998). Unusual chromosomal organization of telomeric sequences and expeditious karyotypic differentiation in the recently evolved *Mus terricolor* complex. *Cytogenetics and Cell Genetics*, **80**, 204–8.

Sharma, T. (1996). Chromosomal and molecular divergence in the Indian pygmy field mice *Mus booduga-terricolor* lineage of the subgenus *Mus*. *Genetica*, **97**, 331–8.

Sharma, T., Bardhan, A., and Bahadur, M. (2002). Increases and decreases of whole-arm heterochromatin in specific chromosomes: an extraordinary situation in hybrids of the *Mus terricolor* complex. *Cytogenetic and Genome Research*, **96**, 244–9.

Sharma, T., Bardhan, A., and Bahadur, M. (2003). Reduced meiotic fitness in hybrids with heterozygosity for heterochromatin in the speciating *Mus terricolor* complex. *Journal of Biosciences*, **28**, 189–98.

Sharma, T., Cheong, N., Sen, P., and Sen, S. (1986). Constitutive heterochromatin and evolutionary divergence of *Mus dunni*, *M. booduga* and *M. musculus*. *Current Topics in Microbiology and Immunology*, **127**, 35–44.

She, J. X., Bonhomme, F., Boursot, P., Thaler, L., and Catzeflis, F. (1990). Molecular phylogenies in the genus *Mus*: comparative analysis of electrophoretic, scnDNA hybridization, and mtDNA RFLP data. *Biological Journal of the Linnean Society*, **41**, 83–103.

Shi, W., Lefevre, L., Yu, Y., *et al.* (2004). Loss-of-imprinting of *Peg1* in mouse interspecies hybrids is correlated with altered growth. *Genesis*, **39**, 65–72.

Shimada, T., Aplin, K., Jenkins, P., and Suzuki, H. (2007). Rediscovery of *Mus nitidulus* Blyth (Rodentia: Muridae), an endemic murine rodent of the central basin of Myanmar. *Zootaxa*, **1498**, 45–68.

Shimada, T., Aplin, K. P., and Suzuki, H. (2010). *Mus lepidoides* (Muridae, Rodentia) of central Burma is a distinct species of potentially great evolutionary and biogeographic significance. *Zoological Science*, **27**, 449–59.

Sicard, B., Navajas y Navarro, M., Jacquart, T., Lachiver, F., and Croset, H. (1985). Métabolisme hydrique de populations de *Mus musculus domesticus* Rutty et de *Mus spretus* Lataste soumises à divers régimes hydriques. *Comptes Rendus de l'Académie des Sciences – Series III – Sciences de la Vie*, **300**: 699–704.

Singh, S. and Sharma, T. (1997). High levels of genetic variation in Indian field and house mice. *Journal of Genetics*, **76**, 189–200.

Suzuki, H., Shimada, T., Terashima, M., Tsuchiya, K., and Aplin, K. (2004). Temporal, spatial, and ecological modes of evolution of Eurasian *Mus* based on mitochondrial and nuclear gene sequences. *Molecular Phylogenetics and Evolution*, **33**, 626–46.

Takada, T., Mita, A., Maeno, A., *et al.* (2008). Mouse inter-subspecific consomic strains for genetic dissection of quantitative complex traits. *Genome Research*, **18**, 500–8.

Tarlington, R., Meers, J., and Young, P. (2008). Biology and evolution of the endogenous koala retrovirus. *Cell Science*, **65**, 3413–21.

Tucker, P. K. (2007). Systematics of the genus *Mus*. In *The Mouse in Biomedical Research, vol. 1: History, Wild Mice, and Genetics*, ed. J. G. Fox, S. W. Barthold, and M. T. Davisson. San Diego, CA: Academic Press/Elsevier, pp. 13–23.

Tucker, P. K., Sandstedt, S. A., and Lundrigan, B. L. (2005). Phylogenetic relationships in the subgenus *Mus* (genus *Mus*, family Muridae, subfamily Murinae): examining gene trees and species trees. *Biological Journal of the Linnean Society*, **84**, 653–62.

Veyrunes, F., Catalan, J., Sicard B., *et al.* (2004). Autosome and sex chromosome diversity among the African pygmy mice, subgenus *Nannomys* (Murinae; *Mus*). *Chromosome Research*, **12**, 369–82.

Veyrunes, F., Catalan, J., Tatard C., *et al.* (2010a). Mitochondrial and chromosomal insights into karyotypic evolution of the pygmy mouse, *Mus minutoides*, in South Africa. *Chromosome Research*, **18**, 563–74.

Veyrunes, F., Chevret, P., Catalan, J., *et al.* (2010b). A novel sex determination system in a close relative of the house mouse. *Proceedings of the Royal Society London B: Biological Sciences*, **277**, 1049–56.

Veyrunes, F., Dobigny, G., Yang, F., *et al.* (2006). Phylogenomics of the genus *Mus* (Rodentia; Muridae): extensive genome repatterning is not restricted to the house mouse. *Proceedings of the Royal Society London B: Biological Sciences*, **273**, 2925–34.

Zechner, U., Hemberger, M., Constancia, M., *et al.* (2002). Proliferation and growth factor expression in abnormally enlarged placentas of mouse interspecific hybrids. *Developmental Dynamics*, **224**: 125–34.

2

Phylogeny and biogeography of the genus *Mus* in Eurasia

HITOSHI SUZUKI AND KEN P. APLIN

Introduction

Mus is a moderately speciose genus of murine rodents with an exceptionally widespread geographic range. Indeed, of the 129 extant genera in the subfamily Murinae, *Mus* alone has a native range that spans not only the full breadth of the Eurasian continent but also the entirety of the African continent (Fig. 2.1). On the present count, there are 41 species recognized within the genus *Mus*, with at least one more awaiting formal description. Ecological diversity is also high, at least in broad terms, with species of *Mus* present in most major biomes from tropical forest to dry steppe and semi-desert. Yet despite this ecological breadth, the ecomorphology of *Mus* species is rather conservative – all species are relatively small (approximate body size range is 5–35 g), all appear to be essentially terrestrial, and all probably feed on grass, seeds, vegetative matter, and insects, in varying proportions.

Although *Mus* obviously presents several features of evolutionary and biogeographical interest, much of the research effort is focused on one particular species – the house mouse, *Mus musculus* (*sensu lato*). First domesticated as 'fancy mice' in Japan (Moriwaki, 1994), this species was adopted into European culture, first as a domestic pet, then as a convenient laboratory animal, and finally as the preeminent 'model organism', with a fundamental and ever-expanding role in biomedicine and biotechnology. However, the recent process of domestication of house mice is only the culmination of a much longer history of close association with people, under the general banner of 'commensalism'. Exactly when and where this developed has long been debated (see Cucchi *et al.*, Chapter 3 in this volume), but it is reasonable to expect that the relationship developed soon after the advent of more settled forms of human life, with the attendant attractions (for mice) of food production and storage, and the construction of semi-permanent artificial shelter. Commensalism also led to co-dispersal, with mice travelling as accidental stowaways on vehicles of all kinds – including in the modern age

Evolution of the House Mouse, ed. Miloš Macholán, Stuart J. E. Baird, Pavel Munclinger, and Jaroslav Piálek. Published by Cambridge University Press.

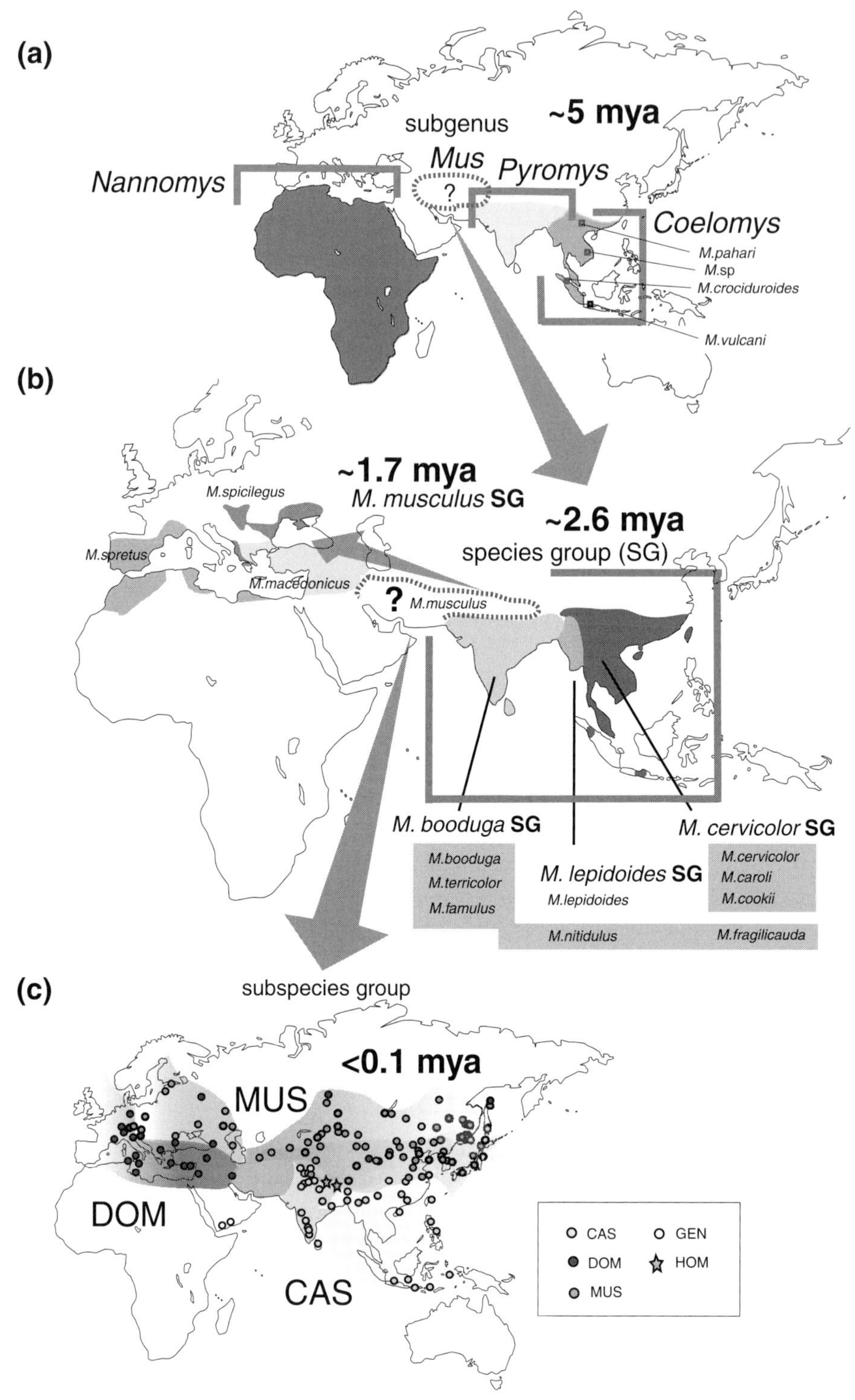

Figure 2.1 Generalized distributions of the Eurasian species of *Mus* (Corbet and Hill, 1992; Mitsainas *et al.*, 2009; IUCN, 2010; M. Macholán, personal communication),

ocean-going ships and aircraft. All of these factors have resulted in significant genetic admixture, so that modern laboratory mouse strains typically contain genetic material originating from multiple original stocks. Unravelling these components has been the impetus for much of the genetic investigation of mice over the last 50 years, with recent attention also turning to wider patterns of genetic diversity in the genus *Mus* as a context for comparative research including both genomics and phenomics (Gondo, 2008).

With these issues in mind, we here review the phylogeny and biogeography of the Eurasian members of the genus *Mus*.

Origins and early diversification of the genus *Mus*

Mus is the type genus of the family Muridae, a group that holds multiple distinctions of its own – the most speciose (by far) mammalian family, with more than 600 species; the only non-volant family to occur naturally on all continents except Antarctica; and the source of almost all of the world's major household and agricultural pests. Within this vast and complex family, *Mus* has its phylogenetic roots within the subfamily Murinae, a group of more than 590 species that occurs naturally on all continents of the Old World, including Australia. As mentioned above, the natural geographical range of the genus *Mus* includes both Eurasia and Africa, while its naturalized range now includes almost every inhabited corner of the Earth (Fig. 2.1c).

Figure 2.1 (*cont.*)
exhibiting the subgenus (a), species groups of the subgenus *Mus* (b), and subspecies groups of *M. musculus* (c). Three of the four subgenera of *Coelomys*, *Pyromys*, and *Nannomys* are based in Southeast Asia, India, and Africa, respectively. The homeland of the remaining subgenus *Mus* is possibly somewhere in the south-central Asian and Indian subcontinent. The four SGs of the subgenus *Mus* are considered to have come from the subtropical region of Eurasia, generating four geographic lineages of India, Myanmar, Southeast Asia, and presumably northern India, respectively, at 2.6 Mya. Subsequently, the northern Indian lineage, adapting to temperate habitat, would have extended the broad ranges from northern India to southern Europe at 1.7 Mya, resulting in three allopatric lineages of *M. musculus*, *M. spicilegus*/*M. macedonicus*, and *M. spretus*. The three lineages that have expanded their ranges throughout Eurasia during the last 100 000 years are those represented by *M. m. castaneus* (CAS), *M. m. domesticus* (DOM), *M. m. musculus* (MUS), and the two locally confined lineages are those of *M. m. gentilulus* (GEN) and *M. m. homoulus* (HOM), which are confined to Yemen (Prager *et al.*, 1998) and Nepal (Terashima *et al.*, 2006), respectively. Geographic distribution of mtDNA haplotypes are shown (Suzuki *et al.*, unpublished). See plate section for a colour version of this figure.

The closest relative of Murinae are members of Gerbillinae and Deomyinae (Michaux *et al.*, 2001; Steppan *et al.*, 2004), with the common ancestry of all three lineages estimated from molecular evidence at around 22 million years ago (Mya) (Steppan *et al.*, 2004). Today, deomyine rodents are restricted to Africa, while living members of Gerbillinae range from Africa to the Indian subcontinent. However, fossil taxa that might be ancestral to one or more of these subfamilies (sometimes called 'primitive dendromurids' (e.g. Flynn *et al.*, 1985)) occur in the late Early Miocene (*ca.* 18 Mya) sediments across a larger geographic range, including Southeast Asia and Europe (López Antoñanzas, 2009). The earliest undisputed fossil murine, *Antemus chinjiensis* from the Siwalik Hills region of northwest India and Pakistan, is significantly younger, with a time range of 13.2–14.3 Mya (Jacobs and Flynn, 2005). However, there is growing evidence to identify members of the genus *Potwarmus* with a stratigraphic range of 18.5–13 Mya, as primitive members of Murinae (López Antoñanzas, 2009), thereby neatly filling the temporal gap.

The Late Miocene was a period of dramatic environmental change, caused in part by global cooling (van Dam *et al.*, 2006), but also by tectonic uplift of the Tibeto-Himalayan massif (Raymo *et al.*, 1988), with intensification of the highly seasonal monsoon climate system in South Asia. These changes triggered an expansion of open habitats including grasslands, which in turn facilitated evolutionary radiation in many groups of organisms (Barry *et al.*, 1985). Among murine rodents, there was considerable diversification across the Asian region, as evidenced both in the fossil record (Barry *et al.*, 1985; Qiu and Li, 2003) and from molecular phylogenies that date the origin of all but one of nine extant tribes to between 12.3 and 10.2 Mya (Rowe *et al.*, 2008; Lecompte *et al.*, 2008). However, local diversification in South Asia was perhaps less important in the long term than geographic range expansion, which is marked by the sudden appearance between 11 and 10 Mya of members of the genus *Progonomys* in the fossil records of China (Qiu and Li, 2003), western Europe (van Dam *et al.*, 2006), and northern Africa (Bernor *et al.*, 1988). With dispersal came further diversification, and by the end of the Miocene, regionally distinctive fossil murine faunas were well established in Europe and Africa (Mein *et al.*, 1993; Winkler, 1992, 2002).

Mus is one of the most isolated of all living murine genera, with its roots set deep in the earliest phase of murine diversification. Lecompte *et al.* (2008) placed *Mus* in its own tribe – the Murini – and together with Rowe *et al.* (2008) observed an ancient affinity between this lineage and the exclusively African tribe Praomyini, with an estimated time of divergence around 10.2 Mya (Lecompte *et al.*, 2008). One taxon that may be phylogenetically closer to *Mus* is *Muriculus imberbis*, the Ethiopian striped mouse. However, this poorly known species has not been examined using molecular methods and shows ambiguous morphological characters (Osgood, 1936); it is probably best treated as *incertae sedis* within Murinae, pending further assessment (Lecompte *et al.*, 2008).

The known fossil record of *Mus*-like murines is consistent with these estimates. Although the earliest definite fossil representative of the *Mus* lineage (*Mus auctor* of the Siwalik sequence; Jacobs, 1978) is dated to *ca.* 6 Mya, each of '*Progonomys*' *debruijni* from *ca.* 8 Mya sediments in the Siwaliks (Jacobs and Pilbeam, 1980) and an as yet unnamed taxon from a *ca.* 9.7 Mya locality in eastern Europe (Mein *et al.*, 1993) are said to have morphological features that presage the distinctive features of *Mus*. These taxa are among the smallest-toothed of currently known early Late Miocene murines, thus it seems likely that small body size evolved early as a characteristic of the *Mus* lineage.

Extant species of *Mus* are divided into four subgenera: *Mus*, *Pyromys*, *Coelomys*, and *Nannomys*. These were established by Marshall (1977) on craniodental criteria but subsequent molecular work largely confirmed each of the subgenera as phylogenetic entities (e.g. Lundrigan *et al.*, 2002). One exception concerned *M. famulus* of western India, which was taken out of *Coelomys* and placed in the subgenus *Mus* based on a range of molecular and morphological evidences (Chevret *et al.*, 2003). One strong biogeographic distinction at this level is between the African subgenus *Nannomys* and the other three subgenera, which are exclusively Eurasian (Fig. 2.1b).

Phylogenetic relationships of the subgenera of *Mus* proved elusive to sequence analysis of single genes and even multiple nuclear genes (Lundrigan *et al.*, 2002; Suzuki *et al.*, 2004; Chevret *et al.*, 2005; Tucker *et al.*, 2005; Veyrunes *et al.*, 2005; Shimada *et al.*, 2007a, 2010). Recently, a convincing solution was obtained through the application of chromosomal painting methods. Using other murine genera as outgroups and a strict cladistic methodology, Veyrunes *et al.* (2006) were able to reconstruct an ancestral karyotype for *Mus* and to infer the history of inferred chromosomal rearrangements. These results unambiguously show *Coelomys* to be the earliest branch of the *Mus* radiation, with a subsequent divergence of *Nannomys* from *Pyromys* and *Mus*. A further inference from this result is that the genus *Mus* underwent some initial lineage differentiation in Eurasia prior to dispersal of the ancestor of the *Nannomys* lineage to Africa. Once again, the molecular and fossil evidence are mutually supportive of this scenario – molecular evidence suggests that diversification amongst subgenera of *Mus* commenced around 6.6 ± 0.7 Mya (Lecompte *et al.*, 2008), whereas the earliest fossil *Mus* from Africa comes from a Kenyan locality dated to *ca.* 4.5 Mya (Winkler, 2002), compared with the *ca.* 6 Mya record of *Mus* in South Asia.

The diversity and phylogeny of Eurasian *Mus* species

In 1993, Musser and Carleton listed 23 species of *Mus* from Eurasia, largely following the revision of Marshall (1977). However, more recent fieldwork in several parts of Eurasia exposed several other well-differentiated species. *Mus fragilicauda* was first detected in northeastern Thailand (Auffray *et al.*, 2003) and reported soon thereafter from Laos (Suzuki *et al.*, 2004). More recently, *M. nitidulus* and *M. lepidoides*, two Myanmar endemics, were resurrected from synonymy of *M. cervicolor* and *M. booduga*, respectively (Shimada *et al.*, 2007a, 2010; see also Auffray and Britton-Davidian, Chapter 1 in this volume). In the Mediterranean region, *M. cypriacus* from Cyprus was recently distinguished from *M. macedonicus*, based on combined evidence from morphology and mtDNA sequences (Bonhomme *et al.*, 2004; Cucchi *et al.*, 2006; Macholán *et al.*, 2007). Finally, fieldwork in Vietnam and Laos by K. P. Aplin, D. Lunde and G. Musser resulted in collection of another entirely new species related to *M. pahari*. This discovery prompted a wider examination of *M. pahari* across its range, the results of which suggest the presence of at least two distinct species in this taxon. From this ongoing discovery of new taxa it is clear that the inventory of wild *Mus* species remains incomplete, with more work required both in the field and in existing museum collections. The endemic Indian species of *Mus* are particularly understudied, with almost no modern genetic investigations or critical morphological assessment since that of Ellerman and Morrison-Scott (1951).

Analysis of phylogenetic relationships within the subgenera has progressed for each of African *Nannomys* (Veyrunes *et al.*, 2005) and the Eurasian subgenus *Mus*. Within the latter group, multiple genetic appraisals using a range of markers consistently indicate three clusters of species (Lundrigan *et al.*, 2002; Chevret *et al.*, 2003, 2005; Suzuki *et al.*, 2004; Veyrunes *et al.*, 2005). Suzuki *et al.* (2004) formalized these evolutionary lineages as species groups (a useful classificatory rank below the subgenus; henceforth SG), namely: (1) the *M. cervicolor* SG for the Southeast Asian species *M. cervicolor*, *M. caroli*, and *M. cookii*; (2) the *M. musculus* SG for the Palaearctic species *M. musculus* (with its subspecies including *domesticus*, *castaneus*, etc.), *M. spretus*, *M. spicilegus*, *M. macedonicus*, and *M. cypriacus*; and (3) the *M. booduga* SG for the Indian species *M. booduga*, *M. terricolor*, and *M. famulus*, and the Southeast Asian species *M. fragilicauda*. Subsequent work showed that *M. nitidulus* from Myanmar belonged to the *M. booduga* SG (Shimada *et al.*, 2007a), but that *M. lepidoides*, also from Myanmar, was not only very distinct from *M. booduga*, but actually fell outside of all the recognized species groups (Shimada *et al.*, 2010), thereby warranting recognition of a fourth unit, the *M. lepidoides* SG (see also Auffray and Britton-Davidian, Chapter 1).

Relationships among the species groups of the subgenus *Mus* are not yet well resolved by available datasets. Incompatibilities among individual gene trees reported by Suzuki *et al.* (2004) and Shimada *et al.* (2010) may be due to random lineage sorting in ancestral lineages or to introgression, possibly in ancient times. To illustrate this point, phylogenetic inference with *Rag1* gene sequences results in strong support for inclusion of *M. caroli* within the *M. cervicolor* SG, whereas in *Irbp* gene trees *M. caroli* is grouped instead with the *M. booduga* SG (Suzuki *et al.*, 2004). In a concatenated dataset *M. caroli* stands apart from both groups (Fig. 2.2). Similarly, with respect to mtDNA, *M. terricolor* is rather independent of the remaining members of the *M. booduga* SG (Fig. 2.3), whereas the nuclear gene trees suggest a closer affinity with this group. Shimada *et al.* (2010) argue from combined molecular and morphological evidence that the newly rediscovered *M. lepidoides* may be the sister taxon of the *M. musculus* SG, but this intriguing possibility requires further testing using wider genomic sampling.

Much less work has been undertaken on the phylogeny of the other Asian subgenera, *Coelomys* and *Pyromys*. In large part, this reflects lack of access to appropriately preserved samples, with some species such as *M.* (*Coelomys*) *mayori* and *M.* (*Pyromys*) *phillipsi* from Sri Lanka and *M.* (*Pyromys*) *shortridgei* from Southeast Asia being known mainly from historical collections. Similarly, an Indonesian mouse, *M.* (*Coelomys*) *vulcani*, an inhabitant of high-altitude forests of Gunug Gede Pangrango in West Java, has not been trapped for several decades, despite several attempts. As mentioned earlier, the genus *Coelomys* probably also contains cryptic species currently hidden within *M. pahari*, while a distinctive new species of this group from Laos and Vietnam lies equidistant from each of Chinese *M. pahari* and *M. crociduroides* from Sumatra (Fig. 2.1a and 2.3). Exploration of phylogenetic affinities within each of *Coelomys* and *Pyromys* will first require clarification of species boundaries.

Phylogeographic patterns among Eurasian *Mus*

Strong phylogeographic signals are present at multiple levels within the genus *Mus*. With the exception of the subgenus *Mus*, which occurs across all of Eurasia, the geographic ranges of the remaining Eurasian subgenera are strongly differentiated – *Coelomys* is most diverse in Southeast Asia but has outliers in northeastern India (*M. pahari* group) and Sri Lanka (*M. mayori*), whereas *Pyromys* is most diverse in India but has one outlier in central Myanmar and Thailand (*M. shortridgei*). This pattern has been interpreted as evidence for allopatric divergence following an early phase of dispersal (Suzuki *et al.*, 2004). Following this logic, it might then be reasonable to suggest a homeland area for the subgenus *Mus* as an area where the other three subgenera are absent – possibly

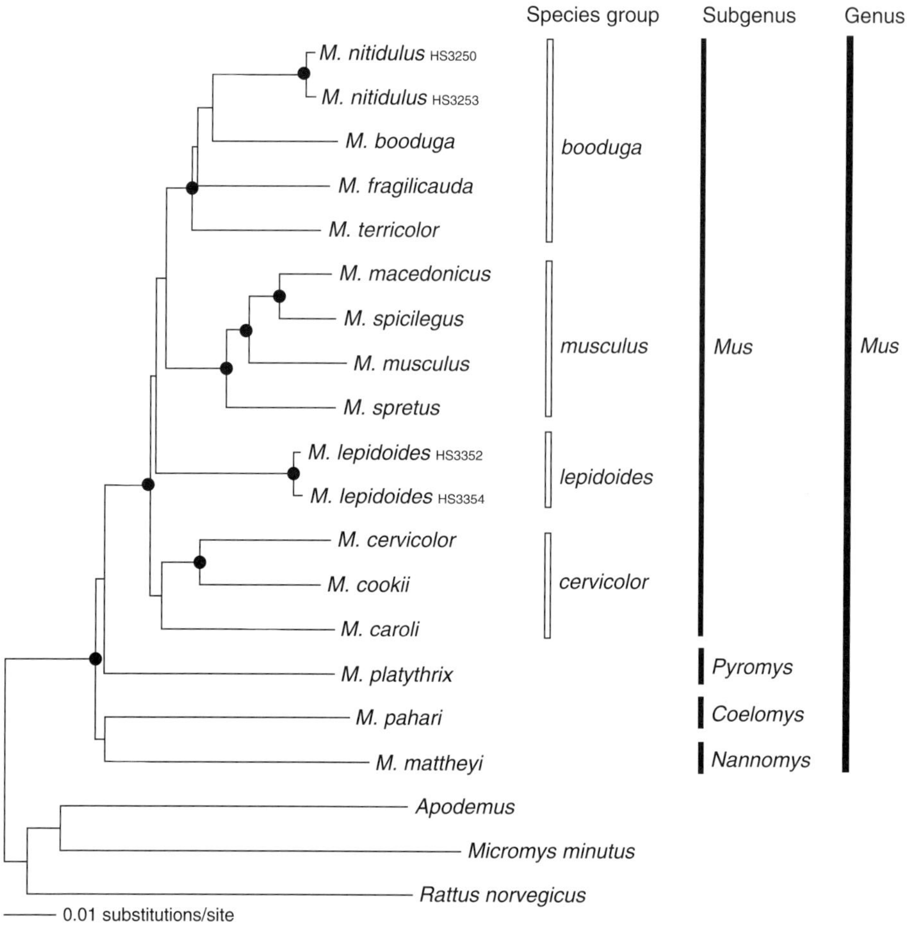

Figure 2.2 A maximum likelihood strict consensus tree based on concatenated sequences of *Cytb* (transversions only), *Irbp*, and *Rag1* (Shimada *et al.*, 2010). Total sequence length is 3294 bp, of which 1140 bp are *Cytb*, 1152 bp are *Irbp*, and 1002 bp are *Rag1*. Nodes denoted with relatively high bootstrap values (>80%) for neighbour-joining (NJ; 1000 pseudoreplicates), maximum parsimony (MP; 1000 pseudoreplicates), and maximum likelihood (ML; 100 pseudoreplicates) analyses are marked with closed circles.

somewhere in western Asia. However, we might also note that species of the two subgenera are also differentiated according to habitat association, with all members of *Coelomys* found in wetter forest habitats (typically evergreen forests) and all members of *Pyromys* found in drier, more seasonal habitats, including semi-arid steppe. In this regard, the presence of *Coelomys* (*mayori*) in evergreen forests in Sri Lanka, and *Pyromys* (*shortridgei*) in the dry dipterocarp forests of Myanmar

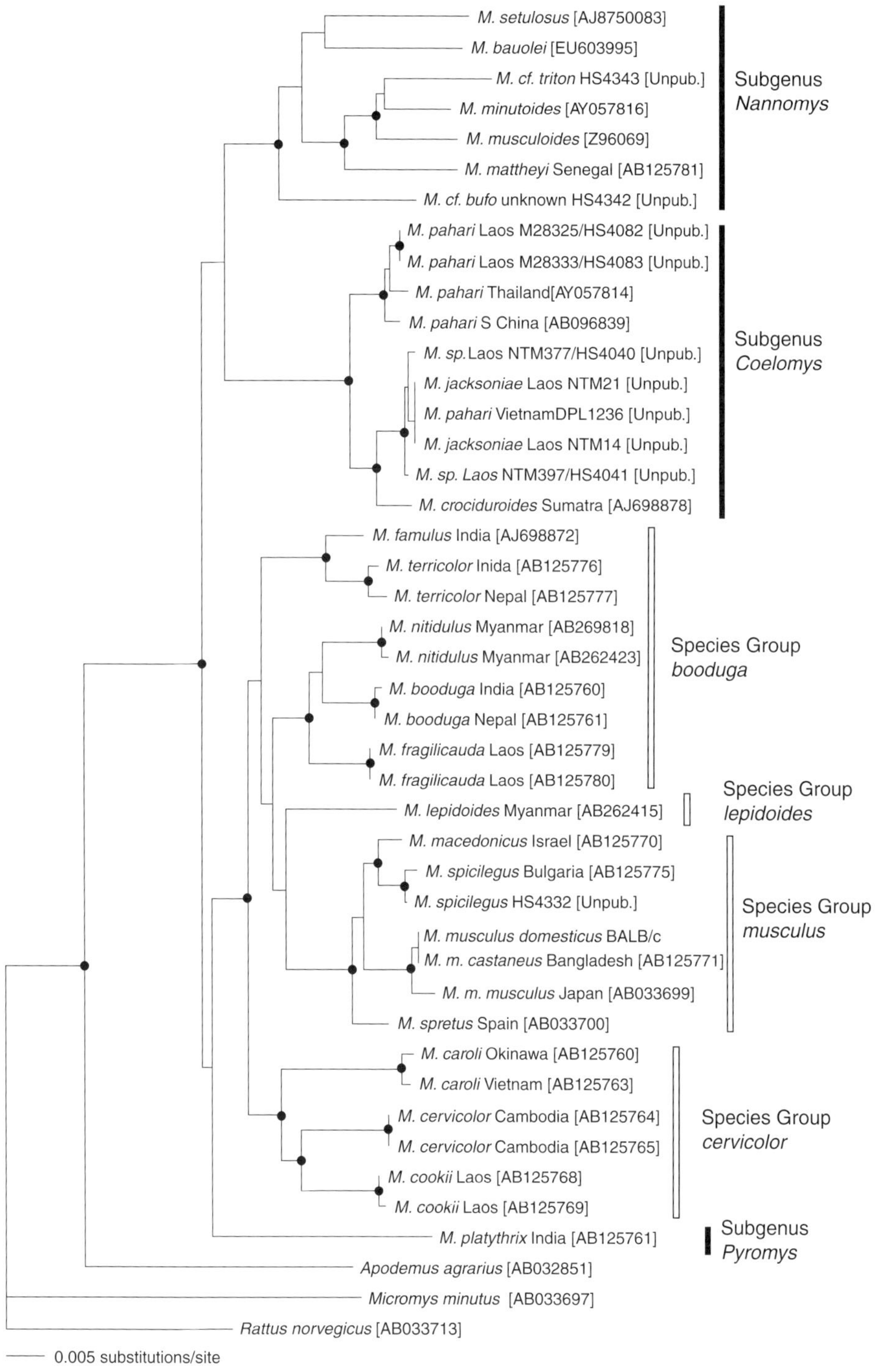

Figure 2.3 Phylogenetic relationships of Eurasian species of *Mus*, based on the mitochondrial *Cytb* gene sequences (transversions only). The phylogenetic trees were constructed using the NJ method with distances inferred from Kimura's two-parameter

and Thailand, may both be best explained as relictual evidence of formerly broader geographic overlap between the two subgenera.

A similar pattern of largely allopatric distributions is also observed within the subgenus *Mus* (Fig. 2.1b). Here, it is SGs that have partially discrete geographic foci – mainland Southeast Asia for the *M. cervicolor* SG; the Indian subcontinent for the *M. booduga* SG; central Myanmar for the monotypic *M. lepidoides* SG; and western Europe to northern India for the *M. musculus* SG. Habitat preferences do not readily explain this geographic patterning. Each of the *M. cervicolor* and *M. booduga* SGs includes species adapted to both semi-evergreen and deciduous forests, and to more open habitats including the margins of wetlands. In contrast, the *M. musculus* SG does show a broader habitat association that includes semi-arid to temperate steppe, which presumably has facilitated its expansion into western Europe. However, once again, a simplistic interpretation in terms of allopatric radiations founders on two anomalous members of the *M. booduga* SG – *M. nitidulus* and *M. fragilicauda*, found in central Myanmar and northeastern Thailand and Laos, respectively (Fig. 2.1). Shimada *et al.* (2010) offered an explanation for *M. nitidulus* in the context of broader evidence for faunal interchange at various intervals during the Pliocene and Pleistocene between India and the 'central dry zone' habitats of Myanmar. However, whether or not this model can be extended to explain the distribution even further to the east of *M. fragilicauda*, remains to be explored.

At an intraregional level, there is a strong phylogeographic pattern within each of *M. caroli*, *M. macedonicus*, *M. spicilegus*, and *M. musculus*. *Mus caroli*, the ricefield mouse, occurs across most of Southeast Asia from Myanmar in the west to Taiwan in the east (Fig. 2.4; Terashima *et al.*, 2003; Shimada *et al.*, 2007b). As its name suggests, *M. caroli* today is best known as an agricultural pest, but it can still be found living in grassland under relatively undisturbed deciduous forest cover. A deep phylogeographic structure is present in mtDNA sequences, with a basal split between Taiwanese populations and those of mainland Southeast Asian populations, and a three-part division of the latter into western, northern, and eastern lineages (Shimada *et al.*, 2007b). Lineage I is the most widespread, with mainland populations spread throughout the lower Mekong

Caption for figure 2.3 (*cont.*)
distances (Kimura, 1980). Bootstrap analyses were performed with NJ and MP methods (1000 replicates for both methods). Nodes denoted with relatively high bootstrap values (>80%) for the NJ and MP methods are marked with closed circles. Taxonomic status of *Mus* sp. from Laos and Vietnam is under investigation (Aplin *et al.*, unpublished).

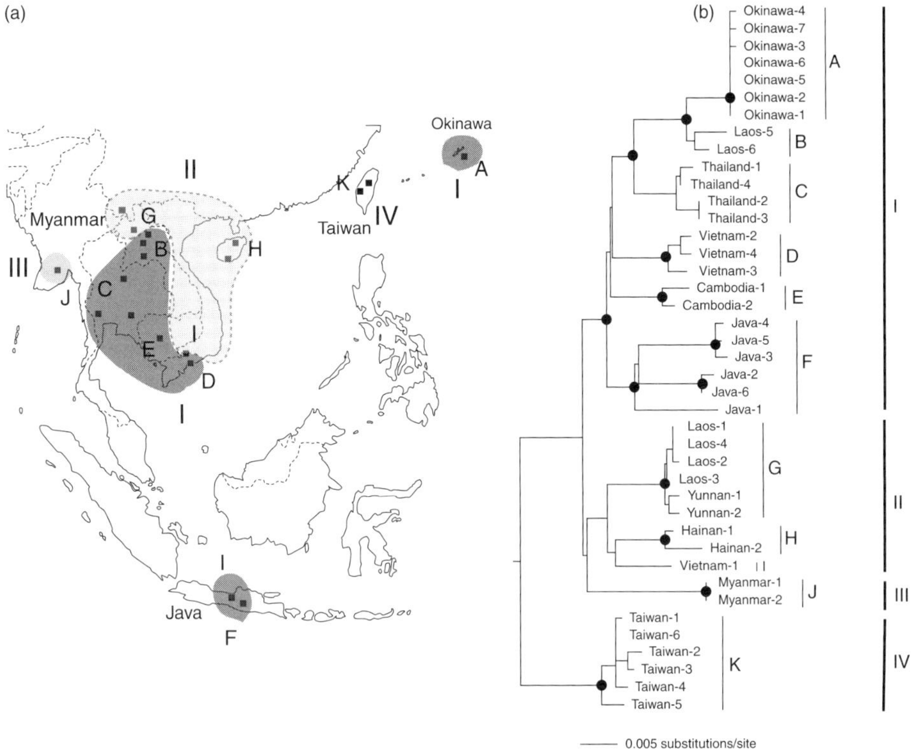

Figure 2.4 (a) Maps showing the geographic context and collection localities of samples. (b) NJ tree constructed for cytochrome *b* gene sequences of *Mus caroli*, using *M. cervicolor* and *M. cookii* as outgroup taxa to root the tree (Terashima *et al.*, 2003; Shimada *et al.*, 2007b). Nodes denoted with relatively high bootstrap values (>80%) for the NJ method are marked with closed circles.

Basin. The northern Lineage II occupies upland regions with basal (i.e. major rivers) elevations in excess of 300 m above sea level (ASL). Lineage III is restricted to the central basin of Myanmar, an area of low basal elevation (<100 m ASL) and limited topographic relief that is enclosed by north–south trending ranges to the west and east and drained by two main rivers, the Ayerawaddy and the Salween. Interestingly enough, highly disjunct island populations of *M. caroli* on Java and Okinawa islands also belong to Lineage I, while the latter is most likely explained by human-assisted dispersals.

Mus macedonicus and *M. spicilegus* are two free-living members of the *M. musculus* SG found in eastern and southeastern Europe and the Middle East. Both show significant phylogeographic structuring. Orth *et al.* (2002) reported deep divergence between populations of *M. macedonicus* in Israel and those from the remainder of the species range in the southern Balkans, Transcaucasia, and Asia Minor (Macholán, 1999). The Israel populations also show some allelic differentiation in nuclear genes and are generally referred to a separate subspecies, *M. m. spretoides*. Macholán *et al.* (2007) explored genetic variation in both subspecies in more detail and found a less complete segregation between European and Asian populations, and evidence for recent population expansions, most likely out of glacial refugia (see Auffray and Britton-Davidian, Chapter 1). Free-living, short-tailed mice on the island of Cyprus were conventionally regarded as a population of *M. macedonicus*, but recent morphological and genetic work favours recognition of this population as a distinct species, *M. cypriacus* (Cucchi *et al.*, 2006). *Mus spicilegus* has a smaller native range that extends from Austria and Slovakia in the north, to Bulgaria in the south, and the Ukraine in the east. There is some geographic variation in morphology, with one subspecies, *M. s. adriaticus*, recently recognized (Kryštufek and Macholán, 1998). Variation in *M. spicilegus* mtDNA is less overall than in *M. macedonicus*, but there is nonetheless a phylogeographic patterning that discriminates all major regional populations (Macholán *et al.*, 2007; Mitsainas *et al.*, 2009; see also Auffray and Britton-Davidian, Chapter 1).

The house mouse, *Mus musculus*, displays extremely complex patterns of both morphological and genetic geographic variation. Schwarz and Schwarz (1943) classified this species into more than ten classical subspecies based on morphological traits, behaviour, and geographical distribution, while Marshall's (1998) careful analysis of all named forms suggested a different arrangement, though still with nine subspecies. Early analyses of molecular markers, including those based on mtDNA RFLP (e.g. Yonekawa *et al.*, 1980, 1981, 1982, 1986; Moriwaki, 1994), suggested a simpler population grouping of Eurasian wild mice into four phylogroups to which they applied subspecies terminology: *M. m. musculus* from the northern part of Eurasia excluding western Europe; *M. m. domesticus* from western Europe and north Africa (also southern Africa, Australia, and the Americas, all as historical introductions); *M. m. castaneus* from Southeast Asia; and *M. m. bactrianus* from Afghanistan and Pakistan (with some evidence that this group of 'central mice' might represent an ancestral population; e.g. Boursot *et al.*, 1996; Din *et al.*, 1996; and review of Tucker, 2007). Subsequent analysis of mtDNA control region sequences found one other highly divergent lineage in populations from Yemen (Prager *et al.*, 1998) and some other areas of prehistoric to recent introduction, including Madagascar (Duplantier *et al.*, 2002). This phylogroup has been designated as *M. m. gentilulus* (Prager *et al.*, 1998). Even more recently, another potentially distinct phylogroup has been identified from individuals from Nepal, which is equally distant from the three major groups of *castaneus*, *domesticus*, and *musculus* in mtDNA cytochrome *b* and control region sequences (Terashima *et al.*, 2006; Suzuki *et al.*, submitted). A major finding of the mtDNA surveys was the presence of both *musculus* and *castaneus* haplotypes in Japanese wild mice, indicating a likely hybrid origin for the 'subspecies' *molossinus* (Yonekawa *et al.*, 1988). This is of particular interest because Japanese 'fancy mice' provided the founder animals for some of the classical laboratory strains. A mixed origin for Japanese mice was strongly supported by later work using a variety of uni- and bi-parentally inherited markers (Terashima *et al.*, 2006; Nunome *et al.*, 2010; for a review, see Yonekawa *et al.*, Chapter 4 in this volume).

The taxonomic status of the mtDNA phylogroups remains unsettled. Contrasting approaches include continued use of subspecies (e.g. Musser and Carleton, 2005), recognition of weakly differentiated 'semispecies' (e.g. Winking *et al.*, 1991), and treatment as full biological species (e.g. Tucker *et al.*, 2005). Relevant information on which to judge the level of genetic isolation between phylogroups includes analyses of interaction at contact zones (e.g. Macholán *et al.*, 2007), analysis of nuclear markers sensitive to detection of hybrids and wider genetic introgression (Abe *et al.*, 2004; Sakai *et al.*, 2005), studies of large sets of single-nucleotide polymorphism (SNP) markers to obtain broader genomic characterization (Frazer *et al.*, 2007), studies of linkage disequilibrium in recombinant haplotypes (Nunome *et al.*, 2010), and investigations of hybrid sterility (Forejt *et al.*, 1991; Trachtulec *et al.*, 1997; for a review, see Forejt *et al.*, Chapter 19 in this volume) and hybrid breakdown in inter-phylogroup crosses (Oka *et al.*, 2004, 2007). As might be expected, these studies provide a wide variety of perspectives on genetic incompatibility among members of what might be best viewed as a *M. musculus* complex. Overall, however, the emerging picture is generally confirmatory of the earlier impression from mtDNA. The 'house mouse' is genetically highly partitioned, consistent with a large natural geographic range and at least partially isolated genetic lineages. Determining the taxonomic status of each genetic unit must now be approached on a pairwise basis, to assess the extent of genetic isolation and incompatibility between divergent populations as they exist in both primary and secondary contexts.

Various recent studies have focused on finer-scale phylogeographic variation in mtDNA of each of the *domesticus* (Searle *et al.*, 2009; Bonhomme *et al.*, 2011; see also Bonhomme and Searle, Chapter 11 in this volume), *musculus*, and *castaneus* (Suzuki *et al.*, submitted) phylogroups. A major challenge in these studies is to distinguish between 'original' patterning within the as yet incompletely understood native ranges of these phylogroups, and the influence of later genetic admixture. An even greater challenge is to extend this approach to nuclear markers where recombination and gene conversion often debar simple interpretation. A recent study using multiple nuclear intron sequences revealed spatial subdivision in the *M. m. musculus* into northern and southern subgroups (Nunome *et al.*, 2010), a pattern that was not observed in earlier analysis of mtDNA markers (Yonekawa *et al.*, 1980, 1981, 1982, 1986) but has emerged with more detailed sampling of eastern Russia and China (Suzuki *et al.*, submitted). Nunome *et al.* (2010) also examined linkage disequilibrium among eight genes on chromosome 8 in Japanese mice. They were able to distinguish ancient from recent recombinant haplotypes, and thus postulate the relative times of introduction to Japan of different genetic stocks.

Diversification, niche transitions and sympatry in the evolution of Eurasian *Mus*

The ecological diversity of the genus *Mus*, coupled with the potentiality conferred by its vast genomic resources, invites detailed examination of the course of its adaptive radiation. A prerequisite to this approach is a robust hypothesis of the history of adaptive shifts in the evolution of the genus. Suzuki *et al.* (2004) reviewed this topic, and it can be updated here in light of new phylogenetic, palaeogeographic, and palaeoecological data.

Early murines like the species of *Potwarmus* evolved in an environment that was dominated by humid evergreen forests across all of low- to mid-latitude Europe and Asia (Nanda and Sehgal, 1993; Hoorn *et al.*, 2000; Liu *et al.*, 2009). At 11.1 Mya a major global cooling episode caused a surge in accumulation of Antarctic glacial ice and a corresponding fall in sea level of around 100 m (Haq *et al.*, 1987; Miller *et al.*, 1991). Exposure of land bridges and creation of new land corridors facilitated both inter- and trans-continental dispersal among many groups of mammals, including the early murine genus *Progonomys*, which appears soon thereafter in eastern Europe and northern Africa. Despite these events, the fossil mammal faunas from Europe, the Near East and the Siwalik Hills show only minor changes in composition across this interval, and for the next 1.4 million years small mammal assemblages continue to be dominated by other groups of rodents, including dormice (Gliridae), hamsters (Cricetidae), and flying squirrels (Pteromyidae). Not surprisingly, fossil murine diversity remains low through this period.

Evidence of major faunal turnover is first observed at *ca.* 9.7 Mya, at which time many of the characteristic mammal lineages of the Middle Miocene disappear from the fossil record of Europe, the Middle East, and South Asia (Agusti *et al.*, 2003; Koufos, 2006; Patnaik, 2010). This period, termed the 'Vallesian Crisis' in European literature (Agusti *et al.*, 2003; Casanovas-Vilar *et al.*, 2005), also saw the first significant pulse in diversification within Murinae, with the appearance of locally endemic genera in each of Europe (Mein *et al.*, 1993; Martin-Suárez and Mein, 1998) and China (Qiu and Li, 2003). Plant macrofossil and pollen evidence from this period suggests a widespread shift towards more seasonal climates, with a replacement of predominantly evergreen forests with predominantly deciduous communities across large areas of Asia and eastern Europe (Agusti *et al.*, 2003). Significant restructuring of global oceanic circulation patterns is also registered at this time (Keller and Barron, 1987; Miller *et al.*, 1991) and changes in global climatic systems seem likely to underpin the changes in terrestrial ecosystems.

The next major change affecting the course of murine evolution including the *Mus* lineage is the widespread vegetation change from a C3 to C4 photosynthetic pathway between 8.3 and 7 Mya, as recorded in sediment isotopic records across large areas of Eurasia, Africa, and North America (Morgan *et al.*, 1994; Cerling *et al.*, 1997; Latorre *et al.*, 1997; Pagani *et al.*, 1999). Grasses are a major group of C4 plants and a global expansion of this group must have created many new niches for Murinae, many of which depend heavily on grass seeds as a seasonal dietary resource (Renaud *et al.*, 2005). According to molecular clock estimates, initial diversification of several of the more speciose murine lineages did indeed take place during this period, including the African Arvicanthini (8.4 ± 0.6 Mya) and Praomyini (7.6 ± 0.6 Mya), and the Asian '*Rattus* sensu lato' (8.4 ± 0.7 Mya). Diversification of extant subgenera of *Mus* appears to narrowly postdate this period, with an estimated age of 6.6 ± 0.6 Mya (Lecompte *et al.*, 2008), but this postdating could well be a product of lineage extinction within this overall less speciose tribe. Many species within *Mus* are in fact adapted to life in seasonal habitats, with grass seeds making up a significant proportion of their diet. However, this is not true for all species and it is therefore instructive to re-examine the pattern of habitat association against the improved phylogenetic framework.

As noted earlier, the earliest cladogenic event among extant lineages of *Mus* appears to have been between the ancestors of subgenus *Coelomys* and a clade comprising all remaining subgenera. This event may also have involved the earliest niche transition in the evolution of the genus. Species of *Coelomys* are currently restricted to South (*C. mayori*) and Southeast Asia (all other species), and there is no known fossil record of this dentally distinctive subgenus outside this region (Chaimanee, 1998). Most *Coelomys* species are inhabitants of evergreen forest types, including montane forests, but several are also known to occur in semi-evergreen forests. The common name of 'shrew mice' for this group underscores a likely significant dietary emphasis on insects, as judged by their cranio-dental morphology (Marshall, 1977). This adaptation contrasts with that among the remaining subgenera which predominantly comprise species of deciduous forests, woodlands, and arid to temperate grassland steppes.

Additionally, all have ecomorphologies that suggest a narrower focus on exploitation of seed resources. The notion of an early niche transition from a more generalized omnivore niche in evergreen to a more specialized seed-eating niche in drier, more seasonal forest types garners some support from the fossil record insofar as the earliest known species of *Mus*, dated to 6.35 Mya and 5.7 Mya in the Siwalik sequence, are associated with fauna suggestive of open, seasonal environments. Moreover, *M. auctor* at 5.7 Mya is not referable to *Coelomys*, even though its subgeneric affinities are not resolved. However, as cautioned by Suzuki *et al.* (2004), the Siwalik assemblages are usually drawn from a range of local habitats (Quade *et al.*, 1989) and the *Mus* species, as rare elements, could well have been living in relictual patches of evergreen forest. Additionally, it is important to note that each of the other subgenera contains at least one taxon that can be found in tropical evergreen forest (i.e. *M. fernandoni* within *Pyromys*; *M. famulus* within the subgenus *Mus*; *M. setulosus* within *Nannomys*). In the case of *Nannomys*, *N. setulosus* occupies the earliest branch among the range of species tested, which may indicate that the earliest immigrant *Mus* into Africa was still a wet-forest-adapted taxon. However, within the subgenus *Mus*, *M. famulus* appears to be embedded phylogenetically within the *M. booduga* SG (Shimada *et al.*, 2009), and thus seems more likely to represent an instance of reinvasion of an ancestral biome. Too little is known about *Pyromys* to speculate on the phylogenetic position of *M. fernandoni*.

Information on more recent evolutionary episodes is largely limited to members of the subgenus *Mus*, with much less known about either of *Pyromys* or *Coelomys*. In the case of the subgenus *Mus*, molecular phylogenies indicate a long period of evolutionary quiescence between the origin of the subgenus (variously dated at 5.0 Mya (Suzuki *et al.*, 2004) and *ca.* 6.0 Mya (Lecompte *et al.*, 2008)) and the more or less simultaneous differentiation of the four SGs at around 2.4–2.7 Mya (Suzuki *et al.*, 2004; Shimada *et al.*, 2010). This quiescent phase spans the Middle to Late Pliocene, a period that experienced warm and stable conditions with globally high sea levels. One significant disruption during this period was caused by a brief but dramatic cooling event at 5.75–5.70 Mya, resulting in a sea level fall variously estimated at −10 m to −50 m (Aharon *et al.*, 1993; Shackleton *et al.*, 1995). In the Mediterranean region, tectonic events interacted with these global changes to produce much more dramatic events: isolation of the Mediterranean basin from the Atlantic Ocean, with eventual drying of the enclosed basin – the so-called 'Messinian salinity crisis' (Hodell *et al.*, 1986; Krijgsman *et al.*, 1999). A phase of inter-continental dispersals at this time is marked by the sudden appearance of various Asian mammals (e.g. hystricid porcupines and leporids; Barry *et al.*, 1985; Winkler, 1994) in the African record, and the appearance in western Europe of various African mammals, including gerbils (Agusti *et al.*, 2006). Chevret (1996) postulated that *Mus* may have entered Africa at this time, and Lecompte *et al.* (2008) noted broader evidence for interchange of murine lineages at that time (e.g. *Occitanomys* from Europe to Africa; and the arvicanthin *Saidomys* from Africa to the Middle East and even to Southeast Asia). In keeping with the lack of evidence for cladogenesis within the subgenus *Mus*, Asian fossil sequences show relatively few changes through this period, despite a phase of accelerated tectonic uplift of the Tibetan Plateau, which led to strengthening of the East Asian monsoon (Quade *et al.*, 1989).

In contrast, evolutionary diversification in the subgenus *Mus* at the end of the Pliocene coincides with a period of growing climatic instability that presaged the full glacial cycling of the Quaternary Period (Ciesielski and Weaver, 1974; Beu and Edwards, 1984; Dwyer *et al.*, 1995). Moreover, it coincides with evidence of dramatic turnover in many regional fossil records, with evidence of trans- and inter- continental dispersals (Barry *et al.*, 1985). Other evidence for elevated rates of speciation around the Pliocene–Pleistocene boundary comes from calibrated molecular phylogenies of diverse groups of mammals, including various other murines (e.g. *Apodemus* (Suzuki *et al.*, 2008); *Rattus*, *Maxomys*, *Niviventer*, and their allies (Pagès *et al.*, 2010)), and mustelids, leporids, and moles (summarized by Suzuki, 2009). As noted earlier, the four SG lineages within the subgenus *Mus* have partially discrete geographic ranges, and it seems likely that initial divergence may have occurred through a process of range expansion and fragmentation followed by local diversification. Similar processes may have underpinned other levels of differentiation within this group, including speciation within each of the SGs and the creation of intra-specific phylogeographic structure such as occurs within the Indochinese *M. caroli* and the more widely distributed members of the *M. musculus* SG. An obvious driver for these population genetic events through the Quaternary is the cyclical climatic swings between glacial and interglacial conditions that began as rapid but low-amplitude perturbations but settled after *ca.* 900 000 years ago into more pronounced and long-lasting oscillations. Across mainland Eurasia, in areas that remained ice-free, these cycles were generally manifested as swings between warm-wet (interglacial) and cold-dry (glacial) conditions, with vegetation communities changing accordingly (Thompson *et al.*, 1997). For the southern part of continental Southeast Asia, represented today by the Sundaic province of the Indo-Malayan Archipelago, the associated rise and fall of global sea level also generated cycles of coalescence and fragmentation of continental shelf islands, and this too would have served to stimulate genetic diversification. In the western part of Eurasia, lineage diversification within the *M. musculus* SG probably commenced around 1.5 Mya (Suzuki *et al.*, 2004), with all of the key speciation events probably over by 700 000 years ago (Macholán *et al.*, 2007).

Taxonomic diversification in the subgenus *Mus* was accompanied by ecological diversification. In each of the Asian *M. cervicolor* and *M. booduga* SGs, a number of lineages have undergone a niche transition from deciduous forests to more open grassland habitats, possibly including the margins of natural wetlands. The circumstances under which these transitions occurred are not obvious, due in part to the fact that many members of these SGs (e.g. *M. caroli*, *M. cervicolor*, *M. booduga*, *M. terricolor*) are today found chiefly in anthropogenic contexts including irrigated rice fields. All of these species are likely to have expanded their geographic ranges in recent times (Aplin and Singleton, 2003) and this may have produced sympatry where it did not formerly occur. In the case of *M. caroli*, long-distance dispersals, presumably with unintended human assistance, have even resulted in the establishment of new island populations (e.g. Ryukyu archipelago; Lesser Sunda islands), but this has not altogether eliminated an original phylogeographic pattern (Shimada *et al.*, 2007a).

All members of the *M. musculus* SG are adapted to life in relatively open habitats, including woodland, shrubland, and grassland steppe, and members of the *M. musculus* species complex have added commensalism to their range of adaptations, thereby allowing them to expand their range well beyond their natural limits of tolerance (Fig. 2.1c). The westward expansion of this SG presumably occurred through a series of sequential niche transitions, some of which may have necessitated the evolution of specific physiological and behavioural traits. For example, *M. spretus* has undergone metabolic changes that allow reduced water requirements (Palomo and Gisbert, 2002), while *M. spicilegus* has developed unique (probably within the genus *Mus*) mound-building and grain-hoarding behaviours that also require communal living (Sokolov *et al.*, 1998). The propensity for commensalism is shared by all members of the *M. musculus* species complex. This is remarkable on two counts. First, it is not otherwise seen in any member of the genus, even where *M. musculus* is absent (e.g. much of Indochina; Aplin *et al.*, 2003). Second, the divergence time among members of the *M. musculus* species complex is invariably estimated in the hundreds of thousands of years (e.g. She *et al.*, 1990; Boursot *et al.*, 1993; Suzuki *et al.*, 2004), which is an order of magnitude older than any form of settled human existence. Clearly, then, commensalism among house mice must be facilitated by some heritable property which it alone possesses within the genus *Mus*, but which evolved in a quite different context to that which involves close interaction with people.

The factors that support sympatry of congeners are of major interest in studies of species diversity. Some of the more obvious factors are differentiation by body size (Brown *et al.*, 1993) or by major niche, as in the case of Eurasian wood mice (*Apodemus* spp.), where geographic range overlap occurs between pairs of ground-dwelling and arboreal species (Suzuki *et al.*, 2008). Where there is no obvious niche or body size segregation between closely related species, it is usually assumed that allopatric speciation has preceded secondary range overlap, with coexistence either maintained by slight niche differentiation or transience, with longer-term outcomes mediated by competition. In the genus *Mus*, all species are terrestrial, similar in body size and, with the possible exception of the members of the subgenus *Coelomys*, not obviously differentiated by diet. Nevertheless, the genus *Mus* seems to have achieved significant levels of regional sympatry. Taking all subgenera into account, the highest intraregional diversity is probably found in parts of Indochina, where five species of *Mus* (*pahari*, *shortridgei*, *cookii*, *cervicolor*, *caroli*; Marshall, 1977) can occur in close proximity. Within the subgenus *Mus*, all three species of the *M. cervicolor* SG are sometimes found within a single locality: *M. cervicolor* and *M. caroli* in the low-lying environments of rice fields, and *M. cookii* in adjacent dry-land habitats. Natural range overlap is also observed within the *M. musculus* SG and between wild house mice and other members of this group. Within the *M. booduga* SG, *M. booduga* and *M. terricolor* occur together in cropping areas and for many years were considered as a single species (Matthey and Petter, 1968). Notably, the two species tend to show slight difference in burrow structure and placement of nests (Singh *et al.*, 2009). Fine-scale niche differentiation thus may have played an important role in the development of sympatry among closely related species of *Mus*.

Adaptive evolution in Eurasian *Mus*

The genome resources available for *M. musculus* provide almost unparalleled opportunities for investigation of many aspects of genetic determination of phenotypes and for studies of natural selection and adaptive evolution within a speciose radiation (Ramm *et al.*, 2008). To date, these capabilities have been little exploited.

In one such study, Shimada *et al.* (2009) examined *Mc1r* gene sequences among ten species of *Mus* selected to represent the four subgenera and each of three SGs within the subgenus *Mus*. The *Mc1r* gene plays a major role in red-yellow (phaeomelanin) and black-brown (eumelanin) melanization (Tamate and Takeuchi, 1981; Klungland *et al.*, 1995; MacDougall-Shackleton *et al.*, 2003), and single-site mutations can sometimes lead to significant changes in phenotype (e.g. Kambe *et al.*, 2011). These properties make the gene particularly susceptible to rapid evolutionary change (Hoekstra and Price, 2004) and have provided convincing examples of natural selection (Mullen and Hoekstra, 2008; Gross *et al.*, 2009). Allele polymorphisms in *Mc1r* are classified either as loss-of-function or gain-of-function mutations, which result in red-yellow or black-brown coat colorations, respectively (Robbins *et al.*, 1993). Even though species of

Mus are not highly variable in coat colour, Shimada *et al.* (2009) found a trend of acceleration of amino acid changes in the ancestral lineage of the subgenus *Mus*. We are currently extending this study with additional species and have found that a similar trend of accelerated amino acid changes is also present in the ancestral lineages of the African subgenus *Nannomys* (*dN*/ *dS* = 0.77) and the Southeast Asian subgenus *Coelomys* (*dN*/ *dS* = 0.61) (Fig. 2.5; T. Shimada *et al.*, unpublished results). Interpretation of long-term evolutionary genetic patterns of this kind presents many new opportunities but also significant challenges in the study of *Mus* species.

Another area of special potential concerns the evolutionary genetics of rodent dental morphology. Recent work on the development of molar crowns, primarily in laboratory mice (*Mus musculus*), has shed light on the genetic and ontogenetic influences that lead to the initiation of dental placodes, the differentiation of the primary and secondary enamel knots, and the formation of accessory crests and basins that create species-specific dental morphologies (Keränen *et al.*, 1998; Jernvall and Jung, 2000; Salazar-Ciudad and Jernvall, 2002; Kassai *et al.*, 2005; Cai *et al.*, 2007; Obara and Lesot, 2007). These studies make it abundantly clear that the murine genome does not contain code that 'individualizes' particular cusps or other dental structures, but rather that these structures emerge as a consequence of developmental cascades involving complex, sequential molecular and cellular events. These findings invite comparative studies using a range of other *Mus* species with contrasting molar morphologies, including differences in overall molar size, molar proportions along the toothrow, overall molar shape, and details of crown complexity, all of which vary between and, in some instances, within species. Insights gained from such a study will assist greatly with interpretation of the fossil record and understanding of patterns of variation in extant species that occupy a variety of habits.

Concluding remarks

Eurasian *Mus* present an exceptional range of research opportunities spanning multiple scientific disciplines. As one of the longest-lived and geographically most widespread lineages within the family Murinae, the evolutionary history of this group can be viewed as a model for small mammal diversification, dispersal, and adaptation, while their regional diversity, local abundance and short generation times make them ideal organisms for studies of ecological interactions, microevolution, and speciation. In addition, the rapidly growing genome resources derived from the special status of the house mouse as a model laboratory organism make them ideal subjects for comparative studies

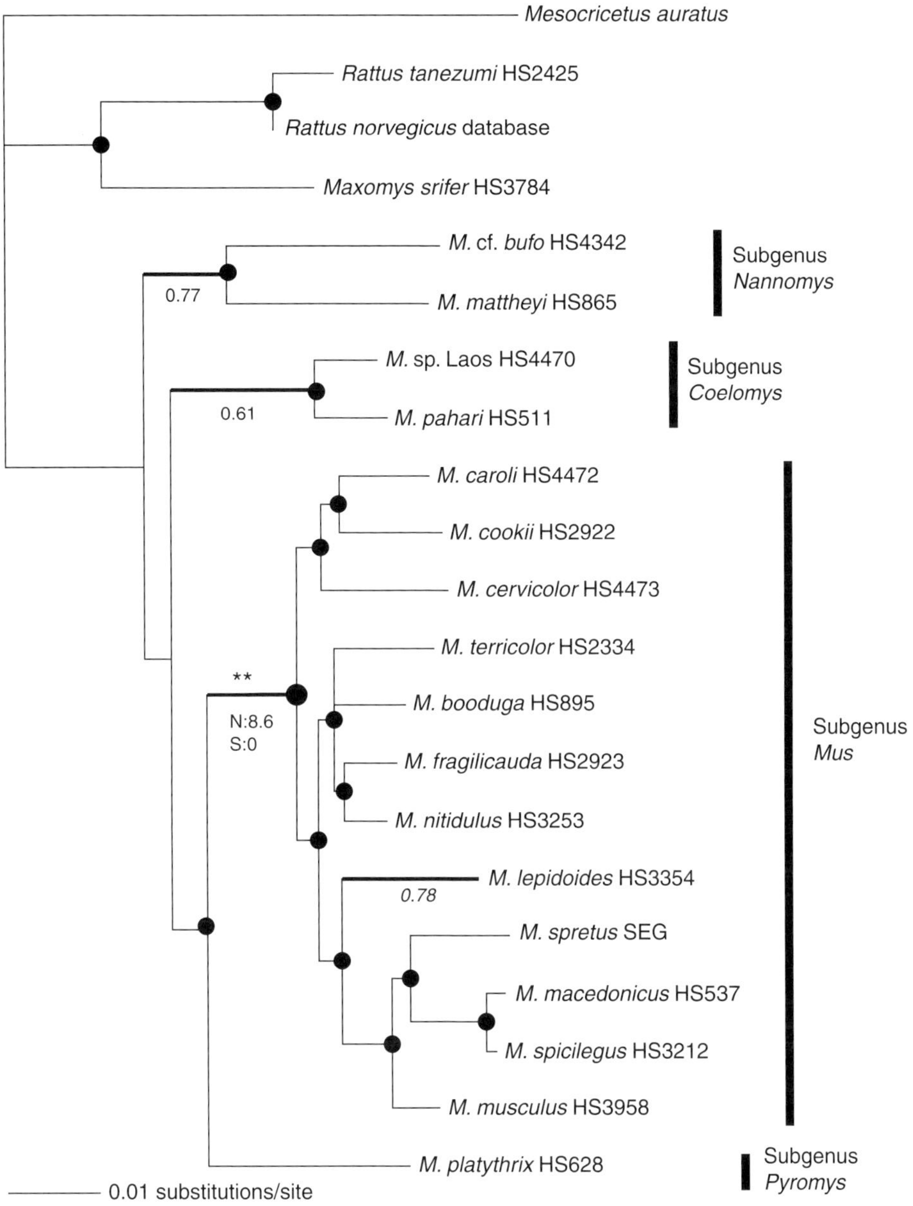

Figure 2.5 NJ tree based on *Mc1r* gene sequences (945 bp, without stop codon) from ten species of the genus *Mus*, with *Rattus norvegicus* included as an outgroup (Shimada *et al.*, 2009; Suzuki *et al.*, unpublished). Nodes denoted with constantly high bootstrap values (>50%) for the NJ method and a maximum parsimony (MP) analysis that yielded a tree with identical topology are marked with closed circles. Non-synonymous (N) and synonymous (S) changes were inferred with codeml in the PAML package (Yang, 1997). The *dN*/ *dS* ratios were calculated with codeml based on the free-ratio model and the codon substitution model 'F3 × 4'. Asterisks indicate when the two-ratio model fitted significantly better than the one-ratio model (**, $p < 0.01$).

of gene evolution across a wide range of evolutionary scales. Success in any of these endeavours in comparative biology relies, of course, on a solid foundation of taxonomic and phylogenetic knowledge. Among Eurasian *Mus*, the foundation is certainly taking shape for the subgenus *Mus*, although even here, recent species discoveries should encourage further close attention in poorly sampled areas. For the other subgenera, *Coelomys* and *Pyromys*, we must first admit that there is still much more gap than substance, and take steps to begin filling the holes.

REFERENCES

Abe, K., Noguchi, H., Tagawa, K., *et al.* (2004). Contribution of Asian mouse subspecies *Mus musculus molossinus* to genomic constitution of strain C57BL/6J, as defined by BAC-end sequence-SKIP analysis. *Genome Research*, **14**, 2439–47.

Agusti, J., Garces, M., and Krijgsman, W. (2006). Evidence for African–Iberian exchanges during the Messinian in the Spanish mammalian record. *Palaeogeography, Palaeoclimatology, Palaeoecology*, **238**, 5–14.

Agusti, J., Sanz de Siria, A., and Garce M. (2003). Explaining the end of the hominoid experiment in Europe. *Journal of Human Evolution*, **45**, 145–53.

Aharon, P., Goldstein, S. L., Wheeler, C. W., and Jacobson, G. (1993). Sea-level events in the South Pacific linked with the Messinian salinity crisis. *Geology*, **21**, 771–5.

Aplin, K. P., Brown, P. R., Jacobs, J., Krebs, C. J., and Singleton, G. R. (2003). *Field Methods for Rodent Studies in Asia and the Indo-Pacific*. Canberra: Australian Center for International Agricultural Research.

Aplin, K. P. and Singleton, G. R. (2003). Balancing rodent management and small mammal conservation in agricultural landscapes: challenges for the present and the future. In *Rats, Mice and People: Rodent Biology and Management*, ed. G. R. Singleton, L. A. Hinds, C. J. Krebs, and D. M. Spratt. ACIAR Technical Report 96, pp. 80–8.

Auffray J.-C., Orth, A., Catalan, J., *et al.* (2003). Phylogenetic position and description of a new species of subgenus *Mus* (Rodentia, Mammalia) from Thailand. *Zoologica Scripta*, **32**, 119–27.

Barry, J. C., Johnson, N. M., Raza, S. M., and Jacobs, L. L. (1985). Neogene mammalian faunal change in southern Asia: correlations with climatic, tectonic, and eustatic events. *Geology*, **13**, 637–40.

Bernor, R. L., Flynn, L. J., Harrison, T., *et al.* (1988). *Dionysopithecus* from southern Pakistan and the biochronology and biogeography of early Eurasian catarrhines. *Journal of Human Evolution*, **17**, 339–58.

Beu, A. G. and Edwards, A. R. (1984). New Zealand Pleistocene and Late Pliocene glacio-eustatic cycles. *Palaeogeography, Palaeoclimatology, Palaeoecology*, **46**, 119–42.

Bonhomme, F., Orth, A., Cucchi, T., *et al.* (2004). Découverte d'une nouvelle espèce de souris sur l'île de Chypre. *Comptes Rendus Biologies*, **347**, 501–7.

Bonhomme, F., Orth, A., Cucchi, T. *et al.* (2011). Genetic differentiation of the house mouse around the Mediterranean basin: matrilineal footprints of early and late colonization. *Proceedings of Royal Society of London B: Biological Sciences*, **278**, 1034–43.

Boursot, P., Auffray, J.-C., Britton-Davidian, J., and Bonhomme, F. (1993). The evolution of house mice. *Annual Review of Ecology and Systematics*, **24**, 119–52.

Boursot, P., Din, W., Anand, R., *et al.* (1996). Origin and radiation of the house mouse: mitochondrial DNA phylogeny. *Journal of Evolutionary Biology*, **9**, 391–415.

Brown, J. H., Marquet, P. A., and Taper, M. L. (1993). Evolution of body size: consequences of an energetic definition of fitness. *The American Naturalist*, **142**, 573–84.

Cai, J., Cho, S.-W., Kim, J.-Y., *et al.* (2007). Patterning the size and number of tooth and its cusps. *Developmental Biology*, **304**, 499–507.

Casanovas-Vilar, I., Moyà -Solà, S., Agustí, J., and Köhler, M. (2005). The geography of a faunal turnover: tracking the Vallesian Crisis. In *Migration of Organisms: Climate, Geography, Ecology*, ed. A. T. Elewa. Heidelberg: Springer-Verlag Publishers, pp. 247–301.

Cerling, T. E., Harris, J. M., MacFadden, B. J., *et al.* (1997). Global vegetation change through the Miocene–Pliocene boundary. *Nature*, **389**, 153–8.

Chaimanee, Y. (1998). Plio-Pleistocene rodents of Thailand. *Thai Studies in Biodiversity*, **3**, 1–303.

Chevret, P. (1996). African Murinae are issued from several waves of immigration: molecular evidence (Abstract). *Mammalia*, **60**, 801.

Chevret, P. D., Jenkins, P., and Catzeflis, F. (2003). Evolutionary systematics of the Indian mouse *Mus famulus* Bonhote, 1898: molecular (DNA/DNA hybridization and 12S rRNA sequences) and morphological evidence. *Zoological Journal of the Linnean Society*, **137**, 385–401.

Chevret, P. D., Veyrunes, F., Britton-Davidian, J. (2005). Molecular phylogeny of the genus *Mus* (Rodentia: Murinae) based on mitochondrial and nuclear data. *Biological Journal of the Linnean Society*, **84**, 417–27.

Ciesielski, P. F. and Weaver, F. M. (1974). Early Pliocene temperature changes in the Antarctic seas. *Geology*, **2**, 511–15.

Corbet, G. B. and Hill, J. E. (1992). *The Mammals of the Indomalayan Region: A Systematic Review*. London: Oxford University Press.

Cucchi, T., Orth, A., Auffray, J.-C., *et al.* (2006). A new endemic species of the subgenus *Mus* (Rodentia, Mammalia) on the island of Cyprus. *Zootaxa*, **1241**, 1–36.

Din, W., Anand, R., Boursot, P., *et al.* (1996). Origin and radiation of the house mouse: clues from nuclear genes. *Journal of Evolutionary Biology* **9**, 519–39.

Duplantier, J. M., Orth, A., Catalan, J., and Bonhomme, F. (2002). Evidence for a mitochondrial lineage originating from the Arabian peninsula in the Madagascar house mouse. *Heredity*, **89**, 154–8.

Dwyer, G. S., Cronin, T. M., Baker, P. A., *et al.* (1995). North Atlantic deepwater temperature change during late Pliocene and late Quaternary climatic cycles. *Science*, **270**, 1347–51.

Ellerman, J. R. and Morrison-Scott, T. C. (1951). *Checklist of Paleartic and Indian Mammals*. London: British Museum (Natural History), pp. 1758–946.

Flynn, L. J., Jacobs, L. L., and Lindsay, E. H. (1985). Problems in muroid phylogeny: relationship to other rodents and origin of major groups. In *Evolutionary Relationships among Rodents*, ed. W. P. Luckett and J. L. Hartenberger. New York: Plenum Press, pp. 589–616.

Forejt, J., Vincek, V., Klein, J., Lehrach, H., and Loudová-Micková, M. (1991). Genetic mapping of the *t*-complex region on mouse chromosome 17 including the hybrid sterility gene, *Mammalian Genome*, **1**, 84–91.

Frazer, K. A., Eskin, E., Kang, H. M., *et al.* (2007). A sequence-based variation map of 8.27 million SNPs in inbred mouse strains. *Nature*, **448**, 1050–3.

Gondo, Y. (2008). Trends in large-scale mouse mutagenesis: from genetics to functional genomics. *Nature Reviews Genetics*, **9**, 803–10.

Gross, J. B., Borowsky, R., and Tabin, C. J. (2009). A novel role for *Mc1r* in the parallel evolution of depigmentation in independent populations of the cavefish *Astyanax mexicanus*. *PLoS Genetics*, **5**, e1000326.

Haq, B. U., Hardenbol, J., and Vail, P. R. (1987). Chronology of fluctuating sea levels since the Triassic. *Science*, **235**, 1156–67.

Hodell, D. A., Elmstrom, K. M., and Kennett, J. P. (1986). Latest Miocene benthic delta^{18}O changes, global ice volume, sea level and the 'Messinian salinity crisis'. *Nature*, **320**, 411–14.

Hoekstra, H. E. and Price, T. (2004). Parallel evolution is in the genes. *Science*, **303**, 1779–80.

Hoorn, C., Ojha, T., and Quade, J. (2000). Palynological evidence for vegetation development and climatic change in the Sub-Himalayan zone (Neogene, Central Nepal). *Palaeogeography, Palaeoclimatology, Palaeoecology*, **163**, 133–61.

IUCN (2010). *IUCN Red List of Threatened Species*, ver. 2010.2. Available at: www.iucnredlist.org.

Jacobs, L. L. (1978). Fossil rodents (Rhizomyidae and Muridae) from Neogene Siwalik deposits, Pakistan. *Museum of Northern Arizona Press, Bulletin Series*, **52**, 1–103.

Jacobs, L. L. and Flynn, L. J. (2005). Of mice *... again: the Siwalik rodent record, murine distribution, and molecular clocks. In *Interpreting the Past: Essays on Human, Primate and Mammal Evolution – In Honor of David Pilbeam*, ed. D. Lieberman, R. Smith, and J. Kelley. Leiden: Brill Academic Publishers, pp. 63–80.

Jacobs, L. L. and Pilbeam, D. (1980). Of mice and men: fossil-based divergence dates and molecular 'clocks'. *Journal of Human Evolution*, **9**, 551–5.

Jernvall, J. and Jung, H.-S. (2000). Genotype, phenotype, and developmental biology of molar tooth characters. *Yearbook of Physical Anthropology*, **43**, 171–90.

Kambe, Y., Tanikawa, T., Matsumoto, Y., *et al.* (2011). Ancient origins of agouti-melanistic polymorphism in wild Black Rats (*Rattus rattus*): evidence from *Mc1r* gene sequences. *Zoological Science*.

Kassai, Y., Munne, P., Hotta, Y., *et al.* (2005). Regulation of mammalian tooth cusp patterning by ectodin. *Science*, **309**, 2067–70.

Keller, G. and Barron, J. A. (1987). Paleodepth distribution of Neogene hiatuses. *Paleoceanography*, **2**, 697–713.

Keränen, S. V., Äberg, T., Kettunen, P., Thesleff, I., and Jernvall, J. (1998). Association of developmental regulatory genes with the development of different molar tooth shapes in two species of rodents. *Development, Genes and Evolution*, **208**, 477–86.

Kimura, M. (1980). A simple method for estimating evolutionary rates of base substitutions through comparative studies of nucleotide sequences. *Journal of Molecular Evolution*, **16**, 111–20.

Klungland, H., Våge, D. I., Gomez-Raya, L., Adalsteinsson, S. and Lein, S. (1995). The role of melanocyte-stimulating hormone (MSH) receptor in bovine coat color determination. *Mammalian Genome*, **6**, 636–9.

Koufos, G. D. (2006). Palaeoecology and chronology of the Vallesian (late Miocene) in the Eastern Mediterranean region. *Palaeogeography, Palaeoclimatology, Palaeoecology*, **234**, 127–45.

Krijgsman, W., Hilgen, F. J., Raffi, I., Sierro, F. J., and Wilson, D. S. (1999). Chronology, causes and progression of the Messinian salinity crisis. *Nature*, **400**, 652–5.

Kryštufek, B. and Macholán, M. (1998). Morphological differentiation in *Mus spicilegus* and the taxonomic status of mound-building mice from the Adriatic coast of Yugoslavia. *Journal of Zoology*, **245**, 185–96.

Latorre, C., Quade, J. and McIntosh, W. C. (1997). The expansion of C_4 grasses and global change in the late Miocene: stable isotope evidence from the Americas. *Earth Planet Science Letter*, **146**, 83–96.

Lecompte, E., Aplin, K., Denys, C., *et al.* (2008). Phylogeny and biogeography of African Murinae based on mitochondrial and nuclear gene sequences, with a new tribal classification of the subfamily. *BMC Evolutionary Biology*, **8**, 199.

Liu, L., Eronen, J. T., and Fortelius, M. (2009). Significant mid-latitude aridity in the middle Miocene of East Asia. *Palaeogeography, Palaeoclimatology, Palaeoecology*, **279**, 201–6.

López Antoñanzas, R. (2009). First *Potwarmus* from the Miocene of Saudi Arabia and the early phylogeny of murines (Rodentia: Muroidea). *Zoological Journal of the Linnean Society*, **156**, 664–79.

Lundrigan, B. L., Jansa, S. A., and Tucker, P. K. (2002). Phylogenetic relationships in the genus *Mus*, based on paternally, maternally, and biparentally inherited characters. *Systematic Biology*, **51**, 410–31.

MacDougall-Shackleton, E. A., Blanchard, L., and Gibbs, L. (2003). Unmelanized plumage patterns in old world leaf warblers do not correspond to sequence variation at melanocortin-1 receptor locus (*MC1R*). *Molecular Biology and Evolution*, **20**, 1675–81.

Macholán, M. (1999). *Mus macedonicus* Petrov & Ružić, 1983. In *The Atlas of European Mammals*, ed. A. J. Mitchell-Jones, J. Amori, W. Bogdanowicz, *et al.* London: Academic Press, pp. 284–5.

Macholán, M., Vyskočilová, M., Bonhomme, F., *et al.* (2007). Genetic variation and phylogeography of free-living mouse species (genus *Mus*) in the Balkans and the Middle East. *Molecular Ecology*, **16**, 4774–88.

Marshall, J. T., Jr (1977). A synopsis of Asian species of *Mus* (Rodentia, Muridae). *Bulletin of the American Museum of Natural History*, **158**, 173–220.

Marshall, J. T. (1998). *Identification and Scientific Names of Eurasian House Mice and Their European Allies, Subgenus Mus (Rodentia: Muridae).* Unpublished report, National Museum of Natural History.

Martín-Suárez, E. and Mein, P. (1998). Revision of the genera *Parapodemus*, *Apodemus*, *Rhagamys*, and *Rhagapodemus* (Rodentia, Mammalia). *Geobios*, **31**, 87–97.

Matthey, R. and Petter, F. (1968). Existence de deux especes distinctes, l'une chromosomiquement polymorphe chez der *Mus* Indiens der groups *booduga*: etude cytogénétique et taxonomique. *Revue Suisse de Zoologie*, **75**, 461–98.

Mein P., Martin-Suarez, E., and Agusti J. (1993). *Progonomys* schaub, 1938 and *Huerzelerimys* gen. nov. (Rodentia); their evolution in western Europe. *Scripta Geologica*, **103**, 41–64.

Michaux, J., Reyes, A., and Catzeflis, F. (2001). Evolutionary history of the most speciose mammals: molecular phylogeny of muroid rodents. *Molecular Biology and Evolution*, **18**, 2017–31.

Miller, K. G., Wright, J. D., and Fairbanks, R. G. (1991). Unlocking the ice house: Oligocene–Miocene oxygen isotopes, eustasy and margin erosion. *Journal of Geophysical Research*, **96B**, 6829–48.

Mitsainas, G. P., Tryfonopoulos G. A., Thanou E. G., *et al.* (2009). New data on the distribution of *Mus spicilegus* Petényi, 1882 (Rodentia, Muridae) and a distinct mtDNA lineage in the southern Balkans. *Mammalian Biology*, **74**, 351–60.

Morgan, M. E., Kingston, J. D., and Marino, B. D. (1994). Carbon isotopic evidence for the emergence of C4 plants in the Neogene from Pakistan and Kenya. *Nature*, **367**, 162–5.

Moriwaki, K. (1994). Wild mouse from a geneticist's viewpoint. In *Genetics in Wild Mice: Its Application to Biomedical Research*, ed. K. Moriwaki, T. Shiroishi, and H. Yonekawa. Basel: S. Karger, pp. xii–xxv.

Mullen, L. M. and Hoekstra, H. E. (2008). Natural selection along an environmental gradient: a classic cline in mouse pigmentation. *Evolution*, **62**, 1555–70.

Musser, G. G. and Carleton, M. D. (2005). Family Muridae. In *Mammal Species of the World: A Taxonomic and Geographic Reference*, ed. D. E. Wilson and D. M. Reeder, 3rd edn. Baltimore, MD: Johns Hopkins University Press, pp. 894–1531.

Nanda, A. C. and Sehgal, R. K. (1993). Siwalik mammalian faunas from Ramnagar (J. K.) and Nurpur (H.P.) and lower limit of hipparion. *Journal of Geological Society of India*, **42**, 115–34.

Nunome, M., Ishimori, C., Aplin, K., *et al.* (2010). Detection of recombinant haplotypes in wild mice (*Mus musculus*) provides new insights into the origin of Japanese mice. *Molecular Ecology*, **19**, 2474–89.

Obara, N. and Lesot, H. (2007). Asymmetrical growth, differential cell proliferation, and dynamic cell rearrangement underlie epithelial morphogenesis in mouse molar development. *Cell and Tissue Research*, **330**, 461–73.

Oka A., Aoto, T., Totsuka, Y., *et al.* (2007). Disruption of genetic interactions between two autosomal regions and the X chromosome causes reproductive isolation between mouse strains derived from different subspecies. *Genetics*, **175**, 185–97.

Oka A., Mita, A., Sakurai-Yamatani, N., *et al.* (2004). Hybrid breakdown caused by substitution of the X chromosome between two mouse subspecies. *Genetics*, **166**, 913–24.

Orth, A., Auffray, J.-C., and Bonhomme, F. (2002). Two deeply divergent mitochondrial clades in the wild mouse *Mus macedonicus* reveal multiple glacial refuges south of Caucasus. *Heredity*, **89**, 353–7.

Osgood, W. H. (1936). New and imperfectly known small mammals of Africa. *Publication of the Field Museum of Natural History, Zoological Series*, **20**, 217–56.

Pagani, M., Freeman, K. H., and Arthur, M. A. (1999). Late Miocene atmospheric CO_2 concentrations and the expansion of C4 grasses. *Science*, **285**, 876–9.

Pagès, M., Chaval, Y., Herbreteau, V., *et al.* (2010). Revisiting the taxonomy of the Rattini tribe: a phylogeny-based delimitation of species boundaries. *BMC Evolutionary Biology*, **10**, 184.

Palomo, L. J. and Gisbert, J. (2002). *Atlas de los Mamíferos Terrestres de España*. Madrid: Organismo Autónomo de Parques y Jardines.

Patnaik, R. (2010). Fossil murine rodents as ancient monsoon indicators of the Indian subcontinent. *Quaternary International*, **229**, 94–104.

Prager, E. M., Orrego, C., and Sage, R. D. (1998). Genetic variation and phylogeography of central Asian and other house mice, including a major new mitochondrial lineage in Yemen. *Genetics*, **150**, 835–61.

Qiu, Z. D. and Li, C. K. (2003). Rodents from the Chinese Neogene: Biogeographic relationships with Europe and North America. *Bulletin of the American Museum of Natural History*, **279**, 586–602.

Quade, J., Cerling, T. E., and Bowman, J. R. (1989). Development of the Asian monsoon revealed by marked ecological shift during the latest Miocene in northern Pakistan. *Nature*, **342**, 163–6.

Ramm, S., Oliver, P., Ponting, C., Stockley, P., and Emes, R. (2008). Sexual selection and the adaptive evolution of mammalian ejaculate proteins. *Molecular Biology and Evolution*, **25**, 207–19.

Raymo, M. E., Ruddiman, W. F., and Froelich, P. N. (1988). Influence of late Cenozoic mountain building on ocean geochemical cycles. *Geology*, **16**, 649–53.

Renaud, S., Michaux, J., Schmidt, D. N., *et al.* (2005). Morphological evolution, ecological diversification and climate change in rodents. *Proceedings of the Royal Society of London B: Biological Science*, **272**, 609–17.

Robbins, L. S., Nadeau, J. H., Johnson, K. R., *et al.* (1993). Pigmentation phenotypes of variant extension locus alleles result from point mutations that alter MSH receptor function. *Cell*, **72**, 827–34.

Rowe, K. C., Reno, M. L., Richmond, D. M., Adkins, R. M., and Steppan, S. J. (2008). Pliocene colonization and adaptive radiations in Australia and New Guinea (Sahul): multilocus systematics of the old endemic rodents (Muroidea: Murinae). *Molecular Phylogenetics and Evolution*, **47**, 84–101.

Sakai, T., Kikkawa, Y., Miura, I., *et al.* (2005). Origins of mouse inbred strains deduced from whole-genome scanning by polymorphic microsatellite loci. *Mammalian Genome*, **16**, 11–19.

Salazar-Ciudad, I. and Jernvall, J. (2002). A gene network model accounting for development and evolution of mammalian teeth. *Proceedings of the National Academy of Science of the United States of America*, **99**, 8116–20.

Schwarz, E. and Schwarz, H. (1943). The wild and commensal stocks of the house mouse, *Mus musculus* Linnaeus. *Journal of Mammalogy*, **24**, 59–72.

Searle, J. B., Jamieson, P. M., Gündüz, İ., *et al.* (2009). The diverse origins of New Zealand house mice. *Proceedings of Royal Society of London B: Biological Science*, **276**, 209–17.

Shackleton, N. J., King Hagelberg, T., and Crowhurst, S. J. (1995). Evaluating the success of astronomical tuning: pitfalls of using coherence as a criterion for assessing pre-Pleistocene timescales. *Paleoceanography*, **10**, 693–7.

She, J. X., Bonhomme, F., Boursot, P., Thaler, L., and Catzeflis, F. (1990). Molecular phylogenies in the genus *Mus*: comparative analysis of electrophoretic, scnDNA hybridization, and mtDNA RFLP data. *Biological Journal of the Linnean Society*, **41**, 83–103.

Shimada, T., Aplin, K. P., Jenkins, P., and Suzuki, H. (2007a). Rediscovery of *Mus nitidulus* Blyth, 1859 (Rodentia, Muridae), an endemic murine rodent of the central basin of Myanmar. *Zootaxa*, **1498**, 45–68.

Shimada, T., Aplin, K. P., Jyogahara, K., *et al.* (2007b). Complex phylogeographic structuring in a continental small mammal from East Asia, the rice field mouse, *Mus caroli* (Rodentia, Muridae). *Mammal Study*, **32**, 49–62.

Shimada, T., Aplin, K. P., and Suzuki, H. (2010). *Mus lepidoides* (Muridae, Rodentia) of Central Burma is a distinct species of potentially great evolutionary and biogeographic significance. *Zoological Science*, **27**, 449–59.

Shimada, T., Sato, J. J., Aplin, K. P., and Suzuki, H. (2009). Comparative analysis of evolutionary modes in *Mc1r* coat color gene in wild mice and mustelids. *Genes and Genetic Systems*, **84**, 225–31.

Singh, S., Cheong, N., Gopeshwar Narayan, G., and Sharma, T. (2009). Burrow characteristics of the co-existing sibling species *Mus booduga* and *Mus terricolor* and the genetic basis of adaptation to hypoxic/hypercapnic stress. *BMC Ecology*, **9**, 6.

Sokolov, V. E., Kotenkova, E. V., and Michailenko, A. G. (1998). *Mus spicilegus*. *Mammalian Species*, **592**, 1–6.

Steppan, S. J., Adkins, R. M., and Anderson, J. (2004). Phylogeny and divergence-date estimates of rapid radiations in muroid rodents based on multiple nuclear genes. *Systematic Biology*, **53**, 533–53.

Suzuki, H. (2009). A molecular phylogenetic view of mammals in the 'three-story museum' of Hokkaido, Honshu, and Ryukyu islands. In *The Wild Mammals of Japan*, ed. S. D. Ohdachi, Y. Ishibashi, M. A. Iwasa, and T. Saitoh. Kyoto: Shoukadoh, pp. 261–3.

Suzuki, H., Filippucci, M. G., Chelomina, G. N., *et al.* (2008). A biogeographic view of *Apodemus* in Asia and Europe inferred from nuclear and mitochondrial gene sequences. *Biochemical Genetics*, **46**, 329–46.

Suzuki, H., Shimada, T., Terashima, M., Tsuchiya, K., and Aplin, K. P. (2004). Temporal, spatial, and ecological modes of evolution of Eurasian *Mus* based on mitochondrial and nuclear gene sequences. *Molecular Phylogenetics and Evolution*, **33**, 626–46.

Tamate, H. B. and Takeuchi, T. (1981). Induction of the shift in melanin synthesis in lethal yellow (A^y/a) mice in vitro. *Developmental Genetics*, **2**, 349–56.

Terashima, M., Furusawa, S., Hanzawa, N., *et al.* (2006). Phylogeographic origin of Hokkaido house mice (*Mus musculus*) as indicated by genetic markers with maternal, paternal and biparental inheritance. *Heredity*, **96**, 128–38.

Terashima, M., Suyanto, A., Tsuchiya, T., *et al.* (2003). Geographic variation of *Mus caroli* from East and Southeast Asia based on mitochondrial cytochrome *b* gene sequences. *Mammal Study*, **28**, 67–72.

Thompson, L. G., Yao, T., Davis, M. E., *et al.* (1997). Tropical climate instability: the last glacial cycle from a Qinghai-Tibetan ice core. *Science*, **276**, 1821–25.

Trachtulec, Z., Mnulova-Fajdelova, M., Hamvas, R. M. J., *et al.* (1997). Isolation of candidate hybrid sterility 1 genes by cDNA selection in a 1.1 megabase pair region on mouse chrosomome 17. *Mammalian Genome*, **8**, 312–16.

Tucker, P. K. (2007). Systematics of the genus *Mus*. In *The Mouse in Biomedical Research, vol. 1. History, Wild Mice, and Genetics*, ed. J. G. Fox, S. W. Barthold, M. T. Davisson, *et al.* San Diego, CA: Academic Press/Elsevier, pp. 13–23.

Tucker, P. K., Sandstedt, S. A., and Lundrigan, B. L. (2005). Phylogenetic relationships in the subgenus *Mus* (genus *Mus*, family Muridae): examining gene trees and species trees. *Biological Journal of the Linnean Society*, **84**, 653–62.

van Dam, J. A., Abdul Aziz, H., Alvarez Sierra, M. A., *et al.* (2006). Long-period astronomical forcing of mammal turnover. *Nature*, **443**, 687–91.

Veyrunes, F., Britton-Davidian, J., Robinson, T. J., *et al.* (2005). Molecular phylogeny of the African pygmy mice, subgenus *Nannomys* (Rodentia, Murinae, *Mus*): implications for chromosomal evolution. *Molecular Phylogenetics and Evolution*, **36**, 358–69.

Veyrunes, F., Dobigny, G., Yang, F., *et al.* (2006). Phylogenomics of the genus Mus (Rodentia; Muridae): extensive genome repatterning is not restricted to the house mouse. *Proceedings of Royal Society of London B: Biological Sciences*, **273**, 2925–34.

Winking, H., Weith, A., Boldyreff, B., *et al.* (1991). Polymorphic HSRs in chromosome 1 of the two semispecies *Mus musculus musculus* and *M. musculus domesticus* have a common origin in an ancestral population, *Chromosoma*, **100**, 147–51.

Winkler A. J. (1992). Systematics and biogeography of middle Miocene rodents from the Muruyur Beds, Baringo District, Kenya. *Journal of Vertebrate Paleontology*, **12**, 236–49.

Winkler, A. J. (1994). The middle/upper Miocene dispersal of major rodent groups between southern Asia and Africa. In *Rodent and Lagomorph Families of Asian Origins and Diversification*, ed. Y. Tomida, C. K. Li, and T. Setoguchi. Tokyo: National Science Museum, pp. 173–84.

Winkler, A. J. (2002). Neogene paleobiogeography and East African paleoenvironments: contributions from the Tugen Hills rodents and lagomorphs. *Journal of Human Evolution*, **42**, 237–56.

Yang, Z. (1997). PAML: a program package for phylogenetic analysis by maximum likelihood. *Computer Applications in BioSciences*, **13**, 555–6.

Yonekawa, H., Moriwaki, K., Gotoh, O., *et al.* (1980). Relationship between laboratory mice and the subspecies *Mus musculus domesticus* based on restriction endonuclease cleavage patterns of mitochondrial DNA. *Japan Journal of Genetics*, **55**, 289–96.

Yonekawa, H., Moriwaki, K., Gotoh, O., *et al.* (1981). Evolutionary relationships among five subspecies *Mus musculus* based on restriction enzyme cleavage patterns of mitochondrial DNA. *Genetics*, **98**, 801–16.

Yonekawa, H., Moriwaki, K., Gotoh, O., *et al.* (1982). Origins of inbred mice deduced from restriction patterns of mitochondrial DNA. *Differentiation*, **22**, 222–6.

Yonekawa, H., Moriwaki, K., Gotoh, O., *et al.* (1988). Hybrid origin of Japanese mice '*Mus musculus molossinus*': evidence from restriction analysis of mitochondrial DNA. *Molecular Biology and Evolution*, **5**, 63–78.

Yonekawa, H., Gotoh, O., Tagashira, Y., *et al.* (1986). A hybrid origin of Japanese mice '*Mus musculus molossinus*'. *Current Topics in Microbiology and Immunology*, **27**, 62–7.

3

On the origin of the house mouse synanthropy and dispersal in the Near East and Europe: zooarchaeological review and perspectives

THOMAS CUCCHI, JEAN-CHRISTOPHE AUFFRAY, AND JEAN-DENIS VIGNE

Introduction

The house mouse (*Mus musculus*) is the most successful and ubiquitous invasive mammal after man (Lowe *et al.*, 2000). This success is due to its commensal interaction with humans, which have allowed it to reach and thrive in places outside its natural bounds. Although the Neolithic revolution – i.e. the birth of a new way of life based on agriculture and husbandry – is accepted as a key factor in the evolution of the house mouse (Tchernov, 1984, 1991a, b, 1993; Auffray *et al.*, 1990c; Cucchi *et al.*, 2005), the components of the interaction between human evolution, the dynamics of change in mouse synanthropy, and the invasion process of the house mouse are still poorly understood (Cucchi and Vigne, 2006). This is mainly because the information documenting the beginning of the process is hidden within the archaeological sites where early farming societies emerge at the beginning of the Holocene (Cucchi *et al.*, 2005). These zooarchaeological remains represent the best source of biological information for understanding the factors in the interaction of cultural and natural systems that led to the increased synanthropy and range of the house mouse. Despite this importance, these remains are still rarely collected and even less often studied.

Being able to date the onset of house mouse synanthropy and invasion has implications for evolutionary biologists working on the diversification of the commensal subspecies, their phylogeny and their hybridization, and the processes shaping the biodiversity we observe today. This is especially obvious since we know that human-induced biological invasion is one of the main factors in biodiversity erosion (Palumbi, 2001). Furthermore, the history of house mouse synanthropy is a key component in the reconstruction of the history of human

Evolution of the House Mouse, ed. Miloš Macholán, Stuart J. E. Baird, Pavel Munclinger, and Jaroslav Piálek. Published by Cambridge University Press.

health. Indeed, commensals like mice and rats, once in contact with humans, brought a brand new array of pathogens into the human environment (Shlyakhov, 1983). Therefore, the onset of rodent synanthropy has played a major role in the epidemiology of human infectious diseases.

Biomolecular markers like mitochondrial DNA (mtDNA) sequences, microsatellites and Y chromosome sequences collected in modern populations have revealed the main trends in the origin and evolution of the commensal mouse lineages (Bonhomme and Searle, Chapter 11 in this volume). However, timings in this history estimated from molecular methods have large intervals of uncertainty. While certainty in timing estimates is sure to increase with advances in technology and methodology, well-dated palaeontological and zooarchaeological occurrences will remain of crucial importance in this field.

This review will summarize our current understanding of the zooarchaeology of the house mouse through the question of the onset of house mouse synanthropy in connection with the Neolithic revolution, one of the main events in recent human history. The current knowledge about the dispersal of *Mus musculus musculus* and *M. m. domesticus* in association with the new Neolithic way of life, migration, and mobility of humans in Europe will then be reviewed. We will finally point towards new research directions in which zooarchaeology will fill crucial gaps in the reconstruction of house mouse evolution.

On the origin of house mouse synanthropy

From an ecological point of view, the relationship between mice and humans is commensal (Schwarz, 1939): mice benefit from this partnership but humans are neither *directly* harmed nor helped by mice. However, humans and commensal rodents also widely interact *indirectly* through a series of other organisms with which they have diverse ecological relationships: mice and rats themselves bear commensals or parasites that can have serious direct or indirect pathological effects on humans (Shlyakhov, 1983); they eat the same plant food as humans, including crops for which they are true competitors with humans, leading to their designation as 'pests' or 'agriculture parasites' (Hopf *et al.*, 1976; Mutze, 1989; Boxall *et al.*, 1997); and their efficiency in competition with other animals in the anthropogenic environment, to which they are very well adapted, may be an important factor in reduction of biodiversity (Lowe *et al.*, 2000). In addition, commensal rodents are also invasive species in the sense of Pascal *et al.* (2006), i.e. they frequently increase their distribution area, which also increases the area where humanity is confronted with their indirect negative effects. On the other hand, humans have taken advantage of this relationship by domesticating the house mouse and making of it the most widely used model for biomedical research (Fox *et al.*, 2007). All these points stress the contradictions with the term commensalism when applied to describing the relationship between mice and men. Indeed, we have demonstrated that mice can both harm (as pests and invasive species) and benefit (as medical model) their host (humans). Therefore, in this chapter, to describe the relationship between mice and men we will rather use the term synanthropy (from the Greek *syn-*, 'together/with' + *anthropos*, 'man'). This term, widely used in ecology, is generally applied to wild animal species which live near, and benefit from, an association with humans and their enviroment. The following section will review our knowledge on the onset of mouse synanthropy.

The Neolithic revolution in the Near East and the onset of house mouse synanthropy

The Neolithic revolution is one of the most profound transformations in the course of human history and laid the foundations for our modern societies. This transition from hunter-gatherer to sedentary societies happened independently in several regions of the world. In the Near East, during the Late Glacial, *ca.* 16 000–13 000 years (kyr) cal. BP, in a vast area including the high Euphrates and Tigris basins and the Jordan valley, some human groups became sedentary, living in small, round, mud and stone houses (Natufian and Khiamian cultures). Three millennia later, around the Holocene climate change, *ca.* 12–11 kyr cal. BP (Pre-pottery Neolithic A or PPNA culture) they began to cultivate wild cereals and legumes (Willcox *et al.*, 2008) and to control large wild-living ungulates (e.g. wild boar) (Vigne *et al.*, 2009; 2011b). The earliest evidence of morphological modifications of plants and animals due to this incipient domestication, however, dates only to the middle of the eleventh millennium BP (early PPNB cultures), and food production through agriculture and husbandry became the main source of subsistence only during the first half of the period 10–9.5 kyr cal. BP (for a recent summary, see Vigne, 2011). From this Near Eastern cradle, domesticated species and farming practices spread to the east (Iranian plateau), to the south (Arabia and North East Africa), to Europe (South Balkans, South Aegean), and to Transcaucasia, starting from 9 kyr BP.

The beginning of this long and complex process in the Near East sets the stage for house mouse synanthropy. The late Eitan Tchernov (Tchernov, 1968, 1984) was the first to provide the material and the theory for our understanding of the origin of this biological interaction between mice and men. He proposed that the shift towards sedentism created a new ecological niche attractive to pre-adapted populations of mice. This new niche would have acted as a barrier against predators and a buffer against temperature variation and food shortage in the natural habitat. This theory relies on the study of small vertebrate turnover in stratified deposits of southern Levant sites, namely the Hayonim cave (Fig. 3.1a) whose stratigraphic

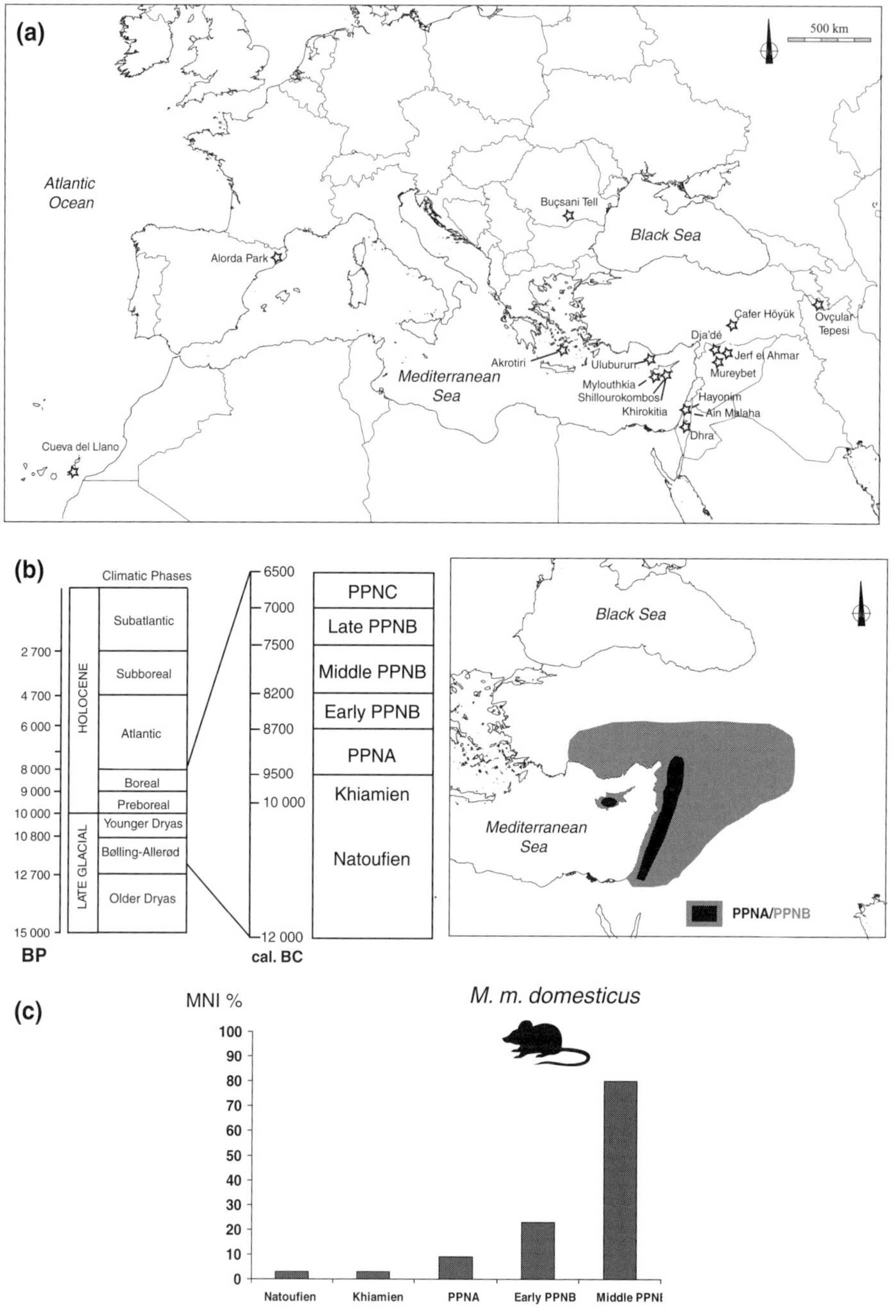

Figure 3.1 (a) Location of the archaeological sites mentioned in the text. (b) Left: chronological and climatic framework of the Near Eastern archaeological sites involved in the house mouse synanthropization; right: geographic extension of PPNA cultures

deposits run from the Middle Palaeolithic up to the Late Neolithic (Tchernov, 1968, 1984). In these deposits, *M. musculus* is present throughout the sequence but shows a massive increase in the layer associated with the Natufian culture (Tchernov, 1984). Based on modern experiments showing how small rural dumps next to human settlement influence the colonization of commensal rodents like *M. musculus* (Courtney and Fenton, 1976), Tchernov built his theory on the correlation between sedentism and the origin of commensalism. He proposed that the high frequency of commensal mice in small vertebrate archaeological assemblages could be used as a bio-indicator of sedentism (Tchernov, 1991a, b).

At the time when Tchernov first published the description of the small mammal fauna of the Hayonim cave (Tchernov, 1968), he didn't know that two different sympatric mouse taxa occurred in Israel and therefore likely also in the Hayonim's fossil deposits: a long-tailed, mostly synanthropic mouse (the western house mouse, *M. m. domesticus*), and a short-tailed, mostly anthropophobic mouse (the Macedonian mouse, *M. macedonicus*) (Auffray *et al.*, 1990a; see also Auffray and Britton-Davidian, Chapter 1 in this volume). The taxonomic status of the mouse remains from Hayonim was assessed by comparison with modern genotyped reference material based on a discriminant skull index (Auffray *et al.*, 1988). The results showed mice remains up to the Natufian layer with skull and teeth measurements within the range of current *M. macedonicus*, while the Natufian layer marks the onset of co-occurrence of both species. For Auffray *et al.* (1988) this was the evidence that house mouse dispersal reached the circum-Mediterranean during the end of the Late Glacial. The interpretation was that the newly created niche related to Natufian dwellings allowed the house mouse to spread in the Fertile Crescent region, avoiding competition with the Macedonian mouse. Indeed, it is established that, today, where the two species occur sympatrically, the house mouse is mainly restricted to indoors while it may be found more frequently outdoors when *M. macedonicus* is absent (Auffray *et al.*, 1990b; see also Auffray and Britton-Davidian, Chapter 1).

For these authors, the biometric attestation of the presence of *M. m. domesticus* in the Natufian from the Levant supported the correlation between sedentism of humans and the origin of commensalism. However, this correlation has been

Figure 3.1 (*cont.*)
involved in the synanthropization process of the house mouse in the Near East.
(c) Histogram displaying relative proportion of *M. m. domesticus* among small mammal accumulation in Neolithic pre-ceramic sites of Syria and Turkey (data from Cucchi, 2005; Haydar, 2004; unpublished PhDs). MNI (minimum number of individuals). Sample sizes per period: Natufian ($n = 33$), Khiamien ($n = 140$), PPNA ($n = 190$), early PPNB ($n = 62$), middle PPNB ($n = 250$).

questioned by Tangri and Wyncoll (1989) and Wyncoll and Tangri (1991). The main criticism concerns the absence of information about the origin of the rodent accumulations in the Natufian caves, which may result from owl pellets (Andrews, 1990) rather than from *in situ* death of commensal mice. Using modern data on strigiformes behaviour, Tangri and Wyncoll (1989) proposed that the changes observed in rodent frequencies throughout the Hayonim stratigraphy instead reflects either a natural variation of rodents in the surrounding habitat or fluctuations in the diet of birds of prey. The same criticism was later put forward regarding *Rattus rattus* remains in Hayonim (Ervynck, 2002). Pending a reappraisal of the circumstances of the small vertebrate accumulation process in the cave sites studied by Tchernov, we take the consensual view based on Auffray *et al.*'s (1988; see also Cauvin, 1977; Ibañez, 2008) parsimonious interpretation which considers that *M. m. domesticus* occurrence in the Natufian layer of the Hayonim cave provide us with the earliest evidence for house mouse dispersal into the circum-Mediterranean. The question of the geographic origin of house mouse synanthropy remains open.

To bypass the inherent problem of interpretation of small mammal accumulations from cave deposits, a study has been undertaken on northern Levant mouse remains collected within the occupation layer of outdoor settlements: Mureybet (Natufian/Khiamian/PPNA – Cauvin, 1977; Ibañez, 2008); Jerf el Ahmar (PPNA/PPNB – Mottram, 1991; Stordeur *et al.*, 1997); Dja'dé (early PPNB – Coquegniot, 1998a, b, 2000); and Çafer Hüyük (middle PPNB – Cauvin and Aurenche, 1982) (Fig. 3.1a). Their *in situ* accumulation within the human dwellings' occupation layers means there is no ambiguity regarding their association with humans (Cucchi, 2005; Cucchi and Vigne, 2006). Located in the cradle of the Near Eastern Neolithic revolution (Fig. 3.1b), these sites cover the transition between the hunter-gatherer subsistence strategies towards sedentary food production, which starts with the Khiamian culture (10 000 BC) but becomes dominant in the PPNA (9500–8700 cal. BC) and early PPNB (8700–8200 cal. BC). This period is characterized by a warmer and more humid climate at the end of the Younger Dryas with the increase of Graminaceae, including cereals like *Triticum beoticum* and *T. dicoccoides* (Sanlaville, 1998) and the onset of the proto-domestication of plants (Hillman, 1996; Willcox, 1996, 1999, 2000) and animals (Vigne, 2000). Several centres of the plant domestication experiment have been observed in different areas of the Near East (Willcox, 2005), but the Upper Euphrates valley is recognized as the earliest and the most important Near Eastern centre (Cauvin, 1997; Guilaine, 2000; Willcox *et al.*, 2008).

Archaeological remains of the genus *Mus* have been identified up to the subspecies level thanks to quantification by geometric morphometrics (Rohlf and Marcus, 1993; Adams *et al.*, 2004) of the molar morphology. Using modern comparative samples unambiguously assigned to the various species and subspecies of mice in Europe and probabilistic approaches, this method allowed identification of the mice remains from the outdoor archaeological sites as *M. m. domesticus* (Cucchi, 2005). Their relative frequency among the other synanthropic small mammals found in the same context shows an exponential increase of house mouse occurrence in Near Eastern Neolithic settlements from the PPNA onward (Fig. 3.1c). House mouse occurrence in northern Levant sedentary hunter-gatherer context (Natufian/Khiamian) is extremely low, being less than 5% of total small mammal remains. From the PPNA onward, house mouse occurrence increases to 10% and reaches 25% by the Early PPNB. By the Middle PPNB it is obvious that the house mouse is thriving in PPN habitats, with almost 80% of synanthropic small mammals collected from the occupation layer being *M. m. domesticus.*

This trend in the increase in the *M. m. domesticus* occurrence among the synanthropes within the human dwellings of the northern Levant from the PPNA between the tenth and ninth millennia cal. BC can be linked to several interrelated factors that increased the carrying capacity and attractiveness of the human settlements and induced/accelerated the synanthropic behavioural shift in mice during the Neolithic revolution of the Near East. The first factor is linked with human demographic increase due to the agricultural system (Bocquet-Appel and Bar-Yosef, 2008) and the exponential increase in settlement size and density of sedentary villages from the PPNA (Kuijt, 2000, 2008). Estimations from excavations in southern Levant suggest that Natufian sedentary sites did not exceed 0.2 ha in area, with a mean community size of 59 people. From the PPNA, community size increased drastically, with village areas larger than 1 ha and mean community size of 332 people (Kuijt, 2000). The settlement areas and community sizes kept on increasing, peaking during the Late PPNB (7500–7000 cal. BC) with settlement areas in southern Levant up to 12 ha and mean community size of almost 4000 people (Kuijt, 2000). This growth in human population and village density from the PPNA onward is due to an increase in birth rate (Guerrero *et al.*, 2008; Kuijt, 2008) and is thought more or less contemporaneous with the development of large-scale food storage in this region. Although it has been argued that sedentary Natufian communities had to rely on large-scale grain storage to reduce seasonal food scarcity risks (Testard, 1981, 1982), so far the only archaeological evidence of Natufian food storage is the paved pits of Ain Mallaha (Perrot, 1966) and the coated pits from the Hayonim terrace (Valla *et al.*, 1989), whose function could have been for funeral or waste disposal purposes rather than as food storage (Valla, 2000). It is undeniable that Natufian communities have heavily exploited their surrounding environment as suggested by the technological investment in the production of stone tools for plant harvesting (sickle blades) and processing (mortars and pestles) (Bar-Yosef, 1998), but the absence of definite archaeological proof of large-scale storage (Bar-Yosef, 1998; Goring-Morris and Belfer-Cohen, 1998) suggests a low level of food storage investment (Kuijt, 2009).

The earliest archaeological evidence for large-scale food storage has been found in PPNA villages of both the northern and southern Levant. In the northern Levant, clear archaeological evidence has been found at the Jerf el Ahmar site (Stordeur *et al.*, 1997) in northern Syria, where direct (bones and droppings) and indirect (gnawing marks on stone vessels) evidence of house mice living within human dwellings has been recorded (Willcox *et al.*, 2008). In Jerf el Ahmar, large-scale food stores have been identified within several community buildings (Fig. 3.2a), which, according to their architectural features and the amount of burnt seed found within, were unequivocally used as granaries, among other collective uses (Stordeur and Willcox, 2009). In the southern Levant, excavation of the PPNA site of Dhra' (Fig. 3.1a), dated to *ca.* 9000 cal. BC and situated next to the Dead Sea has revealed a very sophisticated system of granaries built within the houses and designed with suspended floors (Fig. 3.2b). This would preserve the grain and also protect it from rodent

(a)
(b)

Figure 3.2 (a) Aerial photograph of the community building EA 30 of the PPNA site of Jerf el Ahmar (Syria), where large silos (grain storage) have been identified (arrowed); from the mission to El Kowm-Mureybet of the French Foreign Office, directed by Daniel Stordeur. See plate section for a colour version of this figure. (b) Interpretative reconstruction of the granary structure of Phase 4 in the PPNA site of Dhra' (Jordan); from Kuijt and Finlayson, 2009, illustration by E. Carlson.

parasitism (Kuijt and Finlayson, 2009). In the course of the Neolithic revolution in the Levant, the food storage component appears to have happened simultaneously in north and south between the tenth and ninth millennia cal. BC.

Directly linked with large-scale grain storage, the presence of cultivated fields in the vicinity of the PPNA villages represents another component of Neolithic 'human niche construction' that had a direct effect on the house mouse synanthropy. Although direct evidence of plant domestication is not available, the occurrence of wild cereals and pulses outside their natural range and the presence of weeds of cultivation strongly suggest that cultivation of plants was practised in the northern Levant from the tenth millennium cal. BC (Willcox *et al.*, 2008, 2009). The presence of large and densely inhabited villages with large-scale food stores surrounded by cultivated fields emerging from the PPNA onwards offered a new array of advantages for synanthropic mice compared to the older, sedentary Natufian settlements. This new Neolithic habitat provided a sustained food supply, increased protection against predation and interspecific competition, and perhaps a better buffer against climate variation in the natural habitat. These advantages likely contributed to the behavioural shift of *M. m. domesticus* towards synanthropy that later led to its hegemony among the synanthropic species of the Middle PPNB villages.

If the sedentism is a prerequisite for the beginning of house mouse synanthropy, the increase in farming practices relying on large-scale grain storage since the PPNA created an ecological niche where house mice could really thrive and conquer the human habitat. Once it occupied the villages/houses, this newly synanthropic species could start its conquest of the world following the spread of farming.

Cyprus: earliest house mouse biological invasion

The first zooarchaeological evidence of a successful biological invasion by the house mouse has been documented on Cyprus. Although situated in the Mediterranean Sea, Cyprus does not belong to the European human dispersal of the Neolithic but to the extension of the PPN cultures initiated in the ninth millennium BC from the Levantine corridor. Therefore, the colonization of this island by the house mouse is associated with the Neolithic revolution in the Near East rather than in Europe.

Cyprus has never been connected to the mainland, even during the Pleistocene Glacial maximum (Knapp *et al.*, 1994). Its Pleistocene endemic mammal fauna included dwarf hippos (*Phanourios minutus*) and elephants (*Elephas cypriotes*), a species of genet (*Genetta* cf. *plesictoides*) and one or two unnamed mouse species (Bate, 1903a, b; Boekschoten and Sondaar, 1972). None of these mammals were considered to be the direct ancestors of those described from the Neolithic period on the island (Vigne, 1999). The first wave of anthropogenic transfer of animals onto Cyprus started with maritime foragers (Natufian/Khiamian) coming from the Levant at the beginning of the Holocene. They introduced wild boars (Vigne *et al.*, 2009), probably because the island was devoid of game animals since the extinction of the endemic large mammals during the Younger Dryas. The earliest known Neolithic presence in Cyprus is dated to *ca.* 8700 cal. BC (Manning *et al.*, 2010; Vigne *et al.*, 2011a), but as yet there is no evidence of cultivation or animal husbandry at that time, with the possible exception of the domestic dog. Incipient farming is well attested by, starting from 8400 cal. BC, the introduction of goat, cattle, cereals, and legumes, and later (*ca.* 8000 cal. BC) of sheep, pigs, and the Mesopotamian fallow deer (Peltenburg *et al.*, 2000; Guilaine *et al.*, 2011; Vigne *et al.*, 2011b). We speculate that the house mouse may have arrived with this Neolithic transfer package.

Kissonerga-Mylouthkia (Peltenburg *et al.*, 2000; Peltenburg, 2003), on the western coast of Cyprus, and Shillourokombos (Guilaine *et al.*, 2011) (Fig. 3.1a) are the only Early Neolithic sites of Cyprus that have provided important evidence regarding the beginning of mammal transfer to the island in general (Vigne *et al.*, 2011b) and regarding the arrival of the house mouse in particular (Cucchi *et al.*, 2002). Deep-water wells excavated in Mylouthkia dating to the last centuries of the ninth millennium cal. BC, have acted as pitfall traps for many small vertebrate species living in the vicinity of the early Neolithic village. The majority of these trapped vertebrates were mice. Geometric morphometric analysis of the teeth in these remains indicated two different taxa (Cucchi *et al.*, 2002). Teeth in the first taxon were morphologically close to those of *M. macedonicus*, and significantly different from the second type of tooth (Cucchi *et al.*, 2002). Further morphometric analyses demonstrated similarities between the first taxon and teeth from Pleistocene endemic Cypriot mice (Cucchi, 2005), suggesting that all of them were representatives of a recently described species, the Cypriot mouse, *M. cypriacus* (Bonhomme *et al.*, 2004; Cucchi *et al.*, 2006; see also Auffray and Britton-Davidian, Chapter 1). Genetic characterization of the living 'wild' mice of Cyprus supported this hypothesis (Bonhomme *et al.*, 2004; Cucchi *et al.*, 2006). The second morphotype at Mylouthkia was indistinguishable from those of the house mouse (*M. m. domesticus*), testifying that this species had been introduced to Cyprus *ca.* 8400 cal. BC at the latest as a component of the Neolithic package. It would quickly become the main synanthropic species in the large villages of Late Neolithic Cyprus, like Khirokitia (Cucchi, 2005) (Fig. 3.1a).

These house mice must be seen as unwelcome companions, probably transported as stowaways hidden within the bags of crops and hay, just like modern-day mouse stowaway transport happens in cargoes (Baker, 1994). Therefore, in the absence of direct archaeological evidence of Neolithic maritime vessels, the invasive process of Cyprus by stowaway house mice documents indirectly the seafaring skills of the early Neolithic communities (Vigne and Cucchi, 2005). Presence of boat superstructure such as decks is considered by Vigne (2009) to be a necessary feature to travel unnoticed onboard. In addition, insular populations of the house mouse and rodents in general (Berry, 1964; Britton-Davidian, 1990; Pergams and Ashley, 2001; Millien, 2006) display very fast genetic and morphological drift once isolated from the mainland source. Only a constant migrant flow can prevent these stochastic effects. That is why we suppose that absence of morphological drift observed in Neolithic mice on Cyprus compared to their counterparts of the mainland over several millennia results in a constant migrant flow carried by frequent (several introductions per year) stowaway transports of house mice from the mainland to Cyprus. This migrant flow would have prevented the phenotypic change through neutral drift in the Neolithic house mouse populations of Cyprus either by gene flow or by constant extinction–colonization turnover (Cucchi, 2005; Vigne and Cucchi, 2005). These observations indirectly support the idea that the Early Neolithic seafarers of the eastern Mediterranean were skilled sailors capable of fast and frequent journeys over long distances in unpredictable waters. Modern-day experiments replicating Neolithic seafaring (Tichý, 1999) suggest that the journey to reach Cyprus for Early Neolithic voyagers required crossing at least 70 km of open water. So far, the longest one-day journey in a reconstruction Neolithic rowing boat (monoxylon) is 57 km and took 14 hours and two landings (Tichý, 1999). Voyaging to Cyprus during daylight was probably mandatory for obvious safety reasons, but also for the wellbeing of livestock onboard. These reasons and a lack of favourable currents (Vigne, 2009) suggest that sails were used to speed up the crossing to Cyprus.

Has this human maritime Neolithic dispersal in the Mediterranean, which reached the western end of the basin by the sixth millennium BC, been an important vector of the house mouse invasive process?

The zooarchaeological documentation accumulated on the origin of the house mouse synanthropization so far only concern the *M. musculus domesticus* subspecies. The beginning of the commensalism process for *M. m. musculus* and *M. m. castaneus* is currently unknown since there have been no dedicated zooarchaeology studies for these taxa and therefore no reliable data are available. New data on the synanthropization process for the eastern house mouse (*M. m. musculus*) are just coming out now and will be discussed in the following section.

House mouse invasion of Europe

According to palaeontological sources, the *Mus* genus has been present in Europe since the Middle Pleistocene in the Pannonic basin (Janossy, 1996). After the last glacial period (26 000–20 000 BP), it is absent from the fossil records of mainland Europe and Mediterranean peninsulas and has only been recorded on Mediterranean islands as endemic forms (Auffray *et al.*, 1990c). For these reasons, Pleistocene lineages are not considered to be the direct ancestors of the extant European house mouse lineages (Thaler, 1986; Auffray *et al.*, 1990c; Auffray and Britton-Davidian, 1992). General consensus rather agrees on a dispersal process happening, facilitated by man-mediated dispersal during the mid-Holocene (Boursot *et al.*, 1993; Bonhomme *et al.*, 1994). According to phylogeographic scenarios, the dispersal of the western house mouse (*M. m. domesticus*) followed a Mediterranean route out of its Near Eastern cradle while the eastern house mouse (*M. m. musculus*) followed a route north of the Black Sea and along the Danube (Boursot *et al.*, 1996).

This section will present the latest historical process for the western house mouse dispersal from its Near Eastern cradle towards Europe and beyond, and the latest research state on the origin and dispersal of the eastern house mouse will be discussed.

Transcaucasia

The Transcaucasian isthmus between the Caspian and the Black Sea is a path between Asia and Europe for both man and other animals. As such, it represents a key area to study dispersal of animals and peoples. This path is traversed by mountain chains stretching over 1200 km and which reach 5600 m in altitude, which have acted as both refuges and barriers to the movement of animals, people, goods, and cultures. For archaeologists, Transcaucasia is a key area to understand the evolution of Neolithic culture at the periphery of its Near Eastern cradle. From the seventh millennium cal. BC, cultures bearing most of the Neolithic traits appear in Transcaucasia (Lombard and Chataigner, 2004), but the relative importance of local development versus allochthonous cultural transfer from the Near Eastern cradle via the movement of people, animals, plants, and knowledge needs to be deciphered.

For evolutionary biologists, Transcaucasia is also a key area where hybridization between *musculus* and *domesticus* mouse subspecies occurs (Orth *et al.*, 1996; Mezhzherin *et al.*, 1998). House mice from this region display a unique genetic signature with a high level of ancestral polymorphism. Therefore,

Transcaucasia is interpreted either as a place of divergence between the two lineages or as a secondary contact with a high level of hybridization between them (Milishnikov *et al.*, 2004).

Archaeology and evolutionary biology research fields meet when archaeological remains of mice are collected in the rare known sites of this region. Current archaeological studies led jointly by French and Azeri archaeologists in the Nakhitchevan (region of the Lesser Caucasus) are uncovering important Chalcolithic and Bronze Age occupations dated from 4500 to 2000 cal. BC (Marro and Hauptmann, 2000). In one of these Chalcolithic villages, named Ovçular Tepesi (Fig. 3.1a), deep pit structures for food storage have yielded large amounts of small vertebrate remains (Marro and Hauptmann, 2000). They are mostly from the genus *Mus*, but with other synanthropic species like the common hamster and the white-toothed shrew (Kovacs *et al.*, in prep.). The geometric morphometric study of the mouse remains confirms the occurrence of *M. m. domesticus* in Transcaucasia around 4000 BC, but *domesticus* mice from Ovçular Tepesi display important phenotypic divergence from both modern and ancient *M. m. domesticus* from northern Levant (Kovacs *et al.*, in prep.).

From an archaeological perspective, the presence of *domesticus* as the main synanthrope species in the Chalcolithic villages of the Nakhitchevan region contributes indirectly to the question on the origin of the Transcaucasian Neolithic culture. Given that the earliest commensal *domesticus* populations have been identified in the Upper Euphrates valley between the tenth and ninth millennia cal. BC (Cucchi, 2005; Cucchi *et al.*, 2005; Cucchi and Vigne, 2006), their presence in Ovçular Tepesi by the end of the fifth millennium cal. BC must have been the result of facilitated dispersal from the cultural sphere of northern Mesopotamia. The morphological divergence of the Ovçular Tepesi mice from their Neolithic *domesticus* relatives of northern Levant suggests that this dispersal happened during sporadic contacts without sustained exchanges.

Modern house mouse specimens from the Nachitchevan are currently being studied for nuclear and mtDNA signatures and with morphometrics tools to understand whether these morphological divergences are related to (1) long-distance colonization; (2) several waves of *domesticus* coming from different parts of the Near East; or (3) hybridization between the two lineages.

The history of *M. m. domesticus* invasion in Mediterranean Europe

With the aim of reconstructing the history of the house mouse invasion into the Mediterranean area, the archaeological evidence for occurrences of the species have recently been critically reviewed on the reliability of available data (Cucchi *et al.*, 2005). To be considered as reliable, house mouse archaeological occurrences had to pass statistical (sample size), taxonomic, and archaeological (context, dating, mode of collection) validation thresholds. This review concluded that *M. m. domesticus* remained restricted to the eastern Mediterranean from the ninth to the second millennia BC, but then abruptly spread to the western part of the basin. The colonization of the eastern Mediterranean starts with the early PPNB colonization of Cyprus by the end of the ninth millennium BC as described in the previous section. The invasive process does not seem to have been successful on the other islands of the eastern Mediterranean until the Bronze Age (fourth millennium BC), when house mouse occurrence is documented in the Minoan city of Akrotiri on the Aegean island of Santorini (Papayannis, in press). The role of Bronze Age maritime trade as an efficient vector of house mouse dispersal in the eastern Mediterranean has been directly documented thanks to the thorough excavation of a Late Bronze Age Uluburun shipwreck (Fig. 3.1a) by sub-aquatic archaeologists Cemal Pulak (Texas A&M University) and colleagues. This shipwreck, dated to 1300 cal. BC, was found off the Anatolian shore and became famous due to the richness of its cargo, which represented the wealth of goods exchanged between the main trading ports of the eastern Mediterranean (Pulak, 1996, 1998). Among the sediments encrusted between layers of copper ingots, archaeologists found an entire murine mandible identified as *M. m. domesticus* (Cucchi, 2008). Given the copper incrustation on the mandible due to its long-lasting underwater contact with the oxidizing copper ingots, the association between these house mouse remains and the 44 m deep underwater vessel deposit appears solid. This young house mouse (around three months) didn't display any dental size increase typical of isolated populations on boats (Vigne and Villié, 1995) or small islands (Adler and Levins, 1994; Berry, 1998; Pergams and Ashley, 2001; Millien, 2006), which suggests it was a stowaway transport of a young specimen drowned when the boat sank off the Anatolian coast. Cross-referencing historical and archaeological sources and given the phenotypic similarities between this stowaway mouse and the current population of Syria, it has been suggested that the ship sailed from the city of Ugarit, situated on the modern-day northern Syrian coast (Cucchi, 2008).

The timing and pace of mouse colonization of the western Mediterranean differ from the eastern region. Suddenly, from 1000 BC onward, all the validated western Mediterranean sites display reliable large samples of house mouse. This fast and massive colonization contrasts with the slow colonization process inferred for the eastern Mediterranean and therefore likely implies different processes. Numerous stowaway transport events probably occurred between 8000 and 1000 cal. BC towards the western Mediterranean, but none of these led to successful establishment of permanent mouse populations. Cucchi *et al.* (2005) proposed that the colonization of the western Mediterranean was delayed because maritime trade remained restricted to the main eastern Mediterranean cities and harbours until the last millennium BC. It is only with the commercial and demographic expansions of the Phoenicians and Greeks from the ninth century BC that the Mediterranean became widely open to maritime traffic, with long-distance navigation capable of producing sufficiently long and frequent transport of mice for the species to establish stable populations in the west. It was also proposed that the carrying capacity of the commensal niche in the western Mediterranean could have also played a role in the delayed colonization of that area (Cucchi *et al.*, 2005). Before the last millennium BC, this niche was perhaps not yet ready to sustain viable house mouse populations for two main reasons. First, there may have been a lack of settlements in the western Mediterranean large and stable enough to support commensal mice throughout the Bronze Age: the earliest recorded western Mediterranean cities date to the last millennium BC in the Apennine peninsula and the Etruscan by 800 BC. The second reason is the possible presence of an autochthonous competitor for the western commensal niche, as suggested by the presence of a wood mouse (*Apodemus sylvaticus*) in the deposit of a Late Neolithic granary in the French Jura (Vigne, 1997).

After 1000 BC, sustained mouse migrant flow carried by long-distance and frequent maritime traffic, as well as the multiplication of cities and *oppida* in the northwestern Mediterranean, allowed the successful mouse biological invasion of the whole Mediterranean (Cucchi *et al.*, 2005). On the northeastern Mediterranean shores of Spain the presence of *M. m. domesticus* in the fourth century BC deposit of the Alorda Parc Iron Age fortress (Fig. 3.1a) recently provided the earliest evidence for the successful colonization of the western European commensal niche by the house mouse (Valenzuela-Lamas *et al.*, 2011). Phenetic relationships between these Spanish Iron Age mice and the current Tunisian population that have retained signals of their Punic origin in their mitochondrial genome suggest a dispersal carried by the Cartagena maritime trade network (Valenzuela-Lamas *et al.*, 2011).

What about the Neoltihic dispersal beyond the Straits of Gibraltar? Until recently, the fossil record in the Canary Islands (La Cueva del Llano in Fuerteventura, Fig. 3.1a) suggested dispersal of *M. m. domesticus* via a Neolithic stowaway translocation around 5000 BC (Castillo *et al.*, 2001). This scenario challenged the hypothesis of recent (1000 BC) colonization of the western Mediterranean or suggested another wave of dispersal perhaps along the north African coast. However, although geometric morphometric analysis confirmed the occurrence of *M. m. domesticus* on the archipelago (Michaux *et al.*, 2007), accelerator mass spectrometry (AMS) dating brings the samples date forward to the seventh century BC/third century AD (Alcover *et al.*, 2009). These results are consistent with the Canarian archipelago being colonized by the house mouse following the invasion of the western Mediterranean basin. The precise time and vector of dispersal need to be further investigated.

The *M. m. musculus* colonization of Europe: a fading mystery

Although we now have a rather clear picture of *M. m. domesticus* colonization of the Mediterranean, the history of the biological invasion of eastern and central Europe by *M. m. musculus* has long been unsolved since no archaeological remains were available. The palaeontological and zooarchaeological reappraisal made by Auffray and colleagues at the beginning of the 1990s (Auffray *et al.*, 1990c; Auffray and Britton-Davidian, 1992) suggested that the eastern house mouse colonization of Europe was a recent and arrhythmic process. But it also clearly stated the near absence of zooarchaeological data, and the date of *ca.* 5000 BC that it proposed for the earliest evidence of the house mouse in Belgium has been either invalidated or at least strongly questioned since that time (Cucchi *et al.*, 2005).

More recent modern mtDNA studies suggest that a lineage of *M. m. domesticus* was first to colonize mainland Europe following the Neolithic dispersal of the Linear Pottery culture along the Danube during the sixth and fifth millennia BC. This early dispersal of *M. m. domesticus* would have been outcompeted and pushed backward to its nowadays distribution by a later expansion of *M. m. musculus*. This phylogeographic scenario may find support in the history of the Neolithic expansion throughout Europe. From 6800 BC, Neolithic dispersal in southeastern Europe is thought to be the result of successive human cultural expansions from the northwestern Anatolian peninsula (Marmara zone) (Anthony and Chi, 2010). Given the presence of *M. m. domesticus* as the commensal mouse in Neolithic human dwellings in the Near East and, thus, probably in Anatolia, we could hypothesize that *M. m. domesticus* was the most likely candidate for an early dispersal in southern Europe. Nevertheless, there is so far no archaeological evidence to test this hypothesis (Cucchi *et al.*, 2005).

Recent excavation of Buçsani La Pod Tell (Fig. 3.1a) in the Danube valley of southern Romania has provided a new insight into the house mouse invasion of southern Europe. A Late Neolithic level of the tell (mound) dated between the end of the fifth and the beginning of the fourth millennium BC yielded remains of several houses, burned down along with their entire contents. One house contained a large amount of small mammal remains mainly identified as *Mus*, collected by small-sized wet sieving of the occupation layer of the house and its storage room. Most of these remains display signs of carbonization and calcinations that suggest an *in situ* accumulation of commensal mice killed in their galleries by the intense heat of the fire, and testify to their contemporaneity with the human dwellings. The demographic profile of these mouse remains based on dental age shows the presence of neonates and immature individuals, suggesting a stable synanthropic mouse population. Geometric morphometrics and probabilistic classification of these remains provide highly significant evidence for the occurrence of the eastern house mouse (*M. m. musculus*) as the main synanthropic species within the Late Neolithic human settlements of southern Romania (Cucchi *et al.*, 2011). These unexpected results seem to go against the phylogeographic arguments (above) that suggested *M. m. domesticus* entered southeastern Europe first. Our previous review of archaeological house mouse occurrences in Europe (Cucchi *et al.*, 2005) had already shown that this *domesticus*-first scenario was not supported by the zooarchaeological evidence, since the only validated occurrence of the house mouse from mainland Greece was dated to the Late Bronze Age, around 1000 BC. Furthermore, ongoing research on Aegean small mammal synanthropes have found no validated occurrence of house mice in Neolithic mainland Greece (Papayiannis and van den Hoek Ostende, in press), while finding large quantities of mouse remains with Minoan context on the island of Santorini (Fig. 3.1a) (Papayannis, in press). These current studies suggest that the *M. m. domesticus* colonization of the eastern Mediterranean basin was not driven by Neolithic human dispersal but by the Bronze Age maritime traffic starting around the fourth millennium BC. The discovery of the eastern house mouse in the Late Neolithic context of Romania therefore suggests that *M. m. musculus* was the first commensal subspecies to colonize mainland Europe following the cultural spread within humans of the Neolithic way of life. To address this hypothesis, validated occurrences of the house mouse in the archaeological sites of the Balkans between the seventh and fourth millennia BC must be found.

The presence of *M. m. musculus* as the main synanthropic species of the large Chalcolithic villages of southeastern Europe raises another question: are we dealing with a synanthropic *M. m. musculus* that is native to southeastern Europe and spreads by development of the Neolithic niche or a synanthrope introduced by human dispersal from a source situated further east? Considering the absence of *M. m. musculus* from the fossil deposits of mainland Europe before the Neolithic dispersal (Janossy, 1996), we argue that the *M. m. musculus* lineage is not a native synanthropic species that emerged with neolithization of southeastern Europe. We rather suggest that Buçsani La Pod Tell house mice represent a synanthropic metapopulation settled in southeastern Europe during the westward expansion of the *M. m. musculus* colonization of temperate northern

Europe (Cucchi *et al.*, 2011). The expansion of *M. m. musculus* out of its northeastern Indian cradle may have followed a route through the Iranian plateau and then north of the Caspian Sea into the Pontic steppes (Boursot *et al.*, 1996; Darvish *et al.*, 2006; Rajabi-Maham *et al.*, 2008). Then, *M. m. musculus* may have occurred on the Pontic steppes before the Neolithic revolution in Europe. This natural dispersal of *M. m. musculus* would have come into contact with neolithization of the forest steppe around the sixth millennium BC. The introduction of cultivated plants in the steppes via the Balkans (Monah, 2007) would have triggered the beginning of synanthropy in the *musculus* lineage. Therefore we propose a centre of synanthropization for the *musculus* lineage in eastern Europe north of the Black Sea (see Fig. 3.3). The discovery of mining and metallurgy in southeastern Europe stimulated long-distance trade between the Chalcolithic cultures of southeastern Europe and the Pontic steppes. Balkans copper was traded eastwards into steppe communities and the trade routes from the lower Danube to Moldavia and the Ukraine passed through the towns and villages of the Gumelniţa–Cucuteni–Tripolye settlements. This trading network might have been the vector of *M. m. musculus* expansion from the Pontic steppes towards the Chalcolithic settlements of the lower Danube valleys. Given the history of agriculture in southeastern Europe and along the Danube valleys since the mid-seventh millennium BC, this potential vector would have had good probability to create successful colonies of house mice within the large Chalcolithic settlements of this part of Europe. Long-distance trade and large settlements during the Chalcolithic are regarded as factors contributing to the quick and successful colonization of southeastern Europe by *M. m. musculus* (Cucchi *et al.*, 2011).

Synthesis and perspectives

Zooarchaeology has proven crucial to reconstructing the house mouse history and identifying the factors involved in its synanthropization and dispersal process across Europe, illustrated in Fig. 3.3. The Neolithic revolution starting in the northern Levant between the tenth and ninth millennia cal. BC has been suggested as the driving force at the origin of *M. m. domesticus* synanthropy, creating a fast-increasing commensal niche for this lineage (for a view on mouse synanthropy from outside of the sphere of influence of the Fertile Crescent centre of domestication, see Yonekawa *et al.*, Chapter 4 in this volume). The rise of cereal and legume cultivation and storage during the PPNA, *ca.* 10 000 BC, along with size and density increase of the farming villages from the PPNA, created a new array of advantages allowing commensal house mice to thrive in the Near

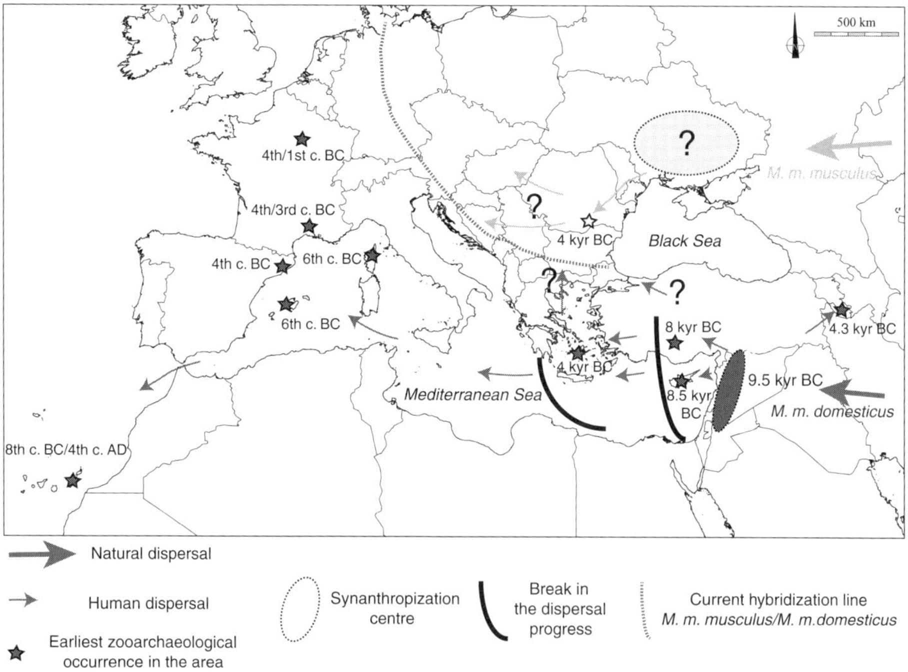

Figure 3.3 Map synthesizing the current knowledge and hypotheses on the history of the house mouse synanthropization and dispersal process in Europe. Kyr = thousand years, c. = century. See plate section for a colour version of this figure.

Eastern Neolithic environment. From this area the *domesticus* lineage was translocated towards Cyprus in the Mediterranean as early as the end of the ninth millennium BC and to Transcaucasia at least by the fourth millennium BC. The pace of *M. m. domesticus* expansion towards the eastern Mediterranean basin seems unrelated with the Mediterranean Neolithic human diaspora. Scarce zooarchaeological evidence rather suggest the colonization of the eastern Mediterranean was not achieved before the Early Bronze Age, driven by maritime trade between the city states and the empires of the eastern Mediterranean. The western Mediterranean dispersal is a later process achieved quickly and on a large scale throughout the western Mediterranean at the very beginning of the last millennium BC. The commercial and demographic expansions of Phoenicians and Greeks from the ninth century BC supported much larger migrant flow throughout the Mediterranean and allowed the successful colonization of the main trading harbours and outposts in the western Mediterranean.

Although *M. m. domesticus* spread is now relatively well understood, the spread of *M. m. musculus* and time of secondary contact between these two lineages in Europe is poorly understood due to the lack of reliable archaeological remains.

The presence of the eastern house mouse (*M. m. musculus*) as the main synanthropic species in a Late Neolithic village of southeastern Romania, *ca.* 3600 BC has only recently been documented (Cucchi *et al.*, 2011). We hypothesize synanthropy in the *musculus* lineage began in eastern Europe north of the Black Sea. The copper trade network between southern and eastern Europe and the large farming settlements of southeastern Europe of the sixth millennium BC along these trade routes have driven its successful invasion of Europe (Cucchi *et al.*, 2011).

This overview of the origin and dispersal of the house mouse in the Near East and Europe have raised several research perspectives that must be undertaken to complete the Holocene history of house mice on the European continent.

The Holocene history of the western house mouse is the most documented, but two parts of its history remain unclear. The first part concerns how the *domesticus* lineage reached the northern Levant from its northern Indian cradle. If human sedentism started in the Natufian culture of the southern Levant, at the western end of the 'Fertile Crescent', then this synanthropic niche cannot explain how the ancestor of the house mouse colonized the Fertile Crescent from the east. Instead, this suggests that the *domesticus* lineage was already present in the Levant when people took up the settled way of life. We propose to test the hypothesis of a natural dispersal of *domesticus* from the northern shores of the Persian Gulf to Mesopotamia, then throughout the Tigris–Euphrates valley and up to the Levant during the warm and moist Interstadial (Bølling/Allerød) at the final stage of the last glacial period, *ca.* 14 700 to 12 700 BP. Current research focused on Pleistocene and Early Holocene mammals in Iran will provide further insights into the natural occurrence of *Mus* in this transition area between the Asian cradle and the putative location of the onset of synanthropy of the *domesticus* lineage in Mesopotamia. The second part concerns the details of the spread of *M. m. domesticus* in southeastern Europe and when it met with *musculus*. Research into the maritime mouse transport across the Aegean and routes across the Balkans looking for reliable evidence of presence/absence of the house mouse in Neolithic settlements between the seventh and fourth millennia BC will hopefully provide definitive evidence to address this question.

The Holocene history of the eastern house mouse is still to be documented, but the recent discovery in southeastern Romania (Cucchi *et al.*, 2005) has highlighted the role of the Neolithic revolution in the spread of *M. m. musculus* in mainland Europe. This finding contrasts with the dispersal of *M. m. domesticus* in Mediterranean Europe, where zooarchaeological clues rather suggest spread related with the Bronze Age maritime trade. We propose to test for a potential synanthropization centre of the *musculus* lineage with a systematic survey of small vertebrates in the Neolithic human dwellings north of the Black Sea between the sixth and fourth millennia BC. To understand the dynamics and factors influencing the spread of *musculus* in central Europe we must stress the crucial need for the systematic small vertebrate collections in reliable zooarchaeological contexts. These research directions must be pursued in conjunction with direct dating of mouse remains to ascertain their chronology.

We would like to finish this perspective with a plea for palaeogenetic studies on zooarchaeological house mouse remains. Although several phylogeographic studies of the house mouse have stressed the use of this taxon as a proxy for past human migration (Searle *et al.*, 2009a, b; Gabriel *et al.*, 2010; Jones *et al.*, 2010; Bonhomme *et al.*, 2011), none of the scenarios can be directly related to the interacting dynamics of human societies and their environments given that the genetic patterns are obtained from extant populations. So far there is no study including ancient house mouse mtDNA sequence and only the Pacific rat (*Rattus exulans*; Matisoo-Smith and Robins, 2004) and the common vole (*Microtus arvalis*; Martínková *et al.*, submitted) have been used for phylogeographic studies on commensal or anthropophilous rodents to track past human migrations.

Ancient DNA sequences from mice collected during archaeological excavations will also provide the information necessary to test evolutionary hypotheses concerning the radiation and diversification processes of the two commensal lineages. In particular, genetic characterization of extant populations is most likely not representative of the successive events of colonization, extinction, and replacement of mouse groups by others. The ancient DNA will allow some of the phases of the complex process of the establishment of the current distributions to be revealed.

Finally, synanthropy is not a property of *M. musculus* (Auffray and Britton-Davidian, Chapter 1; Suzuki and Aplin, Chapter 2). It is obvious that it originated independently for each subspecies, i.e. much after the time of divergence between the subspecies. In this chapter we focused on *M. m. domesticus* and *M. m. musculus* for which numerous data exist. However, the origin of synanthropy in *M. m. castaneus*, as well as dispersion routes for this subspecies, remain poorly documented and require more attention in the future.

REFERENCES

Adams, D. C., Rohlf, F. J., and Slice, D. E. (2004). Geometric morphometrics: ten years of progress following the 'revolution'. *Italian Journal of Zoology*, **71**, 5–16.

Adler, G. H. and Levins, R. (1994). The island syndrome in rodent populations. *The Quarterly Review of Biology*, **64**, 473–90.

Alcover, J. A., Rando, J. C., Garcia-Talavera, F., *et al.* (2009). A reappraisal of the stratigraphy of Cueva del Llano (Fuerteventura) and the chronology of the introduction of the house mouse (*Mus musculus*) into the Canary Islands. *Palaeogeography, Palaeoclimatology, Palaeoecology*, **277**, 184–90.

Andrews, P. (1990). *Owls, Caves and Fossils*. London: British Museum (Natural History).

Anthony, D. W. and Chi, J. (eds) (2010). *The Lost World of Old Europe: The Danube Valley, 5000–3500 BC.* New York: Institute for the Study of the Ancient World.

Auffray, J.-C. and Britton-Davidian, J. (1992). When did the house mouse colonize Europe? *Biological Journal of the Linnean Society*, **45**, 185–90.

Auffray, J.-C., Marshall, J. T., Bonhomme, F., and Thaler, L. (1990a). Focus on the nomenclature of european species of *Mus*. *Mouse Genome*, **88**, 7–8.

Auffray, J.-C., Tchernov, E., Bonhomme, F., *et al.* (1990b). Presence and ecological distribution of *Mus musculus domesticus* and *Mus spretoïdes* in Israel: circum-Mediterranean vicariance in the genus *Mus*. *Zeitschrift für Saügetierkunde*, **55**, 1–10.

Auffray, J.-C., Tchernov, E., and Nevo, E. (1988). Origine du commensalisme de la souris domestique (*Mus musculus domesticus*) vis-à-vis de l'homme. *Comptes Rendus de l'Académie des Sciences – Series III – Sciences de la Vie*, **307**, 517–22.

Auffray, J.-C., Vanlerberghe, F., and Britton-Davidian, J. (1990c). The house mouse progression in Eurasia: a palaeontological and archaeozoological approach. *Biological Journal of the Linnean Society*, **41**, 13–25.

Baker, A. E. M. (1994). Stowaway transport rates of house mice (*Mus domesticus*) and deermice (*Peromyscus maniculatus*). In *Proceedings of the Sixteenth Vertebrate Pest Conference*, ed. W. S. Halverson and A. C. Crabb. Davis, CA: University of California, pp. 106–12.

Bar-Yosef, O. (1998). The Natufian culture in the Levant, threshold to the origins of agriculture. *Evolutionary Anthropology*, **6**, 159–77.

Bate, D. M. A. (1903a). The mammals of Cyprus. *Proceedings of the Zoological Society of London*, **2**, 341–8.

Bate, D. M. A. (1903b). On an extinct species of genet (*Genetta plesictoides*, sp.n.) from the Pleistocene of Cyprus. *Proceedings of the Zoological Society of London*, **2**, 121–4.

Berry, R. J. (1964). The evolution of an island population of the house mouse. *Evolution*, **18**, 468–83.

Berry, R. J. (1998). Evolution of small mammals. In *Evolution on Islands*, ed. P. R. Grant. Oxford: Oxford University Press, pp. 35–50.

Bocquet-Appel, J. P. and Bar-Yosef, O. (eds) (2008). *The Neolithic Demographic Transition and Its Consequences*. New York: Springer.

Boekschoten, G. J. and Sondaar, P. Y. (1972). On the fossil mammalia of Cyprus, I & II. *Proceedings of the Koninklijke Nederlandse Akademie van Wetenschappen (Series B)*, **75**, 306–38.

Bonhomme, F., Anand, R., Darviche, D., Din, W., and Boursot, P. (1994). The house mouse as a ring species? In *Genetics in Wild Mice: Its Application to Biomedical Research*, ed. K. Moriwaki, T. Shiroishi, and H. Yonekawa. Basel/Tokyo: Karger/Japan Scientific Societies Press, pp. 13–23.

Bonhomme, F., Orth, A., Cucchi, T., *et al.* (2004). A new species of wild mice on the island of Cyprus. *Comptes Rendus Biologies*, **327**, 501–7.

Bonhomme, F., Orth, A., Cucchi, T., *et al.* (2011). Genetic differentiation of the house mouse around the Mediterranean basin: matrilineal footprints of early and late colonization. *Proceedings of the Royal Society B: Biological Sciences.*

Boursot, P., Auffray, J.-C., Britton-Davidian, J., and Bonhomme, F. (1993). The evolution of house mouse. *Annual Review of Ecology and Systematics*, **24**, 119–52.

Boursot, P., Din, W., Anand, R., *et al.* (1996). Origin and radiation of the house mouse: mitochondrial DNA phylogeny. *Journal of Evolution Biology*, **9**, 391–415.

Boxall, R., Golob, P., and Taylor, R. (1997). *Pest Management in Farm Granaries*. Chatham: Natural Resources Institute.

Britton-Davidian, J. (1990). Genic differentiation in *M. m. domesticus* populations from Europe, the Middle East and North Africa: geographic patterns and colonization events. *Biological Journal of the Linnean Society*, **41**, 27–45.

Castillo, C., Martin-Gonzales, E., and Coello, J. J. (2001). Small vertebrate taphonomy of La Cueva del Llano, a volcanic cave on Fuerteventura (Canary Islands, Spain): palaeoloecological implications. *Palaeogeography, Palaeoclimatology, Palaeoecology*, **166**, 277–91.

Cauvin, J. (1977). Les fouilles de Mureybet (1971–1974) et leur signification pour les origines de la sédentarisation au Proche-Orient. *Annual of the American School of Oriental Research*, **44**, 19–48.

Cauvin, J. (1997). *Naissance des Divinités Naissance de l'Agriculture*. Paris: Flammarion.

Cauvin, J. and Aurenche, O. (1982). Le néolithique de Cafer Höyük (Malatya-Turquie), fouille 1979–1980. *Cahiers de l'Euphrate*, **3**, 123–38.

Coquegniot, E. (1998a). Dja'de el Mughara 1997. *Chronique archéologiques en Syrie*, **2**, 125–32.

Coquegniot, E. (1998b). Dja'dé el Mughara (Moyen-Euphrate), un village néolithique dans son environnement naturel à la veille de la domestication. In *Espace Naturel, Espace Habité en Syrie du Nord (10e–2e Millénaires av. J.-C.)*, ed. M. Fortin and O. Aurenche. Toronto and Lyon: Canadian Society for Mesopotamian Studies (Bull. 33) et Maison de l'Orient Méditerranéen (TMO 28), pp. 109–14.

Coquegniot, E. (2000). Dja'dé (Syrie), un village à la veille de la domestication (seconde moitié du IXe millénaire av. J.-C.). In *Premiers Paysans du Monde: Naissances des Agricultures*, ed. J. Guilaine. Paris: Errance, pp. 63–79.

Courtney, P. A. and Fenton, M. B. (1976). The effects of a small rural garbage dump on populations of *Peromyscus leucopus* Rafinesque and other small mammals. *Journal of Applied Ecology*, **13**, 413–21.

Cucchi, T. (2005). *Le Commensalisme Murin et les Premières Sociétés Méditerranéennes*. Paris: Muséum National d'Histoire Naturelle.

Cucchi, T. (2008). Uluburun shipwreck stowaway house mouse: molar shape analysis and indirect clues about the vessel's last journey. *Journal of Archaeological Science*, **35**, 2953–9.

Cucchi, T., Bălăşescu, A., Bem, C., *et al.* (2011). First clues on the eastern house mouse (*M. m. musculus*) invasive process of Europe from the burnt houses of Chalcolithic Romania. *The Holocene*, doi: 10.1177/0959683611405233.

Cucchi, T., Orth, A., Auffray, J.-C., *et al.* (2006). A new endemic species of the subgenus *Mus* (Rodentia, Mammalia) on the island of Cyprus. *Zootaxa*, **1241**, 1–36.

Cucchi, T. and Vigne, J. D. (2006). Origin and diffusion of the house mouse in the Mediterranean. *Human Evolution*, **21**, 95–106.

Cucchi, T., Vigne, J. D., and Auffray, J.-C. (2005). First occurrence of the house mouse (*Mus musculus domesticus* Schwarz and Schwarz, 1943) in the Western Mediterranean: a zooarchaeological revision of subfossil occurrences. *Biological Journal of the Linnean Society*, **84**, 429–45.

Cucchi, T., Vigne, J. D., Auffray, J. C., Croft, P., and Peltenburg, E. (2002). Passive transport of the house mouse (*Mus musculus domesticus*) to Cyprus at the Early Preceramic Neolithic (late 9th and 8th millennia cal. BC). *Comptes Rendus Palevol*, **1**, 235–41.

Darvish, J., Orth, A., and Bonhomme, F. (2006). Genetic transition in the house mouse, *Mus musculus* of Eastern Iranian Plateau. *Folia Zoologica*, **55**, 349–57.

Ervynck, A. (2002). Sedentism or urbanism? On the origin of the commensal black rat (*Rattus rattus*). In *Bones and the Man: Studies in Honour of Don Brothwell*, ed. K. Dobney and T. O'Connor. Oxford: Oxbow Books, pp. 95–109.

Fox, J. G., Davisson, M. T., Quimby, F. W., *et al.* (eds) (2007). *The Mouse in Biomedical Research, vol. I. History, Wild Mice, and Genetics*, 2nd edn. San Diego, CA: Academic Press/Elsevier.

Gabriel, S. I., Johannesdottir, F., Jones, E. P., and Searle, J. B. (2010). Colonization, mouse-style. *BMC Biology*, **8**, 131.

Goring-Morris, N. and Belfer-Cohen, A. (1998). The articulation of cultural process and Late Quaternary environmental changes in Cisjordan. *Paléorient*, **23**, 71–93.

Guerrero, E., Naji, S., and Bocquet-Appel, J. P. (2008). The signal of the Neolithic demographic transition in the Levant. In *The Neolithic Demographic Transition and Its Consequences*, ed. J.-P. Bocquet-Appel and O. Bar-Yosef. New York: Springer, pp. 57–80.

Guilaine, J. (ed.) (2000). *Premiers Paysans du Monde:. Naissance des Agricultures*, Paris: Errance.

Guilaine, J., Briois, F., and Vigne, J. D. (2011). *Shillourokombos: Un établissement néolithique pré-céramique à Chypre – Les fouilles du Secteur 1*, Paris: Editions Errance/Ecole Française d'Athènes.

Haydar, A. (2004). *Evolution de l'environnement au Levant nord de l'épipaléolithique au présent à travers l'étude des microfaunes fossiles et actuelles*. Unpublished PhD thesis, Université Paris X, Nanterre.

Hillman, G. (1996). Late Pleistocene changes in wild plant-foods available to hunter-gatherers of the Northern Fertile Crescent: possible preludes to cereal cultivation. In *The Origins and Spread of Agriculture and Pastoralism in Eurasia*, ed. D. Harris. London: University College, pp. 159–203.

Hopf, H. S., Morley, G. E., and Humphries, J. E. (1976). *Rodent Damage to Growing Crops and to Farm and Village Storage in Tropical and Subtropical Regions*. London: Centre for Overseas Pest Research.

Ibañez, J. (ed.) (2008). *Le Site Néolithique de Tell Mureybet (Syrie du Nord), en Hommage à Jacques Cauvin*. Oxford: BAR International Series 1843.

Janossy, D. (1996). Small mammals from the Pleistocene and Holocene of southeast-Europe and the Middle East. *Zeitschrift für Säugetierkunde*, **61**, 242–52.

Jones, E. P., van der Kooij, J., Solheim, R., and Searle, J. B. (2010). Norwegian house mice (*Mus musculus musculus/domesticus*): distributions, routes of colonization and patterns of hybridization. *Molecular Ecology*, **19**, 5252–64.

Knapp, A., Held, S., and Manning, S. (1994). The prehistory of Cyprus: problems and prospects. *Journal of World Prehistory*, **377**, 377–453.

Kuijt, I. (2000). People and space in early agricultural villages: exploring daily lives, community size, and architecture in the Late Pre-Pottery Neolithic. *Journal of Anthropological Archaeology*, **19**, 75–102.

Kuijt, I. (2008). Demography and storage systems during the Southern Levantine Neolithic Demographic Transition. In *The Neolithic Demographic Transition and Its Consequences*, ed. J.-P. Bocquet-Appel and O. Bar-Yosef. New York: Springer, pp. 287–314.

Kuijt, I. (2009). What do we really know about food storage, surplus, and feasting in preagricultural communities? *Current Anthropology*, **50**, 641–4.

Kuijt, I. and Finlayson, B. (2009). Evidence for food storage and predomestication granaries 11,000 years ago in the Jordan Valley. *Proceedings of the National Academy of Sciences of the United States of America*, **106**, 10966–70.

Lombard, P. and Chataigner, C. (2004). Le Néolithique et le Chalcolithique en Transcaucasie: l'exemple des bassins de la Kura et de l'Araxe. In *Aux Marges des Grands Foyers du Néolthique. Périphéries Débitrices ou Créatrices?*, ed. J. Guilaine. Paris: Errance.

Lowe, S., Browne, M., Boudjelas, S., and De Poorter, M. (2000). *100 of the World's Worst Invasive Alien Species Database*. Published by the IUCN Invasive Species Specialist Group (ISSG). Available at: http://www.issg.org/database/species/search.asp?st=100ss.

Manning, S. W., McCartney, C., Kromer, B., and Stewart, S. T. (2010). The earlier Neolithic in Cyprus: recognition and dating of a Pre-Pottery Neolithic A occupation. *Antiquity*, **84**, 693–706.

Marro, C. and Hauptmann, H. (eds) (2000). *Chronologies des Pays du Caucase et de l'Euphrate aux IVème-IIIème millénaires*. Paris: De Boccard.

Martínková, N., Barnett, R., Cucchi, T., *et al.* (Submitted). Islands as arks retaining genetic diversity lost on the mainland. *Nature*.

Matisoo-Smith, E. and Robins, J. H. (2004). Origins and dispersals of Pacific peoples: evidence from mtDNA phylogenies of the Pacific rat. *Proceedings of the National Academy of Sciences of the United States of America*, **101**, 9167–72.

Mezhzherin, S. V., Kotenkova, E. V., and Mikhailenko, A. G. (1998). The house mice, *Mus musculus* s. l., hybrid zone of Transcaucasus. *Zeitschrift für Säugetierkunde*, **63**, 154–68.

Michaux, J., Cucchi, T., Renaud, S., Garcia-Talavera, F., and Hutterer, R. (2007). Evolution of an invasive rodent on an archipelago as revealed by molar shape analysis: the house mouse in the Canary Islands. *Journal of Biogeography*, **34**, 1412–25.

Milishnikov, A. N., Lavrenchenko, L. A., and Lebedev, V. S. (2004). Origin of the house mice (superspecies complex *Mus musculus* sensu lato) from the Transcaucasia region: a new look at dispersal routes and evolution. *Russian Journal of Genetics*, **40**, 1011–26.

Millien, V. (2006). Morphological evolution is accelerated among island mammals. *PLoS Biology*, **4**, 1863–8.

Monah, F. (2007). The spread of cultivated plants in the region between the Carpathians and Dniestr, 6th–4th millenia cal BC. In *The Origins and Spread of Domestic Plants in Southwest Asia and Europe*, ed. S. Colledge and J. Conolly. London: University College London Institute of Archaeology Publications, pp. 111–23.

Mottram, M. (1991). Jerf el Ahmar. *American Journal of Archaeology*, **95**, 218–51.

Mutze, G. L. (1989). Mouse plagues in South Australian cereal growing area: II. Occurence and distribution of plagues from 1900 to 1984. *Australian Wildlife Research*, **16**, 677–83.

Orth, A., Lyapunova, E., Kandaurov, A., *et al.* (1996). L'espèce polytypique *Mus musculus* in Transcaucasia. *Comptes Rendus de l'Acacémie des Sciences de Paris. Serie 3: Sciences de la Vie*, **319**, 435–41.

Palumbi, S. (2001). Humans as the world's greatest evolutionary force. *Science*, **29**, 1786–90.

Papayannis, K. (In press). The microenvironment of urban spaces of the Late Bronze Age: the microfauna from the Late Cycladic I Akrotiri. In *Akrotiri Excavations: 40 Years of Research*, ed. C. Doumas. Athens.

Papayannis, K. and van den Hoek Ostende, L. (In press). The microvertebrates from Theopetra Cave: the Neolithic layers. In *Theopetra Neolithic Layers*, ed. N. Kyparissi-Apostolika.

Pascal, M., Lorvelec, O., and Vigne, J. D. (2006). Holocene turnover of the French vertebrate fauna. *Biological Invasions*, **8**, 395.

Peltenburg, E. (2003). Identifying settlements of the Xth–IXth millennium B.P. in Cyprus from the contents of Kissonerga-Mylouthkia wells. In *Le Néolithique de Chypre (Actes du IIème Colloque International Nicosie, 17–19 Mai 2001)*, ed. J. Guilaine and A. Le Brun. Athens: Actes du Colloque International, organisé par le Département des Antiquités de Chypre et l'Ecole Française d'Athène, pp. 15–34.

Peltenburg, E., Colledge, S., Croft, P., *et al.* (2000). Agro-pastoralist colonization of Cyprus in the 10th millennium BP: initial assessments. *Antiquity*, **74**, 844–53.

Pergams, O. R. W. and Ashley, M. V. (2001). Microevolution in island rodents. *Genetica*, **112–113**, 245–56.

Perrot, J. (1966). Le gisement natoufien de Mallaha (Eynan), Israël. *L'Anthropologie*, **70**, 437–84.

Pulak, C. (1996). A dendrochronological dating of the Uluburun Ship. *INA Quarterly*, **23**, 12–13.

Pulak, C. (1998). The Uluburun shipwreck: an overview. *The International Journal of Nautical Archaeology*, **27**, 188–224.

Rajabi-Maham, H., Orth, A., and Bonhomme, F. (2008). Phylogeography and postglacial expansion of *Mus musculus domesticus* inferred from mitochondrial DNA coalescent, from Iran to Europe. *Molecular Ecology*, **17**, 627–41.

Rohlf, F. J. and Marcus, L. F. (1993). A revolution in morphometrics. *Trends in Ecology and Evolution*, **8**, 129–32.

Sanlaville, P. (1998). Les changements dans l'environnement au Moyen-Orient de 20 000 BP à 6 000 BP. *Paléorient*, **23**, 249–62.

Schwarz, E. (1939). Commensalism and domestication. *The American Naturalist*, **73**, 270–8.

Searle, J. B., Jamieson, P. M., Gündüz, I., *et al.* (2009a). The diverse origins of New Zealand house mice. *Proceedings of the Royal Society B: Biological Sciences*, **276**, 209–17.
Searle, J. B., Jones, C. S., Gündüz, I., *et al.* (2009b). Of mice and (Viking?) men: phylogeography of British and Irish house mice. *Proceedings of the Royal Society B: Biological Sciences*, **276**, 201–7.
Shlyakhov, E. N. (1983). Influence de l'activité humaine sur l'épidémiologie des zooanthroponoses. *Médecine et Maladies Infectieuses*, **13**, 784–7.
Stordeur, D., Helmer, D., and Willcox, G. (1997). Jerf el Ahmar, un nouveau site sur le moyen Euphrate. *Bulletin de la Société Préhistorique Française*, **93**, 1–4.
Stordeur, D. and Willcox, G. (2009). Indices de culture et d'utilisation des céréales à Jerf el Ahmar. In *De Méditerranée et d'ailleurs ... Mélanges offerts à J. Guilaine*, ed. Collectif. Toulouse: Archives d'Ecologie Préhistorique, pp. 693–710.
Tangri, D. and Wyncoll, G. (1989). Of mice and men: is the presence of commensal animals in archaeological sites a positive correlate of sedentism? *Paléorient*, **15**, 85–94.
Tchernov, E. (1968). *Succession of Rodent Faunas during the Upper Pleistocene of Israel (Mammalia Depicta)*. Hamburg and Berlin: Paul Parey.
Tchernov, E. (1984). Commensal animals and human sedentism in the Middle East. In *Animals and Archaeology: 3. Early Herders and Their Flocks*, ed. J. Clutton-Brock and G. Grigson. Oxford: British Archaeological Reports, pp. 91–115.
Tchernov, E. (1991a). Biological evidence for human sedentism in Southwest Asia during the Natufian. In *The Natufian Culture in the Levant*, ed. O. Bar-Yosef and F. Valla. Ann Arbor, MI: International Monographs in Prehistory, pp. 315–40.
Tchernov, E. (1991b). Of mice and men: biological markers for long-term sedentism – a reply. *Paléorient*, **17**, 153–60.
Tchernov, E. (1993). From sedentism to domestication: a preliminary review for Southern Levant. In *Skeletons in Her Cupboard: Festschrift for Juliet Clutton-Brock*, ed. A. T. Clason, S. Payne, and H. P. Uerpmann. Oxford: Oxbow Books, pp. 189–233.
Testard, A. (1981). La conservation des produits végétaux chez les chasseurs cueilleurs. In *Techniques de Conservation à Long Terme*, ed. M. Gast and F. Sigaut. Paris: Maison des Sciences de l'Homme, pp. 181–93.
Testard, A. (1982). The significance of food storage among hunter-gatherers: residence patterns, population densities, and social inequalities. *Current Anthropology*, **23**, 523.
Thaler, L. (1986). Origin and evolution of mice: an appraisal of fossil evidence and morphological traits. *Current Topics in Microbiology and Immunology*, **127**, 1–9.
Tichý, R. (1999). *Monoxylon II. Plavba po 8000 Letech*. Náchod: JB Production.
Valenzuela-Lamas, S., Baylac, S., Cucchi, T., and Vigne, J. D. (2011). House mouse dispersal in Iron Age Spain: a geometric morphometrics appraisal. *Biological Journal of the Linnean Society*, **102**, 483–97.
Valla, F. (2000). La sédentarisation au Proche-Orient: la culture natoufienne. In *Premiers Paysans du Monde. Naissance des Agricultures*, ed. J. Guilaine. Paris: Errance, pp. 13–30.
Valla, F., Plisson, H., and Buxo i Capdevilla, R. (1989). Notes préliminaires sur les fouilles en cours sur la terrasse d'Hayonim. *Paléorient*, **15**, 245–57.

Vigne, J.-D. (1997). Les micromammifères au Néolithique final à Clairvaux-MM et à Chalain 3: contribution à l'histoire du commensalisme. In *Les sites littoraux néolithiques de Clairvaux-les-Lacs et de Chalain (Jura), 3, 3200–2900 av. J.-C*, ed. P. Pétrequin. Paris: Maison des Sciences de l'Homme, pp. 717–22.

Vigne, J.-D. (1999). The large 'true' Mediterranean as a model for the Holocene human impact on the European vertebrate fauna? Recent data and new reflections. In *The Holocene History of the European Vertebrate Fauna. Modern Aspects of Research (Workshop, 6th-9th April 1998, Berlin)*, ed. N. Benecke. Berlin: Deutsches Archäologisches Institut, Eurasian-Abteilung, pp. 295–322.

Vigne, J.-D. (2000). Les début néolithiques de l'élevage des ongulés au Proche Orient et en Méditerranée: acquis récents et questions. In *Premiers Paysans du Monde: Naissance des Agricultures*, ed. J. Guilaine. Paris: Errance, pp. 143–70.

Vigne, J. D. (2009). Introduction et réintroduction de mammifères à Chypres aux 9e et 8e millénaires av. J.-C. (Néolithique précéramique): indices indirects de l'usage de la voile au Néolithique? In *De Méditerranée et d'ailleurs ... Mélanges offerts à J. Guilaine*, ed. B. Miden-Reynes and C. Manen. Toulouse: Archives d'Ecologie Préhistorique, pp. 807–20.

Vigne, J.-D. (2011). The origins of animal domestication and husbandry: a major change in the history of humanity and the biosphere. *Comptes Rendus Biologies*, **334**, 171.

Vigne, J.-D., Briois, F., Zazzo, A., *et al.* (2011a). A new early Pre-Pottery Neolithic site in Cyprus: Ayios Tychonas – Klimonas (ca. 8700 cal. BC). *Neo-lithics*, **1**, 3–18.

Vigne, J. D., Carrère, I., Briois, F., and Guilaine, J. (2011b) The early process of the mammal domestication in the Near East: New Cypriot Pre-Neolithic and Prepottery Neolithic Evidence. *Current Anthropology*, **52**, S255–71.

Vigne, J.-D. and Cucchi, T. (2005). Premières navigations au Proche-Orient: les informations indirectes de Chypre. *Paléorient*, **31**, 195–204.

Vigne, J.-D. and Villié, P. (1995). Une preuve archéologique du transport d'animaux par bateau: le crâne du rat surmulot (*Rattus norvegicus*) de l'épave du 'ça ira'. In *L'homme méditerranéen. Mélanges offerts à Gabriel Camps*, ed. R. Chenorkian. Aix-en-Provence: Publications de l'Université de Provence, pp. 411–16.

Vigne, J.-D., Zazzo, A., Saliege, J. F., *et al.* (2009). Pre-Neolithic wild boar management and introduction to Cyprus more than 11,400 years ago. *Proceedings of the National Academy of Sciences of the United States of America*, **106**, 16135–8.

Willcox, G. (1996). Evidence for plant exploitation and vegetation history from three Early Neolithic pre-pottery sites on the Euphrates (Syria). *Vegetation History and Archaeobotany*, **5**, 143–52.

Willcox, G. (1999). Agrarian change and the beginnings of cultivation in the Near East: evidence from wild progenitors, experimental cultivation and archaeobotanical data. In *Prehistory of Food*, ed. C. Gosden and J. Hather. London: Routledge, pp. 478–500.

Willcox, G. (2000). Nouvelles données sur l'origine de la domestication des plantes au Proche Orient. In *Premiers Paysans du Monde: Naissance des Agricultures*, ed. J. Guilaine. Paris: Errance, pp. 121–40.

Willcox, G. (2005). The distribution, natural habitats and availability of wild cereals in relation to their domestication in the Near East: multiple events, multiple centres. *Vegetation History and Archaeobotany*, **14**, 534–41.
Willcox, G., Buxo, R., and Herveux, L. (2009). Late Pleistocene and early Holocene climate and the beginnings of cultivation in northern Syria. *Holocene*, **19**, 151–8.
Willcox, G., Fornite, S., and Herveux, L. (2008). Early Holocene cultivation before domestication in northern Syria. *Vegetation History and Archaeobotany*, **17**, 313–25.
Wyncoll, G. and Tangri, D. (1991). The origins of commensalism and human sedentism. *Paléorient*, **17**, 157–9.

4

Origin and genetic status of *Mus musculus molossinus*: a typical example of reticulate evolution in the genus *Mus*

HIROMICHI YONEKAWA, JUN J. SATO, HITOSHI SUZUKI, AND KAZUO MORIWAKI

Geographic distribution of the Japanese mouse based on mtDNA polymorphisms

At least five epoch-making events mark the taxonomic and evolutionary study of the house mouse, *Mus musculus*: (1) Schwarz and Schwarz (1943) extensively revised mouse taxonomy, reducing the approximately 133 subspecies of *Mus musculus* in the previous taxonomy to 15. (2) The introduction of biochemical genetics first demonstrated that two taxonomic groups (subspecies) that differ genetically, *M. m. domesticus* and *M. m. musculus*, exist in Europe (Selander *et al.*, 1969). A comprehensive demonstration of genetic and geographical differentiation among mouse subspecies was performed by Bonhomme *et al.* (1984). (3) The discovery of wild mice with Robertsonian (Rb) translocations facilitated an extensive geographic survey of wild mice, particularly European mice (Gropp and Winking, 1981; Riginos and Nachman, 1999). (4) The development of molecular biology, especially the finding that DNA polymorphisms can be detected as restriction fragment-length polymorphisms (RFLPs), enabled subspecies-specific RFLPs to be used as diagnostic markers for geographic populations of mice and permitted more precise analyses of population structures and evolutionary relationships within/among subspecies (or species) of wild mice (Yonekawa *et al.*, 1981). (5) Finally, the introduction of genome science in population genetics and/or in evolutionary genetics enabled analysis of the population structures and dynamics of wild animals at exceptional levels of precision (Frazer *et al.*, 2007; Salcedo *et al.*, 2007).

Before the introduction of mitochondrial DNA (mtDNA) RFLP, taxonomists believed that wild mice distributed from Japan to the northeastern part of China (Manchuria) through the Korean peninsula were an endemic subspecies, termed *M. m. molossinus* (Schwarz and Schwarz, 1943; Marshall, 1981). A survey

Evolution of the House Mouse, ed. Miloš Macholán, Stuart J. E. Baird, Pavel Munclinger, and Jaroslav Piálek. Published by Cambridge University Press.

of Japanese wild mice by Moriwaki *et al.* (1979) and Minezawa *et al.* (1981) using biochemical markers showed that the average heterozygosity of *M. m. molossinus* was as high as that of other subspecies. However, these authors found comparatively little differentiation across geographical populations. Minezawa *et al.* (1981) suggested that this deficit was caused by gene flow from foreign groups of mice that had invaded Japan by different routes. They also suggested that the genetic characteristics of Japanese wild mice were similar to those of the European subspecies *M. m. musculus*. These lines of evidence led us to perform an extensive geographic survey using molecular markers, an approach that had just been developed at that time in the field of molecular biology.

About three decades ago, we surveyed mtDNA RFLPs among the geographically defined populations of wild mice in Japan and its neighbouring countries. These studies produced four important lines of evidence: (1) mtDNA RFLPs can be used to distinguish four major taxonomic groups of *Mus musculus*, and their geographic distributions correspond to the four subspecies described in Schwarz and Schwarz's taxonomy. These subspecies are *M. m. domesticus* (DOM), *M. m. musculus* (MUS), *M. m. castaneus* (CAS) and *M. m. bactrianus* (BAC). This evidence suggests that mtDNA RFLPs could be used in mouse taxonomy as a molecular tool for the diagnosis of the *M. musculus* subspecies. Furthermore, the sequence divergences between each pair of mtDNA haplotypes can easily be estimated by cleavage site analysis based on mtDNA RFLPs, an approach that offers a quantitative mode of analysis in taxonomic and evolutionary genetics (Gotoh *et al.*, 1979). (2) There are no mtDNA haplotypes specific to *M. m. molossinus* in Japanese wild mice. (3) Conversely, two mtDNA haplotypes exist in the population, and they have an unnatural geographic distribution. One haplotype was clustered with MUS and the other with CAS. The population with the *musculus*-type mtDNA haplotypes occupied about two-thirds of Japan, whereas the population with *castaneus*-type mtDNA was distributed throughout the rest of Japan. The two formed geographical border zones of mtDNA haplotypes. (4) Therefore, the distribution of the population with *castaneus*-type mtDNA was disjunct. This split population was located at the northern and southern edges of the Japanese mainland. The population with *musculus*-type mtDNA was located centrally, between these two peripheral regions. Based on these findings, we proposed that *M. m. molossinus* is not an endemic subspecies but has originated by genetic hybridization between MUS and CAS (Yonekawa *et al.*, 1981, 1982, 1986, 1988) (Fig. 4.1).

About two decades ago, we again surveyed the mtDNA polymorphisms. These polymorphisms are based on DNA sequencing of the control region (the D-loop region) of mtDNA, mainly among mouse samples collected from Asia, including Japan, China, and Russia. In this survey we took advantage of the introduction of polymerase chain reaction (PCR) technology in molecular biology. This technology

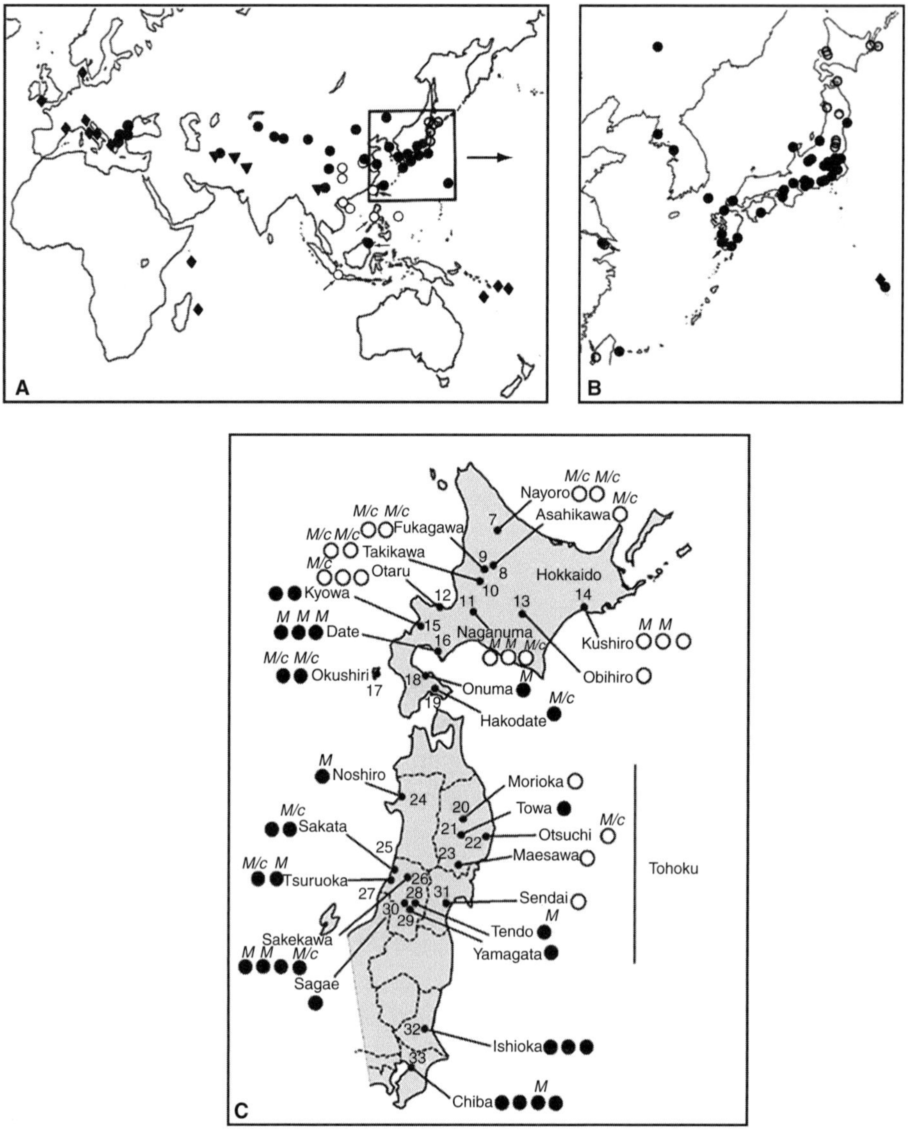

Figure 4.1 (a) Geographic distribution of mtDNA variants on the Eurasian continent. The symbols on the map represent the DOM haplotype (closed rhomboid), MUS haplotype (closed circle), CAS haplotype (open circle), and BAC haplotype (closed triangle). (b) Enlarged view with a focus on Japan and its neighbouring countries; situation in 1988 (Yonekawa *et al.*, 1988). (c) Enlarged view with focus on the northern part of Japan; situation in 2006 (Terashima *et al.*, 2006).

is advantageous because even crude or partially degraded DNA samples can be used for DNA sequencing and because only one pair of primer sets with 15–17 nucleotide lengths, which correspond to two conserved regions on a mouse mtDNA sequence, are needed for PCR amplification. About 70 haplotypes were discovered in the samples surveyed. These haplotypes include several DOM and *M. spretus*. We constructed a neighbour-joining (NJ) tree and found at least three clades. These three clades correspond to the three above-mentioned subspecies, DOM, MUS, and CAS, in the tree. However, no clade corresponding to BAC appeared in the tree. The MUS and the CAS clades can easily be distinguished in the tree. The Japanese wild mice possessed both clades. Based on these lines of experimental evidence, we concluded that the 'hybrid origin hypothesis' is essentially correct (Yonekawa *et al.*, 1994). Furthermore, we surveyed the mtDNA haplotypes for a third time, primarily in the northern Japan population, and we found that the hybrid boundary had moved northward. This finding suggested that invasion of mice with the *musculus*-type haplotype population was occurring in Japan (Terashima *et al.*, 2006) (Fig. 4.1c).

However, mtDNA only provides information on matrilineal structures in wild mouse populations because the maternal inheritance of mtDNA is strict in mice (Kaneda *et al.*, 1995). Although Gyllensten *et al.* (1991) reported an occasional paternal mtDNA contribution based on interspecific hybrids between *M. musculus* and *M. spretus*, we confirmed that the biparental inheritance is restricted only to F_1 hybrid mice (Kaneda *et al.*, 1995; Shitara *et al.*, 1998). Consequently, our genetic survey lacked any patrilineal information. Furthermore, it should be noted that the gene(s) responsible for speciation are encoded in the nuclear genome and not in the mitochondrial genome. Therefore, we shifted our project from a survey of mtDNA polymorphisms to a survey of nuclear genome polymorphisms based mainly on microsatellite polymorphisms.

Major contributor to the nuclear genetic background of Japanese mice, *M. m. molossinus*, is *M. m. musculus* and not *M. m. castaneus*

In the late twentieth and early twenty-first century, we obtained a powerful tool for nuclear genome information. This tool can easily characterize the genetic status of laboratory animals, particularly mice. The tool is the chromosome-specific microsatellite locus. The polymorphisms of microsatellite loci have been widely applied to linkage analysis, quantitative trait loci (QTL) analysis, and positional cloning. This tool was used because the microsatellite loci are universally distributed on all chromosomes. In addition, genetic information regarding more than 9000 loci, as well as high-throughput polymorphism analyses, have been made available for mice, though mainly laboratory mice (Love *et al.*, 1990; Dietrich *et al.*, 1994;

Rhodes *et al.*, 1998). Taking advantage of these genetic properties, several research groups have applied microsatellite polymorphism to distinguish or to trace alleles of interest in wild mouse populations. For example, these polymorphisms were used to test two competing hypotheses about whether a Robertsonian (Rb) translocation had originated independently at multiple times or had arisen once and been spread by long-distance dispersal. The former possibility was supported by Riginos and Nachman (1999). Microsatellite polymorphisms have also been used to disclose population structures in mice with Rb translocations (Dallas *et al.*, 1998), in mice with *t* haplotypes (Dod *et al.*, 2003), or in the gene flow across the subspecies boundary between DOM and MUS (Dallas *et al.*, 1995). These lines of evidence suggest that some chromosome-specific microsatellite loci can be used as diagnostic nuclear markers for subspecies identification.

The origin of laboratory mice has been a topic of considerable discussion. However, from several lines of experimental evidence based on mtDNA haplotypes and polymorphic repetitive sequences on the Y chromosome, it has become evident that the major genetic background of laboratory mice has been derived from that of the European DOM subspecies and that the contribution of Asian mice (e.g. *M. m. molossinus*) is not negligible. To address this issue, we carried out extensive whole-genome scanning using eight common inbred strains of laboratory mice (CIS) and eight wild-derived inbred strains (WIS), including two inbred strains, MSM/Ms and JF1/Ms, derived from Japanese mice. We selected and located 1226 informative loci at 1.2-cM average intervals on all of the chromosomes of the 16 strains, and we compared the polymorphisms of the eight CIS with those from the eight WIS as subspecies representatives. More than 50% of the loci can be identified as WIS (therefore, subspecies-specific) alleles in the CIS genomes. We found that the inbred strains derived from Japanese mice, MSM/Ms and JF1/Ms, which possess *musculus*-type mtDNA, are closely related to the WIS from Korea, and distantly related to the WIS from *M. m. castaneus*. These results suggest that the major contributor to the nuclear genetic background of Japanese mice, *M. m. molossinus*, is *M. m. musculus* and not *M. m. castaneus*, although we did not validate whether the same result holds for the Japanese mice with *castaneus*-type mtDNA (Sakai *et al.*, 2004, 2005). These lines of evidence confirmed that these microsatellite loci are promising nuclear molecular markers for the diagnosis of *M. musculus* subspecies.

Chromosome C-banding patterns of major mouse subspecies groups

As mentioned above, the three major subspecies groups of *M. musculus* (DOM, CAS, and MUS) are genetically and geographically differentiated on the Eurasian continent, a pattern confirmed by recent genomic analyses. From the cytogenetic point of view, the chromosome C-banding pattern is a unique marker that discloses the subspecies relationships and geographical population structures of the Eurasian subspecies. The C-banding method involves either a combined quinacrine and Hoechst 33258 (QH) technique or a Giemsa (GC) technique. These techniques are used to stain centromeric regions of each mouse's chromosomes. The survey of C-banding patterns was initiated more than three decades ago by a number of research groups (Forejt, 1973; Miller *et al.*, 1976), most of which used laboratory mouse strains. To date, only a few comprehensive studies have been reported on populations of wild mice worldwide. In the 1980s, we began to analyse the C-banding patterns of wild mice collected in Eurasia, from western Europe to eastern Asia, including Japan. In our analysis, the modified QH technique (Yoshida *et al.*, 1975) was mainly employed (Fig. 4.2). Each chromosome was identified in relation to previously reported banding patterns (Nesbitt and Francke, 1973). We scored the size of the C-band according to four grades between 0 (null staining) and 3 (large) among the mice, following the method described by Yoshida and Kodama (1983).

The C-band is also a stable feature for discriminating each chromosome. This fact indicates that the C-banding pattern could be a significant genetic marker for identifying each inbred strain of laboratory mice. The C-bands show little diversity among the 20 chromosomes of laboratory mice. They are mostly of medium size, although some chromosomes in certain inbred strains possess an exceptionally strain-specific, small C-band (Yoshida and Kodama, 1983; Moriwaki *et al.*, 1986). The chromosome stretch feature affecting these small C-bands could have been introduced from the MUS group to 'fancy mice' as the progenitors of CIS. A recent bacterial artificial chromosome (BAC)-end sequence analysis of a typical laboratory strain C57BL/6J suggested a possible introduction of approximately 5% of the MUS genome into the major DOM genome, when comparing the genomes between C57BL/6 (DOM strain) and MSM/Ms (MUS strain) (Abe *et al.*, 2004).

Our cytogenetic survey of the subspecies demonstrated that the C-bands of DOM and those of CAS showed less heterogeneity accompanying medium-sized C-banding. Conversely, the C-banding patterns in MUS showed extensive polymorphisms from grade 0 (null staining) to grade 3 (large). Among these patterns, the large bands in chromosome 2 and 18 are quite common in this group from northern China to Korea. The only exception is the MUS group from Bulgaria and Denmark. In *M. m. molossinus* in Japan, the large C-bands on the two chromosomes were common as well, except in the northern part of Honshu and the eastern part of Hokkaido, where the mice exhibited elements of the *castaneus*-type pattern (Table 4.1).

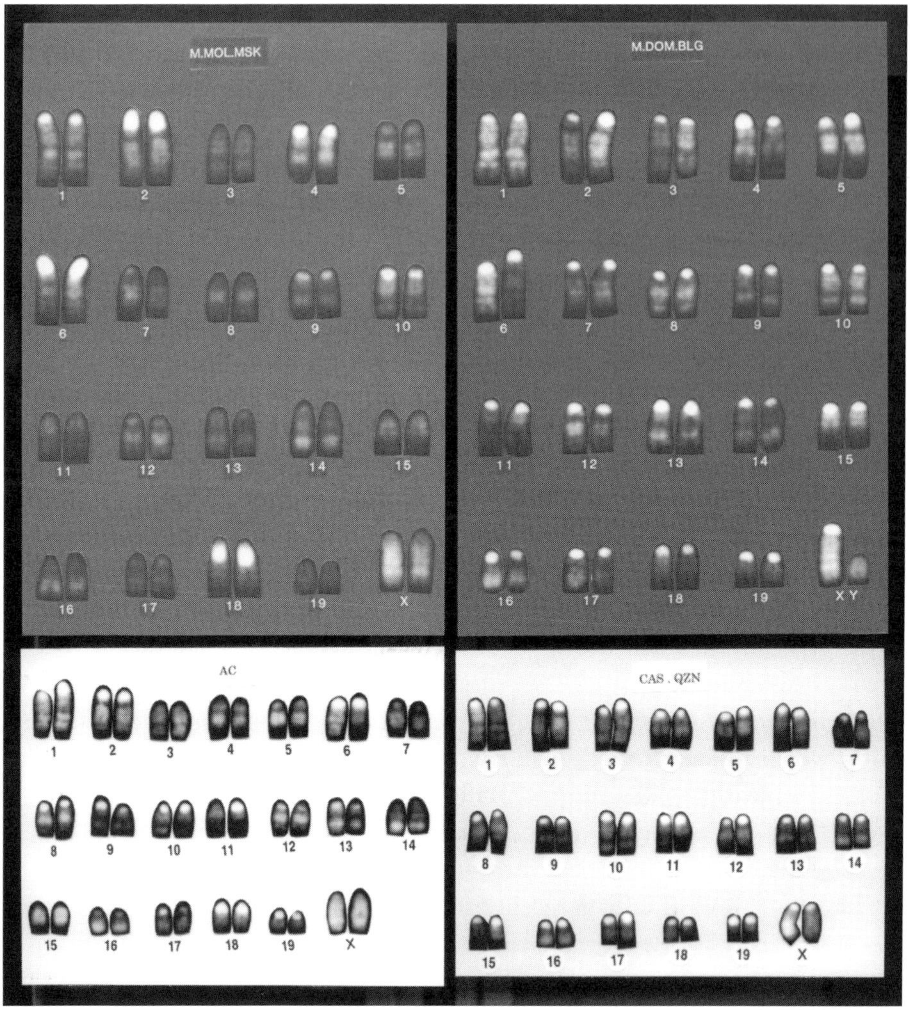

Figure 4.2 Chromosome C-band patterns characteristic of each *Mus musculus* subspecies group. The C-band pattern labelled with M.DOM.BLG was prepared from a Bulgarian mouse belonging to the DOM group. The pattern labelled with M.MOL. MSK is that of *M. m. molossinus* subspecies in the MUS group and was prepared from a Japanese wild mouse in Kanagawa. CAS.QZN means CAS group collected from Quezon City, Philippines. AC was prepared from a Korean mouse belonging to the MUS group.

As judged by the size distribution of the C-band among the DOM, CAS, and MUS groups of mice, DOM is cytogenetically close to CAS, while MUS is distantly related to CAS. However, an extensive sequence analysis using more than 2000 microsatellite loci demonstrated that CAS is closer to MUS than to

Table 4.1 Schematic illustration of chromosome C-band patterns among subspecies of *M. musculus*. Sizes in the C-band staining are indicated as follows: ○ null staining; ▲ small; ● medium; ■ large.

Mouse specimen	C-band size in each chromosome																			
	1	2	3	4	5	6	7	8	9	10	11	12	13	14	15	16	17	18	19	X
Genus *Mus*																				
M. spicilegus	○○	▲▲	▲●	●●	●●	●●	●●	●●	●●	●●	●●	●●	●●	●●	●●	●●	●●	●●	●●	○
M. caroli	●●	▲▲	▲▲	●●	●●	●●	●●	●●	●●	●●	▲●	●●	▲▲	●●	●●	●●	●●	●●	●●	●
M. spretus	●■	●●	●●	■■	●■	●●	●●	●■	●●	●■	▲■	●●	●●	●●	●●	●●	●■	●●	●●	●
DOM subspecies group																				
Dom. Sk. (CAM)	▲▲	●●	▲▲	●●	●●	●●	▲▲	●●	●●	●●	●●	●●	●●	●●	●●	●●	●●	●●	●●	●●
Dom. PW (B)	▲▲	●●	●●	●●	●●	●●	●●	●●	●●	●●	●●	●●	●●	●●	●●	●●	●●	●●	●●	●
Dom. Blg (ZBN)	●●	●●	●●	▲●	●●	▲●	●●	●●	●●	●●	●●	●●	●●	●●	●●	●●	●●	●●	●●	●●
Dom. Pgn.	▲●	●●	●●	▲●	●●	●●	●●	●●	●●	○●	▲●	●●	●●	▲●	●●	●●	●●	●●	●●	●●
Dom. Lak.	●●	▲●	●●	▲●	▲▲	●●	●●	●●	●■	●●	●●	●●	●●	●●	●●	●●	●●	●●	●●	○●
CAS subspecies group																				
Cas. Tch.	▲●	●●	●●	●●	●●	●●	▲▲	●●	●●	●●	●●	●●	●●	●●	●●	●●	▲●	●●	●●	●
Cas. Qzn.	●●	●●	●●	●●	●●	●●	●●	▲●	●●	●●	●●	●●	●●	●●	●●	●●	●●	●●	●●	●●
Bac. Kab.	▲▲	●●	●●	●●	●●	▲●	▲●	●●	●●	●●	▲●	●●	●●	●●	●●	▲●	▲●	▲▲	▲●	▲●
Cas. Shh.	●●	●●	○●	●●	▲▲	▲●	●●	●●	▲▲	●●	○●	▲●	▲●	○○	○●	○●	●●	○●	●●	○○
Cas. Chd	●●	●●	○○	●●	●●	●●	●●	●●	○●	●●	○●	●●	▲●	○○	○○	○○	●●	○○	●●	○
MUS subspecies group																				
Mus. Blg.	○○	○○	●●	●●	●●	●●	●●	●●	●●	●●	●●	●●	●●	▲●	●●	●●	●●	●●	●●	○
Mus. Njl.	●●	●●	●●	●●	●●	●●	●●	●●	●●	●●	●●	●●	●●	●●	●●	●●	●●	●●	●●	○○
Mus. Lzh.	○○	○○	○○	○○	○○	▲▲	○○	▲▲	○▲	▲●	○○	○▲	○▲	○○	○○	○○	■■	■■	○●	○○
Mus. Jyg.	○●	○○	○○	○○	○○	○▲	○●	○▲	○○	○▲	○○	○○	○○	▲▲	○○	○○	■■	■■	●●	○○
Mus. Chc.	●●	○●	○○	○●	●●	●●	●■	●●	●●	■■	●■	○●	○▲	○○	○○	▲▲	▲●	●■	○○	○
Mus. IAS	■■	●●	▲▲	○○	○○	●●	○○	●●	▲●	●●	○○	▲▲	○○	○○	○○	○○	○○	■■	●●	○○
Mus. KJR (KJR)	■■	■■	○○	○○	○○	●●	○○	○○	●●	■■	○○	▲▲	▲▲	○○	○○	○○	●●	■■	●●	○○

Table 4.1 (*cont.*)

Mouse specimen	C-band size in each chromosome																			
	1	2	3	4	5	6	7	8	9	10	11	12	13	14	15	16	17	18	19	X
Mus m. molossinus in MUS Group																				
Mol. Ten.	●●	■■	○○	●●	●●	■■	○○	●●	●●	●●	●●	●●	▲▲	○▲	○○	○○	○○	■■	○○	○○
Mol. Nsb.	●●	●●	○○	●●	○●	▲▲	○○	○○	○●	●●	○▲	●●	○▲	○○	○○	○○	○○	●●	●●	○○
Mol. Mro.	●●	●●	○○	●●	○○	●●	○○	○○	●●	○○	○○	○○	○○	○○	○○	○○	○○	●●	●●	○
Mol. Mmy.	▲●	■■	○○	●●	▲▲	●●	○○	▲▲	▲●	●●	○○	○●	▲▲	○○	○○	○○	○○	■■	○○	○
Mol. Msm.(MSM)	●●	■■	▲▲	●●	▲▲	●●	○○	○○	●●	●●	●●	●●	○○	○○	○○	○○	▲▲	■■	○○	○
Mol. Hkz.	●●	■■	●●	○●	○●	○●	▲▲	○●	●●	●●	○●	○○	○●	▲●	○○	○●	○●	●■	○●	●●

DOM (Sakai *et al.*, 2005). The phylogenetically closer relationship between MUS and CAS was also shown by southern hybridization analysis of genomic DNA with an amplified long genomic sequence (ALGS) probe (Koide *et al.*, 1992). The ALGS is located in the centromeric regions of mouse chromosomes. This phylogenetic discrepancy in the genomic behaviour between the centromeric region and the chromosome long arm region can be explained by the lack of recombinations in the centromeric regions. In other words, these regions characterize mouse ancestral populations.

C-banded karyotypes of the mouse specimens related to in this section will be available through the mouse genome database of the National Institute of Genetics.

Molecular genetic polymorphisms and geographic distribution of the haemoglobin beta chain complex locus

Since extensive collections of wild mice were made in East Asia in the 1970s, many mice have been subjected to morphometric, cytogenetic, biochemical, and genetic analyses, and mice with unique mutations or characteristics have been established as wild-derived inbred strains. These mice have been used widely as live material for biomedical and evolutionary research. As described in the previous section, mtDNA polymorphism analyses have suggested that *M. m. molossinus* is composed of heterogeneous lineages derived from two subspecies, CAS and MUS. We present here the genetic polymorphisms of the haemoglobin beta chain (*Hbb*) locus, their gene structure and their geographic distribution as further supporting evidence for the complex genealogy of *M. m. molossinus*.

The *Hbb* locus constitutes a gene family with two loci, *Hbb-T1* and *Hbb-T2*, located at the distal end of the family with an intervening 14-kb spacer region. Both are expressed in the adult stage (Popp and Bailiff, 1973). Five *Hbb* electromorphs, *s*, *d*, *p*, *w1*, and *w2*, which are identified by protein electrophoresis, have been found widely in the wild mice (Fig. 4.3). The haplotype frequencies are different in each subspecies. For example, the *s*, *d*, and *p* haplotypes are primarily found in DOM, CAS, and MUS, respectively. The newly discovered haplotypes, *w1* and *w2*, are found primarily in the MUS subspecies distributed in northwestern to central China. Subsequent molecular genetic studies using nucleotide sequences of the *Hbb-T1* and *Hbb-T2* subloci demonstrated that the *p* haplotype is a recombinant, carrying the *Hbb-T1* gene of the *d* haplotype and the *Hbb-T2* gene of the *w1* haplotype (Fig. 4.3). Likewise, the *w2* haplotype is the recombinant between the *w1* and unknown haplotypes (Sato *et al.*, 2006, 2008)

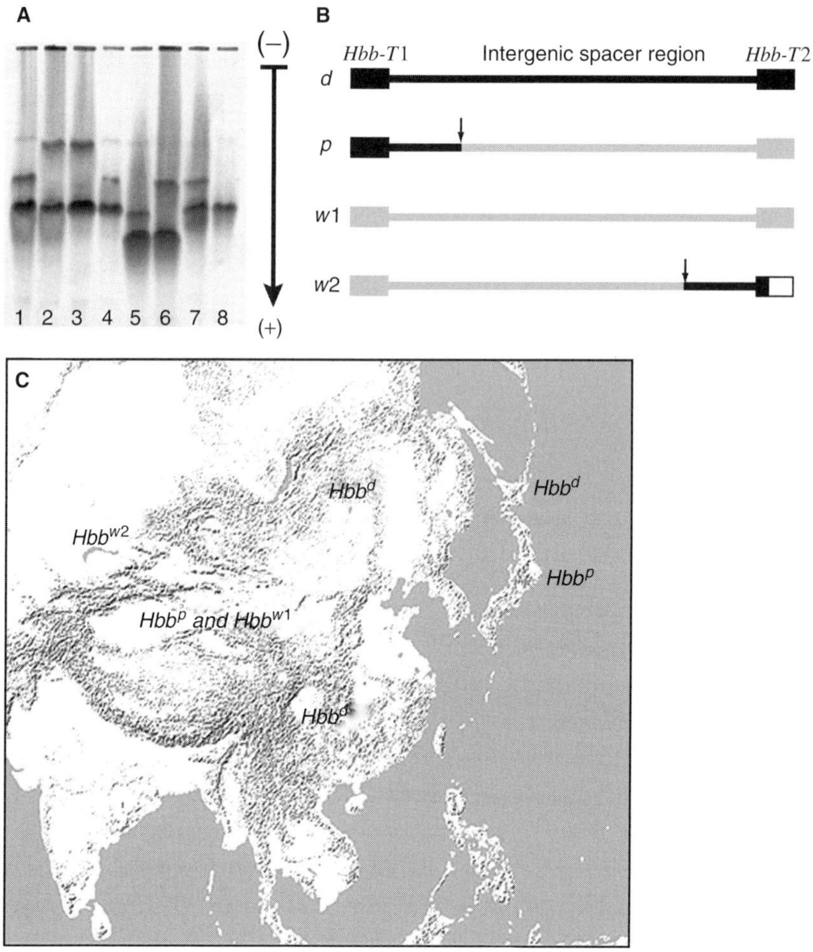

Figure 4.3 (a) Four *Hbb* electromorphs on a cellulose-acetate plate: *d* (1, 5), *p* (2, 6), *w1* (3, 7), and *w2* (4, 8). Samples 5–8 were treated with iodoacetic acid. For more detailed information, see Sato *et al.* (2008). (b) A schematic diagram of four *Hbb* haplotype structures. For detailed information, see Sato *et al.* (2008). (c) Schematic view of the distribution of four *Hbb* haplotypes in East Asia, drawn according to the polymorphisms of the protein electrophoresis examined in Miyashita *et al.* (1994). In fact, several haplotypes were observed in the same location. We show the most common haplotype in each region, with the exception of Hbb^{w2}, which was identified in only one individual.

(Fig. 4.3). These recombination points were also indicated within the intergenic spacer region, a pattern seemingly related to tandem repeat sequences. Both recombinants are supposedly the result of inter-subspecies hybridization. The geographic distribution of each *Hbb* haplotype suggests that reticulate evolution among genetically differentiated subspecies groups may have occurred in continental East Asia.

Other genetic aspects of Japanese wild mice

As shown above, the taxonomic group that includes MUS possesses unique genetic characteristics. For example, Hbb^p and a unique C-band pattern are discussed in detail above. Another example is the $Fv\text{-}4^r$ allele in the *Fv-4* locus. This locus, which endows the mouse with resistance to the Friend leukaemia virus, was first discovered in an outbred strain, ddY, established in Japan. An extensive survey of this allele in wild mice showed that the $Fv\text{-}4^r$ allele was very common in Japanese and South Asian mice. This allele represents a defective provirus of the Friend leukaemia virus (Ikeda *et al.*, 1985; Kai *et al.*, 1986; Odaka *et al.*, 1981). Another recent finding was that the expression-defective mutation of UDP-GlcNAc β-1,6-GlcNAc transferase (GNT) in the kidney-tubular cells of the DBA/1 inbred strain is also specific to this group (Sekine *et al.*, 2001, 2002). Briefly, this mutation is called *Gsl5*, and the nature of *Gsl5* is a *cis*-element in the GNT promoter. The wild-type *cis*-element consists of several blocks with two types of repetitive sequences, whereas the mutated type originally discovered in DBA/1 is largely deleted in the blocks. We surveyed the mutated *cis*-element among wild mice, and found that this mutation exclusively appeared in the MUS group, including *M. m. molossinus*. This suggests that the mutation occurred after the separation of the MUS group from the other subspecies groups.

Because mtDNA and Y-linked genes have no recombination events, they retain the genealogical information of parental populations, mtDNA for matriarchal lineages and Y-linked genes for patriarchal lineages. Therefore, the geographic distribution patterns represented by the use of such uniparentally transmitted genes have advantages over the patterns suggested by nuclear genes with recombination events: namely, the former patterns are always simpler than the latter. Conversely, the simplicity is always associated with some drawbacks. The major drawback is that interaction or gene flow between females and males in the same geographical population cannot be understood, even though both uniparental genes are simultaneously portrayed in a distribution map. To address this issue, it is necessary to develop a new molecular genetic technique, namely a haplotype analysis for a long chromosomal stretch with several molecular diagnostic markers for the subspecies of *M. musculus*. As mentioned above, the ground-breaking and efficient haplotype analysis for wild mice has been performed using *Hbb* haplotypes (Sato *et al.*, 2006, 2008). Recently, we performed an extensive haplotype analysis of wild mice, mainly Japanese wild mice, to estimate the percentage of recombinant-type haplotypes in wild mice. The target is a short but gene-rich stretch (~200 kbp) containing the *Mc1r* locus on mouse chromosome 8 because the *Mc1r* locus was a target of our earlier survey and is embedded in a gene-rich chromosomal stretch. Therefore, this genome situation is thought to facilitate relatively high-resolution haplotype analysis using exon-based PCR primers (Shimada *et al.*, 2009). The eight genes – *Fanca*, *Spire2*, *Tcf25*, *Mc1r*, *Tubb3*, *Def8*, *Afg3l1*, and *Dbndd1* – are encoded on the stretch and are intermittently arranged in this order. Among the 59 mice from throughout Eurasia, we identified the subspecies origin for individual gene alleles and then designated haplotypes for concatenated alleles. We successfully identified all intact haplotypes derived from the subspecies MUS, DOM, CAS, and all recombinant haplotypes (Nunome *et al.*, 2010).

In this report we clearly demonstrate that each of the eight gene regions displays a strong geographic signal within the phyletic clustering. The geographic clusters are also broadly consistent with the subspecies classification of *M. musculus*, as previously defined based on mtDNA and allozyme evidence: MUS in northern Eurasia (eastern Europe, Siberia, Primorye, northern China, Korea, Japan); CAS in southern Asia (India, Southeast Asia, southern China); and DOM in western Eurasia (western Europe, the western part of Iran).

The report also indicates the possible presence of a fourth lineage, distributed in Southeast Asia and India, effectively within the wider CAS territory. This finding is reminiscent of that based on allozyme data (e.g. Din *et al.*, 1996), which identified a fourth genetic component within *M. musculus* – the 'BAC group' from northern India. This finding is also consistent with RFLP evidence for genetic complexity within the CAS group for the nuclear ribosomal RNA gene (Suzuki *et al.*, 1986). Furthermore, the report also identifies additional phylogeographic structures in the MUS group. For example, *Mc1r* alleles comprise two main subgroups whose distribution comprises the northern (eastern Europe, Siberia, Primorye) and southern (Uzbekistan, northern China, Korea, Japan) parts of the MUS territory, respectively. In view of these data, it is tempting to establish a division within MUS between a northern lineage found at high latitudes, from Bulgaria to Sakhalin (MUS-I), and a southern lineage, mostly found at lower latitudes, from Uzbekistan through northern China to Korea and Japan (MUS-II) (Fig. 4.4).

Possible scenario for the establishment of Japanese wild mice

As described above, it is evident that *M. musculus* has experienced a complex evolutionary history involving reticulate evolution among populations of diverse geographic origin. These historical patterns offer hints of a time frame

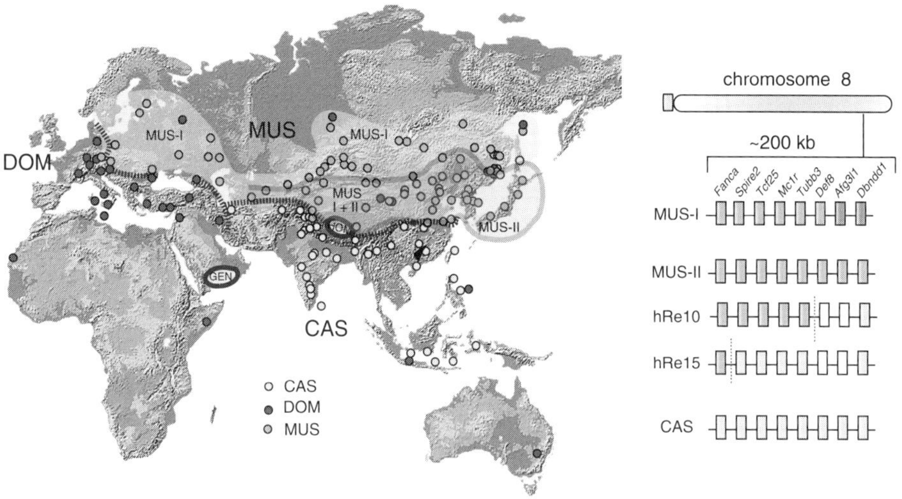

Figure 4.4 Geographic distribution of mtDNA variation in the house mouse *M. musculus*. The three lineages that have expanded in Eurasia are those represented by *M. m. castaneus* (CAS), *M. m. domesticus* (DOM), and *M. m. musculus* (MUS). The two localized lineages are those of *M. m. homoulus* (HOM) and *M. m. gentilulus* (GEN), recovered from Yemen (Prager *et al.*, 1998) and Nepal (Terashima *et al.*, 2006), respectively. The haplotype distributions in the three major lineages are represented by shaded circles (Suzuki *et al.* unpublished results). The spatial structure drawn from the mtDNA analysis is generally consistent with those from nuclear gene analyses, whereas the northern Eurasian group, MUS, can be divided into two subgroups, MUS-I (eastern Europe, Siberia, Primorye, Sakhalin, northern China) and MUS-II (Korea, Japan, northern China), which overlap in northern China, based on nuclear gene sequences (Nunome *et al.*, 2010). The general notion that the two lineages came from the Korean peninsula (MUS-I) and eastern India (CAS) and later intermingled in the Japanese islands is exemplified by a recently conducted haplotype structure analysis of a 200-kb portion of gene region on chromosome 8. This recent analysis recovered recombinant haplotypes (e.g. hRe10, hRe15) between the parental haplotypes of CAS and MUS-II (Nunome *et al.*, 2010). See plate section for a colour version of this figure.

that may predate the geographic expansion of mice as human commensals (Bonhomme *et al.*, 1984; Boursot *et al.*, 1993).

Nunome *et al.* (2010) confirmed that MUS genetic components predominate in Japanese wild mice. Generally, the MUS components of all Japanese wild mice, from Hokkaido to Kyushu, can be ascribed to the 'southern group' within MUS. More specifically, observations reveal a particular allelic affinity to geographically proximate Korean wild mice. Japanese and Korean mice share an allele in the highly polymorphic locus *Afg3l1*, whereas Chinese mice possess different alleles for the MUS components. This finding is concordant with previous data for the Y-specific gene *Sry*, in which mice from Japan and Korea were found to share the same variant sequence (Nagamine *et al.*, 1992; Terashima *et al.*, 2006).

The CAS and DOM components found in Japanese mice present a more complex picture (see Nunome *et al.*, 2010 and references therein for further details). However, the discovery of well-embedded DOM components on chromosome 8 introduces a new element in an already complicated and incompletely documented evolutionary history (Terashima *et al.*, 2006). One possible scenario would have DOM alleles entering Japan in prehistoric times as part of recombinant CAS/DOM haplotypes in mice carrying CAS mtDNA. However, critical evidence in the form of ancient DOM elements in wild mice still resident within the CAS core area has not yet emerged, although large parts of this genetically diverse region remain poorly sampled. Under this scenario, MUS genetic components arrived in Japan more recently, probably via the Korean peninsula. However, recombination frequencies indicate that this MUS component also arrived during prehistoric times, possibly several thousand years ago. As discussed at length by Terashima *et al.* (2006), this scenario implies an intensive exclusion of CAS mtDNA lineages, especially in the western part of Japan, following the introduction of *musculus*-type mice. For nuclear elements, this seems unlikely, unless the interbreeding frequency and/or success was extremely low. However, the alternative scenario posited by Terashima *et al.* (2006), which involves an original *musculus*-type wild mouse population with the later arrival of *castaneus*-type mice in Japan, similarly fails to account for the evidence of the prolonged interaction of CAS and DOM elements, unless this interaction occurred in the Asian region prior to migration. Both scenarios thus imply a complex evolutionary history, much of it significantly predating the period of human-associated long-range movements. Our general finding is thus consistent with that of Bonhomme *et al.* (2007) based on the analysis of minisatellite DNA. Our scenario for the origin of Japanese mice shows several parallels with recent anthropological notions concerning the origin of the Japanese people. First, our conclusion that the earliest immigrant mice were genetically heterogeneous, probably including components of both CAS and DOM lineages, shows parallels with evidence of high genetic diversity in the Palaeolithic Japanese population, as denoted in the mtDNA data from ancient Jomon remains (e.g. Horai *et al.*, 1996; Shinoda and Kanai, 1999). Second, we identify the most recent phase of the dispersal of mice to Japan as occurring through the central route, with northern China as the primary source area and Korea as the final point of departure. This is in accordance with the general notion that modern Japanese are most closely related to Koreans (Horai *et al.*, 1996; Omoto and Saitou, 1997;

Tanaka *et al.*, 2004). These apparent similarities in the history of people and mice might be coincidental. However, they may also point to a far closer and longer association between the two species than is currently credited, with both species making their way eastward across the Eurasian continent during prehistoric times (Searle *et al.*, 2009; Gabriel *et al.*, 2010). Careful attention to the archaeological record is needed to explore the antiquity of the bond that unites mice and people.

Conclusion

The Japanese wild mouse, *Mus musculus molossinus*, has been considered to be an endemic Japanese subspecies. About three decades ago, we surveyed RFLPs of mitochondrial DNA among the geographic populations of Japan and neighbouring countries. We found three important lines of evidence: first, there are no mtDNA haplotypes specific for *M. m. molossinus*. Second, conversely, two mtDNA haplotypes, one clustered with *M. m. musculus* and the other with *M. m. castaneus*, occur in Japan with an unnatural geographic distribution. The population with *musculus*-type mtDNA haplotypes occupies roughly the central two-thirds of Japan, whereas the population with *castaneus*-type mtDNA occupies separate peripheral areas of the country. The two populations form clear hybrid zones. The third line of evidence is that the distribution of the population with *castaneus*-type mtDNA is disjunct; the population is split in two. The intervening population is characterized by *musculus*-type mtDNA. Based on these findings, we propose that *M. m. molossinus* is not an endemic subspecies but has originated by genetic hybridization between *M. m. musculus* and *M. m. castaneus*. Recently, we have surveyed mtDNA haplotypes, mainly among populations in northern Japan. We found that the hybrid zone has moved northward and suggest that invasion by southern populations is still occurring in Japan. Furthermore, a survey based on genome science demonstrates that two distinct subpopulations of Eurasian *M. m. musculus* exist, one eastern and the other western, and that the Japanese population belongs to the former.

REFERENCES

Abe, K., Noguchi, H., Tagawa, K., *et al.* (2004). Contribution of Asian mouse subspecies *Mus musculus molossinus* to genomic constitution of strain C57BL/6J, as defined by BAC-end sequence-SNP analysis. *Genome Research*, **14**, 2439–47.

Bonhomme, F., Catalan, J., Britton-Davidian, J., *et al.* (1984). Biochemical diversity and evolution in the genus *Mus*. *Biochemical Genetics*, **22**, 275–303.

Bonhomme, F., Rivals, E., Orth, A., *et al.* (2007). Species-wide distribution of highly polymorphic minisatellite markers suggests past and present genetic exchanges among house mouse subspecies. *Genome Biology*, **8**, R80.

Boursot, P., Auffray, J.-C., Britton-Davidian, J., and Bonhomme, F. (1993). The evolution of house mice. *Annual Review of Ecology and Systematics*, **24**, 119–52.

Dallas, J. F., Bonhomme, F., Boursot, P., Britton-Davidian, J., and Bauchau, V. (1998). Population genetic structure in a Robertsonian race of house mice: evidence from microsatellite polymorphism. *Heredity*, **80**, 70–7.

Dallas, J. F., Dod, B., Boursot, P., Prager, E. M., and Bonhomme, F. (1995). Population subdivision and gene flow in Danish house mice. *Molecular Ecology*, **4**, 311–20.

Dietrich, W. F., Miller, J. C., Steen, R. G., *et al.* (1994). A genetic map of the mouse with 4,006 simple sequence length polymorphisms. *Nature Genetics*, **7**, 220–45.

Din, W., Anand, R., Boursot, P., *et al.* (1996). Origin and radiation of the house mouse: clues from nuclear genes. *Journal of Evolutionary Biology*, **9**, 519–39.

Dod, B., Litel, C., Makoundou, P., Orth, A., and Boursot, P. (2003). Identification and characterization of t haplotypes in wild mice populations using molecular markers. *Genetical Research*, **81**, 103–14.

Forejt, J. (1973). Centromeric heterochromatin polymorphism in the house mouse: evidence from inbred strains and natural populations. *Chromosoma*, **43**, 187–201.

Frazer, K. A., Eskin, E., Kang, H. M., *et al.* (2007). A sequence-based variation map of 8.27 million SNPs in inbred mouse strains. *Nature*, **448**, 1050–3.

Gabriel, S. I., Johannesdottir, F., Jones, E. P., and Searle, J. B. (2010). Colonization, mouse-style. *BMC Biology*, **8**, 131.

Gotoh, O., Hayashi, J. I., Yonekawa, H., and Tagashira, Y. (1979). An improved method for estimating sequence divergence between related DNAs from changes in restriction endonuclease cleavage sites. *Journal of Molecular Evolution*, **14**, 301–10.

Gropp, A. and Winking, H. (1981). Robertsonian translocations: cytology, meiosis, segregation patterns and biological consequences of heterozygosity. In *Biology of the House Mouse*, ed. R. J. Berry. New York: Academic Press, pp. 141–81.

Gyllensten, U., Wharton, D., Josefsson, A., *et al.* (1991). Paternal inheritance of mitochondrial DNA in mice. *Nature*, **352**, 255–7.

Horai, S., Murayama, K., Hayasaka, K., *et al.* (1996). mtDNA polymorphism in East Asian populations, with special reference to the peopling of Japan. *The American Journal of Human Genetics*, **59**, 579–90.

Ikeda, H., Laigret, F., Martin, M. A., and Repaske, R. (1985). Characterization of a molecularly cloned retroviral sequence associated with Fv-4 resistance. *Journal of Virology*, **55**, 768–77.

Kai, K., Sato, H., and Odaka, T. (1986). Relationship between the cellular resistance to Friend murine leukemia virus infection and the expression of murine leukemia virus-gp70-related glycoprotein on cell surface of BALB/c-Fv-4wr mice. *Virology*, **150**, 509–12.

Kaneda, H., Hayashi, J., Takahama, S., *et al.* (1995). Elimination of paternal mitochondrial DNA in intraspecific crosses during early mouse embryogenesis. *Proceedings of the National Academy of Sciences of the United States of America*, **92**, 4542–6.

Koide, T., Yoshino, M., Niwa, M., *et al.* (1992). The amplified long genomic sequence (ALGS) located in the centromeric regions of mouse chromosomes. *Genomics*, **13**, 1186–91.

Love, J. M., Knight, A. M., McAleer, M. A., and Todd, J. A. (1990). Towards construction of a high resolution map of the mouse genome using PCR-analysed microsatellites. *Nucleic Acids Research*, **18**, 4123–30.

Marshall, J. T. (1981). Taxonomy. In *The Mouse in Biomedical Research*, ed. H. L. Foster, J. D. Small, and J. G. Fox. New York: Academic Press, pp. 17–26.

Miller, D. A., Tantravahi, R., Dev, V. G., and Miller, O. J. (1976). Q- and C-band chromosome markers in inbred strains of *Mus musculus*. *Genetics*, **84**, 67–75.

Minezawa, M., Moriwaki, K., and Kondo, K. (1981). Geographical survey of protein variations in wild populations of Japanese house mouse, *Mus musculus molossinus*. *Japanese Journal of Genetics*, **56**, 27–39.

Miyashita, N., Kawashima, T., Wang, C. H., *et al.* (1994). Genetic polymorphisms of *Hbb* haplotypes in wild mice. In *Genetics in Wild Mice*, ed. K. Moriwaki, T. Shiroishi, and H. Yonekawa. Tokyo and Basel: Japan Scientific Societies Press and Karger, pp. 85–93.

Moriwaki, K., Miyashita, N., Suzuki, H., Kurihara, Y., and Yonekawa, H. (1986). Genetic features of major geographical isolates of *Mus musculus*. *Current Topics in Microbiology and Immunology*, **127**, 55–61.

Moriwaki, K., Shiroishi, T., Minezawa, M., *et al.* (1979). Frequency distribution of histocompatibility-2 antigenic specificities in the Japanese wild mouse genetically remote from the European subspecies. *Journal of Immunogenetics*, **6**, 99–113.

Moriwaki, K., Yonekawa, H., Gotoh, O., *et al.* (1984). Implications of the genetic divergence between European wild mice with Robertsonian translocations from the viewpoint of mitochondrial DNA. *Genetical Research*, **43**, 277–87.

Nagamine, C. M., Nishioka, Y., Moriwaki, K., *et al.* (1992). The *musculus*-type Y chromosome of the laboratory mouse is of Asian origin. *Mammalian Genome*, **3**, 84–91.

Nesbitt, M. N. and Francke, U. (1973). A system of nomenclature for band patterns of mouse chromosomes. *Chromosoma*, **41**, 145–58.

Nunome, M., Ishimori, C., Aplin, K. P., *et al.* (2010). Detection of recombinant haplotypes in wild mice (*Mus musculus*) provides new insights into the origin of Japanese mice. *Molecular Ecology*, **19**, 2474–89.

Odaka, T., Ikeda, H., Yoshikura, H., Moriwaki, K., and Suzuki, S. (1981). *Fv-4*: gene controlling resistance to NB-tropic Friend murine leukemia virus. Distribution in wild mice, introduction into genetic background of BALB/c mice, and mapping of chromosomes. *Journal of the National Cancer Institute*, **67**, 1123–7.

Omoto, K. and Saitou, N. (1997). Genetic origins of the Japanese: a partial support for the dual structure hypothesis. *American Journal of Physical Anthropology*, **102**, 437–46.

Popp, R. A. and Bailiff, E. G. (1973). Sequence of amino acids in the major and minor chains of the diffuse hemoglobin from BALB-c mice. *Biochimica et Biophysica Acta*, **303**, 61–7.

Prager, A., Orrego, C., and Sage, R. D. (1998). Genetic variation and phylogeography of central Asian and other house mice, including a major new mitochondrial lineage in Yemen. *Genetics*, **150**, 835–61.

Rhodes, M., Straw, R., Fernando, S., *et al.* (1998). A high-resolution microsatellite map of the mouse genome. *Genome Research*, **8**, 531–42.

Riginos, C. and Nachman, M. W. (1999). The origin of a Robertsonian chromosomal translocation in house mice inferred from linked microsatellite markers. *Molecular Biology and Evolution*, **16**, 1763–73.

Sakai, T., Kikkawa, Y., Miura, I., *et al.* (2005). Origins of mouse inbred strains deduced from whole-genome scanning by polymorphic microsatellite loci. *Mammalian Genome*, **16**, 11–19.

Sakai, T., Miura, I., Yamada-Ishibashi, S., *et al.* (2004). Update of mouse microsatellite database of Japan (MMDBJ). *Experimental Animals*, **53**, 151–4.

Salcedo, T., Geraldes, A., and Nachman, M. W. (2007). Nucleotide variation in wild and inbred mice. *Genetics*, **177**, 2277–91.

Sato, J. J., Shinohara, A., Miyashita, N., *et al.* (2008). Discovery of a new HBB haplotype *w2* in a wild-derived house mouse, *Mus musculus*. *Mammalian Genome*, **19**, 155–62.

Sato, J. J., Tsuru, Y., Hirai, K., *et al.* (2006). Further evidence for recombination between mouse hemoglobin beta b1 and b2 genes based on the nucleotide sequences of intron, UTR, and intergenic spacer regions. *Genes & Genetic Systems*, **81**, 201–9.

Schwarz, E. and Schwarz, H. K. (1943). The wild and commensal stocks of the house mouse, *Mus musculus* Linnaeus. *Journal of Mammalogy*, **24**, 59–72.

Searle, J. B., Jones, C. S., Gündüz, I., *et al.* (2009). Of mice and (Viking?) men: phylogeography of British and Irish house mice. *Proceedings of the Royal Society B: Biological Sciences*, **276**, 201–7.

Sekine, M., Kikkawa, Y., Takahama, S., *et al.* (2002). Phylogenetic development of a regulatory gene for the core 2 GlcNAc transferase in *Mus musculus*. *The Journal of Biochemistry*, **132**, 387–93.

Sekine, M., Taya, C., Kikkawa, Y., *et al.* (2001). Regulation of mouse kidney tubular epithelial cell-specific expression of core 2 GlcNAc transferase. *European Journal of Biochemistry*, **268**, 1129–35.

Selander, R. K., Hunt, W. G., and Yang, S. Y. (1969). Protein polymorphism and genic heterozygosity in two European subspecies of the house mouse. *Evolution*, **23**, 379–90.

Shimada, T., Sato, J. J., Aplin, K. P., and Suzuki, H. (2009). Comparative analysis of evolutionary modes in Mc1r coat color gene in wild mice and mustelids. *Genes & Genetic Systems*, **84**, 225–31.

Shinoda, K. and Kanai, S. (1999). Intracemetery genetic analysis at the Nakazuma Jomon site in Japan by mitochondrial DNA sequencing. *Anthropological Science*, **107**, 129–40.

Shitara, H., Hayashi, J. I., Takahama, S., Kaneda, H., and Yonekawa, H. (1998). Maternal inheritance of mouse mtDNA in interspecific hybrids: segregation of the leaked paternal mtDNA followed by the prevention of subsequent paternal leakage. *Genetics*, **148**, 851–7.

Suzuki, H., Miyashita, N., Moriwaki, M., *et al.* (1986). Evolutionary implication of heterogeneity of the nontranscribed spacer region of ribosomal DNA repeating units in various subspecies of *Mus musculus. Molecular Biology and Evolution*, **3**, 126–37.

Tanaka, M., Cabrera, V. M., Gonzalez, A. M., *et al.* (2004). Mitochondrial genome variation in eastern Asia and the peopling of Japan. *Genome Research*, **14**, 1832–50.

Terashima, M., Furusawa, S., Hanzawa, N., *et al.* (2006). Phylogeographic origin of Hokkaido house mice (*Mus musculus*) as indicated by genetic markers with maternal, paternal and biparental inheritance. *Heredity*, **96**, 128–38.

Yonekawa, H., Gotoh, O., Tagashira, Y., *et al.* (1986). A hybrid origin of Japanese mice '*Mus musculus molossinus*'. *Current Topics in Microbiology and Immunology*, **127**, 62–7.

Yonekawa, H., Moriwaki, K., Gotoh, O., *et al.* (1981). Evolutionary relationships among five subspecies of *Mus musculus* based on restriction enzyme cleavage patterns of mitochondrial DNA. *Genetics*, **98**, 801–16.

Yonekawa, H., Moriwaki, K., Gotoh, O., *et al.* (1982). Origins of laboratory mice deduced from restriction patterns of mitochondrial DNA. *Differentiation*, **22**, 222–6.

Yonekawa, H., Moriwaki, K., Gotoh, O., *et al.* (1988). Hybrid origin of Japanese mice '*Mus musculus molossinus*': evidence from restriction analysis of mitochondrial DNA. *Molecular Biology and Evolution*, **5**, 63–78.

Yonekawa, H., Takahama, S., Gotoh, O., Miyashita, N., and Moriwaki, K. (1994). Genetic diversity and geographic distributioin of *Mus musculus* subspecies based on the polymorphism of mitochondrial DNA. In *Genetics in Wild Mice*, ed. K. Moriwaki, T. Shiroishi, and H. Yonekawa. Tokyo and Basel: Japan Scientific Societies Press and S. Karger, pp. 25–40.

Yoshida, M. C., Ikeuchi, T., and Sakaki, M. (1975). Differential staining of parental chromosomes in interspecific cell hybrids with a combined quinacrine and 33258 Hoechst technique. *Proceedings of the Japan Academy*, **51**, 184–7.

Yoshida, M. C. and Kodama, Y. (1983). C-band patterns of chromosomes in 17 strains of mice. *Cytogenetics and Cell Genetics*, **35**, 51–6.

5

The complex social environment of female house mice (*Mus domesticus*)

BARBARA KÖNIG AND ANNA K. LINDHOLM

Introduction

Conspecifics are a major environmental factor for most organisms. In sexually reproducing species they are not only mating partners, but also partners in competitive and cooperative interactions. In a social species, females therefore interact with males as potential mating partners and with other females as potential social and cooperation partners. On the other hand, conspecifics are also competitors for limited resources when living in close proximity. Conflicts are thus inevitable when females form groups, despite any benefits of group living (Alexander, 1974; Emlen and Oring, 1977; Sterck *et al.*, 1997). In recent reviews, Clutton-Brock (2009) as well as Stockley and Bro-Jørgensen (2010) emphasize the lack of knowledge about female competitive strategies, despite increasing evidence for intrasexual competition and sexual conflict among females. Female mammals regularly compete among themselves over access to resources such as food, breeding territories, nest sites, and shelter. In species with multi-female groups, they may further compete over breeding (dominance) rank, assistance with offspring care, the number of offspring raised, protection of offspring during babysitting, or social thermoregulation. Females may also compete over access to mating partners in the course of a single breeding cycle, and competitive behaviour among females generally includes aggression, weaponry, alliance formation, cooperation, and the inhibition of other females' reproduction (Rosenqvist and Berglund, 1992; Clutton-Brock, 2009; Stockley and Bro-Jørgensen, 2010). However, few data are available regarding whether differential access to resources (including mating partners) leads to differences in reproductive success, and whether social interactions affect the outcome of such reproductive competition.

The above strategies, which females may use during intrasexual competition, are less conspicuous than those used by males, and this difference has been linked to higher parental investment and lower potential reproductive rates in female

Evolution of the House Mouse, ed. Miloš Macholán, Stuart J. E. Baird, Pavel Munclinger, and Jaroslav Piálek. Published by Cambridge University Press.

mammals (see Clutton-Brock, 2009; Stockley and Bro-Jørgensen, 2010). Nevertheless, any selective influence of reproductive competition will tend to result in the development of traits that allow for more competitive interactions. Such a process has been termed 'social selection' by West-Eberhard (1983, 1987), with sexual selection considered a subset. Social selection refers to the fact that an individual's fitness is not only determined by its own phenotype but also by the phenotype of a social partner. The potential for social selection exists whenever individual fitness varies as a result of social interactions, both in a cooperative and competitive context. In mammals in which females live in social groups, social interactions thus may provide the opportunity for the evolution of traits that improve reproductive competitiveness as well as social partner choice in females, with the prediction that interactants experience variance in fitness. So far, fitness consequences of female social interactions have only been described for primates (Silk *et al.*, 2003) and for house mice (Weidt *et al.*, 2008).

House mice are a tractable species for studies of female social interactions and female intrasexual competition, as well as the potential for reproductive competition to affect a female's social environment. Here, we first review factors that influence female–female social interactions during reproductive cooperation and competition. We then present new data from a free-living population in Switzerland showing seasonal effects on female reproductive competition and social bonding, and analyse whether differences in degree of reproductive competition and in current reproductive state affect social interactions of adult females.

Reproductive competition and cooperation in female house mice

Female house mice (*Mus domesticus*) live in social groups, and field as well as laboratory studies have illustrated their high potential for cooperation and competition over reproduction. House mice have a flexible social structure, but typically they live in small groups consisting of a dominant male, one or several adult females with their litters and several subordinate mice of both sexes (DeLong, 1967; Lidicker, 1976; Bronson, 1979; Berry, 1981; Singleton, 1983; Gray *et al.*, 2000). House mice are plural breeders, with several breeding females per group. Within groups, adult females contribute to territorial defence (Crowcroft, 1966; Latham and Mason, 2004) and cooperate in some kinds of communal care, such as babysitting, social thermoregulation, or defence of pups. The most conspicuous example of cooperation, however, is non-offspring nursing (for a review, see König, 2006). Such non-offspring nursing occurs when two – or, rarely, more – females pool their litters in a communal nest and indiscriminately nurse both own- and non-offspring (Southwick, 1955; Sayler and Salmon, 1971; König, 1989, 1993; Manning *et al.*, 1995).

Kept under standardized conditions in the laboratory or in semi-natural enclosures, non-offspring nursing is an integral part of the reproductive behaviour of female house mice in egalitarian groups. Communal or non-offspring nursing (also referred to as allosuckling or allonursing) has been described for approximately 70 mammalian species across 12 orders, and for reproducing and non-reproducing females; nevertheless, in only 10% of such species were non-offspring nursed as much as own young, as is the case in house mice (reviewed in Packer *et al.*, 1992; Jennions and Macdonald, 1994; König, 1997, 2006; Lewis and Pusey, 1997; Solomon and French, 1997; Hayes, 2000).

Communal nursing in house mice has been studied intensively in the laboratory with direct descendants of wild-caught animals. When two females establish an egalitarian reproductive relationship, communal nursing increases both partners' individual lifetime reproductive success in comparison to females rearing litters solitarily (König, 1993, 1994a). However, the probability for such cooperation is highest when a female shares a nest with a familiar sister to form a low-skew society (König, 1994a, b, 1997, 2006). As a consequence, non-offspring nursing of female house mice in pairs with egalitarian reproduction is adaptive, and involves mutualistic direct fitness benefits. With increasing group size (three adult females per group), on the other hand, reproductive skew increases towards despotic relationships and individual reproductive success decreases below that of a solitary female. The reason why females differ in individual reproductive success as a function of group size and relatedness is that females differ in their probability to reproduce and successfully wean young within their lifetime, due to competition over reproduction, despite communal nesting and communal nursing (König, 1994a, b, 2006).

Individual fitness thus varies as a result of social interactions, and direct benefits of cooperation seem to stabilize non-offspring nursing among female house mice. Spatial proximity and nest sharing usually precede breeding associations in house mice (Manning *et al.*, 1995; Dobson *et al.*, 2000; Hayes, 2000; Rusu and Krackow, 2004), while spatial intolerance and unstable dominance relationships prior to the start of reproduction strongly impair cooperation in communal nests (Rusu and Krackow, 2004). Females may therefore be expected to carefully choose partners for subsequent breeding to avoid fitness loss, and establish social bonds to such preferred partners. In accordance with this hypothesis, female house mice display non-random preferences for social partners when kept in groups of unrelated females in semi-natural enclosures (Weidt *et al.*, 2008). Females that were afterwards allowed to live with previously preferred social partners had a higher probability to reproduce and a significantly higher lifetime reproductive success compared to females living with previously non-preferred partners. This suggests that females generally associate with social partners with whom they have a low potential for reproductive competition (Weidt *et al.*, 2008).

In natural populations and in enclosures, female house mice spatially associate and communally nest with kin (Wilkinson and Baker, 1988; Dobson *et al.*, 2000; Dobson and Baudoin, 2002; Rusu and Krackow, 2004; Rusu *et al.*, 2004). Familiarity during juvenile development (as occurs among siblings) is of importance for such female–female social interactions (König, 1994b; D'Amato, 1997). This is in contrast to female mate choice, where dominance status (Hayashi, 1990; Drickamer, 1992) or genotypic cues (MHC complement – Yamazaki *et al.*, 1976; or *t* haplotype – Lenington and Egid, 1989) are of main relevance. This suggests that for social partner choice direct information about a potential partner's behaviour or physiological status is more important than genetic relatedness (see also König, 1994b, 2006). Nevertheless, choosing a familiar female for social cooperation may result in close association with kin. Incidences of female competition over reproduction, mainly expressed through overt aggression, reproductive inhibition of other females or infanticide of non-offspring, on the other hand, have been typically described for unfamiliar and unrelated females (Hurst, 1987; König, 1994a; Palanza *et al.*, 1996, 2005; Rusu and Krackow, 2004).

The potential for female house mice to establish individualized social bonds among group members is therefore high. House mice mainly use odours (for a review see Stopka *et al.*, Chapter 8 in this volume), but also ultrasonic vocalizations (for review see Musolf and Penn, Chapter 10 in this volume) for intraspecific communication and individual identification, and familiar individuals can recognize each other after a separation period of at least seven days (Hurst, 1990; D'Amato, 1997; D'Amato and Moles, 2001; Nevison *et al.*, 2003).

Seasonal reproduction in house mice

Reproduction in house mice is assumed not to be obligatorily seasonal. Commensal populations usually breed all year, and feral populations breed non-seasonally in some locations, but are seasonal in others (reviewed in Latham and Mason, 2004). In his extensive review of the reproductive ecology of house mice, Bronson (1979) concluded that the interaction between caloric deprivation and the metabolic responses to cold exposure seem to be the most likely candidate for producing a seasonal interruption of the potential of mice for continuous breeding. During cold periods, reproduction is often reduced, although cold-adapted house mice have bred in frozen carcasses in deep freezer houses in the harbour of London (Laurie, 1946). Seasonal variability in commensal barn populations therefore may allow a comparison of periods with high reproductive activity and a potentially high degree of reproductive competition, with periods of relatively low reproductive activity and thus low competition among females.

From free-living populations, however, little information is available on the social life of female house mice and how it is affected by differences in female competition over reproduction and by female bonding. Here, we illustrate the complexity of the social environment of a wild female house mouse, using new data from our long-term study of a free-living population in a barn in Switzerland. We were specifically interested in the following questions:

- Is there evidence that female reproductive competition varies seasonally?
- Does seasonality in reproductive competition affect the social environment of females?
- Does current reproductive state influence female–female social interactions independently of seasonal effects?

Assuming that (1) females display social partner preferences; (2) individual reproductive success is highest in pairs of females and decreases with increasing group size; and (3) high reproductive activity implies high reproductive competition, reproducing females are expected to choose smaller group sizes during periods of high reproductive activity.

Study population of wild house mice

Our study population was initiated in autumn 2002 by Andrea Weidt and Barbara König in a former barn, situated at the border of a forest near Illnau, Kanton Zürich, Switzerland. For a house mouse, a barn is a natural habitat. House mice in Europe occur in anthropogenic habitats, such as grain stores and farm buildings, with feral populations generally restricted to islands (Pocock *et al.*, 2004). This has been the case for a long time. Archaeological and palaeontological evidence shows that house mice were commensal with humans by 8000 BC, coinciding with new farming practices such as large-scale grain storage (Cucchi *et al.*, 2002), and that house mice colonized western Europe, commensal with humans, about 2000 years ago (Cucchi *et al.*, 2005).

Our study population was seeded with 12 wild-caught individuals, six from each of two nearby demes (altogether four males and eight females, caught at two farmhouses situated within 5 km of the barn) in November 2002, and has since August 2003 been open to immigration and emigration (after successful reproduction of the founder individuals). Long-distance dispersal out of the barn occurs, as marked mice have been recovered approximately 1 km away.

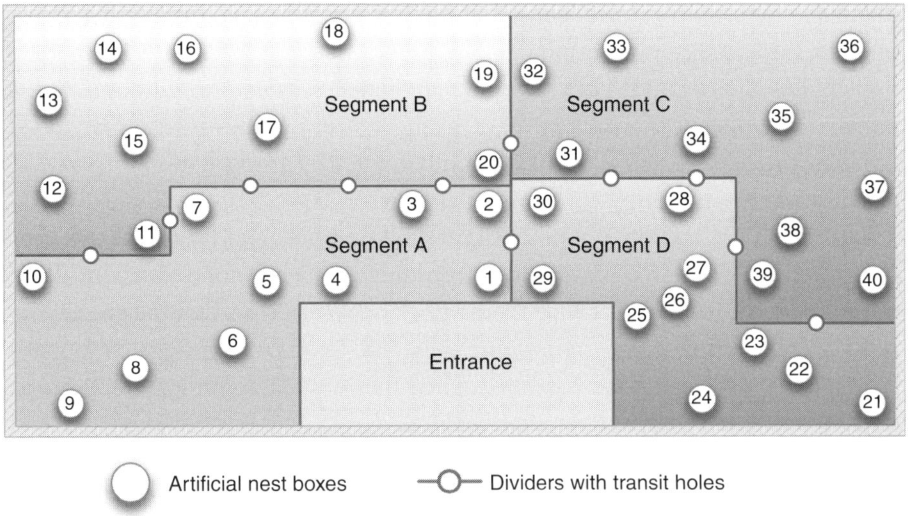

Figure 5.1 Schematic drawing of the floor of the barn (floor space 72 m^2) with 40 nest boxes. Dividing walls are aluminium, 75 cm high, with 11 passages, separating the area into four quarters; next to the entrance door into the barn is a separate area for storage of equipment and for handling the mice – this is also accessible to the animals. Not shown are 2–4 feeding and drinking sites each per quarter, and further structuring of the floor with bricks and smaller wooden and plastic barriers or hides.

The barn is divided into four quarters by aluminium plates 75 cm in height, with 11 small holes allowing the passage of mice (Fig. 5.1). The floor of the barn consists of concrete, and is covered by commercial rodent bedding, and littered with bricks and wooden and plastic barriers to provide internal structure and shelters (not shown in Fig. 5.1). The mice can access all parts of the barn, and can leave it under the roof or through holes in the walls (the walls consist partly of bricks and wood). They nest in 40 artificial nest boxes (ten per quarter) and are provided with straw as nesting material. The interiors of the nest boxes are accessible to us, so offspring can be counted and measured. Food, a 50–50 mixture of oats and commercial rodent food made by Haefliger AG, is provided *ad libitum* at 9–10 feeding trays (at least two per quarter), as is water (3–4 drinking sites per quarter). Availability of food at our study site is less than would be available in grain stores and many barns, but is not limiting. We therefore consider the availability of food within the natural range. The barn itself is free of predators, but not of parasites; mice are exposed to predators, including foxes, badgers, house cats, and birds of prey, whenever they exit the barn.

Male and female mice of minimally 18 g living in the barn are implanted with a transponder (RFID tag; trovan® ID 100, 0.1 g weight, 11.5 mm length, 2.1 mm diameter) by a trained and licensed animal care technician. Implanting transponders into smaller mice results in a high percentage of tags lost. Our method nonetheless allows tagging of all reproducing females, as no pregnant or lactating females weighing less than 19 g have been detected (2006–2010). The transponders provide a unique identity number for each mouse and a means to monitor mice remotely by transponder readers.

In 2003, 2004, and since 2006 until now, all nest boxes and shelters have been monitored weekly for the presence of tagged mice (which is done with a handheld transponder reader placed outside the nest box or a shelter) and for the presence of new litters. During 2005, monitoring was less frequent. The age of pups is estimated, and all litters are measured when they are 13 days of age (range 12–14 days; day of birth of a litter is considered as day 1). We sex the pups and take individual body weight. We refer to the number of pups at day 13 as weaning litter size, because pups are not yet mobile and thus do not mix with other litters by themselves. At 14 days of age, pups open their eyes and are mobile. They begin to eat solid food at 17 days and are weaned when 21–23 days old (dependent on litter size; smaller litters are weaned earlier; König and Markl, 1987). Offspring mortality between days 13 and 17 is almost absent in the laboratory, and we consider litter size at day 13 as a good approximation of weaning litter size also in our barn population.

In addition, at approximately seven-week intervals, comprehensive trapping has been conducted to monitor the adult population (population monitoring). Every mouse is weighed and females are examined for reproductive state (characterized as pregnant and/or lactating according to the swelling of the body and the appearance of the teats), and those adults lacking transponders are tagged. We also monitor the population for remains of deceased mice. Here, we analyse 29 population-monitoring events over a period of four years, from April 2006 until March 2010.

In May 2007 we installed permanent transponder readers in the tunnels that provide entrances to the nest boxes (two antennas per tunnel, which allows us to discriminate between a mouse entering or leaving a nest box). These readers connect to a computer and continuously track movements of tagged mice into and out of nest boxes. This provides 24-hour information on movements and social affiliations of adult mice, and makes breeding females and males easy to locate. Here we include data from 1 January 2008 to 31 December 2009 on tagged females.

Seasonal variability in reproduction

The number of adult males and females in our study population increased from an average (± standard error) of 29.6 ± 4.4 males (range 11–57) and 38.9 ± 3.0 females (range 27–55), respectively, in the 12 months from April 2006, to 69.7 ± 8.3 males (range 55–88) and 70.7 ± 4.8 females (40–100) in the 12

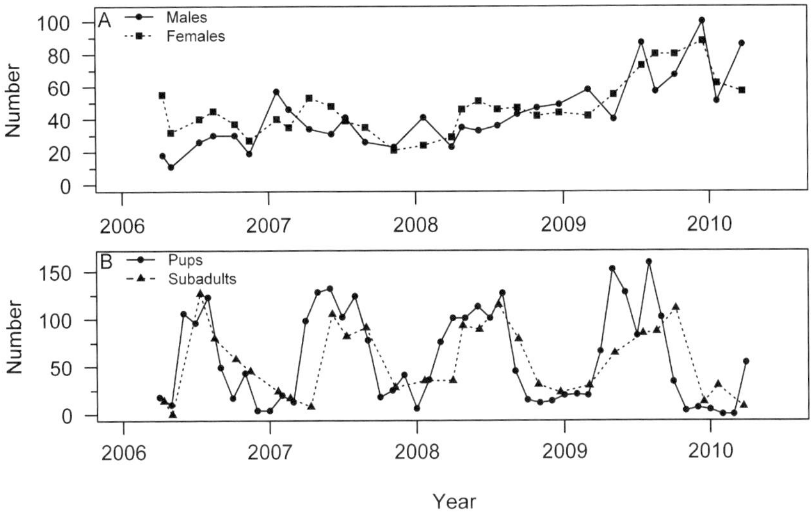

Figure 5.2 (a) Adult males and adult females (body weight at least 18 g) and (b) subadults (weaned offspring of less than 18 g) registered during population-monitoring events over a period of four years (April 2006 to March 2010; $N = 29$); (b) further shows the total number of pups sampled per month at the age of 13 days ($N = 48$).

months from April 2009 (Fig. 5.2a). No seasonal effects on adult sex ratio were obvious; however, during the first three years (2006–2008), the number of adult females was higher than the number of adult males during the summer, with the reverse effect during the winter. The number of pups weaned (13 days of age) and of subadults (weaned individuals up to a body weight of 17.5 g), on the other hand, showed marked seasonal effects, with peaks in reproductive output at 12-month intervals, during late spring/early summer (beginning April/May; Fig. 5.2b). The number of subadults registered during a single population-monitoring event reached values of up to 127 during mid-summer. Although reproduction rarely entirely stopped during the winter, it was typically drastically reduced, beginning in September/October.

Seasonality in reproduction was also reflected in the proportion of reproductive females present in the population (Fig. 5.3a). This measure may reflect female reproductive activity better than pup production, as only pups surviving to the age of 13 days are included in the latter measure. However, it may fail to include non-lactating females in early pregnancy. Generally, the proportion of females pregnant and/or lactating was lower during winter (October to March) than during summer (April to September; binomial GLM, $z_{1,27} = 9.84$, $p < 0.0001$; statistics were performed using R version 2.12.2; R Core Development Team, 2011). This is

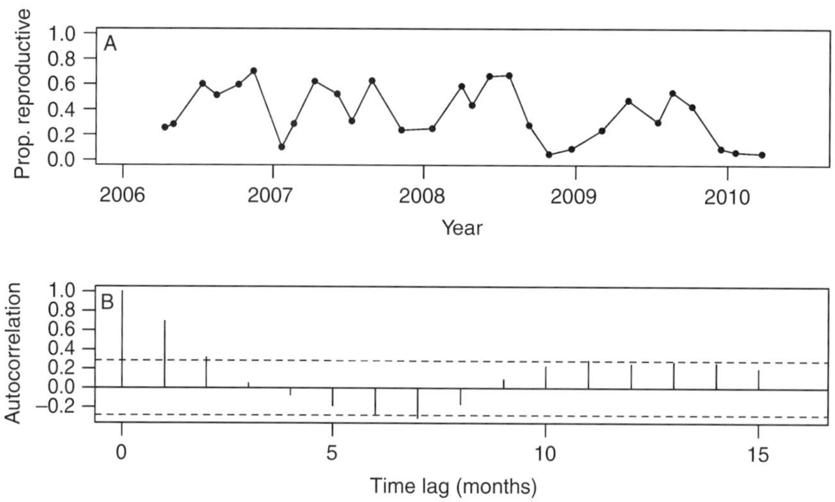

Figure 5.3 Proportion of adult females classified as reproductive (pregnant and/or lactating) during population-monitoring events; (a) over a period of four years (April 2006 to March 2010; $N = 29$); (b) correlogram of temporal autocorrelation. The dashed line gives the 95% confidence interval. Vertical bars crossing the confidence interval indicate significant autocorrelations.

illustrated by the significant negative autocorrelation of the proportion of reproductively active females after a seven-month lag time (Fig. 5.3b), using linear interpolation to estimate missing monthly values (as monitoring sessions were less frequent than once per month). Peaks of positive correlations after 11–14 months indicate an approximate annual cycle in reproduction.

During the study period, the proportion of females showing reproductive activity decreased with time (binomial GLM, $z_{1,27} = 2.10$, $p < 0.05$). As numbers of adult females showed an increase with time (linear regression $t_{1,27} = 4.23$, $p < 0.001$), but numbers of reproductively active females did not (linear regression $t_{1,27} = 0.12$, $p < 0.91$), the decrease in proportion of reproductive females over time could be due to increased female competitive interactions. We used a generalized linear model to test for an effect of the number of adult females present on the proportion of females showing reproductive activity (Fig. 5.4). To remove the effect of seasonality in breeding, we averaged across 12-month intervals, and we log-transformed the average number of females present to avoid over-dispersion of errors. Indeed, reproductive activity significantly declined with increasing numbers of females present (binomial GLM, $z_{1,2} = 5.25$, $p < 0.001$).

To summarize, reproductive skew among adult females was substantial even during periods of favourable breeding conditions during summer. The number

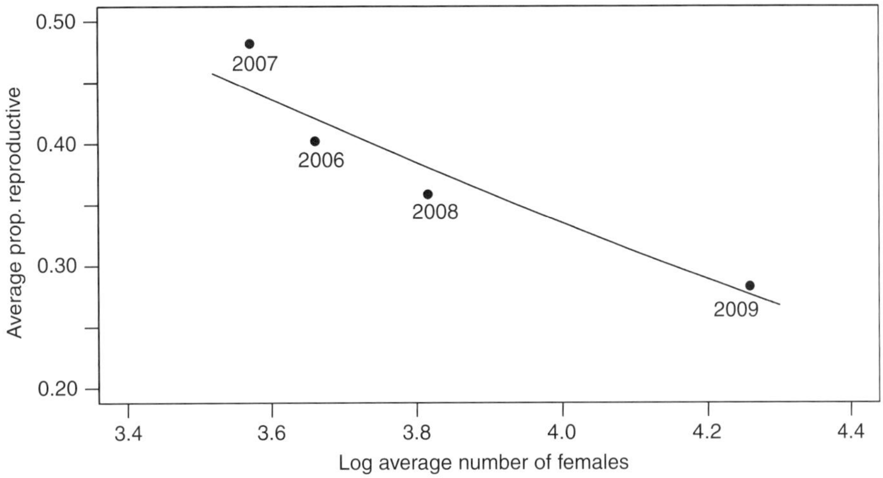

Figure 5.4 Average proportion of females classified as reproductive (pregnant and/or lactating), as a function of the log-transformed average number of females present (points), per year, with generalized linear model prediction (line).

of reproductively active females did not change significantly over the study period, nor did the annual total of pups weaned. However, the number of non-breeding adult females in the population increased over time, suggesting increasing female competition over reproduction, especially during the summer. In winter, however, under conditions of unfavourable, cold temperatures, a female's individual condition or physiology, and not necessarily reproductive competitiveness against other females, might determine whether she is able to reproduce. Reduced food and nesting material availability can be excluded in our study population as causal factors for low breeding during winter, as suggested for other wild populations (Laurie, 1946; Berry, 1968; Randall, 1999). Instead, cold and its ensuing influence on metabolism may be the greatest hazard to a female (see also Lynch, 1992).

Male condition also supports our conclusion that the strength of intrasexual selection varies seasonally. Given that wounds in males are typically inflicted during intrasexual competition over access to females, only 8.8% of adult males examined during population-monitoring events had fresh wounds on their body, head, legs, or tail in winter, but 21% were wounded during summer.

Whatever the cause of the observed seasonality in reproduction, we expect lower female competition during winter than during summer, leading to the next question of seasonal differences in female–female interactions and female bonding.

Seasonal variability in nest box use and in social interactions among females

Female nest box use and social behaviour were analysed from a two-year dataset of transponder readings in the nest boxes. For each tagged female we calculated the following monthly data (beginning with the month of tagging for females implanted after January 2008, and excluding the month during which a female died or disappeared from the barn): cumulative number of nest boxes visited; cumulative number of female partners; and cumulative number and mean duration of interactions. Social interactions and female partners were defined according to meetings in the same nest box. As soon as any two females simultaneously visited the same nest box, they were considered as partners and the meeting as a social interaction. Nest boxes are rather small, with a diameter of 15 cm, and we never observed two separate nests within one box. We therefore assume that individuals have direct contact when simultaneously visiting a box. A total of 226 females provided 1305 monthly records of such data on partners, social interactions, and number of nest boxes used.

Adult females usually met for a rather extended period of time in a nest box. An average social interaction lasted 1296 s (21.6 min), and ranged between 2 s and 4602 s (76.7 min; Fig. 5.5). A short interaction, of less than 1 min, might also be interpreted as a socio-negative or agonistic encounter, resulting in a female being chased out of the box. Substantially longer interactions between females, however, suggest a socio-positive relationship or familiarity with each other, and that

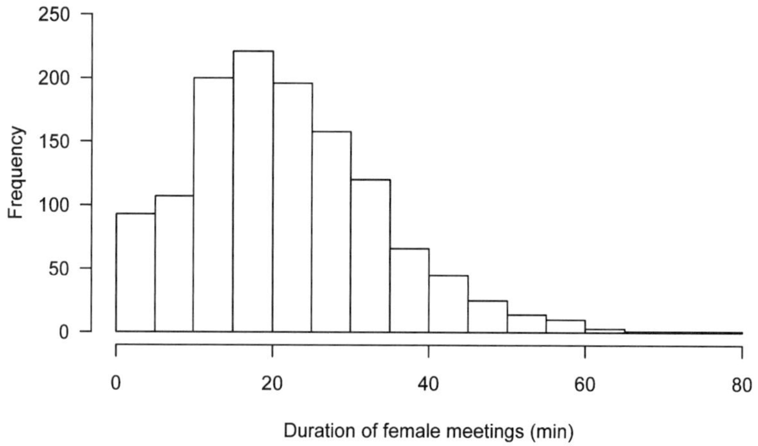

Figure 5.5 Frequency distribution of meeting duration (minutes) of any two tagged females that simultaneously visit a nest box (monthly average of the length of a female's social interactions with any other female; $N = 226$ females).

the females belong to the same social group. In 44 monthly records (5.7%), females used nest boxes without meeting another tagged female.

Influence of female reproductive state on spatial and social behaviour

In 615 monthly records, information was available on female reproductive status during that month from population-monitoring events, and was used for further analyses. We accounted for repeated measures of the same female by implementing linear mixed models in ASReml 3.0 (VSN International) with female identity as a random effect. We tested for independent and multiplicative effects of reproductive status and season on female social interactions. Accounting for repeated measures complicates determination of degrees of freedom; here, denominator degrees of freedom were calculated empirically in ASReml according to Kenward and Roger (1997).

Females differed substantially in the monthly number of nest boxes visited, which ranged between 1 and 33. Reproductively active females used fewer nest boxes than non-reproducing females (Fig. 5.6a). Furthermore, all females generally

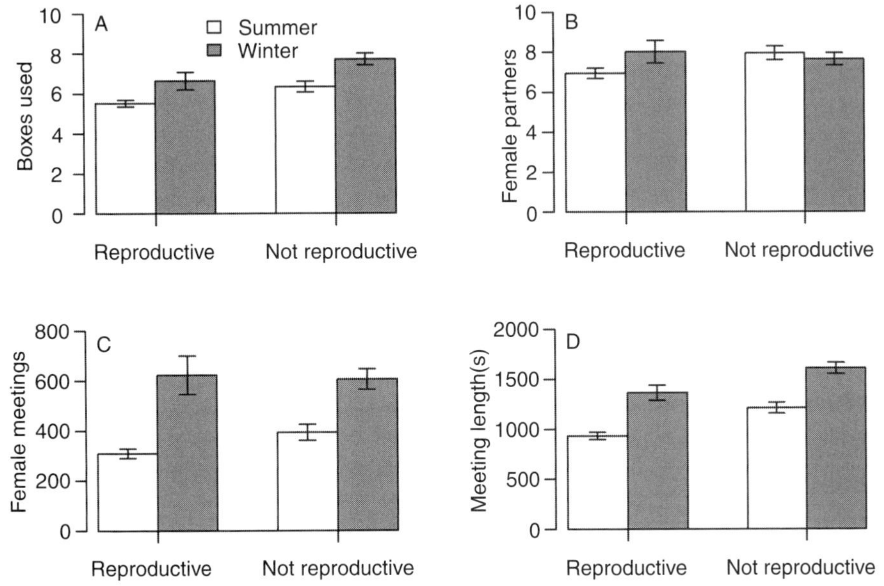

Figure 5.6 Bar plots (mean ± SE) of monthly data on (a) cumulative number of nest boxes visited; (b) cumulative number of female partners; (c) cumulative number of female meetings; and (d) average duration of meetings, according to female reproductive status and season ($N = 615$ monthly records).

used fewer boxes in summer than in winter (effect of reproduction: Wald statistic$_{1,551}$ = 20.0, $p < 0.0001$; season: Wald statistic$_{1,539}$ = 24.9, $p < 0.0001$; interaction: Wald statistic$_{1,545}$ = 0.001, $p < 0.92$). Within nest boxes, females met with 0–19 female partners, without obvious seasonal differences (Fig. 5.6b). Nevertheless, the lowest number of partners was observed for breeding females in summer (on average 6.9 partners), with significantly fewer partners than non-reproducing females in winter (reproduction: Wald statistic$_{1,488}$ = 3.38, $p < 0.07$; season: Wald statistic$_{1,490}$ = 0.38, $p < 0.54$; interaction: Wald statistic$_{1,491}$ = 8.38, $p < 0.005$).

Reproductively active females had fewer (Fig. 5.6c) and shorter meetings (Fig. 5.6d) than non-reproducing females, and during winter meetings generally were more frequent and longer than in summer (number of female meetings: reproduction: Wald statistic$_{1,541}$ = 12.2, $p < 0.0001$; season: Wald statistic$_{1,532}$ = 49.8, $p < 0.0001$; interaction: Wald statistic$_{1,537}$ = 2.26, $p < 0.14$; duration of female meetings: reproduction: Wald statistic$_{1,592}$ = 46.1, $p < 0.0001$; season: Wald statistic$_{1,569}$ = 49.9, $p < 0.0001$; interaction: Wald statistic$_{1,587}$ = 0.06, $p < 0.81$).

In both summer and winter females used only a limited number of nest boxes during a given month (on average 5–8 of 40 available), allowing for the conclusion that such spatial behaviour reflects an individual's home range. We have no information on female behaviour outside the nest boxes, but the rather long duration of the majority of meetings among adult females in a box suggests that partners belong to the same social group, are familiar with each other (with the potential to establish social bonds), and that all group members might contribute to defend nest boxes against non-group members (either through olfactory cues or via aggression during direct interactions; see also Hurst and Nevison, 1994).

Access to nest boxes may be a prerequisite for successful reproduction in females. The vast majority of litters in our population was born and weaned in boxes, presumably because they improve protection from disturbances by conspecifics (see also Crowcroft and Rowe, 1957, 1958; Hurst, 1987; Rusu *et al.*, 2004). Nevertheless, even reproducing females used several nest boxes and sometimes moved litters between boxes. We do not know why females regularly access several boxes even when lactating. Such behaviour can be explained by several factors, such as the option to move when faced with or disturbed by a predator, varying microclimatic needs during different environmental conditions, improved access to feeding and drinking sites, or avoidance of parasites. Since the number of partners significantly increased with the increasing number of nest boxes used (linear regression, $t_{1,1303}$ = 10.7, $p < 0.0001$), at a rate of 0.32 additional partners per additional box used, larger groups may be able to defend larger territories, which again might improve access to resources and/or defence against non-group members.

Reproductive competition nevertheless influenced the spatial and social behaviour of the females. During summer, when a higher proportion of females were breeding, reproducing females used fewer boxes than during winter, and also fewer than non-reproducing females. When radio-tracking female house mice in a feed shed, Wilkinson and Baker (1988) also observed smaller home ranges in lactating females. Reproducing females might access fewer nest boxes because they regularly nurse offspring, keep them warm, and protect them against infanticidal conspecifics, and thus have less time for exploratory behaviour. On the other hand, they are limited in the time spent with the pups, because they have to drastically increase the daily amount of food eaten to allow for milk production (König *et al.*, 1988). Still, we exclude improved access to food as a causal factor to temporarily reduce home range size during breeding in summer. First, we did not observe food storage in nest boxes, and second, in winter, when nutritional and metabolic requirements are expected to be higher than in summer, reproducing females did not show reductions in home range size. Instead, we suggest that reproducing females under conditions suggesting high reproductive competition are more restrictive in access to social partners. As predicted from previous lab studies, they meet with relatively few partners, despite the fact that relatively more other females are breeding. Highly pregnant or lactating females might be more choosy in terms of with whom they meet, as preferences for social partners result in low conflicts over reproduction and improved offspring survival (see Weidt *et al.*, 2008). Furthermore, under experimental lab conditions individual lifetime reproductive success decreases with increasing number of females (König, 2006), and it has been shown for other rodents and larger mammals that female competition increases with increasing group size (reviewed in Silk, 2007).

Adult females rarely stayed in a nest box in the absence of any other adult female, suggesting that meetings in social groups and established social bonds play an important role in reproductive success. Social partners may serve various functions for a reproducing female. If the social partner is also lactating, both can establish a communal nest and gain mutual benefits through cooperative nursing of litters. During summer, on the other hand, relatively many non-reproducing adult females were also observed in our population. Social bonds to a non-reproducing female might allow a lactating female to gain benefits through helping to improve protection of offspring, social thermoregulation, or even allonursing by non-reproducing females.

Future analyses of fitness benefits arising from individual associations between reproducing and non-reproducing females in our study population will help us to understand how the presence of other females influences individual reproductive success. If benefits of communal or cooperative care of young are substantial, we then predict that reproducing females are especially choosy with respect to social partners, and restrict interactions to only a few other reproducing females in order to maximize reproductive success (König, 1994a).

Presumably as a consequence of having fewer partners, reproducing females had fewer meetings during summer. In addition, they had shorter meetings in summer, the period of elevated reproductive competition. Such behaviour might suggest that breeding females alternated in their presence in the nest with other females, and minimized the time of simultaneous nest use. Alternating with other females reduces the time litters are left alone and are thus unprotected during periods of high energetic demand of the mother. Such a benefit would be obvious, especially for communally nursing females, and in captivity wild females have been observed to take turns nursing each other's offspring in communal nests (Wilkinson and Baker, 1988; König, 1989).

In winter, females had more partners, more meetings, and longer meetings than during summer, irrespective of their reproductive condition. Females thus decreased the time outside of nest boxes, presumably to minimize exposure to low temperatures in winter. In addition, larger groups may provide thermoregulatory benefits. Huddling as a social strategy against low temperature is widespread among rodents, and several otherwise solitary species nest communally during the winter, thereby reducing exposure to the environment and consequent heat loss (Martin *et al.*, 1980; Batchelder *et al.*, 1982; McShea, 1990; Hayes, 2000). The absence, or only weak presence, of reproductive competition during winter might have relaxed the reproducing females' selectivity in interactions with other group members, and thus allowed for the observed large number of partners. In African striped mice (*Rhabdomys pumilio*), a strictly seasonal breeder, reproductive competition even favours solitary living. After the breeding season, in the absence of reproductive competition, almost all striped mice live in groups even under very low population densities. During the breeding season, however, mice of both sexes may live solitarily, except under very high population density, when no opportunities for independent breeding exist (Schradin *et al.*, 2010; 2012).

Conclusions

Our free-living study population of house mice in northern Switzerland showed distinct seasonality in female reproductive activity over a period of four years, despite unlimited access to food, water, and protected nesting sites. Such seasonality had interesting consequences for female reproductive competition and the quality of female–female interactions. During summer (April to

September), intrasexual competition among females was pronounced, and might explain why only a limited number of females were able to reproduce in the well-established population. Although population size and the number of adult females increased over the study period, the absolute number of reproductively active females remained unchanged. During winter (October to March), the proportion of reproducing females and the monthly number of offspring weaned were much lower, suggesting relaxed reproductive competition. It is unlikely that female house mice competed strongly over access to mates, since the number of adult males to adult females was rather equal over the entire period. They rather seemed to compete over resource availability, in particular access to nest boxes as a prerequisite to rear and protect a litter. Only when females drastically outnumber males or simultaneously share mate preferences, female competition may also encompass access to mates, as has been suggested by Rusu and Krackow (2004).

Throughout the year, females used a variety of nest boxes where they regularly interacted with several reproducing and adult non-reproducing female conspecifics belonging to the same social group. Despite the fact that analysis of long-term stability in interaction partners is still lacking, such a pattern underlines the significance of same-sex social partners and individualized social groups. Average life expectancy in the study population is 196 days, but adult females can live up to three or four years (Manser *et al.*, 2011). Very interestingly, generation time (average age of reproduction of a female), at nine months, is surprisingly long (Manser *et al.*, 2011), suggesting that especially younger females need to integrate and establish long-term social bonds within a group as a prerequisite to successfully wean offspring. Our long-term data thus propose that females maintain social bonds to other females even during periods of low reproductive competition, as part of a flexible and competitively superior reproductive strategy.

Within groups, social relationships appear to be structured by cooperation and by the existence and resolution of conflicts. During summer, when intrasexual competition was high, reproductively active females had relatively few partners and used few nest boxes. Limitation of lactating females' access to social partners is expected according to laboratory data, since individual reproductive success is highest for communally nursing pairs of females, and decreases with increasing group size (König, 2006). During winter, on the other hand, reproductive competition was low and the benefits of social thermoregulation might outweigh the benefits of reducing the number of social partners.

Theory predicts that the aversive effect of reproductive competition could be offset by kin association. Even when reproductive skew occurs within social groups, the long-term inclusive fitness of interacting females, when related, should then be higher than of females not exhibiting social preferences. Within social groups, female house mice have the option to establish preferences and restrict meetings to fewer partners when reproducing under conditions of high intrasexual competition. Genetic relatedness among interacting partners has still to be analysed for our free-living population, as well as whether individual preferences result in fitness variances. Nevertheless, female social behaviour has to be interpreted in the context of long-term relationships and such relationships are typically complex. Interactions among females thus may be subject to social selection processes, driving the evolution of female traits. Future studies have to analyse whether females choose social partners based on their phenotype, to identify the traits they use for partner preferences, and whether social selection results in assortative traits of social partners.

REFERENCES

Alexander, R. D. (1974). The evolution of social behavior. *Annual Review of Ecology and Systematics*, **5**, 325–83.

Batchelder, P., Kinney, R. O., Demlow, L., and Lynch, C. B. (1982). Effects of temperature and social interactions on huddling behaviour in *Mus musculus. Physiology and Behavior*, **31**, 97–102.

Berry, R. J. (1968). The ecology of an island population of the house mouse. *Journal of Animal Ecology*, **37**, 445–70.

Berry, R. J. (1981). Town mouse, country mouse: adaptation and adaptability in *Mus domesticus* (*M. musculus domesticus*). *Mammal Review*, **11**, 91–136.

Bronson, F. H. (1979). The reproductive ecology of the house mouse. *The Quarterly Review of Biology*, **54**, 265–99.

Clutton-Brock, T. H. (2009). Sexual selection in females. *Animal Behaviour*, **77**, 3–11.

Crowcroft, P. (1966). *Mice All Over*. London: GT Foulis & Co.

Crowcroft, P. and Rowe, F. P. (1957). The growth of confined colonies of the wild house mouse (*Mus musculus* L.). *Proceedings of the Zoological Society of London*, **129**, 359–70.

Crowcroft, P. and Rowe, F. P. (1958). The growth of confined colonies of the wild house-mouse (*Mus musculus* L.): the effect of dispersal on female fecundity. *Proceedings of the Zoological Society of London*, **131**, 357–65.

Cucchi, T., Vigne, J.-D., and Auffray, J.-C. (2005). First occurrence of the house mouse (*Mus musculus domesticus* Schwarz Schwarz, 1943) in the Western Mediterranean: a zooarchaeological revision of subfossil occurrences. *Biological Journal of the Linnean Society*, **84**, 429–45.

Cucchi, T., Vigne, J.-D., Auffray, J.-C., Croft, P., and Peltenburg, E. (2002). Passive transport of the house mouse (*Mus musculus domesticus*) to Cyprus at the Early Preceramic Neolithic (late 9th and 8th millennia cal. BC). *Comptes Rendus Palevol*, **1**, 235–41.

D'Amato, F. R. (1997). Neurobiological and behavioral aspects of recognition in female mice. *Physiology & Behavior*, **62**, 1311–17.

D'Amato, F. R. and Moles, A. (2001). Ultrasonic vocalizations as an index of social memory in female mice. *Behavioral Neuroscience*, **115**, 834–40.
DeLong, K. T. (1967). Population ecology of feral house mice. *Ecology*, **48**, 611–34.
Dobson, F. W. and Baudoin, C. (2002). Experimental tests of spatial association and kinship in monogamous mice (*Mus spicilegus*) and polygynous mice (*Mus musculus domesticus*). *Canadian Journal of Zoology*, **80**, 980–6.
Dobson, F. W., Jacquot, C., and Baudoin, C. (2000). An experimental test of kin association in the house mouse. *Canadian Journal of Zoology*, **78**, 1806–12.
Drickamer, L. C. (1992). Oestrous female house mice discriminate dominant from subordinate males and sons of dominant from sons of subordinate males by odour cues. *Animal Behaviour*, **43**, 868–70.
Emlen, S. T. and Oring, L. W. (1977). Ecology, sexual selection, and evolution of mating systems. *Science*, **197**, 215–23.
Gray, S. J., Jensen, S. P. and Hurst, J. L. (2000). Structural complexity of territories: effects on preference, use of space and territorial defence in commensal house mice (*Mus domesticus*). *Animal Behaviour*, **60**, 765–72.
Hayashi, S. (1990). Social condition influences sexual attractiveness of dominant male mice. *Zoological Science*, **7**, 889–94.
Hayes, L. D. (2000). To nest communally or not to nest communally: a review of rodent communal nesting and nursing. *Animal Behaviour*, **59**, 677–88.
Hurst, J. L. (1987). Behavioural variation in wild house mice *Mus domesticus* Rutty: a quantitative assessment of female social organization. *Animal Behaviour*, **35**, 1846–57.
Hurst, J. L. (1990). Urine marking in populations of wild house mice *Mus domesticus* Rutty: II – communication between females. *Animal Behaviour*, **40**: 223–32.
Hurst, J. L. and Nevison, C. M. (1994). Do female house mice, *Mus domesticus*, regulate their exposure to reproductive priming pheromones? *Animal Behaviour*, **48**, 945–59.
Jennions, M. D. and Macdonald, D. W. (1994). Cooperative breeding in mammals. *Trends in Ecology and Evolution*, **9**, 89–93.
Kenward, M. G. and Roger, J. H. (1997). The precision of fixed effects estimates from restricted maximum likelihood. *Biometrics*, **53**, 983–97.
König, B. (1989). Behavioural ecology of kin recognition in house mice. *Ethology, Ecology Evolution*, **1**, 99–110.
König, B. (1993). Maternal investment of communally nursing female house mice (*Mus musculus domesticus*). *Behavioural Processes*, **30**, 611–74.
König, B. (1994a). Components of lifetime reproductive success in communally and solitarily nursing house mice: a laboratory study. *Behavioral Ecology and Sociobiology*, **34**, 275–83.
König, B. (1994b). Fitness effects of communal rearing in house mice: the role of relatedness and familiarity. *Animal Behaviour*, **48**, 1149–57.
König, B. (1997). Cooperative care of young in mammals. *Naturwissenschaften*, **84**, 95–104.
König, B. (2006). Non-offspring nursing in mammals: general implications from a case study on house mice. In *Cooperation in Primates and Humans: Mechanisms and Evolution*, ed. P. M. Kappeler and van C. P. Schaik. Berlin, Heidelberg: Springer-Verlag, pp. 191–205.

König, B. and Markl, H. (1987). Maternal care in house mice: I – the weaning strategy as a means for parental manipulation of offspring quality. *Behavioral Ecology and Sociobiology*, **20**, 1–9.

König, B., Riester, J., and Markl, H. (1988). Maternal care in house mice (*Mus musculus*): II – the energy cost of lactation as a function of litter size. *Journal of Zoology*, **216**, 195–210.

Latham, N. and Mason, G. (2004). From house mouse to mouse house: the behavioural biology of free-living *Mus musculus* and its implications in the laboratory. *Applied Animal Behaviour Science*, **86**, 261–89.

Laurie, E. M. O. (1946). The reproduction of the house mouse (*Mus musculus*) living in different environmental conditions. *Proceedings of the Royal Society B: Biological Sciences*, **133**, 248–81.

Lenington, S. and Egid, K. (1989). Environmental influences on the preferences of wild female house mice for males of differing *t*-complex genotypes. *Behavior Genetics*, **19**, 257–66.

Lewis, S. E. and Pusey, A. E. (1997). Factors influencing the occurrence of communal care in plural breeding mammals. In *Cooperative Breeding in Mammals*, ed. N. G. Solomon and J. A. French. Cambridge: Cambridge University Press, pp. 335–63.

Lidicker, W. Z., Jr (1976). Social behaviour and density regulation in house mice living in large enclosures. *Journal of Animal Ecology*, **45**, 677–97.

Lynch, C. B. (1992). Clinal variation in cold adaptation in *Mus domesticus*: verification of predictions from laboratory populations. *The American Naturalist*, **139**, 1219–36.

Manning, C. J., Dewsbury, D. A., Wakeland, E. K., and Potts, W. K. (1995). Communal nesting and communal nursing in house mice, *Mus musculus domesticus*. *Animal Behaviour*, **50**, 741–51.

Manser, A., Lindholm, A. K., König, B., and Bagheri, H. (2011). Polyandry and the decrease of a selfish genetic element in a wild house mouse population. *Evolution*, **65**, 2435–47.

Martin, R. A., Fiorentini, M., and Connors, F. (1980). Social facilitation of reduced oxygen consumption in *Mus musculus* and *Meriones unguiculatus*. *Comparative Biochemistry and Physiology*, **65A**, 519–22.

McShea, W. J. (1990). Social tolerance and proximate mechanisms of dispersal among winter groups of meadow voles, *Microtus pennsylvanicus*. *Animal Behaviour*, **39**, 346–51.

Nevison, C. M., Armstrong, S., Beynon, R. J., Humphries, R. E., and Hurst, J. L. (2003). The ownership signature in mouse scent marks is involatile. *Proceedings of the Royal Society B: Biological Sciences*, **270**, 1957–63.

Packer, C., Lewis, S., and Pusey, A. E. (1992). A comparative analysis of non-offspring nursing. *Animal Behaviour*, **43**, 265–82.

Palanza, P., Della Seta, D., Ferrari, P. F., and Parmigiani, S. (2005). Female competition in wild house mice depends upon timing of female/male settlement and kinship between females. *Animal Behaviour*, **69**, 1259–71.

Palanza, P., Re, L., Mainardi, D., Brain, P. F., and Parmigiani, S. (1996). Male and female competitive strategies of wild house mice pairs (*Mus musculus domesticus*) confronted with intruders of different sex and age in artificial territories. *Behaviour*, **133**, 863–82.

Pocock, M. J. O., Searle, J. B., and White, P. C. L. (2004). Adaptations of animals to commensal habitats: population dynamics of house mice *Mus musculus domesticus* on farms. *Journal of Animal Ecology*, **73**, 878–88.

Randall, J. A. (1999). *Vertebrate Pest Management: A Guide for Commercial Applicators*. East Lansing, MI: Michigan State University.

R Core Development Team (2011). *R: A Language and Environment for Statistical Computing*. Vienna, Austria: R Foundation for Statistical Computing.

Rosenqvist, G. and Berglund, A. (1992). Is female sexual behaviour a neglected topic? *Trends in Ecology and Evolution*, **7**, 174–6.

Rusu, A. S., König, B., and Krackow, S. (2004). Pre-reproductive alliance formation in female wild house mice (*Mus domesticus*): the effects of familiarity and age disparity. *Acta Ethologica*, **6**, 53–8.

Rusu, A. S. and Krackow, S. (2004). Kin-preferential cooperation, dominance-dependent reproductive skew, and competition for mates in communally nesting female house mice. *Behavioral Ecology and Sociobiology*, **56**, 298–305.

Sayler, A. and Salmon, M. (1971). An ethological analysis of communal nursing by the house mouse. *Behaviour*, **40**, 60–85.

Schradin, C., König B., and Pillay, N. (2010). Reproductive competition favours solitary living while ecological constraints impose group-living in African striped mice. *Journal of Animal Ecology*, **79**, 515–21.

Schradin, C., Lindholm, A. K., Johannesen, J., *et al.* (2012). Social flexibility and social evolution in mammals: a case study of the African striped mouse *(Rhabdomys pumilio)*. *Molecular Ecology*, **21**, 541–53.

Silk, J. (2007). The adaptive value of sociality in mammalian groups. *Philosophical Transactions of the Royal Society B: Biological Sciences*, **362**, 539–59.

Silk, J., Alberts, S. C., and Altmann, J. (2003). Social bonds of female baboons enhance infant survival. *Science*, **302**, 1231–4.

Singleton, G. R. (1983). The social and genetic structure of a natural colony of house mice, *Mus musculus*, at Healesville Wildlife Sanctuary. *Australian Journal of Zoology*, **31**, 155–66.

Solomon, N. G. and French, J. A. (1997). *Cooperative Breeding in Mammals*. Cambridge: Cambridge University Press.

Southwick, C. H. (1955). Regulatory mechanisms of house mouse populations: social behavior affecting litter survival. *Ecology*, **36**, 627–34.

Sterck, E. H. M., Watts, D. P., and van Schaik, A. P. (1997). The evolution of female social relationships in nonhuman primates. *Behavioral Ecology and Sociobiology*, **41**, 291–309.

Stockley, P. and Bro-Jørgensen, J. (2010). Female competition and its evolutionary consequences in mammals. *Biological Reviews*, **86**, 341–66.

Weidt, A., Hofmann, S. E., and König, B. (2008). Not only mate choice matters: fitness consequences of social partner choice in female house mice. *Animal Behaviour*, **75**, 801–8.

West-Eberhard, M. J. (1983). Sexual selection, social competition, and speciation. *The Quarterly Review of Biology*, **58**, 155–83.

West-Eberhard, M. J. (1987). Phenotypic plasticity and the origins of diversity. *Annual Review of Ecology and Systematics*, **20**, 249–78.

Wilkinson, G. S. and Baker, A. E. M. (1988). Communal nesting among genetically similar house mice. *Ethology*, **77**, 103–14.

Yamazaki, K., Boyse, E. A., Miké, V., *et al.* (1976). Control of mating preferences in mice by genes in the major histocompatibility complex. *Journal of Experimental Medicine*, **144**, 1324–35.

6

Development of the mouse mandible: a model system for complex morphological structures

CHRISTIAN PETER KLINGENBERG AND
NICOLAS NAVARRO

Introduction

Our understanding of evolutionary processes has changed tremendously in the last few decades, most notably by advances in genetics and developmental biology that have transformed the approaches of evolutionary biologists (Amundson, 2005; Klingenberg, 2010). Among other changes, the use of model systems has become more widespread in evolutionary biology. One of those models is the mandible of the house mouse, which is now widely used as a model for complex morphological structures in general (Atchley and Hall, 1991; Atchley, 1993).

The mandible is a complex structure with respect to its composition, developmental origin, and function. It is composed of several anatomical units: the ramus, the coronoid, condylar and angular processes, and the alveolar components where the teeth insert (Moore, 1981; Atchley and Hall, 1991). It is made up of bone, cartilage, and the mineralized tissues of the teeth, as well as other types of connective tissues, bone marrow, nerves, and vascular tissues. The mandible develops from cells of the neural crest and paraxial mesoderm under the influence of various signalling interactions (Depew *et al.*, 2002; Hall, 2005; Chai and Maxson, 2006). Mandibular structure reflects various functional demands for gnawing and chewing, which in turn feed back to development and growth through bone remodelling (Renaud *et al.*, 2010). A key property of the anatomical, developmental, and functional complexity of the mouse mandible, like other morphological structures, is that it is shaped by various interactions and feedbacks among the processes involved, which leads to integration and possible modular organization of its component parts (Klingenberg, 2008).

Evolution of the House Mouse, ed. Miloš Macholán, Stuart J. E. Baird, Pavel Munclinger, and Jaroslav Piálek. Published by Cambridge University Press.

Despite this complexity, the mandible is sufficiently simple so that the effects of the various processes contributing to its development and variation can be identified and studied. Because it is a single bone, the mandible of the mouse and other rodents is suitable for morphometric studies and has been used widely to study intraspecific variation (Klingenberg *et al.*, 2003) as well as evolutionary change (Monteiro *et al.*, 2005; Zelditch *et al.*, 2008). Because the mouse has long been a model organism in genetics, many resources are available for studying the genetic basis of shape variation in the mandible (Klingenberg and Leamy, 2001; Klingenberg *et al.*, 2001). The dentition provides the opportunity for studying a somewhat different kind of complexity (Workman *et al.*, 2002; Macholán, 2006; Laffont *et al.*, 2009).

This chapter provides a review of the development of the mouse mandible, including embryonic patterning and morphogenesis, as well as postembryonic growth and remodelling. We also summarize genetic studies of mandible shape, using either quantitative trait locus (QTL) approaches or classical quantitative genetics. Based on this developmental and genetic perspective, we discuss findings on the evolution of mandible shape in mice and other rodents. Finally, we evaluate what can be learned from the mouse mandible as a model of a complex morphological structure that can be applied to other structures.

Development of the mandible

The mouse mandible consists of several anatomically and functionally differentiated parts, which are also developmentally distinct (Fig. 6.1; Atchley and Hall, 1991). The posterior part of the mandible, the ascending ramus, consists of several parts that serve for the attachment of the mandible to the cranium and for insertion of the jaw musculature. The condylar process articulates with the cranium to form the temporomandibular joint. The most powerful of the jaw muscles is the masseter, which attaches to a large area on the outside of the mandible extending to the angular process, whereas the temporalis muscle, attaching to the coronoid process, is weaker. The anterior part of the mandible supports the teeth and serves for transmission of the forces generated by the jaw muscles to the food or other substrate. This part contains the teeth, which are embedded in alveolar bone that is partly derived from cells of the tooth bud mesenchyme (Diep *et al.*, 2009). The alveolus of the incisor tooth extends far back through the ramus of the mandible, whereas the three molar teeth are grouped together in a region posterior to a wide diastema, a toothless region between the incisor and first molar.

These anatomically distinct components of the mandible also have separate developmental origins, but are integrated into the mandible in a highly controlled

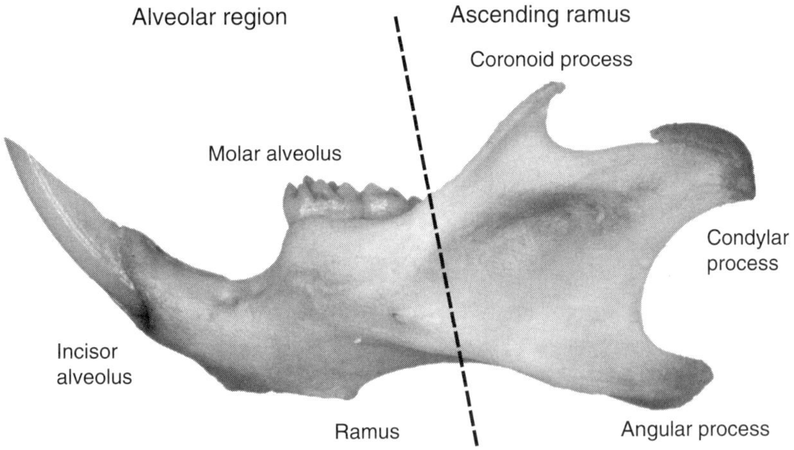

Figure 6.1 The main anatomical parts of the mouse mandible in lateral view. The dashed line is a rough indication of the boundary between the alveolar region and ascending ramus.

manner. While more and more information on the developmental processes involved in the patterning and growth of the mandible is available (Depew *et al.*, 2002; Hall, 2005; Chai and Maxson, 2006; Depew and Simpson, 2006), many questions remain unanswered. In particular, developmental genetic approaches, while powerful for identifying pathways involved in mandible development, are less well suited to investigating how those pathways are integrated and fine-tuned to produce a functional structure and respond to outside stimuli in a coordinated way. Quantitative approaches are more useful for addressing those issues (Klingenberg, 2010). In the end, however, both approaches need to be combined if we are to reach a full understanding of mandible development.

Embryonic development and patterning

Much of the mandible is formed by cells that originate from the cranial neural crest and migrate into the branchial arch region, but cells from the paraxial mesoderm also contribute substantially to the mandible (Chai *et al.*, 2000; Depew *et al.*, 2002; Aggarval *et al.*, 2010). The mandible forms in close association to Meckel's cartilage, which is also substantially derived from neural crest and acts as a sort of scaffold for the mandible (Moore, 1981; Tomo *et al.*, 1997; Depew *et al.*, 2002; Ramaesh and Bard, 2003). Meckel's cartilage originates as a pair of straight rods that grow anteriorly and posteriorly from the molar region of the prospective mandible and fuse together anteriorly. The body of the mandible forms as a mesenchymal cell concentration lateral to Meckel's cartilage and expands to surround it as well as the developing teeth (Ramaesh and Bard, 2003). The body of the mandible undergoes intramembranous ossification, whereas the angular, condylar, and coronoid processes originate from cartilages that derive from separate mesenchymal condensations (Tomo *et al.*, 1997; Ramaesh and Bard, 2003). The anterior part of Meckel's cartilage eventually disappears, partly by resorption and partly by being incorporated in the rostral-most part of the mandible (Ramaesh and Bard, 2003).

The formation of the mandible is accompanied by a variety of signalling and regulatory events that induce and coordinate patterning and morphogenesis and achieve its spatial patterning. Signalling interactions that are critical for mandible formation take place between the mesenchyme and both the oral ectoderm and pharyngeal endoderm (Depew *et al.*, 2002; Chai and Maxson, 2006). These events establish the identity of branchial arches and set up a molecular 'coordinate system' within each branchial arch, which provides the basis for the elaboration of prospective mandibular structures (Depew *et al.*, 2002; Chai and Maxson, 2006). For instance, the *Dlx* genes are important in conveying positional information in the jaws through a '*Dlx* code' (Depew *et al.*, 2005). Signalling interactions and positional specification enable the development of the mandible in a coordinated manner.

The development of teeth has been a focus of particular attention (e.g. Catón and Tucker, 2009; Cobourne and Sharpe, 2010). Teeth are particularly interesting because of their wide range of morphologies in any one species (chisel-shaped incisors and multi-cusped molars in mice) and because of the marked evolution of their number and appearance (Kavanagh *et al.*, 2007; Cobourne and Sharpe, 2010). Tooth development is initiated by signalling interactions between oral epithelium and the underlying mesenchyme, and both epithelium and mesenchyme contribute cells to the developing tooth (Chai *et al.*, 2000; Catón and Tucker, 2009). The mouse dentition is highly reduced by comparison to other mammals: each quadrant of the mouth has only a single incisor and three molars, which are separated by the diastema, an extensive gap without teeth. During embryonic development of mice, tooth primordia that may correspond to the 'missing' premolar teeth form in the region of the prospective diastema, but these primordia are either resorbed or incorporated into the first molar tooth (Prochazka *et al.*, 2010). Signalling between teeth developing in adjacent positions is important for the control of molar number and size, and may have important implications for the potential for evolutionary change (Kangas *et al.*, 2004; Kavanagh *et al.*, 2007).

Overall, the available evidence indicates that mandible development is a highly interactive process that involves many molecular and cellular mechanisms. Accordingly, it is no surprise that knock-out experiments for a large number of genes produce craniofacial abnormalities as part of their phenotypic effects (e.g. review by Depew *et al.*, 2002). In addition, it is to be expected that many additional genes are involved in these processes, even if elimination of their activity does not produce an obvious morphological effect.

Molecular approaches to the study of craniofacial development have mostly focused on the consequences of experimental intervention on patterns of gene expression or on gross phenotypes that were assessed qualitatively. Embryonic development is also increasingly investigated with morphometric methods to capture subtle morphological changes (Young *et al.*, 2007; Boughner *et al.*, 2008; Parsons *et al.*, 2008; Schmidt *et al.*, 2010). These studies used computed tomography to produce scans of embryos, from which landmark positions were recorded for geometric morphometric analyses. This method is promising for the direct study of embryonic development of craniofacial shape, but it also faces considerable methodological challenges, such as substantial artefacts from changes of size and shape during fixation (Schmidt *et al.*, 2010).

Postembryonic growth and remodelling

Development is not terminated at birth; rather, the mouse mandible undergoes considerable change in size and shape during postembryonic growth. There are pronounced changes in the shape of the mandible during postnatal growth, particularly a strong development of the coronoid and angular processes, which leads to an increase in the relative height of the mandible. This is an example of the nearly ubiquitous phenomenon of ontogenetic allometry, the systematic association of changes of size and shape during growth (Cock, 1966; Gould, 1966; Klingenberg, 1996). Even among adult mice, there is clear static allometry, and relation between size and shape within samples of mice of the same age and from the same population. This has been shown by associations of mandible measurements with measures of body size in mice (Atchley *et al.*, 1985b; Bailey, 1985), as well as by the current methods of geometric morphometrics, which show that size and shape of the mandible itself are clearly associated (Fig. 6.2).

Because variation in the extent of growth is an important possible source of variation at a given stage, the connection between static and ontogenetic allometry is potentially important for understanding the origin of phenotypic variation in populations (Cheverud, 1982; Klingenberg and Zimmermann, 1992; Klingenberg, 1998). The pattern of static allometry in the mouse mandible (Fig. 6.2) is qualitatively similar to the main shape changes in postembryonic growth. For other measurements in mice, relationships between static and ontogenetic allometry have also been reported (Leamy and Bradley, 1982), and there are also clear relations to the patterns of evolutionary allometry concerning the divergence among related species (Atchley, 1993). Using the current methods

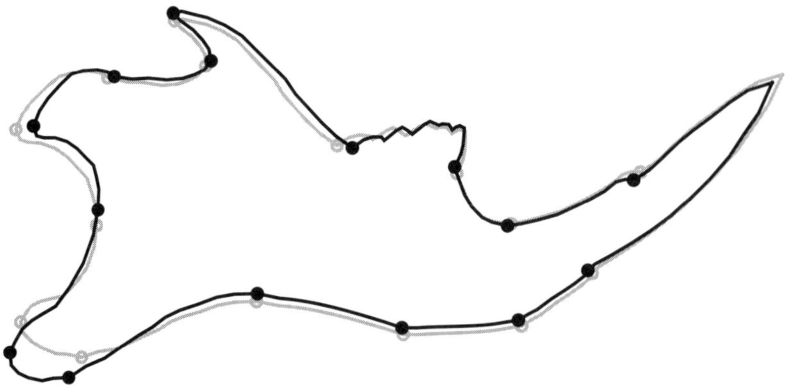

Figure 6.2 Static allometry in the mouse mandible. The diagram shows the change from the average shape (grey outline) to the shape expected for a mandible with a centroid size 5 mm greater than average (an approximately 30% increase). Mandible shape in a heterogeneous stock of mice (Valdar *et al.*, 2006) is characterized by 15 morphological landmarks whose coordinates were subjected to Procrustes analysis (Dryden and Mardia, 1998; Klingenberg, 2010). The allometric change is estimated by multivariate regression of shape on centroid size (Monteiro, 1999; Drake and Klingenberg, 2008).

of geometric morphometrics, such a relationship between static allometry and evolutionary diversification of mandible shape was recently established in a study of mainland and island forms of the house mouse (Renaud and Auffray, 2010). Overall, these studies of allometry highlight the potential for growth processes to provide biases or constraints for evolutionary diversification of mandibular morphology (Arthur, 2001; Klingenberg, 2010).

Bone growth is a dynamic process in which resorption and new deposition of the intercellular matrix in bone tissue create a turnover of material (Enlow and Hans, 1996; Hall, 2005). This remodelling of bone is able to respond to mechanical forces experienced (Herring, 1993), and is therefore a means for developmental processes to respond to environmental stimuli.

A number of studies have documented the effects of the mechanical stimuli on craniofacial morphology in mice with several experimental approaches. Lightfoot and German (1998) used two strains of mice carrying mutations that cause muscular dystrophy and compared the course of growth with a healthy control strain. Mice with the more severe mutation showed considerably reduced growth of various measurements, including the mandible, whereas the more weakly dystrophic strain showed less clear differences. In a different strain of mice with muscular dystrophy, males had larger mandibles than the controls (Renaud *et al.*, 2010). In a strain of mice with overdeveloped musculature due to myostatin deficiency, mandibles were smaller in newborn mice than in controls, but grew larger by the age of 180 days (Vecchione *et al.*, 2010). Mandible shape also responded to these differences in mechanical loading by genetic manipulations of muscle activity (Renaud *et al.*, 2010; Vecchione *et al.*, 2010). Similarly, experimental manipulation of the forces that the mandible experienced by feeding mice either hard or soft food also produced a clear effect on mandible shape (Renaud and Auffray, 2010; Renaud *et al.*, 2010). Another study manipulated the mechanical loading of the mandible by trimming the tip of the mandibular incisor teeth and found a clear response by a lengthening of the condyle (Tagliaro *et al.*, 2009). All these studies show that the mandible can respond by altered growth to various changes in the mechanical forces acting on it. This remodelling enables the mandible to adapt to changes in the environmental conditions and may have considerable evolutionary implications (West-Eberhard, 2003).

Developmental integration and modularity

Morphological structures are integrated so that variation in one part is associated with variation in other parts to maintain overall functionality of the structure as a whole. Development is an important cause of integration, because developmental interactions are the basis for an important component of observable covariation between traits (Klingenberg, 2010).

To investigate the developmental component of integration, it is therefore important to identify covariation of morphological traits that is caused by developmental interactions and to separate it from covariation from other possible origins. An effective way to achieve this is to focus on the covariation in the fluctuating asymmetry among different traits (Klingenberg, 2003, 2005). Because fluctuating asymmetry arises from random variation in developmental processes, correlations between the asymmetries of different traits only occur if there are direct developmental interactions between the developmental processes that produce the traits, so that variation can be transmitted between them. Likewise, the focus on asymmetry also automatically minimizes the effects of confounding factors such as genetic and environmental variation, which can generate covariation among traits without direct developmental interactions between them (for more detailed discussion, see Klingenberg, 2003, 2005).

Integration of asymmetry in the mouse mandible was first reported in a study using traditional morphometric methods (Leamy, 1993), and later confirmed with a geometric morphometric approach (Klingenberg *et al.*, 2003; Klingenberg, 2009). Because the patterns of covariation for fluctuating asymmetry and for the variation among individuals are broadly consistent, it appears likely that a substantial component of the observable phenotypic integration in the mandible originates from direct developmental interactions. A similar study was also conducted for the mandible of deer mice (*Peromyscus maniculatus*) and produced comparable results (Zelditch *et al.*, 2008).

Laffont *et al.* (2009) investigated developmental integration in the molar tooth row of voles (*Microtus arvalis*). They found significant covariation between different teeth for the variation among individuals, but not for fluctuating asymmetry. This contrast suggests that the different molar teeth are developmentally independent of each other to a considerable degree.

Inheritance of mandible shape

To investigate the genetic basis of mandible shape in mice, a range of approaches are available. Because the mouse is a classical model organism in genetics, the genetic tools are particularly well developed. First, the approach of developmental genetics focuses on particular genes of interest and uses loss-of-function mutations to eliminate their function. This approach tends to produce large phenotypic effects that are relatively easy to interpret, and therefore yields unambiguous evidence concerning the gene's involvement in particular processes. But this approach is artificial in a number of ways and therefore is unsuitable for answering many questions, particularly in an evolutionary context. An alternative is to use a QTL approach, to search for genomic regions or genes that are responsible for variation in phenotypic traits. This approach provides useful information about the genetic architecture of a trait, but there are also limitations due to statistical power with manageable sample sizes. Finally, there is the approach of classical quantitative genetics, where information about relatedness among the individuals in the study is used to estimate genetic components of variance and covariance for the traits of interest. This approach is very flexible in that it provides direct information on the total genetic variation and constraints in a population and can be applied in many contexts, as long as genealogical information is available. But this approach does not provide any information about the number, location, or identity of the genes responsible for the observed variation. All three approaches have been used for the study of mouse mandibles. Because the first approach, focusing on particular candidate genes, has mostly been used in developmental genetics, we will not discuss it further here (see discussion above and especially Depew *et al.*, 2002).

Analysis of genetic architecture

Several studies have searched for quantitative trait loci (QTLs) affecting the morphology of the mandible in mice. The most widespread method is to use the F_2 generation from a cross between two inbred lines of mice, and to obtain both the trait measurements and genotype information for genetic markers from a large number of individuals. The QTLs can then be found by correlations between markers and the traits (Lynch and Walsh, 1998).

Cheverud *et al.* (1997) conducted a search for QTLs affecting linear measurements of the mandible in a cross between the Large (LG/J) and Small (SM/J) inbred strains of mice. They found that multiple traits affected by the same QTLs tended to be concentrated either in the alveolar region or in the ascending ramus of the mandible, suggesting a degree of genetic modularity. This impression was reinforced by later analyses (Mezey *et al.*, 2000; Ehrich *et al.*, 2003). Similarly, a QTL analysis for a range of skull measurements in the same cross produced a similar distinction between QTLs affecting mostly measurements of skull parts that are initiated early or late in embryonic development (Leamy *et al.*, 1999).

The methods of geometric morphometrics were applied to analyse landmark data for the same cross of mice (Klingenberg *et al.*, 2001, 2004). The search yielded a substantial number of QTLs, but principal component analysis did not show a clear division between clusters of QTLs whose effects are mostly either in the anterior or posterior region of the mandible. A further analysis suggested that there was a very weak degree of modularity between the two parts of the mandible with respect to the QTL effects (Klingenberg *et al.*, 2004). Moreover, the analysis of QTLs for geometric shape revealed fairly substantial dominance effects. A similar search for QTLs affecting the shape of the molar tooth row produced comparable results (Workman *et al.*, 2002). The QTL study by Burgio *et al.* (2009) used a different design, recombinant congenic strains, to map QTLs affecting cranial shape.

Genetic integration and constraints

To examine the patterns of genetic integration and possible genetic constraints, it is possible to estimate genetic and environmental components of variance and covariance for the traits of interest. This is the approach of classical quantitative genetics, which has recently seen much renewed interest (Falconer and Mackay, 1996; Lynch and Walsh, 1998; Kruuk *et al.*, 2008). This approach has been used extensively for distance measurements of the mouse mandible (e.g. Atchley *et al.*, 1985a, b; Atchley, 1993).

More recently, the methods of geometric morphometrics have been applied in this context (Klingenberg and Leamy, 2001). This study was based on a parent–offspring breeding design and used the ‘animal model’ to estimate variance and covariance components for the shape variables (Lynch and Walsh, 1998; Wilson *et al.*, 2010). The results showed that different shape variables differed considerably in their heritabilities: there was a range from more than 0.7 to near-zero.

Using the multivariate breeder's equation, predictions were made for the response to several hypothetical selection regimes, which suggested that the genetic covariance structure of the mandible deflected the response away from the direction of selection by a considerable angle. Overall, therefore, there is clear evidence for relative constraints, but there is no clear evidence for absolute constraints (see also the discussion in Klingenberg *et al.*, 2010).

The patterns of integration for the genetic and phenotypic covariance matrices of the mouse mandible were similar (Klingenberg and Leamy, 2001) and were also similar to the main patterns of variation among the effects of QTLs obtained by separate studies (Klingenberg *et al.*, 2001, 2004). It therefore appears that there is a consistent pattern of variation at both the phenotypic and genetic levels.

Evolution of mandible shape in mice and other rodents

Studies of variation in the house mouse mandible can provide valuable information that can assist investigators to interpret results from field studies. House mice have diversified considerably, and there are studies that have analysed changes of mandible shape. Renaud and Auffray (2010) compared mandible shape of mice from continental and island populations and considered the possible roles of diet and allometry. Corti and Rohlf (2001) studied mandibles and skulls of different chromosome races of mice, and even found a correlation between mandible shape and aggressive behaviour.

Studies of house mice also provide a basis for analyses in other rodents. Various investigators have studied the evolution of mandible shape in groups such as marmots (Cardini, 2003; Caumul and Polly, 2005), squirrels (Velhagen and Roth, 1997; Swiderski and Zelditch, 2010), or spiny rats (Monteiro *et al.*, 2005; Monteiro and dos Reis, 2005). These studies cover a wide spectrum of approaches and a great diversity of species, but the information derived from studies of house mice provides a useful background for them.

The mouse mandible as a model system

The use of the mouse mandible as a general model system for 'complex morphological structures' (Atchley and Hall, 1991) has inspired many researchers. Accordingly, many of them investigated the mandible in other mammals, such as bats (Monteiro and Nogueira, 2010) and shrews (Young and Badyaev, 2006). This raises the question of how far one can extrapolate the insights gained in this one model system. How simple or how complex can a system be? And how far can the insights be extended to phylogenetically remote taxa?

The mandible is intermediate in anatomical complexity between systems such as insect wings and the mammalian cranium. Insect wings are structurally simple (a single epithelium folded over once) and also presumably have a relatively simple development (although it is known in great detail only for *Drosopila*). In contrast, the cranium is substantially more complex than the mandible, because it contains more structural parts, which also imply many developmental processes that are all highly interactive. So the question is whether the findings from these simpler or more complex systems will be compatible with the intermediate level of the mandible. It is too early to answer this question conclusively. From our personal experience, however, we have the impression that many general patterns extend across these levels. Besides the generality across levels, there is also a good deal of diversity among closely similar systems, which makes any large-scale generalization difficult. But in any case, regardless of the ability to generalize results, the mouse mandible has been a most fruitful model system in that it has produced a large number of pioneering studies that provide inspiration for work in other systems.

REFERENCES

Aggarval, V. S., Carpenter, C., Freyer, L., *et al.* (2010). Mesodermal *Tbx1* is required for patterning the proximal mandible in mice. *Developmental Biology*, **344**, 669–81.

Amundson, R. (2005). *The Changing Role of the Embryo in Evolutionary Thought: Roots of Evo-Devo*. Cambridge: Cambridge University Press.

Arthur, W. (2001). Developmental drive: an important determinant of the direction of phenotypic evolution. *Evolution & Development*, **3**, 271–8.

Atchley, W. R. (1993). Genetic and developmental aspects of variability in the mammalian mandible. In *The Skull*, ed. J. Hanken and B. K. Hall. Chicago, IL: University of Chicago Press, pp. 207–47.

Atchley, W. R. and Hall, B. K. (1991). A model for development and evolution of complex morphological structures. *Biological Reviews*, **66**, 101–57.

Atchley, W. R., Plummer, A. A., and Riska, B. (1985a). Genetics of mandible form in the mouse. *Genetics*, **111**, 555–77.

Atchley, W. R., Plummer, A. A., and Riska, B. (1985b). Genetic analysis of size-scaling patterns in the mouse mandible. *Genetics*, **111**, 579–95.

Bailey, D. W. (1985). Genes that affect the shape of the murine mandible: congenic strain analysis. *Journal of Heredity*, **76**, 107–14.

Boughner, J. C., Wat, S., Diewert, V. M., *et al.* (2008). Short-faced mice and developmental interactions between the brain and the face. *Journal of Anatatomy*, **213**, 646–62.

Burgio, G., Baylac, M., Heyer, E., and Montagutelli, X. (2009). Genetic analysis of skull shape variation and morphological integration in the mouse using interspecific recombinant congenic strains between C57BL/6 and mice of the *Mus spretus* species. *Evolution*, **63**, 2668–86.

Cardini, A. (2003). The geometry of the marmot (Rodentia: Sciuridae) mandible: phylogeny and patterns of morphological evolution. *Systematic Biology*, **52**, 186–205.

Catón, J. and Tucker, A. S. (2009). Current knowledge of tooth development: patterning and mineralization of the murine dentition. *Journal of Anatomy*, **214**, 502–15.

Caumul, R. and Polly, P. D. (2005). Phylogenetic and environmental components of morphological variation: skull, mandible, and molar shape in marmots (*Marmota*, Rodentia). *Evolution*, **59**, 2460–72.

Chai, Y., Jiang, X., Ito, Y., *et al.* (2000). Fate of the mammalian cranial neural crest during tooth and mandibular morphogenesis. *Development*, **127**, 1671–9.

Chai, Y. and Maxson, R. E., Jr. (2006). Recent advances in craniofacial morphogenesis. *Developmental Dynamics*, **235**, 2353–75.

Cheverud, J. M. (1982). Relationships among ontogenetic, static, and evolutionary allometry. *American Journal of Physical Anthropology*, **59**, 139–49.

Cheverud, J. M., Routman, E. J., and Irschick, D. J. (1997). Pleiotropic effects of individual gene loci on mandibular morphology. *Evolution*, **51**, 2006–16.

Cobourne, M. T. and Sharpe, P. T. (2010). Making up the numbers: the molecular control of mammalian dental formula. *Seminars in Cell and Developmental Biology*, **21**, 314–24.

Cock, A. G. (1966). Genetical aspects of metrical growth and form in animals. *The Quarterly Review of Biology*, **41**, 131–90.

Corti, M. and Rohlf, F. J. (2001). Chromosomal speciation and phenotypic evolution in the house mouse. *Biological Journal of the Linnean Society*, **73**, 99–112.

Depew, M. J. and Simpson, C. A. (2006). 21st century neontology and the comparative development of the vertebrate skull. *Developmental Dynamics*, **235**, 1256–91.

Depew, M. J., Simpson, C. A., Morasso, M., and Rubenstein, J. L. R. (2005). Reassessing the *Dlx* code: the genetic regulation of branchial arch skeletal pattern and development. *Journal of Anatomy*, **207**, 501–61.

Depew, M. J., Tucker, A. S., and Sharpe, P. T. (2002). Craniofacial development. In *Mouse Development: Patterning, Morphogenesis, and Organogenesis*, ed. J. Rossant and P. P. L. Tam. San Diego, CA: Academic Press, pp. 421–98.

Diep, L., Matalova, E., Mitsiadis, T. A., and Tucker, A. S. (2009). Contribution of the tooth bud mesenchyme to alveolar bone. *Journal of Experimental Zoology Part B: Molecular and Developmental Evolution*, **312**, 510–17.

Drake, A. G. and Klingenberg, C. P. (2008). The pace of morphological change: historical transformation of skull shape in St. Bernard dogs. *Proceedings of the Royal Society London B: Biological Sciences*, **275**, 71–6.

Dryden, I. L. and Mardia K. V. (1998). *Statistical Shape Analysis*. Chichester: Wiley.

Ehrich, T. H., Vaughn, T. T., Koreishi, S. F., *et al.* (2003). Pleiotropic effects on mandibular morphology I: developmental morphological integration and differential dominance. *Journal of Experimental Zoology Part B: Molecular and Developmental Evolution*, **296**, 58–79.

Enlow, D. H. and Hans, M. G. (1996). *Essentials of Facial Growth*. Philadelphia, PA: W. B. Saunders.

Falconer, D. S. and Mackay, T. F. C. (1996). *Introduction to Quantitative Genetics*. Harlow: Longman.

Gould, S. J. (1966). Allometry and size in ontogeny and phylogeny. *Biological Reviews*, **41**, 587–640.
Hall, B. K. (2005). *Bones and Cartilage: Developmental and Evolutionary Skeletal Biology*. San Diego, CA: Elsevier.
Herring, S. W. (1993). Epigenetic and functional influences on skull growth. In *The Skull*, ed. J. Hanken and B. K. Hall. Chicago, IL: University of Chicago Press, pp. 153–206.
Kangas, A. T., Evans, A. R., Thesleff, I., and Jernvall, J. (2004). Nonindependence of mammalian dental characters. *Nature*, **432**, 211–14.
Kavanagh, K. D., Evans, A. R., and Jernvall, J. (2007). Predicting evolutionary patterns of mammalian teeth from development. *Nature*, **449**, 427–32.
Klingenberg, C. P. (1996). Multivariate allometry. In *Advances in Morphometrics*, ed. L. F. Marcus, M. Corti, A. Loy, G. J. P. Naylor, and D. E. Slice. New York: Plenum Press, pp. 23–49.
Klingenberg, C. P. (1998). Heterochrony and allometry: the analysis of evolutionary change in ontogeny. *Biological Reviews*, **73**, 79–123.
Klingenberg, C. P. (2003). Developmental instability as a research tool: using patterns of fluctuating asymmetry to infer the developmental origins of morphological integration. In *Developmental Instability: Causes and Consequences*, ed. M. Polak. New York: Oxford University Press, pp. 427–42.
Klingenberg, C. P. (2005). Developmental constraints, modules and evolvability. In *Variation: A Central Concept in Biology*, ed. B. Hallgrímsson and B. K. Hall. Burlington, MA: Elsevier, pp. 219–47.
Klingenberg, C. P. (2008). Morphological integration and developmental modularity. *Annual Review of Ecology, Evolution and Systematics*, **39**, 115–32.
Klingenberg, C. P. (2009). Morphometric integration and modularity in configurations of landmarks: tools for evaluating a-priori hypotheses. *Evolution & Development*, **11**, 405–21.
Klingenberg, C. P. (2010). Evolution and development of shape: integrating quantitative approaches. *Nature Reviews Genetics*, **11**, 623–35.
Klingenberg, C. P., Debat, V., and Roff, D. A. (2010). Quantitative genetics of shape in cricket wings: developmental integration in a functional structure. *Evolution*, **64**, 2935–51.
Klingenberg, C. P. and Leamy, L. J. (2001). Quantitative genetics of geometric shape in the mouse mandible. *Evolution*, **55**, 2342–52.
Klingenberg, C. P., Leamy, L. J., and Cheverud, J. M. (2004). Integration and modularity of quantitative trait locus effects on geometric shape in the mouse mandible. *Genetics*, **166**, 1909–21.
Klingenberg, C. P., Leamy, L. J., Routman, E. J., and Cheverud, J. M. (2001). Genetic architecture of mandible shape in mice: effects of quantitative trait loci analyzed by geometric morphometrics. *Genetics*, **157**, 785–802.
Klingenberg, C. P., Mebus, K., and Auffray, J.-C. (2003). Developmental integration in a complex morphological structure: how distinct are the modules in the mouse mandible? *Evolution & Development*, **5**, 522–31.
Klingenberg, C. P. and Zimmermann, M. (1992). Static, ontogenetic, and evolutionary allometry: a multivariate comparison in nine species of water striders. *The American Naturalist*, **140**, 601–20.

Kruuk, L. E. B., Slate, J., and Wilson, A. J. (2008). New answers for old questions: the evolutionary quantitative genetics of wild animal populations. *Annual Review of Ecology, Evolution and Systematics*, **39**, 525–48.

Laffont, R., Renvoisé, E., Navarro, N., Alibert, P., and Montuire, S. (2009). Morphological modularity and assessment of developmental processes within the vole dental row (*Microtus arvalis*, Arvicolinae, Rodentia). *Evolution & Development*, **11**, 302–11.

Leamy, L. (1993). Morphological integration of fluctuating asymmetry in the mouse mandible. *Genetica*, **89**, 139–53.

Leamy, L. and Bradley, D. (1982). Static and growth allometry of morphometric traits in randombred house mice. *Evolution*, **36**, 1200–12.

Leamy, L. J., Routman, E. J., and Cheverud, J. M. (1999). Quantitative trait loci for early- and late-developing skull characters in mice: a test of the genetic independence model of morphological integration. *The American Naturalist*, **153**, 201–14.

Lightfoot, P. S. and German, R. Z. (1998). The effects of muscular dystrophy on craniofacial growth in mice: a study of heterochrony and ontogenetic allometry. *Journal of Morphology*, **235**, 1–16.

Lynch, M. and Walsh, B. (1998). *Genetics and Analysis of Quantitative Traits*. Sunderland, MA: Sinauer.

Macholán, M. (2006). A geometric morphometric analysis of the shape of the first upper molar in mice of the genus *Mus* (Muridae, Rodentia). *Journal of Zoology*, **270**, 672–81.

Mezey, J. G., Cheverud, J. M., and Wagner, G. P. (2000). Is the genotype–phenotype map modular? A statistical approach using mouse quantitative trait loci data. *Genetics*, **156**, 305–11.

Monteiro, L. R. (1999). Multivariate regression models and geometric morphometrics: the search for causal factors in the analysis of shape. *Systematic Biology*, **48**, 192–9.

Monteiro, L. R., Bonato, V., and dos Reis, S. F. (2005). Evolutionary integration and morphological diversification in complex morphological structures: mandible shape divergence in spiny rats (Rodentia, Echimyidae). *Evolution & Development*, **7**, 429–39.

Monteiro, L. R. and dos Reis, S. F. (2005). Morphological evolution in the mandible of spiny rats, genus *Trinomys* (Rodentia: Echimyidae). *Journal of Zoological Systematics and Evolutionary Research*, **43**, 332–8.

Monteiro, L. R. and Nogueira, M. R. (2010). Adaptive radiations, ecological specialization, and the evolutionary integration of complex morphological structures. *Evolution*, **64**, 724–44.

Moore, W. J. (1981). *The Mammalian Skull*. Cambridge: Cambridge University Press.

Parsons, T. E., Kristensen, E., Hornung, L., *et al.* (2008). Phenotypic variability and craniofacial dysmorphology: increased shape variance in a mouse model for cleft lip. *Journal of Anatomy*, **212**, 135–43.

Prochazka, J., Pantalacci, S., Churava, S., *et al.* (2010). Patterning by heritage in mouse molar row development. *Proceedings of the National Academy of Sciences of the United States of America*, **107**, 15497–502.

Ramaesh, T. and Bard, J. B. L. (2003). The growth and morphogenesis of the early mouse mandible: a quantitative analysis. *Journal of Anatomy*, **203**, 213–22.

Renaud, S. and Auffray, J.-C. (2010). Adaptation and plasticity in insular evolution of the house mouse mandible. *Journal of Zoological Systematics and Evolutionary Research*, **48**, 138–50.

Renaud, S., Auffray, J.-C., and de la Porte, S. (2010). Epigenetic effects on the mouse mandible: common features and discrepancies in remodeling due to muscular dystrophy and response to food consistency. *BMC Evolutionary Biology*, **10**, 28.

Schmidt, E. J., Parsons, T. E., Jamniczky, H. A., *et al.* (2010). Micro-computed tomography-based phenotypic approaches in embryology: procedural artifacts on assessments of embryonic craniofacial growth and development. *BMC Developmental Biology*, **10**, 18.

Swiderski, D. L. and Zelditch, M. L. (2010). Morphological diversity despite isometric scaling of lever arms. *Evolutionary Biology*, **37**, 1–18.

Tagliaro, M. L., Mattos de Oliveira, R., Pereira Padilha, D. M., Callegiari-Jacques, S. M., and Jeckel-Neto, E. A. (2009). Morphological changes in the mandible of male mice associated with aging and biomechanical stimulus. *The Anatomical Record*, **292**, 431–8.

Tomo, S., Ogita, M., and Tomo, I. (1997). Development of mandibular cartilages in the rat. *The Anatomical Record*, **249**, 233–9.

Valdar, W., Solberg, L. C., Gaugier, D., *et al.* (2006). Genome-wide genetic association of complex traits in heterogeneous stock mice. *Nature Genetics*, **38**, 879–87.

Vecchione, L., Miller, J., Byron, C., *et al.* (2010). Age-related changes in craniofacial morphology in GDF-8 (myostatin)-deficient mice. *The Anatomical Record*, **293**, 32–41.

Velhagen, W. A. and Roth, V. L. (1997). Scaling of the mandible in squirrels. *Journal of Morphology*, **232**, 107–32.

West-Eberhard, M. J. (2003). *Developmental Plasticity and Evolution*. New York: Oxford University Press.

Wilson, A. J., Réale, D., Clements, M. N., *et al.* (2010). An ecologist's guide to the animal model. *Journal of Animal Ecology*, **79**, 13–26.

Workman, M. S., Leamy, L. J., Routman, E. J., and Cheverud, J. M. (2002). Analysis of quantitative trait locus effects on the size and shape of mandibular molars in mice. *Genetics*, **160**, 1573–86.

Young, N. M., Wat, S., Diewert, V. M., Browder, L. W., and Hallgrímsson, B. (2007). Comparative morphometrics of embryonic facial morphogenesis: implications for cleft-lip etiology. *The Anatomical Record*, **290**, 123–39.

Young, R. L. and Badyaev, A. V. (2006). Evolutionary persistence of phenotypic integration: influence of developmental and functional relationships on complex trait evolution. *Evolution*, **60**, 1291–9.

Zelditch, M. L., Wood, A. R., Bonett, R. M., and Swiderski, D. L. (2008). Modularity of the rodent mandible: integrating bones, muscles, and teeth. *Evolution & Development*, **10**: 756–68

7

Recognition of subspecies status mediated by androgen-binding protein (ABP) in the evolution of incipient reinforcement on the European house mouse hybrid zone

CHRISTINA M. LAUKAITIS AND ROBERT C. KARN

Introduction

Between 5000 and 10 000 years ago, two populations of European house mice, today characterized as the subspecies *Mus musculus domesticus* and *M. m. musculus*, made secondary contact, resulting in formation of the narrow hybrid zone running across Danish Jutland and down through central Europe from the Baltic to the Black Sea. Studies suggest that a mechanism(s) of postzygotic reproductive isolation maintains the European mouse hybrid zone; however, there is also evidence that one or more prezygotic isolation mechanisms may also contribute to the maintenance of this zone of secondary contact. The purpose of this chapter is to review the data that suggest that these two subspecies and the transition of their hybrids across the zone can recognize each other and that this may result in some degree of avoidance of mating with each other. This putative prezygotic reproductive barrier is a behavioural one and a proteinaceous pheromone, mouse salivary androgen-binding protein (ABP) has been proposed to be a signal that mediates subspecies recognition. We review the putative mouse proteinaceous pheromones described thus far and present the state of knowledge about ABP's contribution to mouse communication. In so doing, we explore the relatively recent evidence that ABP-based recognition may be an incipient mechanism of reinforcement operating on the borders of the allopatric subspecies populations.

Chemosensory detection in vertebrates

Vertebrates possess two olfactory systems, the main olfactory system (MOS) and the accessory olfactory system (AOS). The structures and functions

Evolution of the House Mouse, ed. Miloš Macholán, Stuart J. E. Baird, Pavel Munclinger, and Jaroslav Piálek. Published by Cambridge University Press.

of these systems have been the subjects of many reviews (e.g. Dulac and Torello, 2003; Mombaerts, 2004; Baum and Kelliher, 2009; Tirindelli *et al.*, 2009; Touhara and Vosshall, 2009) and will only be summarized here. Both the MOS and AOS are chemosensory systems that transduce chemical sensations into neural activity and both can detect semiochemicals (derived from 'semeon', the Greek for 'signal') – chemical substances that carry messages – either individually or as a mixture. In contrast to the MOS, which detects volatile chemicals, the AOS detects fluid-phase chemicals including proteinaceous pheromones. The main olfactory receptor neurons project to the main olfactory bulb (MOB), whose mitral/tufted cells project in turn to the olfactory cortex. The sensory receptors of the accessory olfactory system are located in the vomeronasal organ (VNO), where receptor neurons innervate the accessory olfactory bulb (AOB), located on the dorsal-posterior portion of the main olfactory bulb. The AOB mitral cells do not project to the cerebral cortex, as in the case of the main olfactory bulb, but instead to hypothalamic areas by way of the medial amygdala, where they influence innate behaviours such as mating and aggression. Induction of major lesions to the main olfactory pathway does not compromise reproductive behaviours, while lesions or genetic inactivation of the VNO and AOB severely compromises sex discrimination and mating behaviours, suggesting that the accessory olfactory pathway is the primary mechanism for detecting pheromonal cues.

The VNO uses a variety of receptor molecules to detect pheromonal signals (for recent reviews, see Dulac and Wagner, 2006; Ferrero and Liberles, 2010). The most common and best-studied receptors in rodents are of the V1R and V2R subfamily of the G-protein coupled seven-transmembrane receptors. V1R-producing neurons respond to low-weight organic molecules and are found in the apical layer of the VNO, whereas V2R-producing cells detect peptide signals and are found in the VNO basal layer. As in the MOB, individual cells express a single vomeronasal receptor. However, unlike the situation in the MOB, a single VNO cell sends projections to multiple glomeruli in the anterior (V1R) or posterior (V2R) portion of the AOB, with projections from cells expressing the same vomeronasal receptor ultimately converging on a few glomeruli. This creates a complex, potential hierarchical, relationship between receptors that appears more conducive to mediating complex discrimination and mating behaviours from complex mixtures of molecules, than to causing a stereotypical response to often-encountered single odours. This assertion is supported by gene knock-out experiments suggesting that V1Rs allow gender discrimination, while V2Rs detect the signals mediating aggression between male mice (Stowers *et al.*, 2002; Dulac and Torello, 2003).

The V1R- and V2R-type receptors are encoded by large, complex gene families. Mice have been reported to have at least 364 V1R genes, of which about half are thought to be psuedogenes (Young *et al.*, 2005). Whereas the V1R genes contain only one coding exon, the more complex structure of the V2R genes has made gene prediction more difficult, but around 120 intact and 160 pseudogenized V2R genes are predicted in mice (Yang *et al.*, 2005; Young and Trask, 2007). The intact genes in each family cluster into more than ten subfamilies with striking amplification since the rat–mouse split. This impressive repertoire of receptor genes is thought to be necessary to detect an ever-changing array of signals in the competition between mice to ensure reproductive success.

Overview of animal communication via proteinaceous pheromones

From an evolutionary perspective, the paramount biological function is reproduction, and mate choice is a fundamental part of that function. The five senses – sight, hearing, taste, smell, and touch – all have a role in mate choice, but one or more of them dominate, depending on the ecological niche of the organism in question. For example, it seems intuitive that sight will have a more important communication role in diurnal animals than in nocturnal animals. Because many rodents are nocturnal, it is not surprising that there are a number of different pheromones that signal various characteristics of a potential mate. One critical component of mate choice is recognizing homospecificity in a potential mate. Mating with the wrong species leads to loss of genetic potential and thus species-specific cues are among the most important signals in communication.

There is a great deal of interest in mammalian communication by pheromones, and much attention has been focused on rodents. This has been especially true of the house mouse, *Mus musculus*, where at least three large gene families encoding proteinaceous pheromones have been described: androgen-binding proteins (ABPs), exocrine gland-secreting peptides (ESPs), and major urinary proteins (MUPs).

ABPs have been shown to mediate assortative mate selection based on subspecies recognition, which potentially limits gene exchange between subspecies where they meet (Laukaitis *et al.*, 1997; Talley *et al.*, 2001). There is evidence that ABP-mediated mate preference across a transect of the European mouse hybrid zone is a case of reproductive character displacement as predicted by reinforcement (Bímová *et al.*, 2005).

ESPs constitute a newly described family of mouse proteinaceous pheromones (Kimoto *et al.*, 2005). Female mice respond to direct facial exposure to an ESP expressed in male exorbital lacrimal glands and released into tear fluid by upregulating *c-Fos* and *egr1* gene expression in vomeronasal sensory neurons

(Kimoto *et al.*, 2007). The same response occurs after close contact with the face or bedding of male mice, and a recombinant ESP protein stimulates electrical activity in an isolated female vomeronasal organ. The male response to similar signals is unremarkable (Kimoto *et al.*, 2005; Kimoto *et al.*, 2007; Kimoto and Touhara, 2005).

The MUPs are a family of lipocalins shown to mediate female recognition of potential mates (for a review, see Hurst, 2009). For the most part, MUP genes are expressed in the liver and the products are passed through the kidneys into the urine. Each adult mouse expresses a pattern of 8–14 different MUP isoforms in its urine, which is determined by its genotype and by its sex, because some MUP genes show sex-limited expression (Hurst, 2009). This individual recognition profile has been likened to a protein 'bar code' (Robertson *et al.*, 1996; Beynon and Hurst, 2003; Armstrong *et al.*, 2005; Cheetham *et al.*, 2007; Logan *et al.*, 2008). MUPs have also been implicated in male–male aggression. Chamero *et al.* (2007) isolated high molecular weight components of male urine that activated dissociated vomeronasal neurons and were sufficient to cause male–male aggressive behaviour when painted onto previously castrated males.

The major histocompatibility complex (MHC) has also been implicated in a mechanism that reduces inbreeding by mediating disassortative mate selection based on MHC genotype. It has been suggested that nine amino acid peptide ligands bind specifically to different MHC proteins and that these ligands are released when the MHC proteins break down (for a review, see Hurst, 2009). The released ligands are filtered through the kidneys and secreted into the urine, where they act as pheromones communicating information about the MHC genotype. Some researchers have pointed to difficulties with this model (Hurst, 2009) and, to date, the genetic basis for the putative peptide pheromones is unknown. Thus, we cannot include the MHC pheromone system in the genomic comparison reported here.

All three families of proteinaceous pheromone genes, *Abp*s, *Esp*s, and *Mup*s, are encoded by large gene clusters in different regions of the mouse genome. The structure of the *Abp* linkage group is complex, in part because the ABP protein is a dimer composed of an alpha subunit disulphide bridged to a beta–gamma subunit, and so two different subunit genes are required to make the subunits for a functional dimer. The *Abp* gene region is comprised of 64 paralogues mapping in a cluster on chromosome 7 (Laukaitis *et al.*, 2008). Of these, 30 are *Abpa* genes, which encode ABP alpha subunits, and 34 are *Abpbg* genes, which encode the beta–gamma subunits. The majority of *Abp* genes occurs in 27 alpha/beta–gamma pairs in a $5'$–$5'$ orientation and numbered 1–27 (Laukaitis *et al.*, 2008). These pairs are designated <*Abpa-Abpbg*> or <*Abpbg-Abpa*> modules, where the arrowheads point in the $3'$ direction.

A group of at least 38 ESP genes, ten of which are putative pseudogenes, map in a cluster on chromosome 17 (Kimoto *et al.*, 2007). The expressed genes encode the monomeric ESP pheromones. The gene family encoding the monomeric MUPs consists of at least 40–42 genes (there are still gaps in this region of the mouse genome), half of which are pseudogenes. These map in a cluster on chromosome 4 (Logan *et al.*, 2008; Mudge *et al.*, 2008). As in the case of the *Abp* region, this cluster of genes is relatively complex in that the centre of the linkage group consists of at least 15 gene pairs, each containing a gene and a pseudogene in 5′–5′ orientation (Logan *et al.*, 2008).

The commonality uniting these gene families, *Abp*, *Esp*, and *Mup*, is that they have all expanded dramatically during house mouse evolutionary history (Kimoto *et al.*, 2007; Laukaitis *et al.*, 2008; Logan *et al.*, 2008).

Discovery and characterization of mouse salivary ABP

The relationship of structure and function is a central theme of biology. Mammalian salivary proteins come in a variety of different forms with diverse functions. In some cases a salivary protein's function has yielded to biochemical studies. For example, salivary amylase digests dietary starches, glycoproteins lubricate hard and soft tissues, and some peptides retard the growth of bacteria and fungi in the oral cavity (reviewed in Karn, 1991). For other salivary proteins, however, the function has remained elusive. Mouse salivary ABP has been one of those enigmatic structures, and its discovery three decades ago was inauspicious compared to the directions taken by the eventual quest for its function. As often happens in science, ABP was discovered during a search for something else. R. C. Karn observed that saliva collected from wild mice contributed by Tønnes Nielsen had a protein with variable migration on electrophoresis gels. There were often 2–4 proteins migrating in a particular area of the gel in the saliva of some mice, but only one in the saliva of others (R. C. Karn, unpublished).

The interest of Karn and his colleagues was initially restricted to documenting a salivary protein polymorphism and it was clear that studying inbred strains of mice would be necessary to reveal the genetic control of the protein in question. A survey of electrophoretic mobility of this protein in inbred strains, especially mice of various ages, led to the surprising conclusion that its mobility could change during the lifetime of a mouse but only in the saliva of males and even then only some inbred strains (Karn *et al.*, 1982). Subsequent studies by this group (Dlouhy and Karn, 1983, 1984) attributed the mobility change to the dominant allele of a gene with sex-limited expression, which they named *Sex-limited salivary protein* (*Ssp*, dominant allele Ssp^S, recessive allele Ssp^F). Extending the observations of Karn (1981), they showed that the action of *Ssp*^{S} was dependent both on increased circulating testosterone (T) levels expressed in males during and following puberty and on the presence of an intact androgen receptor. In addition to identifying a modifier gene, Dlouhy and Karn (1983) observed that the change in electrophoretic mobility from the fast (F) to the slow (S) phenotype appears to be primarily the result of an increase in the molecular weight of the larger subunit, since the isoelectric point changes very little. One possibility was that the protein in the saliva of males possessing *Ssp*^{S} is somehow becoming associated with molecules of testosterone when levels of that hormone rise during puberty. Experiments designed to test this did not support the original hypothesis, but did show that the protein in the saliva of both males and females are capable of binding androgen, even though that did not appreciably alter their migration on electrophoresis gels. Based on that observation, Dlouhy and Karn (1983) named the protein *A*ndrogen-*B*inding *P*rotein (ABP).

Early studies showed that the major tissue source of salivary ABP is the submaxillary gland (Dlouhy and Karn, 1983). Its gene, *androgen-binding protein* (*Abp*), which was subsequently to become a major mouse gene family, was discovered in a second survey of inbred strains performed in Karn's lab, and subsequently confirmed with genetic crosses that revealed two alleles, *Abp*^{a} and *Abp*^{b} (Dlouhy and Karn, 1984). Dlouhy and Karn (1983) had previously shown that the ABP protein is a dimer, composed of two subunits connected by disulphide bridging, but early genetic studies did not identify which subunit (or possibly both) was genetically variant. More extensive genetic studies suggested that there are two dimeric forms of ABP in mouse saliva and that these share a common subunit named alpha (encoded by *Abpa*, later revised to *Abpa27*; see 'Genetic control of ABP' below) but differ in the other subunit, such that one is an alpha–beta dimer and the other an alpha–gamma dimer, where beta is encoded by gene *Abpb* and gamma by gene *Abpg* (Dlouhy *et al.*, 1987). It should be noted, however, that this nomenclature evolved over the next 20 years as genomic studies of the *Abp* region were published (see below and Table 7.1). These two dimers were envisioned as being the products of beta and gamma subunits randomly dimerized with alpha. The same study showed that the original alleles, renamed *Abpa*^{a} and *Abpa*^{b} were variants of the alpha subunit gene.

As interesting as the biochemical genetics of ABP were becoming, it was clear that further characterization of this salivary protein in classical inbred strains of mice was not likely to shed much more light on the function of the protein. Nonetheless, purification of ABP from saliva made it possible to raise an antibody for use in specifically identifying ABP proteins on electroblots of electrophoresis gels (Dlouhy *et al.*, 1986). The ABP phenotypes visualized on anti-ABP antibody-immunostained electroblots of salivas separated by electrophoresis

Table 7.1 *Abp* genes, pseudogenes, nomenclature changes and expression locations (putative expressed genes are in bold typeface).

Original	Emes *et al.*, 2004	Laukaitis *et al.*, 2008	mm9 coordinates		mRNA found?
	ABPa1	*B6_a1ψ*	32039982	32041416	No
	ABPbg1	***B6_bg1***	32048657	32050908	No
Mjal1	*ABPa2*	***B6_a2***	32075543	32076939	Yes
11 kDa	*ABPbg2*	*B6_bg2ψ*	32087682	32089952	Yes
	ABPbg12	*B6_bg3ψ*	32143937	32147206	Yes
		B6_a3	32160507	32161931	Not tested
	ABPbg3	*B6_bg4ψ*	32185931	32188316	No
	ABPa3	*B6_a4ψ*	32194574	32195996	Not tested
	ABPa12	*B6_a28ψ*	32228722	32230136	No
	ABPbg4	*B6_bg5ψ*	32285081	32286475	Not tested
	ABPa4	*B6_a5ψ*	32292340	32293744	Not tested
		B6_bg28ψ	32329941	32331947	Not tested
		B6_bg6ψ	32402639	32404664	Not tested
		B6_a6ψ	32410499	32411928	Not tested
		B6_bg7	32488842	32490875	Not tested
		B6_a7	32497586	32498993	Not tested
		B6_bg8ψ	32563678	32565696	Not tested
		B6_a8ψ	32573178	32574586	Not tested
		B6_bg9ψ	32700132	32702523	Not tested
		B6_a9ψ	32709242	32710665	Not tested
		B6_bg29ψ	32797931	32799937	Not tested
		B6_bg10ψ	32877046	32879077	Not tested
		B6_a10	32885768	32887186	Not tested
		B6_bg30ψ	32931141	32933170	Not tested
		B6_bg11	32994236	32996260	Not tested
		B6_a11	33007756	33009172	Not tested
		B6_bg12	33110343	33112366	Not tested
		B6_a12	33119097	33120516	Not tested
	ABPa13/3	***B6_a29***	33226445	33227857	Yes
		B6_bg13ψ	33292401	33294418	Not tested
		B6_a13ψ	33301910	33303318	Not tested
		B6_bg14ψ	33432547	33434939	Not tested
		B6_a14ψ	33441664	33443088	Not tested
		B6_bg31ψ	33530194	33532200	Not tested
		B6_bg15ψ	33612747	33614778	Not tested

Table 7.1 (*cont.*)

Original	Emes *et al.*, 2004	Laukaitis *et al.*, 2008	mm9 coordinates		mRNA found?
		B6_a15	33621469	33622887	Not tested
		B6_bg16ψ	33659437	33661829	Not tested
		B6_a16ψ	33668554	33669978	Not tested
		B6_bg32ψ	33757093	33759099	Not tested
		B6_bg17ψ	33839646	33841677	Not tested
		B6_a17	33848364	33849782	Not tested
		B6_bg33ψ	33896570	33898596	Not tested
		B6_bg18	33956955	33958979	Not tested
		B6_a18	33970467	33971883	Not tested
		B6_bg19	34063433	34065457	Not tested
		B6_a19	34072206	34073626	Not tested
		B6_bg20	34149406	34151437	Not tested
	ABPa13/1	***B6_a20***	34158154	34159571	Yes
	ABPbg13	*B6_bg34ψ*	34227925	34229954	Not tested
	ABPbg5	***B6_bg21***	34303544	34305567	No
	ABPa5	*B6_a21ψ*	34312275	34313697	Yes
	ABPbg6	*B6_bg22ψ*	34389169	34391191	Not tested
	ABPa6	*B6_a22ψ*	34399532	34400942	Not tested
	ABPbg7	*B6_bg23ψ*	34410209	34412238	Not tested
	ABPa7	*B6_a23ψ*	34420061	34421471	Not tested
	ABPbg8	***B6_bg24***	34522246	34524428	Yes
	ABPa8	***B6_a24***	34528700	34530118	Yes
	ABPbg9	*B6_bg25ψ*	34637923	34639956	No
	ABPa9	*B6_a25ψ*	34652853	34654216	Not tested
Abpg	*ABPbg10*	***B6_bg26***	34728065	34730103	Yes
	ABPa10	*B6_a26ψ*	34743704	34745121	Yes
Abpb	*ABPbg11*	***B6_bg27***	34796983	34799005	Yes
Abpa	*ABPa11*	***B6_a27***	34806467	34807888	Yes
	ABPa14	*B6_a30ψ*	34884647	34885844	No

matched those seen on autoradiographs of ^{14}C-testosterone bound to salivary proteins and subsequently separated by electrophoresis. This advance showed not only that all mouse salivary proteins capable of binding testosterone are some combination of ABP subunits, but also provided an important technical advantage because immunostained electroblots allowed determination of ABP phenotypes without the requirement for first binding the proteins to radioactive androgen.

ABP in the salivas of wild house mice and other murid rodents

Karn employed the electroblotting–immunostaining method to extend ABP typing to the salivas of wild-derived strains maintained in the laboratory of François Bonhomme at the University of Montpellier II in France. These mildly inbred strains had been produced from samples wild-caught in many, widely separated localities in western and eastern Europe, Africa, and Asia. This survey of wild mice produced the surprising discovery that a different allele of the *Abpa* gene was fixed in each of three subspecies of mice occupying distinct geographic ranges in the Old World (Fig. 7.1): *Abpa*^{a} in *M. m. domesticus* (western Europe and the Mediterranean basin); *Abpa*^{b} in *M. m. musculus* (eastern Europe to northern China); and *Abpa*^{c} in *M. m. castaneus* (Southeast Asia and Malaysia) (Karn and Dlouhy, 1991; Karn *et al.*, 2002). Hwang *et al.* (1997) sequenced these three alleles fixed in the subspecies of *M. musculus*, along with those representing three other full species: *M. caroli*, *M. spretus*, and *M. spicilegus*. They concluded that directional selection is a sufficient explanation both for the large ratio of non-synonymous to synonymous substitutions they observed in the coding sequences of *Abpa* and for the unusual population distribution of *Abpa* alleles observed by Karn and Dlouhy (1991). Hwang *et al.* (1997) also observed what appeared to be a high level of homoplasy in the phylogeny of the *Abpa* haplotypes, as shown by the incongruence they found with the canonical

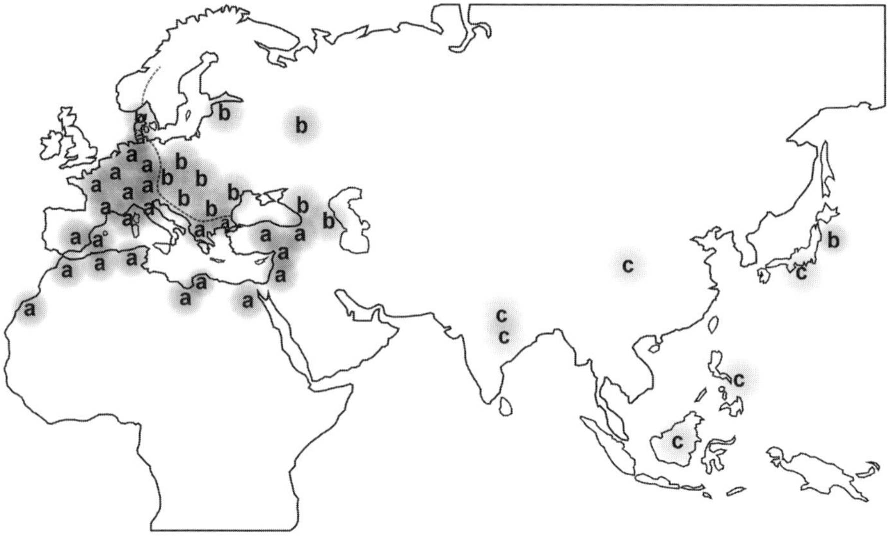

Figure 7.1 The three *Abpa* alleles (*a–c*) found in the peripheral *Mus musculus* subspecies. See plate section for a colour version of this figure.

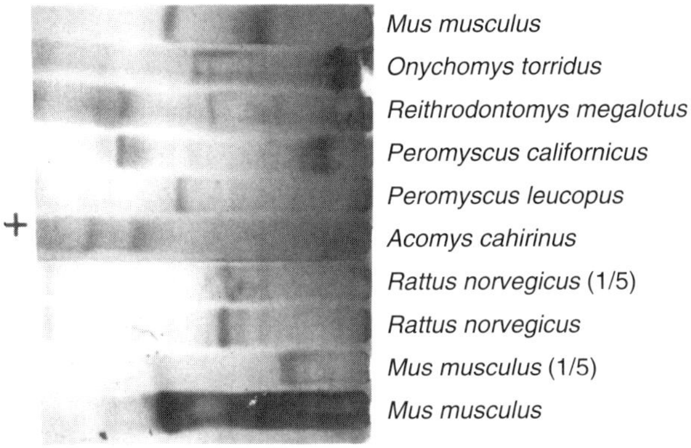

Figure 7.2 Androgen-binding proteins in the saliva of a variety of murid and cricetid rodents (Karn and Dlouhy, 1991). ABP was detected by incubating the saliva with [^{14}C]-testosterone prior to electrophoresis, drying the gels, and producing autoradiographs with X-ray film. The designation 1/5 indicates a dilution of saliva in distilled water before steroid binding.

phylogeny of the genus *Mus*, and suggested the possibility of cyclical selection of certain amino acid variants that became advantageous at some stage in their evolution.

Karn and Dlouhy (1991) also surveyed the saliva of many different murid and cricetid rodents for proteins capable of binding testosterone and found that all salivas sampled have one or more proteins with this characteristic of ABP (Fig. 7.2). This is an important observation because it suggests that the testosterone-binding characteristic must be common to all rodents, rather than one defined narrowly by the ecological niche of the taxon possessing it. Commenting on their observation of unique *Abpa* alleles fixed in the three different subspecies of *M. musculus*, Karn and Dlouhy (1991) suggested that *Abpa* underwent significant microevolution in conjunction with *M. musculus* subspeciation. They discussed two possible explanations: (1) general binding/neutralization of dietary substances (toxins, nutrients, etc.) with structures related to steroids; or (2) binding sex steroids for oral and/or olfactory recognition purposes. Clearly, their finding of a protein(s) with testosterone-binding capability in the salivas of a wide variety of cricetid and murid rodents with very different natural histories augured against the first explanation, and they concluded that the ubiquitous appearance of ABP in rodent saliva and its relative specificity for androgen are particularly interesting in light of the second explanation. Thus was born the concept of a role for ABP in sex and/or subspecies recognition, and this was the first suggestion of a reasonably well-defined role for mouse salivary ABP.

ABP-mediated mate choice

The notion that ABP might serve as a signal of subspecies identity was not without a broader context because there had already been substantial work from a number of different laboratories aimed at defining the interactions of two house mouse subspecies, *M. m. domesticus* and *M. m. musculus*, at the hybrid zone they form in Europe. The question formulated in Karn's laboratory, on which a series of behavioural tests were subsequently predicated, was whether house mice of different subspecies could recognize each other on the basis of their salivary ABP phenotype and, if so, would it cause them to make assortative mate choices. The concept behind the behavioural tests they subsequently developed was a very simplistic analogue to the European hybrid zone because choice was to be between animals representing the allopatric *M. m. domesticus* and *M. m. musculus* subspecies, whereas the parapatric populations in the hybrid zone are actually a transition of hybrid genotypes from one allopatric subspecies population to the other across the zone (Boursot *et al.*, 1993; Sage *et al.*, 1993). Nonetheless, the observations produced with inbred laboratory strains led to testing wild house mice sampled from both allopatric populations and from the hybrid zone itself.

Development of strains of mice congenic for *Abpa* alleles

In order to establish the ABP-based behavioural tests, strains of mice had to be produced that differed essentially only in their *Abpa* genes. Congenic strains are, ideally, two strains with otherwise identical genetic backgrounds that differ only in a gene of interest. In this case, congenic strains would differ only in their *Abpa* alleles, such that one possessed the $Abpa^a$ allele fixed in *M. m. domesticus* populations and the other possessed the $Abpa^b$ fixed in *M. m. musculus* populations. All ABP-based mate preference tests, including those originally run in Karn's laboratory and those later run in collaborative work done between Karn's and Jaroslav Piálek's laboratories, were subsequently designed to take advantage of the power of congenic strains.

The mouse strain congenic for $Abpa^b$ on the inbred C3H strain background was produced by crossing the DBA strain (homozygous for $Abpa^b$) with the C3H strain (homozygous for $Abpa^a$) and then backcrossing the F_1 generation to the C3H strain. The $Abpa^b$ allele was maintained in the line during backcrossing by typing ABP (Karn and Dlouhy, 1991) in offspring and selecting heterozygotes as parents for the next backcross generation. Backcrossing for 16 generations, followed by intercrossing and selecting the $Abpa^b$ homozygotes produced $Abpa^b$ congenic in the C3H strain subsequently named 'b-congenic' (Laukaitis *et al.*,

1997). Estimates have been made of how much heterozygosity remains in congenic strains, depending on the number of backcrosses used to produce them. Naveira and Barbadilla (1992) derived plots of expected combined lengths of heterozygous chromosome segments flanking a selected locus and of the standard deviation. Those plots show that a locus such as *Abpa* that has been selected for 16 generations is expected to have an average combined distance of 12 map units around it, with a standard deviation of eight map units, that is still heterozygous. Thus, while producing the b-congenic line eliminated most of the DBA parental genes, it cannot be said to have eliminated all but $Abpa^b$. Nonetheless, the exact derivation of Naveira and Barbadilla (1992) yields an asymptotic curve and further backcrossing does not much improve the situation.

Testing the effect of ABP on mate choice

Laukaitis *et al.* (1997) provided direct evidence for ABP-assortative mate selection in a laboratory setting: *M. m. domesticus* and *M. m. musculus* female mice recognize and discriminate between the territories of male mice that essentially differ solely in their *Abpa* genotype and, when the males are present, the female prefers to mate with the one of her own ABP type (Table 7.2). The observation that females could differentiate between the territories of the two males when those

Table 7.2 Choice of inbred, wild-derived and wild mice for congenic-derived targets (Laukaitis *et al.*, 1997).

Test type	Strain and gender tested	Target type	*Abpa* choice	*Abpb* choice	*P* value	Reference
Marked chamber, irrevocable choice	C3H females	Congenic mice	68	24	<0.001	
	Czech II females	Congenic mice	39	66	0.004	
Marked territory, female free-roaming	C3H females	Congenic mice	23	8	0.007	Laukaitis *et al.* (1997)
Free-roaming female, tethered male	C3H females	Congenic mice	9	2	<0.02	
Reproduction of females exposed to free-roaming males	C3H females	Congenic mice	7	1	<0.02	

Table 7.3 Choice of inbred, wild-derived and wild mice for congenic-derived targets in Y-mazes.

Cohort tested	Target type	*Abpa* choice	*Abpb* choice	*P* value (first choice)	Reference
C3H females	Congenic saliva	61	39	0.01	
C3H females	Congenic urine	54	46	0.21	
Wild-derived *M. m. domesticus* females	Congenic saliva	36	17	0.005	Talley *et al.* (2001)
Wild-derived *M. m. musculus* females	Congenic saliva	11	25	0.01	
Allopatric populations*	Congenic saliva	6	2	0.08‡	Bímová *et al.* (2005)
Parapatric populations*	Congenic saliva	21	9	0.014‡	

Note: *Number of tested *M. m. domesticus* and *M. m. musculus* populations making a consubspecific choice (in *Abpa* column) or heterosubspecific choice (in *Abpb* column). Males and females and subspecies origin of populations are all pooled.

‡Corrected from Bímová *et al.* (2005) for one-tailed Chi-square test (Laukaitis *et al.*, 1997; Talley *et al.*, 2001).

mice were absent suggests that the males marked their territories with ABP. In their study, they also detected ABP on the pelts of male mice and in their environment. They reasoned that it is likely that the animals apply the protein to their pelts by licking and that it is then deposited in their environment. They suggested that females of the two subspecies are able to discriminate between males of those subspecies on the basis of this protein molecule and that ABP-based discrimination acts as a kind of prezygotic isolation mechanism. One of their most important conclusions was that preference ratio increased as testing conditions more closely mimicked natural conditions (Table 7.2).

Talley *et al.* (2001) studied the effects of a single genetic change in two-way preference tests (Y-mazes) using C3H and b-congenic male mouse saliva and urine as targets. Females exhibited a preference for investigating salivas of males of their own genetic type of ABP but not for urines of either type of male (Table 7.3). In addition, this preference behaviour is consistent when test subjects were females of wild derived, mildly inbred strains developed from a variety of geographically widely separated populations on either side of the *M. musculus* hybrid zone in Europe (Table 7.3). These results are consistent with the hypothesis that ABP is a signal for subspecies recognition because it is secreted into saliva but is not found in urine (Dlouhy *et al.*, 1986). The work reported here complements earlier work (Laukaitis *et al.*, 1997) insofar as they suggest that saliva is the source of ABP-based mate selection behaviour. The results of Talley *et al.* (2001) also show that the response of females from wild populations is consistent with the response of laboratory strains. The results of both studies provide an explanation for the observation that this gene is evolving under strong selection (Karn and Nachman, 1999).

Tests of ABP-based preference in wild mouse populations have supported our previous results with classical strains of inbred mice and wild-derived inbred mice reported by Laukaitis *et al.* (1997) and Talley *et al.* (2001), contrary to the claim of Ganem in this book (Chapter 15). These studies began with a collaboration between Karn's laboratory in the United States and Piálek's laboratory in the Czech Republic, which was subsequently funded by a COBASE grant from the US National Research Council. Karn and Laukaitis provided the b-congenic strain for the behavioural testing conducted by Barbora Bímová in the course of her PhD thesis work in Piálek's laboratory. The first of two studies of ABP-based preference in populations of wild mice extended the available information on salivary ABP preferences in mice by testing wild populations of the two European *M. musculus* subspecies in a two-way choice system using Y-mazes. In this work, Bímová *et al.* (2005) tested the hypothesis that ABP is a signal for subspecies recognition in animals from wild allopatric and parapatric populations sampled on the Czech–Bavarian transect across the *M. m. domesticus* and *M. m. musculus* hybrid zone. While the study did not reveal a consistent statistically significant trend of homosubspecific preferences in individual allopatric and parapatric populations, nonetheless the whole pattern of preference was skewed towards homosubspecific preference (Table 7.3). The 21: 9 preference ratio ($P = 0.014$; one-tailed Chi-square test) is consistent with the 2: 1 ratio of preference for homosubspecific targets in large test chambers (Laukaitis *et al.*, 1997) or Y-mazes (Talley *et al.*, 2001). Even with a one-tailed Chi-square test, a minimum population size of $N = 27$ is required for a significant deviation from the null hypothesis. Ganem and Karn (cited by Ganem as her personal observations) tested members of wild-caught populations from a hybrid zone in Danish Jutland using Y-mazes and congenic saliva targets provided by Karn. However, the disproportionate numbers of localities sampled on the *M. m. domesticus* and *M. m. musculus* sides of the hybrid zone and their small population sizes did not allow a conclusion on which both collaborators could agree. In summary, it is likely that the 2: 1 consubspecific preference ratio obtained in Y-maze tests reflects the fact that mice do not encounter saliva spotted on filter paper in the wild, and this is only one of the artificialities of such tests. In that context it is worth reiterating the observations of Laukaitis *et al.* (1997), that preference ratios increased as testing conditions more closely mimicked natural conditions.

The pattern of homosubspecific preferences observed by Bímová *et al.* (2005) in wild mice was stronger for the time spent sniffing than it was for the first choice of the signal, perhaps reflecting a somewhat different response in animals sampled directly from the wild compared to that of inbred strains. Thus, in this first study of preference in wild house mice, overall preference was nearly the same 2:1 ratio seen in male-marked territory tests (Laukaitis *et al.*, 1997) and Y-maze tests of laboratory inbred strains and wild-derived inbred strains (Talley *et al.*, 2001). Perhaps the most interesting finding of this study was that it provided the first hint that a more complex picture of ABP-based preference might occur in this area of secondary contact of both subspecies, where preferences are predicted to be stronger (reproductive character displacement; Fig. 7.3) by the theory of reinforcement (reviewed in Howard, 1993).

The second study of wild house mouse preference focused especially on the issue of reinforcement, because behavioural isolation may lead to complete speciation when partial postzygotic isolation acts in the presence of divergent specific mate recognition systems, conditions that exist where *M. m. domesticus* and *M. m. musculus* come into contact and hybridize. Vošlajerová Bímová *et al.* (2011) studied two mate recognition signal systems, salivary ABP and urines, across the Czech–Bavarian portion of the European hybrid zone. One of these, a study of introgression of the *Abp* gene region on chromosome 7, was relatively straightforward because different alleles of all the numerous *Abp* genes in this large region (3 Mb, see below) appear to be fixed in the two subspecies (Karn and Laukaitis, unpublished observations). On the other hand, it was necessary to study introgression of proxy markers for the region on chromosome 4 encoding the major urinary proteins (MUPs) because no MUP has been identified with different alleles fixed in the two subspecies of *M. musculus*. Introgression for these two gene regions was compared to introgression at loci assumed to be nearly neutral, as well as those under selection against hybridization. Clinal analyses confirmed nearly neutral introgression of an *Abp* locus, confirming the clinal pattern reported earlier (Dod *et al.*, 2005), as well as two loci closely linked to the *Abp* gene cluster while, by contrast, two markers flanking the *Mup* gene region revealed massive introgression.

Vošlajerová Bímová *et al.* (2011) developed a model for the analysis of the transition of traits under reinforcement selection, considering a hierarchy of nested hypotheses designed to summarize the key features of the preference

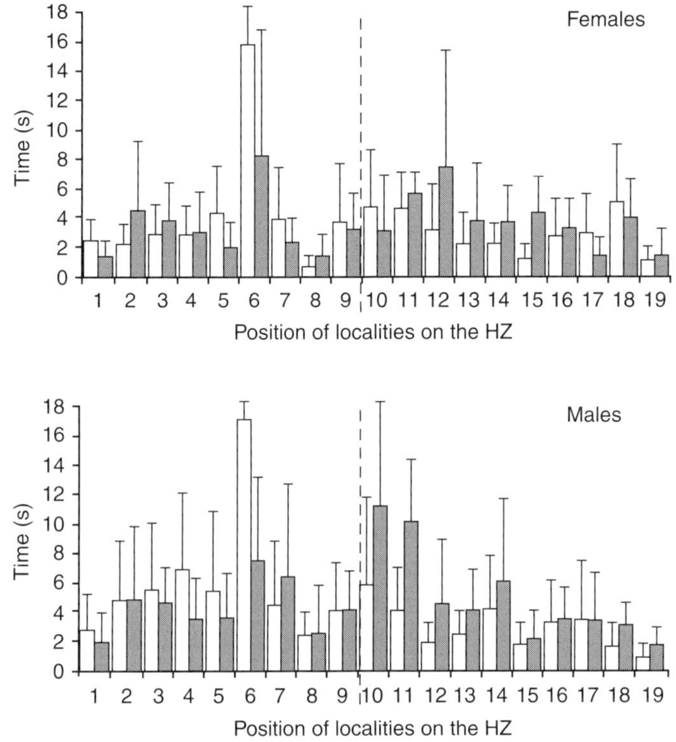

Figure 7.3 Mean time spent sniffing the signals with the *Abpa*^{a} (white columns) and *Abpa*^{b} (black columns) alleles within populations across the hybrid zone transect for females (top) and males (bottom) (Bímová *et al.*, 2005). The lines above each bar show the standard error of the mean. The populations are ordered according to their relative position along a transect from pure *M. m. domesticus* populations (left), to pure *M. m. musculus* populations (right). The centre of the zone is indicated by a dashed line.

clines and accepting additional parameters only when they significantly increased the likelihood of the data conditioned on the model. In models where there was no reinforcement (parameter R constrained at 1), cline width estimates were not robust to details of the model (this may be seen as an indication that the non-reinforcement models in general were poor). By contrast, reinforcement ($R > 1$) was strongly supported over all scenarios considered and steepened preference clines by a factor of 2–3, making them between 25 km and 30 km wide. The scale of change in the underlying hybrid index on which reinforcement is proposed to have acted is also robust to model details, with estimated width around 60 km. For both tested signals, ABPs and urines, and sexes, models of behavioural clines including a reinforcement parameter showed significantly better fits than sigmoid cline models. From their results, they concluded that recognition between the two house mouse subspecies is a complex system involving several signals and associated preferences, where the latter are shaped by reinforcing selection, and that behavioural barriers are an important component of a mosaic of reproductive isolation between house mouse taxa.

These various experimental tests of the idea that mouse salivary ABP influences mate selection behaviour have gone a long way to developing a picture of a possible pheromonal function for ABP. Recognition of one's own species versus a foreign one is an extremely important role in the evolutionary scheme of things. At least two additional developments will be necessary to pursue this putative function further. The first of these is to seek a better understanding of the natural conditions under which ABP is recognized. Saliva targets in Y-maze tests are convenient for testing large numbers of subjects but, as discussed above, early work (Laukaitis *et al.*, 1997) strongly suggests that testing conditions that more closely mimic natural conditions increase the power of the test. More realistic test conditions may lead not only to higher preference ratios but also to a better understanding of how ABP influences the behaviour of mice. The second development will be identification of the receptor(s) that recognize ABP as a signal of species and/or subspecies identity. Numerous reports have suggested that the receptor is the vomeronasal organ (VNO; see, for example, Karn and Russell, 1993; Karn, 1994; Laukaitis *et al.*, 1997) but this has yet to be confirmed. Using a purely genetic and evolutionary approach, Karn *et al.* (2010) have identified a candidate V1R receptor for ABP, Vmn1r67 (see details below), but have called for direct electrophysiological experiments that will corroborate the genetic identification of a VNO receptor for the ABP molecules proposed to advertise subspecies identity in house mice, and possibly species identity in other rodents of the genus *Mus*.

Genetic control of ABP

Using the putative amino acid sequence of the ABP alpha subunit reported by Karn and Russell (1993) as a search string, Karn (1994) discovered significant identity (50%) between the alpha subunit of mouse salivary ABP and Chain 1 of cat allergen Fel dI. In addition, he reported significant identity with two other proteins that share identity with Chain 1 of Fel dI: rabbit uteroglobin (27% identity with ABP alpha) and human lung Clara 10 (27% identity with ABP alpha). The secondary structure predicted for the mouse ABP alpha subunit is a very good fit with the secondary structure determined by X-ray crystallography for rabbit uteroglobin, a protein that shares with mouse ABP the capability of binding steroid. Fel dI is found in cat saliva, sebaceous glands, and pelt. Its function is not known, but it has been proposed to be involved in protecting dry epithelia, a parallel to uteroglobin protecting wet epithelia. Since mice, like cats, lick themselves and each other extensively, coating their pelts with ABP may be part of this or another biological function. This comparison was extended by Dominguez (1995), and the group of related proteins was eventually named the secretoglobin family of proteins (Mukherjee and Chilton, 2000; Mukherjee *et al.*, 2007).

The assignment of individual gene status to the three genes – *Abpa*, *Abpb*, and *Abpg* – encoding ABP subunits that randomly associate to produce alpha–beta and alpha–gamma dimers predicted that three different genes would eventually be found (Dlouhy *et al.*, 1987). Recombinant inbred (RI) strains allowed *Abpa* to be assigned to mouse chromosome 7, and preliminary evidence suggested that *Abpg* was also polymorphic and was closely linked to *Abpa* (Dlouhy *et al.*, 1987). However, the 'three-gene hypothesis' went untested for 16 years until, using chromosome 7 contigs, Laukaitis *et al.* (2003) sequenced *Abpa*, *Abpb*, and *Abpg* and mapped them near the proximal end of chromosome 7. In producing a map of the *Abp* gene region, they noted that a previously reported gene for a lacrimal secretoglobin mapped proximal to the *Abp* genes. They also identified the structures of the three *Abp* genes and described a 5′–5′ orientation of *Abpa* and *Abpb* that was shown eventually to be a common arrangement for many pairs of *Abp* genes.

Emes *et al.* (2004) conducted the first investigation of the mouse sequence that contains the three *Abpa*/*Abpb*/*Abpg* genes, as well as orthologous regions in rat, human, and chimpanzee genomes; they reported extensive remodelling of this region. Instead of finding only the three *Abp* genes reported by Laukaitis *et al.* (2003), they found 14 *Abpa*-like genes and 13 *Abpb-*/*Abpg*-like genes closely linked near the proximal end of mouse chromosome 7 (Fig. 7.4a). Because that region of the draft mouse genome was not free of gaps, they surmised that there might be more unidentified *Abp* paralogues lurking in the gaps. Clearly, a new *Abp* gene nomenclature was required (Table 7.1), and they extended the original nomenclature by numbering the series of alpha-like and beta-/gamma-like paralogues as *Abpa1*, *Abpa2* ... *Abpa14*, and *Abpbg1*, *Abpbg2* ... *Abpbg13*, respectively (Table 7.1). In the new system, numbering was sequential according to the order in which paired genes appear in the mouse or rat genome. The original *Abpa* was renamed as *Abpa11*, *Abpb* as *Abpbg11*, and *Abpg* as *Abpbg10*, but they did not propose to change the previous designations of *Abpa*, *Abpb*, and *Abpg*, which proved useful in putting their observations in a historical context.

Emes *et al.* (2004) found that duplication rates of *Abpa*-like and *Abpbg*-like genes in mice are more than two orders of magnitude higher than the average rate for all mouse genes; synonymous nucleotide substitution rates are two-fold

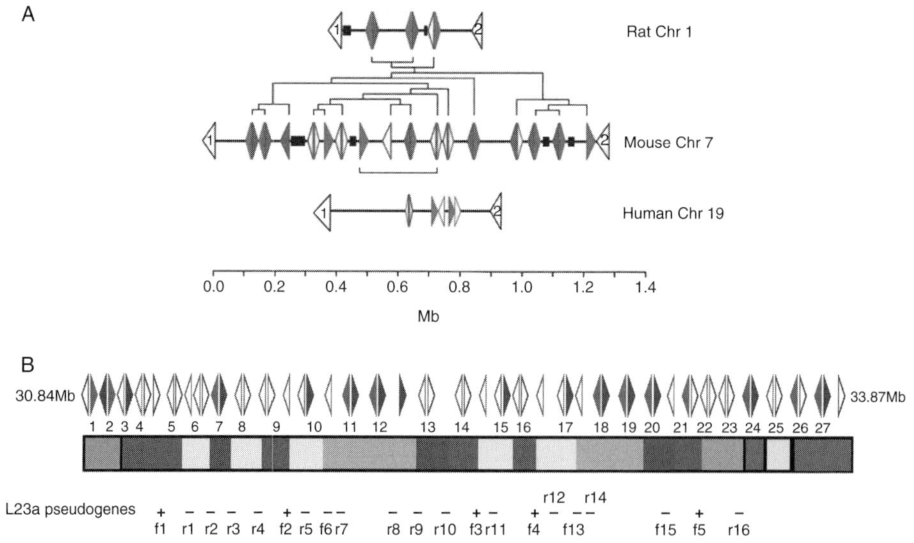

Figure 7.4 The *Abp* gene family linkage groups reported by Emes *et al.* (2004) and Laukaitis *et al.* (2008). (a) A graphical representation of the relative position and transcriptional orientation of the *Abpa*- and *Abpbg*-like genes and pseudogenes located on *Rattus norvegicus* chromosome 1, *Mus musculus* chromosome 7, and *Homo sapiens* chromosome 19. Coordinates are taken from genome releases rn3 (Baylor RGSC v3.1), mm3 (NCBI build 30), and hg16 (NCBI build 34), respectively. The $5'$–$3'$ orientations of the genes are shown by the direction of the arrowheads. *Scn1b* and *Uble1b* genes, which lie in orthologous genomic regions in all three species, are numbered 1 and 2, respectively. *Abpa*-like genes are shown in blue, *Abpbg*-like genes in red (see the version of this figure in the plate section), and primate *SCGB4A1–4(P)* genes in green. Filled arrowheads represent predicted functional genes; open arrowheads denote predicted pseudogenes. The sequence of duplication events among rodent genes inferred from phylogenetic trees is implied from the dendrogram, shown in black. Blue and red lines in this dendrogram represent branches that are not supported by the phylogenetic trees shown in Emes *et al.* (2004: figure. 7.3). The scale bar shows approximate genomic distance in megabases. Gaps (>5 kb) in the genomic assembly of each species are shown as black boxes.
(b) Predicted genes and pseudogenes are indicated as filled and unfilled triangles, respectively. *Abpa* paralogues are shown in blue and *Abpbg* paralogues are shown in red. The original paralogues described by Emes *et al.* (2004) are faded, while the new paralogues described by Laukaitis *et al.* (2008) are in bright colours. Ribosomal protein L23a pseudogenes on the forward and reverse strands are shown as '+' and '–' symbols, respectively, and are numbered at the bottom. The five *M. musculus Abpa* clades, each predicted to have originated from a single gene in the last common ancestor with *M. pahari*, are indicated by rectangles outlined in black on the bar beneath the paralogue symbols. Each of the five rectangles is annotated with the *Abpa–Abpbg* gene pairs (namely 1–2, 3–23, 24, 25, and 26–27) that arose from these single common ancestors. More recent *Mus*-lineage-specific duplications are represented by smaller coloured rectangles embedded within the outlined rectangles. See plate section for a colour version of this figure.

higher; and the *Abp* genomic region has expanded nearly three-fold since divergence of the rodents. During this time, one in six amino acid sites in ABP beta-/gamma-like proteins appear to have been subject to positive selection; they surmised that these may constitute a site of interaction with receptors or ligands. Greater adaptive variation among *Abpbg*-like sequences than among *Abpa*-like sequences suggests that assortative mating preferences are more influenced by variation in *Abpbg*-like genes. They proposed a role for ABP alpha/beta/gamma proteins as pheromones, or in modulating odourant detection, suggesting that this would account for the extraordinary adaptive evolution of these genes, and surrounding genomic regions, in murid rodents. Laukaitis and Karn (2005) subsequently updated the phylogeny of the secretoglobins, adding new secretoglobin protein sequences and discussing their potential evolutionary relationships. They presented a comparative genomic picture of the secretoglobin genes in three species of mammals: mouse, rat, and human, suggesting that the secretoglobins can be divided into at least five subfamilies. In rodent and human genomes, these gene families are found in two main clusters that share an evolutionary history (Fig. 7.5).

Laukaitis *et al.* (2008) interrogated a finished mouse (*Mus musculus*) genome sequence assembly (mm8) to show that the *Abp* gene repertoire is, in fact, twice as large as reported by Emes *et al.* (2004) for mm2. The completed *Abp* gene cluster in the mouse genome contains 30 *Abpa* and 34 *Abpbg* genes and pseudogenes in a 3 Mb region of mouse chromosome 7 (Fig. 7.4b; Table 7.1). Sequencing *Abpa* homologues from species within the genus *Mus* suggests that this burst of gene duplication occurred very recently, within the past seven million years. These paralogues occur in <*Abpa-Abpbg*> pairs (the arrows point in the $3'$ directions) and appear to have arisen primarily by independent bursts of tandem duplication from approximately five gene pairs found in the last common ancestor of the subgenus *Mus*. Moreover, these five ancestral *Abp* gene pairs in *Mus* have arisen since the last common ancestor with rat (*Rattus norvegicus*), because the dramatic expansion in *Mus* (27 <*Abpa-Abpbg*> gene pairs and three *Abpa* and seven *Abpbg* singletons) and the more modest one in *Rattus* (three <*Abpa-Abpbg*> gene pairs) appear to have occurred independently (Fig. 7.6) (Emes *et al.*, 2004; Laukaitis *et al.*, 2008).

Laukaitis *et al.* (2008) surveyed *Abp* orthologues in genomes from across the mammalian clade and showed that bursts of *Abp* gene duplications are not unique to the murid rodents; they have also occurred recently in the lagomorph (rabbit, *Oryctolagus cuniculus*) and ruminant (cattle, *Bos taurus*) lineages (Fig. 7.7). It appears that *Abp* genes have undergone repeated bursts of gene duplication and adaptive sequence diversification, perhaps driven by these genes' participation in chemosensation and/or sexual identification. At least one *Abpa* and one *Abpbg*

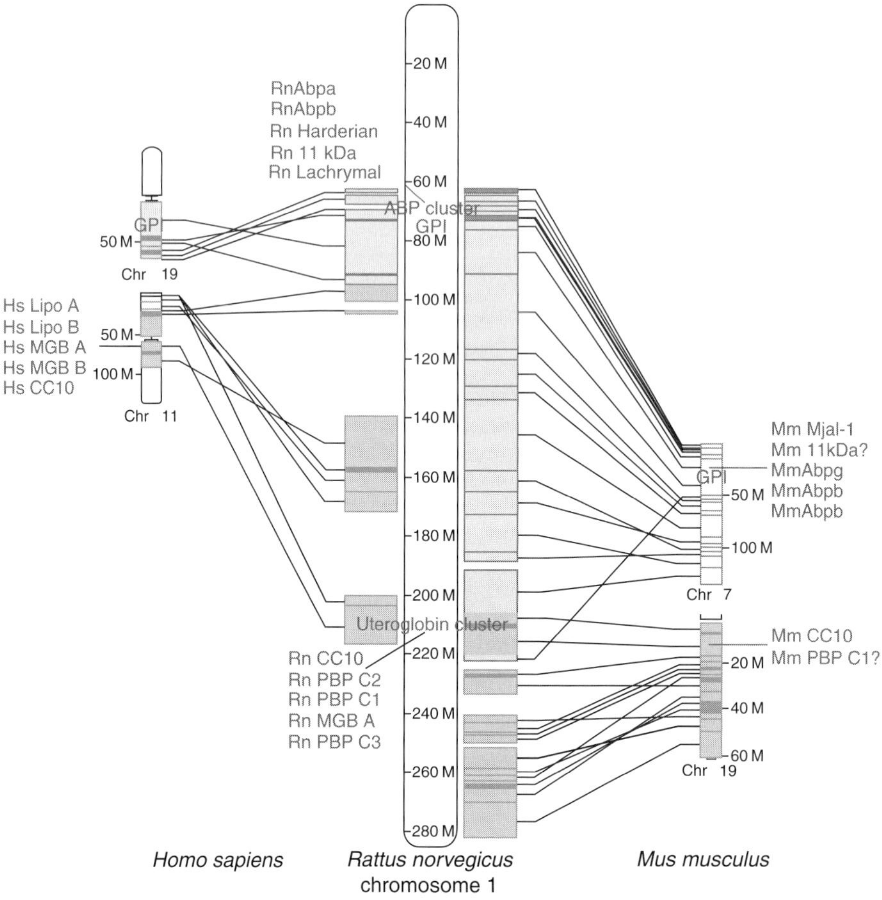

Figure 7.5 Syntenic relationships between secretoglobins expressed in mice, rats, and humans (Laukaitis and Karn, 2005). Rat secretoglobins are located in two clusters on chromosome 1 and these are mapped in the middle chromosome of the figure. Syntenic relationships on mouse chromosome 7 for B and E secretoglobins and chromosome 19 for A, C, and D secretoglobins are shown on the right of the figure. Syntenic relationships on human chromosome 11 for A, C, and D secretoglobins are shown on the left of the figure. Human GPI is shown on chromosome 19, but no B or E secretoglobins have yet been identified. Genes are listed in their approximate, predicted order on the chromosome. The base illustrations and synteny assignments are modified from Ensembl. See plate section for a colour version of this figure.

gene was present in the last common ancestor of eutherians and metatherians, approximately 180 million years ago, and this single <*Abpa-Abpbg*> gene pair has been conserved in active form in most mammalian lineages, but was silenced by pseudogenization in primates. By contrast, *Abp* evolutionary history in *Mus*,

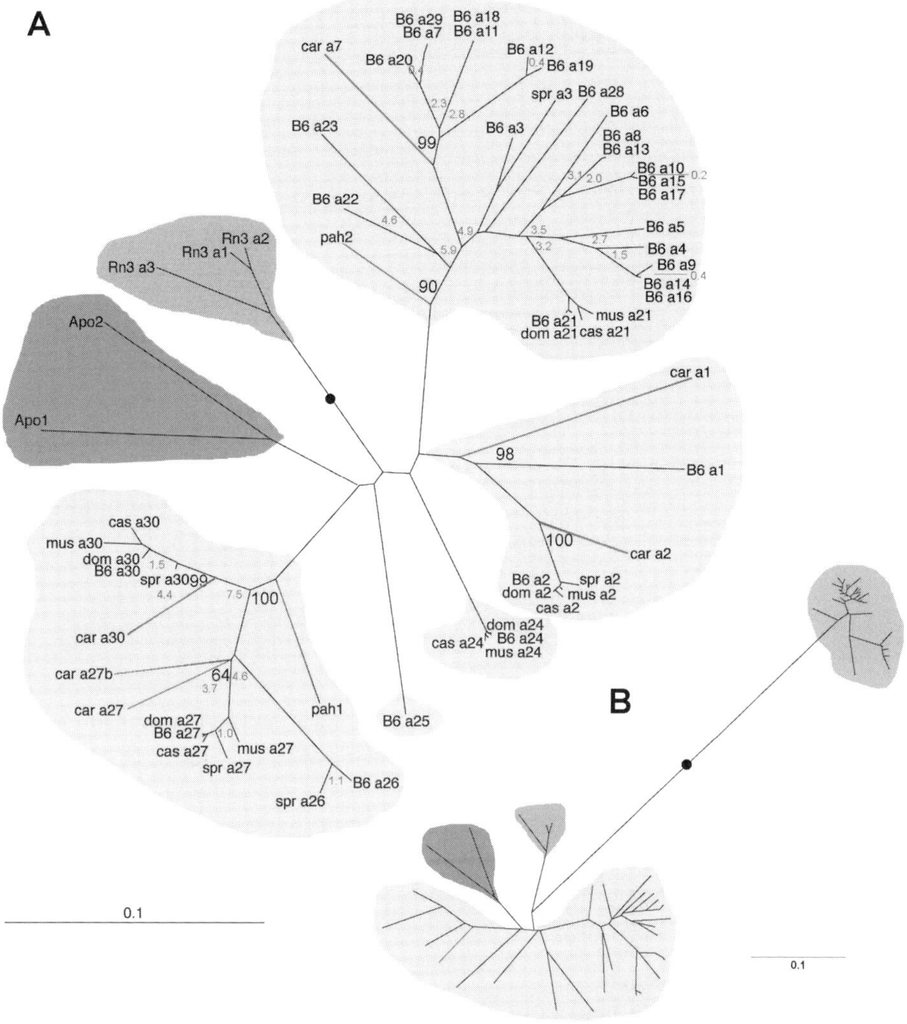

Figure 7.6 Evolutionary relationship between *Abpa* paralogues in rodents and the rabbit (Laukaitis *et al.*, 2008). (a) Neighbour-joining (NJ) phylogeny of intron 2 from rodent *Abpa* genes. Values from a superimposable bootstrap tree were used to annotate a phylogenetic tree calculated with NJ methodology. The 30 *Abpa* genes found in mm8 are shown as B6_*a1*–*a30*. Corresponding genes found in subgenus *Mus* taxa are abbreviated: dom (*M. m. domesticus*), mus (*M. m. musculus*), cas (*M. m. castaneus*), spr (*M. spretus*) and car (*M. caroli*). *Abpa* genes found in other murid taxa are abbreviated: pah (*M. pahari*), Apo (*Apodemus*) and Rn3 (2004 *Rattus* genome release). Subgenus *Mus Abpa* clades are shaded in yellow (see the version of this figure in the plate section). In the case of two of these clades, an *M. pahari* paralogue appears as an outgroup (branch shaded in red). *M. caroli* paralogues serve as outgroups to *M. musculus* clades (branches shaded in blue). The *Apodemus* clade is shaded purple and the rat clade blue. The bootstrap values for all internal nodes except two exceed 60%; key bootstrap values are shown in black typeface. Divergence times in millions of years (MY) are shown in red typeface. (b) NJ phylogeny of intron 2 from rabbit and rodent *Abpa* genes. The rabbit clade is shown with a green background; the rodent clades are shaded as in (a). In both panels, the black dots represent the probable roots of these phylogenies, the locations of which are supported by the locations of rat (a) and rabbit (b) genes which serve as outgroups. See plate section for a colour version of this figure.

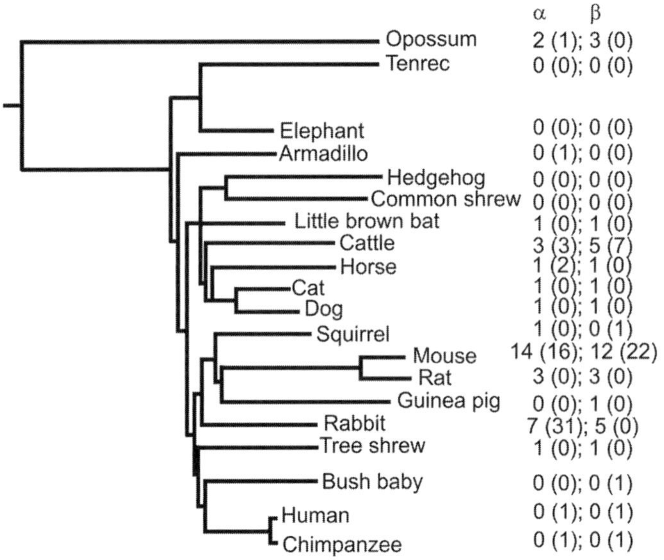

Figure 7.7 Predicted minimum *Abp* gene number per species for mammals with available genome sequences (Laukaitis *et al.*, 2008). Genes are indicated below their alpha or beta subunit heading. Pseudogenes are shown in parentheses; see Laukaitis *et al.* (2008) for species names and data sources. The phylogeny is modified from Margulies *et al.* (2005). Results for macaque and platypus are not shown owing to the absence of evidence for *Abp* genes in their current genome assemblies.

Rattus, *Apodemus*, rabbit, and cattle lineages is characterized by rapid and independent expansions of these gene families by gene duplication, with some of the paralogues remaining active and some likely becoming silenced. *Abp* genes have diverged widely through evolutionary time, both in sequence and in gene copy number, and remain under the influence of positive selection.

Characterizing the dramatic *Abp* gene expansion in the genus *Mus*

Laukaitis *et al.* (2008) considered, and eventually discarded, the possibility that non-allelic gene conversion has caused the homogeneity of *Abpa* genes they observed. They were able to discount this explanation by inspecting the phylogenetic tree of ribosomal protein L23a pseudogenes, which frequently appear to have co-duplicated with *Abpa–Abpbg* gene pairs (Fig. 7.4b). These duplications are of recent origin since similar sequences are absent from the syntenic region in the rat. In a more recent analysis of the most closely related *Abp* sequences, Karn and Laukaitis (2009) pursued the possibility that gene conversion occurred at a smaller scale by using the program GENECONV to look for evidence of short gene conversion tracks. GENECONV seeks aligned DNA or protein segments for which a pair of sequences is sufficiently similar to suggest that gene conversion occurred. These are classified as inner or outer fragments. Inner fragments are evidence of a possible gene conversion event between ancestors of two sequences in the alignment. Outer fragments are runs of unique sites that may be evidence of past gene conversion events that originated from outside the alignment or else from within the alignment but such that evidence of the source has been destroyed by later mutation or gene conversion. Their GENECONV analysis of *Abpa* paralogues identified no inner and no outer fragments that were globally significant, suggesting that there is no compelling evidence of gene conversion in *Abpa* paralogues. In the case of the *Abpbg* paralogues, the analysis identified only one inner (*Abpbg26* and *Abpbg34*) and two outer fragments (*Abpbg5p* and *Abpbg19*) that were globally significant. Karn and Laukaitis (2009) also calculated the GC content of the *Abp* gene region because sequences undergoing frequent gene conversion, either ectopic or allelic, are expected to become GC rich (Galtier *et al.*, 2001, 2008). Karn and Laukaitis (2009) found that the average GC content in the *Abp* gene region is low, about 41–42%, compared with genes undergoing gene conversion, such as ribosomal operons and transfer RNAs, which have much higher GC contents (Galtier *et al.*, 2001). Thus, it appears that gene conversion has made a minimal, but not non-existent, contribution to the evolutionary history of the *Abp* gene family. It certainly has not been significant enough to have confounded the phylogenetic inference presented by Laukaitis *et al.* (2008), and it should not have adversely affected the analysis of recently duplicated products described by Karn and Laukaitis (2009).

Karn and Laukaitis (2009) found that copy number variation (CNV) is common in at least part of the *Abp* gene region of the house mouse genome. Their evaluation of the pattern of *Abp* paralogues and interspersed repeated elements suggests that the mechanism for this was duplication of blocks containing combinations of <*Abpa-Abpbg*> modules and single *Abpbg* paralogues. This mechanism is consistent with parts of the *Abp* region of the mouse genome duplicating as low-copy repeats (LCRs) (Lupski, 1998; Stankiewicz and Lupski, 2002) by non-allelic homologous recombination (NAHR) (Shaffer and Lupski, 2000; Stankiewicz and Lupski, 2002). It is evident, however, that the massive expansion of the *Abp* gene region in the genus *Mus* did not start out this way. Nor does it explain the origin of the single <*Abpa-Abpbg*> modules seen in the genomes of most mammals. Katju and Lynch (2003) found that two-thirds of tandem gene duplicates are in inverse orientation with respect to one another and the inverse (i.e. 5′–5′) orientation of the original pair of *Abpa* and *Abpbg* genes (an <*Abpa-Abpbg*> module) fits this description. Karn and Laukaitis (2009) proposed that the original duplication of a single ancestral *Abp* gene in an early mammalian ancestor produced two paralogues in inverse adjacent order and these evolved into the original <*Abpa-Abpbg*> pair, consistent with the most widespread *Abp* gene configuration in mammals (Laukaitis *et al.*, 2008). It also seems likely that the first duplication of an <*Abpa-Abpbg*> module occurred in this way in the ancestor of the genus *Mus* because the <*Abpa1-Abpbg1*> . . . <*Abpa2-Abpbg2*> set, constituting one of the oldest clades, is in the inverse orientation to all the other *Abp* paralogues. One of the original duplicated modules was likely the ancestor of <*a1-bg1*> . . . <*a2-bg2*>, and the other was the ancestor of <*bg26-a26p*> . . . <*bg27-a27*> . . . <*a30* (unpaired) and possibly <*bg24-a24*> and <*bg25p-a25p*> because these sets of modules are the oldest in the *Abp* gene group, comprising four of the five ancestral clades of genes (Laukaitis *et al.*, 2008), and designated with coloured bars at the top of Fig. 7.8 (see colour version in the plate section). Very likely, the large fifth clade described by Laukaitis *et al.* (2008) arose later, from a module that duplicated from one of these and is now represented by <*bg3p-a3*> and/or <*bg21-a21p*>.

Something apparently changed following this scenario, however, because the most recent *Abp* duplicates in the centre of this large group of genes appear in direct, not inverse, order with respect to all members of their clade, and with respect to three of the four other clades. Rather than the relatively slower mechanism of duplication of single <*Abpa-Abpbg*> modules by primer-slipping during DNA replication (Chen *et al.*, 2005), blocks of multiple modules, sometimes including unpaired *Abpa* and *Abpbg* paralogues, were duplicated by NAHR, at least in the largest, most volatile *Abp* clade. In this scenario, duplication of blocks of genes pushed the ancestral gene sets apart, leaving the more diverged sequences on the flanks, reminiscent of the mechanism proposed by Achaz *et al.* (2000, 2001). Duplication of blocks of *Abp* modules on the interior of the *Abp* gene cluster by NAHR helps explain the very high rate of duplication (dubbed the 'snowball effect' by Kondrashov and Kondrashov, 2006). The direct repeat nature of the most recent *Abp* duplicates is also consistent with duplication by NAHR, which relies on LCRs flanking the duplicating region and commonly produces direct orientation. Thus, Karn and Laukaitis (2009) argued that duplication of blocks of *Abp* modules by NAHR is the most likely scenario, at least in the youngest, largest, and most volatile clade of *Abp* genes. They reinforced this view with observations using the Mouse Paralogy Browser, which suggest that gene birth and death accelerated by the ability to duplicate and delete numerous paralogues in large blocks is common in the *Abp* gene region.

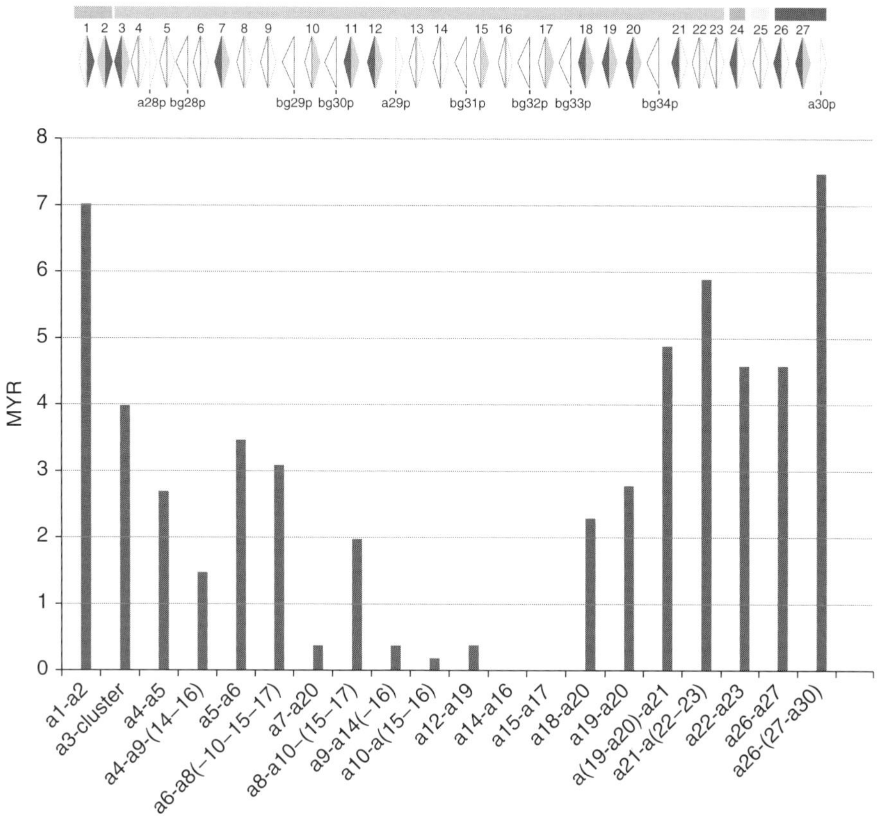

Figure 7.8 Ages of *Abpa* paralogues in millions of years (Myr) estimated by Laukaitis *et al.* (2008) were plotted below the physical map of the *Abp* gene region. Older paralogues plot closer to the left and right boundaries, whereas younger paralogues are in the interior of the region. Coloured bars (see plate section) above the linkage group represent the five clades described by Laukaitis *et al.* (2008). See plate section for a colour version of this figure.

The role of selection in the evolution of ABP subunits

Several lines of evidence have supported the idea that mouse *Abp* genes have evolved under positive selection: (1) the overall ratio of non-synonymous/synonymous substitution (*Ka/Ks*) in *Abpa27*; (2) the failure of a test of the predictions of a model of neutral molecular evolution applied to *Abpa27*; (3) a divergence in coding regions of *Abpbg26* and *Abpbg27* genes that is actually greater than in their non-coding regions; (4) codeml analysis that predicts numerous codons ('ω+ sites') that have been subjected to positive selection in *Abpa27*, *Abpbg27*, and *Abpbg26* genes; and (5) assessment of the entire *Abp* gene region with the Mouse Paralogy Browser, which suggests that substantial volatility has occurred there. The details are reviewed below.

Hwang *et al.* (1997) reported what appears to be a high level of homoplasy in the phylogeny of the *Abpa* haplotypes, as shown by the incongruence with the canonical phylogeny of the genus *Mus*. They suggested the possibility of cyclical selection of certain amino acid variants that became advantageous at some stage in their evolution. Investigating patterns of DNA sequence variation at *Abpa27* within and between species of mice, Karn and Nachman (1999) tested several predictions of a model of neutral molecular evolution. They compared intra-specific variation among ten *M. m. domesticus* alleles with divergence between *M. m. domesticus* and *M. caroli* for *Abpa27* and two X-linked genes, *Glra2* and *Amg*. They observed reduced variability in *M. m. domesticus* at *Abpa27* relative to *Amg* or *Glra2*, consistent with a recent adaptive fixation (a selective sweep) at or near *Abpa27*. They also made interspecific comparisons among *M. m. domesticus*, *M. m. musculus*, *M. m. castaneus*, *M. spretus*, *M. spicilegus*, and *M. caroli*, which showed that the ratio of non-synonymous substitutions to synonymous substitutions on a per-site basis (*Ka*/*Ks*) was generally greater than one. The combined observations of no variation at *Abpa27* within *M. m. domesticus* and uniformly high *Ka*/*Ks* values between species suggest that positive directional selection has acted recently at this locus. Karn *et al.* (2002) characterized the exon and intron sequence variations of *Abpa27* in house mouse populations from central Eurasia, a region not previously sampled and thought to be close to the cradle of the radiation of the subspecies. They confirmed the general pattern of rapid evolution by essentially non-synonymous substitutions, both inter- and intraspecifically, supporting the idea that Darwinian selection has driven the evolution of this gene. Their observation of a single 998-bp *Abpa27*^{b} sequence, very different from that of *Abpa27*^{a}, in five samples of the *M. m. musculus* populations south of the Caucasus reinforced the hypothesis advanced by Karn and Nachman (1999) that allelic sweeps could account for the significantly reduced *Abpa27* intron polymorphism in peripheral subspecies.

Karn and Laukaitis (2003) reported the first observations of elevated *Ka*/*Ks* in *Abpbg27* (formerly *Abpb*) and *Abpbg26* (formerly *Abpg*) in a pairwise comparison of cDNAs representing the three subspecies of *Mus musculus*. They concluded that *Abpbg27* and *Abpbg26* have also evolved under positive selection. Using a different approach, Laukaitis *et al.* (2003) found supporting evidence for this conclusion; they calculated and compared distances in terms of nucleotide divergences in coding and non-coding regions of *Abpbg27* and *Abpbg26*. In most proteins, the divergence in the coding regions is expected to be less than that in the non-coding regions because of purifying (negative) selection acting on the amino acid sequence of the gene product (thus, there is less constraint expected on non-coding regions, i.e. introns (Hughes and Yeager, 1997)). Positive selection for changes in protein sequence, on the other hand, could create a situation in which the divergence in the coding regions is actually greater than in the non-coding regions. This was in fact the case with *Abpbg26* and *Abpbg27*, the two *Abpbg* genes studied by Laukaitis *et al.* (2003). They observed no divergence at all in the coding regions for the signal peptides. The divergences for the coding regions of Exons 2 and 3, on the other hand, are significantly higher than those for either intron. This suggests that positive selection has acted strongly on the coding regions for the secreted proteins, whereas perhaps the coding regions for the signal peptides have been under purifying selection. An alternative explanation, that gene conversion was responsible for the conservation of Exon 1 (5′ UTR and signal peptide), has been found to be unlikely in light of later work (see above).

Emes *et al.* (2004) used codeml to predict codons ('ω+ sites') that have been subjected to positive selection. Five and 28 such ω+ sites were identified for mouse *Abpa*-like or *Abpbg*-like homologues, respectively, demonstrating that adaptation has been the dominant evolutionary force during recent divergence of these rodent homologues. They mapped these ω+ sites to the crystal structure of the cat Fel dI heterodimer (Fig. 7.9). Although predicted ω+ sites appear to be relatively uniformly distributed along the protein sequence, a clear pattern is revealed when the sites are examined in relation to the predicted three-dimensional structure: they are overrepresented among exposed amino acids and they form a pronounced cluster on one face of the heterodimer (Fig. 7.9). Thus, the majority of these ω+ sites are available to participate in binding interactions. Moreover, they are entirely absent from the dimer's internal cavity and are distinct from sites in uteroglobin thought to bind internal ligands (Callebaut *et al.*, 2000; Kaiser *et al.*, 2003; Karn and Laukaitis, 2003). Several secretoglobins are known to bind hydrophobic ligands within an interior cavity formed at the interface of two homologous chains (Beato and Baier, 1975; Parker *et al.*, 1982; Dlouhy and Karn, 1983; Callebaut *et al.*, 2000), and the cat Fel dI crystal structure contains strong electron density within this cavity, indicating the presence of a bound ligand of unknown type (Kaiser *et al.*, 2003). Our results suggest that positive selection has not acted strongly on sites that confer specificity on binding internal ligands. Rather, it has constantly remodelled a single face of the heterodimer that consists of all but one of the five alpha-helices of *Abpbg*-like homologues.

A very different perspective of rapid evolution of the *Abp* region of the mouse genome comes from the observations of Karn and Laukaitis (2009), who used the Mouse Paralogy Browser to evaluate the nature of the evolution of the three mouse pheromonal gene regions, *Abp*, *ESP*, and *MUP*. They observed that

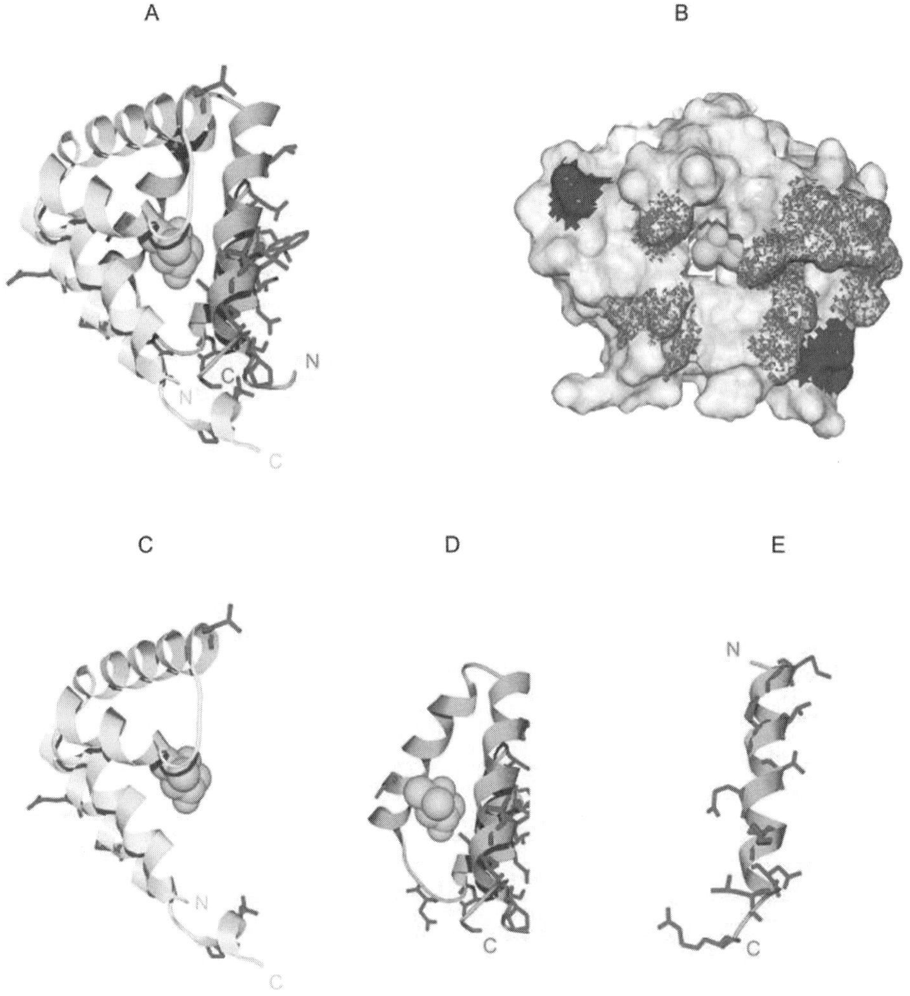

Figure 7.9 Site-specific K_A/K_S analysis of *Abpa*-like and *Abpbg*-like genes mapped to the tertiary structure of cat Fel dI (Emes *et al.*, 2004). ω+ codons that are predicted to be under positive selection for mouse paralogues are mapped to a ribbon representation of the tertiary structure of feline major allergen Fel dI (PDB 1POU; Kaiser *et al.*, 2003). (a) Ribbon diagram of the Fel dI dimer. Side chains of mapped predicted ω+ sites are highlighted in red (see plate section). A single ligand molecule (2-methyl-2,4-pentanediol) present in the crystal structure is coloured orange. (b) A representation of the molecular surface of the Fel dI chain 1 and 2 heterodimer. Mapped ABPα ω+ sites are shown in blue, and ABPβγ ω+ sites are shown in red. (c) Mapped ω+ sites predicted for the *Abpa*-like mouse paralogues. (d) Mapped ω+ sites predicted for the *Abpbg*-like mouse paralogues. (e) Prediction of the ABPβγ terminal extension and the mapped location of identified ω+ sites. The ABPβγ terminal extension is predicted to form a helical structure of 24 residues and is shown to scale. Swiss-PDBviewer (http://www.expasy.org/spdbv; Guex *et al.*, 1999) was used for all structural manipulations, and POVRAY (http://www.povray.org) was used to generate images. See plate section for a colour version of this figure.

evolution of the *Abp* region has been volatile, resulting in extensive gene birth and death, probably accelerated by the ability to duplicate and delete numerous paralogues in large blocks (see above). By contrast, they did not find evidence for any significant divergence of the *ESP* or *MUP* gene regions in any one of the three *M. musculus* subspecies in the data they obtained from the Mouse Paralogy Browser. This led them to conclude that there has been much less volatility in terms of recent gene birth and death in the evolutionary histories of these two families of pheromone genes than in the *Abp* family. This is consistent with the communication attributes of the three families of pheromones, because ABP is the only one for which a role in mate selection based on subspecies recognition has been suggested (Laukaitis *et al.*, 1997; Talley *et al.*, 2001; Bímová *et al.*, 2005). Karn and Laukaitis (2009) contended that the contrast in volatility between the *Abp* gene region and the other two pheromone gene regions supports the idea of a very different communication role for mouse ABPs, one involving intersubspecific and perhaps interspecific recognition.

Expression of *Abp* paralogues

Laukaitis *et al.* (2005) designed primers for the *Abpa* and *Abpbg* paralogues described by Emes *et al.* (2004) as likely to be expressed, and used their primer sets to search for transcripts in cDNA libraries of various organs and glands of C3H mice. They observed expression of six *Abpa* paralogues and five *Abpbg* paralogues in ten glands and other organs located predominantly in the head and neck (olfactory lobe of the brain, three salivary glands, lacrimal gland, Harderian gland, vomeronasal organ, and major olfactory epithelium) (Fig. 7.10). These *Abp* paralogues differed dramatically in their specific expression in these different glands and in their sexual dimorphism of expression. They also studied the appearance of expression in both late-stage embryos and postnatal animals prior to puberty and found significantly different timing of the onset of expression among the various paralogues. The developmental sequence of expression of these genes provides clues that help further narrow the choices of functional role for ABP. *Abpa/Abpbg* expression begins in the head of the embryo as early as 13 days of gestation, and comprises a relatively simple pattern of *Abpa11/Abpbg11* and *Abpbg10* (now *Abpa27/Abpbg27* and *Abpbg26*; see Table 7.1). By the first day of life, this has changed little, but by six days many other paralogues are expressed in the submaxillary and lacrimal glands. Expression proceeds to 15 days of life without yet revealing the sexually dimorphic patterns seen in several glands in adult mice. The transition to sexually dimorphic expression following puberty appears incompatible with roles for these in host immune defence. If the variety of proteins produced by expression of the mouse *Abp* paralogues act as components

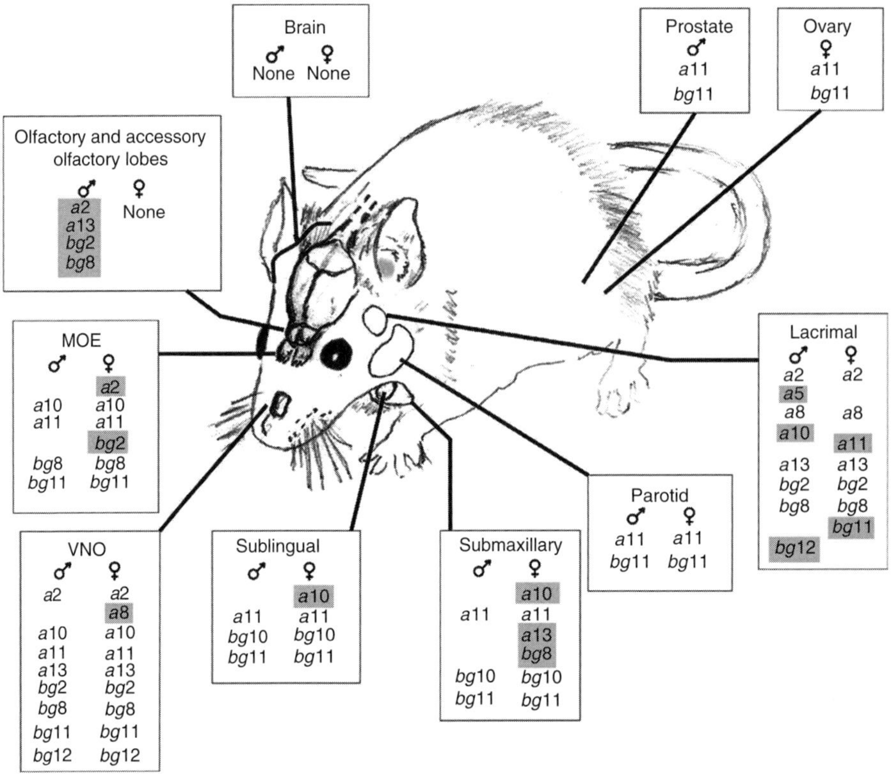

Figure 7.10 *Abp* paralogue expression shown on a diagram of the glands/organs of a mouse, with expression in males and females differentiated (Laukaitis and Karn, 2005). Paralogues which exhibited sexually dimorphic expression are shaded in grey.

of a complex pheromone system, then the secretions of the lacrimal gland and salivary glands provide a constellation of olfactory information, such as an animal's age and sex. Indeed, it has been proposed previously that polymorphism in salivary ABP communicates the species/subspecies of the animal (Karn and Dlouhy, 1991; Hwang *et al.*, 1997; Laukaitis *et al.*, 1997; Talley *et al.*, 2001), and Laukaitis *et al.* (2008) suggested that the strong evidence for adaptive evolution of *Abp* paralogous genes is consistent with a pheromonal role for their protein products.

Future directions

A number of studies have identified the mouse VNO as the tissue containing receptors for pheromones (reviewed in Baum and Kelliher, 2009; Keller *et al.*, 2009). Karn *et al.* (2010) recently used a genetic approach to look for

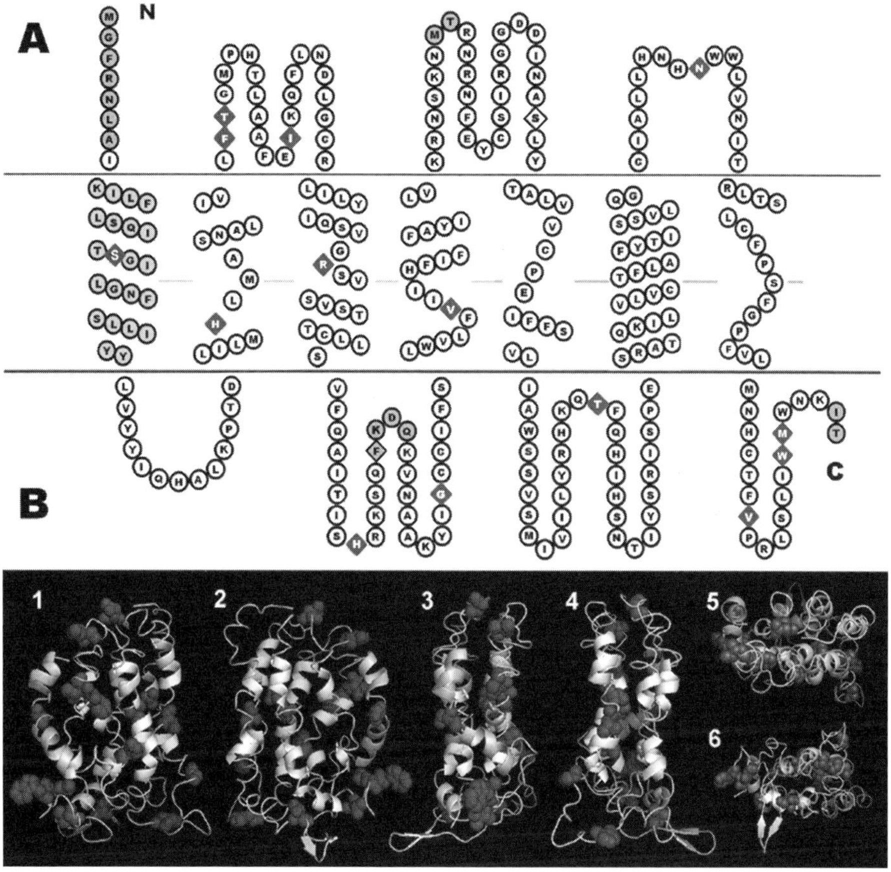

Figure 7.11 (a) Diagrammatic representation of the amino acid sequence of *Vmn1r67* and (b) the three-dimensional structure on which it was threaded. Amino acids in (a) that were not threaded on the *d1ln6a* model in (b) are shaded in grey (Karn *et al.*, 2010). The amino acid positions affected by 15 non-synonymous differences fixed between *M. m. domesticus* and *M. m. musculus* are shown as diamonds. Fourteen non-synonymous sites detected by a test for homoplasy are coloured red (see plate section). For orientation to the three-dimensional models in (b), the first helix is coloured light orange. Panel B views are (1) front, (2) rear, (3) left side, (4) right side, (5) top and (6) bottom. See plate section for a colour version of this figure.

a candidate receptor(s) for ABP. They studied the mouse *V1R* genes, which encode a large family of receptors in the VNO, by screening Perlegen SNP data and identified one, *Vmn1r67*, with 24 fixed SNP differences, most of which (15 of 24) are non-synonymous nucleotide substitutions between *M. m. domesticus* and *M. m. musculus* (Fig. 7.11). They observed substantial linkage disequilibrium (LD) between *Vmn1r67* and *Abpa27*, a mouse salivary androgen-binding-protein gene that encodes a proteinaceous pheromone (ABP) capable of mediating assortative mating, perhaps in conjunction with its bound small lipophilic ligand (Fig. 7.12). The LD they observed is likely a case of association rather than residual physical linkage from a very recent selective sweep, because an intervening gene, *Vmn1r71*, shows significant intra(sub)specific polymorphism but no inter(sub)specific divergence in its nucleotide sequence. Karn *et al.* (2010) discuss alternative explanations of these observations – for example that *Abpa27* and *Vmn1r67* are coevolving as signal and receptor to reinforce subspecies hybridization barriers or that the unusually divergent *Vmn1r67* allele was not a product of fast positive selection, but was derived from an introgressed allele, possibly from *Mus spretus*.

While their approach was an indirect one, it had the efficacy of identifying candidate receptor genes that can then be subjected to other studies, including experiments with mouse strains in which either the signal gene (e.g. *Abpa27*) and/or the receptor gene (*Vmn1r67* in the case of their study) has been knocked out. Additional or alternative experiments involving mate-choice behaviour, using subject mice with recombinant *Vmn1r67–Abpa27* haplotypes may not only reinforce the conclusion that the *Vmn1r67* receptor recognizes the ABP signal, but may also reveal more details about the interaction of the receptor and the signal. In any event, they concluded that future work should also involve direct electrophysiological experiments that will corroborate the genetic identification of a VNO receptor for the ABP molecules proposed to advertise subspecies identity in house mice, and possibly species identity in other rodents of the genus *Mus*.

Conclusions

The various mouse pheromones identified to date are capable of communicating different kinds of information. But what is required for identification of a pheromone capable of communicating subspecies status, resulting in assortative mate choice that directly restricts gene exchange between the two subspecies? To answer this question, one must consider how behavioural tests are designed and what they ultimately test. Features of behavioural testing systems based on odours include the stimulus or signal, the response and the testing conditions or environment.

The stimulus is a source of the odour, which might be a live animal, a complex undefined mixture of odours such as urine or soiled bedding, or a chemically defined molecule or molecules. In some cases, fluids such as saliva or urine from genetically defined sources such as strains congenic for a particular gene on an otherwise homogeneous genetic background allow specification of the genetic source of the odour to the exclusion of other odours. The response is the behaviour being measured and is defined by the design of the experiments. For

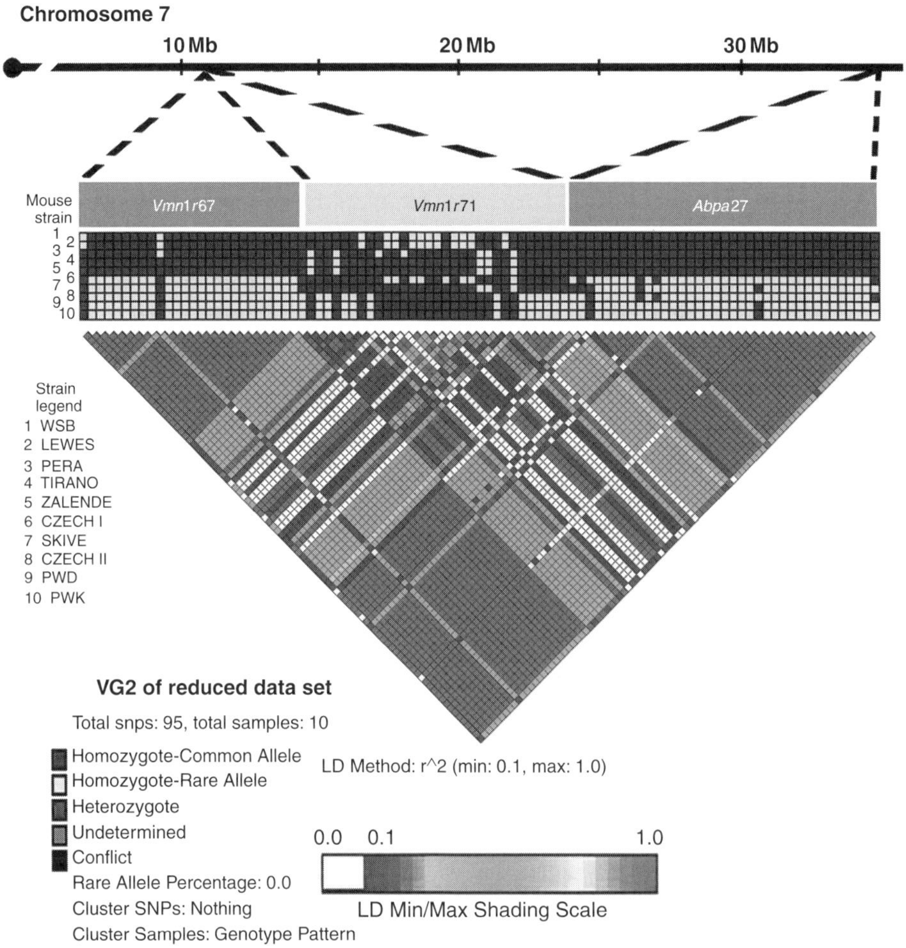

Figure 7.12 Linkage disequilibrium (LD) between *Vmn1r67* and *Abpa27*, with the gene *Vmn1r71* mapping between them (Karn *et al.*, 2010). The proximal end of mouse chromosome 7 shown at the top of the figure represents the relative positions of the three genes, which appear diagrammed as grey-shaded blocks below the chromosome and over their portion of the LD diagram. A comparison of the degree of LD, represented as r^2 values, appears in the triangle at the bottom of the figure and a shading scale for LD minimum to maximum appears under the LD diagram. The wild-derived inbred mouse strains used in this study are numbered on the left side of the LD diagram with a strain legend immediately below. The genes *Vmn1r67* and *Abpa27* show strong LD, while *Vmn1r71* shows no significant LD with either of them. See plate section for a colour version of this figure.

example, the response tested in a Y-maze is the preference between either of two odours. Other experiments are designed to test specific behaviours such as aggression, shown by tail rattling or mating behaviour such as mounting and lordosis. The conditions of the test are defined by the experimental design. Some examples are: testing under lights versus testing in darkened conditions (e.g. under red light); testing in enclosures set up in the wild versus open arenas or relatively confined Y-mazes in the laboratory, etc.

An enormous number of rodent behavioural studies have relied on complex undefined mixtures of odours such as urine or bedding. Informative as these studies might have been regarding the response of the subjects, the characteristics of the stimulus or stimuli (i.e. the chemical signal or signals) remain a mystery. The response could depend on either or both large (e.g. proteins) and small (e.g. steroids) molecules acting as signals, either collectively or independently. For example, a protein may be a carrier for a small pheromone molecule, binding it for transport and then releasing it when coupled to a receptor molecule. Genetic variability of particular amino acid sites may be important in some aspects of the protein's function in this relationship or the protein might be no more than a passive vehicle in terms of its contribution. There might even be antagonism in the effects of more than one molecule in the odour source. None of these things are revealed when the source is an undefined mixture of molecules.

One means of bringing some definition into such studies uses the power of genetics to isolate the stimulus when it involves the protein product of a gene. The requirement in such an instance is that two forms of the gene product can be differentiated in terms of their effect on the animal's behaviour. As described earlier, different alleles of *Abp* are fixed in each of the three subspecies of *Mus musculus*, and two of these have been made congenic in the same strain (C3H) to provide alternative signals in a number of different behavioural tests. Breeding strategies to develop such congenic strains represent a relatively traditional approach, while similar strains can now be produced with transgenic technology. In either case, the advantages are that (1) the signal is relatively well defined in relation to the genetic background, which is identical in the two strains; and (2) the potentially confounding effect of small molecules is essentially excluded from the tests.

Another means of exploring the chemical nature of putative signals in complex mixtures is to analyse their components through one or more types of biochemical analysis. Perhaps the simplest approach is to first determine whether the response is triggered by large molecules or small ones, two groups which are easily separated by a number of straightforward biochemical techniques (e.g. spin-column dialysis). If neither fraction elicits the response, remixing them may restore the effect, suggesting that the signal is a combination of small and large molecules. Protein components of the source can be analysed by one or more of several different mass spectrometer-based proteomics analyses. Relatively small amounts of protein are required for such analyses. Evaluation of the protein complement of the complex odour source may provide clues to the most likely candidates for molecules involved in communication. Modern biochemical separation techniques such as high-performance liquid chromatography and analysis options require extremely small amounts of material, usually in the nanomole or picomole range. These kinds of approaches make possible the isolation and characterization of putative signalling components of complex sources of odours, which may have provided interesting preliminary results in behavioural experiments.

The methods described above are particularly applicable to defining the components of odour sources proposed to mediate recognition of taxon status at a zone of secondary contact between two different taxa such as the house mouse hybrid zone in Europe. Ultimately, differentiation of at least one characteristic of a chemical signal such that a different form is possessed by each of the two taxa in contact is necessary to communicate their status. Behavioural studies relying on urine or bedding as odour sources may be a reasonable way to start a study aimed at identifying signals involved in subspecies recognition in mice, but without eventual definition of the signal responsible, they will be of limited use. The only chemical cue that has been shown to differentiate the subspecies of *M. m. domesticus* and *M. m. musculus* is ABP. Although urine has been proposed to have a role in mediating subspecies recognition in the European house mouse hybrid zone, no component with molecular characteristics that potentially could differentiate the two subspecies has yet been described.

Finally, conclusions about the significance of a particular response elicited by a putative chemical cue must be either obvious from the response or justified by additional observations that further define the response. For example, a response in a male–male aggression test may obviously relate to some aspect of aggression. By contrast there is no guarantee that preference for an odour in a Y-maze would lead to preferential mating between the subject and the animal that provided the signal unless that has been demonstrated in some other way. Y-maze tests have the advantage of being relatively simple in design and therefore allowing rapid accumulation of preference data compared to more complex tests, which otherwise more closely simulate the actual mate-choice environment. Yet a more complex test that more closely simulates the natural environment may enhance the preference ratio substantially. More importantly, only those tests in which both males and females are present are capable of demonstrating that a decision to mate with the animal that provided the preferred odour source is the actual outcome of preference (see the description of various types of test in the case of ABP preference above and Table 7.2). Claims that a signal mediates assortative mating in an encounter between subspecies must be validated by observations that the female subject chooses to mate with the consubspecific male as an outcome of preference for the signal he possesses. Otherwise, there is no assurance that preference for an odour will directly result in restricting gene exchange between the two subspecies.

ABP is the only communication system thus far described that meets all these criteria for mediating recognition of subspecies status. Here we have summarized the recent evidence that it is involved in the evolution of incipient reinforcement on the European house mouse hybrid zone.

REFERENCES

Achaz, G., Coissac, E., Viari, A., and Netter, P. (2000). Analysis of intrachromosomal duplications in yeast *Saccharomyces cerevisiae*: a possible model for their origin. *Molecular Biology and Evolution*, **17**, 1268–75.

Achaz, G., Netter, P., and Coissac, E. (2001). Study of intrachromosomal duplications among the eukaryotic genomes. *Molecular Biology and Evolution*, **18**, 2280–8.

Armstrong, S., Robertson, D., Cheetham, S., Hurst, J., and Beynon, R. (2005). Structural and functional differences in isoforms of the major urinary proteins: a male-specific protein that preferentially binds a male pheromone. *Biochemical Journal*, **391**, 343–50.

Baum, M. and Kelliher, K. (2009). Complementary roles of the main and accessory olfactory systems in mammalian mate recognition. *Annual Review of Physiology*, **71**, 141–60.

Beato, M. and Baier, R. (1975). Binding of progesterone to the proteins of the uterine luminal fluid. Identification of uteroglobin as the binding protein. *Biochimica et Biophysica Acta*, **392**, 346–54.

Beynon, R. and Hurst, J. (2003). Multiple roles of major urinary proteins in the house mouse, *Mus domesticus*. *Biochemical Society Transactions*, **31**, 142–6.

Bímová, B., Karn, R., and Piálek, J. (2005). The role of salivary androgen-binding protein in reproductive isolation between two subspecies of house mouse: *Mus musculus musculus* and *Mus musculus domesticus*. *Biological Journal of the Linnean Society*, **84**, 349–61.

Boursot, P., Auffray, J.-C., Britton-Davidian, J., and Bonhomme, F. (1993). The evolution of house mice. *Annual Review of Ecology and Systematics*, **24**, 119–52.

Callebaut, I., Poupon, A., Bally, R., *et al.* (2000). The uteroglobin fold. *Annals of the New York Academy of Sciences*, **923**, 90–112.

Chamero, P., Marton, T., Logan, D., *et al.* (2007). Identification of protein pheromones that promote aggressive behaviour. *Nature*, **450**, 899–902.

Cheetham, S., Thom, M., Jury, F., *et al.* (2007). The genetic basis of individual recognition signals in the mouse. *Current Biology*, **17**, 1771–7.

Chen, J. M., Chuzhanova, N., Stenson, P. D., Ferec, C., and Cooper, D. N. (2005). Meta-analysis of gross insertions causing human genetic disease: novel mutational mechanisms and the role of replication slippage. *Human Mutation*, **25**, 207–21.

Dlouhy, S. R. and Karn, R. C. (1983). The tissue source and cellular control of the apparent size of androgen binding protein (Abp), a mouse salivary protein whose electrophoretic mobility is under the control of sex-limited saliva pattern (Ssp). *Biochemical Genetics*, **21**, 1057–70.

Dlouhy, S. R. and Karn, R. C. (1984). Multiple gene action determining a mouse salivary protein phenotype: identification of the structural gene for androgen binding protein (Abp). *Biochemical Genetics*, **22**, 657–67.

Dlouhy, S. R., Nichols, W. C., and Karn, R. C. (1986). Production of an antibody to mouse salivary androgen binding protein (ABP) and its use in identifying a prostate protein produced by a gene distinct from Abp. *Biochemical Genetics*, **24**, 743–63.

Dlouhy, S. R., Taylor, B. A., and Karn, R. C. (1987). The genes for mouse salivary androgen-binding protein (ABP) subunits alpha and gamma are located on chromosome 7. *Genetics*, **115**, 535–43.

Dod, B., Smadja, C., Karn, R., and Boursot, P. (2005). Testing for selection on the androgen-binding protein in the Danish mouse hybrid zone. *Biological Journal of the Linnean Society*, **84**, 447–59.

Dominguez, P. (1995). Cloning of a Syrian hamster cDNA related to sexual dimorphism: establishment of a new family of proteins. *FEBS Letters*, **376**, 257–61.

Dulac, C. and Torello, A. (2003). Molecular detection of pheromone signals in mammals: from genes to behaviour. *Nature Reviews Neuroscience*, **4**, 551–62.

Dulac, C. and Wagner, S. (2006). Genetic analysis of brain circuits underlying pheromone signaling. *Annual Review of Genetics*, **40**, 449–67.

Emes, R. D., Riley, M. C., Laukaitis, C. M., *et al.* (2004). Comparative evolutionary genomics of androgen-binding protein genes. *Genome Research*, **14**, 1516–29.

Ferrero, D. M. and Liberles, S. D. (2010). The secret codes of mammalian scents. *Wiley Interdisciplinary Reviews: Systems Biology and Medicine*, **2**, 23–33.

Galtier, N., Duret, L., Glemin, S., and Ranwez, V. (2008). GC-biased gene conversion promotes the fixation of deleterious amino acid changes in primates. *Trends in Genetics*, **25**, 1–5.

Galtier, N., Piganeau, G., Mouchiroud, D., and Duret, L. (2001). GC-content evolution in mammalian genomes: the biased gene conversion hypothesis. *Genetics*, **159**, 907–11.

Guex, N., Diemand, A., and Peitsch, M. C. (1999). Protein modelling for all. *Trends in Biochemical Sciences*, **24**, 364–7.

Howard, D. (1993). Reinforcement: the origin, dynamics, and fate of an evolutionary hypothesis. In *Hybrid Zones and the Evolutionary Process*, ed. R. Harrison. New York: Oxford University Press, pp. 46–69.

Hughes, A. and Yeager, M. (1997). Comparative evolutionary rates of introns and exons in murine rodents. *Journal of Molecular Evolution*, **45**, 125–30.

Hurst, J. L. (2009). Female recognition and assessment of males through scent. *Behavioural Brain Research*, **200**, 295–303.

Hwang, J. M., Hofstetter, J. R., Bonhomme, F., and Karn, R. C. (1997). The microevolution of mouse salivary androgen-binding protein (ABP) paralleled subspeciation of *Mus musculus*. *Journal of Heredity*, **88**, 93–7.

Kaiser, L., Gronlund, H., Sandalova, T., *et al.* (2003). The crystal structure of the major cat allergen Fel dI, a member of the secretoglobin family. *The Journal of Biological Chemistry*, **278**, 37730–5.

Karn, R. C. (1981). Sex-limited effects of the expression of the *db* gene in mice during puberty. *Biochemical Genetics*, **19**, 355–63.

Karn, R. C. (1991). Genetic control of mammalian salivary proteins. In *Aspects of Oral Molecular Biology*, ed. D. Ferguson. Basel: S. Karger.

Karn, R. C. (1994). The mouse salivary androgen-binding protein (ABP) alpha subunit closely resembles chain 1 of the cat allergen Fel dI. *Biochemical Genetics*, **32**, 271–7.

Karn, R. C. and Dlouhy, S. R. (1991). Salivary androgen-binding protein variation in *Mus* and other rodents. *Journal of Heredity*, **82**, 453–8.

Karn, R. C., Dlouhy, S. R., Springer, K. R., Hjorth, J. P., and Nielsen, J. T. (1982). Sex-limited genetic variation in a mouse salivary protein. *Biochemical Genetics*, **20**, 493–504.

Karn, R. C. and Laukaitis, C. M. (2003). Characterization of two forms of mouse salivary androgen-binding protein (ABP): implications for evolutionary relationships and ligand-binding function. *Biochemistry*, **42**, 7162–70.

Karn, R. C. and Laukaitis, C. M. (2009). The mechanism of expansion and the volatility it created in three pheromone gene clusters in the mouse (*Mus musculus*) genome. *Genome Biology and Evolution*, **1**, 494–503.

Karn, R. C. and Nachman, M. W. (1999). Reduced nucleotide variability at an androgen-binding protein locus (*Abpa*) in house mice: evidence for positive natural selection. *Molecular Biology and Evolution*, **16**, 1192–7.

Karn, R. C., Orth, A., Bonhomme, F., and Boursot, P. (2002). The complex history of a gene proposed to participate in a sexual isolation mechanism in house mice. *Molecular Biology and Evolution*, **19**, 462–71.

Karn, R. C. and Russell, R. (1993). The amino acid sequence of the alpha subunit of mouse salivary androgen-binding protein (ABP), with a comparison to the partial sequence of the beta subunit and to other ligand-binding proteins. *Biochemical Genetics*, **31**, 307–19.

Karn, R. C., Young, J., and Laukaitis, C. M. (2010). A candidate subspecies discrimination system involving a vomeronasal receptor gene with different alleles fixed in *M. m. domesticus* and *M. m. musculus*. *PLoS ONE*, **5**, e12638.

Katju, V. and Lynch, M. (2003). The structure and early evolution of recently arisen gene duplicates in the *Caenorhabditis elegans* genome. *Genetics*, **165**, 1793–803.

Keller, M., Baum, M., Brock, O., Pa, B., and Bakker, J. (2009). The main and the accessory olfactory systems interact in the control of mate recognition and sexual behaviour. *Behavioural Brain Research*, **200**, 268–76.

Kimoto, H., Haga, S., Sato, K., and Touhara, K. (2005). Sex-specific peptides from exocrine glands stimulate mouse vomeronasal sensory neurons. *Nature*, **437**, 898–901.

Kimoto, H., Sato, K., Nodari, F., *et al.* (2007). Sex- and strain-specific expression and vomeronasal activity of mouse ESP family peptides. *Current Biology*, **17**, 1879–84.

Kimoto, H. and Touhara, K. (2005). Induction of c-Fos expression in mouse vomeronasal neurons by sex-specific non-volatile pheromone(s). *Chemical Senses*, **30** (Suppl. 1), i146–7.

Kondrashov, F. and Kondrashov, A. (2006). Role of selection in fixation of gene duplications. *Journal of Theoretical Biology*, **239**, 141–51.

Laukaitis, C. M., Critser, E. S., and Karn, R. C. (1997). Salivary androgen-binding protein mediates sexual isolation in *Mus musculus*. *Evolution*, **51**, 2000–5.

Laukaitis, C. M., Dlouhy, S. R., Emes, R. D., Ponting, C. P., and Karn, R. C. (2005). Diverse spatial, temporal, and sexual expression of recently duplicated androgen-binding protein genes in *Mus musculus*. *BMC Evolutionary Biology*, **5**, 40.

Laukaitis, C. M., Dlouhy, S. R., and Karn, R. C. (2003). The mouse salivary androgen-binding protein (ABP) gene cluster on chromosomes 7: characterization and evolutionary relationships. *Mammalian Genome*, **14**, 679–91.

Laukaitis, C. M., Heger, A., Blakley, T. D., *et al.* (2008). Rapid bursts of androgen-binding protein (Abp) gene duplication occurred independently in diverse mammals. *BMC Evolutionary Biology*, **8**, 46–62.

Laukaitis, C. M. and Karn, R. C. (2005). Evolution of the secretoglobins: a genomic and proteomic view. *Biological Journal of the Linnean Society*, **84**, 493–501.

Logan, D., Marton, T., and Stowers, L. (2008). Species specificity in major urinary proteins by parallel evolution. *PLoS ONE*, **3**, e3280.

Lupski, J. (1998). Genomic disorders: structural features of the genome can lead to DNA rearrangements and human disease traits. *Trends in Genetics*, **14**, 417–22.

Margulies, E., Maduro, V., Thomas, P., *et al.* (2005). Comparative sequencing provides insights about the structure and conservation of marsupial and monotreme genomes. *Proceedings of the National Academy of Sciences of the United States of America*, **102**, 3354–9.

Mombaerts, P. (2004). Genes and ligands for odourant, vomeronasal and taste receptors. *Nature Reviews Neuroscience*, **5**, 263–78.

Mudge, J., Armstrong, S., McLaren, K., Beynon, R., and Hurst, J. (2008). Species specificity in major urinary proteins by parallel evolution. *Genome Biology*, **9**, R91.

Mukherjee, A. B. and Chilton, B. S. (2000). The uteroglobin/Clara cell protein family: Nomenclature Committee Report. *Annals of the New York Academy of Sciences*, **923**, 348–56.

Mukherjee, A. B., Zhang, Z., and Chilton, B. S. (2007). Uteroglobin: a steroid-inducible immunomodulatory protein that founded the secretoglobin superfamily. *Endocrine Reviews*, **28**, 707–25.

Naveira, H. and Barbadilla, A. (1992). The theoretical distribution of lengths of intact chromosome segments around a locus held heterozygous with backcrossing in a diploid species. *Genetics*, **130**, 205–9.

Parker, M., Needham, M., and White, R. (1982). Prostatic steroid binding protein: gene duplication and steroid binding. *Nature*, **298**, 92–4.

Robertson, D., Cox, K., Gaskell, S., Evershed, R., and Beynon, R. (1996). Molecular heterogeneity in the major urinary proteins of the house mouse *Mus musculus*. *Biochemical Journal*, **316**, 265–72.

Sage, R., Atchley, W., and Capanna, E. (1993). House mice as models in systematic biology. *Systematic Biology*, **42**, 523–61.

Shaffer, L. and Lupski, J. (2000). Molecular mechanisms for constitutional chromosomal rearrangements in humans. *Annual Review of Genetics*, **34**, 297–329.

Stankiewicz, P. and Lupski, J. (2002). Genome architecture, rearrangements and genomic disorders. *Trends in Genetics*, **18**, 74–82.

Stowers, L., Holy, T. E., Meister, M., Dulac, C., and Koentges, G. (2002). Loss of sex discrimination and male–male aggression in mice deficient for TRP2. *Science*, **295**, 1493–500.

Talley, H. M., Laukaitis, C. M., and Karn, R. C. (2001). Female preference for male saliva: implications for sexual isolation of *Mus musculus* subspecies. *Evolution*, **55**, 631–4.

Tirindelli, R., Dibattista, M., Pifferi, S., and Menini, A. (2009). From pheromones to behaviour. *Physiological Reviews*, **89**, 921–56.

Touhara, K. and Vosshall, L. (2009). Sensing odourants and pheromones with chemosensory receptors. *Annual Review of Physiology*, **71**, 307–32.

Vošlajerová Bímová, B., Macholán, M., Baird, S., *et al.* (2011). Reinforcement selection acting on the European house mouse hybrid zone. *Molecular Ecology*, **20**, 2403–24.

Yang, H., Shi, P., Zhang, Y. P., and Zhang, J. (2005). Composition and evolution of the V2r vomeronasal receptor gene repertoire in mice and rats. *Genomics*, **86**, 306–15.

Young, J. M., Kambere, M., Trask, B. J., and Lane, R. P. (2005). Divergent V1R repertoires in five species: amplification in rodents, decimation in primates, and a surprisingly small repertoire in dogs. *Genome Research*, **15**, 231–40.

Young, J. M. and Trask, B. J. (2007). V2R gene families degenerated in primates, dog and cow, but expanded in opossum. *Trends in Genetics*, **23**, 212–15.

8

Mechanisms of chemical communication

PAVEL STOPKA, ROMANA STOPKOVÁ, AND
KATEŘINA JANOTOVÁ

Introduction

The house mouse is a standard model in human biomedical science because many of its genes have almost identical functions in mouse and man. This functional similarity stems from common ancestry within the clade Euarchontoglires. Despite evolutionary divergence in numerous traits, gene function is similar even in these distant euarchontoglirid lineages. Because most of the functional studies in mice had primary goals in medicine, there are areas of rodent biology that have been heavily under-studied. Among such areas, chemical communication was a prime example until it was noted that humans may synchronize their menstrual cycles in a similar fashion to rats (McClintock, 1984). Since then, most of the research on chemical communication has been performed on laboratory strains of mice and rats. Therefore, to fully understand and appreciate the various mechanisms of chemical communication in the contexts of mouse behavioural ecology and evolution it is necessary to study wild house mice in their natural habitats and contexts: the biology of inbred lines is insufficient, although it still remains a good initial scaffold.

There may be other reasons why the importance of mouse chemical communication has escaped earlier human attention. Mice have four times more olfactory G-protein coupled receptor (GPCR) genes than humans. Human investigators have little chance of imagining the importance of smell to such organisms. However, with senses dominated by vision, we can imagine brightness of signal depending on the signal volume and signal hues representing pheromonal blends.

Chemical signals, odorants, and pheromones

The olfactory system enables recognition of a wide spectrum of information mediated by various chemical substances of different origin and structure.

Evolution of the House Mouse, ed. Miloš Macholán, Stuart J. E. Baird, Pavel Munclinger, and Jaroslav Piálek. Published by Cambridge University Press.

These substances are often distinguished as *general odorants* and *chemical signals*. John Maynard Smith and David Harper (2003) defined signal as any structure or act that alters behaviour in other organisms and which evolved because of that effect and is effective because the receiver's response has also evolved. Odorants are produced by all living organisms, including plants, microbes, and animals – and even by inorganic materials – and provide important information about environment, food, potential danger, or even the health condition of an individual. The term *pheromone* is reserved for chemical signals that trigger specific evolved behavioural responses. According to Karlson and Lüscher (1959), a pheromone is a chemical substance that is secreted by one individual and received by another individual of the same species, inducing endocrine changes and stimulating innate behavioural responses. However, there are ambiguous aspects in the definition of pheromones (species specificity, volatility, innate response, and chemical composition – one substance versus pheromonal blend) that make this definition unclear, so it requires further clarification. For example, the display of instinctive behaviour along with the physiological changes can be elicited also by odorant molecules, which stimulate similar olfactory and neuronal pathways (Sam *et al.*, 2001). The former theories postulating that odorants are perceived specifically through olfactory receptors of the main olfactory epithelium (MOE), while receptors of the vomeronasal organ (VNO) are stimulated entirely by pheromones, are false simply because many studies have now provided evidence that the two systems overlap to some extent in the detection of both types of molecules (Trinh and Storm, 2003, 2004; Restrepo *et al.*, 2004; Wang *et al.*, 2006; Xu *et al.*, 2005; Brennan and Zufall, 2006; Marchlewska-Koj *et al.*, 2000). On the other hand, non-volatile proteinaceous substances with pheromonal function seemed to activate only V2 receptors in the VNO (Krieger *et al.*, 1999). Therefore, the main benefit of the VNO may be that animals are able to recognize even large molecules as chemical signals; however, this requires direct physical contact with the source of such information.

It is now obvious that in chemical communication animals use a variety of chemical signals that do not fit the original definition of 'pheromone' (Doty, 2003). Perhaps most important among these are so-called signature odours that are used to distinguish identity, kin, or species of signal releaser (Wyatt, 2005). On the basis of such information, animals choose an appropriate reproductive or social partner. Moreover, some responses to pheromones are dependent on such information. A classic example is the Bruce effect in which a female's early pregnancy is terminated by a strange male (or his odour) with whom she did not mate (Bruce, 1959). This effect requires that a particular female remembers the identity of her mating partner and recognizes him among others (Parkes and Bruce, 1961).

Chemical communication and behaviour

The key to unravelling complex interactions between individual mice is to understand the particular roles of emitted molecules in the modulation of the receiver's physiology and behaviour. In their natural environment, mice use a system of chemical signals and signature odours that reflect the intensity of space use and the composition of a social group. The most studied source of chemical signals and signature odours in mice is urine, deposited in the form of scent marks (Novotny *et al.*, 1980; Hurst, 1987; Drickamer, 1989). Dominant males mark the whole territory to advertise their ownership (Desjardins *et al.*, 1973; Hurst, 1990). All members of the group, however, deposit scent marks. Subordinates may use this to advertise their kin and/or lower social status to gain social tolerance from dominant individuals. The complex odour of 'home substrate' is, therefore, enriched by each individual, thus yielding social tolerance and stability of the group (Hurst *et al.*, 1993).

In a harmonious 'mouse world' any representative of an inappropriate odour can easily be detected and is usually ejected or eradicated because scent marks also (if not primarily) serve as input information during mate choice, having enormous modulatory consequences on the reproductive physiology of individuals of both sexes (Drickamer, 1992; Novotny *et al.*, 1999b; Daev and Dukelskaya, 2003). To an observer, it may seem difficult to imagine mice have any rules of decision making during mating because they easily mate (even) with relatives under laboratory conditions, thus providing science with numerous inbred lines. Still, one can imagine at least three levels of resolution of an optimum signal that generates fitness benefits for a particular individual. It is likely that social status is of prime importance to ensure high-quality offspring. For the sake of energy saving, it is important to quickly detect individuals of the relevant (opposite) sex. Males save effort by quick detection of female temporal reproductive condition (i.e. the receptive phase of the oestrous cycle). Females that quickly detect and identify interested males may have the opportunity to avoid non-preferred matings. Last, but not least, mice avoiding heterospecific or heterosubspecific interbreeding may avoid production of hybrids with lower viability, reduced fertility, or even sterility (Ptacek, 2000; Johansson and Jones, 2007).

Chemical signals and pheromones involved at each level of mate choice may be invariable or contiguous rather than mutually exclusive (Ryan and Rand, 1993). Thus, the same compound may be used to transmit information facilitating decisions at two or even all three levels. Accordingly, several volatiles produced by male mice have been shown to provide both mate assessment and mate recognition signals – see Table 8.1. Their production is known to be associated with social status

Table 8.1 Pheromones and chemical signals of the house mouse (Beynon and Hurst, 2003; Novotny, 2003; Lin *et al.*, 2005; Spehr *et al.*, 2006).

Structure	Name	Bound to MUPs	Detection system	Source	Sex	Function
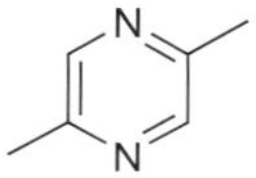	2,5-dimethylpyrazine	Yes	VNO/ MOE	Urine	♀	Puberty delay, impaired spermiogenesis
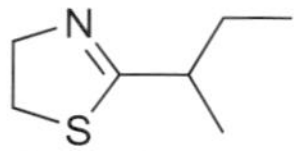	2-*sec*-butyl-4,5-dihydrothiazol	Yes	VNO	Urine	♂	Puberty acceleration, oestrus synchronization, male aggression, female attraction
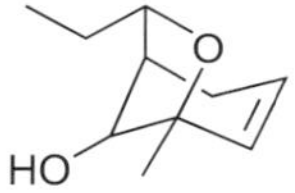	2,3-dehydro-exo-brevicomin	Yes	VNO	Urine	♂	Puberty acceleration, oestrus synchronization, male aggression
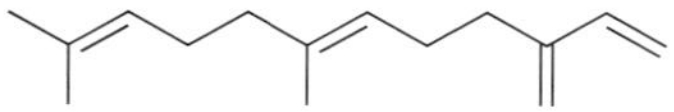	E-b-farnesene and	Yes	VNO	Preputial gland	♂	Oestrus synchronization, puberty acceleration, male aggression
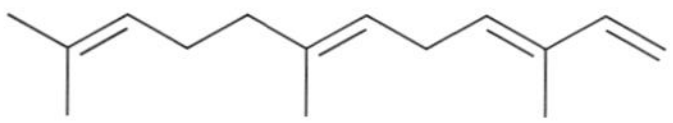	E,E-a-farnesene					
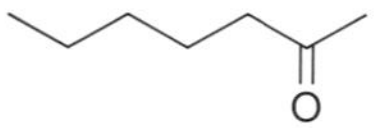	2-heptanone	Yes	VNO/ MOE	Urine	♀	Oestrus synchronization
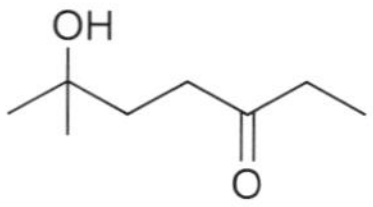	6-hydroxy-6-methyl-3-heptanone	Yes	VNO/ MOE	Urine	♂	Puberty acceleration
	methylthiomethanethiol (MTMT)	No	MOE	Urine	♂	Female attraction
e.g. SYFPEITHI; AAPDNRETF	MHC peptides	No	VNO/ MOE	Urine	♀/♂	Signature odour (Bruce effect)
	MUP	No	VNO	Urine, saliva and other secretions	♀/♂	Male aggression, influences on reproductive physiology, signature odour

and promotes competitive behaviour between males, thus allowing particular females to assess mate qualities (Novotny *et al.*, 1985, 1990). Simultaneously, these compounds act as female attractants and elicit profound changes in their reproductive physiology, which lead to increased reproductive receptivity (Jemiolo *et al.*, 1985; Novotny *et al.*, 1999a). Solid evidence for quantitative variation of volatiles has been reported in two subspecies of the house mouse, *M. m. musculus* and *M. m. domesticus*, which may mean that they represent subspecific recognition signals (Mucignat-Caretta *et al.*, 2010). Therefore, mate choice can be seen as a continuous rather than step-by-step evaluation process which results in finding the best of all potential partners (Johansson and Jones, 2007).

Mating is a very complex process that involves various levels of signal filtering, resulting in the production of offspring of which the quality depends at least to some extent on the innate level of discrimination skills of particular mating partners. Fitness benefits might have driven the evolution of the variety of sex pheromones that are detected during mating and which have additive modulatory functions upon the receiver's physiology (Novotny *et al.*, 1999b). Such effects are mostly demonstrated on females since reproduction is more costly for females than for males, who often provide only sperm (Davies, 1991). Therefore, females must modulate their reproductive physiology at each step of reproduction to prevent potential wasted investment from the very beginning of the process. For example, the cycle of female field mice (*Apodemus sylvaticus*) is shortened and oestrus prolonged when caged with males, but the effect is absent when caged alone and reversed when caged with other females (Stopka and Macdonald, 1998). In various microtine species males may even induce abortion in particular females (Stehn and Jannett, 1981). These effects are commonly and to various extents facilitated by many rodents, thus suggesting that sexual selection favours the synchronization of mates due to energy saving.

House mouse reproductive biology has been explored using inbred line experiments. Male signals have been repeatedly demonstrated to have major impacts on the onset, and variation in length, of the female receptive phase. While the presence of a male or his chemical signals shortens the female oestrous cycle and prolongs the oestrus phase, an all-female environment causes extended duration of the oestrous cycle with a very short phase of receptivity. These effects are called the Whitten and Lee–Boot effects and can be viewed as two aspects of one phenomenon that allows switching on and off receptivity and fertilizing capacity in females in relation to the probability of mating (Whitten, 1958; Lee and Boot, 1955). In juvenile mice, sexual maturation is accelerated (the Vandenbergh effect) or delayed in response to the same stimuli (Vandenbergh, 1969; Massey and Vandenbergh, 1980). The search for a pheromone responsible for the Whitten effect has shown that a number of male-derived signals have the potential to stimulate female physiology, including pheromone-carrying major urinary proteins (MUPs) (Jemiolo *et al.*, 1986; Ma *et al.*, 1999; More, 2006). Similarly, volatile signals that are responsible for the Lee–Boot effect are present in female urine, especially when housed in crowded conditions (Novotny *et al.*, 1986).

The abortion and rapid restoration of female receptivity by male pheromones (the Bruce effect) may be induced in the beginning of pregnancy. It is supposed that pregnancy blocking is a strategy for avoiding investment in offspring that have a high probability of infanticide by an unfamiliar non-sire male (Bruce, 1959). However, pregnancy failure may also be elicited by other stress factors, indicating that it may be a general response to a stressful environment rather than a specific response to the presence of unfamiliar non-sire males. Moreover, pregnancy block has been studied only in laboratory conditions; under natural conditions in wild house mice, where females usually encounter many males, the occurrence of this effect remains unclear (Thom and Beynon, 2006).

Chemical signals involved in pregnancy blocking have been studied using congenic strains of mice that differed in their major histocompatibility complex (MHC) (Yamazaki *et al.*, 1983). Later studies have supported that MHC underlies recognition of identity in the context of the Bruce effect. The authors showed that prototypical ligands of MHC class 1 molecules elicit pregnancy block when added to the urine of a sire male and that this response is mediated by vomeronasal neurons expressing V2R receptors (Leinders-Zufall *et al.*, 2004). Conversely, another study has found no overlap between the vomeronasal neurons activated by synthetic MHC class 1 peptides and those activated by the whole mouse urine (He *et al.*, 2008); moreover, the presence of MHC class 1 peptides in the urine has not been proved yet (Singh *et al.*, 1987; Hurst, 2009). The involvement of MHC in pregnancy blocking is, however, still possible through MHC-associated odours (Yamazaki *et al.*, 1983; Brennan and Peele, 2003). Interestingly, other results indicate that MUPs, another potential source of signature odours in male urine, are not involved in triggering the Bruce effect. Instead, it appears to be associated with volatiles contained in the urine itself (Bruce, 1965; Peele *et al.*, 2003). Moreover, the efficiency of pregnancy blocking appears to depend on testosterone levels (Bruce, 1965) and, therefore, it is unlikely that MHC class 1 peptides alone are able to trigger it. These data suggest that the Bruce effect can be caused by two separate signals, one responsible for identity information and the other carrying a signal on male status (Brennan and Peele, 2003).

The olfactory acuity of mice allows them to discriminate between odour stimuli that differ only in the expression of one or a few genes (Yamazaki *et al.*, 1983; Laukaitis *et al.*, 1997; Hurst *et al.*, 2001). It is, therefore, not surprising that olfactory cues contribute to behavioural isolation between *M. m. muculus* and *M. m. domesticus* (Christophe and Baudoin, 1998; Munclinger and Frynta, 2000;

Smadja and Ganem, 2002; see also Laukaitis and Karn, Chapter 7, and Ganem, Chapter 15 in this volume), which are known to interbreed in the unimodal hybrid zone occurring in Europe (Boursot *et al.*, 1993; see also Baird and Macholán, Chapter 14). A number of studies have been aimed at describing mating preferences in these two subspecies and their hybrids (Smadja and Ganem, 2002; Bímová *et al.*, 2005, 2009; Ganem *et al.*, 2005) and at identifying differences in their chemical profiles. However, little attention has been paid to the divergence in chemosensory organs that could lead to differential perception of chemical signals and thus reproductive isolation. As described later in this chapter, to date, there is just a single paper that shows that chemosensory receptors may be tuned to consubspecific signals (Karn *et al.*, 2010).

The search for subspecific chemical signals revealed divergent production of quite a few urinary chemical components between *M. m. musculus* and *M. m. domesticus*. These include urinary volatiles as well as MUPs (Stopková *et al.*, 2007; Mucignat-Caretta *et al.*, 2010). The profile of *M. m. domesticus* volatiles seems to be quantitatively richer than that of *M. m. musculus*. Moreover, F_1 hybrids of these subspecies reveal quite complex profiles of urinary volatiles that are not directly related to profiles of any of the parental species (Mucignat-Caretta *et al.*, 2010). Therefore, it is likely that there are differences in regulatory pathways that control the production of volatile compounds in each subspecies. Moreover, these results suggest that the subspecies recognition signature is not provided by the presence or absence of single components, but rather by a variation in their relative proportion in particular odour profiles (Mucignat-Caretta *et al.*, 2010).

The divergence in MUP expression in *M. m. musculus* and *M. m. domesticus* is both qualitative and quantitative. Qualitative differences in MUP profiles are apparent mainly in males, which express a more complex pattern of isoforms than females (P. Stopka *et al.*, unpublished data; Fig. 8.1). Quantitatively, the highest level of expression is in males of *M. m. musculus*. Females of both subspecies produce less MUPs than males (Wicks, 1941). However, for females, higher expression has been reported in *M. m. domesticus*. These results are corroborated also on the level of protein concentration in the urine (Stopková *et al.*, 2007). Interestingly, species specificity in MUP quantity has been demonstrated among various species of the genus *Mus*, with both subspecies of the house mouse having the highest values. Moreover, the level of sexual dimorphism differed in each of the tested species (Sampsell and Held, 1985; Beynon *et al.*, 2008). The results of these studies indicate that the concentration of MUPs reflects specific requirements of chemical signalling that might be different in each species with respect to its social structure.

It has been repeatedly demonstrated that in *M. m. domesticus* (in contrast to *M. m. musculus*) preferences for consubspecific individuals are much less pronounced or even absent (Christophe and Baudoin, 1998; Smadja and Ganem, 2002; Smadja

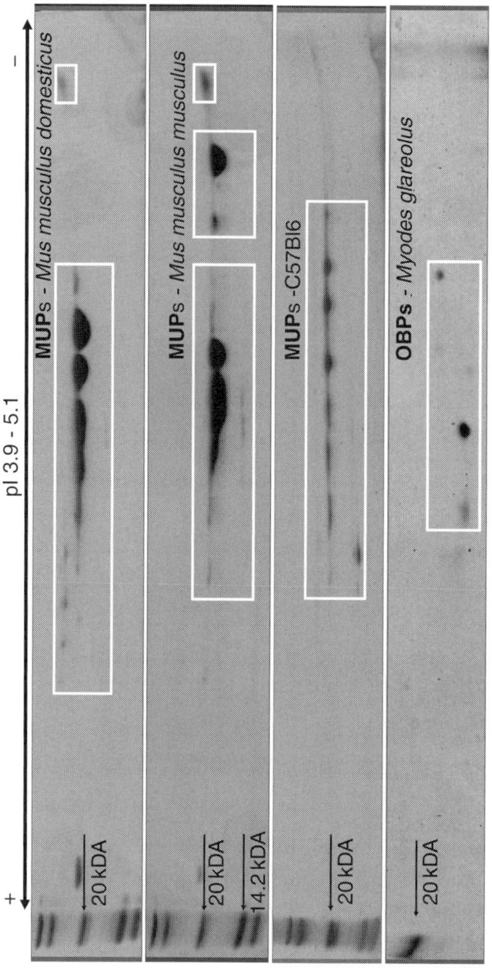

Figure 8.1 Urinary lipocalins: visualization of 2-DE analysis of urinary lipocalins reveals that mice of the genus *Mus* have evolved enhanced capacity to protect and transfer pheromones via major urinary proteins by the means of a high number of isoforms (i.e. gene variants) and high levels of expression. Note that particular isoforms (protein spots) in wild *Mus* have different levels of expression. On the other hand, some other rodent species such as *Myodes glareolus* use related lipocalins – aphrodisin-like odorant binding proteins (OBPs) for this purpose; however, the number of urinary OBP isoforms and the level of their expression are lower. The 2-DE gels were prepared and the proteins (in white boxes) identified with mass spectrometry techniques as described in Stopková *et al.* (2010).

et al., 2004). This asymmetrical pattern of divergence indicates that mating preferences have been influenced by specific selective pressures and that the cost of hybridization has differed in the two subspecies (Smadja and Ganem, 2005, 2008). Differential selection may be associated with the different levels of aggression observed in *M. m. musculus* and *M. m. domesticus* (Thuesen, 1977; van Zegeren and van Oortmerssen, 1981). Interestingly, in more aggressive *M. m. domesticus* higher production of several volatiles that are known to be associated with presentation of male dominance has been reported (Novotny *et al.*, 1990; Mucignat-Caretta *et al.*, 2010). On the other hand, the expression level of MUPs which bind and release these volatiles is lower in males belonging to this subspecies when compared to *M. m. musculus* males (Stopková *et al.*, 2007). Therefore, it is conceivable that a different MUP-to-volatile ratio may contribute to subspecies recognition, further supporting the hypothesis that quantitative variation rather than the production of individually specific compounds is important in signalling the subspecies specificity.

Another set of compounds that have been proposed to be involved in assortative mate choice in *M. m. musculus* and *M. m. domesticus* are the androgen-binding proteins (ABPs; for review see Laukaitis and Karn, Chapter 7). This hypothesis was developed on the basis of the fact that each subspecies has a fixed variant of the *Abpa* gene that codes for the alpha subunit of the protein (Karn and Dlouhy, 1991). While there is some supporting evidence (Laukaitis *et al.*, 1997; Talley *et al.*, 2001), other data do not confirm a recognition signalling role of ABPs in natural populations (Bímová *et al.*, 2005), although this is open to interpretation (see Laukaitis and Karn, Chapter 7). Moreover, results of several studies strongly suggest that the primary source of a subspecies recognition signal is urine, in which ABPs are not present (Smadja and Ganem, 2002; Bímová *et al.*, 2009). Other results suggest that even in salivary glands the original function of ABPs may be transport of steroids (Dlouhy *et al.*, 1986, 1987; Karn, 1998). Recently, Zhou *et al.* (2009) have demonstrated that major sex steroids, including testosterone, oestradiol, and progesterone play important regulatory roles in the olfactory chemosensory system. They have provided evidence that the lateral nasal gland, known to produce the highest level of odorant binding proteins (see below), contains also a higher volume of testosterone than all other glands and epithelia in oral and nasal cavity. In addition, it expresses as much as 50-fold higher volume of ABPs than the remaining olfactory tissues (Zhou *et al.*, 2009). Salivary ABPs may also transport steroids from males to females during a course of mouth-to-mouth contact, as with wild boar in which males trigger oestrus with saliva that contains high volumes of pheromaxein – a transporter that primarily binds and transports androstendione (Marchese *et al.*, 1998). These data thus suggest that ABPs represent multifunctional transporters which have various functions in chemical communication, different from those of urine.

Candidate gene approach

Olfactory and vomeronasal receptors

The best-known example of chemical communication, which has since been extended to mice by the use of the candidate gene approach, was discovered by Nobel Prize winners Linda Buck and Richard Axel (Buck and Axel, 1991; Buck, 2000, 2005): the odorant-binding G-protein coupled receptors known as olfactory GPCR. They were first to realize that the mammalian olfactory system can recognize and discriminate a large number of different odorant molecules via numerous odorant GPCR on dendritic cilia of chemosensory neurons (Buck and Axel, 1991). Furthermore, the fact that mammals can recognize various chemical substances has been emphasized by numerous observations that small differences in the structure of a chemical signal often lead to profound changes in the perception of given odour. Therefore, Buck and Axel (1991) postulated a model of perception based on three basic facts: (1) the odorant receptors are likely to belong to the superfamily of receptors similar to rhodopsin (i.e. transducing intracellular signals by coupling to GTP-binding proteins); (2) due to the high diversity of odours that mammals can perceive, this family of receptors themselves should exhibit enormous diversity; and (3) the expression of odorant receptors should be restricted to the main and/or accessory olfactory epithelium (Buck and Axel, 1991). In their experiments each of these proposals was experimentally demonstrated. Since their original research on rats widened general understanding of olfactory perception and was rewarded by the Nobel Prize in 2004, several other model organisms have been studied in detail.

The house mouse genome contains approximately 1400 genes for olfactory GPCR, of which, however, about 24% represent non-coding pseudogenes (Zhang and Firestein, 2002). On the basis of phylogenetic analyses, olfactory receptors can be divided into 228 protein families, each of which contains 1–50 genes. These families are further divided into Class I olfactory receptors (families 1–42) and Class II olfactory receptors (families numbered from 101–286) (Zhang and Firestein, 2002).

Apart from the olfactory receptors located in the main olfactory epithelium, there is another large group of GPCR receptors expressed in the accessory olfactory VNO. These vomeronasal receptors (VR) in the mouse are divided into the V1R group, which are expressed on the apical end of the VNO, and the V2R group of receptors found on the basal part of the VNO (Dulac and Axel, 1995; Herrada and Dulac, 1997; Thompson *et al.*, 2007; Jia and Halpern, 1996). A later study (Krieger *et al.*, 1999) showed that G protein (Gi) coexpressed with the V1 receptor was activated by lipophilic volatile substances, while the proteinaceous structure (alpha2u globulin) activated G protein (Go) coexpressed with V2 receptors. Contrary to V1R, V2R receptors contain an extremely large extracellular region of approximately 600 amino acids and, therefore, it is possible that they are able to recognize high molecular weight, non-volatile ligands (Krieger *et al.*, 1999). The number of vomeronasal receptors is lower than OR, with approximately 187 V1R and 121 V2R intact genes (Shi and Zhang, 2007). In species where olfaction has been displaced as the dominant sense (e.g. by vision or the auditory system), the fraction of OR and VR pseudogenes increases; for example, humans have more than 50% OR pseudogenes and even 90% VR pseudogenes (Rodriguez *et al.*, 2000).

The comparison of mouse and rat receptors shows that vomeronasal receptors (V1R) are more species-specific than are the OR of the main olfactory epithelium (Zhang *et al.*, 2007). This may provide evidence for the idea that vomeronasal receptors detect species or subspecies-specific pheromonal substances (Karn *et al.*, 2010). The variety of chemical substances and their receptors are studied with regard to speciation using two mouse subspecies, *M. m. musculus* and *M. m. domesticus*, as a model system. A recent study using five wild-derived strains of *M. m. musculus* and *M. m. domesticus* identified the first vomeronasal receptor (Vmn1r67) to show a species-specific (diagnostic) pattern in sequence structure (24 SNPs, more than half of which are non-synonymous) and which therefore may be essential in the perception of signals for interspecific diversity (Karn *et al.*, 2010).

Lipocalins: candidate transporters of chemical signals

Molecular mechanisms underlying chemical communication involve a system of signal input (olfactory receptors) and a system of signal output (signal transporters). Proteins that belong to the family of lipocalins are candidate molecules for involvement in the transport of biologically active chemical substances. This is a family of 18–22 kDa globular proteins with the capacity to bind a wide spectrum of volatile ligands. Despite their low sequence similarity, all lipocalins share highly conserved exon/intron organization (see Fig. 8.2) and, therefore, have similar tertiary organization, forming a typically eight-stranded beta barrel composed of a characteristic number of antiparallel beta sheets (Flower *et al.*, 2000; Salier, 2000). Structurally conserved regions of lipocalins generally have stable motifs in their amino acid sequences (Flower, 1996). The main lipocalin signature is the sequence GxW, present (with some exceptions) in all lipocalins on the second exon (Salier, 2000). At least, the two conservative cystein residues exist in all lipocalins, forming a characteristic intramolecular disulphidic bond. Other sequence motifs described in Flower (1996) split lipocalins into two main subgroups named 'kernel' and 'outlier' lipocalins.

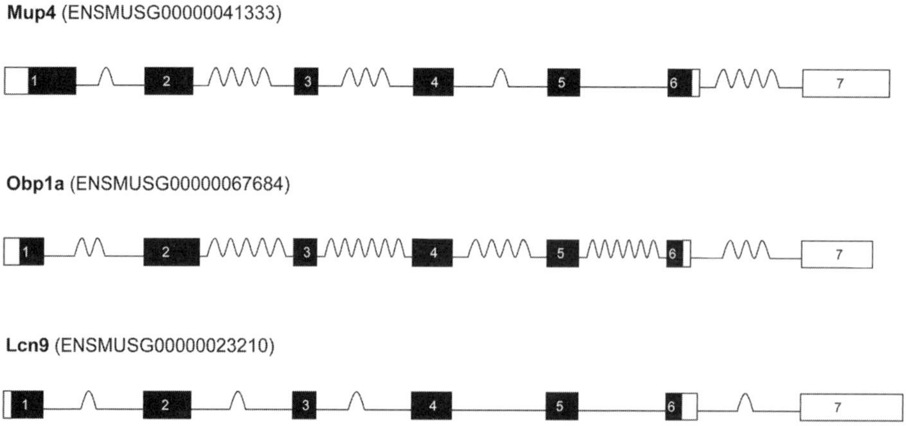

Figure 8.2 Schematic exon/intron organization of three mouse lipocalins: major urinary protein (*MupIV*), odorant binding protein (*Obp1a*), and lipocalin 9 (*Lcn9*). All mammalian lipocalins share a typical array of seven exons and six introns, where the first exon is partially coding and is variable in its length. Exons 2–5, coding the beta sheets, are highly conservative in their length. Contrary to the exons, intronic sequences length is very variable and is symbolized by wavy lines (the more twirls the more bases). The stop codon is in the sixth exon and the seventh exon is therefore untranslated (Salier, 2000).

Lipocalins are abundant proteins in both prokaryotic and eukaryotic cells and they cover a large range of functions, including antibacterial function (Strong, 2006); immune system response (Flower, 1994; Fournier *et al.*, 2000); prostaglandin synthesis (Igarashi *et al.*, 1992); and various aspects of reproduction, like maintaining pregnancy (Crosset *et al.*, 1996), fertilization process (Oehninger *et al.*, 1995; Seppälä *et al.*, 2002), or sperm maturation (Flower, 1994). Some of the lipocalin members are already known to act as signal transporters or even as signal molecules themselves (i.e. MUPs) but the function of many others is still unclear.

Major urinary proteins

MUPs are 20 kDa kernel lipocalins encoded by a large number of genes, which form an approximately 2 Mb cluster on mouse chromosome 4 (Hastie *et al.*, 1979). This large segment contains 21 intact genes and the same number of pseudogenes (Logan *et al.*, 2008; Stopková *et al.*, 2009) According to phylogenetic analyses, *Mup* genes can be divided into two groups (Logan *et al.*, 2008): ancestral group A, which includes *Mup*s known as MUPIII, IV and V (in fact it is a group of six genes – *Mup1*, *Mup2*, *Mup18*, *Mup19*, *Mup20*, and *Mup21* – see Fig. 8.3); and a recently reduplicated massive central part of the cluster forming a second group B, which consists of the 15 remaining genes known as MUPI and

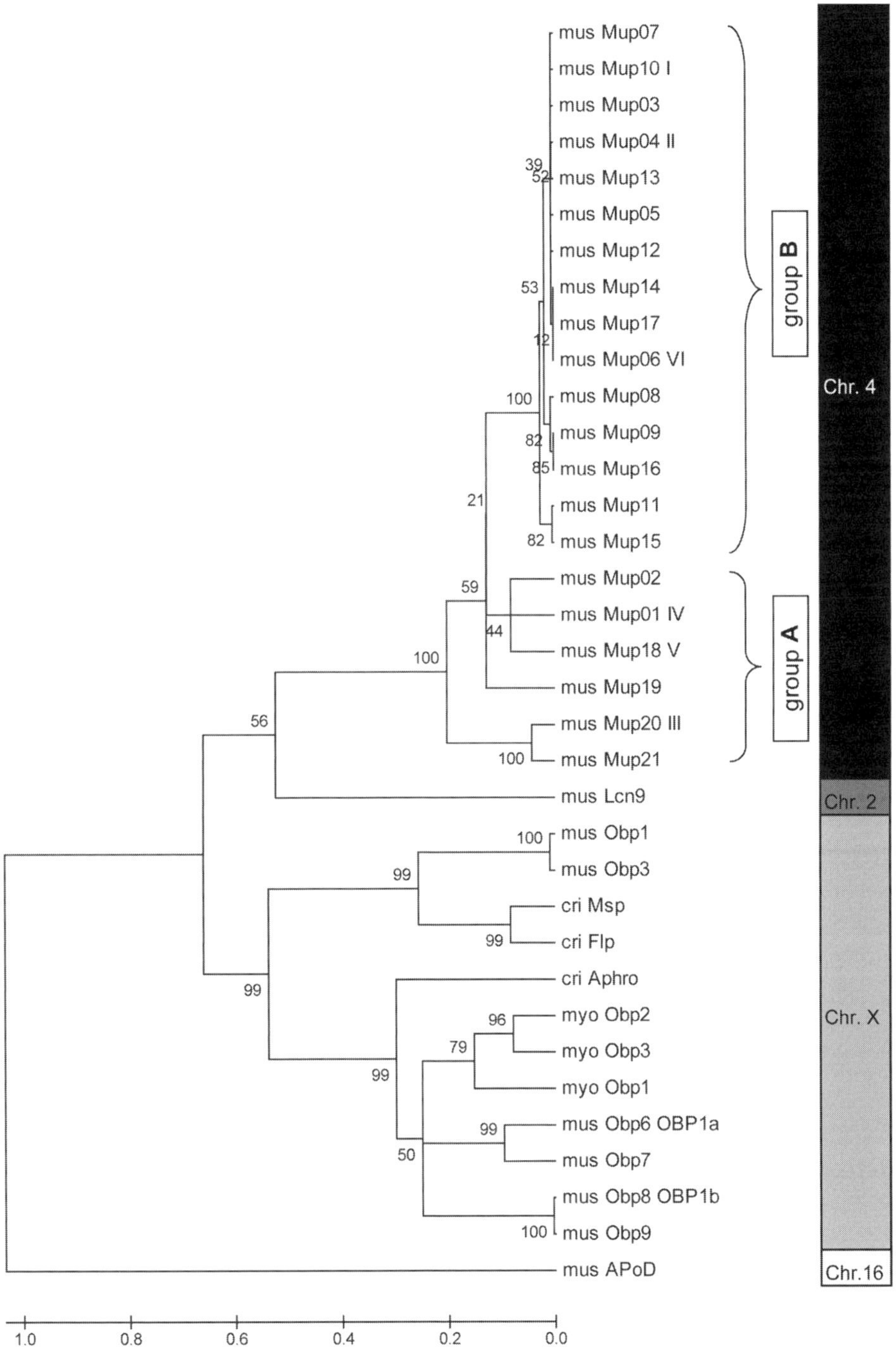

Figure 8.3 Evolutionary relationships of selected lipocalins. Genes used for phylogenetic analysis are mouse (*M. m. domesticus*) genes for major urinary proteins (mus *Mup1–21*), odorant binding proteins (mus *Obp1–9*), lipocalin 9 (mus *Lcn9*), and ancestral lipocalin apolipoprotein D (mus *ApoD*); hamster (*Mesocricetus auratus*) gene for aphrodisin (cri *Aphro*), male-specific salivary protein (cri *Msp*), and female lacrimal protein (cri *Flp*); bank vole (*Myodes glareolus*) genes for odorant binding proteins (myo *Obp1–3*).

MUPII types, depicted in a cladogram in Fig. 8.3. The characteristic design of group A is their sense–antisense organization of genes and their respective pseudogenes and very high sequence similarity (98%). All pseudogenes have the same G/T substitution that encodes the premature stop codon, so they produce only the initial hexapeptide EEAARS (Finlayson *et al.*, 1974; Clark *et al.*, 1985), which some authors consider as a biologically active substance eliciting physiological and behavioural responses (Mucignat-Caretta *et al.*, 1995; More, 2006). However, these studies include only synthetic products and there is as yet no evidence that such natural hexapeptides are secreted into urine.

Function and regulation of major urinary proteins

In the social life of the house mouse (*M. musculus*), MUPs represent an essential tool for the deposition and slow release of pheromones. Only once this slow release is complete does the original mark need to be overmarked. The longevity of such marking is in each individual's interest as the presence of a particular mouse is manifested even if it is temporarily absent from its territory (Bacchini *et al.*, 1992; Hurst *et al.*, 1998). The logic of such individual presentation, therefore, suggests that strong selective forces will have driven the evolution of MUP gene expression to an evolutionary optimum. This situation has been achieved through two evolutionary processes, gene duplication (Logan *et al.*, 2008; Stopková *et al.*, 2009) and steroid-dependent expression, by which an individual's status is honestly manifested according to its particular hormonal levels.

The main site of MUP expression is the liver, from where they are filtered through the kidney into the urine – the most important scent-marking secretion in mice. However, MUPs are also expressed in other secretory glands including

Caption for Figure 8.3 (*cont.*)
Associated numbers in mouse *Mup*s and *Obp*s correspond to a consecutive gene location on the DNA strand. Pseudogenes are not included in the analysis. Major urinary proteins form an ancestral group A and recently duplicated group B. Roman numbers mark the MUP types. Mouse odorant binding proteins cluster into three groups, each containing a pair of genes; the first group is *Obp1a* genes (*Obp6* and *Obp7*), the second is *Obp1b* genes (*Obp8* and *Obp9*); the third is formed by *Obp1* and *Obp3* sequences, which are located at the beginning of the *Obp* gene cluster on the X chromosome. Interestingly, hamster aphrodisin together with bank vole *Obp*s are more similar to *Obp1a* and *Obp1b* genes, while other hamster lipocalins with *Obp* structure (*Msp* and *Flp*) are similar to the mouse *Obp1* and *Obp3*. The evolutionary history was inferred using the neighbour-joining method in the MEGA5 software (Tamura *et al.*, 2007) as in Stopková *et al.* (2010).

salivary, lachrymal, and mammary glands (Finlayson *et al.*, 1965; Shaw *et al.*, 1983; Shahan and Derman, 1984). Their expression is tissue-specific, which results in quantitatively and qualitatively specific MUP phenotypes in each secretion (Hastie *et al.*, 1979; Shaw *et al.*, 1983).

The regulation of expression of urinary MUPs has been repeatedly demonstrated in experiments in which growth hormone, thyroid hormone, or steroid hormones were peritoneally administered in mice, thus showing that the levels of expression in castrated males may be re-established to normal and lower levels of MUP expression, and in females can be upregulated to the levels typical for males (Rümke and Thung, 1964; Knopf *et al.*, 1983; Clissold *et al.*, 1984; Johnson *et al.*, 1995). However, there is little supporting evidence that such changes in the regulation of expression actually occur under different social and environmental stimuli in wild mice. In the laboratory mouse strain C57BL/6, it has been shown that levels of MUP expression significantly correlate with different stages of the oestrous cycle (Stopka *et al.*, 2007). Precise calculation of different cellular types from vaginal smears occurring in different stages of the oestrous cycle provided a continuous oestrous curve that has been cross-correlated with varying levels of MUP expression. This analysis revealed that the level of MUP expression reaches its maximum just one day before oestrus, a period during which females also reach maximum beta-2-oestradiol levels. Because of their increased exploratory activity, this phase is called behavioural oestrus. This observation, therefore, represents further evidence that MUPs release honest signals on the reproductive status of female signallers.

The expression of MUPs is sexually dimorphic, with the level of sexual dimorphism varying between species of the genus *Mus* (Sampsell and Held, 1985). Males produce 3–10 times more MUPs than females, while individuals of each sex alter the levels of MUP expression according to specific social structure. In the laboratory strain C57BL/6, males and females upregulate MUP expression when paired in a cage dividing them by a metal grid (Stopka *et al.*, 2007). The evolutionary advantages of this dynamic variation are straightforward. Males stimulate an onset of female oestrus by an increase in MUP production, which in turn stimulates changes in male reproductive physiology. MUP expression, therefore, leads to more pheromones being released and thus drives extensive changes in a receiver's physiology. At the same time, the receiver influences changes in the physiology of a signalling individual. Therefore, chemical signals have modulatory effects via dynamic production of MUPs while their production is also modulated by another individual (Stopka *et al.*, 2007).

The expression of MUPs may also be regulated at the intrasexual level. In an experiment with wild-derived house mice (*M. m. musculus*) Janotová and Stopka (2011) demonstrated that males downregulate the expression of MUPs as a direct response to paired housing in which two males are separated by a grid. This phenomenon could have arisen due to at least two potential selective forces. First, downregulation of MUPs represents a mechanism by which males decrease the amount of volatiles that may stimulate attack-related behaviours in the receiver. Thus other males would tolerate an individual that develops such expression dynamics and this mechanism would be selectively advantageous. However, if the main function of this mechanism were to modulate reproductive physiology of a receiver, downregulation of MUPs in such intrasexual context would be an energy-saving mechanism that would also be supported by natural selection. Furthermore, because the modulation of chemical communication is to some extent tuned to social contexts, the adaptive value of this mechanism would be further enforced if such individuals escape predation during fruitless searching when no females are around.

Profiles of MUPs consist of up to 15 MUP isoforms that are products of multiple *Mup* loci (Beynon and Hurst, 2003). In *M. m. domesticus* from Great Britain and Switzerland, remarkable heterogeneity of MUP profiles has been observed (Robertson *et al.*, 1997; Pes *et al.*, 1999; Payne *et al.*, 2001; Rusu *et al.*, 2008). Since this variation in MUP profiles is based on differences in heritable genotypes, it has been suggested that it provides information about a variety of genetically determined individual characteristics, like identity (Hurst *et al.*, 2001; Cheetham *et al.*, 2007), kinship (Sherborne *et al.*, 2007), gender (Armstrong *et al.*, 2005; Roberts *et al.*, 2010), and heterozygosity (Thom *et al.*, 2008). Indeed, it has been shown that mice are able to recognize differences in MUP profiles and use them to discriminate between conspecifics (Hurst *et al.*, 2001), at least for *M. m. domesticus* on which the majority of behavioural observations and analyses have been performed. However, other components of mouse chemical systems, for example the MHC, have been reported to carry this type of information (Brennan and Kendrick, 2006). Although a few attempts to assess the role of MUPs when compared to MHC and vice versa have been made (Brennan, 2001; Sherborne *et al.*, 2007; Thom *et al.*, 2008), it still remains unclear to what extent the communication of particular genetically determined characteristics depends on each of these complexes.

Another issue concerns the transmission process of the information contained in MUP profiles. Recently, considerable progress has been made demonstrating that male MUPs can directly, without their pheromonal ligands, induce response in specific receptors in the VNO (Chamero *et al.*, 2007). The ability to respond to ligand-depleted MUPs has also been corroborated on both behavioural and physiological levels. In males, MUPs have been shown to trigger aggressive behaviour (Chamero *et al.*, 2007), while in females held in crowded living conditions they enhanced the rate of ovulation (More, 2006). Moreover, a recent study by Papes *et al.* (2010) revealed that MUP orthologs derived from cats or rats promote aversive behaviour in mice via the VNO. They also demonstrated that cat and rat MUPs activate specific vomeronasal neurons different from those that have been shown to be activated by mouse MUPs.

The importance and exclusivity of information carried by MUPs themselves is evident from different mouse responses, either to direct contact with scent marks (i.e. protein with its ligands) or when the contact with the mark was prevented such that mice perceived only the volatile component. In females, only direct contact with the scent mark stimulates attraction and thus may be important in mate choice for either mate assessment or mate recognition (Ramm *et al.*, 2008). Other studies have explored marking behaviour in males, which are known to advertise their competitive ability and territory ownership by intense scent marking of the whole territory and quick countermarking of all the scent marks of competitors (Desjardins *et al.*, 1973). Interestingly, when different fractions of urine that contained either both non-volatile and volatile components or just urinary volatiles were tested, the fraction with non-volatile proteins stimulated significantly more countermarking than that without these proteins (Humphries *et al.*, 1999). Moreover, when direct access to the scent mark was prevented, males failed to show countermarking responses. Therefore, although males investigated all the scent marks, they evaluated as competitive only those in which they could detect the non-volatile protein component (Nevison *et al.*, 2003). Since countermarking is important not just for territory defence but also as a measure of competitive abilities of a male, a measure females frequently use in mate choice (Rich and Hurst, 1998), these results imply that MUPs play an important role in advertising male qualities to both males and females.

Odorant-binding proteins (OBPs)

The role of MUPs in chemical communication has long been studied. However, many mammals have lost the MUP-encoding genes and use different proteins for the same purpose. In hamsters, OBPs play almost identical roles. However, these proteins are also present in mice (*Mus*), but their role is still insufficiently studied in this genus.

OBPs are small, extracellular proteins, belonging to the group of outlier lipocalins and sharing typical lipocalin features, including specific sequence motifs, exon/intron organization depicted in Fig. 8.2, and a characteristic beta barrel structure. Mouse *Obp* genes are located on the X chromosome but there is still a lack of information about the exact number of gene variants and their expression in respective tissues. An NCBI search revealed that there are seven intact genes and one pseudogene which is oriented in the opposite direction (Stopková *et al.*, 2009). However, according to recent data from the Ensembl database, there are six intact genes and three pseudogenes in the region of 75.110 to 75.491 Mb on the X chromosome. In contrast to the DNA level, only two OBP variants on the protein level (named OBP1a and OBP1b) have been identified in mouse nasal mucosa (Pes *et al.*, 1992). Binding capacity analysis revealed that the most common ligand with a capacity to specifically bind to OBP is the 'bell pepper' odorant with pyrazine structure: IMBP 2-isobutyl-3-methoxypyrazine (Bignetti *et al.*, 1985; Pes *et al.*, 1992; Burova *et al.*, 1999; Briand *et al.*, 2000b). Further studies revealed that there exists an even wider spectrum of volatile molecules with different structures that are able to form complexes with OBPs (Pevsner *et al.*, 1990; Lobel *et al.*, 2002).

Obp orthologues are found in many different animal taxa, mainly insects, but including mammalian species such as cow (Bignetti *et al.*, 1985), rat (Pevsner *et al.*, 1988b; Briand *et al.*, 2000b), pig (Spinelli *et al.*, 1998), rabbit (Dal Monte *et al.*, 1991; Lazar *et al.*, 2002), elephant (Lazar *et al.*, 2002), porcupine (Henzel *et al.*, 1988; Felicioli *et al.*, 1993), hamster (Henzel *et al.*, 1988), and have recently been discovered in the bank vole (Stopková *et al.*, 2010) and the dwarf hamster (Turton *et al.*, 2010). This is in contrast with MUPs which, while abundant in mice, rats, cats, and, as recently discovered, in *Mastomys coucha* (Daniszová *et al.*, 2009), remain unconfirmed in other species. Moreover, it seems that in some species where MUPs are missing the OBP becomes the dominant protein component of urine, as in the case of the bank vole (Stopková *et al.*, 2010).

All theories that emphasize the function of OBPs are based on the fact that they are specifically expressed in the nasal mucosa. Therefore, many authors have suggested that OBPs act as odorant transporters presenting volatile molecules to the olfactory receptor cells (Pevsner *et al.*, 1988a, 1990; Steinbrecht, 1998; Tegoni *et al.*, 2000). The fact that volatile substances may act upon receptor cells faster without binding to OBP than OBP/ligand complexes led some authors to present a hypothesis that the role of OBP is rather in deactivating odorants after receptor stimulation (Lazar *et al.*, 2002) or in enhancing the scent signalling by keeping the odorant molecules close to the receptor cells (Taylor *et al.*, 2008; Vidic *et al.*, 2008). However, data from various species presented evidence that apart from the nasal mucosa, OBPs are also excreted into the urine (Stopková *et al.*, 2010; Turton *et al.*, 2010), saliva (Stopková *et al.*, 2010), and/or vagina (Kruhoffer *et al.*, 1997; Briand *et al.*, 2004b) and, therefore, it is possible that the role of OBPs is much wider than originally thought.

The first evidence that OBPs are involved in chemical communication was provided by the group of Alan Singer, who discovered that the particular OBP variant called aphrodisin has a pheromonal effect. They identified a lipocalin in the vaginal discharge of female hamsters (*Mesocricetus auratus*) as a molecule that stimulates copulatory behaviour in males that were exposed to the fraction of the vaginal discharge dominated by this protein (Macrides *et al.*, 1984; Singer *et al.*, 1986, 1987). Because of its aphrodisiacal effect they named this protein aphrodisin. Nevertheless, structurally it belongs to the OBPs, being most similar to the mouse *Obp1a*-like genes depicted in Fig 8.3. Recombinant aphrodisin showed high affinity for 2-isobutyl-3-methoxypyrazine – IMBP, and for other substances presented in vaginal secretions, namely dimethyl disulphide – DMDS and methylthiobutyrate – MTB (Briand *et al.*, 2000a). However, the analysis of natural aphrodisin revealed five more ligands specifically bound to this protein; some of these molecules have been previously described as parts of insect pheromonal blends (Briand *et al.*, 2004a). After ligand depletion, aphrodisin seems to retain the same effect as the proteinaceous fraction of the vaginal discharge (Singer *et al.*, 1986); therefore, this finding contributed to the theory that aphrodisin also acts as a non-volatile pheromone. However, another study showed that recombinant aphrodisin did not induce the same stimulation effects on copulatory behaviour in males (Singer and Macrides, 1990).

Chemical signals involved in reproduction often show sexual dimorphism in their production. This is very well documented in the case of MUPs, in which the dimorphism is biased toward males (Janotová and Stopka, 2011; Stopková *et al.*, 2007, 2009), and conversely in hamsters, in which the expression of aphrodisin is specifically attributed to females (Magert *et al.*, 1995, 1999). In the case of mouse OBPs the balance of sexual dimorphic expression remains to be determined. The similarity of some OBP variants with other described lipocalins selectively inhibited by male- and/or female-specific steroids (De, 1996; Ranganathan *et al.*, 1999) suggest that the hypothesis that OBPs play some role in sexually specific signalling may be relevant.

Conclusion

The studies surveyed above in the context of chemical communication in the house mouse provide experimental evidence that priming effects by pheromones are maintained, enhanced, and regulated by protein transporters belonging to the lipocalin family of proteins. The house mouse is a model in which this system was extensively studied, and the most recent papers show that these proteins have huge impacts on the reproductive physiology of potential receivers (Stopka *et al.*, 2007; Janotová and Stopka, 2009). On the other hand, most of this research was focused on MUPs at the expense of other related proteins with almost identical structure and capacity to bind pheromones – a fact evident in other rodent species where MUPs are absent or non-functional.

This situation is interesting because it demonstrates that selection operates on the level of a particular function and structure. Tertiary structure and binding capacity have been selected regardless of the particular gene – a scenario that is commonly observed in rodents. Furthermore, some species facilitate MUPs (e.g. *Mus*) while others (e.g. *Cricetus*, *Phodopus*, *Myodes*) use OBPs.

Lipocalins are proteins that have always played the role of transporters and retained this function during a transition from uni-cellular to complex multi-cellular organisms. The house mouse genome (or at least that of an inbred strain) contains at least 54 genes for lipocalins with different functions ascribed to them (Stopková *et al.*, 2009), but with one common feature: they all have a tertiary structure with an eight-stranded beta barrel that has the capacity to transport various hydrophobic ligands, including steroids, bilins, odorants, and pheromones. Some experiments have even shown that these transporters are heavily involved in removing free iron, free radicals, and other potentially toxic compounds that regularly emerge during metabolism (Petrak *et al.*, 2007). This fact opens a new era of mouse research as the house mouse appears to have retained the plesiomorphic profile of lipocalins, which has to some extent been lost in other rodents, hares, and rabbits, and apes within the larger taxon Euarchontoglires.

REFERENCES

Armstrong, S. D., Robertson, D. H. L., Cheetham, S. A., Hurst, J. L., and Beynon, R. J. (2005). Structural and functional differences in isoforms of mouse major urinary proteins: a male-specific protein that preferentially binds a male pheromone. *Biochemical Journal*, **391**, 343–50.

Bacchini, A., Gaetani, E., and Cavaggioni, A. (1992). Pheromone binding proteins of the mouse, *Mus musculus*. *Experientia*, **48**, 419–21.

Beynon, R. J. and Hurst, J. L. (2003). Multiple roles of major urinary proteins in the house mouse, *Mus domesticus*. *Biochemical Society Transactions*, **31**, 142–6.

Beynon, R. J., Hurst, J. L., Turton, M. J., *et al.* (2008). Urinary lipocalins in Rodentia: is there a generic model? In *Chemical Signals in Vertebrates*, ed. J. L. Hurst, R. J. Beynon, S. C. Roberts, and T. D. Wyatt. New York: Springer, pp. 37–50.

Bignetti, E., Cavaggioni, A., Pelosi, P., *et al.* (1985). Purification and characterisation of an odorant-binding protein from cow nasal tissue. *European Journal of Biochemistry*, **149**, 227–31.

Bímová, B., Albrecht, T., Macholán, M., and Piálek, J. (2009). Signalling components of mate recognition system in the house mouse. *Behavioural Processes*, **80**, 20–7.

Bímová, B., Karn, R. C., and Piálek, J. (2005). The role of salivary androgen-binding protein in reproductive isolation between two subspecies of house mouse: *Mus musculus musculus* and *Mus musculus domesticus*. *Biological Journal of the Linnean Society*, **84**, 349–61.

Boursot, P., Auffray, J.-C., Britton-Davidian, J., and Bonhomme, F. (1993). The evolution of house mice. *Annual Review of Ecology and Systematics*, **24**, 119–52.

Brennan, P. (2001). How mice make their mark. *Nature*, **414**, 590–1.

Brennan, P. A. and Kendrick, K. M. (2006). Mammalian social odours: attraction and individual recognition. *Philosophical Transactions of the Royal Society B: Biological Sciences*, **361**, 2061–78.

Brennan, P. A. and Peele, P. (2003). Towards an understanding of the pregnancy-blocking urinary chemosignals of mice. *Biochemical Society Transactions*, **31**, 152–5.

Brennan, P. A. and Zufall, F. (2006). Pheromonal communication in vertebrates. *Nature*, **444**, 308–15.

Briand, L., Blon, F., Trotier, D., and Pernollet, J.-C. (2004a). Natural ligands of hamster aphrodisin. *Chemical Senses*, **29**, 425–30.

Briand, L., Huet, J.-C., Perez, V., *et al.* (2000a). Odorant and pheromone binding by aphrodisin, a hamster aphrodisiac protein. *FEBS Letters*, **476**, 179–85.

Briand, L., Nespoulos, C., Perez, V., *et al.* (2000b). Ligand-binding properties and structural characterization of a novel rat odorant-binding protein variant. *European Journal of Biochemistry*, **267**, 3079–89.

Briand, L., Trotier, D., and Pernollet, J.-C. (2004b). Aphrodisin, and aphrodisiac lipocalin secreted in hamster vaginal secretions. *Peptides*, **25**, 1545–52.

Bruce, H. M. (1959). An exteroreceptive block to pregnancy in the mouse. *Nature*, **184**, 105.

Bruce, H. M. (1965). Effect of castration on the reproductive pheromones of male mice. *Journal of Reproduction & Fertility*, **10**, 141–3.

Buck, L. B. (2000). The molecular architecture of odor and pheromone sensing in mammals. *Cell*, **100**, 611–18.

Buck, L. B. (2005). Unraveling the sense of smell. *Angewandte Chemie International Edition*, **44**, 6128–40.

Buck, L. and Axel, R. (1991). A novel multigene family may encode odorant receptors: a molecular basis for odor recognition. *Cell*, **65**, 175–87.

Burova, T., Choiset, Y., Jankowski, C. K., and Haertle, T. (1999). Conformational stability and binding properties of porcine odorant binding protein. *Biochemistry*, **38**, 15043–51.

Chamero, P., Marton, T. F., Logan, D. W., *et al.* (2007). Identification of protein pheromones that promote aggressive behaviour. *Nature*, **450**, 899–903.

Cheetham, S. A., Thom, M. D., Jury, F., *et al.* (2007). The genetic basis of individual-recognition signals in the mouse. *Current Biology*, **17**, 1771–7.

Christophe, N. and Baudoin, C. (1998). Olfactory preferences in two strains of wild mice, *Mus musculus musculus* and *Mus musculus domesticus*, and their hybrids. *Animal Behaviour*, **56**, 365–9.

Clark, A. J., Ghazal, P., Bingham, R. W., Barrett, D., and Bishop, J. O. (1985). Sequence structures of a mouse major urinary protein gene and pseudogene compared. *EMBO Journal*, **4**, 3159–65.

Clissold, P. M., Hainey, S., and Bishop, J. O. (1984). Messenger RNAs coding for mouse urinary proteins are differentially induced by testosterone. *Biochemical Genetics*, **22**, 379–87.

Crosset, B., Allen, W. R., and Stewart, F. (1996). A 19 kDa protein secreted by the endometrium of the mare is a novel member of the lipocalin family. *Biochemical Journal*, **320**, 137–43.

Daev, E. V. and Dukelskaya, A. V. (2003). The female pheromone 2,5-dimethylpyrazine induces sperm-head abnormalities in male CBA mice. *Russian Journal of Genetics*, **39**, 969–74.

Dal Monte, M., Andreini, I., Revoltella, R., and Pelosi, P. (1991). Purification and characterization of two odorant-binding proteins from nasal tissue of rabbit and pig. *Comparative Biochemistry and Physiology: Part B – Comparative Biochemistry*, **99**, 445–51.

Daniszová, K., Janotová, K., Jedelský, P., and Stopka, P. (2009). Urinary lipocalins in *Mastomys coucha*. *Folia Zoologica*, **58**, 56–64.

Davies, N. B. (1991). Mating systems. In *Behavioural Ecology: An Evolutionary Approach*, ed. J. R. Krebs and N. B. Davies. London: Blackwell Scientific Publications, pp. 263–94.

De, P. K. (1996). Sex-hormonal regulation of 20.5 and 24 kDa major male-specific proteins in Syrian hamster submandibular gland. *The Journal of Steroid Biochemistry and Molecular Biology*, **58**, 183–7.

Desjardins, C., Maruniak, J. A., and Bronson, F. H. (1973). Social rank in house mice: differentiation revealed by ultraviolet visualization of urinary marking patterns. *Science*, **182**, 939–41.

Dlouhy, S. R., Nichols, W. C., and Karn, R. C. (1986). Production of an antibody to mouse salivary androgen-binding protein (ABP) and its use in identifying a prostate protein produced by a gene distinct from *Abp*. *Biochemical Genetics*, **24**, 9–10.

Dlouhy, S. R., Taylor, B. A., and Karn, R. C. (1987). The genes for mouse salivary androgen-binding protein (ABP) subunits alpha and gamma are located on chromosome 7. *Genetics*, **115**, 535–43.

Doty, R. L. (2003). Mammalian pheromones: fact or fantasy? In *Handbook of Olfaction and Gustation*, ed. R. L. Doty. New York: Marcel Dekker, pp. 571–631.

Drickamer, L. C. (1989). Patterns of deposition of urine containing chemosignals that affect puberty and reproduction by wild stock male and female house mice. *Journal of Chemical Ecology*, **15**, 1407–21.

Drickamer, L. C. (1992). Oestrus female house mice discriminate dominant from subordinate males and sons of dominant from sons of subordinate males by odour cues. *Animal Behaviour*, **43**, 868–70.

Dulac, C. and Axel, R. (1995). A novel family of genes encoding putative pheromone receptors in mammals. *Cell*, **83**, 195–206.

Felicioli, A., Gann, M., Garibotti, M., and Pelosi, P. (1993). Multiple types and forms of odorant-binding proteins in the Old-World porcupine *Hystrix cristata*. *Comparative Biochemistry and Physiology: Part B – Biochemistry & Molecular Biology*, **105**, 775–84.

Finlayson, J. S., Asofsky, R., Potter, M., and Runner, C. C. (1965). Major urinary protein complex of normal mice: origin. *Science*, **149**, 981–2.

Finlayson, J. S., Potter, M., Shinnick, C. S., and Smithies, O. (1974). Components of the major urinary protein complex of inbred mice: determination of N-terminal sequences and comparison with homologous components from wild mice. *Biochemical Genetics*, **11**, 325–35.

Flower, D. R. (1994). The lipocalin protein family: a role in cell regulation. *FEBS Letters*, **354**, 7–11.

Flower, D. R. (1996). The lipocalin protein family: structure and function. *Biochemical Journal*, **318**, 1–14.

Flower, D. R., North, A. C. T., and Sansom, C. E. (2000). The lipocalin protein family: the structural and sequence overview. *Biochimica et Biophysica Acta*, **1482**, 9–24.

Fournier, T., Medjoubi, N., and Porquet, D. (2000). Alpha-1-acid glykoprotein. *Biochimica et Biophysica Acta*, **1482**, 157–71.

Ganem, G., Ginane, C., Ostrowski, M. F., and Orth, A. (2005). Assessment of mate preference in the house mouse with reference to investigations on assortative mating. *Biological Journal of the Linnean Society*, **84**, 461–71.

Hastie, N. D., Held, W. A., and Toole, J. J. (1979). Multiple genes coding for the androgen-regulated major urinary proteins of the mouse. *Cell*, **17**, 449–57.

He, J., Ma, L., Kim, S. S., Nakai, J., and Yu, C. R. (2008). Encoding gender and individual information in the mouse vomeronasal organ. *Science*, **25**, 535–8.

Henzel, W. J., Rodriguez, H., Singer, A. G., *et al.* (1988). The primary structure of aphrodisin. *Journal of Biological Chemistry*, **263**, 16682–7.

Herrada, G. and Dulac, C. (1997). A novel family of putative pheromone receptors in mammals with a topographically organized and sexually dimorphic distribution. *Cell*, **90**, 763–73.

Humphries, R. E., Robertson, D. H. L., Beynon, R. J., and Hurst, J. L. (1999). Unravelling the chemical basis of competitive scent marking in house mice. *Animal Behaviour*, **58**, 1177–90.

Hurst, J. L. (1987). The functions of urinary marking in a free-living population of house mice, *Mus domesticus* Rutty. *Animal Behaviour*, **35**, 1433–42.

Hurst, J. L. (1990). Urine marking in populations of wild mice *Mus domesticus* Rutty: I – communication between males. *Animal Behaviour*, **40**, 209–22.

Hurst, J. L. (2009). Female recognition and assessment of males through scent. *Behavioural Brain Research*, **200**, 295–303.

Hurst, J. L., Fang, J., and Barnard, C. J. (1993). The role of substrate odours in maintaining social tolerance between male house mice, *Mus musculus domesticus*. *Animal Behaviour*, **45**, 997–1006.

Hurst, J. L., Payne, C. E., Nevison, C. M., *et al.* (2001). Individual recognition in mice mediated by major urinary proteins. *Nature*, **414**, 631–4.

Hurst, J. L., Robertson, D. H. L., Tolladay, U., and Beynon, R. J. (1998). Proteins in urine scent marks of male house mice extend the longevity of olfactory signals. *Animal Behaviour*, **55**, 1289–97.

Igarashi, M., Nagata, A., Toh, H., Urade, Y., and Hayashi, O. (1992). Structural organization of the gene for prostaglandin D synthase in the rat brain. *Proceedings of the National Academy of Sciences of the United States of America*, **89**, 5376–80.

Janotová, K. and Stopka, P. (2009). Mechanisms of chemical communication: the role of major urinary proteins. *Folia Zoologica*, **58** (Suppl. 1), 41–55.

Janotová, K. and Stopka, P. (2011). The level of major urinary proteins is socially regulated in wild *Mus musculus musculus*. *Journal of Chemical Ecology*, doi.org/10.1007/s10886–011–9966–8.

Jemiolo, B., Alberts, J., Sochinski-Wiggins, S., Harvey, S., and Novotny, M. V. (1985). Behavioural and endocrine responses of female mice to synthetic analogs of volatile compounds in male urine. *Animal Behaviour*, **33**, 1114–18.

Jemiolo, B., Harvey, S., and Novotny, M. (1986). Promotion of the Whitten effect in female mice by synthetic analogs of male urinary constituents. *Proceedings of the National Academy of Sciences of the United States of America*, **83**, 4576–9.

Jia, C. and Halpern, M. (1996). Subclasses of vomeronasal receptor neurons: differential expression of G proteins ($G_{i\alpha 2}$ and $G_{o\alpha}$) and segregated projections to the accessory olfactory bulb. *Brain Research*, **719**, 117–28.

Johansson, B. G. and Jones, T. M. (2007). The role of chemical communication in mate choice. *Biological Reviews*, **82**, 265–89.

Johnson, D., Al-Shawi, R., and Bishop, J. O. (1995). Sexual dimorphism and growth hormone induction of murine phero-binding proteins. *Journal of Molecular Endocrinology*, **14**, 21–34.

Karlson, P. and Lüscher, M. (1959). Pheromones: a new term for a class of biologically active substances. *Nature*, **183**, 55–6.

Karn, R. C. (1998). Steroid binding by mouse salivary proteins. *Biochemical Genetics*, **36**, 105–17.

Karn, R. C. and Dlouhy, S. R. (1991). Salivary androgen-binding protein variation in *Mus* and other rodents. *Journal of Heredity*, **82**, 453–8.

Karn, R. C., Young, D. A., and Laukaitis, C. M. (2010). A candidate subspecies discrimination system involving a vomeronasal receptor gene with different alleles fixed in *M. m. domesticus* and *M. m. musculus*. *PloS ONE*, **5**, e12638.

Knopf, J. L., Gallagher, J. F., and Held, W. A. (1983). Differential, multihormonal regulation of the mouse major urinary protein gene family in the liver. *Molecular and Cellular Biology*, **3**, 2232–40.

Krieger, J., Schmitt, A., Lobel, D., *et al.* (1999). Selective activation of G protein subtypes in the vomeronasal organ upon stimulation with urine-derived compounds. *Journal of Biological Chemistry*, **274**, 4655–62.

Kruhoffer, M., Bub, A., Cieslak, A., *et al.* (1997). Gene expression of aphrodisin in female hamster genital tract segments. *Cell and Tissue Research*, **287**, 153–60.

Laukaitis, C. M., Critser, E. S., and Karn, R. C. (1997). Salivary androgen-binding protein (ABP) mediates sexual isolation in *Mus musculus*. *Evolution*, **51**, 2000–5.

Lazar, J., Greenwood, D. R., Rasmussen, L. E. L., and Prestwich, G. D. (2002). Molecular and functional characterization of an odorant binding protein of the Asian elephant, *Elephas maximus*: implications for the role of lipocalins in mammalian olfaction. *Biochemistry*, **41**, 11786–94.

Lee, S. V. D. and Boot, L. M. (1955). Spontaneous pseudopregnancy in mice. *Acta Physiologica et Pharmacologica Neerlandica*, **4**, 442–3.

Leinders-Zufall, T., Brennan, P., Widmayer, P., *et al.* (2004). MHC Class I peptides as chemosensory signals in the vomeronasal organ. *Science*, **306**, 1003–37.

Lin, D. Y., Zhang, S.-Z., Block, E., and Katz, L. C. (2005). Encoding social signals in the mouse main olfactory bulb. *Nature*, **434**, 470–7.

Lobel, D., Jacob, M., Volkner, M., and Breer, H. (2002). Odorants of different chemical classes interact with distinct odorant binding protein subtypes. *Chemical Senses*, **27**, 39–44.

Logan, D. W., Marton, T. F., and Stowers, L. (2008). Species specificity in Major Urinary Proteins by parallel evolution. *PLoS ONE*, **3**, e3280.

Ma, W., Miao, Z., and Novotny, M. V. (1999). Induction of estrus in grouped female mice (*Mus domesticus*) by synthetic analogues of preputial gland constituents. *Chemical Senses*, **24**, 289–93.

Macrides, F., Clancy, A. N., Singer, A. G., and Agosta, W. C. (1984). Male hamster investigatory and copulatory responses to vaginal discharge: an attempt to impart sexual significance to an arbitrary chemosensory stimulus. *Physiology & Behavior*, **33**, 627–32.

Magert, H. J., Cieslak, A., Alkan, O., *et al.* (1999). The golden hamster aphrodisin gene. *Journal of Biological Chemistry*, **274**, 444–50.

Magert, H. J., Hadrys, T., Cieslak, A., *et al.* (1995). cDNA sequence and expression pattern of the putative pheromone carrier aphrodisin. *Proceedings of the National Academy of Sciences of the United States of America*, **92**, 2091–5.

Marchese, S., Pes, D., Scaloni, A., Carbone, V., and Pelosi, P. (1998). Lipocalins of boar salivary glands binding odours and pheromones. *European Journal of Biochemistry*, **252**, 563–8.

Marchlewska-Koj, A., Cavaggioni, A., Mucignat-Caretta, C., and Olejniczak, P. (2000). Stimulation of estrus in female mice by male urinary proteins. *Journal of Chemical Ecology*, **26**, 2355–66.

Massey, A. and Vandenbergh, J. G. (1980). Puberty delay by a urinary cue from female house mice in feral populations. *Science*, **209**, 821–2.

Maynard Smith, J. and Harper, D. (2003). Signal and cues. In *Animal Signals*, ed. P. H. Harvey and R. M. May. New York: Oxford University Press, pp. 3–9.

McClintock, M. K. (1984). Group mating in the domestic rat as a context for sexual selection: consequences for the analysis of sexual behavior and neuroendocrine responses. *Advances in the Study of Behavior*, **14**, 1–50.

More, L. (2006). Mouse major urinary proteins trigger ovulation via the vomeronasal organ. *Chemical Senses*, **31**, 393–401.

Mucignat-Caretta, C., Caretta, A., and Cavaggioni, A. (1995). Acceleration of puberty onset in female mice by male urinary proteins. *Journal of Physiology*, **486**, 517–22.

Mucignat-Caretta, C., Redaelli, M., Orsetti, A., *et al.* (2010). Urinary volatile molecules vary in males of the 2 European subspecies of the house mouse and their hybrids. *Chemical Senses*, **35**, 647–54.

Munclinger, P. and Frynta, D. (2000). Social interactions within and between two distant populations of house mouse. *Folia Zoologica*, **49**, 1–6.

Nevison, C. M., Armstrong, S., Beynon, R., Humphries, R., and Hurst, J. (2003). The ownership signature in mouse scent marks is involatile. *Proceedings of the Royal Society B: Biological Sciences*, **270**, 1957–63.

Novotny, M. V. (2003). Pheromones, binding proteins and receptor responses in rodents. *Biochemical Society*, **31**, 117–22.

Novotny, M., Harvey, S., and Jemiolo, B. (1990). Chemistry of male dominance in the house mouse, *Mus domesticus*. *Experientia*, **46**, 109–13.

Novotny, M. V., Harvey, S., Jemiolo, B., and Alberts, J. (1985). Synthetic pheromones that promote inter-male aggression in mice. *Proceedings of the National Academy of Sciences of the United States of America*, **82**, 2059–61.

Novotny, M. V., Jemiolo, B., Harvey, S., Wiesler, D., and Marchlewska-Koj, A. (1986). Adrenal-mediated endogeneous metabolites inhibit puberty in female mice. *Science*, **231**, 722–5.

Novotny, M., Jorgenson, J. W., Carmack, M., *et al.* (1980). Chemical studies of the primer mouse pheromones. In *Chemical Signals: Vertebrates and Aquatic Invertebrates*, ed. D. Muller-Schwarze and R. M. Silverstein. New York: Plenum Press, pp. 377–90.

Novotny, M. V., Ma, W., Wiesler, D., and Žídek, L. (1999a). Positive identification of the puberty-accelerating pheromone of the house mouse: the volatile ligands associating with the major urinary protein. *Proceedings of the Royal Society B: Biological Sciences*, **266**, 2017–22.

Novotny, M. V., Ma, W., Žídek, L., and Daev, E. (1999b). Recent biochemical insights into puberty acceleration, estrus induction, and puberty delay in the house mouse. In *Advances in Chemical Communication in Vertebrates*, ed. R. E. Johnston, D. Muller-Schwarze, and P. Sorensen. New York: Plenum Press, pp. 99–116.

Oehninger, S., Coddington, C. C., Hodgen, G. D., and Seppala, M. (1995). Factors affecting fertilization: endometrial placental protein 14 reduces the capacity of human spermatozoa to bind to the human zona pellucida. *Fertility and Sterility*, **63**, 377–83.

Papes, F., Logan, D. W., and Stowers, L. (2010). The vomeronasal organ mediates inter-species defensive behaviors through detection of protein pheromone homologs. *Cell*, **141**, 692–703.

Parkes, A. S. and Bruce, H. M. (1961). Olfactory stimuli in mammalian reproduction. *Science*, **134**, 1049–54.

Payne, C. E., Malone, N., Humphries, R., *et al.* (2001). Heterogeneity of major urinary proteins in house mice: population and sex differences. In *Chemical Signals in Vertebrates, vol. 9*, ed. A. Marchlewska-Koj, J. J. Lepri, and D. Muller-Schwarze. New York: Kluwer Academic/Plenum Publishers, pp. 233–40.

Peele, P., Salazar, I., Mimmack, M., Keverne, E. B., and Brennan, P. A. (2003). Low molecular weight constituents of male mouse urine mediate the pregnancy block effect and convey information about the identity of the mating male. *European Journal of Neuroscience*, **18**, 622–8.

Pes, D., Dal Monte, M., Gianni, M., and Pelosi, P. (1992). Isolation of two odorant-binding proteins from mouse nasal tissue. *Comparative Biochemistry and Physiology*, **103B**, 1011–17.

Pes, D., Robertson, D. H. L., Hurst, J. L., Gaskell, S., and Beynon, R. J. (1999). How many major urinary proteins are produced by the house mouse *Mus domesticus*? In *Advances in Chemical Communication in Vertebrates*, ed. R. E. Johnston, D. Muller-Schwarze, and P. Sorensen. New York: Plenum Press, pp. 149–62.

Petrak, J., Myslivcova, D., Halada, P., *et al.* (2007). Iron-independent specific protein expression pattern in the liver of HFE-deficient mice. *International Journal of Biochemistry and Cell Biology*, **39**, 1006–15.

Pevsner, J., Hou, V., Snowman, A. M., and Snyder, S. H. (1990). Odorant-binding protein. *Journal of Biological Chemistry*, **265**, 6118–25.

Pevsner, J., Hwang, P. M., Sklar, P. B., Venable, J. C., and Snyder, S. H. (1988a). Odorant-binding protein and its mRNA are localized to lateral nasal gland implying a carrier function. *Proceedings of the National Academy of Sciences of the United States of America*, **85**, 2383–7.

Pevsner, J., Reed, R. R., Feinstein, P., and Snyder, S. H. (1988b). Molecular cloning of odorant-binding protein: member of a ligand carrier family. *Science*, **241**, 336–9.

Ptacek, M. B. (2000). The role of mating preferences in shaping interspecific divergence in mating signals in vertebrates. *Behavioural Processes*, **51**, 111–34.

Ramm, S. A., Cheetham, S. A., and Hurst, J. L. (2008). Encoding choosiness: female attraction requires prior physical contact with individual male scents in mice. *Proceedings of the Royal Society: Biological Sciences*, **275**, 1727–35.

Ranganathan, V., Jana, N. R., and De, P. K. (1999). Hormonal effects on hamster lacrimal gland female-specific major 20 kDa secretory protein and its immunological similarity with submandibular gland major male-specific proteins. *Journal of Steroid Biochemistry and Molecular Biology*, **70**, 151–8.

Restrepo, D., Arellano, J., Oliva, A. M., Schaeffer, M. L., and Lin, W. (2004). Emerging views on the distinct but related roles of the main and accessory olfactory systems in responsiveness to chemosensory signals in mice. *Hormones and Behavior*, **46**, 247–56.

Rich, T. J. and Hurst, J. L. (1998). Scent marks as reliable signals of the competitive ability of mates. *Animal Behaviour*, **56**, 727–35.

Roberts, S. A., Simpson, D. M., Armstrong, S. D., *et al.* (2010). Darcin: a male pheromone that stimulates female memory and sexual attraction to an individual male's odour. *BMC Biology*, **8**, 75.

Robertson, D. H. L., Hurst, J. L., Bolgar, M. S., Gaskell, S. J., and Beynon, R. J. (1997). Molecular heterogeneity of urinary proteins in wild house mouse populations. *Rapid Communications in Mass Spectrometry*, **11**, 786–90.

Rodriguez, I., Greer, C. A., Mok, M. Y., and Mombaerts, P. (2000). A putative pheromone receptor gene expressed in human olfactory mucosa. *Nature Genetics*, **26**, 18–19.

Rümke, P. and Thung, P. J. (1964). Immunological studies on the sex-dependent prealbumin in mouse urine and on its occurence in the serum. *Acta Endocrinologica*, **47**, 156–64.

Rusu, A. S., Krackow, S., Jedelský, P. L., Stopka, P., and König, B. (2008). A qualitative investigation of major urinary proteins in relation to the onset of aggressive behavior and dispersive motivation in male wild house mice (*Mus musculus domesticus*). *Journal of Ethology*, **26**, 127–35.

Ryan, M. J. and Rand, A. S. (1993). Species recognition and sexual selection as a unitary problem in animal communication. *Evolution*, **47**, 647–57.

Salier, J. (2000). Chromosomal location, exon/intron organization and evolution of lipocalin genes. *Biochimica et Biophysica Acta*, **1482**, 25–34.

Sam, M., Vora, S., Malic, B., *et al.* (2001). Odorants may arouse instinctive behaviours. *Nature*, **412**, 142.

Sampsell, B. and Held, W. (1985). Variation in the major urinary protein multigene family in wild-derived mice. *Genetics*, **109**, 549–68.

Seppälä, M., Taylor, R. N., Koistinen, H., and Milgrom, E. (2002). Glycodelin: a major lipocalin protein of the reproductive axis with diverse actions in cell recognition and differentiation. *Endocrine Reviews*, **23**, 401–30.

Shahan, K. and Derman, E. (1984). Tissue-specific expression of major urinary protein (MUP) genes in mice: characterization of MUP mRNAs by restriction mapping of cDNA and by in vitro translation. *Molecular and Cellular Biology*, **4**, 2259–65.

Shaw, P. H., Held, W. A., and Hastie, N. D. (1983). The gene family for major urinary proteins: expression in several secretory tissues of the mouse. *Cell*, **32**, 755–61.

Sherborne, A. L., Thom, M. D., Paterson, S., *et al.* (2007). The genetic basis of inbreeding avoidance in house mice. *Current Biology*, **17**, 2061–6.

Shi, P. and Zhang, J. (2007). Comparative genomic analysis identifies an evolutionary shift of vomeronasal receptor gene repertoires in the vertebrate transition from water to land. *Genome Research*, **17**, 166–74.

Singer, A. G., Agosta, W. C., Clancy, A. N., and Macrides, F. (1987). The chemistry of vomeronasally detected pheromones: characterization of an aphrodisiac protein. *Annals of the New York Academy of Sciences*, **519**, 287–98.

Singer, A. and Macrides, F. (1990). Aphrodisin: pheromone or transducer? *Chemical Senses*, **15**, 199–203.

Singer, A., Macrides, F., Clancy, A. N., and Agosta, W. C. (1986). Purification and analysis of a proteinaceous aphrodisiac pheromone from hamster vaginal discharge. *Journal of Biological Chemistry*, **261**, 13323–6.

Singh, P. B., Brown, R. E., and Roser, B. (1987). MHC antigens in urine as olfactory recognition cues. *Nature*, **327**, 161–4.

Smadja, C., Catalan, J., and Ganem, G. (2004). Strong premating divergence in a unimodal hybrid zone between two subspecies of the house mouse. *Journal of Evolutionary Biology*, **17**, 165–76.

Smadja, C. and Ganem, G. (2002). Subspecies recognition in the house mouse: a study of two populations from the border of a hybrid zone. *Behavioral Ecology*, **13**, 312–20.

Smadja, C. and Ganem, G. (2005). Asymmetrical reproductive character displacement in the house mouse. *Journal of Evolutionary Biology*, **18**, 1485–93.

Smadja, C. and Ganem, G. (2008). Divergence of odorant signals within and between the two European subspecies of the house mouse. *Behavioral Ecology*, **19**, 223–30.

Spehr, M., Kelliher, K. R., Li, X. H., *et al.* (2006). Essential role of the main olfactory system in social recognition of major histocompatibility complex peptide ligands. *The Journal of Neuroscience*, **26**, 1961–70.

Spinelli, S., Ramoni, R., Grolli, S., *et al.* (1998). The structure of the monomeric porcine odorant binding protein sheds light on the domain swapping mechanism. *Biochemistry*, **37**, 7913–18.

Stehn, R. A. and Jannett, F. J. (1981). Male-induced abortion in various microtine rodents. *Journal of Mammalogy*, **62**, 369–72.

Steinbrecht, R. A. (1998). Odorant-binding proteins: expression and function. *Annals of the New York Academy of Sciences*, **855**, 323–32.

Stopka, P., Janotová, K., and Heyrovský, D. (2007). The advertisement role of major urinary proteins in mice. *Physiology & Behavior*, **91**, 667–70.

Stopka, P. and Macdonald, D. W. (1998). Signal interchange during mating in the wood mouse (*Apodemus sylvaticus*): the concept of active and passive signalling. *Behaviour*, **135**, 231–49.
Stopková, R., Hladovcová, D., Kokavec, J., *et al.* (2009). Multiple roles of secretory lipocalins (MUP, OBP) in mice. *Folia Zoologica*, **58**, 29–40.
Stopková, R., Stopka, P., Janotová, K., and Jedelský, P. L. (2007). Species-specific expression of major urinary proteins in the house mice (*Mus musculus musculus* and *Mus musculus domesticus*). *Journal of Chemical Ecology*, **33**, 861–9.
Stopková, R., Zdráhal, Z., Ryba, S., *et al.* (2010). Novel OBP genes similar to hamster Aphrodisin in the bank vole, *Myodes glareolus*. *BMC Genomics*, **11**, 45.
Strong, R. K. (2006). Siderocalins. In *Lipocalins*, ed. B. Akestrom, N. Borregaard, D. R. Flower and J. P. Salier. Georgetown, TX: Landes Bioscience, pp. 83–98.
Talley, H. M., Laukaitis, C. M., and Karn, R. C. (2001). Female preference for male saliva: implications for sexual isolation of *Mus musculus* preferences. *Evolution*, **55**, 631–4.
Tamura, K., Dudley, J., Nei, M., and Kumar, S. (2007). MEGA4: Molecular Evolutionary Genetics analysis (MEGA) software version 4.0. *Molecular Biology and Evolution*, **24**, 1596–9.
Taylor, A. J., Cook, D. J., and Scott, D. J. (2008). Role of odorant binding proteins: comparing hypothetical mechanisms with experimental data. *Chemosensory Perception*, **1**, 153–62.
Tegoni, M., Pelosi, P., Vincent, F., *et al.* (2000). Mammalian odorant binding proteins. *Biochimica et Biophysica Acta*, **1482**, 229–40.
Thom, M. D. and Beynon, R. J. (2006). The role of the Major Histocompatibility Complex in scent communication. In *Chemical Signals in Vertebrates*, ed. R. T. Mason, M. P. LeMaster, and D. Müller-Schwarze. New York: Springer, pp. 173–82.
Thom, M. D., Stockley, P., Jury, F., *et al.* (2008). The direct assessment of genetic heterozygosity through scent in the mouse. *Current Biology*, **18**, 619–23.
Thompson, R. N., Napier, A., and Wekesa, K. S. (2007). Chemosensory cues from the lacrimal and preputial glands stimulate production of IP3 in the vomeronasal organ and aggression in male mice. *Physiology & Behavior*, **90**, 797–802.
Thuesen, P. (1977). A comparison of the agonistic behaviour of the *Mus musculus* L. and *Mus musculus domesticus* Rutty (Mammalia and Rodentia). *Videnskabelige Meddelelser fra Dansk naturhistorisk Forening i Kobenhavn*, **140**, 117–28.
Trinh, K. and Storm, D. R. (2003). Vomeronasal organ detects odorants in absence of signaling through main olfactory epithelium. *Nature Neuroscience*, **6**, 519–25.
Trinh, K. and Storm, D. R. (2004). Detection of odorants through the main olfactory epithelium and vomeronasal organ of mice. *Nutrition Reviews*, **62**, S189–92.
Turton, M. J., Robertson, D. H. L., Smith, J. R., Hurst, J. L., and Beynon, R. J. (2010). Roborovskin, a lipocalin in the urine of the Roborovski hamster, *Phodopus roborovskii*. *Chemical Senses*, **35**, 675–84.
van Zegeren, K. and van Oortmerssen, G. A. (1981). Frontier disputes between the West-and East-European house mouse in Schleswig-Holstein, West Germany. *Zeitschrift für Säugetierkunde*, **46**, 363–9.
Vandenbergh, J. G. (1969). Male odor accelerates female sexual maturation in mice. *Endocrinology*, **84**, 658–60.

Vidic, J., Grosclaude, J., Momnnerie, R., *et al.* (2008). On a chip demonstration of a functional role for Odorant Binding Protein in the preservation of olfactory receptor activity at high odorant concentration. *Lab Chip*, **8**, 678–88.

Wang, Z., Sindreu, C. B., Li, V., *et al.* (2006). Pheromone detection in male mice depends on signaling through the type 3 adenylyl cyclase in the main olfactory epithelium. *The Journal of Neuroscience*, **26**, 7375–9.

Whitten, W. K. (1958). Modification of the oestrous cycle of the mouse by external stimuli associated with the male: changes in the oestrous cycle determined by vaginal smears. *Journal of Endocrinology*, **17**, 307–13.

Wicks, L. F. (1941). Sex and proteinuria of mice. *Proceedings of the Society for Experimental Biology and Medicine*, **48**, 395–400.

Wyatt, T. D. (2005). Pheromones: convergence and contrasts in insects and vertebrates. In *Chemical Signal in Vertebrates 10*, ed. R. T. Mason, M. P. LeMaster, and D. Müller-Schwarze. New York: Springer, pp. 7–20.

Xu, F., Schaefer, M., Kida, I., *et al.* (2005). Simultaneous activation of mouse main and accessory olfactory bulbs by odors or pheromones. *The Journal of Comparative Neurology*, **489**, 491–500.

Yamazaki, K., Beauchamp, G. K., Wysocki, C. J., *et al.* (1983). Recognition of H-2 types in relation to the blocking of pregnancy in mice. *Science*, **221**, 186–8.

Zhang, X. H., Zhang, X. M., and Firestein, S. (2007). Comparative genomics of odorant and pheromone receptor genes in rodents. *Genomics*, **89**, 441–50.

Zhang, X. M. and Firestein, S. (2002). The olfactory receptor gene superfamily of the mouse. *Nature Neuroscience*, **5**, 124–33.

Zhou, X., Zhang, X. L., Weng, Y., *et al.* (2009). High abundance of testosterone and salivary androgen-binding protein in the lateral nasal gland of male mice. *Journal of Steroid Biochemistry and Molecular Biology*, **117**, 81–6.

9

The evolution of MHC diversity in house mice

DUSTIN J. PENN AND KERSTIN MUSOLF

Introduction

The existence within natural populations of large amounts of genetic variation in molecules and morphology presents an evolutionary problem. (Clarke, 1979)

The genes of the major histocompatibility complex (MHC) appear to be the most polymorphic loci in vertebrates, and explaining how natural selection maintains such genetic diversity ('population allelic richness') is a major unsolved problem in evolutionary biology. There is accumulating evidence that MHC polymorphisms are maintained by balancing selection (Apanius *et al.*, 1997; Spurgin and Richardson, 2010); however, the nature of this selection is still unclear. MHC genes encode cell-surface glycoproteins (class I and II molecules) that present peptide antigens to T cells, and thereby play an important role in the development of the T cell receptor (TCR) repertoire, immunological self/non-self recognition, and resistance to pathogens and parasites. Many MHC alleles increase susceptibility to infectious and autoimmune diseases, which makes MHC polymorphisms especially puzzling, as these harmful alleles should be eliminated by natural selection. The evolutionary *origin* of MHC diversity is generated by mutation, recombination, and gene conversion (Martinsohn *et al.*, 1999); however, these mechanisms do not explain how selection maintains polymorphisms. In this chapter, we review theoretical and empirical studies on the evolution of MHC diversity, with a particular focus on house mice (*Mus musculus*).

There are two general hypotheses proposed to explain how selection maintains MHC polymorphisms. One hypothesis suggests that MHC polymorphisms are maintained by selection from pathogens and parasites (pathogen-mediated selection (PMS)) (Apanius *et al.*, 1997; Spurgin and Richardson, 2010). This idea is the most likely explanation, given the role MHC molecules play in immune recognition of pathogens. The most viable models for PMS include negative frequency-dependent

Evolution of the House Mouse, ed. Miloš Macholán, Stuart J. E. Baird, Pavel Munclinger, and Jaroslav Piálek. Published by Cambridge University Press.

selection (FDS) and fluctuating selection (FS) (Hill *et al.*, 1991; Hedrick, 2002), though direct evidence for these hypotheses is lacking. If pathogens are the *only* factor driving MHC diversity, however, then most other immunological receptor loci should also be highly polymorphic (Penn and Potts, 1999). As MHC genes appear to have exceptional polymorphism, it is crucial to consider other sources of selection.

A second general hypothesis suggests that MHC diversity is maintained through sexual selection (non-random mating) (SS). Studies in house mice, and an increasing number of other species, indicate that MHC genes influence odour and mating preferences (Penn and Potts, 1999; Penn, 2002; Yamazaki and Beauchamp, 2007; Radwan *et al.*, 2008; Schwensow *et al.*, 2008; for humans, see Havlicek and Roberts, 2009). MHC-disassortative mating preferences are sufficient to drive MHC polymorphisms (Hedrick, 1992), and there are several possible reasons why such preferences evolved. MHC-dependent mating preferences may function to enhance offspring resistance to pathogens, avoid inbreeding, or both (Penn and Potts, 1999; Penn *et al.*, 2002; Piertney and Oliver, 2006). In addition to increasing offspring MHC and overall heterozygosity, MHC-disassortative preferences may function to optimize offspring MHC heterozygosity or provide a 'moving target' to protect offspring against rapidly evolving pathogens (Penn and Potts, 1999). The PMS and SS hypotheses are sometimes mistakenly assumed to be mutually exclusive alternatives; however, these ideas are completely compatible. In fact, there are several reasons to expect PMS to selectively favour the evolution of MHC-dependent mating preferences (Penn and Potts, 1999).

In this chapter we provide an integrative overview of the evolution of MHC diversity (for other reviews, see Apanius *et al.*, 1997; Edwards and Hedrick, 1998; Penn and Potts, 1999; Meyer and Thomson, 2001; Penn, 2002; Bernatchez and Landry, 2003; Garrigan and Hedrick, 2003; Sommer, 2005; Milinski, 2006; Piertney and Oliver, 2006; Spurgin and Richardson, 2010). Recent reviews on this topic focus on 'non-model' species, whereas here we focus mainly on house mice (*M. musculus*), an important model species for MHC research (Box 9.1). We summarize surveys of MHC polymorphisms, and we consider a related problem, the evolution of duplications. We consider both PMS and SS hypotheses, and address misunderstandings about the SS hypothesis. Finally, we provide ideas for the future and explain why more studies are needed on selection on MHC genes in more natural ecological conditions.

MHC polymorphisms

This degree of polymorphism is unprecedented; there is no other locus whose polymorphism comes anywhere near that of the HLA-A,B and H-2K,D loci. (Klein, 1987)

Box 9.1 House mice as a model organism in MHC research

Domesticated house mice are the premier model species for biomedical research, and have long played an important role in MHC research (Klein, 1975, 1986; Melvold, 2001; Penn and Ilmonen, 2005). MHC genes were first discovered in research on tissue rejection using domesticated strains of house mice. In the 1930s, Peter Gorer found mouse blood group antigens, which he called 'antigens I–IV' and George Snell and his colleagues later bred the first congenic strains of mice and showed that tissue rejection is controlled by multiple closely linked loci, rather than a single locus (MHC genes in mice are often called '*H2*' genes, a term that originated by combining Snell's 'histocompatibility (H) genes' with Gorer's 'antigen II', whereas in humans they are often called 'human leucocyte antigen' (HLA)). Subsequent studies with house mice have led to some of the most important discoveries in immunology, including the finding that MHC genes control immune responses to antigens (McDevitt and Benacerraf, 1969; McDevitt and Chinitz, 1969) and the development of the TCR (Zinkernagel and Doherty, 1975, 1997). The unusually high levels of polymorphisms were first described in wild populations of house mice (Klein, 1970; Klein and Bailey, 1971). In more recent years, the development of mutant, transgenic, and other strains of mice have enabled researchers to examine how MHC genes affect disease resistance, including immune resistance to infection, tumour surveillance, autoimmunity, and other forms of immunopathology. Genetic techniques that allow disruption or insertion of particular MHC and other loci (so-called genetic 'knock-outs' and 'knock-ins') provide state-of-the-art methods (Shastry, 1995; Wolfer *et al.*, 2002). An amazing number of MHC congenic, transgenic, and other mouse strains are available from JAX labs and other commercial suppliers, and these mice have become central tools for biomedical research (Crawley, 1999, 2000). Unfortunately, however, there has been comparatively little work conducted on MHC – or other genes – in populations of wild house mice to understand how natural selection maintains the diversity of these genes.

There are five closely linked classical, antigen-presenting, class I and II MHC loci in house mice (Box 9.2), and population surveys find extensive genetic polymorphisms at these loci, with over 100 major alleles per locus (Klein, 1970; Duncan *et al.*, 1979a; Klein and Figueroa, 1981). A typical house mouse population will carry more than a dozen alleles per locus, as with humans and many other species. However, levels of variation vary between loci of different classes and among the different chains (α and β) of class II – e.g. Eα shows reduced polymorphisms (Duncan *et al.*, 1979b) – and MHC alleles or their frequencies differ among neighbouring populations in spite of a social structure that would tend to reduce heterozygosity (Duncan *et al.*, 1979a). In addition to being polymorphic, MHC genes also show unusually even distributions of allelic frequencies (indicating balancing selection) (Nadeau *et al.*, 1988) and large sequence divergences among alleles (suggesting ancient origins) (Apanius *et al.*, 1997; Hughes and Yeager, 1998; Klein *et al.*, 2007).

It has been suggested that MHC polymorphisms can be used for estimating the size of the founding population of new species and analysing the long-term population demographics of phylogenetic lineages (Klein *et al.*, 2007). However, to distinguish demographic from selective processes, one must use many unlinked loci, because a bottleneck will affect the entire genome, whereas a selective sweep affects some loci more than others (Galtier *et al.*, 2000).

Box 9.2 Genetic architecture/genomic organization (H2, haplotypes, and strain nomenclature)

The MHC as an organizational unit originated during the evolution of jawed vertebrates (Kelley *et al.*, 2005). Its genomic organization appears conserved in many species, as linkage between the class I and class II regions is present from cartilaginous fishes to humans, with the exception of bony fishes (Kulski *et al.*, 2002). In mice, MHC (also known as '*H2*') genes are located on chromosome 17. On the basis of distinct structural and functional characteristics, MHC genes are divided between class I and II molecules (so-called class III genes are not shown in Fig. 9.1). Adjacent to the classical class I and II MHC molecules, other homologous genes are non-classical MHC genes, though their functions are not as well understood. The classical H2 complex (see Box 9.1) in mice is small, with three class I loci (K, D, and L) and two types of class II loci (A and E), each further divided into their respective α and β chains. Recombination within is relatively rare, and therefore a set of alleles is usually transmitted together as a unit called a haplotype. This is indicated in laboratory strain nomenclature by superscript designations (e.g. $H2^b$, $H2^d$, $H2^k$). Within one haplotype, individual alleles have been correspondingly designated with respect to their original locus (e.g. K^d, $A\beta^d$, $A\alpha^d$, $E\beta^d$, $E\alpha^d$, D^d, and L^d). Through recombination events, haplotypes occur with alleles derived from different original sources (see Table 1 in Stuart, 2010). Interestingly, a recent study comparing two closely linked MHC class II genes across the European house mouse hybrid zone demonstrated that despite their tight linkage, both loci showed disparate evolutionary patterns (Čížková *et al.*, 2011), indicating the need for multi-locus analyses.

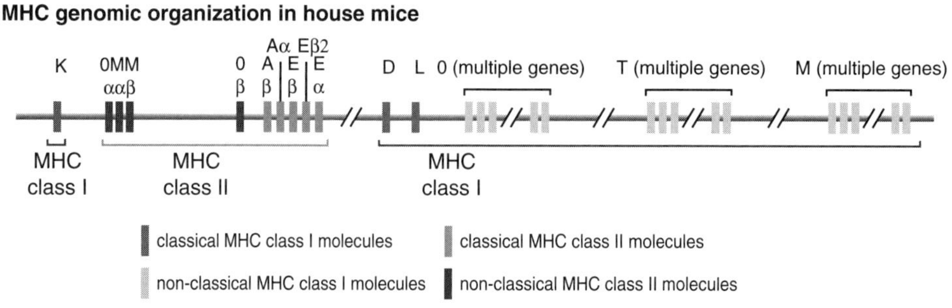

Figure 9.1 MHC genomic organization in house mice (modified with permission from DeFranco *et al.*, 2007). See plate section for a colour version of this figure.

Klein and his colleagues (1987, 2007) surveyed MHC diversity in wild house mice and discovered surprisingly poor consistency between the phylogeny of MHC alleles and phylogenetic trees that reveal the evolutionary history of different *Mus* species. One expects that MHC alleles would be more closely related to alleles within than among species, and yet they often show just the opposite pattern. To explain this odd finding, Klein and his colleagues suggested that MHC allelic lineages are ancient and predate speciation (cladogenesis) events. Such 'trans-species polymorphism' is defined as the passage of allelic lineages from ancestral to descendant species (Klein, 1987; Figueroa *et al.*, 1988; Klein *et al.*, 2007). Subsequent findings in rodents are mixed, with evidence for ancestral polymorphism of MHC class II genes in the genus *Mus* (Edwards *et al.*, 1997) and *Apodemus* (Musolf *et al.*, 2004), but not in *Peromyscus* (Richman *et al.*, 2003), suggesting that variable selection pressures have contributed to the evolution of MHC in rodents. For cases of trans-species polymorphisms it appears as if selection has maintained MHC allelic lineages for tens of millions of years, which seems improbable. Another possibility is that these patterns may be due to gene conversion (segmental exchanges of sequence motives between alleles of the same or different loci; Martinsohn *et al.*, 1999). Results from comparative phylogenetic analyses have suggested that interspecific allele sharing is generated by a recent origin and independent convergent evolution from similar selection pressures from pathogens, rather than common ancestry (Yeager and Hughes, 1999). If so, this would require widespread independent convergence in DNA sequences among many different species. It is as if MHC genes are horizontally transferred among closely related hosts, which might be feasible given that MHC sequences are horizontally transferred from hosts to pathogens (Imase *et al.*, 2001). However, we are unaware of studies that have considered this idea. These various hypotheses constructed to explain the evolutionary origins of MHC diversity are not mutually exclusive, and they will be difficult to distinguish (Sette *et al.*, 2003).

Some species lack polymorphic MHC loci, and several hypotheses have been proposed to explain why: (1) selective sweeps driven by infectious diseases; (2) genetic bottlenecks and drift; (3) reduced pathogens in solitary species and aquatic mammals (Slade, 1992; Bowen *et al.*, 2004); and (4) limited gene flow among populations. Island populations are likely to have reduced MHC diversity, as found in Australian bush rats (*Rattus fuscipes*) (Seddon and Baverstock, 1999). As more MHC studies are published on different species, comparative approaches will help understand the evolutionary origins and maintenance of MHC diversity. They can be used, for example, to test whether MHC polymorphisms are elevated among species exposed to more infectious diseases (or greater fluctuations) (Nevo and Beiles, 1992), as predicted by PMS, or species showing MHC-dependent mating preferences, as predicted by SS (Penn and

Potts, 1999). Inter- and intraspecific variation in MHC polymorphisms may also be influenced by social structure (e.g. social versus solitary) (Faulkes *et al.*, 1990; Hambuch and Lacey, 2002). Comparative approaches must take into account phylogeny, population size, and historical bottlenecks, which can be a challenge. It would be of practical importance to know whether low levels of MHC diversity increases the risk of extinction, as this question has implications for endangered species and conservation (O'Brien and Evermann, 1988; Hughes and Nei, 1992; Radwan *et al.*, 2010).

Non-classical MHC genes (Box 9.2) are monomorphic in laboratory mouse strains; however, there are surprisingly few studies on these genes in wild mice, as most studies of the MHC region have focused on the immunological functions of classical class I and II molecules (Ohtsuka *et al.*, 2008). As we explain below, non-classical MHC genes appear to evolve from classical MHC genes, and the latter duplicate with a high frequency (Hughes and Nei, 1989). Non-classical MHC genes are often silenced and non-functional, though some perform various functions in the immune system and reproduction (e.g. non-classical MHC genes are expressed on the placenta). Recent studies discovered that non-classical Ib MHC genes in mice regulate the expression of receptors of the vomeronasal organ (VNO), which controls pheromone detection (reviewed in Hedge, 2003; Ishii *et al.*, 2003; Loconto *et al.*, 2003). Studies are needed to examine polymorphism of non-classical MHC loci, and determine how they evolve new functions.

Early studies on the evolution of MHC genes relied mainly on genetic data, coalescence theory, and comparisons of synonymous *versus* non-synonymous substitutions to test for selection over macro-evolutionary timescales (Hughes and Nei, 1988; Takahata and Nei, 1990) and to examine their evolutionary origins. Bioinformatic approaches using the large amounts of genomic data increasingly available will surely continue along these lines. These approaches are important for understanding the history of MHC genes; however, they shed little light on the nature of selection on MHC genes (Apanius *et al.*, 1997; Garrigan and Hedrick, 2003; modelling approach in Ejsmond *et al.*, 2010; Spurgin and Richardson, 2010). Understanding how selection maintains MHC diversity requires studying fitness of animals living in the wild or semi-natural conditions. Before considering how selection maintains MHC polymorphisms, however, we examine a closely related aspect of diversity, the evolution of MHC gene duplications.

The evolution of MHC gene duplications

Until around 1990, most multigene families were thought to be subject to concerted evolution, in which all member genes of a family evolve as a unit in concert. However, phylogenetic analysis of MHC and other immune system genes showed a quite different evolutionary pattern, and a new model called birth-and-death evolution was proposed. (Nei and Rooney, 2005)

MHC genes are a multiple-gene family and the evolution of gene duplications has also been difficult to explain (Innan and Kondrashov, 2010). The number of MHC loci can vary greatly among different species, and it is unclear why. Although duplicate genes may only rarely evolve new functions, the stochastic silencing of such genes may play a significant role in the passive origin of new species (Lynch and Conery, 2000). Generation of new alleles can be achieved via gene conversion events or via birth-and-death evolution (Nei and Rooney, 2005). Birth-and-death evolution produces new orthologous genes by duplication, the duplicates diverging by accumulating mutations over time. Some of these duplicate genes persist in the genome, and some are deleted or become pseudogenes (Nei and Rooney, 2005). Duplicated MHC genes can be found in a wide range of species (e.g. Miller and Lambert, 2004; Miska *et al.*, 2004; Reusch *et al.*, 2004; Harf and Sommer, 2005; Baker *et al.*, 2006; Bryja *et al.*, 2006; Schwensow *et al.*, 2007) and support the idea that gene duplication is an important process in MHC evolution.

Many species show copy number variation (CNV) for MHC loci, though to our knowledge there has been little work on this area in house mice. A comparison of two different inbred strains of mice suggested that the number of class I genes vary by up to 20% among different haplotypes (Rogers *et al.*, 1985). Interestingly, a recent study found that *M. m. musculus/domesticus* hybrids show increased CNV in the genome (Scavetta and Tautz, 2010).

Understanding the evolution of MHC gene duplication and maintenance of CNV should help shed light on the selective maintenance of MHC polymorphisms – and vice versa. Heterozygote advantage, for example, is suspected to provide positive selection for new gene duplications, as well as promoting polymorphisms, since MHC heterozygosity often enhances disease-resistance and fitness (Penn *et al.*, 2002) (see below). The strength of selection for duplications should be similar to the amount favouring heterozygotes (Otto and Yong, 2002). Natural selection may favour individuals with an intermediate rather than maximum number of MHC molecules for immune function ('optimal MHC heterozygosity' hypothesis) (reviewed in Penn and Potts, 1999; Woelfing *et al.*, 2009). Evidence for optimizing selection on individual allele number is insufficient to explain the evolution of MHC polymorphisms (Hedrick, 2004); however, it shows *stabilizing selection* on the number of MHC loci. Studies are needed to test whether the optimum number of individual allelic diversity varies in space or time due to fluctuating selection (see below). Mating preferences for MHC heterozygotes or individuals with high allelic diversity (Griggio *et al.*, 2011;

Thoß *et al.*, 2011) may also provide positive or fluctuating selection for new gene duplications. Sexual selection may reinforce PMS, favouring new disease-resistant mutations at new duplications and stabilizing selection for an intermediate or optimum number of loci.

To understand the evolution of duplications, it is necessary to determine the various functions of MHC molecules. Although MHC genes are generally assumed to function primarily or exclusively for antigen presentation to T cells, studies with house mice show that MHC genes are also involved in many other – and completely unexpected – functions, including individual odour (Yamazaki *et al.*, 1990; reviewed in Penn and Potts, 1998b), commensal microflora communities (Lanyon *et al.*, 2007), pregnancy rejection (Rülicke *et al.*, 1998), iron metabolism (Cardoso *et al.*, 2002), pheromone detection (reviewed in Hedge, 2003; Ishii *et al.*, 2003; Loconto *et al.*, 2003), and the development and function of cortical neurons in the brain (Zohar *et al.*, 2008). Some, though not all, of these studies have ruled out potential confounding effects due to genes linked to the MHC – further work is needed here. Nevertheless, these findings have important implications for understanding the evolution and the consequences of MHC diversity and gene duplication. Moreover, it has been suggested that *promiscuous functions* of an existing protein (also known as 'cross-reactivity' or 'moon-lighting activity') can provide a selective advantage in changing environments, allow for the evolution of new gene functions, and enhance evolvability (Otto and Yong, 2002; Aharoni *et al.*, 2005). Since MHC genes are functional for more than antigen presentation, it is necessary to consider how these other functions influence selection. Before considering the possible role of sexual selection, however, we first examine selection from pathogens.

Pathogen-mediated selection

[T]he struggle against disease, and particularly infectious disease, has been a very influential evolutionary agent, and some of its results have been rather unlike those of the struggle against natural forces, hunger, and predators, or with members of the same species . . .. The most that the average species can achieve is to dodge its minute enemies by constantly producing new genotypes. (Haldane, 1949)

There is much evidence that MHC genes influence resistance to pathogens and parasites, especially in laboratory mice (Apanius *et al.*, 1997; Penn, 2002), though surprisingly little evidence that pathogens impose balancing selection in wild populations (also see Goüy de Bellocq *et al.*, Chapter 18 in this volume, for recent studies challenging the hypothesis that *M. m. musculus/M. m. domesticus* hybrids are susceptible to parasites). Several ideas have been proposed for how selection from pathogens might drive MHC diversity; however, it will not be easy to distinguish between these non-mutually exclusive hypotheses (Spurgin and Richardson, 2010).

Group selection

H-2 polymorphism prevents [extinction of the species] .... It assures the existence in a population of at least some individuals with the right H-2 alleles and the right T-cell repertoire to enable activation of defense reactions to any pathogen. (Klein, 1979)

It was once widely assumed that MHC polymorphisms evolved as a way to protect the species or populations from extinction, but it is increasingly recognized that such group selection is inadequate to explain MHC diversity. Group selection may favour populations and species with greater MHC diversity; however, it cannot overcome the process of individual or genetic selection eroding diversity within groups. If selection within populations promotes MHC diversity, group selection is unnecessary. This argument does not mean that group selection does not play a role, and should be discounted. Group selection might contribute to diversification (e.g. such multi-level, balancing selection as has been used to explain the evolutionary maintenance of T alleles in house mice), but it is not a sufficient explanation. Thus, the main goal is to determine how selection operates on MHC diversity within populations.

Heterozygote advantage

This idea of heterozygote advantage – or overdominance – is still central to MHC research. (Meyer and Thomson, 2001)

It is often suggested that MHC diversity is maintained by heterozygote advantage (Doherty and Zinkernagel, 1975; Hughes and Nei, 1988; Takahata and Nei, 1990). This hypothesis assumes that MHC heterozygotes (*ab*) have higher fitness than either of the parental homozygotes (*aa* or *bb*). To avoid confusion, this synergistic effect is better labelled 'heterozygote superiority' or 'overdominant selection', as explained below. This idea is based on the expectation that MHC heterozygotes present a greater diversity of antigens to the immune system and cover more immunological 'blind spots' than homozygotes (Doherty and Zinkernagel, 1975). Immunological studies, primarily with house mice, however, paint a more complicated picture. MHC heterozygosity may broaden the array of antigens for TCR; however, it may also have negative effects that favour intermediate rather than maximal levels of MHC heterozygosity ('optimal MHC heterozygosity' hypothesis) (reviewed in Penn and Potts, 1999; Woelfing *et al.*,

2009). MHC genes influence the development of the TCR repertoire, and increasing the number of different MHC alleles expressed may eventually have negative effects on many aspects of immunity (TCR diversity, regulatory T cells, immunopathology, etc.). The development of the TCR repertoire is influenced by both MHC and background genes (Vukusic *et al.*, 1995), and it appears to depend mainly on the diversity of antigens presented by MHC molecules (Matsutani *et al.*, 2011). Therefore, the benefits of MHC heterozygosity may decline with increasing levels of background or overall heterozygosity in the genome (Ilmonen *et al.*, 2007; Thoß *et al.*, 2011).

Observational and experimental studies of mice show that MHC heterozygosity enhances host resistance to most pathogens and parasites; however, contrary to what is often assumed, these studies provide few examples of MHC *heterozygote superiority* (Williams *et al.*, 1978; McLeod *et al.*, 1989; reviewed in Penn, 2002). Most experimental studies show that MHC-dependent disease resistance is generally *dominant*, rather than overdominant (i.e. heterozygotes are as resistant as the best parental homozygote, but not more so; see Fig. 9.2). MHC-dependent resistance can also be co-dominant for some pathogens, which provides no HA (Penn, 2002; Wedekind *et al.*, 2005, 2006), and recessive, which can result in a heterozygote *disadvantage* (Penn, 2002; Ilmonen *et al.*, 2007). Dominance is the most common pattern, and it is sufficient to provide heterozygotes an advantage compared to the average parental homozygote (Penn, 2002). Therefore, dominance may explain cases of HA in observational studies with wild, outbred species, indicating a need for allele-specific measures in future studies (Penn, 2002; Lipsitch *et al.*, 2003). Such 'heterozygote advantage through dominance' will not generate the balancing selection required to maintain MHC diversity, although it could provide a selective advantage for MHC-disassortative mating preferences (Penn, 2002). When resistance is dominant, overdominance may emerge during multiple infections, especially with combinations of pathogens showing reciprocal resistance profiles (Apanius *et al.*, 1997; Penn and Potts, 1999). One experiment found support for this idea (McClelland *et al.*, 2003), but more studies are needed to compare the relative *fitness* of MHC genotypes as well as pathogen clearance. For some infectious diseases, MHC heterozygotes can be high immune responders, but are less likely to survive the infection compared to homozygotes (due to immunopathology) (Doherty and Zinkernagel, 1975).

Studies measuring fitness of wild-derived house mice indicate that MHC heterozygosity can increase or reduce fitness depending upon exposure to infectious agents and genetic background. First, a study on congenic strains of mice found that MHC heterozygosity enhanced pathogen clearance and host survival (against multiple strains of *Salmonella*) (Penn *et al.*, 2002). HA was due to resistance being dominant, and not overdominant, as in most previous studies.

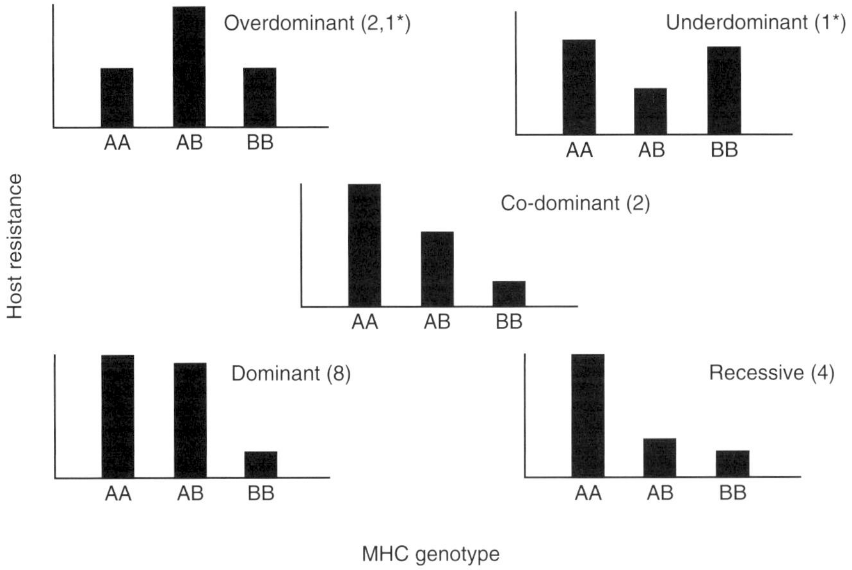

Figure 9.2 Experimental infection studies with laboratory mice show that MHC-dependent resistance to infection is generally (a) dominant (eight studies), rather than (b) overdominant (one or two studies), (c) co-dominant (two studies), (d) recessive (four studies), or (e) under-dominant (either zero or one study) (Penn, 2002). The study usually cited in support of overdominant selection found that MHC heterozygous mice present more antigens to T cells than homozygotes; however, the heterozygous mice died in the experiment, apparently from over-responsiveness (Doherty and Zinkernagel, 1975). Subsequent studies have found that MHC-dependent resistance to infection can be dominant (Penn *et al.*, 2002; McClelland *et al.*, 2003) or co-dominant (Wedekind *et al.*, 2006), and a study with semi-wild mice found that resistance to infection was recessive, which resulted in heterozygote disadvantage (Ilmonen *et al.*, 2007).

Similar results were found when antiviral defences were examined in *bm* mutant mice, which only differ genetically from controls by four amino acid substitutions in the peptide-binding region of class I molecules (Messaoudi *et al.*, 2002). Second, a study on semi-wild mice (wild × MHC-congenic crosses) found no evidence that MHC heterozygosity enhanced resistance to *Salmonella* infection (resistance was recessive); on the contrary, MHC-heterozygous females had reduced numbers of offspring compared with homozygotes (Ilmonen *et al.*, 2007). This result may have been due to the genetically heterozygous background or testing animals with maximal levels of MHC heterozygosity (optimal heterozygosity hypothesis). Third, a recent study on wild mice (F_2 from wild-caught *M. m. musculus*) living in large population enclosures found that MHC heterozygosity enhanced individual mating and reproductive success (Thoß *et al.*, 2011). This finding was not confounded by rare MHC alleles (there were none) or inbreeding (close inbreeding was controlled through experimental outbreeding of the mice and confirmed with genetic markers). It is crucial to control these factors when studying outbred animals since MHC heterozygosity is often correlated with genome-wide or background heterozygosity. This study found an interaction between MHC and background heterozygosity, such that the benefits of MHC heterozygosity became more variable and diminished with increasing levels of background heterozygosity (perhaps due to changes in the TCR repertoire by increased antigens presented) (Thoß *et al.*, 2011). Therefore, it is important to study how MHC affects host fitness as a whole, not only survival and resistance to one or a few pathogens, but also including reproductive success in wild, outbred animals living in natural or semi-natural conditions (Box 9.3).

Thus, overdominance rarely occurs, and while dominance can result in HA, this mechanism is insufficient to generate balancing selection by itself (Penn, 2002). Moreover, the models that support overdominant selection require that all heterozygotes and homozygotes have equal fitness (Takahata and Nei, 1990). When this unrealistic assumption is not met, the chances of maintaining polymorphism are reduced and only a few alleles can be maintained (Lewontin *et al.*, 1978; Hedrick and Kim, 2000). Intermediate rather than maximal heterozygosity (or individual allele number) increases resistance to multiple parasites (Wegner *et al.*, 2003), but this form of selection is insufficient to maintain MHC polymorphisms (Hedrick, 2004). On the other hand, HA through dominance or overdominance can still play an important role in contributing to the evolution of MHC polymorphisms, especially through fluctuating selection (see below).

Rare allele advantage (negative frequency-dependent selection)

I wish to suggest that the selection of rare biochemical genotypes has been an important agent not only in keeping species variable, but also in speciation. (Haldane, 1949)

A rather intriguing explanation for how pathogens might drive MHC diversity is through a process of host–pathogen coevolution that results in a 'rare allele advantage' or 'negative frequency-dependent selection' (FDS) (Bodmer, 1972; Clarke, 1979; Takahata and Nei, 1990; Borghans *et al.*, 2004). Since viruses and other pathogens often have much shorter generation times than their hosts, they can potentially evolve adaptations around their hosts' immune defences. If pathogens tend to adapt most rapidly to common MHC genotypes, which seems a reasonable assumption, this process will tend to provide a selective advantage for rare genotypes. Consequently, hosts with rare, resistant MHC

Box 9.3 The importance of studying wild house mice in natural or semi-natural conditions

Although studies on domesticated house mice in the laboratory have been extremely useful for understanding how MHC genes influence immune resistance to pathogens, odour, and other phenotypes, it is unclear whether and how these findings can be extrapolated to wild house mice. First, domesticated, laboratory strains of house mice kept in conventional conditions suffer from many health problems (Martin *et al.*, 2010), and there are significant differences in their behaviour and physiology compared to their wild, outbred counterparts (Crawley *et al.*, 1997; Miller *et al.*, 2002). Second, MHC genes can have different effects depending on the genetic background (G × G interactions), and whenever inconsistent results are found among strains, it is impossible to generalize without testing outbred animals. Third, MHC effects might only become apparent, for example, when studied on a homozygous background, when they are the only loci in the genome that differ among individuals. Additionally, determining how natural selection maintains MHC diversity in wild populations requires studying the survival and reproductive success (Darwinian fitness) of animals living in natural or semi-natural ecological conditions. For these reasons, researchers have studied MHC effects on fitness of mice living in semi-natural conditions, not only with congenic strains (Penn *et al.*, 2002), but also with 'semi-wild' (from domesticated × wild crosses) (Potts *et al.*, 1991; Penn and Potts, 1999; Ilmonen *et al.*, 2007) and 'wild-derived' house mice (recently trapped in the wild and reared in the laboratory) (Sherborne *et al.*, 2007; Thoß *et al.*, 2011). We know of no studies that have addressed how MHC genes influence fitness in wild, free-ranging house mice, whose survival and reproduction is not influenced by researchers, and such studies might prove to be most informative. We would emphasize, however, that since studies on free-ranging animals also involve human observers, trapping, marking, sampling, they are 'semi-natural' as well, and differ only in the relative degree or type of human interference. Perhaps the most compelling reason for conducting studies in natural or semi-natural conditions is that fitness effects sometimes only become apparent in competitive settings outside of the laboratory, e.g. inbreeding depression (Meagher *et al.*, 2000; Ilmonen *et al.*, 2008); selection against segregation distorters (t alleles; Carroll *et al.*, 2004); and effects of MHC heterozygosity on reproductive success (Ilmonen *et al.*, 2007; Thoß *et al.*, 2011).

alleles evolve to a higher frequency over time, which then would provide a selective advantage for pathogens to adapt to the new common MHC allele. This process of reciprocal adaptation can theoretically result in continual coevolutionary cycles, and maintain genetic polymorphisms in both hosts and pathogens (models reviewed in Apanius *et al.*, 1997; Hedrick and Kim, 2000; Tellier and Brown, 2007). Host–pathogen coevolution can easily account for polymorphisms of more than 50 alleles per MHC locus (Borghans *et al.*, 2004).

There is evidence for FDS on MHC alleles in human HIV studies (Moore *et al.*, 2002; Trachtenberg *et al.*, 2003); however, there is no direct evidence in house mice or any other species to our knowledge. Many studies, mainly of laboratory mice, have discovered a variety of mechanisms through which pathogens can adapt to host MHC genotypes, such as by escaping antigen presentation (Potts and Slev, 1995). Some pathogen adaptations that allow evasion of MHC-dependent recognition, such as downregulating of MHC expression, may not cause selection on the antigen-binding site, though they might still play a role in driving MHC diversity. However, to our knowledge, there is no evidence for genotype-specific resistance and infectivity between host MHC and pathogen alleles, as required for FDS, or models showing that reciprocal coevolution can occur even though host–parasite interactions are generally 'diffuse' (Thompson, 1989). FDS will be difficult to test and to distinguish from other models, as they share several predications (Apanius *et al.*, 1997; Hedrick, 2002; Spurgin and Richardson, 2010).

Fluctuating selection over time or space

We know virtually nothing about the role of spatial and temporal heterogeneity of environment in evolution and it is here that the most fruitful and most difficult area of research lies. (Lewontin, 1964, cited in Hedrick and Kim, 2000)

The most realistic hypothesis for how pathogens drive MHC diversity is probably through temporal or spatial fluctuations in selection (FS) (Hill *et al.*, 1991; Hedrick, 2002; for general overview of models see Gillespie, 1991; Thompson, 2005). If parasites vary geographically – either in species presence, density, or genetically – over their hosts' range, then in such a mosaic landscape different MHC alleles will be selectively favoured in different populations. MHC diversity can become reduced within populations as a result of selective sweeps, but MHC diversity among host populations should be maintained. Similarly, if parasites vary over time, then fluctuating temporal selection can also maintain MHC diversity (even without FDS or other forms of host–pathogen coevolution). FS in time or space provides broader conditions for maintaining polymorphisms than in constant environments. The conditions favouring diversity are generally more robust for spatial than temporal variation (Hedrick, 2006), though both are likely to occur and may interact (pathogens in some locations may change more over time than others).

Hedrick (2002) examined a model that assumes different MHC alleles confer resistance to different pathogens (specific resistance alleles), resistance to infection is dominant, and that pathogens fluctuate over time in terms of their presence or

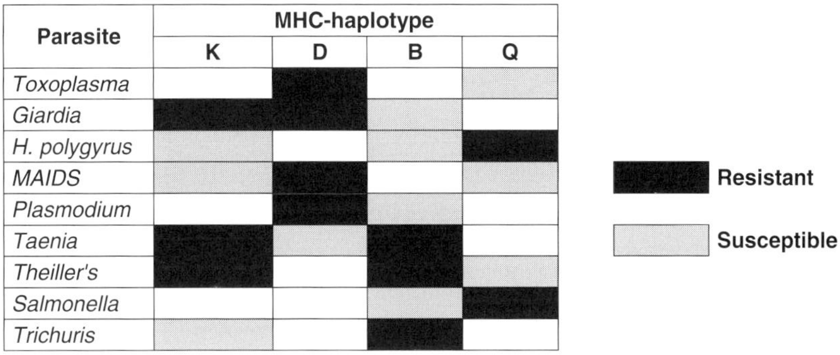

Figure 9.3 Studies on house mice show that MHC haplotypes that confer resistance to one parasite often increase susceptibility to other parasites, which supports the idea that parasites impose balancing selection on MHC genes (see references in Apanius *et al.*, 1997, Table 3).

absence. This model indicates that fluctuating temporal selection can maintain high levels of MHC polymorphisms, and requires no FDS or 'intrinsic heterozygote advantage' (overdominance). It is unclear whether fluctuations in pathogens are sufficient to explain MHC diversity in house mice, though studies on laboratory house mice provide support for the assumptions in this model. First, a survey of experimental studies on MHC-dependent resistance to infectious diseases found that MHC haplotypes that enhance resistance to one pathogen in one study often increase susceptibility to other pathogens in other studies (allele-specific resistance/susceptibility tradeoffs) (Fig. 9.3) (Penn and Potts, 1999; Penn *et al.*, 2002). Such tradeoffs were confirmed with experimental infections (McClelland *et al.*, 2003). Second, experimental studies with mice show that MHC-dependent disease resistance is generally dominant (Penn, 2002; Penn *et al.*, 2002; McClelland *et al.*, 2003) (Fig. 9.2). More work is needed, however, to examine whether parasites show sufficient spatial or temporal fluctuations to drive MHC diversity (see review of the evidence in Piertney and Oliver, 2006).

The finding that MHC alleles confer resistance tradeoffs with different pathogens (Penn and Potts, 1999; Penn, 2002; Penn *et al.*, 2002; McClelland *et al.*, 2003) provides evidence for balancing selection on MHC diversity. Such balancing selection is not sufficient, however, to maintain high levels of MHC diversity under realistic conditions (unequal fitness among homozygotes and among heterozygotes). Some other factor is necessary, such as fluctuations in the presence/absence of pathogens (though environmental fluctuations do not solve the problem by generating G × E interactions, contrary to what is often suggested (see Hedrick, 2006)).

The various hypotheses for PMS are not mutually exclusive and there are reasons to expect that they may interact and reinforce each other. For example, spatial and temporal fluctuations can stabilize host–parasite coevolutionary dynamics and help to maintain genetic diversity in hosts and pathogens (Sasaki *et al.*, 2002). It is still unclear whether PMS is adequate to explain high levels of MHC diversity, especially since other immune recognition loci do not appear to be highly polymorphic. Increasingly, researchers are considering the possibility that sexual selection may also play a role in driving MHC diversity, and in the next section we review the evidence in house mice.

Sexual selection

These mating preferences could in natural populations serve the purpose of increasing the representation of particular H-2 haplotypes or of maintaining heterozygosity of genes in the region of H-2. (Yamazaki *et al.*, 1976)

Studies on house mice have found that MHC genes influence odour (reviewed in Penn and Potts, 1998b; Penn, 2002; Beauchamp and Yamazaki, 2003; Yamazaki and Beauchamp, 2007; Kwak *et al.*, 2010) and disassortative mating preferences (reviewed in Penn and Potts, 1999; Penn, 2002; Yamazaki and Beauchamp, 2007; Roberts, 2009). MHC disassortative preferences can potentially drive MHC polymorphisms, though there is still no direct evidence from populations of wild house mice. It has been suggested that MHC-dependent mating preferences in mice 'could not be much more than a quirk of nature without general significance' (Klein *et al.*, 1993). However, several recent studies have found evidence for MHC-dependent mating preferences in other taxa, including fish, reptiles, and even birds (Zelano and Edwards, 2002; Bernatchez and Landry, 2003; Milinski, 2006; Havlicek and Roberts, 2009). The most interesting implication of these studies is that they provide a potential example of how mating preferences can generate selection and evolution of genes that control immune recognition of pathogens and parasites.

Mating preferences

Yamazaki and his colleagues (1976, 1978) performed the pioneering studies that first found evidence that MHC genes influence odour and mating preferences. They studied MHC-congenic strains and found that males preferred to mate with females from MHC-dissimilar strains. However, since only some strains – and only males – showed such mating preferences, it remained unclear whether this behaviour even occurs in wild mice. Brown and his colleagues found that female laboratory mice displayed MHC-disassortative odour and mating preferences (Egid and Brown, 1989), but found no evidence with males (Eklund *et al.*, 1991). Potts and his colleagues (1991) studied populations of (semi-wild) mice living in semi-natural conditions and found that half the litters contained offspring sired by males other than the dominant, territorial male. Among these litters, MHC loci were more heterozygous than expected if females had mated only with the territorial male. This finding suggested that females sought extra-pair copulations with more-MHC-disparate males.

Yamazaki *et al.* (1988) found that males' MHC-mediated mating preferences could be reversed by rearing them with MHC-dissimilar families (cross-fostering). This effect was also found in female mate preferences in laboratory strains (Eklund, 1997a) and to a much lesser extent in wild female mice (Eklund, 1997b). This finding indicated that mice learn the MHC-identity of their family ('familial imprinting') and avoid mating with partners that have familial MHC genes. Penn and Potts (1998c) tested this idea in (semi-wild) female mice. Behavioural observations and genotyping offspring indicated that females avoid mating with MHC-similar males, unless they were cross-fostered at birth; then they avoided males similar to their family. These findings provide direct evidence for MHC-dependent mating preferences in females living in semi-natural conditions, and support the familial imprinting hypothesis.

Two recent studies raise questions about whether MHC genes influence mating biases in wild populations of house mice. Roberts and Gosling (2003) found that female mice are attracted to the scent marks of MHC-dissimilar males, but only when scent-marking rate was similar between stimulus males (and females' preference was better explained by males' marking rate than by MHC sharing). These findings provide evidence that females are attracted to the scent of MHC-dissimilar males, but also raise the possibility that such odour biases may not affect mating patterns in natural conditions where males vary in their scent marking. Another study compared the relative importance of MUP (major urinary proteins) versus MHC genes for inbreeding avoidance in wild-derived mice in semi-natural enclosures (Sherborne *et al.*, 2007). The mice could choose between full versus half-siblings (no option to avoid inbreeding), and genetic analyses of offspring were used to assess mating preferences. The authors found a deficit in successful matings (based on number of offspring) between mice sharing MUPs, but not MHC haplotypes, and concluded that inbreeding avoidance is based on MUP rather than MHC sharing. This conclusion is premature for several reasons, however. First, the results indicate that there was no inbreeding avoidance in this study (no bias to mate with half versus full-sibs), which is necessary to conclude that MUPs but not MHC genes influence inbreeding avoidance. Second, the mice may have avoided mating with familial

MHC genotypes (Yamazaki *et al.*, 1988; Penn and Potts, 1998c), but familial imprinting was not considered. Third, before being released into the enclosures, the mice were reared since birth together in cages with individuals carrying all the MHC haplotypes later encountered in the enclosures. Consequently, the mice were forced to choose their mates among individuals that potentially had already been classified as kin through familial imprinting (which might explain why there was no inbreeding avoidance) (Ruff *et al.*, 2011). Although MUPs play an important role in chemical communication (Hurst, 2009), there are only a few attempts to assess their role compared to MHC (Brennan, 2001; Sherborne *et al.*, 2007; Thom *et al.*, 2008; see also Stopka *et al.*, Chapter 8).

Cryptic female choice

MHC may also play a role in post-copulatory (cryptic) female choice in house mice and other species (Wedekind, 1994). There is mixed evidence for non-random fertilization success of gametes of different MHC haplotypes, though this effect appears to depend on a female's health or infection status (Wedekind *et al.*, 1996; Rülicke *et al.*, 1998). The mechanism may be controlled by egg–sperm interactions or perhaps modulated by olfactory receptor (OR) genes (Ziegler *et al.*, 2002) or other MHC-linked genes. MHC genes do not appear to be expressed at this stage, and yet in humans selection maintains a linkage disequilibrium between OR and MHC loci (Santos *et al.*, 2010). The sperm receptor selection hypothesis (reviewed in Ziegler *et al.*, 2005, 2010) proposes that during spermatogenesis the expression of chemoreceptors (such as olfactory receptors or VNO receptors) is modulated through elimination of self-reactive receptors, resulting in the generation of sperm that are sensitive only to non-self-MHC antigens. Ligand binding affects sperm motility and thereby fertilization potential. While the egg does not express MHC molecules itself, the surrounding granulose cells do, thereby shedding antigens into the follicular fluid (Dohr *et al.*, 1987), which could interact with approaching sperm.

Not all studies find MHC-dependent mating (pre- or post-copulatory) preferences, and it would be interesting to know if there are differences among geographical populations maintained by fluctuating selection. Also, individual preferences can change as mating preferences are often conditional, and expressed only under certain circumstances. For example, females may pay attention to the MHC of potential mates when seeking extra-pair mates, and especially when their own mates are genetically similar to themselves. One of the main problems with the sexual selection hypothesis, however, is that it is difficult to understand how MHC genes influence odour (reviewed in Penn and Potts, 1998b; Penn, 2002; Beauchamp and Yamazaki, 2003; Yamazaki and Beauchamp,

2007; Kwak *et al.*, 2010). MHC genes clearly influence odour, but how remains a mystery (see Stopka *et al.*, Chapter 8). In addition to determining the mechanisms underlying MHC-mediated mate choice, there is increasing interest in determining the underlying functions.

Evolutionary functions of MHC-dependent mating preferences

Two main types of hypotheses have been suggested to explain the potential functions of MHC-dependent mating preferences. MHC-disassortative mating preferences may be selectively favoured to: (1) enhance offspring resistance to pathogens through several possible mechanisms; (2) facilitate inbreeding avoidance; or both (reviewed in Penn and Potts, 1999; Penn, 2002). These hypotheses are not mutually exclusive, as inbreeding avoidance may also enhance offspring resistance to infectious diseases (Keller and Waller, 2002; Ilmonen *et al.*, 2009).

There are two main ways that MHC-disassortative mating preferences have been suggested to enhance offspring resistance to pathogens. First, MHC-disassortative mating preferences might function to produce disease-resistant, MHC heterozygous offspring (Penn and Potts, 1999; Penn *et al.*, 2002). As addressed above, the evidence for this idea shows that MHC heterozygosity often (but not always) enhances disease-resistance and fitness, and more studies are needed with wild or wild-derived house mice (Thoß *et al.*, 2011; see also Goüy de Bellocq *et al.*, Chapter 18). Similarly, MHC-dependent mating preferences might help optimize rather than maximize offspring MHC heterozygosity (Penn and Potts, 1999; Penn *et al.*, 2002) or individual allelic diversity (Reusch *et al.*, 2001; Aeschlimann *et al.*, 2003). Second, MHC-disassortative mating preferences may provide a 'moving target' against pathogens adapted to the parental genotypes (Penn and Potts, 1999). This 'moving target' hypothesis is supported by a computer simulation (Howard and Lively, 2002), and by an experiment with congenic house mice, which found that MHC heterozygotes can more effectively resolve co-infections of pathogens that are resistant to the immune defences of parental MHC genotypes (McClelland *et al.*, 2003).

MHC-disassortative mating preferences might function to help avoid kin matings (inbreeding avoidance) (Apanius *et al.*, 1997; Penn, 2002). Some studies find that house mice avoid kin matings (Winn and Vestal, 1986; Krackow and Matuschak, 1991), and there can be surprisingly strong selection against inbreeding (Barnard and Fitzsimons, 1989; Meagher *et al.*, 2000; Ilmonen *et al.*, 2008). The benefits from inbreeding avoidance appear to be greater than those from enhancing offspring MHC-heterozygosity (Potts *et al.*, 1988). The inbreeding avoidance hypothesis suggests that MHC genes play a role in kin recognition, though only a few studies have tested this idea with mice. Female mice sometimes nest communally, and a study with semi-wild mice found that females nest with MHC-similar females when siblings are unavailable (Manning *et al.*, 1992). Another study, however, found no evidence for MHC-similar odour preferences among congenic strains of female mice (Ehman and Scott, 2001). Yamazaki *et al.* (2000) studied parent–offspring recognition with MHC congenic mice, and found that females are more likely to retrieve pups when they are MHC-similar, and pups are attracted to the odours of mothers that are MHC-similar. This study provides direct evidence that MHC genes play a role in kin recognition, but there have been no such studies on wild mice to our knowledge.

There are other ways that SS can potentially play a role in promoting MHC diversity, besides disassortative mating preferences. For example, females often prefer to mate with disease-resistant males (Hamilton and Zuk, 1982), which will likely carry disease-resistant MHC alleles ('good genes' sexual selection) (von Schantz *et al.*, 1989; Penn, 2002; Eizaguirre *et al.*, 2009). Such directional mating preference for health or condition should enhance offspring fitness and reinforce PMS (FDS or FS) on MHC diversity. This idea has not been tested in house mice, but it is consistent with evidence that females prefer the scent of healthy versus infected males (Kavaliers and Colwell, 1995; Zala *et al.*, 2004) (reviewed in Penn and Potts, 1998a) and genetically resistant versus susceptible males (Zala *et al.*, 2008). Furthermore, females may be attracted to disease-resistant, MHC-heterozygous males, and there is some evidence for this idea in house mice (Thoß *et al.*, 2011). It is unclear whether preferences for MHC heterozygotes confer genetic benefits, though such preferences would reinforce PMS favouring heterozygotes.

Thus, the hypothesis that SS plays a role in shaping MHC diversity has been gaining support, but the evidence is still mixed in house mice and other species. The SS hypothesis has been controversial and sometimes treated with much scepticism, though sometimes due to misunderstandings (Box 9.4).

Outlook: functional evolutionary genomics and integrating PMS and SS

Because of the richness of data for this system and the apparent importance of a number of evolutionary factors affecting variation at [MHC loci], we feel that [the MHC] has become an exemplary system for understanding evolutionary genetics. (Hedrick *et al.*, 1987)

There is growing interest in trying to understand the evolutionary origins and selective maintenance of MHC genetic diversity (polymorphisms and duplications). It is unclear why this intriguing problem has not attracted more attention

Box 9.4 Why the sexual selection hypothesis is often treated with scepticism

The idea that SS might help explain the evolution of MHC polymorphisms has been controversial, but some of the scepticism is based on misunderstandings. For example, sometimes SS studies are misinterpreted as claims that mate choice is controlled solely by MHC genes. This misunderstanding may help explain why human studies have been controversial (reviewed in Roberts, 2009). Also, as we previously pointed out, the SS hypothesis is sometimes mistakenly assumed to be a mutually exclusive alternative to PMS, although these ideas are completely compatible (Penn and Potts, 1999).

Additionally, the SS hypothesis is often perceived as far-fetched. It is not always intuitive that mating preferences can function to produce disease-resistance offspring or provide an important source of selection, driving evolutionary genetic changes. Nevertheless, theoretical models support the idea that pathogens can drive the evolution of disassortative mating and other forms of non-random mating preferences (Howard and Lively, 2003; Nuismer *et al.*, 2008). In fact, the leading explanation for the evolutionary function of sexual reproduction itself is to allow hosts to 'keep up' in a never-ending co-evolutionary race with rapidly evolving pathogens and parasites (Red Queen hypothesis) (Hamilton, 2001). There is increasing evidence that mate choice in many species, including house mice, can function to provide genetic benefits for offspring (reviewed in Hettyey *et al.*, 2010), including resistance to infectious agents (Hamilton and Zuk, 1982). Thus, from an evolutionary perspective, the SS hypothesis is not as extraordinary as some assume.

Regardless of evolutionary theory and empirical evidence, the SS hypothesis will continue to be viewed as far-fetched by some researchers until the underlying mechanisms are better resolved. There is much evidence that MHC genes influence odour, and though there are several potential mechanisms, it remains a major challenge to determine how this occurs (reviewed in Penn and Potts, 1998b; Penn, 2002; Beauchamp and Yamazaki, 2003; Yamazaki and Beauchamp, 2007; Kwak *et al.*, 2010; see also Stopka *et al.*, Chapter 8). It is feasible that genes in the MHC region *other than class I or II loci* have a more important role in influencing odour and mating preferences, but there has been little work on this alternative. Nonetheless, we cannot rule out mating preferences just because the underlying chemosensory mechanisms are unresolved, no matter how far-fetched this hypothesis might seem. The general emphasis on the immunological functions of MHC genes can blind us to other possibilities, and it has led some to promote group selection and overdominant selection explanations, despite the fact that theoretical (Lewontin *et al.*, 1978; Hedrick and Kim, 2000) and empirical evidence (Apanius *et al.*, 1997; Penn, 2002) have long failed to support these ideas.

in the past – especially given its important consequences for human health – e.g. tissue rejection and resistance to infectious and autoimmune diseases. Biomedical research places a greater emphasis on mechanistic than evolutionary questions, and consequently, the vast majority of studies on MHC genes have been conducted on immunological mechanisms with domesticated strains of laboratory mice (Box 9.1). This approach has proven useful for unravelling immunological mechanisms controlling resistance to infectious diseases; however, as we previously stressed, determining how natural selection maintains MHC diversity requires studying Darwinian fitness of animals living in natural or semi-natural ecological conditions (functional ecological genomics) (Box 9.3). Testing PMS in the field is a major challenge, especially since it requires tracking changes in host pathogens – and their fitness effects on hosts – over time and space. Future studies on MHC polymorphisms in wild populations would benefit by considering hypotheses about sexual selection, as well as pathogen-mediated selection. One of the most important reasons is that MHC heterozygosity can have significant effects on mating and reproductive success (Darwinian fitness) *even without any discernable effects on viability or survival* in the laboratory (Thoß *et al.*, 2011).

The development of high-throughput DNA-sequencing technologies is exciting and it has prompted an increasing number of studies on MHC polymorphisms in non-model organisms (see Wegner, 2009; Babik, 2010 and references therein). The availability of genomic technologies and bioinformatic analyses will undoubtedly provide new insights into MHC diversity, but there are many caveats and limitations. As Hedrick (2006: 81) points out:

> There appear to be many genomic regions with a history of adaptive selection, but only a small proportion of these indicate a genetic signal consistent with balancing selection. However, the smaller region thought to exhibit a balancing selection signal, compared with a selective sweep, may make detection beyond the resolution of most genomic scans. Further, alleles identified under directional selection may be part of balancing selection at a gene, and further examination of specific alleles may be necessary to clarify the nature of selection.

Perhaps the greatest challenge for studies on the evolution of MHC diversity is obtaining sufficiently *large sample sizes* to detect small selection coefficients. It is ironic that the large diversity of alleles that makes MHC genes so interesting also makes these genes difficult to study due to the large sample sizes necessary to compare fitness of different genotypes. This is problematic because even minimal selection coefficients may nonetheless drive evolutionary changes in MHC polymorphism (Apanius *et al.*, 1997; Meyer and Thomson, 2001).

REFERENCES

Aeschlimann, P. B., Haeberli, M. A., Reusch, T. B. H., Boehm, T., and Milinski, M. (2003). Female sticklebacks *Gasterosteus aculeatus* use self-reference to optimize MHC allele number during mate selection. *Behavioral Ecology and Sociobiology*, **54**, 119–26.

Aharoni, A., Gaidukov, L., Khersonsky, O., *et al.* (2005). The 'evolvability' of promiscuous protein functions. *Nature Genetics*, **37**, 73–6.

Apanius, V., Penn, D. J., Slev, P. R., Ruff, L. R., and Potts, W. K. (1997). The nature of selection on the major histocompatibility complex. *Critical Reviews in Immunology*, **17**, 179–224.

Babik, W. (2010). Methods for MHC genotyping in non-model vertebrates. *Molecular Ecology Resources*, **10**, 237–51.

Baker, C. S., Vant, M. D., Dalebout, M. L., *et al.* (2006). Diversity and duplication of DQB and DRB-like genes of the MHC in baleen whales (suborder: Mysticeti). *Immunogenetics*, **58**, 283–96.

Barnard, C. J. and Fitzsimons, J. (1989). Kin recognition and mate choice in mice: fitness consequences of mating with kin. *Animal Behaviour*, **38**, 35–40.

Beauchamp, G. K. and Yamazaki, K. (2003). Chemical signalling in mice. *Biochemical Society Transactions*, **31**, 147–51.

Bernatchez, L. and Landry, C. (2003). MHC studies in nonmodel vertebrates: what have we learned about natural selection in 15 years? *Journal of Evolutionary Biology*, **16**, 363–77.

Bodmer, W. F. (1972). Evolutionary significance of the HL-A system. *Nature*, **237**, 139–45.

Borghans, J. A. M., Beltman, J. B. and De Boer, R. J. (2004). MHC polymorphism under host–pathogen coevolution. *Immunogenetics*, **55**, 732–9.

Bowen, L., Aldridge, B. M., Gulland, F., *et al.* (2004). Class II multiformity generated by variable *MHC-DRB* region configurations in the California sea lion (*Zalophus californianus*). *Immunogenetics*, **56**, 12–27.

Brennan, P. (2001). How mice make their mark. *Nature*, **414**, 590–1.

Bryja, J., Galan, M., Charbonnel, N., and Cosson, J. F. (2006). Duplication, balancing selection and trans-species evolution explain the high levels of polymorphism of the *DQA* MHC class II gene in voles (Arvicolinae). *Immunogenetics*, **58**, 191–202.

Cardoso, E. M., Macedo, M. G., Rohrlich, P., *et al.* (2002). Increased hepatic iron in mice lacking classical MHC class I molecules. *Blood*, **100**, 4239–41.

Carroll, L. S., Meagher, S., Morrison, L., Penn, D. J., and Potts, W. K. (2004). Fitness effects of a selfish gene (the *Mus t* complex) are revealed in an ecological context. *Evolution*, **58**, 1318–28.

Čížková, D., Goüy de Bellocq, J. G., Baird, S. J., Piálek, J., and Bryja, J. (2011). Genetic structure and contrasting selection pattern at two major histocompatibility complex genes in wild house mouse populations. *Heredity*, **106**, 727–40.

Clarke, B. C. (1979). The evolution of genetic diversity. *Proceedings of the Royal Society of London Series B: Biological Sciences*, **205**, 453–74.

Crawley, J. N. (1999). Behavioral phenotyping of transgenic and knockout mice: experimental design and evaluation of general health, sensory functions, motor abilities, and specific behavioral tests. *Brain Research*, **835**, 18–26.

Crawley, J. N. (2000). *What's Wrong with My Mouse? Behavioral Phenotyping of Transgenic and Knockout Mice*. New York: Wiley.

Crawley, J. N., Belknap, J. K., Collins, A., *et al.* (1997). Behavioral phenotypes of inbred mouse strains: implications and recommendations for molecular studies. *Psychopharmacology*, **132**, 107–24.

DeFranco, A. L., Locksley, R. M., and Robertson, M. (2007). *Immunity: The Immune Response in Infectious and Inflammatory Disease*. Oxford: Oxford University Press.

Doherty, P. C. and Zinkernagel, R. M. (1975). Enhanced immunological surveillance in mice heterozygous at the H-2 gene complex. *Nature*, **256**, 50–2.

Dohr, G. A., Motter, W., Leitinger, S., *et al.* (1987). Lack of expression of HLA class I and class II molecules in the human oocyte. *Journal of Immunology*, **138**, 3766–70.

Duncan, W. R., Wakeland, E. K., and Klein, J. (1979a). Heterozygosity of H-2 loci in wild mice. *Nature*, **281**, 603–5.

Duncan, W. R., Wakeland, E. K., and Klein, J. (1979b). Histocompatibility-2 system in wild mice: VIII – frequencies of H-2 and Ia antigens in wild mice from Texas. *Immunogenetics*, **9**, 261–272.

Edwards, S. V., Chesnut, K., Satta, Y., and Wakeland, E. K. (1997). Ancestral polymorphism of MHC class II genes in mice: implications for balancing selection and the mammalian molecular clock. *Genetics*, **146**, 655–68.

Edwards, S. V. and Hedrick, P. W. (1998). Evolution and ecology of MHC molecules: from genomics to sexual selection. *Trends in Ecology and Evolution*, **13**, 305–11.

Egid, K. and Brown, J. L. (1989). The major histocompatibility complex and female mating preferences in mice. *Animal Behaviour*, **38**, 548–50.

Ehman, K. D. and Scott, M. E. (2001). Urinary odour preferences of MHC congenic female mice, *Mus domesticus*: implications for kin recognition and detection of parasitized males. *Animal Behaviour*, **62**, 781–9.

Eizaguirre, C., Yeates, S. E., Lenz, T. L., Kalbe, M., and Milinski, M. (2009). MHC-based mate choice combines good genes and maintenance of MHC polymorphism. *Molecular Ecology*, **18**, 3316–29.

Ejsmond, M. J., Babik, W., and Radwan, J. (2010). MHC allele frequency distributions under parasite-driven selection: a simulation model. *BMC Evolutionary Biology*, **10**, 332.

Eklund, A. C. (1997a). The effect of early experience on MHC-based mate preferences in two B10W strains of mice (*Mus domesticus*). *Behavior Genetics*, **27**, 223–9.

Eklund, A. C. (1997b). The major histocompatibility complex and mating preference in wild house mice (*Mus domesticus*). *Behavioral Ecology*, **8**, 630–4.

Eklund, A. C., Egid, K., and Brown, J. L. (1991). The major histocompatibility complex and mating preferences of male mice. *Animal Behaviour*, **42**, 693–4.

Faulkes, C. G., Abbott, D. H., and Mellor, A. L. (1990). Investigation of genetic diversity in wild colonies of naked mole-rats (*Heterocephalus glaber*) by DNA fingerprinting. *Journal of Zoology*, **221**, 87–97.

Figueroa, F., Gunther, E., and Klein, J. (1988). MHC polymorphism pre-dating speciation. *Nature*, **335**, 265–7.

Galtier, N., Depaulis, F., and Barton, N. H. (2000). Detecting bottlenecks and selective sweeps from DNA sequence polymorphism. *Genetics*, **155**, 981–7.

Garrigan, D. and Hedrick, P. W. (2003). Perspective: detecting adaptive molecular polymorphism: lessons from the MHC. *Evolution*, **57**, 1707–22.

Gillespie, J. H. (1991). *The Causes of Molecular Evolution*. New York: Oxford University Press.

Griggio, M., Biard, C., Penn, D. J., and Hoi, H. (2011). Female house sparrows 'count on' male genes: experimental evidence for MHC-dependent mate preference in birds. *BMC Evolutionary Biology*, **11**, 44.

Haldane, J. B. S. (1949). Disease and evolution. *La Ricerca Scientifica Supplement*, **19**, 68–76.

Hambuch, T. M. and Lacey, E. A. (2002). Enhanced selection for MHC diversity in social Tuco-Tucos. *Evolution*, **56**, 841–5.

Hamilton, W. D. (2001). *Narrow Roads of Gene Land: The Collected Papers of W. D. Hamilton: Volume 2: The Evolution of Sex*. Oxford: Oxford University Press.

Hamilton, W. D. and Zuk, M. (1982). Heritable true fitness and bright birds: a role for parasites? *Science*, **218**, 384–7.

Harf, R. and Sommer, S. (2005). Association between major histocompatibility complex class II DRB alleles and parasite load in the hairy-footed gerbil, *Gerbillurus paeba*, in the southern Kalahari. *Molecular Ecology*, **14**, 85–91.

Havlicek, J. and Roberts, S. C. (2009). MHC-correlated mate choice in humans: a review. *Psychoneuroendocrinology*, **34**, 497–512.

Hedge, A. N. (2003). MHC molecules in the vomeronasal organ: contributors to pheromonal discrimination? *Trends in Neurosciences*, **26**, 646–50.

Hedrick, P. W. (1992). Female choice and variation in the major histocompatibility complex. *Genetics*, **132**, 575–81.

Hedrick, P. W. (2002). Pathogen resistance and genetic variation at MHC loci. *Evolution*, **56**, 1902–8.

Hedrick, P. W. (2004). Comment on 'Parasite selection for immunogenetic optimality'. *Science*, **303**, 957.

Hedrick, P. W. (2006). Genetic polymorphism in heterogeneous environments: the age of genomics. *Annual Review of Ecology Evolution and Systematics*, **37**, 67–93.

Hedrick, P. W. and Kim, T. J. (2000). Genetics of complex polymorphisms: parasites and maintenance of the histocompatibility complex variation. In *Evolutionary Genetics: From Molecules to Morphology*, ed. R. S. Singh and C. B. Krimbas. Cambridge: Cambridge University Press, pp. 204–34.

Hedrick, P. W., Thomson, G., and William, K. (1987). Evolutionary genetics and HLA: another classic example. *Biological Journal of the Linnean Society*, **31**, 311–31.

Hettyey, A., Hegyi, G., Puurtinen, M., *et al.* (2010). Mate choice for genetic benefits: time to put the pieces together. *Ethology*, **116**, 1–9.

Hill, A. V., Allsopp, C. E., Kwiatkowski, D., *et al.* (1991). Common west African HLA antigens are associated with protection from severe malaria. *Nature*, **352**, 595–600.

Howard, R. S., and Lively, C. M. (2002). The Ratchet and the Red Queen: the maintenance of sex in parasites. *Journal of Evolutionary Biology*, **15**, 648–56.

Howard, R. S. and Lively, C. M. (2003). Opposites attract? Mate choice for parasite evasion and the evolutionary stability of sex. *Journal of Evolutionary Biology*, **16**, 681–9.

Hughes, A. L. and Nei, M. (1988). Pattern of nucleotide substitution at major histocompatibility complex class I loci reveals overdominant selection. *Nature*, **335**, 167–70.

Hughes, A. L. and Nei, M. (1989). Evolution of the major histocompatibility complex: independent origin of nonclassical class I genes in different groups of mammals. *Molecular Biology and Evolution*, **6**, 559–79.

Hughes, A. L. and Nei, M. (1992). Maintenance of MHC polymorphism. *Nature*, **355**, 402–3.

Hughes, A. L. and Yeager, M. (1998). Natural selection at major histocompatibility complex loci of vertebrates. *Annual Review of Genetics*, **32**, 415–35.

Hurst, J. L. (2009). Female recognition and assessment of males through scent. *Behavioural Brain Research*, **200**, 295–303.

Ilmonen, P., Penn, D. J., Damjanovich, K., *et al.* (2007). Major histocompatibility complex heterozygosity reduces fitness in experimentally infected mice. *Genetics*, **176**, 2501–8.

Ilmonen, P., Penn, D. J., Damjanovich, K., *et al.* (2008). Experimental infection magnifies inbreeding depression in house mice. *Journal of Evolutionary Biology*, **21**, 834–41.

Ilmonen, P., Stundner, G., Thoß, M., and Penn, D. J. (2009). Females prefer the scent of outbred males: good-genes-as-heterozygosity? *BMC Evolutionary Biology*, **9**, 104.

Imase, A., Kobayashi, K., Ohmae, H., Matsuda, H., and Iwamura, Y. (2001). Horizontal and vertical transmission of mouse class I MHC sequence in *Schistosoma mansoni*. *Parasitology*, **123**, 163–8.

Innan, H. and Kondrashov, F. (2010). The evolution of gene duplications: classifying and distinguishing between models. *Nature Reviews Genetics*, **11**, 97–108.

Ishii, T., Hirota, J., and Mombaerts, P. (2003). Combinatorial coexpression of neural and immune multigene families in mouse vomeronasal sensory neurons. *Current Biology*, **13**, 394–400.

Kavaliers, M. and Colwell, D. D. (1995). Discrimination by female mice between the odours of parasitized and non-parasitized males. *Proceedings of the Royal Society of London Series B: Biological Sciences*, **261**, 31–5.

Keller, L. F. and Waller, D. M. (2002). Inbreeding effects in wild populations. *Trends in Ecology and Evolution*, **17**, 230–41.

Kelley, J., Walter, L., and Trowsdale, J. (2005). Comparative genomics of major histocompatibility complexes. *Immunogenetics*, **56**, 683–95.

Klein, J. (1970). Histocompatibility-2 (H-2) polymorphism in wild mice. *Science*, **168**, 1362–4.

Klein, J. (1975). *Biology of the Mouse Histocompatibility-2 Complex: Principles of Immunogenetics Applied to a Single System*. New York: Springer.

Klein, J. (1979). The major histocompatibility complex of the mouse. *Science*, **203**, 516–21.

Klein, J. (1986). *Natural History of the Major Histocompatibility Complex*. New York: Wiley.

Klein, J. (1987). Origin of major histocompatibility complex polymorphism: the trans-species hypothesis. *Human Immunology*, **19**, 155–62.

Klein, J. and Bailey, D. W. (1971). Histocompatibility differences in wild mice: further evidence for the existence of deme structure in natural populations of the house mouse. *Genetics*, **68**, 287–97.

Klein, J. and Figueroa, F. (1981). Polymorphism of the mouse H-2 loci. *Immunological Reviews*, **60**, 24.

Klein, J., Sato, A., and Nikolaidis, N. (2007). MHC, TSP, and the origin of species: from immunogenetics to evolutionary genetics. *Annual Review of Genetics*, **41**, 281–304.

Klein, J., Satta, Y., O'hUigin, C., and Takahata, N. (1993). The molecular descent of the major histocompatibility complex. *Annual Review of Immunology*, **11**, 269.

Krackow, S. and Matuschak, B. (1991). Mate choice for non-siblings in wild house mice: evidence from a choice test and a reproductive test. *Ethology*, **88**, 99–108.

Kulski, J. K., Shiina, T., Anzai, T., Kohara, S., and Inoko, H. (2002). Comparative genomic analysis of the MHC: the evolution of class I duplication blocks, diversity and complexity from shark to man. *Immunological Reviews*, **190**, 95–122.

Kwak, J., Willse, A., Preti, G., Yamazaki, K., and Beauchamp, G. K. (2010). In search of the chemical basis for MHC odourtypes. *Proceedings of the Royal Society of London Series B: Biological Sciences*, **277**, 2417–25.

Lanyon, C. V., Rushton, S. P., O'Donnel, A. G., *et al.* (2007). Murine scent mark microbial communities are genetically determined. *FEMS Microbiology Ecology*, **59**, 576–83.

Lewontin, R. C. (1964). The role of linkage in natural selection. *Genetics Today: Proceedings of the XI International Congress of Genetics*, 517–25.

Lewontin, R. C., Ginzburg, L. R., and Tuljapurkar, S. (1978). Heterosis as an explanation for large amounts of genic polymorphism. *Genetics*, **88**, 149–70.

Lipsitch, M., Bergstrom, C. T., and Antia, R. (2003). Effect of human leukocyte antigen heterozygosity on infectious disease outcome: the need for allele-specific measures. *BMC Medical Genetics*, **4**, 1–9.

Loconto, J., Papes, F., Chang, E., *et al.* (2003). Functional expression of murine V2R pheromone receptors involves selective association with the M10 and M1 families of MHC class Ib molecules. *Cell*, **112**, 607–18.

Lynch, M. and Conery, J. S. (2000). The evolutionary fate and consequences of duplicate genes. *Science*, **290**, 1151–5.

Manning, C. J., Wakeland, E. K., and Potts, W. K. (1992). Communal nesting patterns in mice implicate MHC genes in kin recognition. *Nature*, **360**, 581–3.

Martin, B., Ji, S., Maudsley, S., and Mattson, M. P. (2010). 'Control' laboratory rodents are metabolically morbid: why it matters. *Proceedings of the National Academy of Sciences of the United States of America*, **107**, 6127–33.

Martinsohn, J. T., Sousa, A. B., Guethlein, L. A., and Howard, J. C. (1999). The gene conversion hypothesis of MHC evolution: a review. *Immunogenetics*, **50**, 168–200.

Matsutani, T., Ogata, M., Fujii, Y., *et al.* (2011). Shortening of complementarity determining region 3 of the T cell receptor α chain during thymocyte development. *Molecular Immunology*, **48**, 623–9.

McClelland, E. E., Penn, D. J., and Potts, W. K. (2003). Major histocompatibility complex heterozygote superiority during coinfection. *Infection and Immunity*, **71**, 2079–86.

McDevitt, H. O. and Benacerraf, B. (1969). Genetic control of specific immune responses. *Advances in Immunology*, **11**, 31–74.

McDevitt, H. O. and Chinitz, A. (1969). Genetic control of the antibody response: relationship between immune response and histocompatibility (H-2) type. *Science*, **163**, 1207–8.

McLeod, R., Skamene, E., Brown, C. R., Eisenhauer, P. B., and Mack, D. G. (1989). Genetic regulation of early survival and cyst number after peroral *Toxoplasma gondii* infection

Rogers, M. J., Germain, R. N., Hare, J., Long, E., and Singer, D. S. (1985). Comparison of MHC genes among distantly related members of the genus *Mus*. *Journal of Immunology*, **134**, 630–6.

Ruff, J. S., Nelson, A. C., Kubinak, J. L., and Potts, W. K. (2011). MHC signaling during social communication. In *Self and Non-Self*, ed. C. López-Larrea. Landes Bioscience and Springer Science+Business Media.

Rülicke, T., Chapuisat, M., Homberger, F. R., Macas, E., and Wedekind, C. (1998). MHC-genotype of progeny influenced by parental infection. *Proceedings of the Royal Society of London Series B: Biological Sciences*, **265**, 711–16.

Santos, P. S. C., Seki Uehara, C. J., Ziegler, A., Uchanska-Ziegler, B., and Bicalho, M. d. G. (2010). Variation and linkage disequilibrium within odorant receptor gene clusters linked to the human major histocompatibility complex. *Human Immunology*, **71**, 843–50.

Sasaki, A., Hamilton, W. D., and Ubeda, F. (2002). Clone mixtures and a pacemaker: new facets of Red-Queen theory and ecology. *Proceedings of the Royal Society of London Series B: Biological Sciences*, **269**, 761–72.

Scavetta, R. J. and Tautz, D. (2010). Copy number changes of CNV regions in intersubspecific crosses of the house mouse. *Molecular Biology and Evolution*, **27**, 1845–56.

Schwensow, N., Eberle, M., and Sommer, S. (2008). Compatibility counts: MHC-associated mate choice in a wild promiscuous primate. *Proceedings of the Royal Society of London Series B: Biological Sciences*, **275**, 555–64.

Schwensow, N., Fietz, J., Dausmann, K. H., and Sommer, S. (2007). Neutral versus adaptive genetic variation in parasite resistance: importance of major histocompatibility complex supertypes in a free-ranging primate. *Heredity*, **99**, 265–77.

Seddon, J. M. and Baverstock, P. R. (1999). Variation on islands: major histocompatibility complex (*MHC*) polymorphism in populations of the Australian bush rat. *Molecular Ecology*, **8**, 2071–9.

Sette, A., Sidney, J., Livingston, B. D., *et al.* (2003). Class I molecules with similar peptide-binding specificities are the result of both common ancestry and convergent evolution. *Immunogenetics*, **54**, 830–41.

Shastry, B. S. (1995). Genetic knockouts in mice: an update. *Experientia*, **51**, 1028–39.

Sherborne, A. L., Thom, M. D., Paterson, S., *et al.* (2007). The genetic basis of inbreeding avoidance in house mice. *Current Biology*, **17**, 1–6.

Slade, R. W. (1992). Limited MHC polymorphism in the southern elephant seal: implications for MHC evolution and marine mammal population biology. *Proceedings of the Royal Society of London Series B: Biological Sciences*, **249**, 163–71.

Sommer, S. (2005). The importance of immune gene variability (MHC) in evolutionary ecology and conservation. *Frontiers in Zoology*, **2**, 16.

Spurgin, L. G. and Richardson, D. S. (2010). How pathogens drive genetic diversity: MHC, mechanisms and misunderstandings. *Proceedings of the Royal Society of London Series B: Biological Sciences*, **277**, 979–88.

Stuart, P. M. (2010). Major histocompatibility complex (MHC): mouse. In *Encyclopedia of Life Sciences (ELS)*. Chichester: John Wiley & Sons.

Takahata, N. and Nei, M. (1990). Allelic genealogy under overdominant and frequency-dependent selection and polymorphism of major histocompatibility complex loci. *Genetics*, **124**, 967–78.

Tellier, A. and Brown, J. K. M. (2007). Stability of genetic polymorphism in host–parasite interactions. *Proceedings of the Royal Society of London Series B: Biological Sciences*, **274**, 809–17.

Thom, M. D., Stockley, P., Jury, F., *et al.* (2008). The direct assessment of genetic heterozygosity through scent in the mouse. *Current Biology*, **18**, 619–23.

Thompson, J. N. (1989). Concepts of coevolution. *Trends in Ecology and Evolution*, **4**, 179–83.

Thompson, J. N. (2005). *The Geographic Mosaic of Coevolution*. Chicago, IL: University of Chicago Press.

Thoß, M., Ilmonen, P., Musolf, K., and Penn, D. J. (2011). Major histocompatibility complex heterozygosity enhances reproductive success. *Molecular Ecology*, **20**, 1546–57.

Trachtenberg, E., Korber, B., Sollars, C., *et al.* (2003). Advantage of rare HLA supertype in HIV disease progression. *Nature Medicine*, **9**, 928–35.

von Schantz, T., Göransson, G., Andersson, G., *et al.* (1989). Female choice selects for a viability-based mate trait in pheasants. *Nature*, **337**, 166–9.

Vukusic, B., Poplonski, L., Phillips, L., *et al.* (1995). Both MHC and background gene heterozygosity alter T cell receptor repertoire selection in an antigen-specific response. *Molecular Immunology*, **32**, 1355–67.

Wedekind, C. (1994). Mate choice and maternal selection for specific parasite resistances before, during and after fertilization. *Philosophical Transactions of the Royal Society of London Series B: Biological Sciences*, **346**, 303–11.

Wedekind, C., Chapuisat, M., Macas, E., and Rülicke, T. (1996). Non-random fertilization in mice correlates with the MHC and something else. *Heredity*, **77**, 400–9.

Wedekind, C., Walker, M., and Little, T. J. (2005). The course of malaria in mice: major histocompatibility complex (MHC) effects, but no general MHC heterozygote advantage in single-strain infections. *Genetics*, **170**, 1427–30.

Wedekind, C., Walker, M., and Little, T. J. (2006). The separate and combined effects of MHC genotype, parasite clone, and host gender on the course of malaria in mice. *BMC Genetics*, **7**, 55.

Wegner, K. M. (2009). Massive parallel MHC genotyping: titanium that shines. *Molecular Ecology*, **18**, 1818–20.

Wegner, K. M., Kalbe, M., Kurtz, J., Reusch, T. B., and Milinski, M. (2003). Parasite selection for immunogenetic optimality. *Science*, **301**, 1343.

Williams, D. M., Grumet, F. C., and Remington, J. S. (1978). Genetic control of murine resistance to Toxoplasma gondii. *Infection and Immunity*, **19**, 416–20.

Winn, B. E., and Vestal, B. M. (1986). Kin recognition and choice of males by wild female house mice (*Mus musculus*). *Journal of Comparative Psychology*, **100**, 72–5.

Woelfing, B., Traulsen, A., Milinski, M., and Boehm, T. (2009). Does intra-individual major histocompatibility complex diversity keep a golden mean? *Philosophical Transactions of the Royal Society of London Series B: Biological Sciences*, **364**, 117–28.

Wolfer, D. P., Crusio, W. E., and Lipp, H.-P. (2002). Knockout mice: simple solutions to the problems of genetic background and flanking genes. *Trends in Neurosciences*, **25**, 336–40.

Yamazaki, K. and Beauchamp, G. K. (2007). Genetic basis for MHC-dependent mate choice. *Advances in Genetics*, **59**, 129–45.

Yamazaki, K., Beauchamp, G. K., Curran, M., Bard, J., and Boyse, E. A. (2000). Parent–progeny recognition as a function of MHC odortype identity. *Proceedings of the National Academy of Sciences of the United States of America*, **97**, 10500–2.

Yamazaki, K., Beauchamp, G. K., Imai Y., *et al.* (1990). Odor types determined by the major histocompatibility complex in germfree mice. *Proceedings of the National Academy of Sciences of the United States of America*, **87**, 8413–16.

Yamazaki, K., Beauchamp, G. K., Kupniewski, D., *et al.* (1988). Familial imprinting determines H-2 selective mating preferences. *Science*, **240**, 1331–2.

Yamazaki, K., Boyse, E. A., Mike, V., *et al.* (1976). Control of mating preferences in mice by genes in the major histocompatibility complex. *Journal of Experimental Medicine*, **144**, 1324–35.

Yamazaki, K., Yamaguchi, M., Andrews, P. W., Peake, B., and Boyse, E. A. (1978). Mating preferences of F2 segregants of crosses between MHC-congenic mouse strains. *Immunogenetics*, **6**, 253–9.

Yeager, M. and Hughes, A. L. (1999). Evolution of the mammalian MHC: natural selection, recombination, and convergent evolution. *Immunological Reviews*, **167**, 45–58.

Zala, S. M., Chan, B. K., Bilbo, S. D., *et al.* (2008). Genetic resistance to infection influences a male's sexual attractiveness and modulation of testosterone. *Brain, Behavior, and Immunity*, **22**, 381–7.

Zala, S. M., Potts, W. K., and Penn, D. J. (2004). Scent-marking displays provide honest signals of health and infection. *Behavioral Ecology*, **15**, 338–44.

Zelano, B. and Edwards, S. V. (2002). An *MHC* component to kin recognition and mate choice in birds: predictions, progress, and prospects. *The American Naturalist*, **160** (Suppl. 6), S225–37.

Ziegler, A., Dohr, G., and Uchanska-Ziegler, B. (2002). Possible roles for products of polymorphic MHC and linked olfactory receptor genes during selection processes in reproduction. *American Journal of Reproductive Immunology*, **48**, 34–42.

Ziegler, A., Kentenich, H., and Uchanska-Ziegler, B. (2005). Female choice and the MHC. *Trends in Immunology*, **26**, 496–502.

Ziegler, A., Santos, P. S. C., Kellermann, T., and Uchanska-Ziegler, B. (2010). Self/Nonself perception, reproduction and the extended MHC. *Landes Bioscience*, **1**, 176–91.

Zinkernagel, R. M. and Doherty, P. C. (1975). *H-2* compatability requirement for T-cell-mediated lysis of target cells infected with lymphocytic choriomeningitis virus: different cytotoxic T-cell specificities are associated with structures coded for in *H-2K* or *H-2D*. *Journal of Experimental Medicine*, **141**, 1427–36.

Zinkernagel, R. M. and Doherty, P. C. (1997). The discovery of MHC restriction. *Immunology Today*, **18**, 14–17.

Zohar, O., Reiter, Y., Bennink, J. R., *et al.* (2008). MHC class I-Ly49 interaction regulates neuronal function. *Journal of Immunology*, **180**, 6447–51.

10

Ultrasonic vocalizations in house mice: a cryptic mode of acoustic communication

KERSTIN MUSOLF AND DUSTIN J. PENN

Introduction

House mice (*Mus musculus*) are thought to communicate primarily through chemosensory signals (see Stopka *et al.*, Chapter 8 in this volume), and yet, like many rodents, house mice also communicate using ultrasonic vocalizations (USVs) (Box 10.1). Zippelius and Schleidt (1956) first studied USVs in house mice when they developed a recording device sufficient to detect ultrasound from pups. Since then, many studies have investigated rodent USVs, concentrating on the occurrence of the calls and the underlying proximate mechanisms. In house mice, USVs are emitted by pups as distress calls (Ehret, 2005) and by adults during courtship (primarily males) (Nyby, 1983) and same-sex social interactions (primarily females) (Costantini and D'Amato, 2006). More than one century ago, song-like features were reported from rare individual house mice, which sang at audible frequencies (Farr, 1857; Coburn, 1912; reviewed in Dice, 1932), and it was predicted that singing might be common in *M. musculus* and would occur at ultrasonic frequencies (Sidebotham, 1877; Dice, 1932). In the 1970s, Gillian Sales, Günter Ehret, and John Nyby and colleagues independently pioneered work on the complexity, proximate control and evolutionary origins of USVs in pups and adult house mice (Sales, 1972b; Ehret, 1975; Nyby *et al.*, 1976). More recently, a detailed, qualitative spectrographic analysis of laboratory mice concluded that male USVs fulfil the criteria for *song* (i.e. a series of syllables, generally more than one type, uttered in succession to form a recognizable sequence or pattern in time (Holy and Guo, 2005)). Amazingly, the playbacks of males' courtship USVs, when slowed down for human hearing, sound like songbirds (http://www.plosbiology.org/article/info:doi/10.1371/journal.pbio.0030386, Audio S1–S5)! We recently found evidence that the courtship USVs of male house mice provide a secondary sexual display (Hoffmann *et al.*, 2009; Musolf *et al.*, 2010) along with chemosensory signals, which appears to be functionally analogous to extravagant ornaments of peacocks and other birds. By

Evolution of the House Mouse, ed. Miloš Macholán, Stuart J. E. Baird, Pavel Munclinger, and Jaroslav Piálek. Published by Cambridge University Press.

Box 10.1 What is ultrasound? Bioacoustics of USVs

Sound is caused by the detection of a travelling wave; an oscillation of pressure is transmitted through a solid, liquid, or gas and the vibrations are perceived in hearing organs. Ultrasound *per se* is not different from ordinary sound except it is produced at frequencies that exceed the range of human hearing. Humans can perceive frequencies between 20 Hz and 20 kHz, with peak sensitivity at 2 kHz, the so-called 'acoustic frequency' range (Fig. 10.1). Infrasound (frequencies <20 Hz) is used in elephant communication, for example. Ultrasound (frequencies >20 kHz) is used in a variety of animals, such as bats, marine mammals, and rodents (Sales and Pye, 1974).

Most audible sound is produced by vocal fold vibrations under the influence of lung pressure. Due to their pure tone and dependence on a functioning larynx, USVs are thought to be produced by a whistle mechanism in the vocal tract, a hypothesis which has been supported by direct observations of non-vibrating vocal folds during production of these calls (Sanders *et al.*, 2001).

A study of auditory sensitivity in laboratory mouse strains found that the highest hearing sensitivity is around 15 kHz, and a second, subsidiary peak occurs at about 50 kHz (Ehret, 1974). This finding, and further adaptation in sound processing in the brain, indicate that mice have evolved hearing abilities tuned to both their sonic and ultrasonic vocalizations and even specialized their reception of specific ultrasonic calls (Ehret, 2001). Measurements of calling frequency vary among different laboratory strains – it remains to be tested whether these differences are reflected in hearing sensitivity; some laboratory strains do not hear well in the ultrasonic range or are even deaf as adults (Ralls, 1967). Deaf and knock-out mouse strains are useful tools for research on acoustic communication pathways.

Ultrasonic vocalizations, because of their short range, are suitable for communication both inside and outside the burrow. The attenuation inside the burrow depends on wall surfaces – the rougher the structure the greater the attenuation (Fletcher, 1992), whereas outside the burrow the habitat (grass *vs* wood) influences the vocalization range (Smith, 1979). Besides predator evasion through shifting communication in ultrasonic ranges, the ability to perceive ultrasound also might be helpful for mice to detect their enemies. For example, footsteps generate vibrations as well as ultrasonic signatures (Ekimov and Sabatier, 2006, 2008); therefore, mice might use ultrasound for detecting potential predators.

focusing mainly on chemical communication in house mice, researchers have been overlooking an important mode of communication with possible functions in mate choice and social recognition.

The discovery that male house mice emit complex courtship song is stimulating much new research on USVs, and such studies could prove to be as fruitful as the remarkable work on bird song (Marler and Slabbekoorn, 2004). Genetic analyses of USVs are particularly tractable due to the availability of a large number of well-characterized strains of laboratory mice and large amounts of readily available

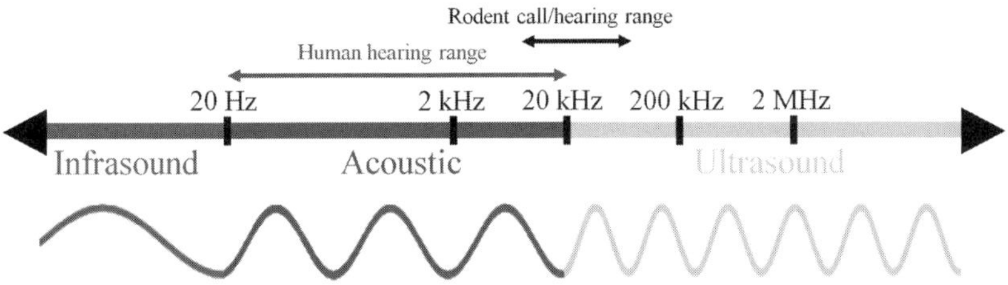

Figure 10. 1 Schematic overview of sound categories defined by frequencies. The human hearing range and the rodent call and hearing range are indicated in the text. See the plate section for a colour version of this figure.

sequence information. USVs of approximately 20 different inbred strains have been studied (reviewed in Hahn and Lavooy, 2005), including MHC-congenic strains (Brown *et al.*, 1999) and hybrids of strain crosses. The differences found among laboratory strains, together with the finding that hybrids (both pups and adults) call more frequently than the pure strains (Maggio and Whitney, 1986; Hahn *et al.*, 1997) indicate that USVs have a genetic basis. The precise role of genetics on USVs can be investigated in greater detail with the availability of an increasing number of mutant, congenic, and transgenic strains (Box 10.2). Combined with behavioural phenotyping, laboratory strains will provide a useful tool to identify and investigate the influence of specific genes on the development and control of USV communication (Crawley, 2000). Such candidate genes could be used to help in understanding the evolutionary origins and functions of USVs in wild house mice and to investigate whether they have a common genetic basis with other rodent species. Almost everything currently known about house mouse USVs comes from studies on inbred laboratory strains, however, and research on USVs in wild or wild-derived mice has only just begun (i.e. Hoffmann *et al.*, 2009, 2012a, b; Musolf *et al.*, 2010).

In this chapter we provide a broad and integrated ethological overview of USVs in house mice and highlight our current knowledge of both the mechanistic (ontogeny and physiology) and evolutionary (functional and phylogenetic) features of this behaviour.

Pup vocalizations

In many rodent species, pups are born blind and deaf, and are incapable of feeding themselves or thermoregulating during their first one or two postnatal weeks. However, at this stage they already produce a variety of sounds, mostly pure ultrasounds (Ehret and Bernecker, 1986; Hahn and Lavooy, 2005) that function to elicit maternal care (Branchi *et al.*, 1998; Hahn and Lavooy, 2005).

Box 10.2 Genetics of USVs in house mice

Our current knowledge on the genetic basis of USVs comes from studies on laboratory mice. Different inbred laboratory strains show distinctive USV patterns. Strains vary in their rates of USV production (low emitters: NZB and C57BL/6J; high emitters: BALB/c, DBA, A/J, BTBR, FVB/NJ), call duration, sound frequency, and bandwidth. The genetic basis of USV emission is corroborated by studies on knock-out and other transgenic mice, which have shown that deletion or insertion of selected genes markedly modifies vocal behaviour (Winslow *et al.*, 2000; Weller *et al.*, 2003; Moles *et al.*, 2004; Shu *et al.*, 2005; Scattoni *et al.*, 2008a, b). A cross-fostering study on laboratory mice pups revealed that call features were primarily dependent on the genotype of the pup, and call amplitude was the only feature dependent on the maternal genotype, i.e. on early environmental factors (Wöhr *et al.*, 2008). Another cross-fostering study with BALB and B6 mice, which show strain-specific call types, revealed that adult courtship calls of fostered males remained similar to calls of their genetic parental strain, and did not copy their fostered parents (Kikusui *et al.*, 2011). This study confirms a genetic influence on USVs and indicates that the strain-specific call types are not socially learned.

Studies investigating ultrasonic vocalization in crosses of mouse strains agree that hybrids always call at higher rates than individuals of the respective parental strain, indicating heterosis and a directional dominance towards specific call parameters (e.g. higher rate of calling, longer calls) (in pups: Hahn *et al.*, 1997, 1998; in adults: Maggio and Whitney, 1986), which may have important fitness values.

Foxp2 is a candidate locus for vocalization and influences sound production in humans, birds, and mice (Teramitsu and White, 2008). It is unknown, however, whether *Foxp2* or a linked receptor influences variation in USVs of adult mice. In contrast to early published reports (Shu *et al.*, 2005), *Foxp2* is not essential for ultrasonic vocalization in mice – recessive homozygotes generate harmonically structured audible calls and complex ultrasounds only under conditions of elevated stress (Fisher and Scharff, 2009). Mutations at *Foxp2* are known to lead to severe reductions in cerebellar growth and postnatal weight gain in mice pups (Groszer *et al.*, 2008), factors that can influence natural behaviours.

A mutant mouse was recently discovered which sang at audible frequencies (A. Uchimura, Osaka University's Graduate School of Frontier Biosciences). Further research with this mutant might also help in understanding the evolution of song and language.

The rate of USV production in house mice increases during the first 6–7 days of life, peaks around day 8, and then starts to decrease until the total disappearance of these calls after 14–16 days (Nyby *et al.*, 1976; Elwood and Keeling, 1982), when pups have grown fur and begin to thermoregulate independently. Pup USVs change during this period both in call length and frequency characteristics (Hahn *et al.*, 1998). Pup and adult calls fall into distinct spectral and temporal categories

(Liu *et al.*, 2003). A recent study followed individual mouse USV development from pup to adult in the CBA/CaJ strain and found that although most (9 of 11) syllable types were already present in the pup repertoire, the proportion and characteristics of each type changed developmentally and the complexity of vocal bouts increased with age (Grimsley *et al.*, 2011). Further studies are needed to identify how these developmental changes occur.

Proximate mechanisms of USV production: ontogeny and physiology

USVs in infant rodents are triggered as part of a general stress reaction, such as cold stress, and are produced through laryngeal braking forced by an abdominal compression reaction aimed at increasing blood flow to the heart (Blumberg and Alberts, 1990). House mouse pups are deaf until the age of 10 days (Ehret, 1983), and variation in call parameters during their development may be influenced by the developmental stage of individual pups, as well as their general condition. Vocalizations are thought to be modulated by emotional/motivational mechanisms governed by the limbic system (Ehret, 2005). Receptor systems of the brain are the target for modulating neurotransmitters, such as serotonin and dopamine, and other neuromodulators, such as opioids and cholecystokinin, which can both facilitate and inhibit USV production (Barron and Gilbertson, 2005; Ehret, 2005; Hofer, 2010). Interestingly, the profile of neurotransmitters involved in separation vocalization in infant pups is very similar to the neurotransmitter pattern that mediates human anxiety and is also thought to influence separation anxiety in children (Hofer, 2010). USVs of infant mice have become an important phenotypic marker for the characterization of mouse strains for neurodevelopmental disorders, such as autism (Scattoni *et al.*, 2009). The genetic mutant TSC2 affects learning and memory, and also mother–offspring communication and subsequently pup USVs (Young *et al.*, 2010), but a detailed understanding of the link between USV production and neurological development requires further investigation. It remains to be determined whether pups learn from their parents' ultrasonic calls (familial imprinting) or develop dialects, as do many songbirds (Boughman and Moss, 2003).

USV functions in parent–offspring communication

In young house mice USVs have an important function in maternal–offspring communication. Pups emit USVs under potentially threatening conditions, such as isolation from the nest, low temperature, unfamiliar odours, and during rough handling (Branchi *et al.*, 1998), and the USVs of pups initiate specific maternal behaviours, such as searching and retrieval of pups (Hahn and Lavooy, 2005). In addition to USVs, mouse pups have other calls in the acoustic range, such as a 'wriggling call', which they emit when they are nursed to initiate grooming and other maternal behaviours (Ehret and Bernecker, 1986), and 'pain calls', which they emit as a reaction to being bitten, pinched, or injured, and postpartum sounds when the mother is cleaning them after birth (Ehret, 1975). Interestingly, pups adjust their USV calling rate in different social contexts: calling increases 2–3-fold when an isolated pup has a brief encounter with its dam and gets re-isolated (Shair *et al.*, 2005); however, pups cease emitting USVs when facing unfamiliar – and potentially infanticidal – males (Branchi *et al.*, 1998). In addition to the influence of conspecific interactions on USV production, predation may also play a role in facultative changes in call production. Small mammalian predators (e.g. weasel, mink, cat, and dog) of house mice can detect ultrasonic sounds to some extent (Heffner and Heffner, 1985; Powell and Zielinski, 1989), suggesting that high rates of call production may increase predation risk.

Dams show a stronger retrieval response to their own pups compared to alien individuals (D'Amato *et al.*, 2005), which indicates that USVs play a role in offspring recognition. This finding demonstrates kin recognition in the broad sense (the ability to differentiate kin from non-kin), as well as kin-biased behaviour, but it is unclear if males are able to recognize their own offspring via their USV calls (genetic similarity detection or kin recognition in the narrow sense; Penn and Frommen, 2010). USVs alone are sufficient to stimulate maternal searching behaviour towards the sound source, and female mice prefer USVs and synthesized calls with typical acoustic properties of their own species (reviewed in Ehret, 2005). Interestingly, the sex and developmental stage of offspring influences maternal retrieval behaviours (male pups receive more maternal care than females, and more-developed pups are retrieved sooner than less-developed individuals) (Hahn and Lavooy, 2005), something that may be explained by differences in the calling rate of infant male and female mice (Hahn *et al.*, 1998). Interestingly, maternal responsiveness of a mouse strain is negatively correlated with pup USV calling rate, indicating a close linkage of these behaviours in lab strains (D'Amato *et al.*, 2005; Wöhr *et al.*, 2008). This interaction of maternal response rates and offspring USV production suggests the potential for parent–offspring conflicts, which could be a fruitful area for future research (Box 10.3).

As USV call production is closely associated with thermal regulation, pups' distress USVs may have initially evolved as sounds produced as a by-product of their response to cold, which were evolutionarily co-opted and modified for communication functions. Also, pup USVs may have evolved before USVs in

Box 10.3 Evolution of animal communication

Communication occurs when the action of (or cues given by) one animal influence the behaviour of another (Endler and Basolo, 1998). Both natural and sexual selection can influence both the signal design and the behaviour of the receiver. Sometimes the communication signal evolves based on a pre-existing sensory property of the receiver, which evolved for reasons unrelated to communication. This possibility is referred to as 'sensory exploitation' (Ryan, 1998). As in most signalling systems, information transfer is vulnerable to deception and manipulation (Dawkins and Krebs, 1978) by the sender, though the receiver is expected to attempt to ensure the accuracy of information transfer. Communication is best viewed not as a harmonious exchange of information, but as the focus of an arms race between signallers as manipulators and receivers as 'mind-readers' (Dawkins and Krebs, 1978). This is also true when looking at communication between the sexes. It is no longer assumed that courtship and mating are only cooperative interactions. There is increasing evidence for evolutionary conflicts between the sexes (Holland and Rice, 1998; Arnqvist and Rowe, 2005) as males are typically under stronger selection to maximize mate acquisition and can benefit by feigning higher quality to potential mates. In contrast, females are under stronger selection to mate with superior partners and consequently benefit most from reliable assessment (Trivers, 1972; Clutton-Brock, 2007). When signals are energetically expensive to produce, according to the 'handicap principle' (Zahavi and Zahavi, 1997), they should provide honest information about the signaller's quality.

Sound as a communicating medium is used by several classes of vertebrates and non-vertebrates (Bradbury and Vehrencamp, 1998). Vocalizations are used in intra- and interspecific interactions and have evolved to communicate information about individuals' condition, social state, and quality to potential rivals and mates, as well as to inform listeners about objects (e.g. predators) and events in the environment (e.g. group movement) (Bradbury and Vehrencamp, 1998; Owings and Morton, 1998).

adults, and the sensitivity of females' ears to pups' USVs may have been exploited by males, facilitating the evolution of the courtship USVs in adult males (sensory exploitation) (see 'The evolutionary origins of USVs in house mice' below).

Adult vocalizations

Adult house mice emit USVs to other adult conspecifics (Nyby and Whitney, 1978; Nyby, 1983; Maggio and Whitney, 1985; White *et al.*, 1998; Gourbal *et al.*, 2004). In sexual encounters, male house mice can produce USV calls at high rates, while females emit few or no calls (largely defensive squeaks in the acoustic frequency range) (Wang *et al.*, 2008). USVs can be stimulated by using anesthetized partners or respective urinary cues (Nyby *et al.*, 1977b; Maggio and Whitney, 1985; Warburton *et al.*, 1989), and the sex of the adult stimulus animal is critical for the elicitation of USVs: male house mice emit USVs when presented with female mice or their urinary scent, and emit few, if any, USVs when presented with male scent (Nyby *et al.*, 1976; Musolf *et al.*, 2010). This sex-specific stimulus response is precisely the opposite for females, but female–female pairs produce USVs at comparable rates (at least during the first minutes of social interaction) (Maggio and Whitney, 1985; Gourbal *et al.*, 2004; Costantini and D'Amato, 2006). Spectral differences of USVs between the sexes have been found in wild *Peromyscus boylii* (Kalcounis-Rueppell, personal communication) and wild-caught singing mice (*Scotinomys* and *Baiomys*, Miller and Engstrom, 2007), though we know of no such comparisons in house mice.

Male USVs: courtship calls

Proximate mechanisms of USV production

Most research on USVs in house mice has been aimed at understanding the influence of hormones and pheromonal regulation in call production. Male ultrasound production is only found in individuals at sexual maturity (*ca.* 55 days) (Whitney *et al.*, 1973; Nyby *et al.*, 1976), when the concentrations of circulating androgens begin to increase, and males begin emitting 70-kHz ultrasounds in response to females, in association with other courtship behaviours (Sales, 1972b; Whitney *et al.*, 1973). Castration reduces adult male calling (Dizinno and Whitney, 1977) and androgen replacement restores it (Nunez *et al.*, 1978; Nunez and Tan, 1984; Warburton *et al.*, 1989), indicating a strong link between the levels of circulating sex hormones and call production. USV calling rate can be used as an indicator of sexual arousal leading to 'reflexive testosterone release', which increases the male's copulatory behaviours (James and Nyby, 2002; James *et al.*, 2006). Recent studies on a receptor knock-out strain suggest that muscarinic receptors involved in neuronal interactions are required for male USV production during male–female interactions, perhaps via their roles in dopamine activation, which imply a potential role in reward, motivation, and cognition (Wang *et al.*, 2008). However, molecular and genetic control of USV production is largely undetermined.

Exposure to female house mice, or female odorants including urine, saliva, and vaginal fluids, provide the most potent sensory cue found so far to elicit males' USVs (Whitney *et al.*, 1974; Nyby *et al.*, 1977b; Guo and Holy, 2007), whereas male urine and control chemicals are relatively ineffective (laboratory mice: Whitney *et al.*, 1974; Nyby *et al.*, 1977b; wild-derived mice: Musolf *et al.*, 2010). Interestingly, males' courtship USVs are a species-specific response: male laboratory mice elicit USVs to fresh female urine of their own strain and a reduced response to fresh female urine of other strains of *M. domesticus*, but USV production is strangely attenuated when mice are presented with either female rat or human urine (Wang *et al.*, 2008).

Two female pheromones have been identified that elicit male USVs (Sipos *et al.*, 1992). These compounds are only present in female urine after sexual maturity, explaining the lack of male response to the urine of immature females (Whitney and Nyby, 1979; Musolf *et al.*, 2010). One of these pheromones is 'potent' in freshly voided urine, as it elicits USVs from males regardless of males' former social or sexual experience (Sipos *et al.*, 1992); however, this compound is also ephemeral (after 15–18 hours its activity is destroyed by oxidation) (Sipos *et al.*, 1993, 1995). The second pheromone is stable for at least 30 days (Nyby and Zakeski, 1980), but requires male social or sexual experience in order to trigger USVs; naïve males do not respond to this urinary signal. The pheromones in female mouse urine that elicit male USVs may be the same compounds that produce testosterone surges, as they are both non-volatile, ovarian-independent, and regulated directly by the pituitary gland (Nyby *et al.*, 1979; Johnston and Bronson, 1982). Whereas males do not habituate in their USV response when presented with fresh urine from females (eight trials within 23 days; Sipos *et al.*, 1992), aged urine leads to habituation after three or four exposures (Dizinno *et al.*, 1978) and short-term freezing of urine decreases males' USV responses (Hoffmann *et al.*, 2009). These findings provide examples of how males' USVs can be used to examine their ability to detect females' pheromones and sexual attractiveness to different females.

A recent study on laboratory mice (strain B6D2F1, which is an F_1 hybrid between C57Bl/6 and DBA2/J) found that male USVs show features of birdsong (Holy and Guo, 2005). The study used a narrow, *sensu stricto* definition for 'song' (Broughton, 1963), and though this label is sometimes debated, this finding opens up many possibilities for future research, which should complement the extensive body of research on bird song. We recently conducted a detailed description of specific call structures and elements, which indicates striking differences between laboratory versus wild-derived house mice (Hoffmann *et al.*, 2012b). Although calls of both groups could be divided into two spectral categories, the calls of wild-derived mice (65 kHz and 103 kHz) appear to be shifted to higher frequencies than those documented in laboratory mice (40 kHz and 67 kHz) (Barthelemy *et al.*, 2004). Our findings support other evidence for genetic influences on USVs, though the evolutionary origin of the differences between domesticated and wild mice remains to be elucidated.

To our knowledge only one study has examined individual variability (or consistency) of males' USVs (Hoffmann *et al.*, 2012a). The USVs of individual, wild-derived male mice were consistent in their spectrographic features, enabling call classification into respective kin groups and beyond that to the individual level, suggesting that these calls may provide information useful for individual recognition.

It has been suggested that social experience, such as prior physical and sensory interactions with conspecifics, is a prerequisite to elicit males' USV responses to (aged) female urine (Sipos *et al.*, 1992), and that it enhances males' USV responses to female scent (but see Musolf *et al.*, 2010). Exposing males to artificially odorized females results in males subsequently emitting USVs to the artificial fragrance on its own (Nyby *et al.*, 1978). Interestingly, social experience in females has the opposite effect, as increased experience decreases their USV response to other females (D'Amato and Moles, 2001). Although these studies suggest that social learning may play a role in the development and expression of USVs, it is important to note that social deprivation used in these studies (using isolated individuals as controls) could have produced laboratory artefacts, and future studies using animals held under more natural controls are needed. Genetics appear to be another source of individual variation in USVs (Box 10.2). Identifying genes that influence USVs could help to better understand the evolution of USVs, as well as the underlying mechanisms.

Functions of USVs in mate choice

Why do male house mice produce USVs? Early studies concluded that male USVs function to facilitate or coordinate mating, but they did not suggest why females are attracted to these vocalizations. Similarly, male USV have been suggested to reduce females' aggression when the male is approaching the female, though this effect appears to be inconsistent (Bean *et al.*, 1986). Since USVs in house mice are sexually dimorphic with respect to calling rate, androgen-dependent, and produced chiefly by males during mating (triggered by female pheromones), they can be considered to be *secondary sexual characters*, which, like scent-marking (Zala *et al.*, 2004), evolve through sexual selection (Musolf *et al.*, 2010). Males' USVs may evolve through intrasexual (direct male–male competition) or intersexual selection (female choice) or a combination of both. There is as yet no direct evidence that USVs enhance male mating success (through either type of selection), and studies are needed on wild or wild-derived mice in natural or semi-natural conditions to determine the costs as well as the benefits of producing USVs, something which cannot be investigated under laboratory conditions. With the development of improved technologies for measuring USVs, an increasing number of USV studies on wild rodents have been undertaken (*Peromyscus* mice: Kalcounis-Rueppell *et al.*, 2006, 2010; *Scotinomys* and *Baiomys*: Miller and Engstrom, 2007). Previous studies on laboratory mice, and the few studies conducted so far on wild-derived house mice, provide some insights into the functions of USVs.

As emphasized above, USVs during courtship and mating are produced primarily by males (White *et al.*, 1998), and recent studies suggest these vocalizations play a role in female choice. Females are more attracted to vocalizing than non-vocalizing males (muted by surgical abduction), and male ultrasonic vocalizations prolong the time females spend near potential mating partners (Pomerantz *et al.*, 1983). Studies using recorded USV playback also indicate that females are attracted to males' USVs. In both laboratory (Hammerschmidt *et al.*, 2009) and wild-derived (Musolf *et al.*, 2010) mice, females spent more time near a speaker playing recorded male USVs than a speaker playing only background noise. However, females rapidly habituated to this experimental setting and showed no further preferences when tested in a second trial (Hammerschmidt *et al.*, 2009; Musolf *et al.*, 2010). While females appear to show a preference for males producing USVs, female house mice still copulate with devocalized males when housed in mating pairs (White *et al.*, 1998), indicating that male USVs are not essential for reproduction (at least in cages). Male ultrasonic vocalizations may affect female behaviours during copulation and extend copulation duration, a result documented in hamsters (*Mesocricetus auratus*; Floody and Pfaff, 1977b).

Do USVs provide information about males' quality or compatibility? Males' USVs likely play a role in *sex* and *species recognition* (allowing hybridization avoidance), but these are not likely to be their only functions, as highly complex calls would not seem to be necessary for such roles (simple sex- and species-specific chirps would presumably suffice). Males' courtship USVs might also play a role in *individual* and *kin recognition* (Musolf *et al.*, 2010), and females may use these vocalizations to assess male compatibility, to obtain direct or indirect genetic benefits for their offspring. We have recently shown that wild-derived females prefer USV playback from non-kin males to those produced by their male siblings, providing evidence that USVs alone are sufficient for kin discrimination (Musolf *et al.*, 2010).

USVs may also enable females to assess males' quality, such as their social status (see below) and condition (e.g. health and resistance to infectious diseases), as do chemosensory signals (reviewed in Penn and Potts, 1998). There is much evidence to support the handicap hypothesis of costly signalling, even if signals do not necessarily need to be costly to be honest (Maynard Smith and Harper, 2003). Ultrasonic calling is assumed to be energetically costly (Nyby and Whitney, 1978), and there are several reasons to suspect that males' USVs may be costly to produce in terms of time as well as energy, though we know of no direct evidence in house mice. Food-deprived mice produce USVs at lower rates than those produced by well-nourished individuals (Moles and D'Amato,

2000). Similarly, malnutrition leads to a reduction of specific call types in rat pups (Barron and Gilbertson, 2005). The production of USVs may be particularly costly for food-restricted animals, a hypothesis suggested in several empirical studies. The production of USV may also reduce the time available for foraging and thus lead to courtship/foraging tradeoffs, a pattern found in birds (reviewed in Gil and Gahr, 2002). The investigation of such tradeoffs requires studies on wild or semi-natural populations of mice. If USV production can be shown to impose time and energy tradeoffs, USVs might provide honest signals indicating the caller's condition, signals that could be used in mate-choice decisions.

While there is generally a negative relationship between body size and the frequency of sound production (low frequencies require large body structures) (Pye, 1979), this relationship does not appear to hold for USV production in rodents. Although some studies support a relationship between body size and call frequency (e.g. USVs emitted during male–male encounters: calls of field voles (*Microtus agrestis*) have higher fundamental frequencies than those of smaller common voles (*Microtus arvalis*) and bank voles (*Myodes glareolus*); Kapusta *et al.*, 2007), others do not (e.g. exploratory ultrasonic calls of both male and female wood mice (*Apodemus sylvaticus*) are higher in frequency than those of the larger yellow-necked mouse (*A. flavicollis*); Hoffmeyer and Sales, 1977). No studies to our knowledge have yet investigated intraspecific variation in body size and its relationship to USV production.

Males' courtship USVs do not necessarily function to provide *information* to females, as much as to persuade and manipulate them (as with all advertisements), and future research should examine their potential role from a sexual conflict perspective (Box 10.3).

Neither laboratory (Nyby *et al.*, 1976) nor wild-derived males (Musolf *et al.*, 2010) appear to use USVs in intrasexual communication, suggesting that male USVs are not currently under intrasexual selection. This pattern stands in striking contrast to the pattern observed in birds, where song plays a key role in territorial defence and direct male–male competition, and songs are displayed regularly if not exclusively to other males (Collins, 2004). However, this sexual bias does not necessarily mean that USVs in mice do not play a role in male–male competition. Nyby *et al.* (1976) found that subordinates utter fewer USVs than dominants in the presence of a female, whereas D'Amato (1991) found the opposite: subordinates called more in the temporary absence of the dominant male, though subordinates' calling rate was inhibited by exposure to urine of the dominant males interacting with a female. Social defeat has also been shown to reduce males' USVs (Lumley *et al.*, 1999). Thus, USVs appear to provide information on a male's social status, as found with other mammal vocalizations (Fischer *et al.*, 2004; Behr *et al.*, 2006; Vannoni and McElligott, 2008), and even humans (Sell *et al.*, 2010), and thus affect the behaviour and physiology of other males (e.g. differences between dominant and subordinate male mice: scent marking behaviour (Hurst, 1990), urine excretion (Drickamer, 1995), urine chemistry (Novotny *et al.*, 1990), and aggression (Hurst *et al.*, 1993)). Future studies are needed to determine whether and how male USVs might play a role in male–male interactions.

Female USVs

Proximate mechanisms

There have been relatively few studies of female house mouse USVs and they remain relatively poorly studied. The few studies so far suggest that female USVs are also hormonally modulated, as their production varies during the oestrous cycle, and sexually receptive females emit fewer USVs than non-receptive individuals in the presence of a female intruder (Moles *et al.*, 2007). In contrast to males, the control of USVs in females does not appear to be androgen-dependent. Ovariectomized females began to call to stimulus females at similar rates as long-term gonadectomized males when receiving testosterone treatment (Nyby *et al.*, 1977a). However, USVs were accompanied by mounting bouts in this study, suggesting that females demonstrated inappropriate male-typical behaviour, which does not take place in natural female–female encounters. Further studies are needed to understand the development and proximate mechanisms controlling sex differences in USVs in house mice.

Functions of female USVs

As female USVs have received so little attention, their adaptive function remains a mystery. In general, sexual ornaments in females have often been considered to be a by-product of selective pressures on male ornamentation (an analogue of males' nipples); but evidence suggests that female ornaments have often evolved independently and may have important functions in competitive situations (Amundsen, 2000). Since female ultrasonic vocalization appears to occur principally during female–female social investigations (Moles *et al.*, 2007), this behaviour may play a role in the establishment of social dominance among females. The role of female USVs has not yet been investigated in courtship or male mate choice. It has been suggested that female USVs are important for sex recognition (Wysocki *et al.*, 1982); these calls may also play a role in species, group, kin, and individual recognition (Musolf *et al.*, 2010). Indeed, it has been shown that female USVs can be used to indicate social recognition: resident females recognize familiar individuals intruding into their territory, and the rate of USVs rapidly decreases in such situations, a pattern that differs from that observed with novel females, whereas their olfactory behaviour did not differ in these situations (D'Amato and Moles, 2001). Social recognition ability is known to depend on the time interval between encounters (the ability disappears after 60 min; D'Amato and Moles, 2001), and recognition also depends on the age of the female (older females having shorter memory spans; Moles *et al.*, 2007).

Besides individual recognition, female USVs may function to communicate additional condition-dependent information. Moles and D'Amato (2000) found that females emitted a higher number of USVs during the first minute of social interaction when the intruder had previously fed on a palatable food in contrast to one fed on unpalatable food. Whereas the results of these few studies are interesting, a more systematic approach to the study of female USVs is needed in order to understand their potential function in social communication.

The evolutionary origins of USVs in house mice

There have been no comparative studies on the evolutionary origins of USVs in mammals to our knowledge, though USVs have been detected in at least 50 species in 30 genera, all from two suborders, the Sciuromorpha (two families) and Myomorpha (eight families) (see overview in Sales, 2010). There are no single species-specific frequency bands or patterns that would permit species recognition (Bradbury and Vehrencamp, 1998), but there are some differences in the form and frequency of USVs produced in different groups. Infant house mice and other members of the Muridae appear to call at higher frequencies than pups of the genera *Rattus* and *Thamnomys* (Sales, 2010). The pattern of frequency modulation, which also appears to vary among taxa, seems to be more common in murid rodents than in cricetids. Harmonic elements (i.e. additional signal(s) whose frequency(ies) are integral multiples of the reference signal) have not been detected in all species, but these differences may, to a certain extent, reflect the methods of recording and reporting of the calls (Sales, 2010).

Studies on USVs in Norway rats (*Rattus norvegicus*) show interesting similarities and differences compared to house mice. In rats, adult male USV calls are emitted during reproductive and agonistic situations, and adolescent rats call during play activities and even in response to human 'tickling' (Brudzynski and Ociepa, 1992), in addition to which they also have specific alarm calls (Brudzynski, 2007). Both male and female rats emit ultrasonic calls during sexual encounters (White and Barfield, 1990), as with golden hamsters (Floody and Pfaff, 1977a), collared lemmings (Brooks and Banks, 1973), and different vole species: *Microtus arvalis*, *M. agrestis*, *M. oeconomus*, *Myodes glareolus* (Kapusta *et al.*, 1999, 2007; Kapusta and Sales, 2009). In rats, male pre-ejaculatory vocalizations solicit female darting, possibly coordinating male and female reproductive behaviour (Thomas *et al.*, 1982), and facilitate lordosis in females (White and Barfield, 1990). Male and female voles elicit USVs both in intra- and intersexual encounters (Kapusta *et al.*, 1999, 2007; Kapusta and Sales, 2009). Therefore, compared with rats and other rodents, house mice USVs appear to be produced under a more restrictive set of social conditions.

Two hypotheses have been developed to explain the evolutionary origins of USVs in rodents. The first of these, the *sensory exploitation hypothesis*, suggests that USV communication originally evolved in pups, where it plays a role in mother–offspring interactions, and was later co-opted and modified for communication in adults (Ehret, 2005). Infants of virtually every species of mammal, as well as birds, vocalize when separated from their home nest, conspecifics, and most clearly their mothers. Newborn mouse pups need their mothers not only for protection and food provisioning, but also for thermoregulation. It has been suggested that laryngeal braking associated with thermoregulations may be responsible for USV production in pups, suggesting that USVs may have arisen originally as an acoustic by-product of physiological adaptations to recover from hypothermia in juvenile rodents and that the role of these vocalizations in maternal–offspring communication may have followed their initial role in thermoregulation.

Both male and female rodents produce ultrasounds in various situations (e.g. Sales, 1972a; Nyby, 1983; Maggio and Whitney, 1985; Brudzynski *et al.*, 1993), and though the acoustic properties of adult USVs may differ from those of pups in several respects (Liu *et al.*, 2003), they seem to reflect states of arousal and motivation/emotions similar to those of infant rodents and may have evolved from infant calls (Ehret, 2005). It is interesting to consider the possibility that once USVs became established in maternal offspring communication, this sensory adaptation may have opened an opportunity for these cues to be co-opted during courtship, with males producing USVs at high rates being favoured by females (sensory drive or sensory exploitation). Although the opposite evolutionary sequence is also possible, with pups exploiting the pre-existing USV detection ability of their mothers, this scenario seems less probable since pups are deaf for their first days of life (Ehret, 1983).

A second – though not mutually exclusive – hypothesis suggests that the evolution of USVs may have been mediated by *predator avoidance*. Many predators (and parasitoids) eavesdrop on the communication signals of their prey, and USVs might have initially evolved, or been reinforced, as a way to evade predator detection (Hofer, 2010). The ability to detect USVs requires complex morphological changes in the inner ear bones and additional sensory hair cells. While ultrasonic calls are not entirely cryptic to predators, many predators cannot detect the full range of USV and USV alarm calls that are utilized by some rodents (i.e. rats; Brudzynski, 2007). Further, particular physical characteristics of ultrasounds (greater directionality, greater attenuation, greater scattering, decreased localizability and suitability for communication in underground burrows) make ultrasound a superior alternative to sonic vocalization for short-range communication, particularly in critical situations.

Conclusions: potential future directions

USVs have been studied in laboratory rodents since the late 1950s, and the recent discovery that USVs of male house mice have features of song (Holy and Guo, 2005) is sure to stimulate even more research into this communication mode that is cryptic to humans' senses. To understand the proximate mechanisms controlling USV communication and their ontogeny, more neuro-ethological studies are needed. For example, to determine how female pheromones activate male courtship USVs, we will need to understand how olfactory signals can be processed via receptors and transferred to the brain for USV production.

Vocal communication is not the only information channel animals rely on, of course, and it would be particularly interesting to know how mice integrate information from multiple sensory modalities, such as olfactory and acoustic signals. Such *multimodal sensory integration* is a challenging area of research and more accurately reflects natural situations in which animals receive information on different sensory channels. Neuro-ethological and other studies on mechanisms of USV production could benefit by considering more ecological contexts, as some behavioural paradigms developed in the laboratory are non-existent in the wild (Wolff, 2003).

While the structure and function of adult male USV has been studied in detail, a lack of comparable studies on adult female house mice means we still have a relatively poor understanding of female USVs. Future studies (e.g. comparing intersexual differences in call characteristics) are needed to improve our knowledge of female ultrasonic communication.

To understand the functions of USVs, studies on wild house mice are crucial, especially in an ecological context (behavioural ecology; ecological genomics). Our research on the USVs of wild-derived house mice provide a new perspective, as they indicate that males produce USVs as part of their courtship display to attract females (Musolf *et al.*, 2010). USVs are individually consistent and distinguishable (Hoffmann *et al.*, 2011a) and females could use them for kin recognition (Musolf *et al.*, 2010). Moreover, as previously mentioned, domesticated strains of mice have low individual variability and show many differences from their wild counterparts in behaviour (Crawley *et al.*, 1997) and life history traits (Miller *et al.*, 2002) due to inbreeding and hybridization of different *M. musculus* subspecies (Wade and Daly, 2005). Further, rearing house mice in

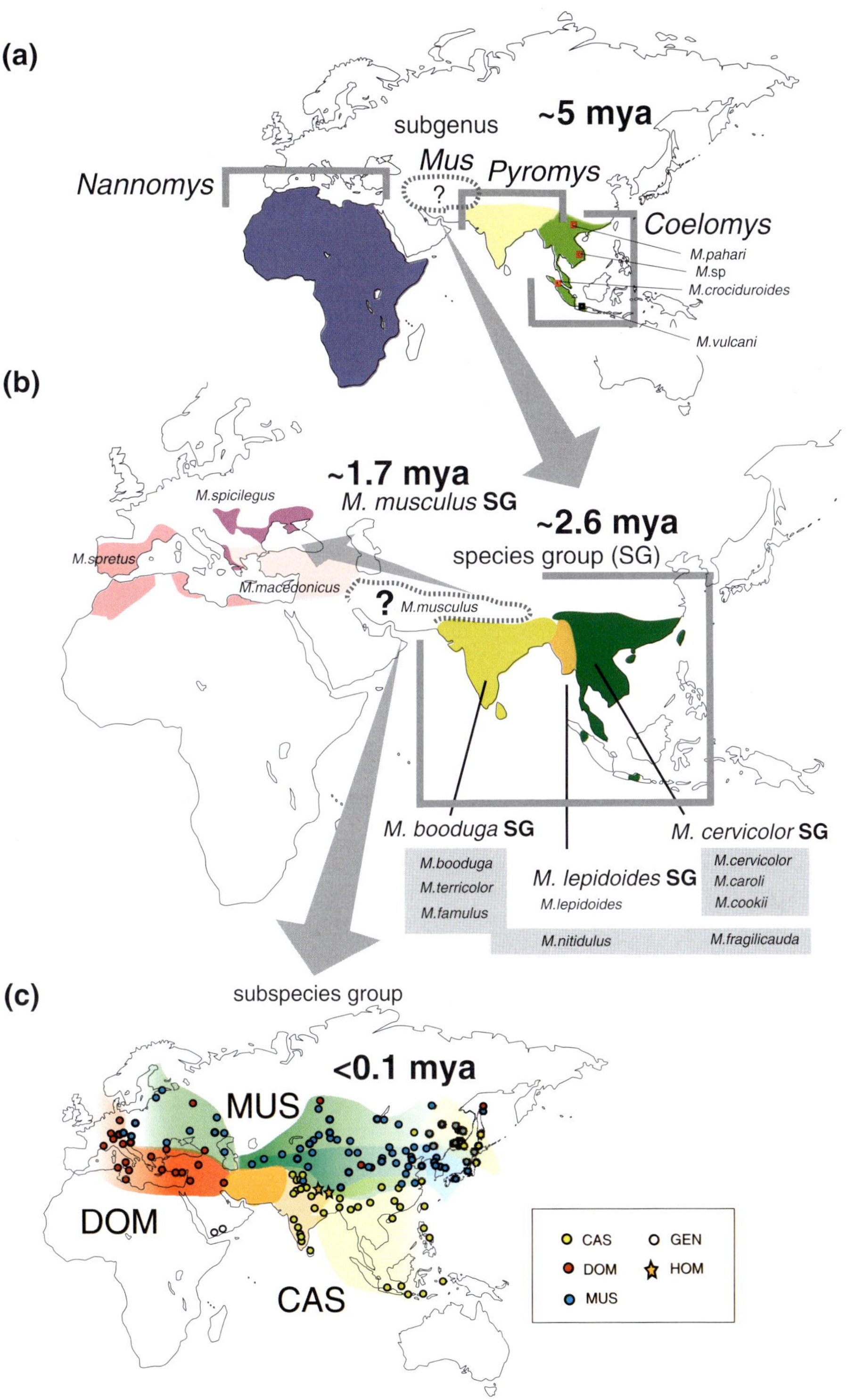

Figure 2.1 Generalized distributions of the Eurasian species of *Mus* (Corbet and Hill, 1992; Mitsainas *et al.*, 2009; IUCN, 2010; M. Macholán, personal communication), exhibiting the subgenus (a), species groups of the subgenus *Mus* (b), and subspecies groups of *M. musculus* (c). Three of the four subgenera of *Coelomys*, *Pyromys*, and *Nannomys* are based in Southeast Asia, India, and Africa, respectively. (For the full caption text see Fig. 2.1 on p. 36.)

Figure 3.2 Aerial photograph of the community building EA 30 of the PPNA site of Jerf el Ahmar (Syria), where large silos (grain storage) have been identified (arrowed); from the mission to El Kowm-Mureybet of the French Foreign Office, directed by Daniel Stordeur.

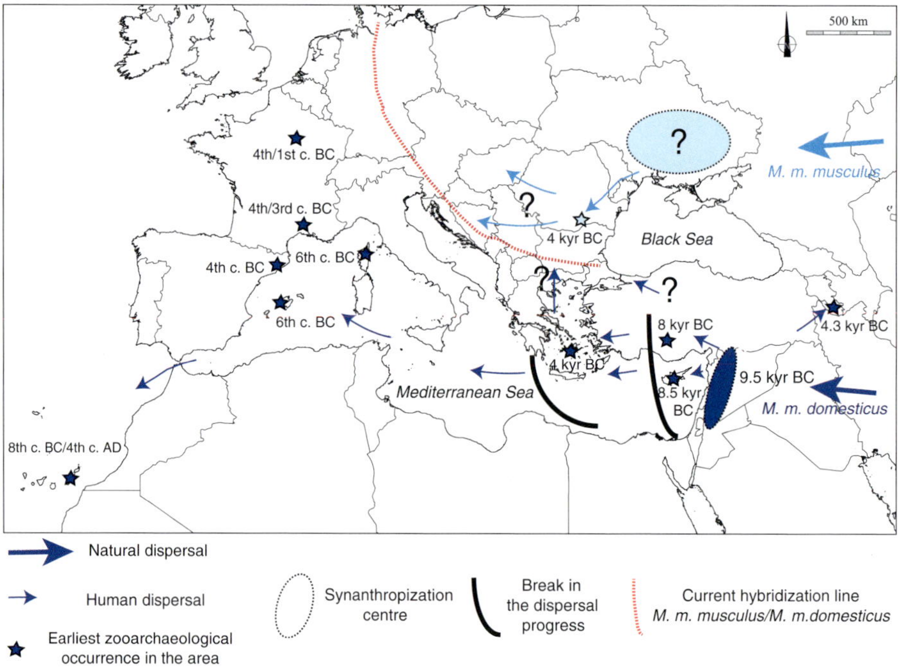

Figure 3.3 Map synthesizing the current knowledge and hypotheses on the history of the house mouse synanthropization and dispersal process in Europe. Kyr = thousand years, c. = century.

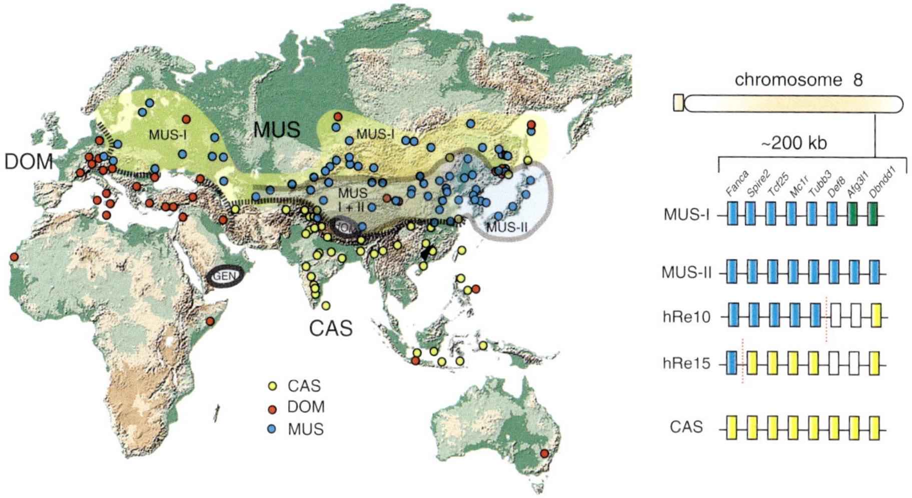

Figure 4.4 Geographic distribution of mtDNA variation in the house mouse *M. musculus*. CAS, DOM, and MUS stand for *M. m. castaneus*, *M. m. domesticus*, and *M. m. musculus*, respectively, and their haplotype distributions are shown with coloured circles (Suzuki *et al.*, unpublished). MUS can be divided into two subgroups, MUS-I and MUS-II, which overlap in northern China. Two lineages coming from the Korean Peninsula (MUS-I) and eastern India (CAS), respectively, are mingled in the Japanese islands (Nunome *et al.*, 2010). (For the full caption text see Fig. 4.4 on p. 107.)

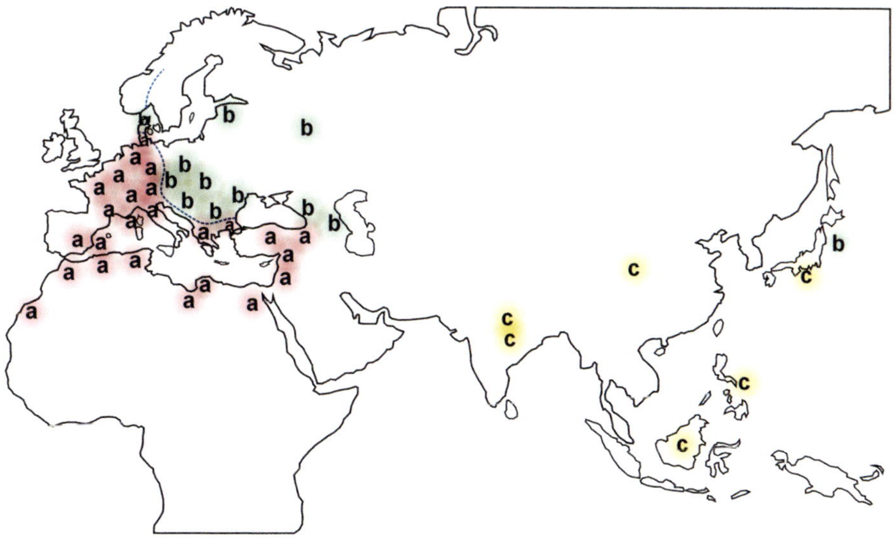

Figure 7.1 The three *Abpa* alleles (*a–c*) found in the peripheral *Mus musculus* subspecies.

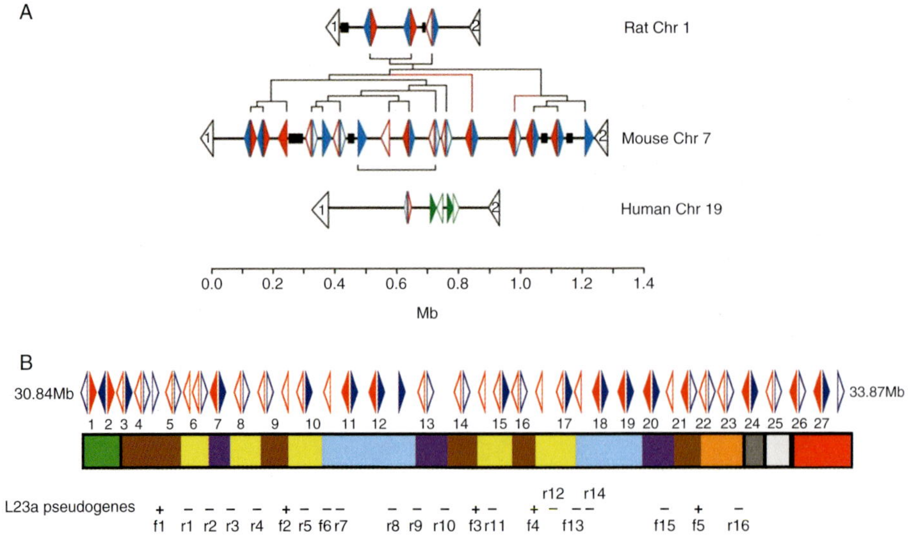

Figure 7.4 The *Abp* gene family. (a) blue: *Abpa*; red: *Abpbg*; green: primate SCGBs. Filled/open arrowheads: genes/pseudogenes; black boxes: genome gaps. (b) Filled/unfilled triangles: completed *M. musculus* genome predicted genes/pseudogenes; '+' and '–': ribosomal protein L23a pseudogenes. (For the full caption text see Fig. 7.4 on p. 168.)

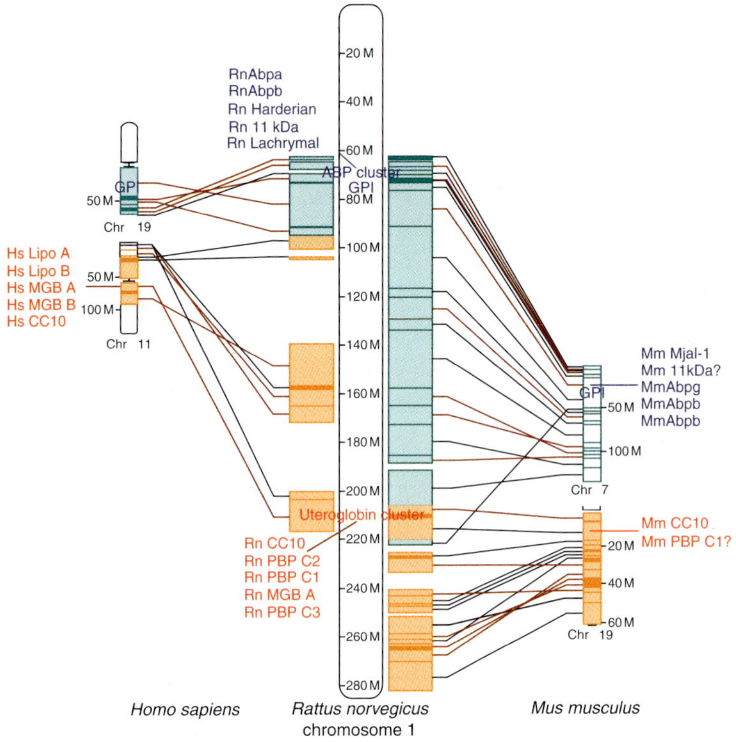

Figure 7.5 Syntenic relationships between secretoglobins expressed in mice, rats, and humans (Laukaitis and Karn, 2005). Syntenic relationships: rat, two clusters on chromosome 1; mouse, chromosomes 7 and 19; human, chromosome 11. (For the full caption text see Fig. 7.5 on p. 170.)

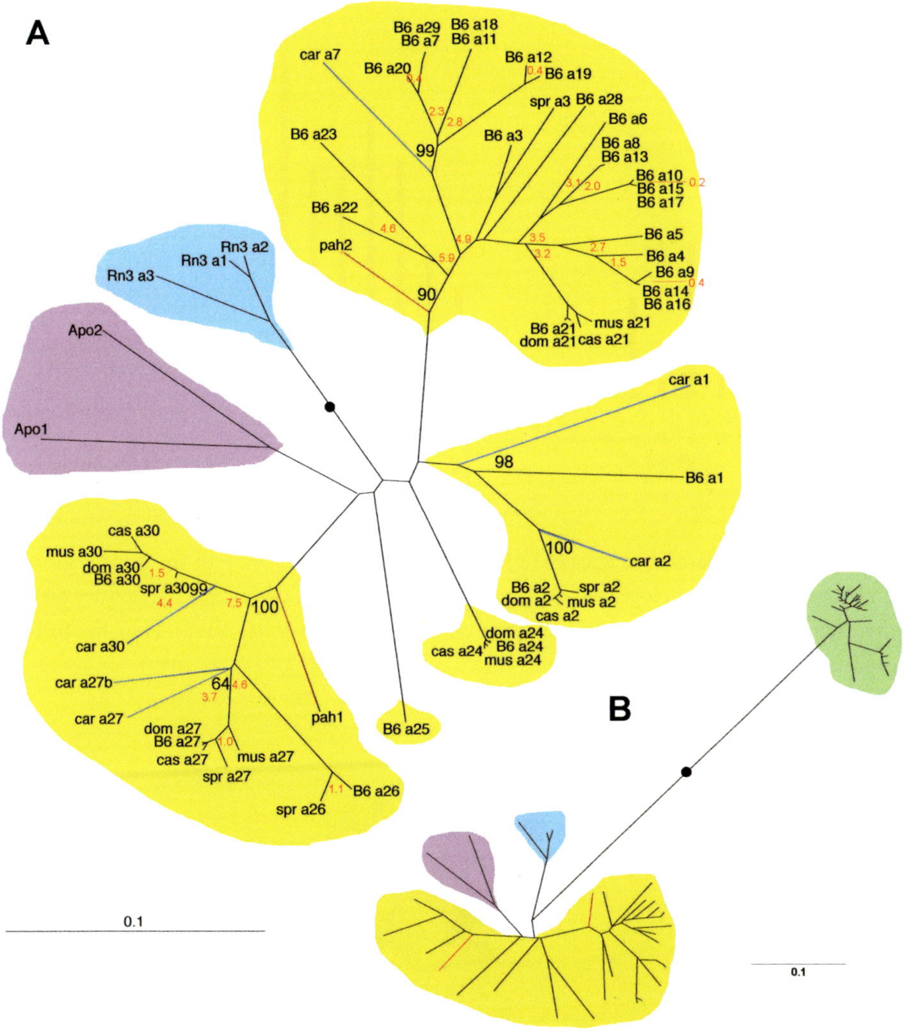

Figure 7.6 *Abpa* evolution. Intron 2 NJ phylogenies: (a) Yellow: *Mus*; purple: *Apodemus*; blue: rat; red typeface: divergence (MY). (b) Green: rabbit; as in A: rodent; black dots: probable roots with rat (a) and rabbit (b) outgroups. (For the full caption text sec Fig. 7.6 on p. 171.)

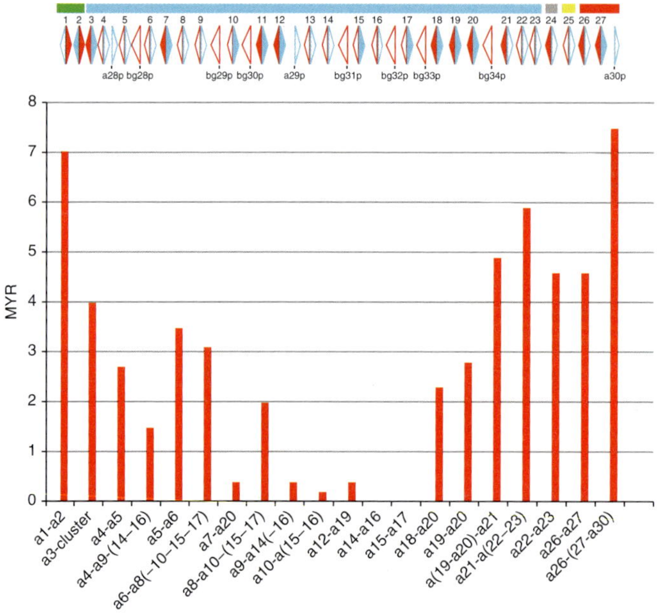

Figure 7.8 *Abpa* paralogue ages (MY) below the gene region map of the gene region. Older paralogues plot closer to the left and right boundaries, younger paralogues are in the interior. Coloured bars (top): the five clades (Laukaitis *et al.*, 2008). (For the full caption text see Fig. 7.8 on p. 175.)

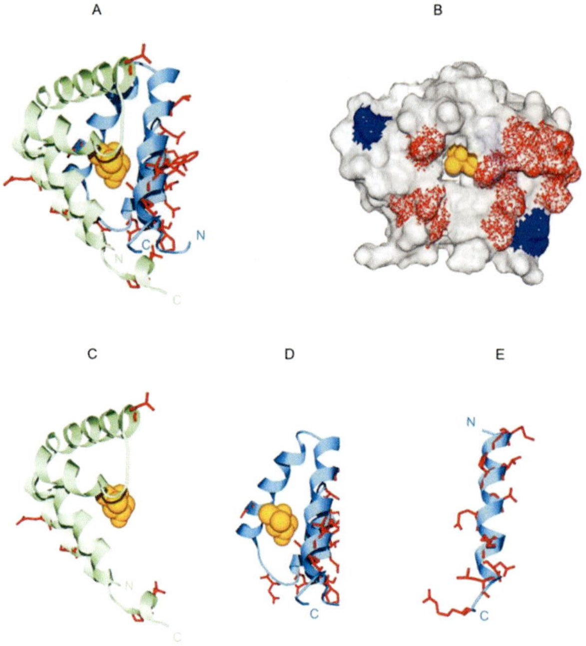

Figure 7.9 (*KA*/*KS*) analysis of *Abpa*-like and *Abpbg*-like genes mapped to PDB 1POU (Emes *et al.*, 2004). ω+ sites: (a) red with ligand orange; (b) blue: ABPA, red: ABPBG; (c) *Abpa* (d) *Abpbg* (e) ABPBG terminal extension. (For the full caption text see Fig. 7.9 on p. 178.)

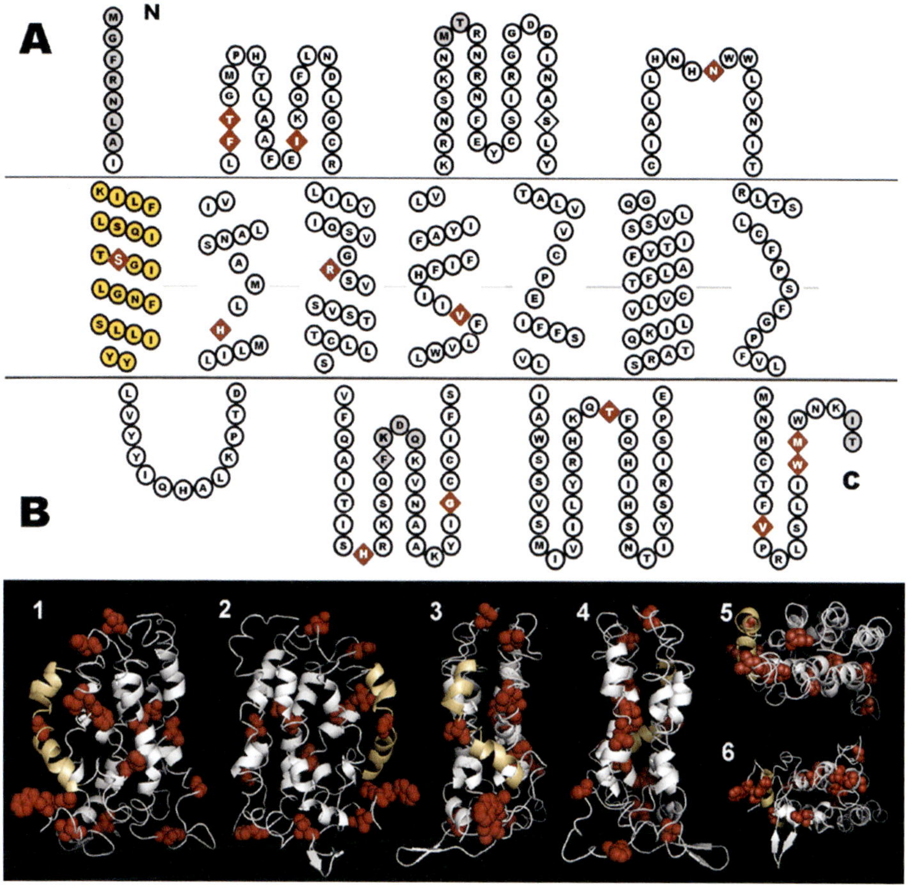

Figure 7.11 (a) *Vmn1r67* amino acid sequence diagram and (b) threaded three-dimensional structure. Amino acids in (a) that were not threaded on the *d1ln6a* model in (b) are shaded in grey (Karn *et al.*, 2010). (For the full caption text see Fig. 7.11 on p. 181.)

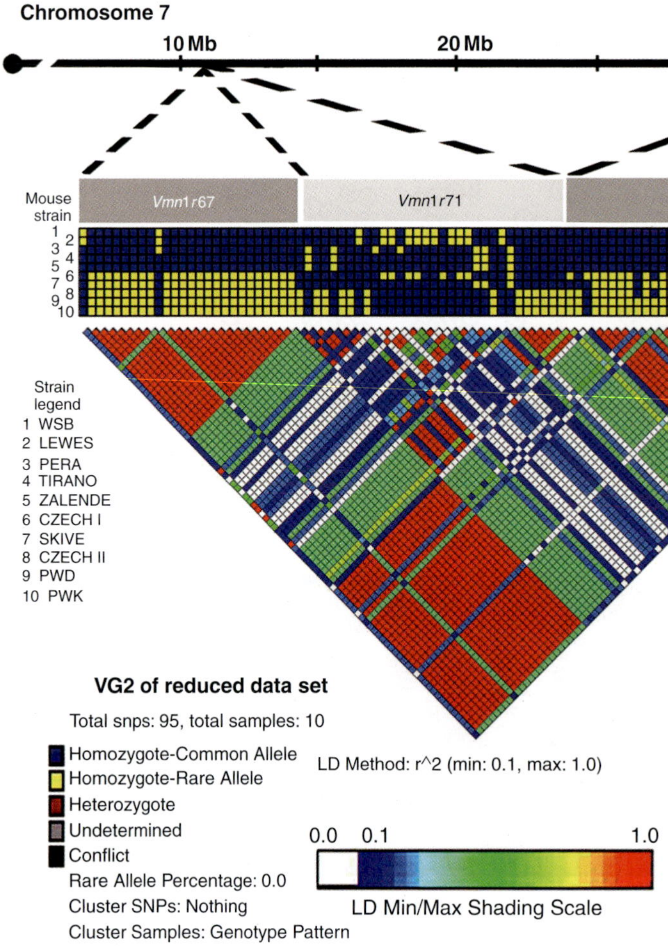

Figure 7.12 LD between *Vmn1r67* and *Abpa27*, *Vmn1r71* mapping between (grey blocks; Karn *et al.*, 2010). LD (*r 2*) is plotted in the triangle under the SNPs; an LD shading scale appears under the LD diagram. (For the full caption text see Fig. 7.12 on p. 183.)

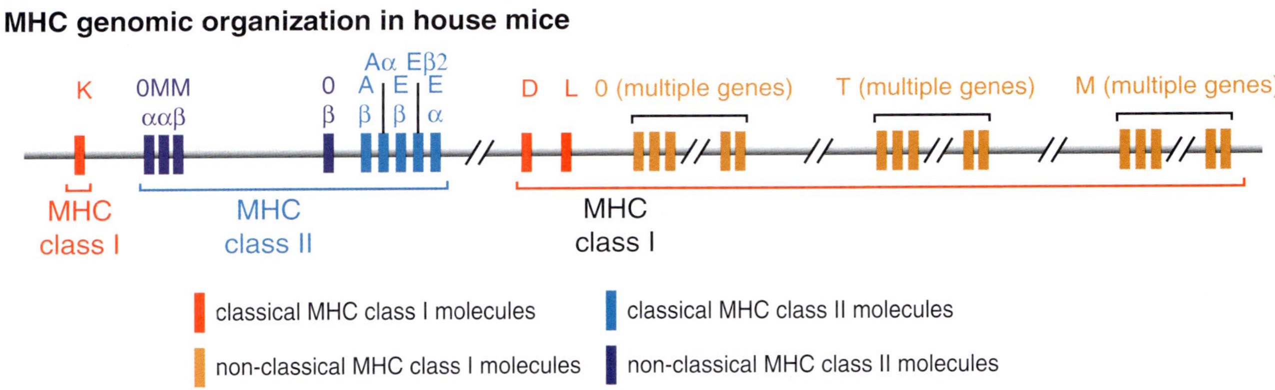

Figure 9.1 MHC genomic organization in house mice (modified with permission from DeFranco *et al.*, 2007).

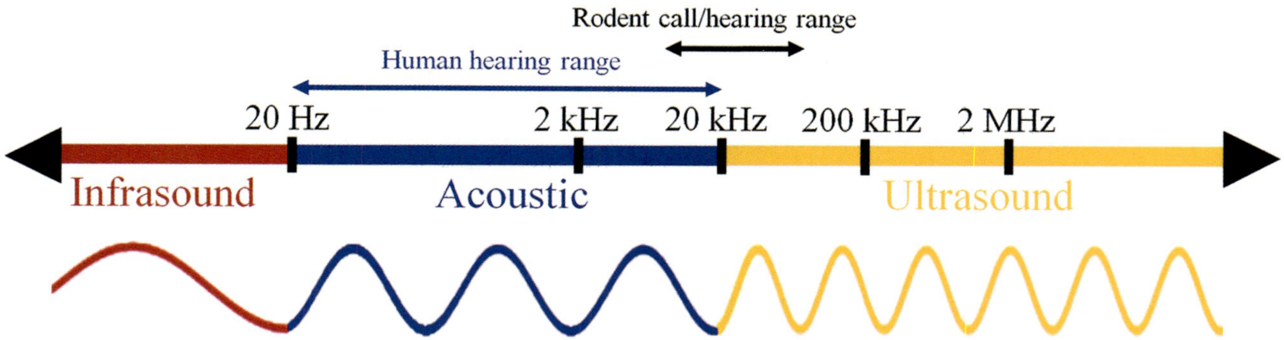

Figure 10.1 Schematic overview of sound categories defined by frequencies. The human hearing range and the rodent call and hearing range are indicated in the text.

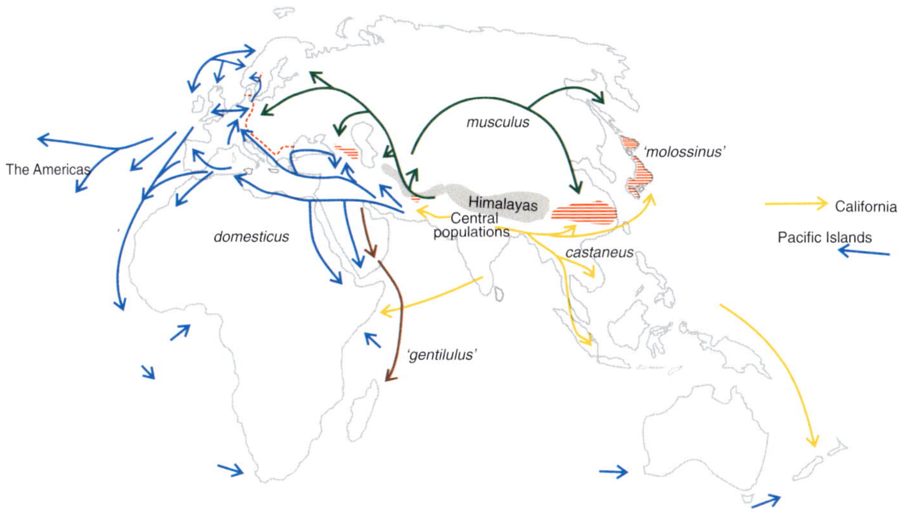

Figure 11.1 House mouse expansion routes inferred from mitochondrial variation. Blue: *M. m. domesticus*, green: *M. m. musculus*, yellow: *M. m. castaneus*. (For the full caption text see Fig. 11.1 on p. 280.)

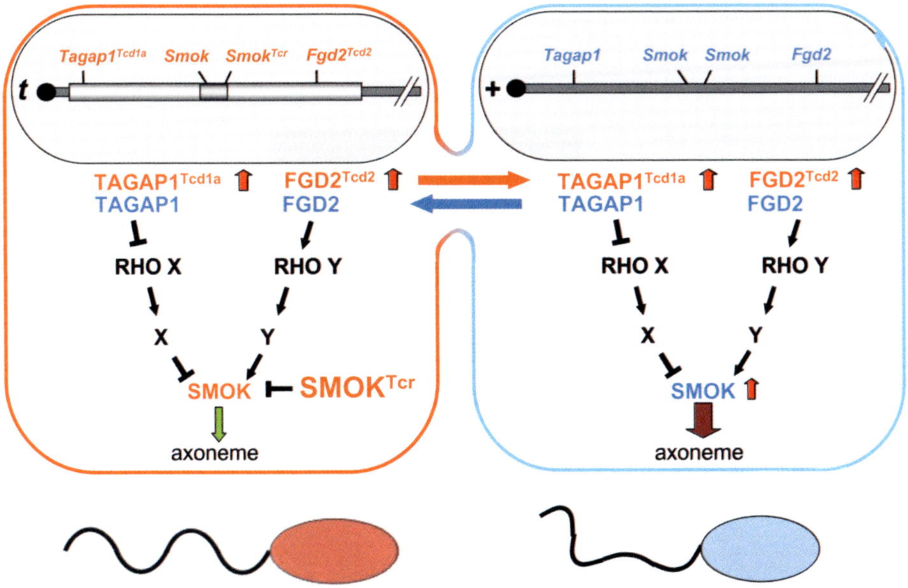

Figure 12.5 Model of the signalling mechanisms controlling sperm motility and causing TRD. Red upward arrows indicate enhanced activity; red downward arrow impaired, green arrow rescued control of the axoneme resulting in aberrant or normal sperm motility. (For the full caption text see Fig. 12.5 on p. 306.)

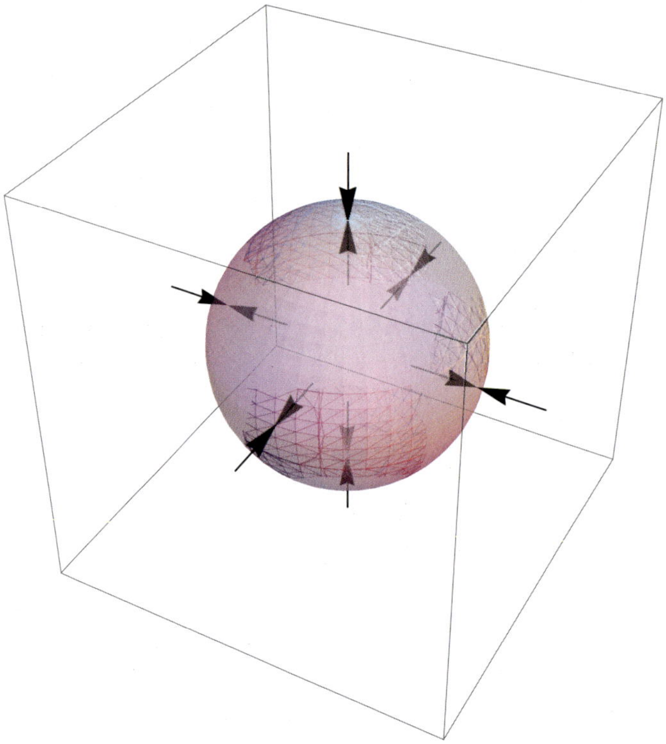

Figure 14.1 Tension zones behave like soap bubble surfaces: wobbling over the landscape they settle where the pressure of gene flow is equal from either side and the extent of the zone is minimized, smoothing the path of the zone centre towards a straight line. (For the full caption text see Fig. 14.1 on p. 336.)

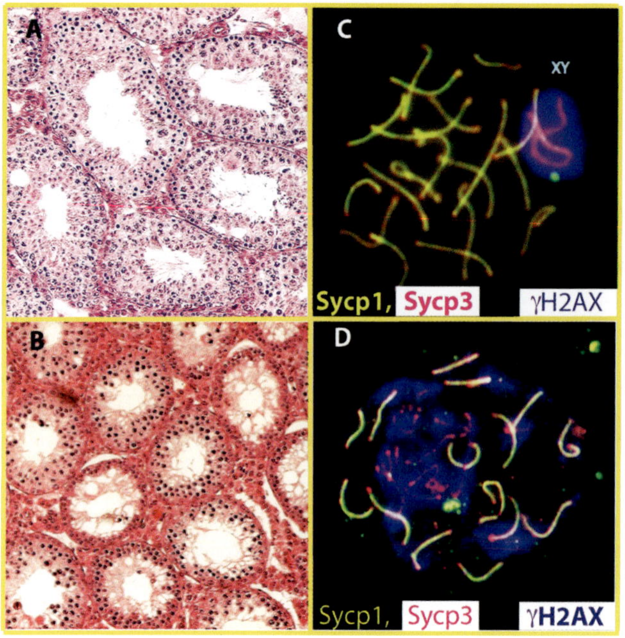

Figure 19.1 Spermatogenesis of intersubspecific hybrids. (a, b) Histological sections and (c, d) pachytene spreads from fertile and sterile hybrids. Pachynemas from sterile males display unsynapsed autosomes (SYCP3+, SYCP1–) with phosphorylated histone γH2AX but lack the sex (XY) body. (For the full caption text see Fig. 19.1 on p. 486.)

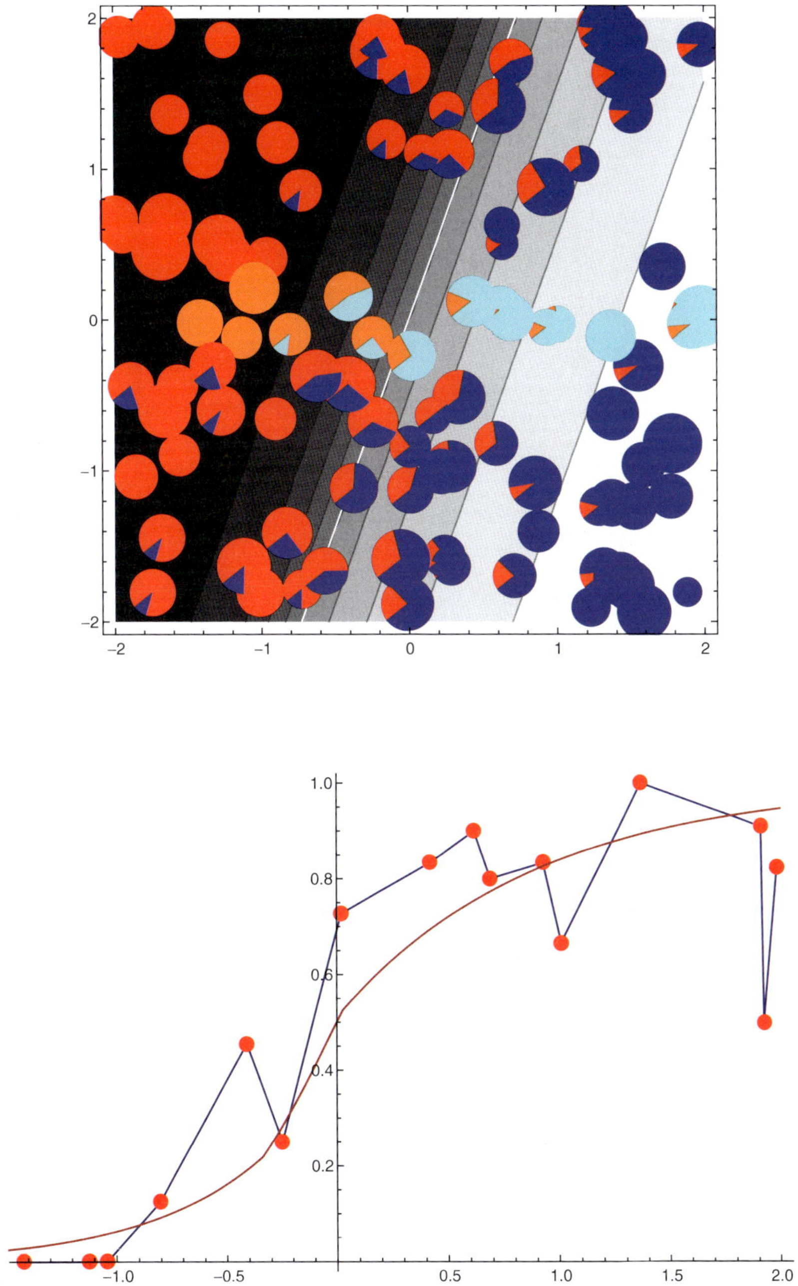

Figure 14.4 (a) Allele frequency samples from an underlying cline (shades of grey); orange and cyan: sites close to a horizontal linear transect. (b) The shape of the underlying cline compared to allele frequencies from samples along the transect: sample noise displaces the estimated cline centre to the left or overinflates its width. (For the full caption text see Fig. 14.4 on p. 343.)

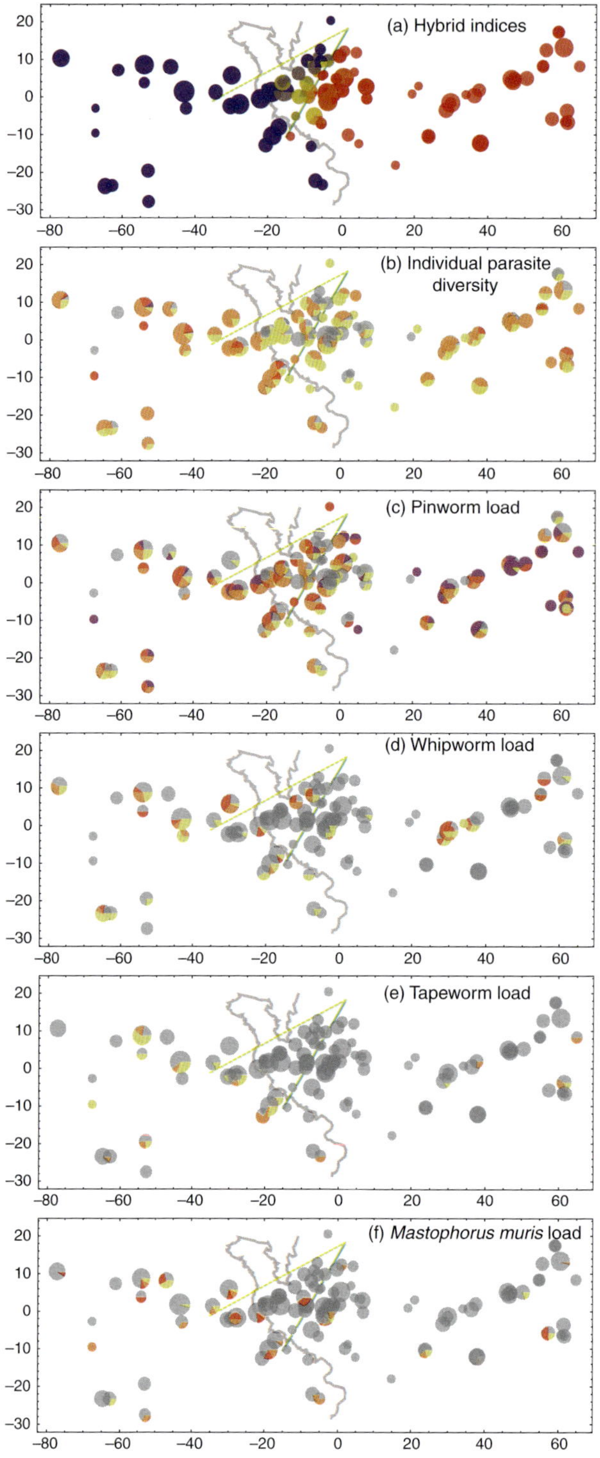

Figure 18.2 Hybrids and parasites across the Bavarian–Bohemian border (grey line) with zone centre (solid green), Y introgression (dashed), and pies proportional to log sample sizes. (a) Individuals' hybrid indices (*HI*s) within localities. Colours blend from blue (*domesticus*) through yellow (*HI* = 0.5) to red (*musculus*). (b) Parasite diversity over individuals within localities. Grey: 0; yellow, orange, red, purple indicate increasing diversity. (c–f) Individuals' parasite loads for localities. Grey: 0; yellow through purple, increasing load. (For the full caption text see Fig. 18.2 on p. 465.)

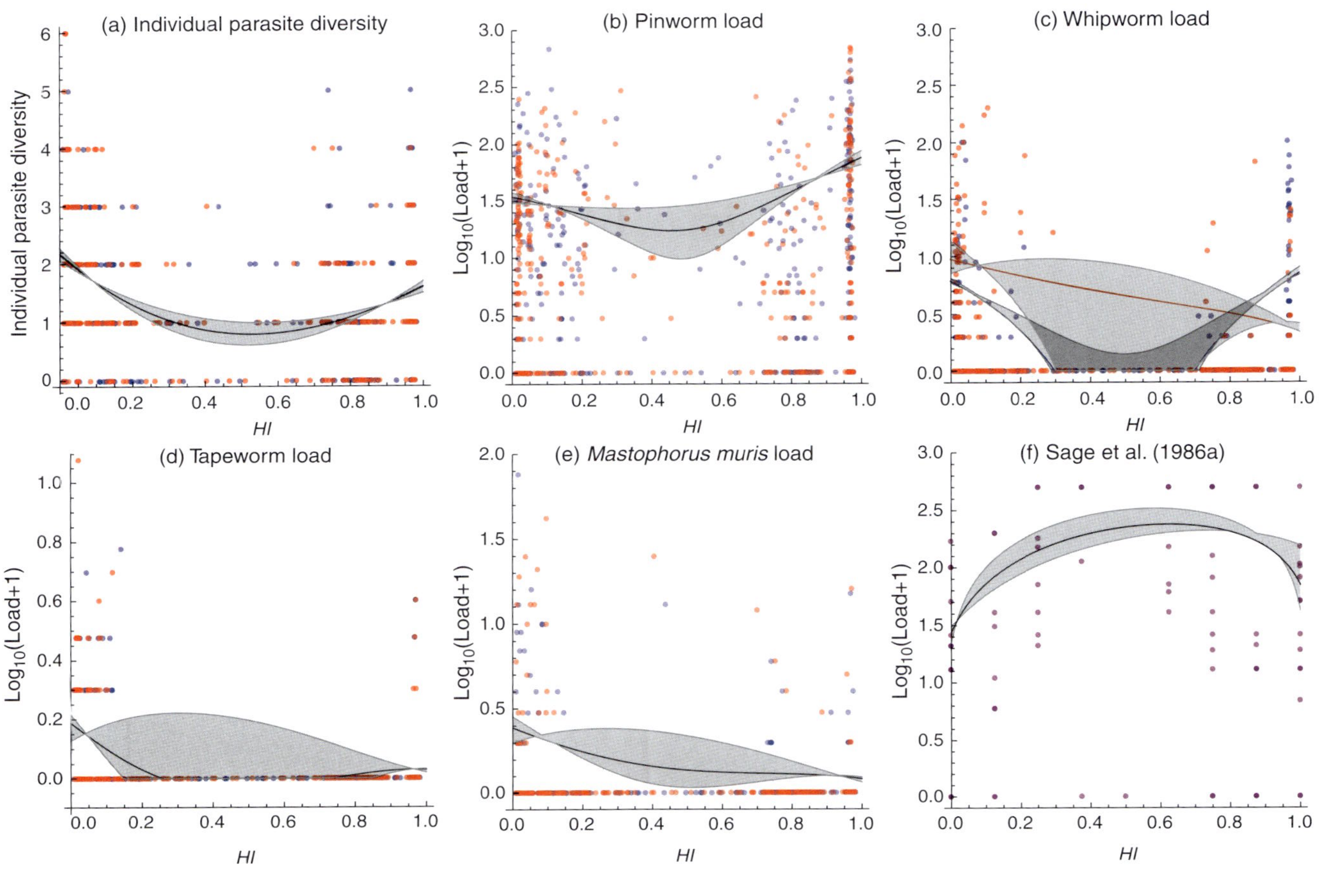

Figure 18.3 Individual parasite diversity and mean load compared to best models. *HI* ranges from 0 (*domesticus*) to 1 (*musculus*). Blue: males, red: females, purple: sex unknown. The scale of the y-axis is individual parasite diversity (a) and $\log_{10}$ (load + 1) (b–f). (For the full caption text see Fig. 18.3 on p. 466.)

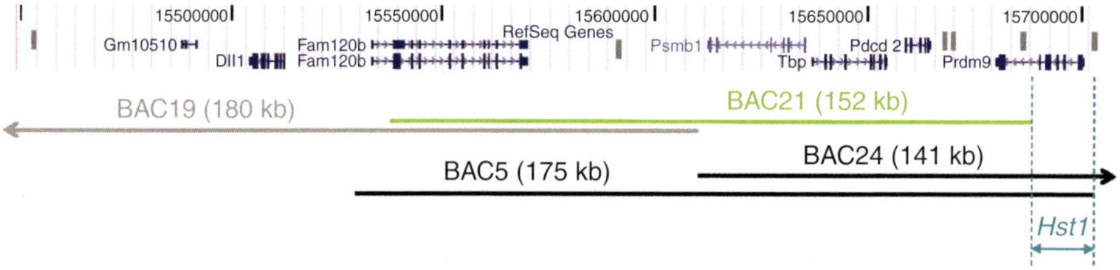

Figure 19.2 Transgenic rescue of *Hst1* sterility. The *Hst1* co-segregating region (252 kb) as a UCSC Genome Browser track (mm9): boxes, exons (pseudogenes in grey); black and green horizontal lines, BAC transgenes rescuing and not rescuing hybrid sterility, respectively; *Hst1*, region necessary for the rescue. (For the full caption text see Fig. 19.2 on p. 490.)

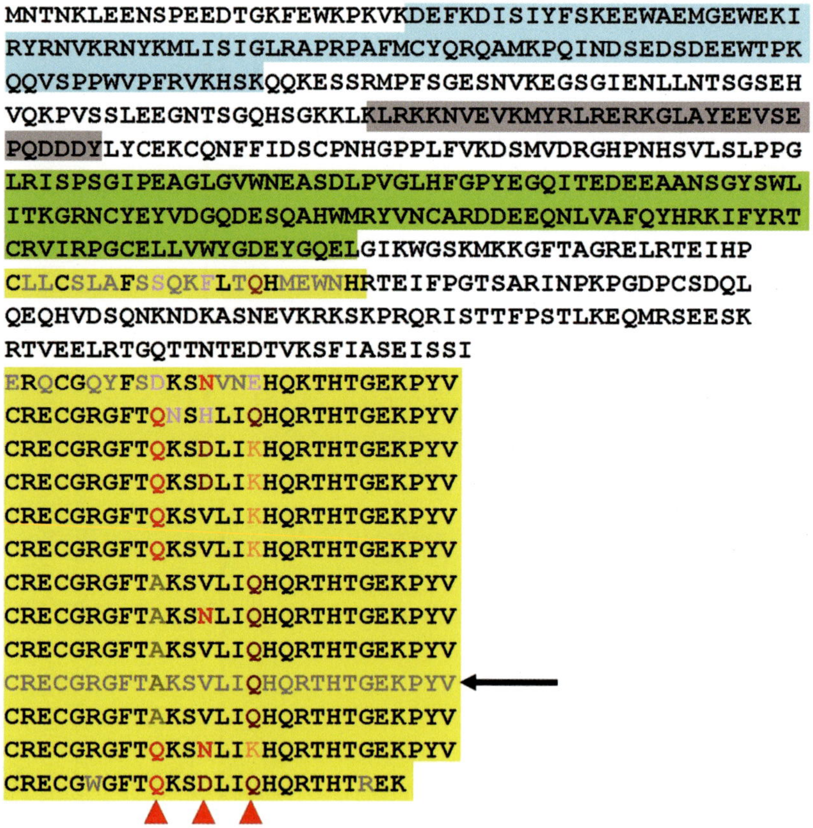

Figure 19.3 The protein sequence of the C3H allele of PRDM9. The domains are marked by colours: KRAB, blue; SSXRD, grey; PR/SET, green; C_2H_2 zinc finger, yellow. The arrow indicates the additional C_2H_2 domain of C3H allele compared with B6. (For the full caption text see Fig. 19.3 on p. 493.)

standard laboratory conditions reduces the diversity of their USVs compared to rearing in socially and environmentally enriched conditions (mixed-genders, nest boxes, shelters) (Portfors, 2007), and therefore more such studies are needed to investigate the effect of rearing conditions and laboratory artefacts. Nevertheless, such findings strongly underscore the importance of studying USVs in outbred, wild mice, and also under wild, or at least semi-natural, conditions.

Wild mice likely exhibit greater diversity and complexity in their USVs than laboratory mice, as has been demonstrated in *Peromyscus californicus* (Kalcounis-Rueppell *et al.*, 2010), and studies on wild populations combining USV characteristics with genetic analyses might help us to understand the evolution of these courtship displays. More detailed analyses on male courtship USVs should help to assign call characteristics to specific functions, which will be needed in order to determine what kinds of information are conveyed in male USVs. Moreover, future studies are needed to examine whether sexual conflict plays a role in USV communication (e.g. males' USVs may alter females' sexual maturity, reproductive cycles, and pregnancy maintenance, as with chemical signals). Manipulation of call parameters and playback experiments make it possible to experimentally test such functional hypotheses. Future studies are needed to investigate *receivers*, as well as senders, both for sensory (hearing) mechanisms, and the selective pressures to detect USVs. Such studies have already been conducted in some detail for mother–offspring communication in laboratory mice both on females' perceptional variation (Ehret and Schmid, 2009) and mother–pup interactions (Wöhr *et al.*, 2010), though not for male courtship calls and detection by females.

Finally, as we have previously emphasized, molecular genetic tools provide great opportunities for determining the genetic basis of USVs and the evolutionary origins of these calls (Box 10.2). Such genetic dissection would complement studies on other mammals and even songbirds. As more studies begin to use USV behaviour to characterize phenotypic effects in knock-out strains (Scattoni *et al.*, 2009), such studies will provide new insights into the genetic mechanisms and pathways controlling USV behaviour. Although there is already some evidence that USVs are heritable in laboratory strains, virtually nothing is known about the genetics or evolution of USVs in wild house mice. Integrating studies on well-characterized strains of laboratory mice with detailed investigations of the functions of these vocalizations under ecologically relevant situations should continue to provide new insights into this exceptional form of auditory communication.

REFERENCES

Amundsen, T. (2000). Why are female birds ornamented? *Trends in Ecology and Evolution*, **15**, 149–55.

Arnqvist, G. and Rowe, L. (2005). *Sexual Conflict*. Princeton, NJ: Princeton University Press.

Barron, S. and Gilbertson, R. (2005). Neonatal ethanol exposure but not neonatal cocaine selectively reduces specific isolation-induced vocalization waveforms in rats. *Behaviour Genetics*, **35**, 93–102.

Barthelemy, M., Gourbal, B. E. F., Gabrion, C., and Petit, G. (2004). Influence of the female sexual cycle on BALB/c mouse calling behaviour during mating. *Naturwissenschaften*, **91**, 135–8.

Bean, N. J., Nyby, J., Kerchner, M., and Dahinden, Z. (1986). Hormonal regulation of chemosignal-stimulated precopulatory behaviours in male housemice (*Mus musculus*). *Hormones and Behaviour*, **20**, 390–404.

Behr, O., von Helversen, O., Heckel, G., *et al.* (2006). Territorial songs indicate male quality in the sac-winged bat *Saccopteryx bilineata* (Chiroptera, Emballonuridae). *Behavioural Ecology*, **17**, 810–17.

Blumberg, M. S. and Alberts, J. R. (1990). Ultrasonic vocalizations by rat pups in the cold: an acoustic by-product of laryngeal braking? *Behavioural Neuroscience*, **104**, 808–17.

Boughman, J. W. and Moss, C. F. (2003). Social sounds: vocal learning and development of mammal and bird calls. In *Acoustic Communication*, ed. A. Megela Simmons, A. N. Popper, and R. Fay. New York: Springer Verlag, pp. 138–224.

Bradbury, J. W. and Vehrencamp, S. L. (1998). *Principles of Animal Communication*. Sunderland, MA: Sinauer Associates.

Branchi, I., Santucci, D., Vitale, A., and Alleva, E. (1998). Ultrasonic vocalizations by infant laboratory mice: a preliminary spectrographic characterization under different conditions. *Developmental Psychobiology*, **33**, 249–56.

Brooks, R. J. and Banks, E. M. (1973). Behavioural biology of the collared lemming (*Dycrostonyx groenlandicus*, Traill): an analysis of acoustic communication. *Animal Behaviour Monographs*, **6**, 1–83.

Broughton, W. P. (1963). Glossary. In *Acoustic Behaviour of Animals*, ed. R. G. Burnel. Boston, MA: Elsevier, p. 883.

Brown, R. E., Schellinck, H. M., and Jagosh, J. (1999). Behavioural studies of MHC-congenic mice. *Genetica*, **104**, 249–57.

Brudzynski, S. M. (2007). Ultrasonic calls of rats as indicator variables of negative or positive states: acetylcholine–dopamine interaction and acoustic coding. *Behavioural Brain Research*, **182**, 261–73.

Brudzynski, S. M., Bihari, F., Ociepa, D., and Fu, X.-W. (1993). Analysis of 22 kHz ultrasonic vocalization in laboratory rats: long and short calls. *Physiology & Behaviour*, **54**, 215–21.

Brudzynski, S. M. and Ociepa, D. (1992). Ultrasonic vocalization of laboratory rats in response to handling and touch. *Physiology & Behaviour*, **52**, 655–60.

Clutton-Brock, T. (2007). Sexual selection in males and females. *Science*, **318**, 1882–5.
Coburn, C. A. (1912). Singing mice. *Journal of Animal Behaviour*, **2**, 364–6.
Collins, S. (2004). Vocal fighting and vocal flirting: the functions of birdsong. In *Nature's Music*, ed. P. Marler and H. Slaberkoorn. London: Elsevier Academic Press, pp. 39–80.
Costantini, F. and D'Amato, F. R. (2006). Ultrasonic vocalizations in mice and rats: social contexts and functions. *Acta Zoologica Sinica*, **52**, 619–33.
Crawley, J. N. (2000). *What's Wrong with My Mouse? Behavioural Phenotyping of Transgenic and Knockout Mice*. New York: John Wiley & Sons.
Crawley, J. N., Belknap, J. K., Collins, A., *et al.* (1997). Behavioural phenotypes of inbred mouse strains: implications and recommendations for molecular studies. *Psychopharmacology*, **132**, 107–24.
D'Amato, F. R. (1991). Courtship ultrasonic vocalizations and social status in mice. *Animal Behaviour*, **41**, 875–85.
D'Amato, F. R. and Moles, A. (2001). Ultrasonic vocalizations as an index of social memory in female mice. *Behavioural Neuroscience*, **115**, 834–40.
D'Amato, F. R., Scalera, E., Sarli, C., and Moles, A. (2005). Pups call, mothers rush: does maternal responsiveness affect the amount of ultrasonic vocalizations in mouse pups? *Behaviour Genetics*, **35**, 103–12.
Dawkins, R. and Krebs, J. R. (1978). Animal signals: information or manipulation? In *Behavioural Ecology: An Evolutionary Approach*, ed. J. R. Krebs and N. B. Davies. Oxford: Blackwell Science, pp. 282–309.
Dice, L. R. (1932). The songs of mice. *Journal of Mammalogy*, **13**, 187–96.
Dizinno, G. and Whitney, G. (1977). Androgen influence on male mouse ultrasounds during courtship. *Hormones and Behaviour*, **8**, 188–92.
Dizinno, G., Whitney, G., and Nyby, J. (1978). Ultrasonic vocalizations by male mice to female sex pheromone: experiential determinants. *Behavioural Biology*, **22**, 104–13.
Drickamer, L. C. (1995). Rates of urine excretion by house mouse (*Mus domesticus*): differences by age, sex, social status, and reproductive condition. *Journal of Chemical Ecology*, **21**, 1481–93.
Ehret, G. (1974). Age-dependent hearing loss in normal hearing mice. *Naturwissenschaften*, **61**, 506–7.
Ehret, G. (1975). Schallsignale der Hausmaus (*Mus musculus*). *Behaviour*, **52**, 38–56.
Ehret, G. (1983). Auditory processing and perception of ultrasound in house mice. In *Advances in Vertebrate Neuroethology*, ed. J. P. Ewert, R. R. Capranica, and D. J. Ingle. New York: Plenum Press, pp. 911–18.
Ehret, G. (2001). Adaptation in the mouse auditory system for perception of ultrasonic communication calls. *Journal of Evolutionary Biochemistry and Physiology*, **37**, 562–8.
Ehret, G. (2005). Infant rodent ultrasounds: a gate to the understanding of sound communication. *Behaviour Genetics*, **35**, 19–29.
Ehret, G. and Bernecker, C. (1986). Low-frequency sound communication by mouse pups (*Mus musculus*): wriggling calls release maternal behaviour. *Animal Behaviour*, **34**, 821–30.
Ehret, G. and Schmid, C. (2009). Reproductive cycle-dependent plasticity of perception of acoustic meaning in mice. *Physiology & Behaviour*, **96**, 428–33.

Ekimov, A. and Sabatier, J. M. (2006). Vibration and sound signatures of human footsteps in buildings. *Journal of the Acoustical Society of America*, **120**, 762–8.

Ekimov, A. and Sabatier, J. M. (2008). Human motion analyses using footstep ultrasound and Doppler ultrasound. *Journal of the Acoustical Society of America*, **123**, EL149–54.

Elwood, R. W. and Keeling, F. (1982). Temporal organization of ultrasonic vocalizations in infant mice. *Developmental Psychobiology*, **15**, 221–7.

Endler, J. A. and Basolo, A. L. (1998). Sensory ecology, receiver biases and sexual selection. *Trends in Ecology and Evolution*, **13**, 415–20.

Farr, J. (1857). Singing-mouse. *Zoologist*, **15**, 5591.

Fischer, J., Kitchen, D. M., Seyfarth, R. M., and Cheney, D. L. (2004). Baboon loud calls advertise male quality: acoustic features and their relation to rank, age, and exhaustion. *Behavioural Ecology and Sociobiology*, **56**, 140–8.

Fisher, S. E. and Scharff, C. (2009). FOXP2 as a molecular window into speech and language. *Trends in Genetics*, **25**, 166–77.

Fletcher, N. H. (1992). *Acoustic Systems in Biology*. New York: Oxford University Press.

Floody, O. R. and Pfaff, D. W. (1977a). Communication among hamsters by high-frequency acoustic signals: II. Determinants of calling by males and females. *Journal of Comparative & Physiological Psychology*, **91**, 807–19.

Floody, O. R. and Pfaff, D. W. (1977b). Communication among hamsters by high-frequency signals: III. Responses evoked by natural and synthetic ultrasound. *Journal of Comparative & Physiological Psychology*, **91**, 820–9.

Gil, D. and Gahr, M. (2002). The honesty of bird song: multiple constraints for multiple traits. *Trends in Ecology and Evolution*, **17**, 133–41.

Gourbal, B. E. F., Barthelemy, M., Petit, G., and Gabrion, C. (2004). Spectrographic analysis of the ultrasonic vocalizations of adult male and female BALB/c mice. *Naturwissenschaften*, **91**, 381–5.

Grimsley, J. M. S., Monaghan, J. J. M., and Wenstrup J. J. (2011). Development of social vocalizations in mice. *PLoS ONE*, **6**, e17460.

Groszer, M., Keays, D. A., Deacon, R. M. J., *et al.* (2008). Impaired synaptic plasticity and motor learning in mice with a point mutation implicated in human speech deficits. *Current Biology*, **18**, 354–62.

Guo, Z. and Holy, T. E. (2007). Sex selectivity of mouse ultrasonic songs. *Chemical Senses*, **32**, 463–73.

Hahn, M. E., Hewitt, J. K., Schanz, N., Weinreb, L., and Henry, A. (1997). Genetic and developmental influences on infant mouse ultrasonic calling: I. A diallel analysis of the calls of 3-day olds. *Behaviour Genetics*, **27**, 133–43.

Hahn, M. E., Karkowski, L., Weinreb, L., *et al.* (1998). Genetic and developmental influences on infant mouse ultrasonic calling: II. Developmental patterns in the calls of mice 2–12 days of age. *Behaviour Genetics*, **28**, 315–25.

Hahn, M. E. and Lavooy, M. J. (2005). A review of the methods of studies on infant ultrasound production and maternal retrieval in small rodents. *Behaviour Genetics*, **35**, 31–52.

Hammerschmidt, K., Radyushkin, K., Ehrenreich, H., and Fischer, J. (2009). Female mice respond to male ultrasonic 'songs' with approach behaviour. *Biology Letters*, **5**, 589–92.

Heffner, R. S. and Heffner, H. E. (1985). Hearing in mammals: the least weasel. *Journal of Mammalogy*, **66**, 745–55.

Hofer, M. A. (2010). Evolution of the infant separation call: rodent ultrasonic vocalization. In *Handbook of Mammalian Vocalization: An Integrative Neuroscience Approach*, ed. S. M. Brudzynski. Amsterdam/Boston, MA: Elsevier/Academic Press, pp. 29–35.

Hoffmann, F., Musolf, K., and Penn, D. J. (2009). Freezing urine reduces its efficacy for eliciting ultrasonic vocalizations from male mice. *Physiology & Behaviour*, **96**, 602–5.

Hoffmann, F., Musolf, K., and Penn, D. J. (2012a). Spectrographic analyses reveal signals of individuality and kinship in the ultrasonic courtship vocalizations of wild house mice. *Physiology and Behavior*, **105**, 766–71.

Hoffmann, F., Musolf, K., and Penn, D. J. (2012b). Ultrasonic courtship vocalizations in wild house mice: spectrographic analyses. *Journal of Ethology*, DOI 10.1007/s10164-011-0312-y

Hoffmeyer, I. and Sales, G. D. (1977). Ultrasonic behaviour of *Apodemus sylvaticus* and *A. flavicollis*. *Oikos*, **29**, 67–77.

Holland, B. and Rice, W. R. (1998). Chase-away sexual selection: antagonistic seduction versus resistance. *Evolution*, **52**, 1–7.

Holy, T. E. and Guo, Z. (2005). Ultrasonic songs of male mice. *PLoS Biology*, **3**, e386.

Hurst, J. L. (1990). Urine marking in populations of wild house mice, *Mus domesticus* Rutty: I. Communication between males. *Animal Behaviour*, **40**, 209–22.

Hurst, J. L., Fang, J., and Barnard, C. J. (1993). The role of substrate odours in maintaining social tolerance between male house mice, *Mus musculus domesticus*. *Animal Behaviour*, **45**, 997–1006.

James, P. J. and Nyby, J. G. (2002). Testosterone rapidly affects the expression of copulatory behaviour in house mice (*Mus musculus*). *Physiology & Behaviour*, **75**, 287–94.

James, P. J., Nyby, J. G., and Saviolakis, G. A. (2006). Sexually stimulated testosterone release in male mice (*Mus musculus*): roles of genotype and sexual arousal. *Hormones and Behaviour*, **50**, 424–31.

Johnston, R. E. and Bronson, F. (1982). Endocrine control of female mouse odors that elicit luteinizing hormone surges and attraction in males. *Biology of Reproduction*, **27**, 1174–80.

Kalcounis-Rueppell, M. C., Metheny, J. D., and Vonhof, M. J. (2006). Production of ultrasonic vocalizations by *Peromyscus* mice in the wild. *Frontiers in Zoology*, **3**, 3.

Kalcounis-Rueppell, M. C., Petric, R., Briggs, J. R., *et al.* (2010). Differences in ultrasonic vocalizations between wild and laboratory California mice (*Peromyscus californicus*). *PLoS ONE*, **5**, e9705.

Kapusta, J., Pachinger, K., and Marchlewska-Koj, A. (1999). Behavioural variation in two populations of root voles. *Acta Theriologica*. **44**, 337–43.

Kapusta, J. and Sales G. D. (2009). Male–female interactions and ultrasonic vocalization in three sympatric species of voles during conspecific and heterospecific encounters. *Behaviour*, **146**, 939–62.

Kapusta, J., Sales, G. D., and Czuchnowski, R. (2007). Aggression and vocalization behaviour of three sympatric vole species during conspecific and heterospecific same-sex encounters. *Behaviour*, **144**, 283–305.

Kikusui, T., Nakanishi, K., Nakagawa, R., *et al.* (2011). Cross fostering experiments suggest that mice songs are innate. *PLoS ONE*, **6**, e17721.

Liu, R. C., Miller, K. D., Merzenich, M. M., and Schreiner, C. E. (2003). Acoustic variability and distinguishability among mouse ultrasound vocalizations. *Journal of the Acoustic Society of America*, **114**, 3412–22.

Lumley, L. A., Sipos, M. L., Charles, R. C., Charles, R. F., and Meyerhoff, J. L. (1999). Social stress effects on territorial marking and ultrasonic vocalizations in mice. *Physiology and Behavior*, **67**, 769–75.

Maggio, J. C. and Whitney, G. (1985). Ultrasonic vocalizing by adult female mice (*Mus musculus*). *Journal of Comparative Psychology*, **99**, 420–36.

Maggio, J. C. and Whitney, G. (1986). Heterosis of adult mouse (*Mus musculus*) ultrasonic vocalizing. *Behaviour Genetics*, **16**, 493–506.

Marler, P. R. and Slabbekoorn, H. (2004). *Nature's Music: The Science of Birdsong*. London/ San Diego, CA: Elsevier/Academic Press.

Maynard Smith, J. and Harper, D. (2003). *Animal Signals*. New York: Oxford University Press.

Miller, J. R. and Engstrom, M. D. (2007). Vocal stereotypy and singing behaviour in Baiomyine mice. *Journal of Mammalogy*, **88**, 1447–65.

Miller, R. A., Harper, J. M., Dysko, R. C., Durkee, S. J., and Austad, S. N. (2002). Longer life spans and delayed maturation in wild-derived mice. *Experimental Biology and Medicine*, **227**, 500–8.

Moles, A., Costantini, F., Garbugino, L., Zanettini, C., and D'Amato, F. R. (2007). Ultrasonic vocalizations emitted during dyadic interactions in female mice: a possible index of sociability? *Behavioural Brain Research*, **182**, 223–30.

Moles, A. and D'Amato, F. R. (2000). Ultrasonic vocalization by female mice in the presence of a conspecific carrying food cues. *Animal Behaviour*, **60**, 689–94.

Moles, A., Kieffer, B. L., and D'Amato, F. R. (2004). Deficit in attachment behaviour in mice lacking the μ-opioid receptor gene. *Science*, **304**, 1983–6.

Musolf, K., Hoffmann, F., and Penn D. J. (2010). Ultrasonic courtship vocalizations in wild house mice, *Mus musculus*. *Animal Behaviour*, **79**, 757–64.

Novotny, M., Harvey, S., and Jemiolo, B. (1990). Chemistry of male dominance in the house mouse, *Mus domesticus*. *Experientia*, **46**, 109–13.

Nunez, A. A., Nyby, J., and Whitney, G. (1978). The effects of testosterone, estradiol, and dihydrotestosterone on male mouse (*Mus musculus*) ultrasonic vocalizations. *Hormones and Behaviour*, **11**, 264–72.

Nunez, A. A. and Tan, D. T. (1984). Courtship ultrasonic vocalizations in male Swiss-Webster mice: effects of hormones and sexual experience. *Physiology & Behaviour*, **32**, 717–21.

Nyby, J. (1983). Ultrasonic vocalizations during sex behaviour of male house mice (*Mus musculus*): a description. *Behavioural and Neural Biology*, **39**, 128–34.

Nyby, J., Dizinno, G., and Whitney, G. (1976). Social status and ultrasonic vocalizations of male mice. *Behavioural Biology*, **18**, 285–9.

Nyby, J., Dizinno, G., and Whitney, G. (1977a). Sexual dimorphism in ultrasonic vocalizations of mice (*Mus musculus*): gonadal hormone regulation. *Journal of Comparative & Physiological Psychology*, **91**, 1424–31.

Nyby, J. and Whitney, G. (1978). Ultrasonic communication of adult myomorph rodents. *Neuroscience & Biobehavioural Reviews*, **2**, 1–14.

Nyby, J., Whitney, G., Schmitz, S., and Dizinno, G. (1978). Postpubertal experience establishes signal value of mammalian sex odor. *Behavioural Biology*, **22**, 545–52.

Nyby, J., Wysocki, C. J., Whitney, G., and Dizinno, G. (1977b). Pheromonal regulation of male mouse ultrasonic courtship (*Mus musculus*). *Animal Behaviour*, **25**, 333–41.

Nyby, J., Wysocki, C. J., Whitney, G., Dizinno, G., and Schneider, J. (1979). Elicitation of male mouse (*Mus musculus*) ultrasonic vocalizations: I. Urinary cues. *Journal of Comparative & Physiological Psychology*, **93**, 957–75.

Nyby, J. and Zakeski, D. (1980). Elicitation of male mouse ultrasounds: bladder urine and aged urine from females. *Physiology & Behaviour*, **24**, 737–40.

Owings, D. H. and Morton, E. S. (1998). *Animal Vocal Communication: A New Approach*. Cambridge: Cambridge University Press.

Penn, D. J. and Frommen, J. G. (2010). Kin recognition: an overview of conceptual issues, mechanisms and evolutionary theory. In *Behaviour: Mechanisms and Functions*, ed. P. Kappeler. Berlin: Springer-Verlag, pp. 55–86.

Penn, D. J. and Potts, W. K. (1998). Chemical signals and parasite-mediated sexual selection. *Trends in Ecology and Evolution*, **13**, 391–6.

Pomerantz, S. M., Nunez, A. A., and Bean, N. J. (1983). Female behaviour is affected by male ultrasonic vocalizations in house mice. *Physiology & Behaviour*, **31**, 91–6.

Portfors, C. V. (2007). Types and functions of ultrasonic vocalizations in laboratory rats and mice. *Journal of the American Association for Laboratory Animal Science*, **46**, 28–34.

Powell, R. A. and Zielinski, W. J. (1989). Mink response to ultrasound in the range emitted by prey. *Journal of Mammalogy*, **70**, 637–8.

Pye, J. D. (1979). Why ultrasound? *Endeavour*, **3**, 57–62.

Ralls, K. (1967). Auditory sensitivity in mice, *Peromyscus* and *Mus musculus*. *Animal Behaviour*, **15**, 123–8.

Ryan, M. J. (1998). Sexual selection, receiver biases, and the evolution of sex differences. *Science*, **281**, 1999–2003.

Sales, G. D. (1972a). Ultrasound and aggressive behaviour in rats and other small mammals. *Animal Behaviour*, **20**, 88–100.

Sales, G. D. (1972b). Ultrasound and mating behaviour in rodents with some observation on other behavioural situations. *Journal of Zoology*, **168**, 149–64.

Sales, G. D. (2010). Ultrasonic calls of wild and wild-type rodents. In *Handbook of Mammalian Vocalization: An Integrative Neuroscience Approach*, ed. S. M. Brudzynski. Amsterdam/Boston, MA: Elsevier/Academic Press, pp. 77–88.

Sales, G. D. and Pye, J. D. (1974). *Ultrasonic Communication by Animals*. London: Chapman and Hall.

Sanders, I., Weisz, D. J., Yang, B. Y., Fung, K., and Amirali, A. (2001). The mechanism of ultrasonic vocalization in the rat. *Society for Neuroscience Abstracts*, **27**, 19.

Scattoni, M. L., Crawley, J., and Ricceri, L. (2009). Ultrasonic vocalizations: a tool for behavioural phenotyping of mouse models of neurodevelopmental disorders. *Neuroscience & Biobehavioural Reviews*, **33**: 508–15.

Scattoni, M. L., Gandhy, S. U., Ricceri, L., and Crawley, J. N. (2008a). Unusual repertoire of vocalizations in the BTBR T+tf/J mouse model of autism. *PLoS ONE*, **3**, e3067.

Scattoni, M. L., McFarlane, H. G., Zhodzishsky, V., *et al.* (2008b). Reduced ultrasonic vocalizations in vasopressin 1b knockout mice. *Behavioural Brain Research*, **187**, 371–8.

Sell, A., Bryant, G. A., Cosmides, L., *et al.* (2010). Adaptations in humans for assessing physical strength from the voice. *Proceedings of the Royal Society London B: Biological Sciences*, **277**, 3509–18.

Shair, H. N., Brunelli, S. A., and Hofer, M. A. (2005). Lack of evidence for mu-opioid regulation of a socially mediated separation response. *Physiology & Behaviour*, **83**, 767–77.

Shu, W., Cho, J. Y., Jiang, Y., *et al.* (2005). Altered ultrasonic vocalization in mice with a disruption in the *Foxp2* gene. *Proceedings of the National Academy of Sciences of the United States of America*, **102**, 9643–8.

Sidebotham, J. (1877). Singing mice. *Nature*, **17**, 29.

Sipos, M. L., Kerchner, M., and Nyby, J. G. (1992). An ephemeral sex pheromone in the urine of female house mice (*Mus domesticus*). *Behavioural and Neural Biology*, **58**, 138–43.

Sipos, M. L., Nyby, J. G., and Serran, M. F. (1993). An ephemeral sex pheromone of female house mice (*Mus domesticus*): pheromone fade-out time. *Physiology & Behaviour*, **54**, 171–4.

Sipos, M. L., Wysocki, C. J., Nyby, J. G., Wysocki, L., and Nemura, T. A. (1995). An ephemeral pheromone of female house mice: perception via the main and accessory olfactory systems. *Physiology & Behaviour*, **58**, 529–34.

Smith, J. C. (1979). Factors affecting the transmission of rodent ultrasounds in natural environments. *American Zoologist*, **19**, 432–42.

Teramitsu, I. and White, S. A. (2008). Motor learning: the FoxP2 puzzle piece. *Current Biology*, **18**, R335–7.

Thomas, D. A., Howard, S. B., and Barfield, R. J. (1982). Male-produced postejaculatory 22-kHz vocalizations and the mating behaviour of estrous female rats. *Behavioural and Neural Biology*, **36**, 403–10.

Trivers, R. L. (1972). Parental investment and sexual selection. In *Sexual Selection and the Descent of Man 1871–1971*, ed. B. Campbell. Chicago, IL: Aldine, pp. 136–79.

Vannoni, E. and McElligott, A. G. (2008). Low frequency groans indicate larger and more dominant fallow deer (*Dama dama*) males. *PLoS ONE*, **3**, e3113.

Wade, C. M. and Daly, M. J. (2005). Genetic variation in laboratory mice. *Nature Genetics*, **37**, 1175–80.

Wang, H., Liang, S., Burgdorf, J., Wess, J., and Yeomans, J. (2008). Ultrasonic vocalizations induced by sex and amphetamine in M2, M4, M5 muscarinic and D2 dopamine receptor knockout mice. *PLoS ONE*, **3**, e1893.

Warburton, V. L., Sales, G. D., and Milligan, S. R. (1989). The emission and elicitation of mouse ultrasonic vocalizations: the effects of age, sex and gonadal status. *Physiology & Behaviour*, **45**, 41–7.

Weller, A., Leguisamo, A. C., Towns, L., *et al.* (2003). Maternal effects in infant and adult phenotypes of 5HT1A and 5HT1B receptor knockout mice. *Developmental Psychobiology*, **42**, 194–205.

White, N. R. and Barfield, R. J. (1990). Effects of male pre-ejaculatory vocalizations on female receptive behaviour in the rat (*Rattus norvegicus*). *Journal of Comparative Psychology*, **104**, 140–6.

White, N. R., Prasad, M., Barfield, R. J., and Nyby, J. G. (1998). 40- and 70-kHz vocalizations of mice (*Mus musculus*) during copulation. *Physiology & Behaviour*, **63**, 467–73.

Whitney, G., Alpern, M., Dizinno, G., and Horowitz, G. (1974). Female odors evoke ultrasounds from male mice. *Animal Learning & Behaviour*, **2**, 13–18.

Whitney, G., Coble, J. R., Stockton, M. D., and Tilson, E. F. (1973). Ultrasonic emissions: do they facilitate courtship of mice? *Journal of Comparative & Physiological Psychology*, **84**, 445–52.

Whitney, G. and Nyby, J. (1979). Cues that elicit ultrasound from adult male mice. *American Zoology*, **19**, 457–63.

Winslow, J. T., Hearn, E. F., Ferguson, J., *et al.* (2000). Infant vocalization, adult aggression, and fear behaviour of an oxytocin null mutant mouse. *Hormones and Behaviour*, **37**, 145–55.

Wöhr, M., Dahlhoff, M., Wolf, E., *et al.* (2008). Effects of genetic background, gender, and early environmental factors on isolation-induced ultrasonic calling in mouse pups: an embryo-transfer study. *Behaviour Genetics*, **38**, 579–95.

Wöhr, M., Oddi, D., and D'Amato, F. R. (2010). Effect of altricial pup ultrasonic vocalization on maternal behaviour. In *Handbook of Mammalian Vocalization: An Integrative Neuroscience Approach*, ed. S. M. Brudzynski. Amsterdam/Boston, MA: Elsevier/Academic Press, pp. 159–66.

Wolff, J. O. (2003). Laboratory studies with rodents: facts or artifacts? *Bioscience*, **53**, 421–7.

Wysocki, C. J., Nyby, J., Whitney, G., Beauchamp, G. K., and Katz, Y. (1982). The vomeronasal organ: primary role in mouse chemosensory gender recognition. *Physiology & Behaviour*, **29**, 315–27.

Young, D. M., Schenk, A. K., Yang, S.-B., Jan, Y. N., and Jan, L. Y. (2010). Altered ultrasonic vocalizations in a tuberous sclerosis mouse model of autism. *Proceedings of the National Academy of Sciences of the United States of America*, **107**, 11074–9.

Zahavi, A. and Zahavi, A. (1997). *The Handicap Principle: A Missing Piece of Darwin's Puzzle*. New York: Oxford University Press.

Zala, S. M., Potts, W. K., and Penn, D. J. (2004). Scent-marking displays provide honest signals of health and infection. *Behavioural Ecology*, **15**, 338–44.

Zippelius, H.-M. and Schleidt, W. M. (1956). Ultraschall-Laute bei jungen Mäusen. *Naturwissenschaften*, **43**, 502.

11

House mouse phylogeography

FRANÇOIS BONHOMME AND JEREMY B. SEARLE

Introduction

'Phylogeography' is a field of evolutionary biology involving the use of molecular markers to explain the extant genetic variation over a species range in a temporal and spatial context (Avise, 2000). This geographic pattern of variation reflects range expansions or contractions through time, and therefore phylogeography aims at elucidating the colonization/contraction history of organisms by using molecular tools. Gene lineages (or 'phylogroups') are defined and the primary analysis involves geographical mapping of these phylogroups, sublineages, and individual haplotypes, determining genetic diversities in phylogeographic units and geographic areas, and using methods to age phylogeographic units and demographic events. Much phylogeographic analysis has been directed at elucidating the natural colonization history of organisms since the Last Glacial Maximum, when the ranges of temperate species were restricted to glacial refugia from which range expansions occurred with the amelioration of climate (Hewitt, 2000). By inferring what happened to separate lineages within species, it is possible to build up a picture of the species-wide colonization history, from the location of probable glacial refugia to the limits of the range expansions. Of course, the molecular records tend to be more complex when they include the impact of several successive glacial events.

The phylogeography of most members of the genus *Mus* fits into this typical framework, although few species other than the house mouse have been under phylogeographic scrutiny over their complete range. We can cite as exceptions *Mus spretus* and *M. macedonicus* (Boursot *et al.*, 1984; Macholán *et al.*, 2007). For most other species in the genus, available information primarily concerns phylogeny and broad distribution (see, for example, Suzuki and Aplin, Chapter 2 in this volume) but hardly more. However, the house mouse (*M. musculus*) has a more complex phylogeography because of the involvement of humans in its

Evolution of the House Mouse, ed. Miloš Macholán, Stuart J. E. Baird, Pavel Munclinger, and Jaroslav Piálek. Published by Cambridge University Press.

colonization history. There has been much work directed at this history of the house mouse in conjunction with humans and we will focus this chapter on that aspect. However, it is important to see this in the context of the earlier pre-human phase, which we will review first.

Early phylogeography of house mouse subspecies

For historical reasons, *M. musculus* was first described and characterized from rather peripheral parts of its range. This includes *M. m. musculus* described by Linnaeus from Scandinavia, *M. m. domesticus* from western Europe, *M. m. molossinus* in Japan, and *M. m. castaneus* from Southeast Asia. So, for a long time the origins and history of these subspecies, as well as their interrelationships, remained obscure, until sampling included more central parts of the species distribution occurring in parallel with improved methodologies, including the development of suitable molecular techniques. Among those, electrophoretically detected protein polymorphism was historically the first technique to yield valuable pieces of information, although it was more of use in a systematic rather than phylogeographic context. Up to now, it is thus primarily the study of mitochondrial DNA that has most contributed to shedding light on the phylogeography of the species. It is believed that *M. musculus* and its ancestors thrived naturally in a temperate steppic climate (Berry and Jakobson, 1975; Brothwell, 1981; Berry, 1986), a fact that would have prevented its natural expansion across deserts, mountain ranges, or dense temperate/tropical forest. Although our understanding is far from complete, a picture emerges that suggests the Indo-Pak subcontinent and neighbouring Afghanistan and Iran as the area where *M. musculus* originated and initially differentiated (Fig. 11.1).

This highly dissected terrain, with mountain barriers and intervening lowlands, would have promoted separation of gene pools each time climatic conditions allowed natural migration of the species into new territory. This process probably occurred several times during the Quaternary climatic oscillations, as indicated by the divergence times of the major *M. musculus* subspecies, estimated from the coalescence of their mitochondrial genomes. There are five distinct phylogroups or clades that can be retrieved from cytochrome *b* (Cyt-*b*) sequences available in GenBank as of May 2011, which are almost equidistant from each other. The average third codon position divergence over 844 nucleotide sites among the three lineages referred to as *castaneus* amounted to 8.45%, 10.57% between these and *domesticus*, 9.61% between these and *musculus*, and 9.61% between *domesticus* and *musculus*. The sixth phylogroup of the house mouse, known as *gentilulus*, hasn't been studied for the Cyt-*b* gene yet, but appears from the mitochondrial control region as almost equidivergent from the preceding five. It is helpful that Nabholz

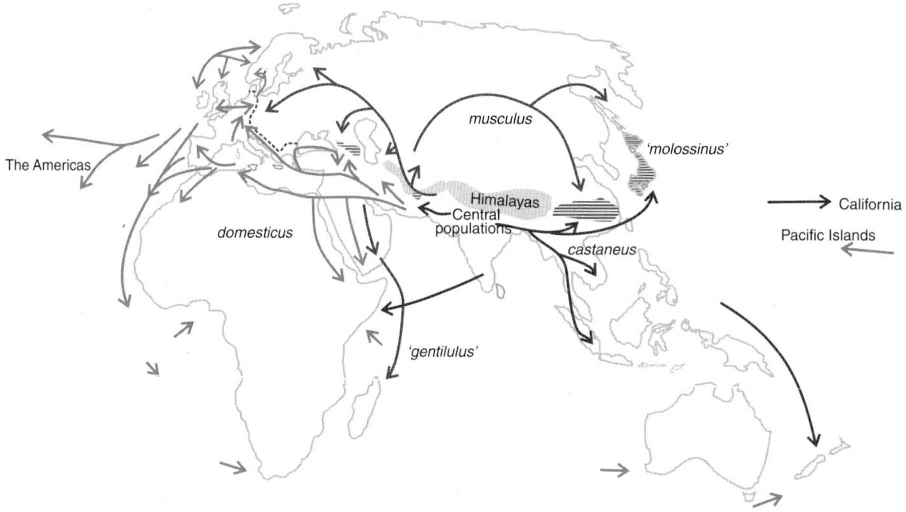

Figure 11.1 Summary of the inferred colonization history of the house mouse (*Mus musculus*) starting with an origin and differentiation in the vicinity of the Indo-Pak subcontinent and with arrows indicating the colonizing movements of *castaneus*, *domesticus*, and *musculus* subspecies and the *'gentilulus'* lineage. These inferences are based primarily on mitochondrial DNA studies. Dashing and hatching indicate regions of hybridization between the subspecies, including the generation of the *'molossinus'* form in Japan. For maps showing more detail of the inferred colonization history in Europe, see Gabriel *et al.* (2010). See the plate section for a colour version of this figure.

et al. (2008) have re-evaluated the substitution rate for the third codon position of Cyt-*b* in many mammalian orders and genera, including *Mus*, with a substitution rate of 0.22 substitution/third codon position/million years (Myr) in *M. indutus*. If this rate also holds for *M. musculus*, the values observed would correspond roughly to a 250 000-year-old divergence, both between the so-called *castaneus* lineages, which do not necessarily appear monophyletic with respect to *domesticus* and *musculus*, and between *domesticus* and *musculus* (Rajabi-Maham *et al.*, submitted). Therefore, the initial differentiation of the various present-day subspecies possibly started two major interglacial periods ago, somewhere in the area south of the Himalayas long occupied by the genus *Mus* (see the review of the genus fossil record in Auffray *et al.*, 1990; see also Auffray and Britton-Davidian, Chapter 1). Iran, Afghanistan, Pakistan, and India constitute the part of the world with the greatest genetic diversity in *M. musculus*, where all the above-mentioned matrilines can be found, together with a high diversity of Y chromosome patrilines (Rajabi-Maham *et al.*, submitted). The *castaneus* clades are found in various proportions in these four countries, but identifying a primary place of differentiation is hampered by much secondary movement, and will require additional sampling and molecular data. The *domesticus* phylogroup is found in southwest Iran, which probably constitutes – together with the part of the Fertile Crescent located in Turkey, Syria, and Iraq – the area of origin for this subspecies (Rajabi-Maham *et al.*, 2008; but see also Prager *et al.*, 1998). The *musculus* phylogroup is found in the northeast and northwest corners of Iran, along the shore of the Caspian Sea and south of the Caucasus in Georgia and Azerbaijan (Spiridonova *et al.*, 1990; Orth *et al.*, 1998; Milishnikov *et al.*, 2004; Darvish *et al.*, 2006). Which of these locations is the point of origin of this subspecies remains uncertain.

Range expansion in association with humans

Compared with the earlier period of major differentiation in the house mouse, association with humans has been more recent and has occasioned a spectacular period of range expansion. The house mouse has been a commensal of humans for about 12 000 years, since the beginnings of agriculture and the creation of the first grain stores (Tchernov, 1984; Cucchi *et al.*, 2005; see also Cucchi *et al.*, Chapter 3). Subsequently, house mice have exploited humans and spread with them around the world (Boursot *et al.*, 1993; Pocock *et al.*, 2005). The house mouse is one of the most successful invasive mammals and it is intriguing to know how it attained its cosmopolitan range (Searle, 2008). Also, because natural populations of the house mouse are increasingly being studied to elucidate fundamental aspects of evolutionary genetics (e.g. Ihle *et al.*, 2006; Baines and Harr, 2007; Geraldes *et al.*, 2008; see also Laukaitis and Karn *et al.*, Chapter 7; Teschke *et al.*, Chapter 13), it is important to know the recent history of these populations. Because humans may transport them frequently and over long distances, house mice in a particular area may have a mixed genome; this can either be viewed as a confounding factor or an exciting research opportunity, depending on one's interest (Searle, 2008; Yang *et al.*, 2011).

The association of humans and house mice makes the phylogeography of the latter of exceptional interest for another reason: study of the colonization history of house mice is inevitably informative about the movements of the humans that transported the mice. Therefore, the phylogeography of the house mouse can act as a proxy for trading links and movement history of humans (Searle *et al.*, 2009a, 2009b, 2009c). Obviously, archaeologists and historians can use human artefacts and human DNA to gain knowledge about this, but the phylogeography of house mice may be viewed as an interesting and informative additional research tool. Over the period when mice may be useful (i.e. the 12 000 years since they have been commensal), phylogeographic interpretation based on human DNA can be difficult (Soares *et al.*, 2010), and so it may be very valuable to have mice as a different line of contemporary DNA evidence.

Mitochondrial DNA (mtDNA) is particularly good as a marker for range expansion of house mice, and sequences have been collected over the last 20 years (Prager *et al.*, 1993). MtDNA is maternally inherited and the control region (D-loop), which is the region that is generally sequenced, is highly variable. There is experimental genetic evidence that female house mice do not readily penetrate existing populations of house mice and thus the mtDNA sequences in a certain area are likely to represent the founding females. Thus, when mice of both sexes were introduced from one small Scottish island to another (Eday onto the Isle of May), introduced nuclear markers spread throughout the colonized island, but introduced mtDNA did not spread away from the release point, which suggests that males successfully spread their genes on secondary colonization but females do not (Jones *et al.*, 1995). The mitochondrial and nuclear data for house mice on Madeira are most easily interpreted in these terms (see below; Förster *et al.*, 2009). The large dataset from the Kerguelen archipelago provides a further example of impenetrability of established house mouse populations to infiltration by new mtDNA sequences (Hardouin *et al.*, 2010) for a relatively long period of time.

So, the phylogeography of the house mouse, particularly based on mtDNA, is likely to often reflect the first colonizing individuals, and therefore may be an unusually useful marker of human first arrival in an area if that arrival was within a period of time that humans have been transporting mice. Of course, the signal will not always be as 'clean' as that, because mice may go extinct in a particular area (for whatever reason), there could occasionally be particularly large influxes of secondary colonists or there may be selection events involving mtDNA. Different parts of an archipelago or even a single large island may have a different mouse history (e.g. Cyprus is characterized by an astonishing seven mtDNA clades: Bonhomme *et al.*, 2011). All of these things (and more) could confuse the phylogeographic pattern. Late-colonized regions are self-evidently much less likely to show complex patterns than more ancient ones, as is the case for much of the Near East, for instance, but also in other places like Germany (Rajabi-Maham *et al.*, 2008). However, with those caveats, the phylogeographic findings so far in the house mouse have indicated a remarkable fit with human history (which can be complex, too!), which bodes well for its future value to historians and archaeologists, as we will illustrate in the following case studies.

Case studies of phylogeography in the house mouse

Mus musculus domesticus: Near East and Mediterranean

As described above, mtDNA has been instrumental in helping to decipher the geographical site of origin of *M. m. domesticus* and its spread thereafter.

Several clues suggest that the Fertile Crescent *sensu lato* (from the tip of the Arabo-Persian Gulf to south Anatolia through the former Mesopotamia; see Cucchi *et al.*, Chapter 3, for a detailed account) was the area from which the human-mediated range expansion of *M. m. domesticus* originated. First, the diversity of mtDNA matrilines and the mismatch distribution in samples from this area fit with an ancient sudden expansion (Rajabi-Maham *et al.*, 2008), supported by a high diversity for X chromosome markers (Baines and Harr, 2007). Second, the available zooarchaeological data also designate this region and the Levant as the first places of association between Neolithic humans and mice as soon as there was stable sedentism and granaries were established (Cucchi *et al.*, 2005). Note that this geographical area is also occupied by *M. macedonicus* (Macholán *et al.*, 2007), but interestingly this species did not become commensal even though *M. musculus* did, even though *M. macedonicus* was found in close proximity to early Neolithic settlements (Auffray *et al.*, 1988). The association of humans and house mice (i.e. *M. musculus*) took place immediately after the first post-glacial temperature rebound during the Bølling/Allerød (12 700–10 700 BC) warm period (Cucchi and Vigne, 2006), and initiated the subsequent house mouse expansion when Neolithic farming practices expanded.

The molecular traces of this expansion can be seen from the D-loop phylogeny, which is our prime source of information for the time being, although the internal nodes of this phylogeny are poorly resolved. In what follows, we refer to the nomenclature used in the most comprehensive and latest publications and do not discuss this point again. Two haplogroups (Hg) appear to be associated primarily with the Fertile Crescent and the Levant (Hg6 and Hg10; Bonhomme *et al.*, 2011), their split being tentatively linked with the cold Younger Dryas period (between 10 800 and 9500 BC), which may have occasioned a retreat of Neolithic storage practice and a split between the southern and northern Levant. Another haplogroup (Hg3) that appears to be specific to Asia Minor and the north of the Balkan Peninsula (Gündüz *et al.*, 2005; Rajabi-Maham *et al.*, 2008) may correspond to an early expansion taken by the early Neolithic farmers of the so-called Ribbon (or banded) ceramic culture that subsequently followed a Danubian route. A further set of haplogroups apparently took a northern Mediterranean route in a westward expansion following the southern edge of Asia Minor, Cyprus, and the Greek islands in association with the so-called Cardial ceramic Neolithic culture, supported by archaeological findings of house mice (Cucchi *et al.*, 2005, see also Cucchi *et al.*, Chapter 3). The arrival of house mice in western Europe would not have been complete until the Late Bronze Age/Iron Age, as shown by these same archaeological data, when maritime trade intensified, with decked vessels that were more likely to carry house mice, and large human settlements became available for colonization by them (Cucchi *et al.*,

2005). As to the other (southern) side of the Mediterranean, its occupation by house mice apparently followed a similar scheme, with the western part of north Africa probably colonized later on, like western Europe with the same set of haplogroups, while the shore of the eastern basin could have been colonized earlier following a stepwise progression of agriculture through Egypt and Libya to Tunisia (Bonhomme *et al.*, 2011). However, archaeological findings of mice are almost absent from the southern Mediterranean (Cucchi *et al.*, Chapter 3), and the D-loop data have not provided robust dating. At any rate, further molecular data encompassing nuclear genes on larger geographical samplings will allow tests of these Mediterranean colonization scenarios for house mice.

Phylogeography of Atlantic western Europe

There has been a particular focus on the phylogeography of the house mouse in Atlantic western Europe, i.e. the regions and countries that border the Atlantic Ocean and the oceanic islands. These areas were peripheral in terms of human history at the time of major movement of house mice in the Late Bronze Age/Iron Age (Cucchi *et al.*, 2005), when the Mediterranean and central Europe were the influential areas in terms of political power, trade, and commerce. However, the Atlantic coastal zone in western Europe became very important as a region of human activity in medieval and post-medieval times and, not surprisingly, was the source area of human colonization of many of the oceanic islands of the northeast Atlantic. It is to be expected that house mouse phylogeography in this area would somehow reflect human history.

Portugal is the southernmost country in Atlantic western Europe and the house mouse has been extensively surveyed there (Förster *et al.*, 2009). Not surprisingly, Portuguese mice primarily have affinity with those from the Mediterranean and the mtDNA clades termed B and C by Jones *et al.* (2011b)[1] are the best represented.

It would be expected that geographical areas closely affiliated with Portugal within the last 3000 years would show the same genetic characteristics in their house mice as Portugal itself. Madeira is a dependency of Portugal that, in the historical literature, was discovered by Henry the Navigator of Portugal in 1419 and has had close links with the country ever since. So, there is an expectation that the genetics of Madeira mice would reflect this Portuguese connection, and indeed both allozymes and microsatellites do (Britton-Davidian *et al.*, 2007; Förster *et al.*, in prep.). However, despite the extensive sampling on both Madeira and Portugal, there is no overlap in mtDNA, apart from a single individual on Madeira that has a Portuguese mtDNA type (Gündüz *et al.*, 2001; Förster *et al.*, 2009). Instead, Madeira has a remarkable similarity in terms of mtDNA with northern Germany and Scandinavia, with four haplotypes in common; the two most widespread and abundant haplotypes in Madeira are also widespread and abundant in northern Germany and Scandinavia (Gündüz *et al.*, 2001; Förster *et al.*, 2009; Jones *et al.*, 2010, 2011a, 2011b). These haplotypes shared between Madeira and northern Europe are all closely related to each other (subclade D1, Hg8); indeed, the star phylogeny obtained from the Madeiran mtDNA sequences indicate colonization by a small number of individuals from one location, with new haplotypes generated through time by the accumulation of new mutations.

Although other explanations are possible (Förster *et al.*, 2009), most parsimoniously these nuclear and mitochondrial data for Madeira suggest an initial (unexpected) colonization from northern Europe (reflected in the mtDNA sequences) and repeated (expected) secondary colonization from Portugal, by males at least (reflected in the nuclear markers). The initial colonization can only reasonably have occurred before the Portuguese link was established by Henry the Navigator and the only likely seafarers to have inadvertently visited Madeira from northern Europe are the Danish Vikings. They made trips to the Mediterranean in the ninth century and although visits to Madeira are not recorded in the historical record, they are perfectly possible. Any introduced mice would have been able to survive in the outdoor environment in the absence of competitors, as has occurred on other oceanic islands (Ruscoe and Murphy, 2005). In relation to these Madeira data, Gündüz *et al.*'s (2001) initial assertion that house mice may be a novel marker for Viking movements has been borne out by subsequent studies in northwest Europe (see below).

These data from Madeira show without a doubt the resistance of mtDNA to penetration on secondary colonization. Certainly, thousands of mice would have been inadvertently transported from Portugal to the Madeiran archipelago, and yet there is no mitochondrial signal of these movements. Likewise between Madeira and Porto Santo within the Madeiran archipelago, where there is a small but consistent difference in the mtDNA sequence (Gündüz *et al.*, 2001; Förster *et al.*, 2009).

Also within Macaronesia, the mice in the Canary Islands have been analysed by Bonhomme *et al.* (2011). These give a picture of later colonization than Madeira, based on levels of genetic variation. Thus, mice of clade D1 (Hg8) and C1 (Hg1) colonized, likely from Madeira and elsewhere in Mediterranean Europe, respectively (Bonhomme *et al.*, 2011). The first Europeans to establish a settlement on the Canary Islands were Portuguese from Madeira (Fregel *et al.*, 2009), and it appears that house mice also colonized then. That the mice on the Canary Islands should be more recent colonists than on Madeira may seem surprising because there is archaeological evidence that house mice have been on the Canary Islands for at least 1600 years (Alcover *et al.*, 2009). It appears that the descendants of these earliest colonists became extinct for some reason, and the present-day mouse populations are the product of more recent arrivals.

While the house mice from the Atlantic seaboard of continental Iberia show clear affinities with the west Mediterranean (Förster *et al.*, 2009), further up the continental coast this affinity disappears. In particular, the coast of north France has been well-sampled (Jones *et al.*, 2011b) and the mtDNA clade there is 'E', which is basal in the phylogeny and yet barely found among mice from the Mediterranean area (Bonhomme *et al.*, 2011; Jones *et al.*, 2011b). Clade E is predominant on the island of Britain (except for the extreme north of Scotland: Jones *et al.*, 2011b), in the Netherlands (Gabriel *et al.*, 2011), and extends into Germany. There is archaeological evidence that house mice arrived in this area in the Iron Age (Cucchi *et al.*, 2005; Jones *et al.*, 2011b), and this distribution fits well with the realm of interaction of Iron Age people in northern Europe (Cucchi *et al.*, 2005; Searle *et al.*, 2009b; Jones *et al.*, 2011b). The fact that clade E is not well represented in central Europe is interesting and suggests that this clade did not arrive with people overland. Instead, it can be suggested that it came via a maritime route from the source area of *M. m. domesticus* in the Near East, perhaps transported by a chain of seafarers through the Mediterranean and along the Atlantic coast (Jones *et al.*, 2011b). The Phoenicians were Late Bronze Age/Iron Age seafarers who could have taken the mice through the length of the Mediterranean to its contact with the Atlantic (Bonhomme *et al.*, 2011; Jones *et al.*, 2011b).

There is another clade in *M. m. domesticus* in northern Europe that did apparently penetrate overland through central Europe: clade D, with its derivative D1 at the apex of the expansion (Jones *et al.*, 2011b). This 'overland' transport of mice may actually have largely happened by boat along rivers. The Rhône corridor is an important route by which Iron Age people travelled up from the Mediterranean into central Europe and it is reasonable to suggest that mice of clade D took this route as stowaways (Jones *et al.*, 2011b). In terms of the Atlantic coast, it is subclade D1 that is important. Sequences of this clade are common in a region encompassing northern Germany and Scandinavia. It has been suggested that Danish Vikings may have taken *M. m. domesticus* with mtDNA of this clade to Madeira from the southern part of this region, as described above. However, the D1 clade is also found in northern Denmark, Sweden, and Finland, where the mice are *M. m. musculus* in terms of nuclear markers but with *M. m. domesticus* mtDNA (Prager *et al.*, 1993; Jones *et al.*, 2010). It is tempting to think that it was the Vikings that were also involved in this colonization process. The Vikings were very important in urbanizing Scandinavia (Skre, 2008), and for *M. m. domesticus* at least, urban centres appear to be critically important for the subspecies to spread under competition from *Apodemus sylvaticus* (Cucchi *et al.*, 2005). *A. sylvaticus* occurs throughout Germany, Denmark, and southern Sweden (Mitchell-Jones *et al.*, 1999), and so may have blocked the colonization of northern Denmark, southern Sweden, and further north until Viking urbanization. However, the subspecies that was colonizing was *M. m. musculus* and it is not certain that urbanization is so important for this subspecies (but see Ganem, Chapter 15) and therefore it may have been able to colonize Scandinavia sooner than the Viking Age (Jones *et al.*, 2010).

It appears that the colonization of Sweden and northern Denmark involved mice derived from the hybrid zone between *M. m. musculus* and *M. m. domesticus* in northern Germany (Prager *et al.*, 1993), and it was for this reason that they had hybrid nuclear/mitochondrial characteristics. The colonizing mice spread the length of Sweden, into Finland (Prager *et al.*, 1993). They also got into Norway, and clade D1 is found all the way to the Norwegian coast in southwest Norway (Jones *et al.*, 2010). However, the west Norwegian mice primarily have a *M. m. domesticus* nuclear genome. Although the rather complicated situation in Norway is being studied further (E. P. Jones, personal communication), it may have involved an initial spread throughout Norway of the *M. m. musculus* with clade D1 mtDNA from Sweden (Jones *et al.*, 2010). Thus, initially all Norwegian house mice may have had this same nuclear–mitochondrial hybrid genome. However, it appears that *M. m. domesticus* was transported by sea to the coast of Norway and replaced *M. m. musculus* over much of the west of the country. There currently is a hybrid zone between the two subspecies near Oslo (Jones *et al.*, 2010).

The *M. m. domesticus* along coastal Norway are generally characterized by clade F (Jones *et al.*, 2010). It appears that *M. m. domesticus* (nuclear) + *domesticus* clade F (mtDNA) colonizing eastwards from the west coast of Norway met *M. m. musculus* (nuclear) + *domesticus* clade D1 (mtDNA). Interestingly, in southwest Norway the mice are *M. m. domesticus* (nuclear) + *domesticus* clade D1 (mtDNA). Therefore, it can be suggested that *M. m. musculus* 'captured' *domesticus* clade D1 mtDNA from *M. m. domesticus* in the hybrid zone in northern Germany, and that *M. m. domesticus* 're-captured' that same D1 mtDNA in western Norway.

Interestingly, it is the *M. m. domesticus* (nuclear) + *domesticus* clade D1 (mtDNA) that colonized the Faroes (Jones *et al.*, 2011a). There are historic data which suggest that the human colonizers of the Faroe Islands were Vikings from southwest Norway, and their livestock appear to have come from there (Jones *et al.*, 2011a). It appears that the mice were taken there after the generation of the new nuclear/mitochondrial combination in southwest Norway.

How did clade F arrive in Norway? Clade F is found through much of the Norwegian Viking realm of influence in the British Isles and Norway (Searle *et al.*, 2009b). The clade is basal in *M. m. domesticus* and is not represented in central continental Europe, and therefore, like clade E, may have reached the Atlantic coast by mice being carried along a maritime route from the source area of the *M. m. domesticus* expansion in the Near East (Jones *et al.*, 2011b). Mice belonging to clade F presumably got to a locality where they were picked up by a Viking boat, and thereafter the lineage would have been spread around the region occupied by Norwegian Vikings (Searle *et al.*, 2009b).

Concluding this section we can say that there have been a series of studies targeted at the phylogeography of mice along the Atlantic seaboard and that these have highlighted the Vikings as influential human agents. This finding is of interest. Zooarchaeological studies have shown that house mice reached much of western Europe considerably earlier than the Viking Age. However, as with the colonizing movements of mice to the west Mediterranean, for many parts of the Atlantic coast there was a delay until master seafarers were available to move the mice. In the case of the west Mediterranean the mice were available to be moved already in Neolithic times, but it apparently required the Late Bronze Age/Iron Age Phoenicians to take them there. In the case of the Atlantic coast, the mice were available to be moved in the Iron Age, but it was predominantly the Vikings who were responsible for many of the important colonizing movements.

M. m. musculus and *M. m. castaneus*: eastern Asia

The history of association of house mice with humans does not only concern its western subspecies, but also *M. m. musculus* in Central Asia and *M. m. castaneus* in Southeast Asia. Much fewer data are available regarding these subspecies and we still lack precise indications of where and when *musculus* and *castaneus* first became commensal. Concerning *musculus*, as recounted above, there is still some debate concerning the site of initial differentiation of the subspecies (during one of the last interglacial cycles), i.e. whether it occurred to the south of the Caucasus (Kotenkova and Bulatova, 1994; Mezhzherin *et al.*, 1998; Milishnikov *et al.*, 2004) or to the north (Darvish *et al.*, 2006). But wherever it was, these animals cannot have colonized by their own means the vast expanses of land north of a line starting with the Alps in the west, passing along the Balkans, the Black Sea, the Caucasus, the Elbourz, the Kopet Dag, and finally the Himalayas. Instead, their association with humans permitted this colonization, and allowed them to occupy areas that are too cold for them to overwinter outside. One can therefore suggest that, once agriculture progressed into these territories, at some stage *M. m. musculus* was able to spread throughout the area. One of the few zooarchaeological records available indicates an already strong association with humans around 5000 years ago north of the Danube in Romania (Cucchi *et al.*, Chapter 3). Now this subspecies is firmly established from eastern

Europe to Siberia and northern China, and there have been reports of a certain degree of genetic differentiation within this range (e.g. Frisman *et al.*, 1990; Usuda *et al.*, 2009; Nunome *et al.*, 2010).

As to *M. m. castaneus*, it also became commensal when humans became sedentary and started to provide a suitable supply of resources and shelter for mice. When and where this occurred is still a matter for research, and it is hoped that a zooarchaeological analysis of early Neolithic settlements will provide the necessary information. Preliminary phylogeographic information gathered by Rajabi-Maham *et al.* (submitted) reveals that, among the differentiated matrilines that constitute the so-called *castaneus* subspecies, the haplogroup predominant in India is also the one that has been primarily exported from there (for example, to Kenya, China, Indonesia, and even New Zealand – see Fig. 11.1). What is certain, however, is that house mice are not able to survive in a free-living form over most of the tropical environment of Southeast Asia, although they occur there in human dwellings. Indeed, they face strong competition with many kinds of rodents, including several other species of mice (see the summary of the phylogeny of the genus *Mus* in Asia provided by Suzuki and Aplin (Chapter 2 and literature cited therein)) better adapted than themselves to the wet, tropical environment. Humans have thus undoubtedly been the dispersal agent for *M. m. castaneus*, bringing them into contact with *M. m. musculus* within China and Japan. A hybrid zone between these two subspecies can be seen along the Yangtze River in China (Kawashima *et al.*, 1995) and the interaction between the subspecies has been well studied in Japan (Yonekawa *et al.*, 1986, 1988; see also Yonekawa *et al.*, Chapter 4), where they were possibly brought in by humans over several waves (Terashima *et al.*, 2006). House mice are now classically considered as forming a hybrid taxon in Japan called '*M. m. molossinus*' (Fig. 11.1), with varied levels of admixture and distribution of matrilines inside the archipelago reflecting certain aspects of human history (Yonekawa *et al.*, Chapter 4, and literature cited therein).

All subspecies: long-distance colonization

Unlike the mouse colonization of Eurasia and north Africa considered thus far, long-distance colonization of the rest of the world is a recent phenomenon, with mice taken to points from Greenland to the sub-Antarctic islands (Berry *et al.*, 1978; Jakobson, 1981; Hardouin *et al.*, 2010 and references therein). *M. m. domesticus* is the subspecies that has been transported the most widely by humans, having colonized the Americas and Australia, for example (e.g. Gabriel *et al.*, 2011), but it is interesting to note that in some areas colonization by multiple subspecies has occurred, especially around the Pacific and Indian Oceans for *domesticus* and *castaneus*. This is the case for many islands like Chichi (Bonhomme *et al.*, 1989), Hawaii (Wheeler and Selander, 1972; Moriwaki *et al.*, 1986), and others, but also the eastern side of Africa, where matrilines of both ancestry can be found in Kenya (Bonhomme *et al.*, 2011), as well as California (Orth *et al.*, 1998). In the case of the large island of Madagascar, the nuclear ancestry is mostly *castaneus*, while the matrilines derive from the Arabian peninsula (the so-called '*gentilulus*' type: Fig. 11.1 and Duplantier *et al.*, 2002). A recent report (Searle *et al.*, 2009a) has even demonstrated a three-way contribution in New Zealand, where the three subspecies have left footprints in various proportions. Moreover, in many large cities or harbours it is common to find traces of multiple origins as reflected by the presence of various haplogroups often pertaining to more than one subspecies. At the very least, these findings demonstrate that the three main subspecies are not tightly genetically isolated, and that introgression and intermixture occur when the conditions are favourable, despite the existence of hybrid zones between the subspecies in other situations.

Conclusions

Phylogeography of the house mouse as a tool in investigation of human history

One of the key interests of phylogeography in the house mouse is the use of results to make inferences about human history. House mice have been associated with humans for the last 12 000 years (Cucchi *et al.*, 2005; see also Cucchi *et al.*, Chapter 3) and over that period they have the potential to reveal historical movements. House mice readily live in anthropogenic habitats and exploit food intended for humans or livestock. They are pests that are easily missed and therefore can readily be loaded into a boat or other means of transport of goods without being noticed, and they are not necessarily easily caught even when they are. As such, house mice have usually been taken wherever humans have settled or traded. Applying molecular markers to house mice to match a colonized area with possible source areas means that a movement of people can be grounded, and this evidence can be added to other historical or archaeological data to infer human colonizations or commercial interactions.

There have been various efforts to match mouse history and human history, particularly in Europe (Gündüz *et al.*, 2001; Rajabi-Maham *et al.*, 2008; Förster *et al*, 2009; Searle *et al.*, 2009b; Gabriel *et al.*, 2010; Jones *et al.*, 2010, 2011a, b; Bonhomme *et al.*, 2011), with the mtDNA control region as a particularly good marker of mouse colonization (see above). To date these studies have generally confirmed what we already know about human history, but they do act as a 'proof of principle' that mouse phylogeography may tell us about human colonization and trade. Mouse phylogeography will never replace historical documents or artefacts as a source of knowledge about the human past. However, it is another piece of evidence that archaeologists and historians may use. An example of this is the finding that, in terms of mtDNA, house mice on Madeira are more closely related to mice in northern Europe than those in Portugal, contrary to the written history of Madeira. To explain this result, it has been suggested that the mice were accidentally introduced onto Madeira by Danish Vikings (see above; Förster *et al.*, 2009; Gündüz *et al.*, 2001). At present, there is no evidence from historical writings or artefacts to support this finding. However, the molecular work has created a new historical hypothesis that can be explored further by historians and archaeologists. It is not the only possible explanation for the molecular data, but it is currently the most reasonable one. In this way the mice have added a new perspective to our understanding of the journeys of the Danish Vikings to the southern parts of Europe.

Outlook with new molecular tools

In the phylogeographic analysis of the house mouse we are currently at a stage where a small number of molecular markers have been used on a large sample of individuals over a very wide geographic area. Our knowledge is still largely based on allozyme surveys (e.g. Selander and Yang, 1969; Minezawa *et al.*, 1981; Sage, 1981; Kotenkova and Bulatova, 1994; Din *et al.*, 1996), mitochondrial RFLP analysis (Vanlerberghe *et al.*, 1988; Prager *et al.*, 1993; Boursot *et al.*, 1996), and D-loop or Cyt-*b* sequencing (references cited in the present chapter), and to a lesser extent microsatellites (Thomas *et al.*, 2005; Ihle *et al.*, 2006; Teschke *et al.*, 2008; Hardouin *et al.*, 2010; Jones *et al.*, 2011a) and Y chromosome analysis (Nishioka and Lamothe, 1987; Tucker *et al.*, 1989; Boissinot and Boursot, 1997; Prager *et al.*, 1997; Nachman, 1998; Terashima *et al.*, 2006). There has been little work with other markers (but see Bonhomme *et al.*, 2007). However, new tools – such as single nucleotide polymorphism (SNP) arrays (Yang *et al.*, 2011) and resequencing of various kinds – are now becoming available, with much higher resolution than the old markers. These new techniques should provide a much more complete view of migration and expansion in the house mouse. We may expect that the overall perspective provided by the classical markers will be retained, but genome-wide studies will be rich in complementary information about secondary admixture and selective phenomenon, with a shift of emphasis from 'where do the animals originally come from?' to 'what can be said about the movements of the genetic variants inside the species gene pool?'.

Note

1. An almost exact correspondence between the clades defined by Jones *et al.* (2011b: letters) and the haplogroups coined by Bonhomme *et al.* (2011: numbers) is as follows: A–10; B–3, 5, 9; C–2; C1–1; D–7; D1–8; E–11; F–4; abs–6.

REFERENCES

Alcover, J. A., Rando, J. C., Garcia-Talavera, F., *et al.* (2009). A reappraisal of the stratigraphy of Cueva del Llano (Fuerteventura) and the chronology of the introduction of the house mouse (*Mus musculus*) into the Canary Islands. *Palaeogeography, Palaeoclimatology, Palaeoecology*, **277**, 184–90.

Auffray, J. -C., Tchernov, E., and Nevo, E. (1988). Origin of commensalism of the house mouse (*Mus musculus domesticus*) in relation to man. *Comptes Rendus de l'Académie des Sciences – Series III – Sciences de la Vie*, **307**, 517–22.

Auffray, J. -C., Vanlerberghe, F., and Britton-Davidian, J. (1990). The house mouse progression in Eurasia: a palaeontological and archaeozoological approach. *Biological Journal of the Linnean Society*, **41**, 13–25.

Avise, J. C. (2000). *Phylogeography: The History and Formation of Species*. Cambridge, MA: Harvard University Press.

Baines, J. F. and Harr, B. (2007). Reduced X-linked diversity in derived populations of house mice. *Genetics*, **175**, 1911–21.

Berry, R. J. (1986). Genetical processes in wild mouse populations: past myth and present knowledge. *Current Topics in Microbiology and Immunology*, **127**, 86–94.

Berry, R. J. and Jakobson, M. E. (1975). Adaptation and adaptability in wild living house mice. *Journal of Zoology*, **176**, 391–402.

Berry, R. J., Peters, J., and van Aarde, R. J (1978). Sub-Antarctic house mice: colonization, survival and selection. *Journal of Zoology*, **184**, 127–41.

Boissinot, S. and Boursot, P. (1997). Discordant phylogeographic patterns between the Y chromosome and mitochondrial DNA in the house mouse: selection on the Y chromosome? *Genetics*, **146**, 1019–34.

Bonhomme, F., Miyashita, N., Boursot, P., Catalan, J., and Moriwaki, K. (1989). Genetical variation and polyphyletic origin in Japanese *Mus musculus*. *Heredity*, **63**, 299–308.

Bonhomme, F., Orth, A., Cucchi, T., *et al.* (2011). Genetic differentiation of the house mouse around the Mediterranean basin: matrilineal footprints of early and late colonization. *Proceedings of the Royal Society B: Biological Sciences*, **278**, 1034–43.

Bonhomme, F., Rivals, E., Orth, A., *et al.* (2007). Species-wide distribution of highly polymorphic minisatellite markers suggests past and present genetic exchanges among house mouse subspecies. *Genome Biology*, **8**, R80.1–R80.15.

Boursot, P., Auffray, J. -C., Britton-Davidian, J., and Bonhomme, F. (1993). The evolution of house mice. *Annual Review of Ecology and Systematics*, **24**, 119–52.

Boursot, P., Bonhomme, F., Britton-Davidian, J., *et al.* (1984). Introgression différentielle des génomes nucléaires et mitochondriaux chez deux semi-espèces de souris. *Comptes Rendus de l'Académie des Sciences – Series III – Sciences de la Vie*, **299**, 365–70.

Boursot, P., Din, W., Anand, R., *et al.* (1996). Origin and radiation of the house mouse: mitochondrial DNA phylogeny. *Journal of Evolutionary Biology*, **9**, 391–415.

Britton-Davidian, J., Catalan, J., Lopez, J., *et al.* (2007). Patterns of genic diversity and structure in a species undergoing rapid chromosomal radiation: an allozyme analysis of house mice from the Madeira archipelago. *Heredity*, **99**, 432–42.

Brothwell, D. (1981). The Pleistocene and Holocene archaeology of the house mouse and related species. *Symposium of the Zoological Society of London*, **47**, 1–13.

Cucchi, T. and Vigne, J. -D. (2006). Origin and diffusion of the house mouse in the Mediterranean. *Human Evolution*, **21**, 95–106.

Cucchi, T., Vigne, J. -D., and Auffray, J. -C. (2005). First occurrence of the house mouse (*Mus musculus domesticus* Schwarz & Schwarz, 1943) in the western Mediterranean: a zooarchaeological revision of subfossil occurrences. *Biological Journal of the Linnean Society*, **84**, 429–45.

Darvish, D., Orth, A., and Bonhomme, F. (2006). Genetic transition in the house mouse *Mus musculus* of eastern Iranian Plateau. *Folia Zoologica*, **55**, 349–57.

Din, W., Anand, R., Boursot, P., *et al.* (1996). Origin and radiation of the house mouse: clues from nuclear genes. *Journal of Evolutionary Biology*, **9**, 519–39.

Duplantier, J. -M., Orth, A., Catalan, J., and Bonhomme, F. (2002). Evidence for a mitochondrial lineage originating from the Arabian peninsula in the Madagascar house mouse (*Mus musculus*). *Heredity*, **89**, 154–8.

Förster, D. W., Gündüz, İ., Nunes, A. C., *et al.* (2009). Molecular insights into the colonization and chromosomal diversification of Madeiran house mice. *Molecular Ecology*, **18**, 4477–94.

Fregel, R., Gomes, V., Gusmao, L., *et al.* (2009). Demographic history of Canary Islands male gene-pool: replacement of native lineages by European. *BMC Evolutionary Biology*, **9**, 181.

Frisman, L. V., Korobitsina, K. V., Yakimenko, L. V., Bokshtein, F. M., and Muntyanu, A. I. (1990). Genetic differentiation of the U.S.S.R. house mice: electrophoretic study of proteins. *Biological Journal of the Linnean Society*, **41**, 65–72.

Gabriel, S. I., Jóhannesdóttir, F., Jones, E. P., and Searle, J. B. (2010). Colonization, mouse-style. *BMC Biology*, **8**, 131.

Gabriel, S. I., Stevens, M. I., Mathias, M. L., and Searle, J. B. (2011). Of mice and 'convicts': origin of the Australian house mouse, *Mus musculus*. *PLoS ONE*, **6**, e28622.

Geraldes, A., Basset, P., Gibson, B., *et al.* (2008). Inferring the history of speciation in house mice from autosomal, X-linked, Y-linked and mitochondrial genes. *Molecular Ecology*, **17**, 5349–63.

Gündüz, İ., Auffray, J.-C., Britton-Davidian, J., *et al.* (2001). Molecular studies on the colonization of the Madeiran archipelago by house mice. *Molecular Ecology*, **10**, 2023–9.

Gündüz, İ., Rambau. R. V., Tez, C., and Searle, J. B. (2005). Mitochondrial DNA variation in the western house mouse (*Mus musculus domesticus*) close to its site of origin: studies in Turkey. *Biological Journal of the Linnean Society*, **84**, 473–85.

Hardouin, E., Chapuis, J., Stevens, M., *et al.* (2010). House mouse colonization patterns on the sub-Antarctic Kerguelen Archipelago suggest singular primary invasions and resilience against re-invasion *BMC Evolutionary Biology*, **10**, 325.

Hewitt, G. M. (2000). The genetic legacy of the Quaternary ice ages. *Nature*, **405**, 907–13.

Ihle, S., Ravaoarimanana, I., Thomas, M., and Tautz, D. (2006). An analysis of signatures of selective sweeps in natural populations of the house mouse. *Molecular Biology and Evolution*, **23**, 790–7.

Jakobson, M. E. (1981). Physiological adaptability: the response of the house mouse to variation in the environment. *Symposium of the Zoological Society of London*, **47**, 301–35.

Jones, C. S., Noble, L. R., Jones, J. S., *et al.* (1995). Differential male genetic success determines gene flow in an experimentally manipulated mouse population. *Proceedings of the Royal Society B: Biological Sciences*, **260**, 251–6.

Jones, E. P., Jensen, J.-K., Magnussen, E., *et al.* (2011a). A molecular characterization of the charismatic Faroe house mouse. *Biological Journal of the Linnean Society*, **102**, 471–82.

Jones, E. P., Jóhannesdóttir, F., Gündüz, İ., Richards, M. B., and Searle, J. B. (2011b). The expansion of the house mouse into north-western Europe. *Journal of Zoology*, **283**, 257–68.

Jones, E. P., van der Kooij, J., Solheim, R., and Searle, J. B. (2010). Norwegian house mice (*Mus musculus musculus/domesticus*): distributions, routes of colonization and patterns of hybridization. *Molecular Ecology*, **19**, 5252–64.

Kawashima, T., Miyashita, N., Tsuchiya, K., *et al.* (1995). Geographical distribution of the Hbb haplotypes in the *Mus musculus* subspecies in eastern Asia. *Japanese Journal of Genetics*, **70**, 17–23.

Kotenkova, E. V. and Bulatova, N. S. (1994). *The House Mouse: Origin, Distribution, Systematics, Behaviour*. Moscow: Russian Academy of Sciences.

Macholán, M., Vyskočilová, M., Bonhomme, F., *et al.* (2007). Genetic variation and phylogeography of free-living mouse species (genus *Mus*) in the Balkans and the Middle East. *Molecular Ecology*, **16**, 4774–88.

Mezhzherin, S. V., Kotenkova, E. V., and Mikhailenko, A. G. (1998). The house mice, *Mus musculus* s.l., hybrid zone of Transcaucasus. *Zeitschrift für Säugetierkunde*, **63**, 154–68.

Milishnikov, A. N., Lavrenchenko, L. A., and Lebedev, V. S. (2004). Origin of the house mice (superspecies complex *Mus musculus* sensu lato) from the Transcaucasia region: a new look at dispersal routes and evolution. *Russian Journal of Genetics*, **40**, 1011–26.

Minezawa, M., Moriwaki, K., and Kondo, K. (1981). Geographical survey of protein variations in wild populations of Japanese mouse *Mus musculus molossinus*. *Japanese Journal of Genetics*, **56**, 27–39.

Mitchell-Jones, A. J., Amori, G., Bogdanowicz, W., *et al.* (1999). *The Atlas of European Mammals*. London: Poyser.

Moriwaki, K., Miyashita, N., Suzuki, H., Kurihara, Y., and Yonekawa, H. (1986). Genetic features of major geographical isolates of *Mus musculus*. *Current Topics in Microbiology and Immunology*, **127**, 55–61.

Nabholz, B., Glemin, S., and Galtier, N. (2008). Strong variations of mitochondrial mutation rate across mammals: the longevity hypothesis. *Molecular Biology and Evolution*, **25**, 120–30.

Nachman, M. W. (1998). Y chromosome variation of mice and men. *Molecular Biology and Evolution*, **15**, 1744–50.

Nishioka, Y. and Lamothe, E. (1987). The *Mus musculus musculus* type Y chromosome predominates in Asian house mice. *Genetical Research*, **50**, 195–8.

Nunome, M., Ishimori, C., Aplin, K. P., *et al.* (2010). Detection of recombinant haplotypes in wild mice (*Mus musculus*) provides new insights into the origin of Japanese mice. *Molecular Ecology*, **19**, 2474–89.

Orth, A., Adama, T., Din, W., and Bonhomme, F. (1998). Hybridation naturelle entre deux sous espèces de souris domestique *Mus musculus domesticus* et *Mus musculus castaneus* près de Lake Casitas (Californie). *Genome*, **41**, 104–10.

Pocock, M. J. O., Hauffe, H. C., and Searle, J. B. (2005). Dispersal in house mice. *Biological Journal of the Linnean Society*, **84**, 565–83.

Prager, E., Boursot, P., and Sage, R. (1997). New assays for Y chromosome and p53 pseudogene clines among East Holstein house mice. *Mammalian Genome*, **8**, 279–81.

Prager, E. M., Orrego, C., and Sage, R. D. (1998). Genetic variation and phylogeography of central Asian and other house mice, including a major new mitochondrial lineage in Yemen. *Genetics*, **150**, 835–61.

Prager, E. M., Sage, R. D., Gyllensten, U., *et al.* (1993). Mitochondrial DNA sequence diversity and the colonization of Scandinavia by house mice from East Holstein. *Biological Journal of the Linnean Society*, **50**, 85–122.

Rajabi-Maham, H., Orth, A., and Bonhomme, F. (2008). Phylogeography and postglacial expansion of *Mus musculus domesticus* inferred from mitochondrial DNA coalescent, from Iran to Europe. *Molecular Ecology*, **17**, 627–41.

Rajabi-Maham, H., Orth, A., Siahsarvie, R., *et al.* (Submitted). The south-eastern house mouse *Mus musculus castaneus* (Rodentia: Muridae) is a polyphyletic subspecies. *Biological Journal of the Linnean Society*.

Ruscoe, W. A. and Murphy, E. C. (2005). House mouse. In *The Handbook of New Zealand Mammals*, 2nd edn, ed. C. M. King. Melbourne: Oxford University Press, pp. 204–21.

Sage, R. D. (1981). Wild mice. In *The Mouse in Biomedical Research*, ed. H. L. Foster, J. D. Small, and J. G. Fox. New York: Academic Press, pp. 40–90.

Searle, J. B. (2008). The genetics of mammalian invasions: a review. *Wildlife Research*, **35**, 185–92.

Searle, J. B., Jamieson, P. M., Gündüz, İ., *et al.* (2009a). The diverse origins of New Zealand house mice. *Proceedings of the Royal Society B: Biological Sciences*, **276**, 209–17.

Searle, J. B., Jones, C. S., Gündüz, İ., *et al.* (2009b). Of mice and (Viking?) men: phylogeography of British and Irish house mice. *Proceedings of the Royal Society B: Biological Sciences*, **276**, 201–7.

Searle, J. B., Kotlík, P., Rambau, R. V., *et al.* (2009c). The Celtic fringe of Britain: insights from small mammal phylogeography. *Proceedings of the Royal Society B: Biological Sciences*, **276**, 4287–94.

Selander, R. K. and Yang, S. Y. (1969). Protein polymorphism and genic heterozygosity in a wild population of the house mouse (*Mus musculus*). *Genetics*, **63**, 653–67.

Skre, D. (2008). The development of urbanism in Scandinavia. In *The Viking World*, ed. S. Brink and N. Price. Oxford: Routledge, pp. 83–93.

Soares, P., Achilli, A., Semino, O., *et al.* (2010). The archaeogenetics of Europe. *Current Biology*, **20**, R174–83.

Spiridonova, L. N., Chelomina, G. N., Moriwaki, K., Yonekawa, H., and Bogdanov, S. (1990). Genetic and taxonomic diversity of the house mouse *Mus musculus* from the Asian part of the Former Soviet Union. *Systematic Zoology*, **39**, 40–59.

Tchernov, E. (1984). Commensal animals and human sedentism in the Middle East. In *Animals and Archaeology: 3. Early Herders and Their Flocks*, ed. J. Clutton-Brock and G. Grigson. Oxford: British Archaeological Reports, pp. 91–115.

Terashima, M., Furusawa, S., Hanzawa, N., *et al.* (2006). Phylogeographic origin of Hokkaido house mice (*Mus musculus*) as indicated by genetic markers with maternal, paternal and biparental inheritance. *Heredity*, **96**, 128–38.

Teschke, M., Mukabayire, O., Wiehe, T., and Tautz, D. (2008). Identification of selective sweeps in closely related populations of the house mouse based on microsatellite scans. *Genetics*, **180**, 1537–45.

Thomas, M., Ihle, S., Ravaoarimanana, I., *et al.* (2005). Microsatellite variability in wild populations of the house mouse is not influenced by differences in chromosomal recombination rates. *Biological Journal of the Linnean Society*, **84**, 629–35.

Tucker, P. K., Lee, B. K., and Eicher, E. M. (1989). Y chromosome evolution in the subgenus *Mus* (Genus *Mus*). *Genetics*, **122**, 169–79.

Usuda, D., Takada, T., Moriwaki, K., *et al.* (2009). Phylogeographic study of wild house mice in Eurasia. *Genes & Genetic Systems*, **84**, 445.

Vanlerberghe, F., Boursot, P., Nielsen, J. T., and Bonhomme, F. (1988). A steep cline for mitochondrial DNA in Danish mice. *Genetical Research*, **52**, 185–93.

Wheeler, L. L. and Selander, R. K. (1972). Genetic variation in populations of the house mouse *Mus musculus* in the Hawaiian islands. *Studies in Genetics VII*. Austin: University of Texas.

Yang, H., Wang, J., Didion, J., *et al.* (2011). Subspecific origin and haplotype diversity in the laboratory mouse. *Nature Genetics*, **43**, 648–55.

Yonekawa, H., Gotoh, O., Tagashira, Y., *et al.* (1986). A hybrid origin of Japanese mice '*Mus musculus molossinus*'. *Current Topics in Microbiology and Immunology*, **127**, 62–7.

Yonekawa, H., Moriwaki, K., Gotoh, O., *et al.* (1988). Hybrid origin of Japanese mice '*Mus musculus molossinus*': evidence from restriction analysis of mitochondrial DNA. *Molecular Biology and Evolution*, **5**, 63–78.

12

The mouse *t*-haplotype: a selfish chromosome – genetics, molecular mechanism, and evolution

BERNHARD G. HERRMANN AND HERMANN BAUER

Discovery and genetic dissection of the *t*-haplotype

In 1927 Nadine Dobrovolskaia-Zavadskaia, in one of the first mutagenesis screens of the mouse, discovered a short-tailed offspring in animals treated with X-rays (reviewed in Korzh and Grunwald, 2001). She found that the phenotype had a genetic cause and named the gene Brachyury (short tail; gene symbol *T*). To identify new alleles of *T* she crossed *T*/+ mice with mice caught in the wild and isolated tailless individuals from their offspring. Tailless mice produced short-tailed and normal-tailed offspring. The latter apparently carried a weak allele of *T*, which produced no phenotype on its own, but enhanced the Brachyury phenotype. Therefore it was regarded as a recessive allele at the *T* locus, abbreviated *t*. Tailless mice produced only tailless litters upon intercrossing, but neither short-tailed nor normal-tailed pups. Inspection of the embryos from such crosses showed that about half of the embryos died in utero. They showed two distinct phenotypes, caused by homozygosity of *T* and *t*, respectively (Chesley and Dunn, 1936). Later, several *t*-alleles were discovered, which showed different embryonic lethal phenotypes in the homozygous condition. They were all initially regarded as recessive alleles of *T*, and some authors suggested they might encode subunits of a multimeric protein complex involved in controlling embryonic development (Bennett, 1975). It was at first called the *T*-locus, later the *T/t*-complex or just *t*-complex, and mapped to proximal chromosome 17.

Transmission ratio distortion

In 1936 Chesley and Dunn reported the observation that males but not females of the genotype T/t^0 produced far more normal-tailed than short-tailed offspring when crossed with wild-type mice (Chesley and Dunn, 1936). This breach

Evolution of the House Mouse, ed. Miloš Macholán, Stuart J. E. Baird, Pavel Munclinger, and Jaroslav Piálek. Published by Cambridge University Press.

of Mendel's law was associated with the *t*-allele, and confirmed by several researchers to occur with various *t*-alleles. Dunn was the first to refer to this phenomenon as 'distortion of transmission' (cited in Bennett, 1977). According to Mary Lyon, 'the term transmission ratio distortion (TRD) refers to instances in which the two alleles carried by a heterozygote are transmitted unequally to the zygote at the time of fertilisation, a phenomenon also sometimes known as meiotic drive' (Lyon, 2003).

In early studies the (*T*) tail interaction effect (*tct*), embryonic lethality, and TRD co-segregated and were thought to be caused by one pleiotropic locus, but over the years *t*-alleles that had lost either the recessive-lethal phenotype or *tct* and showed reduced transmission ratios were isolated in various laboratories (Bennett, 1975; Lyon and Mason, 1977). These observations suggested that *tct* and the different recessive lethals resided at separate loci. The identification of the tufted (*tf*) gene, located 8 cM distal to *T* on the wild-type chromosome, finally provided a tool for genetic dissection of the *T* locus (Lyon, 1956). Lyon generated a series of '*t*-alleles' (later termed partial *t*-haplotypes), which derived from rare recombination between the wild-type chromosome and the *t*-chromosome in the region between *T* and *tf* (Lyon and Meredith, 1964a, 1964b, 1964c). She tested them for viability of homozygotes, segregation ratio of *T*:*t*, presence of *tct*, fertility of homozygotes, and recombination frequency with the wild-type chromosome. These experiments allowed the mapping of *T*, recessive embryonic lethality and sterility (*LS*), and the factor responsible for TRD – the *A* factor, now known as *Tcr* – to different subregions of the *t*-chromosome, and demonstrated that the different factors can be separated but retain their genetic properties (Lyon and Mason, 1977; Fig. 12.1a). Lyon also showed that the *LS* locus is involved in TRD and suggested a third, proximal factor, which was later experimentally confirmed by Hammerberg (1981, 1982). In the follow-up to these reports, Lyon demonstrated that TRD is due to the cumulative action of multiple distorter genes (*Tcd1* to *Tcd3*) on the responder (*Tcr*) (Lyon, 1984). Two more distorters (*Tcd4* and *Tcd5*) were reported later, but *Tcd5* could not be independently confirmed (Silver and Remis, 1987; Silver, 1989). Importantly, by genetically separating the distorters from *Tcr*, Lyon (1984) showed that the distorters act in *trans*, while the single responder acts in *cis* (Fig. 12.2b). High TRD is obtained only in the presence of all distorters (Fig. 12.2a); fewer distorters reduce the TRD effect, and in the absence of all distorters the *t*-haplotype shows a low transmission ratio – far below 50% (Fig. 12.2c; Lyon, 1984).

Male sterility

The genetic basis of male sterility associated with the *t*-haplotype remained unclear until Lyon demonstrated that the distorters additively contribute to this phenotype (Lyon, 1986). A single distorter had no effect, while a

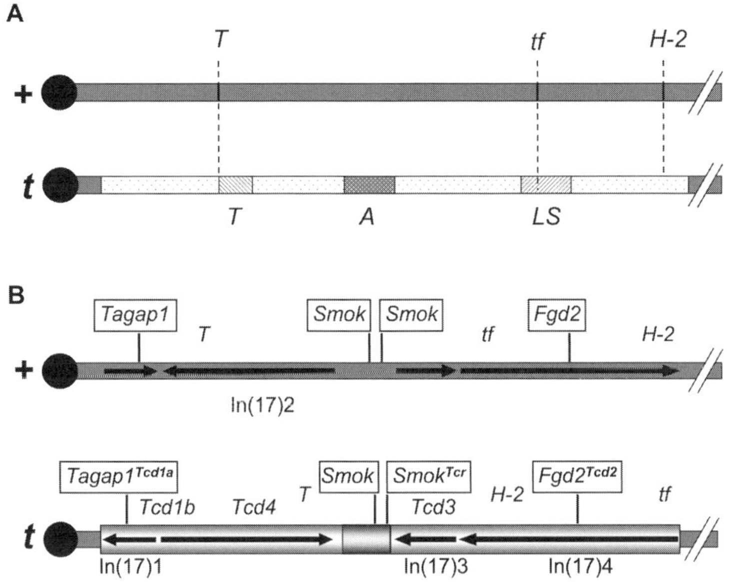

Figure 12.1 Structure of the *t*-haplotype in comparison to the wild-type *t*-complex. (a) The structure according to Lyon (Lyon and Mason, 1977), the *T* gene, *A* locus (later renamed *Tcr*) responsible for TRD, and the region carrying lethality and sterility loci (*LS*) are indicated. (b) The present-day view of the structure of the *t*-haplotype, indicating the four inversions In(17)1 to In(17)4, the marker genes *T*, *tf*, and the *H-2* complex used for genetic mapping, and the known and isolated factors (boxed gene symbols) involved in TRD: the distorters *Tcd1b*, *Tcd3*, *Tcd4*, $Tagap^{Tcd1a}$, $Fgd2^{Tcd2}$, wild-type *Smok*, and the responder $Smok^{Tcr}$ are shown with respect to their position on the *t*-haplotype or wild-type *t*-complex. Abbr.: +, wild-type *t*-complex; *t*, *t*-haplotype. Arrows indicate orientation of genomic DNA in the wild-type *t*-complex with respect to the *t*-haplotype.

combination of *Tcd1* and *Tcd2* produced semi-sterility (Fig. 12.2d). Increasing the distorter gene dosage (*Tcd1* or *Tcd2*) results in complete sterility (Fig. 12.2e). In the presence of *Tcr*, fertility can be maintained but only until a strong 'harmful' level of distorter activity is reached (Fig. 12.2b, f). Lyon proposed that the distorters act harmfully on the wild-type form of the responder, while *Tcr* is partially resistant to this effect. But in males homozygous for *Tcd* loci, the resistance of *Tcr* is overcome by the additive harmful effect of the distorters, resulting in sterility (Lyon, 1986). Thus, the presence of the distorter loci, while promoting the successful transmission of the *t*-haplotype, can lead to sterility (Lyon, 1986). Since the distorters in Lyon's studies were associated with partial *t*-haplotypes comprising large chromosomal regions rather than single genes, some authors doubted that distorters and

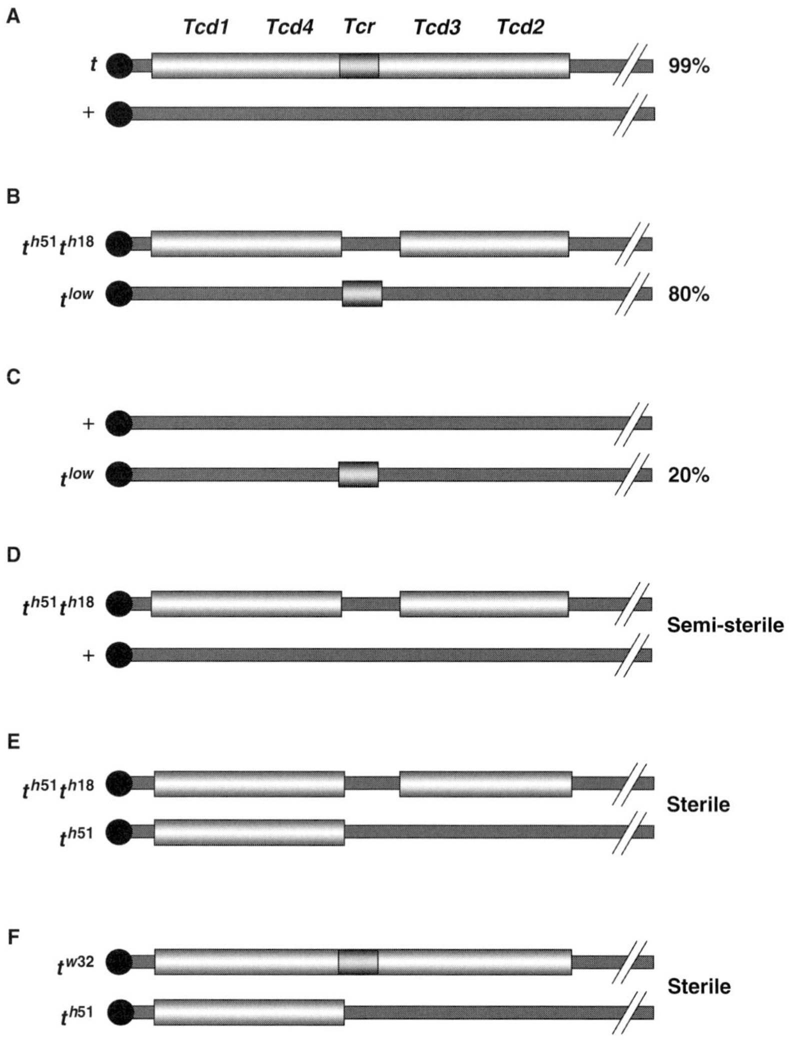

Figure 12.2 Genetic dissection of transmission ratio distortion and sterility phenotypes. Pairs of *t*-haplotype and wild-type chromosomes are shown along with the TRD or sterility phenotypes of the carrier males. The percentage on the right indicates the transmission rate of the neighbouring chromosome. (a) The complete *t*-haplotype is transmitted at up to 99% to the offspring of a heterozygous male. (b) The distorters (*Tcd*) act in *trans* on the responder (*Tcr*), which is transmitted at high ratio; however, an isolated responder is transmitted at a low ratio (c). (d) In the absence of *Tcr* the distorters can cause semi-sterility of the carrier male, which is enhanced to a complete sterility phenotype when the distorter dosage is further increased (e). (f) This occurs even in the presence of *Tcr*.

sterility loci were identical. Deletion analysis and functional genetic studies provided evidence for at least one sterility gene that does not act as a distorter (Planchart *et al.*, 2000; Schimenti *et al.*, 2005). However, since the *t*-haplotype causes homozygous male sterility, such mutations might have actually been picked up secondarily as 'passenger mutations', a term taken from cancer genetics indicating that the genes have no functional relevance. Thus, Lyon's original unifying hypothesis that male sterility is caused by the distorters remains valid.

Structure of the *t*-haplotype

Genetic mapping data provide strong evidence for extended meiotic recombination suppression between the wild-type and *t*-form of the *t*-complex. The actual cause of this phenomenon was ascribed to suppression of meiotic chiasmata formation in the *t*-complex region (Forejt, 1972; Forejt *et al.*, 1980), yet its molecular mechanism remained elusive for some time, leading to much speculation about its molecular nature (reviewed in Bennett, 1975). However, recombination between *t*-haplotypes was shown to occur freely over the *t*-chromatin. The chance occurrence of a tufted mutation in the t^{w12} haplotype and novel genomic probes derived from the major histocompatibility complex (*H2*) finally provided genetic tools for unravelling this mystery. Genetic mapping of these markers showed an inverse order along the *t*-chromosome relative to the wild-type, providing the first proof of a large inversion in the region marked by these loci (Artzt *et al.*, 1982). Additional molecular (polypeptide and DNA) markers allowed for the discovery of three more inversions, comprising almost all of the *t*-haplotype (Herrmann *et al.*, 1986; Hammer *et al.*, 1989). These data finally showed that several chromosomal rearrangements are the cause of recombination suppression between the *t*-haplotypes and the wild-type *t*-complex (Fig. 12.1b).

Physiological basis of TRD and male sterility

The fact that TRD is confined to *t*-carrying males that are sterile in the homozygous state suggests that the *t*-haplotype affects sperm cells. Spermatogenesis was found to proceed normally in *t*/+ heterozygotes, and ejaculate from these males contains *t*-sperm and +-sperm in equal numbers (Silver and Olds-Clarke, 1984). Because +-sperm is not degenerated in *t*/+ males, TRD must take place by the competition of sperm for egg cells. Indeed, an earlier report had shown a bimodal distribution of the beat frequency in spermatozoa from T/t^{12} mice (Katz *et al.*, 1979).

These data brought forth the question of whether *t*-sperm are superior to +-sperm derived from the same male. The answer was provided by sperm mixing followed by artificial insemination. Upon mixing of *t*/+-derived with +/+-derived sperm at 1 : 1 ratio, the transmission ratio of the *t*-haplotype dropped from 95% to 22%, demonstrating that *t*-sperm is no better at fertilization than the +-sperm derived from +/+ males (Olds-Clarke and Peitz, 1985) This fostered the idea that factors expressed by the *t*-haplotype might render wild-type sperm dysfunctional in a *t*/+ environment. One year after this work appeared, Lyon published her genetic model suggesting that the distorters act strongly and harmfully on the wild-type allele of the responder and weakly on *Tcr*, leading to distortion in favour of *Tcr*, but that homozygosity of distorters is also harmful for *Tcr*, leading to male sterility (Lyon, 1986). In combination, the two publications suggested that distorter genes express products which impair sperm motility, and that *Tcr* provides resistance to this harmful effect only in *t*-sperm, leading to ratio distortion in favour of *t*-sperm (Fig. 12.3).

Several years after their initial sperm studies, Patricia Olds-Clarke and colleagues investigated the flagellar behaviour of sperm derived from *t*/+

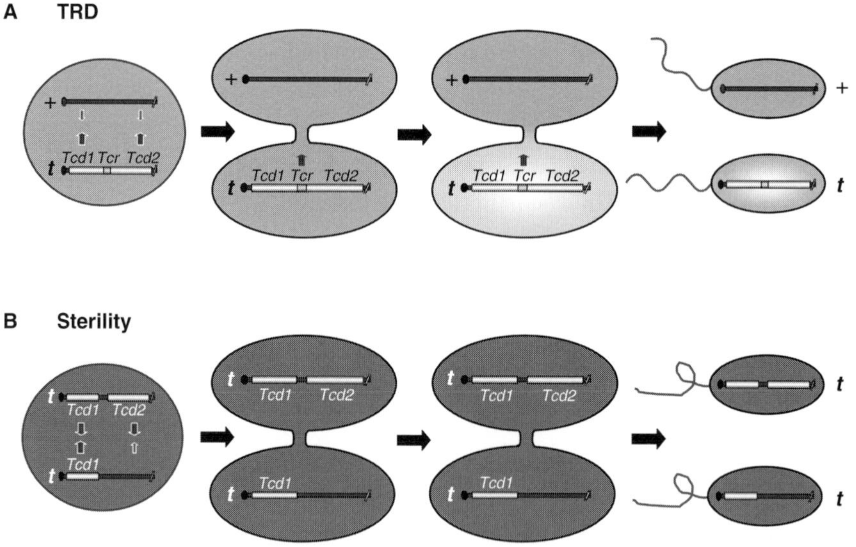

Figure 12.3 Physiological effects of *Tcd*s and *Tcr* on sperm motility. Stages of spermatogenesis proceed from diploid to haploid (left to right; spermatids are connected by inter-cellular bridges). (a) The distorters (*Tcd1*, *Tcd2*), which act in *trans*, cause impairment of sperm motility in all sperm cells (indicated by shading). In contrast, *Tcr*, which acts in *cis*, rescues sperm motility, but only in the spermatozoa carrying the gene (indicated by light grey). (b) A high dosage of the distorters (indicated by dark grey shading) results in failure of sperm motility and sterility of the carrier male. Arrows adjacent to *Tcd1*, *Tcd2*, and *Tcr* indicate gene activity; the shape of the sperm tail indicates flagellar behaviour.

heterozygous males in more detail (Olds-Clarke and Johnson, 1993). They used congenic wild-type mice as well as males carrying two complementing *t*-haplotypes (t^x/t^y) as controls and demonstrated that *t*-haplotypes compromise sperm flagellar function. Two sperm motility parameters, curvilinear velocity – a measure of sperm swimming speed – and linearity were found to differ between the three classes of sperm (+/+, t^{w5}/+ or t^{w32}/+, and t^{w5}/t^{w32}) (Olds-Clarke and Johnson, 1993). Sperm derived from t^{w5}/t^{w32} males showed a curlicue bending of the flagellum and severe motility abnormalities, while the *t*/+-derived sperm showed milder defects. Statistical analysis suggested that sperm derived from *t*/+ males is not homogeneous, but rather comprises one population behaving like sperm derived from wild-type controls and another population with impaired swimming behaviour. This work confirmed that the *t*-haplotype affects sperm motility and suggested that factors expressed by *t*-carrying sperm are harmful to their wild-type meiotic partners (Fig. 12.3).

The molecular basis of TRD

In the early 1980s the group of Hans Lehrach began work on the mouse *t*-complex as a model system for developing a positional cloning approach in order to isolate genes identified by interesting genetic traits, such as Brachyury. Inspired by approaches developed in *Drosophila*, they derived DNA fragments from the proximal half of mouse chromosome 17 using micro-dissection of metaphase chromosomes (Rohme *et al.*, 1984). The DNA markers were used as genomic DNA probes in the search for restriction fragment length variations, which served as genetic markers in chromosomal mapping. Several informative DNA markers were identified and localized along the *t*-complex (Rohme *et al.*, 1984; Fox *et al.*, 1985), allowing mapping of DNA markers to genetic traits located on particular partial *t*-haplotypes. This work provided the first molecular map of the *t*-haplotype and identified starting points for the isolation of genes causing the various phenotypes associated with the *t*-haplotype. The subsequent detailed structure analysis identified an inversion in the proximal region of the *t*-complex and culminated in the isolation of the Brachyury gene (Herrmann *et al.*, 1986, 1987, 1990).

Isolation of the *Tcr* gene

Attempts to isolate genes involved in TRD soon resulted in the discovery of a candidate gene for *Tcr* (Schimenti *et al.*, 1988). This was facilitated by the fact that *Tcr* could be mapped to a chromosomal candidate region of just 150–220 kb, which was shown to be part of an amplified genomic region with a complex structure (Rosen *et al.*, 1990). The gene was analysed in great detail, but ultimately turned out to be false (Ewulonu *et al.*, 1996). More than a decade after the first *Tcr* candidate was reported, we published the isolation of a novel protein kinase family, designated sperm-motility-kinase (*Smok*) (Herrmann *et al.*, 1999). One member of this gene family, which localized to the *Tcr* genomic region, was shown to differ from all other members by structure, amino acid sequence, and kinase activity. It had arisen by rearrangement, producing a fusion between a *Smok* gene family member and one copy of a triplicated ribosomal S6 kinase gene, *Rsk3*. The fusion contained the promoter and coding exon of *Smok* and converted the *Rsk3* gene into the 3′-untranslated region (Fig. 12.4a). Examination of the amino acid sequence of this fusion protein showed many alterations compared to other members of the family, and the kinase activity was shown to be reduced by one order of magnitude, suggesting that it may act as a dominant-negative enzyme. Importantly, in transgenic models this altered kinase was able to distort the Mendelian ratio. In one mouse line the transgene fortuitously co-segregated with the Y chromosome and caused sex ratio distortion in the presence of several distorters, wherein twice as many male than female offspring were produced from males carrying the transgene. This data provided proof that the altered, presumably dominant-negative, kinase is the true *t*-complex-responder (gene symbol $Smok^{Tcr}$). Interestingly, this study also demonstrated that *Tcr* can act outside the *t*-complex on any chromosome (Herrmann *et al.*, 1999). Further analysis of staged testes samples showed that *Tcr* expression commences postmeiotically, in the haploid phase of spermatogenesis. Altogether, these data provide proof at the molecular level for the genetic model put forward by Lyon more than a decade earlier. The resistance of *Tcr* to the harmful action of the distorters could be explained by the presumed dominant-negative kinase $\mathrm{SMOK}^{\mathrm{Tcr}}$, and the action of *Tcr* in *cis* was consistent with $Smok^{Tcr}$ expression in haploid sperm cells. But what were the harmful effects of the distorters on the wild-type allele of $Smok^{Tcr}$? Herrmann proposed that the distorters encode components of signalling cascades regulating wild-type SMOK, which controls sperm motility, and enhance SMOK activity in all sperm cells leading to abnormal flagellar movement. $\mathrm{SMOK}^{\mathrm{Tcr}}$ counterbalances the resulting hyper-activation of SMOK by the distorters, leading to rescue of sperm motility, but only in sperm carrying the gene. This would result in rescued *t*-sperm showing normal motility and wild-type sperm with impaired motility, and thus TRD in favour of *t*-sperm. This model remains valid and has been further confirmed with regards to the function of the distorters (see below and Fig. 12.5).

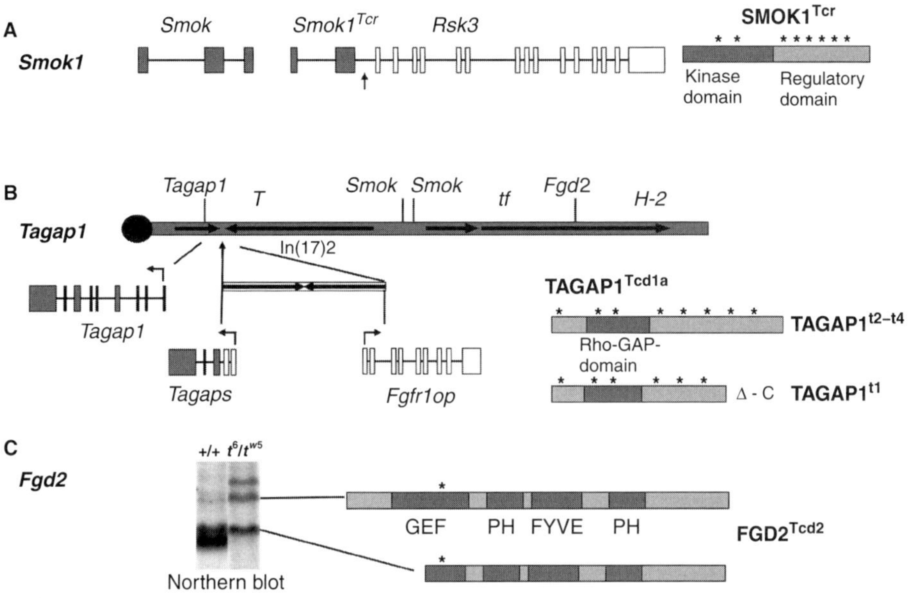

Figure 12.4 Molecular characterization of the responder *Smok*^{Tcr} and the distorters *Tagap1*^{Tcd1a} and *Fgd2*^{Tcd2}. (a) *Smok*^{Tcr} is formed by a rearrangement fusing *Smok* with *Rsk3*; a neighbouring *Smok* gene is shown on the left for comparison (arrow indicates point of gene fusion). A subset of amino acid changes in SMOK^{Tcr}, some leading to down-regulation of the Smok kinase activity, are indicated by asterisks on the schematic structure of the protein (right). (b) *Tagap1* was identified as a full-length copy of *Tagaps*, which was formed by an inverted duplication transposing the 5′-region of *Fgfr1op* to a centromer-proximal position, and thereby truncating and inactivating one *Tagap1* gene. *Tagaps* forms the junction between the first inversion (In(17)1) and second inversion (In(17)2) of the *t*-complex, as indicated by an arrow. The protein structure of the full-length TAGAP1 and a C-terminally truncated variant, both expressed from *Tagap*^{Tcd1a} are shown on the right. Changes in amino acid residues identified in the predicted *t*-proteins are indicated by asterisks. (c) Northern blot hybridization of testis RNA showing the expression of the full-length and a smaller *Fgd2* transcript encoding a truncated, possibly dominant-negative protein (schematic on the right). Expression of the full-length transcript is increased and expression of the small transcript reduced from the *t*-haplotype compared to the wild-type allele (northern blot; left panel). The asterisk indicates a *t*-specific amino acid change in the catalytic domain.

Isolation of distorter genes

Unlike the localization of *Tcr* to a relatively small genomic region, which strongly facilitated its cloning, the distorters could only be localized to *t*-complex subregions of several megabases in size. Positional cloning approaches alone were

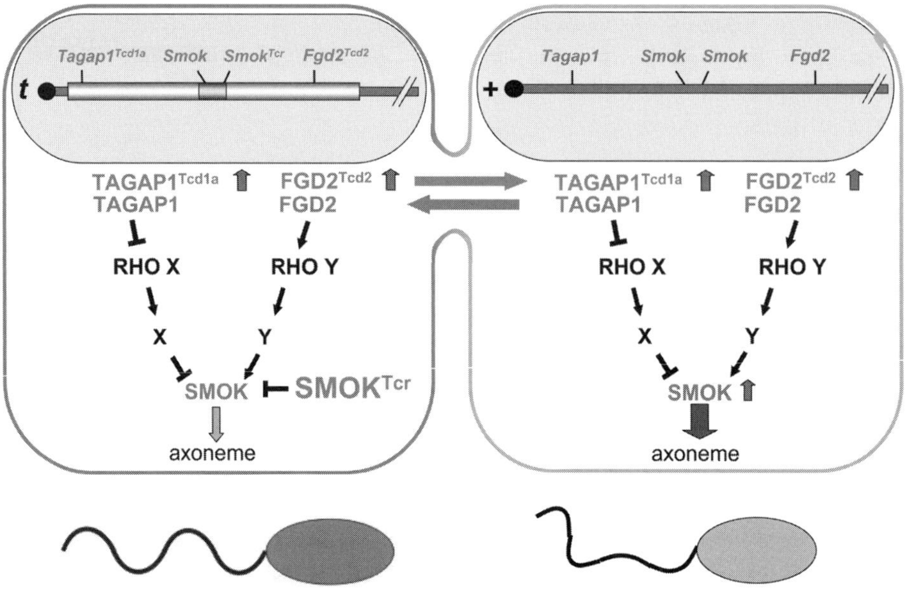

Figure 12.5 Model of the signalling mechanisms controlling sperm motility and causing TRD. Haploid sperm cells are connected by inter-cellular bridges allowing passage of the distorter gene products (indicated by arrows) between cells carrying the *t*-haplotype or the wild-type *t*-complex. TAGAP1 inhibits a Rho small GTPase (RHO X) acting as repressor of SMOK; in contrast, FGD2 is an activator of another Rho small G protein (RHO Y), which activates SMOK. Both *Tagap1*^{Tcd1a} and *Fgd2*^{Tcd2} are hypermorphs expressing excess activity of TAGAP1 and FGD2, thus both distorters are able to contribute additively to hyper-activation of SMOK, causing impairment of flagellar motility in all cells. However, SMOK^{Tcr} can counteract this harmful effect of the distorters and rescue sperm motility, but only in the cells carrying the gene (left half of panel). Thus, the haploid-restricted rescue activity of SMOK^{Tcr} promotes the transmission of the *t*-haplotype to the next generation. Dark grey upward arrows indicate enhanced activity; dark grey downward arrow impaired; light grey arrow rescued control of the axoneme resulting in aberrant or normal sperm motility. See the plate section for a colour version of this figure.

not adequate to attempt isolation of these genes since the subregions could not readily be reduced by genetic means. Therefore, several groups identified candidates on the basis of three main criteria: (1) the gene is localized to the *t*-complex; (2) the gene is expressed in testis; and (3) the gene exhibits molecular differences between the wild-type and the *t*-allele. The most promising candidates reported encoded proteins involved in flagellar function, such as TCTEX1 (Lader *et al.*, 1989), a dynein light chain protein, or the axonemal dynein heavy chain, DNAHC8 (Fossella *et al.*, 2000). However, none of these candidates were proven to act as a distorter in transgenic mouse models.

The molecular nature of distorters actually turned out to be quite different (Bauer *et al.*, 2005). Since structural rearrangements often affect gene integrity or function, and since a number of such rearrangements had occurred in the *t*-complex, we pursued the idea that distorter loci might be associated with chromosomal rearrangements in the *t*-complex. We searched for genes located at the proximal breakpoint of the second inversion of the *t*-complex, which coincides with the breakpoint of a large inverted duplication (Herrmann *et al.*, 1987; Fig. 12.4b). From this region we isolated a fusion gene consisting of the promoter and first exon of *Fgfr1op* (Fgfr1 oncogene partner), which had been duplicated and translocated to the proximal end of the inverted duplication and fused with the 3′-region of a gene that could not be initially identified (Fig. 12.4b). This fusion gene, later designated *Tagaps* (T-cell activation Rho GTPase-activating protein short), is highly expressed in the testis of wild-type mice but not in *t*-haplotypes, and was thus regarded as a distorter candidate. Knock-out analysis of *Tagaps* in transgenic mice did not reveal distorter activity (Bauer *et al.*, 2005; and unpublished data), but genomic Southern blot analysis of *Tagaps* suggested the presence of a related gene located proximal to *Tagaps* within the *t*-complex. Isolation of this gene identified it as *Tagap1*, encoding a GTPase-activating protein for Rho small G proteins. This gene fits the hypothesis put forward earlier that distorters are involved in signalling cascades acting upstream of SMOK (Herrmann *et al.*, 1999). The *t*-haplotype turned out to contain four copies of *Tagap1*, producing more transcript than the single wild-type gene. Over-expressing the wild-type gene *Tagap1* in transgenic mice increased the transmission rate of a partial *t*-haplotype, and knock-out of the wild-type allele reduced the transmission of the *t*-chromosome, providing proof that *Tagap1* is a distorter (Bauer *et al.*, 2005). According to its position within the *t*-complex, *Tagap1* was assigned to *Tcd1*. Lyon had previously reported that *Tcd1* can be phenocopied by a deletion in the wild-type chromosome (Lyon, 1992); therefore, Lyon's *Tcd1* could not be *Tagap1*. However, a more recent report suggested that *Tcd1* actually consisted of two separate loci, *Tcd1a* and *Tcd1b* (Fig. 12.1b; Lyon *et al.*, 2000). According to its map position, *Tagap1* is located in the region of *Tcd1a*, and thus its *t*-allele is designated $Tagap1^{Tcd1a}$, whereas *Tcd1b* has not yet been reported.

The isolation of $Tagap1^{Tcd1a}$ led directly to the isolation of *Tcd2*, which is located in the distal region of the *t*-haplotype (Bauer *et al.*, 2007). This region contains *Fgd2* (Faciogenital dysplasia 2), encoding a guanine nucleotide exchange factor acting on Rho small G proteins. Expression of this gene in the testes of males homozygous for the *t*-haplotype was found to be much higher than in wild-type males, suggesting that the *t*-allele of *Fgd2*, similarly to $Tagap1^{Tcd1a}$, is a hypermorph (Fig. 12.4c). If $Fgd2^{t}$ acts as a distorter, enhanced gene activity should increase the transmission rate of a *t*-haplotype containing *Tcr*, while lowering *Fgd2* gene activity should lead to a reduction of the transmission rate. Knock-out analysis indeed demonstrated that a reduction of the *Fgd2* gene dosage by half reduced the transmission rate of the *t*-haplotype t^{h49} significantly. This proved that the *t*-allele of *Fgd2* is *Tcd2* (gene symbol $Fgd2^{Tcd2}$) and that it acts as a hypermorph (Bauer *et al.*, 2007). Both *Tagap1* and *Fgd2* are expressed in diploid spermatogenic cells, permitting the action of the distorters in all spermatozoa. This is in line with the finding of Lyon that the distorters act in *trans* (Lyon, 1984; Fig. 12.2b, 12.3).

The mechanism for creation of $Fgd2^{Tcd2}$ is quite different from that resulting in $Tagap1^{Tcd1a}$. While $Tagap1^{Tcd1a}$ originated by gene amplification, $Fgd2^{Tcd2}$ is most likely the result of an intragenic switch between two alternative promoters. *Fgd2* expresses two transcripts in the testis, a full length and a short transcript (Fig. 12.4c). The latter encodes a truncated protein believed to act as a dominant-negative guanine nucleotide exchange factor (GEF). Expression of the full-length transcript of $Fgd2^{Tcd2}$ is increased at the expense of the short transcript, suggesting that the relative activities of the corresponding promoters have switched in comparison to the wild-type allele, which shows the inverse situation, more of the shorter transcript and less of the full-length transcript (Bauer *et al.*, 2007; Fig 12.4c). The molecular mechanism for this switch, however, remains unresolved.

The finding that both $Tagap1^{Tcd1a}$ and $Fgd2^{Tcd2}$ are hypermorphs, but encode proteins with opposing functions, suggested that the two proteins act on different Rho small G proteins, which have opposing effects on SMOK (Bauer *et al.*, 2007; Fig. 12.5). We therefore proposed that TAGAP1 acts in a repressive pathway, while FGD2 is involved in an activating pathway. Up-regulation of TAGAP1 activity by $Tagap1^{Tcd1a}$ will result in down-regulation of the repressive pathway and thus an increase of SMOK activity. Higher levels of FGD2 protein provided by $Fgd2^{Tcd2}$ would likewise up-regulate SMOK activity. SMOK can therefore be hyper-activated by the additive action of both pathways and trigger the impairment of flagellar motility, the 'harmful' effect of the distorters suggested in Lyon's model. This harmful effect of the distorters is counterbalanced and rescued by the dominant-negative kinase $SMOK^{Tcr}$, but since the latter acts in a haploid cell-restricted manner, only the sperm cells expressing $Smok^{Tcr}$ profit and gain an advantage in the competition for egg cells (Fig. 12.5).

Mechanism for haploid cell-restricted action of $Smok^{Tcr}$

Haploid sperm cells are connected in syncytia and are supposed to be phenotypically equal due to exchange of RNA and protein products. $Smok^{Tcr}$ does not follow this rule. We have recently shown that $Smok^{Tcr}$-derived transcripts are retained in cells carrying the gene and are not shared with neighbouring cells devoid of $Smok^{Tcr}$, most likely by tethering of the transcripts (Véron *et al.*, 2009). Translation of $Smok^{Tcr}$ transcripts is also considerably delayed. The protein is first detectable in the flagella of spermatozoa located in the lumen of seminiferous tubules. Immunohistochemical detection of the protein on ultra-thin sections of sperm using electron microscopy techniques followed by statistical analysis showed that males carrying a $Smok^{Tcr}$ transgene indeed produce two types of sperm: half of the sperm contains $\mathrm{SMOK}^{\mathrm{Tcr}}$ protein and the other half is devoid of it (Véron *et al.*, 2009). This clearly demonstrates that $Smok^{Tcr}$ can make sperm phenotypically unequal. In fact, the wild-type *Smok1* gene behaves the same way. Therefore, $Smok^{Tcr}$ is not a peculiarity, which evolved just once and there may be more genes which behave similarly.

How might the *t*-haplotype have evolved?

All complete *t*-haplotypes isolated from the wild are structurally identical and show very few nucleotide changes, suggesting that they derive from a single ancestor, which arose about two to three million years ago. This is based on analysis of various *t*-complex-derived molecular markers (Willison *et al.*, 1986; Morita *et al.*, 1992; Hammer and Silver, 1993). The high molecular and structural similarity of the *t*-haplotypes is contrasted by striking variability in *t*-lethal mutations. At least 16 different recessive lethals have been identified on *t*-haplotypes isolated from various geographic regions (Klein *et al.*, 1984), suggesting that the *t*-lethals arose independently, either in parallel with or after spreading of the *t*-haplotype through the *Mus musculus* subspecies. To explain this observation some authors have suggested that the *t*-haplotype may have evolved in a geographically isolated mouse population and later introgressed into the *M. musculus* subspecies in which it was found (Silver *et al.*, 1987; Morita *et al.*, 1992). This introgression may have begun as recently as 7000–9000 years ago, coincident with dissemination of agriculture and trade routes, which might have facilitated escape of the *t*-haplotype from the geographic region in which it evolved (Willison *et al.*, 1986; Morita *et al.*, 1992).

This plausible scenario implies that, prior to introgression, the ancestral *t*-haplotype coevolved with the wild-type chromosome 17 in an isolated mouse population (Fig. 12.6). We propose that the evolutionary process leading to the present-day *t*-haplotype was initiated by the formation of the dominant-negative *Smok* homologue $Smok^{Tcr}$. This mutation resulted in a decrease of SMOK activity in the sperm carrying the gene, and thus created a phenotypic difference between $Smok^{Tcr}$-positive and $Smok^{Tcr}$-negative sperm, directly leading to TRD. Our experimental data derived from transgenic mice expressing $Smok^{Tcr}$ on the wild-type background suggest that sperm expressing $Smok^{Tcr}$ are superior to their wild-type meiotic partners (B. G. Herrmann, unpublished data). This observation contradicts the low-ratio phenotype of the t^{low}-haplotype described by Lyon

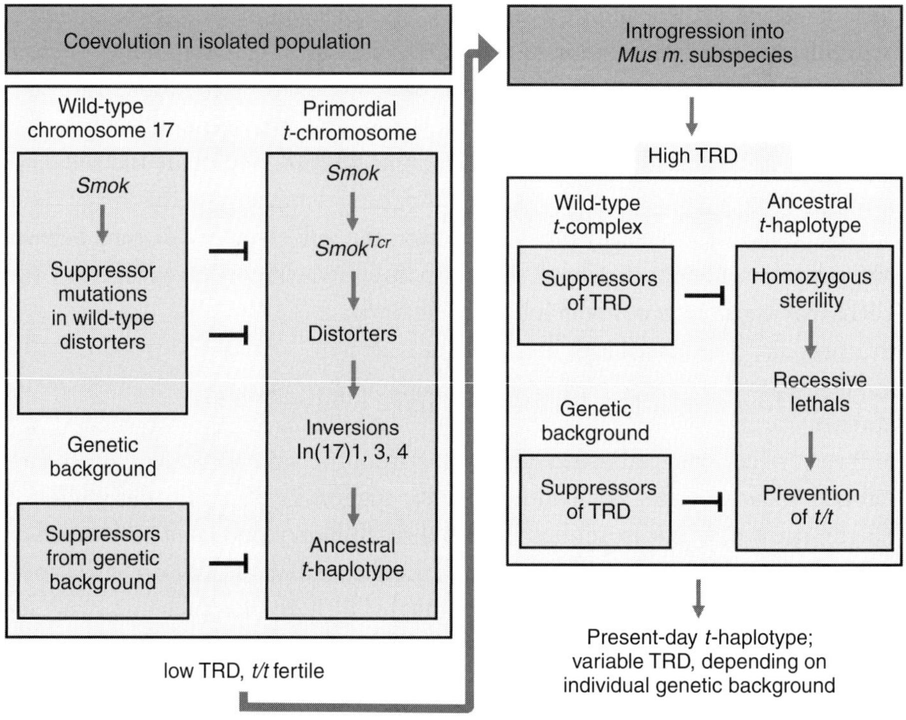

Figure 12.6 Model for the evolution of the *t*-haplotype. For details see text. Arrows indicate progression over evolutionary time; blocked lines indicate genetic mechanisms counteracting TRD in favour of the *t*-haplotype.

(Fig. 12.2c); however, the latter may have evolved as an adaptation to the improvement of the distorters, which needed to be counterbalanced by strongly reducing total SMOK activity in order to keep sperm motility optimal. Evidence for the latter comes from the t^{low}-phenotype itself, which clearly indicates that the SMOK activity of t^{low}-sperm, which carry only *t*-haplotype-encoded *Smok* genes and $Smok^{Tcr}$ but no distorters, is compromised, leading to strong ratio distortion in favour of wild-type meiotic partners. We propose that the formation of $Smok^{Tcr}$ created a selective advantage for the $Smok^{Tcr}$-carrying sperm sufficient to outcompete the sister chromosome. But why did it not become genetically fixed? Why was the primordial *t*-chromosome able to accumulate multiple mutations over an extended period of time, which created several distorters and three inversions – In(17)1, In(17)3, and In(17)4 – preventing segregation of the distorters from each other and from $Smok^{Tcr}$ by meiotic recombination, and incrementally improved the primordial *t*-haplotype towards the ancestral *t*-haplotype (Fig. 12.6)? We propose that improvements in TRD by the evolving *t*-haplotype were counterbalanced by parallel mutations in wild-type distorter genes and in the genetic background (i.e. in genes involved in the Rho signalling pathways controlling SMOK), which promoted the transmission of the wild-type chromosome. In line with these arguments we also propose that the ancestral *t*-haplotype did not cause significant TRD or sterility in its population of origin due to suppressors controlling both effects. In other words, the evolution of the ancestral *t*-haplotype was only possible through coevolution of TRD *and* suppressors of TRD.

However, once the *t*-haplotype escaped its region of origin, the *Mus* subspecies which had not coevolved suppressive mechanisms were not able to counteract the TRD effect, and thus were rapidly invaded by the *t*-haplotype. The strong TRD was most likely associated with homozygous male sterility, which would have led to decimation of the invaded populations, accompanied by a risk of complete extinction. However, it is likely that the genetic pool of wild mouse populations was sufficiently diverse to allow for rapid resistance to TRD. The occurrence of recessive lethals on the *t*-haplotype, and the selective advantage of suppressors of TRD provided by distorter variants in the wild-type *t*-complex and by the genetic background, would ultimately result in the present-day situation.

The assumption that suppressors of TRD exist in the *Mus* subspecies invaded by the *t*-haplotype is supported by the finding that the genetic background of laboratory mouse strains has a considerable influence on the strength of TRD (Bennett *et al.*, 1983). For instance, the *t*-haplotype t^{12} when maintained on the BTBRTF/Nev strain background is reported to be transmitted at 52%, while on the C3H/DiSn background a transmission rate of 99% is observed (Gummere *et al.*, 1986). Further support comes from the molecular analysis of *Tagap1* and *Fgd2*, which shows considerable variability in RNA expression levels between different wild-type strains.

Conclusion

To date, the mouse *t*-haplotype is among the best characterized and best understood of selfish genetic elements, at both the genetic and molecular levels. The detailed analysis of two distorters has revealed a mechanism for how the cumulative impact of several modifiers on the cellular signalling network can trigger cellular malfunction and impaired sperm motility, resulting, in the most extreme outcome, in male sterility. We have shown that TRD is due to the haploid cell-restricted rescue of the sperm motility defect that is caused by the distorters (modifiers), by a kinase variant, $SMOK^{Tcr}$. Thus, the *t*-haplotype provides a striking example of how the accrued effects of several genetic variants can trigger a disease phenotype, while others may instead provide protection against it.

We have also shown that the haploid cell-restricted action of *Smok*^{Tcr} is not a peculiarity of a mutant gene but also applies to its wild-type homologue, *Smok*. *Smok* and *Smok*^{Tcr} have revealed a mechanism that promotes phenotypic differences between sperm cells. It is very likely that more genes act in this manner, and thus have a direct impact on the inheritance of the chromosomal segment on which they are located. We presume that this mechanism may also occur in other species, including man. The identification and analysis of genes showing haploid cell-restricted action in sperm thus promise new insights into the mechanisms of inheritance and evolution.

REFERENCES

Artzt, K., Shin, H. S., and Bennett, D. (1982). Gene mapping within the *T/t* complex of the mouse: II. Anomalous position of the H-2 complex in *t* haplotypes. *Cell*, **28**, 471–6.

Bauer, H., Veron, N., Willert, J., and Herrmann, B. G. (2007). The *t*-complex-encoded guanine nucleotide exchange factor *Fgd2* reveals that two opposing signaling pathways promote transmission ratio distortion in the mouse. *Genes & Development*, **21**, 143–7.

Bauer, H., Willert, J. R., Koschorz, B., and Herrmann, B. G. (2005). The *t* complex-encoded GTPase-activating protein *Tagap1* acts as a transmission ratio distorter in mice. *Nature Genetics*, **37**, 969–73.

Bennett, D. (1975). *T*-locus of mouse. *Cell*, **6**, 441–54.

Bennett, D. (1977). L. C. Dunn and his contribution to *T*-locus genetics. *Annual Review of Genetics*, **11**, 1–12.

Bennett, D., Alton, A. K., and Artzt, K. (1983). Genetic analysis of transmission ratio distortion by t-haplotypes in the mouse. *Genetical Research*, **41**, 29–45.

Chesley, P. and Dunn, L. C. (1936). The inheritance of taillessness (anury) in the house mouse. *Genetics*, **21**, 525–36.

Ewulonu, U. K., Schimenti, K., Kuemerle, B., Magnuson, T., and Schimenti, J. (1996). Targeted mutagenesis of a candidate *t* complex responder gene in mouse *t* haplotypes does not eliminate transmission ratio distortion. *Genetics*, **144**, 785–92.

Forejt, J. (1972). Chiasmata and crossing-over in the male mouse (*Mus musculus*): suppression of recombination and chiasma frequencies in the ninth linkage group. *Folia Biologica*, **18**, 161–70.

Forejt, J., Čapková, J., and Gregorová, S. (1980). T(16: 17)43H translocation as a tool in analysis of the proximal part of chromosome 17 (including *T-t* gene complex) of the mouse. *Genetical Research*, **35**, 165–77.

Fossella, J., Samant, S. A., Silver, L. M., *et al.* (2000). An axonemal dynein at the *Hybrid Sterility 6* locus: implications for t haplotype-specific male sterility and the evolution of species barriers. *Mammalian Genome*, **11**, 8–15.

Fox, H. S., Martin, G. R., Lyon, M. F., *et al.* (1985). Molecular probes define different regions of the mouse t complex. *Cell*, **40**, 63–9.

Gummere, G. R., McCormick, P. J., and Bennett, D. (1986). The influence of genetic background and the homologous chromosome 17 on *t*-haplotype transmission ratio distortion in mice. *Genetics*, **114**, 235–45.

Hammer, M. F., Schimenti, J., and Silver, L. M. (1989). Evolution of mouse chromosome 17 and the origin of inversions associated with *t* haplotypes. *Proceedings of the National Academy of Sciences of the United States of America*, **86**, 3261–5.

Hammer, M. F. and Silver L. M. (1993). Phylogenetic analysis of the alpha-globin pseudogene-4 (*Hba-ps4*) locus in the house mouse species complex reveals a stepwise evolution of *t* haplotypes. *Molecular Biology and Evolution*, **10**, 971–1001.

Hammerberg, C. (1981). The influence of *TOrl* upon male fertility in *t*-bearing mice. *Genetical Research*, **37**, 71–7.

Hammerberg, C. (1982). The effects of the *t*-complex upon male reproduction are due to complex interactions between its several regions. *Genetical Research*, **39**, 219–26.

Herrmann, B. G., Barlow, D. P., and Lehrach, H. (1987). A large inverted duplication allows homologous recombination between chromosomes heterozygous for the proximal *t* complex inversion. *Cell*, **48**, 813–25.

Herrmann, B. G., Bucan, M., Mains, P. E., *et al.* (1986). Genetic analysis of the proximal portion of the mouse *t* complex: evidence for a second inversion within *t* haplotypes. *Cell*, **44**, 469–76.

Herrmann, B. G., Koschorz, B., Wertz, K., *et al.* (1999). A protein kinase encoded by the *t* complex responder gene causes non-mendelian inheritance. *Nature*, **402**, 141–6.

Herrmann, B. G., Labeit, S., Poustka, A., King, T. R., and Lehrach, H. (1990). Cloning of the *T* gene required in mesoderm formation in the mouse. *Nature*, **343**, 617–22.

Katz, D. F., Erickson, R. P., and Nathanson, M. (1979). Beat frequency is bimodally distributed in spermatozoa from *T/t12* mice. *Journal of Experimental Zoology*, **210**, 529–35.

Klein, J., Sipos, P., and Figueroa, F. (1984). Polymorphism of *t*-complex genes in European wild mice. *Genetical Research*, **44**, 39–46.

Korzh, V. and Grunwald, D. (2001). Nadine Dobrovolskaia-Zavadskaia and the dawn of developmental genetics. *Bioessays*, **23**, 365–71.

Lader, E., Ha, H. S., O'Neill, M., Artzt, K., and Bennett, D. (1989) *Tctex-1:* a candidate gene family for a mouse *T*-complex sterility locus. *Cell*, **58**, 969–79.

Lyon, M. F. (1956). Hereditary hair loss in the tufted mutant of the house mouse. *Journal of Heredity*, **47**, 101–3.

Lyon, M. F. (1984). Transmission ratio distortion in mouse *t*-haplotypes is due to multiple distorter genes acting on a responder locus. *Cell*, **37**, 621–8.

Lyon, M. F. (1986). Male sterility of the mouse *t*-complex is due to homozygosity of the distorter genes. *Cell*, **44**, 357–63.

Lyon, M. F. (1992). Deletion of mouse *t-complex distorter-1* produces an effect like that of the *t*-form of the distorter. *Genetical Research*, **59**, 27–33.

Lyon, M. F. (2003). Transmission ratio distortion in mice. *Annual Review of Genetics*, **37**, 393–408.

Lyon, M. F. and Mason, I. (1977). Information on the nature of *t*-haplotypes from the interaction of mutant haplotypes in male fertility and segregation ratio. *Genetical Research*, **29**, 255–66.

Lyon, M. F. and Meredith, R. (1964a). Investigations of the nature of *t*-alleles in the mouse: I. Genetic analysis of a series of mutants derived from a lethal allele. *Heredity*, **19**, 301–12.

Lyon, M. F. and Meredith, R. (1964b). Investigations of the nature of *t*-alleles in the mouse: II. Genetic analysis of an unusual mutant allele and its derivatives. *Heredity*, **19**, 313–25.

Lyon, M. F. and Meredith, R. (1964c). Investigations of the nature of *t*-alleles in the mouse: III. Short tests of some further mutant alleles. *Heredity*, **19**, 327–30.

Lyon, M. F., Schimenti, J. C., and Evans, E. P. (2000). Narrowing the critical regions for mouse *t* complex transmission ratio distortion factors by use of deletions. *Genetics*, **155**, 793–801.

Morita, T., Kubota, H., Murata, K., *et al.* (1992). Evolution of the mouse *t* haplotype: recent and worldwide introgression to *Mus musculus*. *Proceedings of the National Academy of Sciences of the United States of America*, **89**, 6851–5.

Olds-Clarke, P. and Johnson, L. R. (1993). *t* haplotypes in the mouse compromise sperm flagellar function. *Developmental Biology*, **155**, 14–25.

Olds-Clarke, P. and Peitz, B. (1985). Fertility of sperm from *t*/+ mice: evidence that +-bearing sperm are dysfunctional. *Genetical Research*, **47**, 49–52.

Planchart, A., You, Y., and Schimenti, J. C. (2000). Physical mapping of male fertility and meiotic drive quantitative trait loci in the mouse *t* complex using chromosome deficiencies. *Genetics*, **155**, 803–12.

Rohme, D., Fox, H., Herrmann, B., *et al.* (1984). Molecular clones of the mouse *t* complex derived from microdissected metaphase chromosomes. *Cell*, **36**, 783–8.

Rosen, L. L., Bullard, D. C., Silver, L. M., and Schimenti, J. C. (1990). Molecular cloning of the t complex responder genetic locus. *Genomics*, **8**, 134–40.

Schimenti, J., Cebra-Thomas, J. A., Decker, C. L., *et al.* (1988). A candidate gene family for the mouse *t* complex responder (*Tcr*) locus responsible for haploid effects on sperm function. *Cell*, **55**, 71–8.

Schimenti, J. C., Reynolds, J. L., and Planchart, A. (2005). Mutations in *Serac1* or *Synj2* cause proximal *t* haplotype-mediated male mouse sterility but not transmission ratio distortion. *Proceedings of the National Academy of Sciences of the United States of America*, **102**, 3342–7.

Silver, L. M. (1989). Gene dosage effects on transmission ratio distortion and fertility in mice that carry *t* haplotypes. *Genetical Research*, **54**, 221–5.

Silver, L. M., Hammer, M., Fox, H., *et al.* (1987). Molecular evidence for the rapid propagation of mouse *t* haplotypes from a single, recent, ancestral chromosome. *Molecular Biology and Evolution*, **4**, 473–82.

Silver, L. M. and Olds-Clarke, P. (1984). Transmission ratio distortion of mouse *t* haplotypes is not a consequence of wild-type sperm degeneration. *Developmental Biology*, **105**, 250–2.

Silver, L. M. and Remis, D. (1987). Five of the nine genetically defined regions of mouse *t* haplotypes are involved in transmission ratio distortion. *Genetical Research*, **49**, 51–6.

Véron, N., Bauer, H., Weisse, A. Y., *et al.* (2009). Retention of gene products in syncytial spermatids promotes non-Mendelian inheritance as revealed by the *t* complex responder. *Genes & Development*, **23**, 2705–10.

Willison, K. R., Dudley, K., and Potter, J. (1986). Molecular cloning and sequence analysis of a haploid expressed gene encoding *t* complex polypeptide 1. *Cell*, **44**, 727–38.

13

Tracing recent adaptations in natural populations of the house mouse

MEIKE TESCHKE, ANNA BÜNTGE, AND DIETHARD TAUTZ

Natural selection: a short introduction

Selection is one of the principal forces shaping phenotypic and genotypic variation within populations. To understand the consequences of selection at the DNA level, it is necessary to distinguish between positive and negative selection. Negative selection removes disadvantageous alleles and thus leads to relative sequence conservation. Positive or directional selection acts upon new advantageous mutations, fixes them in populations, and thus leads to sequence divergence. But sequence divergence can also be caused by random fixation of neutral mutations, and it is therefore necessary to take the neutral expectation as the null model if one wants to trace positive selection. Although there is persuasive evidence for phenotypic adaptations caused by natural selection, comparatively few genes have been identified so far that are directly involved. This is for two major reasons: (1) the majority of DNA sequence changes are due to neutral evolution, implying that most mutations that can be traced by direct sequencing approaches are not caused by selection; (2) adaptations affect mainly adult phenotypes, while modern genetic approaches are heavily focused on cell biological and embryonic phenotypes. Accordingly, little is known about gene functions involved in adult morphology, physiology, and behaviour, which would be expected to be the prime target of positive selection.

Thus, it is a general goal in contemporary evolutionary biology to learn more about genes, or genomic regions, that have been the target of positive selection (Tautz and Schmid, 1998; Sabeti *et al.*, 2007; Oleksyk *et al.*, 2010). One of the main effects of recent selection is to modify the levels of variability within and between species, thereby leaving characteristic signatures in the polymorphism level of the genomes in populations. Tracing these signals is of central importance to understanding how natural selection shapes genetic variation among species.

Evolution of the House Mouse, ed. Miloš Macholán, Stuart J. E. Baird, Pavel Munclinger, and Jaroslav Piálek.
Published by Cambridge University Press.

The relative role of positive selection in causing evolutionary divergence has long been debated. Haldane (1957) was the first to argue that positive selection is associated with a cost – namely, loss of individuals which do not share the new advantageous allele – and that this cost of selection would place an upper limit on the rate of evolution. He estimated that the limit of selection within a diploid population was one gene substitution per 300 generations. This was later contested by Maynard Smith (1968) and Moran (1970), who showed that interactions in genes need to be taken into account when considering this question and that much higher frequencies of positive selection could occur in natural populations.

Kimura proposed in his neutral theory of molecular evolution (1983) that most mutations would have no effect on fitness, and consequently have their fate dictated by chance alone. It is now clear that neutral mechanisms can indeed explain most polymorphism patterns found in populations, but it has also long been known that the number of observed polymorphisms in various species cannot be explained by neutrality alone (Lewontin, 1974). While the neutral theory predicts a linear correlation between population size and heterozygosity, the actual measurements of heterozygosity usually arrive at non-linear relationships and an upper limit. It is likely that associated selection is the major force generating these patterns in natural populations, but there is still an ongoing debate over how much of this is due to negative selection (or background selection; Charlesworth *et al.*, 1993) or to positive selection (selective sweeps). Maynard Smith and Haigh (1974) showed that in large populations the joint fixation of variants linked to selected sites can reduce neutral diversity much more than random genetic drift (Barton, 2000). Gillespie (2000) has formalized this in an extended theory and has called the process 'genetic draft'.

Assessing the frequency of natural selection that shapes genetic variation is one step towards clarifying the interaction of natural selection and random drift at the molecular level (Orr, 2005). Since all mutations, even those that are advantageous, are at risk of being lost by random genetic drift, drift and natural selection do not act independently in natural populations. However, the degree to which alleles are affected by drift or selection varies – the stronger the advantage of a mutation, the less likely it is lost (Eyre-Walker and Keightley, 2007). Hence, not only the frequency of positive selection events, but also the selection coefficients are of relevance. But little is known about both of these parameters so far.

The house mouse as a model

Two main features characterize the house mouse as a particularly suitable model organism for investigating the genetic basis of adaptations. First, its evolutionary history is well known since mouse phylogeny and history have been intensely studied (Boursot *et al.*, 1993; Guénet and Bonhomme, 2003). Second, the nearly complete genome sequence has been available since 2002 (Mouse Genome Sequencing Consortium, 2002; Church *et al.*, 2009), and excellent genetic and genomic resources are available.

The house mouse is a recent phylogenetic offshoot of the genus *Mus* (Auffray and Britton-Davidian, Chapter 1 in this volume; Suzuki and Aplin, Chapter 2). It evolved on the Indian subcontinent, from where it radiated in several directions to the well-described peripheral subspecies (*Mus musculus musculus*, *M. m. domesticus*, and *M. m. castaneus*) (Din *et al.*, 1996). As commensals to humans the species spread all over the world in historic times.

These days the subspecies *M. m. musculus* is found all over northern Asia, as well as in eastern, central, and Scandinavian Europe. *M. m. domesticus* has its contemporary range in western Europe, the Near East, north Africa, and was recently introduced by humans into the New World, sub-Saharan Africa, and Australia. The third subspecies, *M. m. castaneus*, is spread all over Southeast Asia (Boursot *et al.*, 1993). This complex history of repeated successful colonization and the concomitant adaptations to new environments highlight this species as a perfect model system for evolutionary research.

The fairly large effective population sizes of mice in their ancestral habitats (in the order of 4.4×10^5 to 2.1×10^6; Baines and Harr, 2007) as well as the short generation time (three generations per year; Karn *et al.*, 2002) are additional factors that are of relevance for evolutionary studies. Drift effects are small in large populations, allowing selection to act more effectively. Thus, even small selection coefficients can shape the genetic pattern. Compared to humans, mice have up to 50-fold more generations per time unit. Thus, even during the recent historic times for which we know patterns of human movements, as well as ecological conditions, mice had a large potential to evolve. For example, the genetic divergence between mouse populations that have split only 3000 years ago can be as much as the divergence between humans and Neanderthals.

In Fig. 13.1 we give an overview of a number of studies on various populations worldwide to approach the question of occupation of new environments and adaptations in house mice. The figure depicts examples of adaptations from morphological, physiological, behavioural, and genetic perspectives.

The work by Berry and colleagues in particular presents a large amount of phenotypic data on possibly adaptive changes, which have been observed on island populations. Mice tend to be large and carry increased amounts of brown fat as a response to cold temperatures on South Georgia and Macquarie (Berry and Peters, 1975; Berry *et al.*, 1979). On various sub-Antarctic islands Berry and colleagues observed opposing selection at different life stages (Berry and Peters, 1975; Berry *et al.*, 1978, 1979). Various new food resources have been exploited by

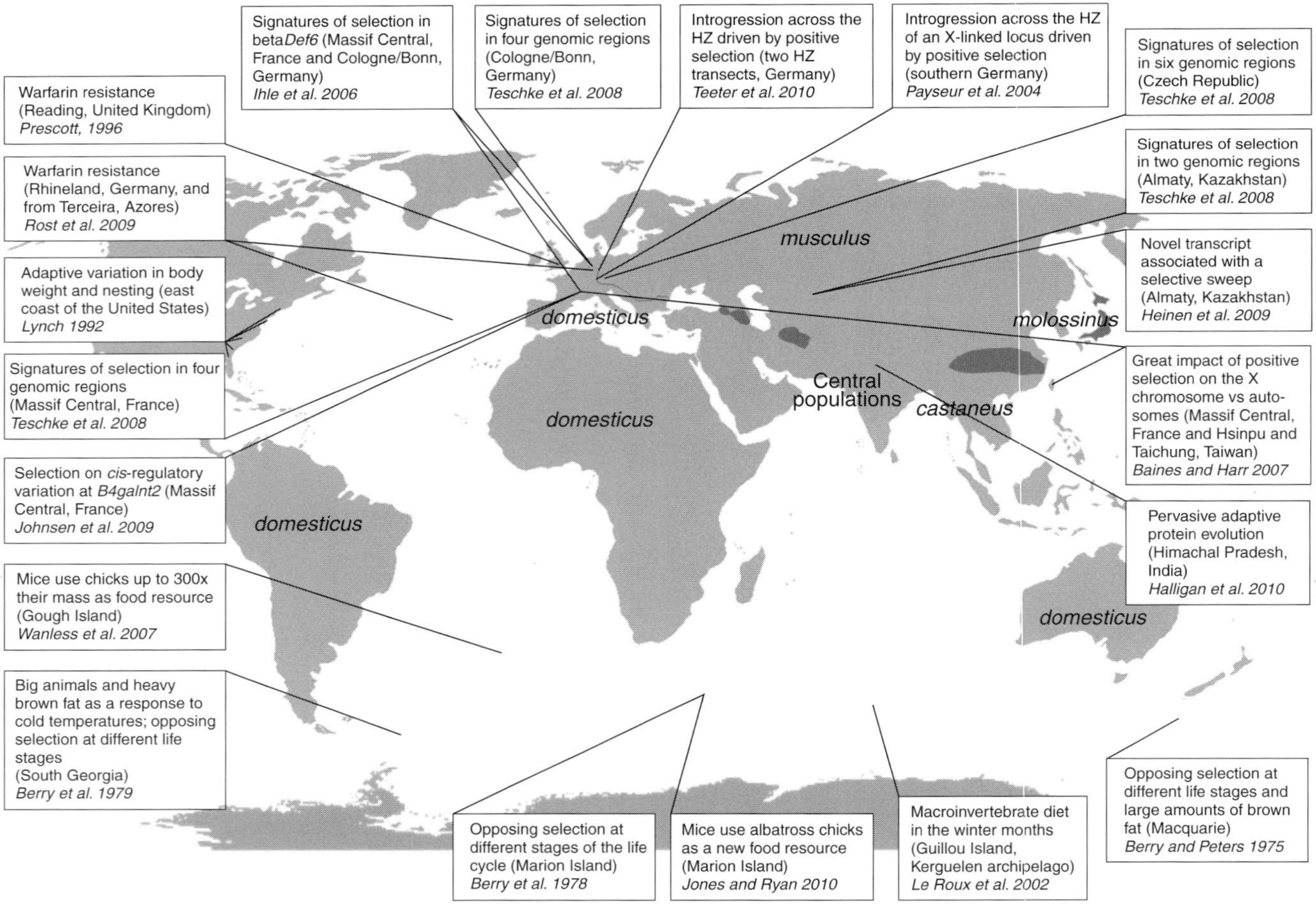

Figure 13.1 Depiction of studies on adaptations in house mouse populations. The picture is only meant to give an impression of the range of studies conducted so far, and does not cover all relevant studies.

mice on islands. The most spectacular example is mice eating bird chicks of body mass multiple times their own (Wanless *et al.*, 2007; Jones *et al.*, 2010). Another example is the macroinvertebrate diet of mice on the Kerguelen Islands, which represents a large behavioural change as a response to extreme climatic circumstances (Le Roux *et al.*, 2002). Also, under less extreme conditions, such as clinal variation along the east coast of the United States, behavioural and morphological adaptations can be observed (Lynch, 1992).

There are also several examples of individual genes being identified as involved in adaptations. This ranges from studies on the evolution of resistance against warfarin, a poison to eradicate rats and mice (Prescott, 1996; Rost *et al.*, 2009), to the description of the appearance of apparently advantageous novel transcripts (Harr *et al.*, 2006; Heinen *et al.*, 2009). The introgression characteristics of a locus across a hybrid zone can also be used to infer whether positive selection is involved, and Payseur *et al.* (2004) and Teeter *et al.* (2010) have shown that this is indeed the case for the mouse hybrid zones. Variation has been maintained at specific loci for much longer than expected under neutrality (Ferguson *et al.*, 2008; Johnsen *et al.*, 2009). In general, mice exhibit pervasive adaptive protein evolution, which has been shown for many genes on the X chromosome (Baines and Harr, 2007) and across the entire genome (Halligan *et al.*, 2010; Fraser *et al.*, 2011).

Levels of population differentiation

To systematically study frequencies of selection, it is necessary to have a good knowledge of the history and level of population differentiation. As discussed above, this is available for the house mouse, but for operational purposes it is useful to distinguish different levels. We propose to distinguish three such levels of divergence, which allow asking different questions.

Level 1: comparison between subspecies. This refers to comparisons between *M. m. musculus*, *M. m. domesticus*, and *M. m. castaneus*, assumed to have split 0.5–1 million years ago (She *et al.*, 1990; Geraldes *et al.*, 2008). These comparisons are particularly useful for studying the genetic basis of the onset of reproductive isolation and speciation processes, including hybrid zone studies in areas of secondary contact. These topics are further discussed in Chapter 14 (Baird and Macholán). However, they are less well suited to studying recent adaptations, since the time since the split would have been sufficient for multiple successive rounds of positive selection at any given locus. Also, the details of population history, such as bottlenecks, migration, and hybridization cannot easily be reconstructed.

Level 2: ancestral and derived populations of *M. m. domesticus*. The ancestral populations of *M. m. domesticus* occur in the Middle East, most notably in Iran and Iraq, and are probably descended from the mice that have developed commensalism in conjunction with the rise of agricultural civilizations in these parts of the world, up to 10 000 years ago. From there they moved westwards and became established in central Europe around 3000 years ago (Cucchi *et al.*, 2005). Comparisons between such populations are thus within time frames in which single or few rounds of positive selection at a given locus can be expected and in which population history parameters can be reasonably inferred. These comparisons are therefore particularly useful to estimate frequencies and strength of selection events, as further detailed below.

Level 3: newly established populations. Beginning less than 500 years ago, *M. m. domesticus* populations from western Europe have started to colonize the remaining continents and oceanic islands in the wake of inadvertent transport on ships. This allows looking at population dynamic processes shortly after establishment in a new area, including the first wave of local adaptations to new climates. This is particularly useful to better understand the very first steps of adaptation after arriving in a new environment (Gabriel *et al.*, 2010; Hardouin *et al.*, 2010). However, since these colonizations were usually accompanied by bottlenecks, polymorphism levels are generally low and it will be difficult to tease apart founder effects from adaptive changes. Also, since the most recently established populations have not yet reached new population genetic equilibria, it is more difficult to assess them with respect to models that are based on equilibrium assumptions. Hence, such populations are currently of less relevance for deriving general selection parameters.

Appropriate population comparisons require the sampling of natural populations, since the polymorphisms that are present in inbred strains will always be biased, even if their origin is well known. But to capture the allelic diversity in natural populations, it is necessary to adjust the sampling scheme to the specific behavioural pattern of mice. It is known that related mice breed together and that females show communal nesting (Southwick, 1955; Berry and Bronson, 1992). Dispersal occurs mainly via males, and generally dispersal distances tend to be very small, mostly within a few metres (Pocock *et al.*, 2005). We have devised a sampling scheme which takes these factors into account and aims to obtain usually only one mouse per location and then move about 1 km further to capture the next mouse (Ihle *et al.*, 2006). As a local population sample we consider mice that are caught within a radius of about 50 km (Fig. 13.2). When analysing expected and observed heterozygosities in such samples, we find strong indications for a deficit of heterozygous animals (Table 13.1), confirming the notion that there is an underlying deme structure.

Table 13.1 Average heterozygosity values based on 204 microsatellite loci genotyped in population samples of 57–68 individuals for the four mouse populations studied in Ihle *et al.* (2006).

Population	Observed heterozygosity	Expected heterozygosity
Cameroon	0.45	0.60
France	0.53	0.72
Germany	0.49	0.73
Kazakhstan	0.60	0.75

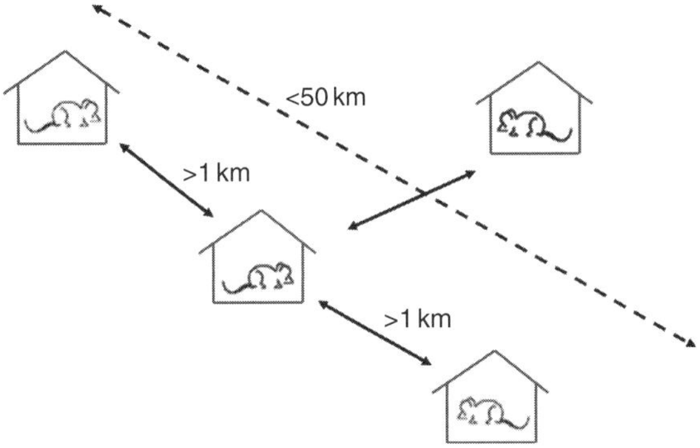

Figure 13.2 Sampling scheme for mouse populations to capture variability of a given area. Only single mice are collected at given locations in order to avoid catching related mice. The different locations are at least 1 km apart and the whole area has a diameter of about 50 km.

Signatures of selection

When a favourable mutation arises, the affected allele rises in frequency in the population at a speed that depends on the selection coefficient. This process will alter the frequencies of alleles at closely linked loci. In the most extreme case, a single favourable mutation arises at a site, which is completely linked to a neutral polymorphism. Fixation of the beneficial allele in the population will be accompanied by the fixation of the linked neutral variant. Thus, positive selection will result in reduction of heterozygosity of neutral polymorphisms to an extent which varies with map distance from the substituted locus (Maynard Smith and Haigh, 1974).

The change in allele frequencies due to fixation of associated advantageous mutations is called 'genetic hitchhiking', and the 'footprint' that is left in the genome is referred to as a 'selective sweep' (Fig. 13.3). Screening for signatures of

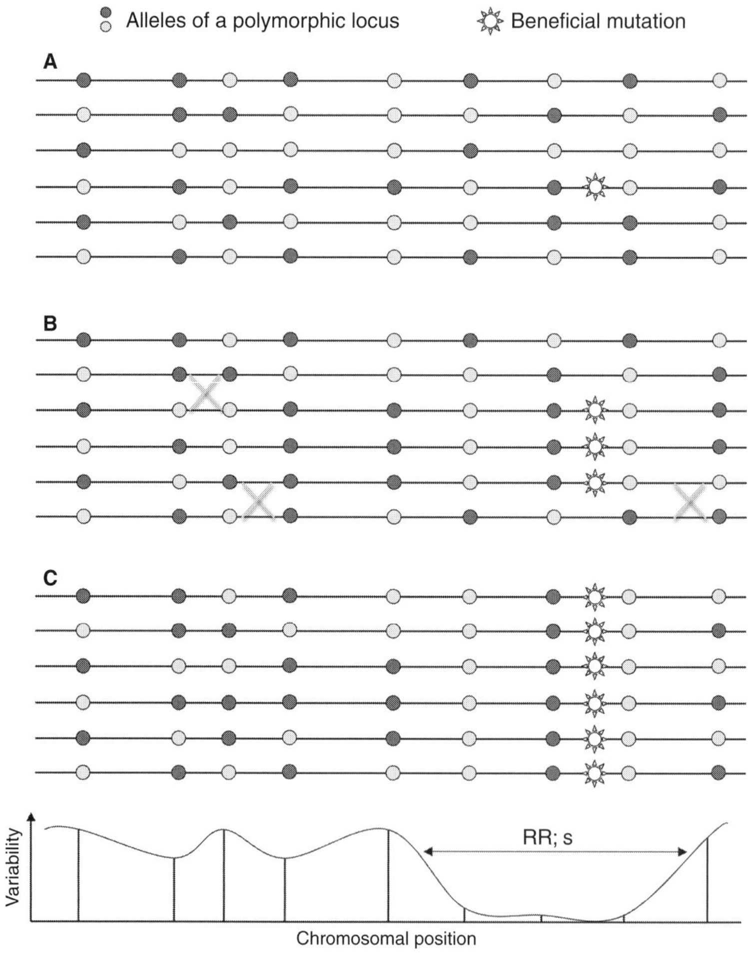

Figure 13.3 Scheme for the signature of a selective sweep. The lines represent single chromosomes in a population, circles represent polymorphisms. (a) Starting situation where an advantageous mutation arises within a polymorphic background. (b) The chromosomes carrying the mutation increase in frequency. The sites close to the selected site hitchhike with this site, while the more distant ones become separated by recombination events (symbolized by crosses). (c) The advantageous mutation is fixed, together with the neighbouring neutral sites. The variability pattern across the population is depicted at the bottom. The width of the sweep window is dependent on the recombination rate (RR) and the selection coefficient (s).

selective sweeps by comparing variability levels between populations has been called 'hitchhiking mapping' (Harr *et al.*, 2002; Schlötterer, 2003).

The pattern and shape of a selective sweep within the genome is mainly determined by the local recombination rate and the selection coefficient (Maynard Smith and Haigh, 1974), the former being negatively associated with the width, the latter positively. Thus, in regions of high recombination, the hitchhiking pattern will be more disrupted by recombined variation (Fay and Wu, 2000).

After the beneficial allele reaches fixation within the population, the footprint is gradually lost by acquisition of new mutations. This 'recovery' pattern is characterized by an excess of new mutations at low frequencies. Thus, the time frame in which the pattern of positive selection will be observable in a hitchhiking mapping approach depends on the mutation rate of the investigated neutral marker (Slatkin, 1995; Wiehe, 1998). Single nucleotide polymorphisms (SNPs) and microsatellites (short, tandemly repeated sequences of 1–6 bp in length) are commonly used as such markers. While the mutation rates for SNPs are relatively constant – although they may differ between regions of the genome (Wolfe *et al.*, 1989): about 2.5×10^{-8} in humans (Nachman and Crowell, 2000), 2.1×10^{-8} in mice (Nachman, 1997), and 3×10^{-9} in insects (Andolfatto and Przeworski, 2000) – the mutation rate of microsatellites is highly variable. It is mainly determined by the repeat pattern and the number of repeat units, whereby it is expected that the mutation rate rises with repeat number (Ellegren, 2004). The average mutation rates of microsatellites are estimated to be several orders of magnitude higher than those of SNPs. Hence, a single beneficial substitution can only be detected in polymorphism data as long as it occurred recently. Przeworski (2002) estimates that the signature of selection can be inferred up to about 10^4 generations in humans and about 10^6 in *Drosophila melanogaster* based on SNPs. Concerning microsatellites, sweep patterns are expected to be blurred much more quickly by new mutations due to their high mutation rates, only allowing a focus on loci that have only very recently been relevant for adaptations.

Another important aspect of interpreting sweep patterns is that due to the heterogeneity of microsatellite mutation rates, different loci are not comparable with each other. To circumvent this problem, selective changes are deduced by comparing the same microsatellite locus among different populations or species (Schlötterer, 2002). In such comparisons two key parameters can be calculated: the variance in repeat number (V) as a measure of variability at the locus (Goldstein and Clark, 1995), and the expected heterozygosity (H) (Nei, 1978). It has been shown that under neutrality the logarithm of the ratio of Θ, ($\ln(\Theta_{\text{locus1,pop1}}/\Theta_{\text{locus1,pop2}})$; whereby Θ determines the level of variation under the neutral model (Gillespie, 2004), which can be deduced from either V or H ($\ln RV$ and $\ln RH$) follows a normal distribution (Schlötterer, 2002; Kauer *et al.*, 2003).

Hence, this allows detection of loci that depart from the null hypothesis with a defined probability. Thereby, outlier loci are identified with respect to differences in polymorphism between the two compared populations, which are likely linked to sites under selection. In contrast to using the variance, the heterozygosity measure seems to have a higher power to detect such loci due to robustness against variance caused by large differences in repeat number between alleles (Kauer *et al.*, 2003).

The advantages and drawbacks of the hitchhiking mapping approach are well understood (Teshima *et al.*, 2006; Thornton *et al.*, 2007). The problem that is considered the most severe is that demographic effects, such as a dramatic, recent population expansion, can produce patterns in the genome that closely mimic the patterns of selective sweeps. Since bottlenecks readily lead to large variances in the genealogical (coalescent) history of samples from different loci along a chromosome (Hermisson, 2009), their traces can severely impact the inference of selection (Jensen *et al.*, 2005). Hence, for 'hitchhiking mapping' it is important to take the demographic history of the populations into account, which is best described for the level 2 comparisons among mouse populations, as discussed above.

Genome scans for recent selection in the house mouse

The analysis of microsatellite allele patterns has the best potential to focus on recent sweeps and to catch adaptation in action. Although there are several studies that have used microsatellite scans in other systems, the house mouse is probably currently the best one for focusing on recent effects in a reasonably well-defined historical setting. We have therefore developed this approach during the past years. The main focal populations for these studies were mice from southern France (Massif Central) and mice from western Germany (Cologne/Bonn area). The maximum divergence time between these populations is 3000 years, based on the assumption that they have quickly spread across western Europe after the initial colonization and have then not further mixed to large extents. Indeed, the two populations carry rather different mitochondrial haplotypes and are well differentiated between each other (Ihle *et al.*, 2006). In the overall comparison there is also no indication for a major bottleneck after the initial colonization, i.e. it is not expected that demographic factors would have a major impact on allele patterns (Ihle *et al.*, 2006; Teschke *et al.*, 2008).

A second population pair that was assessed in parallel are mice from the Czech Republic and from Kazakhstan. These represent *M. m. musculus* populations, but their detailed history is less well known.

The first study focused on the relative frequency of sweep patterns between candidate genes and randomly chosen genes (Ihle *et al.*, 2006). Candidate genes were retrieved based on known functions in immune defence or hybrid sterility. Approximately 200 microsatellite loci were typed for up to 60 unrelated individuals from each population and evaluated for signs of selective sweeps on the basis of Schlötterer's ln*RV* and ln*RH* statistics (Schlötterer, 2002; Kauer *et al.*, 2003). This study showed only slightly more signs of selective sweeps than would have been expected by chance alone and also no differences between candidate genes and randomly chosen genes. Interestingly, one locus showed a clear sweep pattern that distinguished the two *M. m. domesticus* populations versus the two *M. m. musculus* populations, although this comparison should have been beyond the time horizon captured by the microsatellite scan. It turned out that the microsatellite locus that displayed the sweep pattern was, in this case, a locus with interruptions in the repeat sequence, which is expected to reduce the mutation rate. This confirmed the notion that the time horizon that can be assessed with the microsatellite screen depends on the mutation rate of the respective loci.

The second study (Teschke *et al.*, 2008) used a prescreening approach that allowed an efficient assessment of a larger number of loci (Thomas *et al.*, 2007). A total of 915 microsatellite loci were chosen from the vicinity of annotated genes and initially assessed for a reduction of variability in pooled DNA from the individuals of the respective populations. This was followed by individual typing of loci, which showed a reduction in the overall variability profile in specific populations. For each population pair, eight loci with a significant sweep signature were further characterized by typing flanking microsatellite loci to assess the size of the sweep windows. This allowed assessing whether the flanking loci also showed signs of selective sweeps, even if they were not significant in the statistical tests. Within the sweep window, it was possible to identify 'recovery-patterns', characterized by one major allele flanked by minor alleles one or two repeat units away from the major allele of the respective locus. Such recovery patterns are expected to arise when a locus with a fixed microsatellite allele starts to accumulate new mutations which are, in the case of microsatellites, mostly only single repeat units away from the source allele (Fig. 13.4). Finding such recovery patterns significantly strengthens the notion that the locus was subject to a recent sweep. Furthermore, determining the sweep window size allows us also to apply a second statistical procedure (linked locus statistics; Wiehe *et al.*, 2007). Sweep window sizes were found to be usually confined to one or two genes and it was estimated that the most recent sweep between the French and the German populations had started about 400–600 years ago, with an estimated selection coefficient of $s = 0.01$ (Teschke *et al.*, 2008).

In the third study we explored a 454 sequencing approach to assess systematically a whole contiguous chromosome (chromosome 19) for patterns of selective

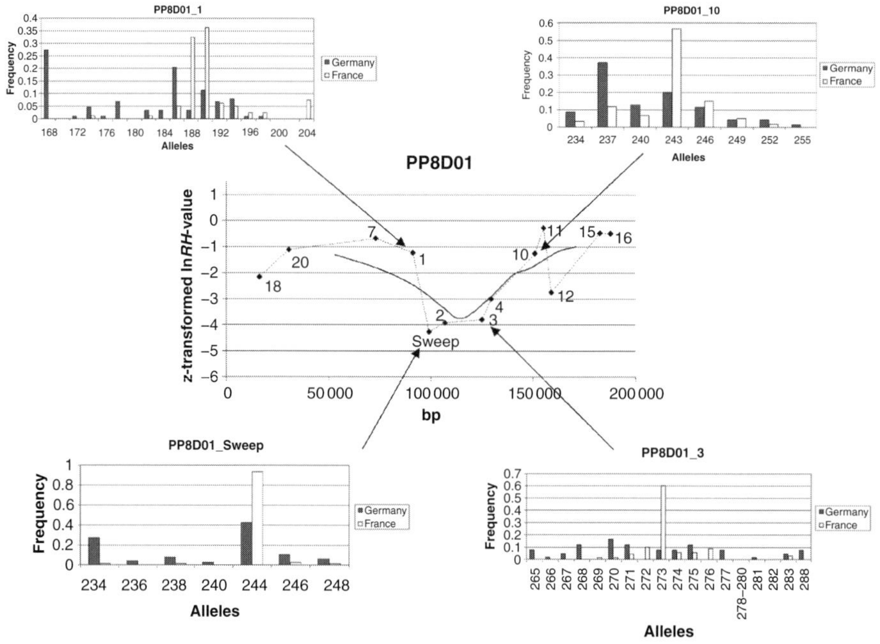

Figure 13.4 Example of a selective sweep, taken from Teschke *et al.* (2008). The centre depicts a 200-kb region within which an initial locus was found that was identified as a sweep candidate ('Sweep'). Additional microsatellite loci were then typed around this one to assess the width of the sweep (symbolized by dots and numbers). The allele spectra are depicted for four of the loci. The 'Sweep' locus is characterized by a near fixation of one allele in the population from France. A neighbouring locus ('3') also has a high-frequency allele in the same population, but additional alleles surrounding the high-frequency one, which can be interpreted as single step mutations that have arisen after fixation. The same pattern is evident for locus '10'. Locus '1' on the other hand, also has additional alleles that are further away from the sweep allele, which can be interpreted to have accumulated through recombination.

sweeps (Büntge, 2010). In this case microsatellites were chosen solely on the basis of a more or less even distribution along the chromosome, independent of the vicinity to genes and hence covering also the gene deserts (which can be larger than 1 Mb in size). Hence, this locus selection was unbiased and therefore suitable to potentially identifying large sweep regions that could be associated with high selection coefficients and/or low recombination rates. Assessment of approximately 760 loci with an average spacing of about 80 kb revealed a comparable number of candidate sweep loci as in our second study (including one in a gene desert), but provided no evidence for extended sweep regions larger than 200 kb (Büntge, 2010).

These three major screens (Ihle *et al.*, 2006; Teschke *et al.*, 2008; Büntge, 2010) yielded at least 20 loci that showed credible patterns of selective sweeps. This should allow us to derive some inferences regarding which kinds of genes are preferentially affected by sweeps. However, no clear picture has emerged as yet. Both highly conserved and highly divergent genes are among the sweep candidates, genes involved in signalling or regulation, structural genes or orphan genes, and even non-coding regions. In a follow-up study for one of the loci, β-defensin 6, no amino acid change in the variant that was subject to a sweep was found, although β-defensins are known to evolve quickly. Instead, it appeared that a regulatory change could have caused the sweep (Ihle *et al.*, 2006). This suggested that regulatory mutations might more frequently be subject to positive selection than protein changes. However, in a dedicated study to assess this question, Staubach *et al.* (2010) found that most regulatory changes between the mouse subspecies appeared to follow the principles of neutral evolution. Deeper inferences into the reasons why particular loci have been subject to positive selection will therefore have to await further gene-by-gene analysis.

Estimates of the frequency of adaptations

Given that there is a reasonably well supported time frame in which the German and French populations were established, the above genome scans allow a rough estimate of the frequency of positive selection events in natural populations of the house mouse. Although the core window size of selective sweeps ranges over 20–100 kb (Teschke *et al.*, 2008), the size of the sweep that is detectable with the initial stringent screening procedure is closer to 20 kb. Based on this, one can calculate that the 915-locus screen has covered about 1.8×10^7 bp, or roughly 1% of the mouse genome. Given that five significant regions were detected across both populations, one would anticipate that 500 detectable sweeps have occurred in the whole genome. Scaling this by divergence time (3000 years, approximately 9000 generations in each lineage) suggests approximately one sweep per 40 generations. The genome scan of chromosome 19 allows estimation of a similar average frequency (Büntge, 2010).

However, for various reasons even this relatively high frequency is an underestimate. The sweeps were identified using the ln*RH* statistics, which detects significant signatures only if the pairwise differences of heterozygosities vary extremely. In other words, a sweep is only detected immediately after one allele was driven to very high frequency in one population and new mutations have not yet accumulated to a certain extent. Thus, the time period within which a selective event can be identified is fairly small and the time a sweep pattern will be maintained depends very much on the mutation rate of the microsatellite. This is mainly determined by the number of repeat units, its motif, and the motif length. If a short allele is fixed through selection, the sweep pattern persists for a longer time, whereas long repeat stretches recover variability within a relatively short time frame. In fact, in many sweep signatures a short allele was fixed, indicating that cases are missing where longer alleles were fixed, but have since then accumulated new mutations. It is currently not possible to estimate how many sweeps were not detected in the current screens, but it is clear that the observed frequency must be an underestimate, probably by at least a factor of two.

Another reason for underestimating the true rate is that a sweep is not necessarily caused by a new mutation, but could be based on an existing allele in the standing variation (Pennings and Hermisson, 2006). In this case, the selected mutation will be flanked by more than one allele of a neighbouring microsatellite locus and the signature of selection will also remain undetected (Przeworski *et al.*, 2005). Similarly, if recombination occurs in the early stage of a sweep event, two different alleles will be linked to the selected site and the sweep pattern will be shifted towards multiple frequent alleles instead of a single one (Fay and Wu, 2000).

Previous estimates of the frequency of selection events have been made by comparison of genome data in *Drosophila* (Smith and Eyre-Walker, 2002; Andolfatto, 2005). Depending on the consideration of non-coding DNA estimates of amino acid substitution driven by natural selection, they greatly vary between 0.02 and 10 in 100 generations. Since many experimental approaches suffer from only detecting relatively strongly selected mutations, estimating the strength of selection is still of central importance to revealing the exact frequency of natural selection (Wright and Andolfatto, 2008). Many attempts try to answer this question by using amino acid substitution models (Chamary *et al.*, 2006; Eyre-Walker and Keightley, 2007; Yang and Nielsen, 2008). However, it has become clear that it is also necessary to incorporate highly conserved non-coding sequences into such analyses, although their functional role is still only poorly understood (Siepel *et al.*, 2005; Wright and Andolfatto, 2008).

Conclusion

Using the house mouse as a model to study the genetic basis of positive selection in natural populations has only just begun. But it is already clear that it is an excellent model that provides many new perspectives, in particular since there are so many local populations that are expected to harbour specific adaptations. The sweep screens have shown that interesting loci can readily be detected and the challenge currently is to actually identify the mutations that have caused the positive selection. Once candidate mutations are identified, the next challenge will be to functionally test them. Since there are many tools available for manipulating the mouse genome, it should be possible to compare the function of a newly selected allele with a non-selected allele in the context of the whole organism. However, since the phenotypic consequences can be expected to be small, it will be necessary to assess the role of such alleles under semi-natural population conditions – but even this can be done in the mouse system. Hence, although it will still be a long way to understand the genetics of adaptation in the mouse, the relevant tools are all at hand.

REFERENCES

Andolfatto, P. (2005). Adaptive evolution of non-coding DNA in *Drosophila*. *Nature*, **437**, 1149–52.

Andolfatto, P. and Przeworski, M. (2000). A genome-wide departure from the standard neutral model in natural populations of *Drosophila*. *Genetics*, **156**, 257–68.

Baines, J. F. and Harr, B. (2007). Reduced X-linked diversity in derived populations of house mice. *Genetics*, **175**, 1911–21.

Barton, N. H. (2000). Genetic hitchhiking. *Philosophical Transactions of the Royal Society B: Biological Sciences*, **355**, 1553–62.

Berry, R. J., Bonner, W. N., and Peters, J. (1979). Natural selection in house mice (*Mus musculus*) from South Georgia (South Atlantic Ocean). *Journal of Zoology*, **189**, 385–98.

Berry, R. J. and Bronson, F. H. (1992). Life history and bioeconomy of the house mouse. *Biological Reviews of the Cambridge Philosophical Society (London)*, **67**, 519–50.

Berry, R. J. and Peters, J. (1975). Macquarie Island house mice: a genetical isolate on a sub-Antarctic island. *Journal of Zoology*, **176**, 375–89.

Berry, R. J., Peters, J., and Van Aarde, R. J. (1978). Sub-antarctic house mice: colonization, survival and selection. *Journal of Zoology*, **184**, 127–41.

Boursot, P., Auffray, J. -C., Britton-Davidian, J., and Bonhomme, F. (1993). The evolution of house mice. *Annual Review of Ecology and Systematics*, **24**, 119–52.

Büntge, A. (2010). *Tracing Signatures of Positive Selection in Natural Populations of the House Mouse*. Unpublished PhD thesis, University of Kiel.

Chamary, J. V., Parmley, J. L., and Hurst, L. D. (2006). Hearing silence: non-neutral evolution at synonymous sites in mammals. *Nature Reviews Genetics*, **7**, 98–108.

Charlesworth, B., Morgan, M. T., and Charlesworth, D. (1993). The effect of deleterious mutations on neutral molecular variation. *Genetics*, **134**, 1289–303.

Church, D. M., Goodstadt, L., Hillier, L. W., et al. (2009). Lineage-specific biology revealed by a finished genome assembly of the mouse. *PLoS Biology* **7**(5), e1000112.

Cucchi, T., Vigne, J. -D., and Auffray, J. -C. (2005). First occurrence of the house mouse (*Mus musculus domesticus* Schwarz & Schwarz, 1943) in the Western Mediterranean: a

zooarchaeological revision of subfossil occurrences. *Biological Journal of the Linnean Society*, **84**, 429–46.
Din, W., Anand, R., Darviche, D., *et al.* (1996). Origin and radiation of the house mouse: mitochondrial DNA phylogeny. *Journal of Evolutionary Biology*, **9**, 391–415.
Ellegren, H. (2004). Microsatellites: simple sequences with complex evolution. *Nature Reviews Genetics*, **5**, 435–45.
Eyre-Walker, A. and Keightley, P. D. (2007). The distribution of fitness effects of new mutations. *Nature Reviews Genetics*, **8**, 610–18.
Fay, J. C. and Wu, C. I. (2000). Hitchhiking under positive Darwinian selection. *Genetics*, **155**, 1405–13.
Ferguson, W., Dvora, S., Gallo, J., Orth, A., and Boissinot, S. (2008). Long-term balancing selection at the West Nile virus resistance gene, *Oas1b*, maintains transspecific polymorphisms in the house mouse. *Molecular Biology and Evolution*, **25**, 1609–18.
Fraser, H. B., Babak, T., Tsang, J., *et al.* (2011). Systematic detection of polygenic *cis*-regulatory evolution. *PLoS Genetics*, **7**, e1002023.
Gabriel, S. I., Jóhannesdóttir, F., Jones, E. P., and Searle, J. B. (2010). Colonization, mouse-style. *BMC Biology*, **8**, 131.
Geraldes, A., Basset, P., Gibson, B., et al. (2008). Inferring the history of speciation in house mice from autosomal, X-linked, Y-linked and mitochondrial genes. *Molecular Ecology*, **17**, 5349–63.
Gillespie, J. H. (2000). Genetic drift in an infinite population: the pseudohitchhiking model. *Genetics*, **155**, 909–19.
Gillespie, J. H. (2004). *Population Genetics: A Concise Guide*. Baltimore, MD: Johns Hopkins University Press.
Goldstein, D. B. and Clark, A. G. (1995). Microsatellite variation in North American populations of *Drosophila melanogaster*. *Nucleic Acids Research*, **23**, 3882–6.
Guénet, J. and Bonhomme, F. (2003). Wild mice: an ever-increasing contribution to a popular mammalian model. *Trends in Genetics*, **19**, 24–31.
Haldane, J. B. S. (1957). The cost of natural selection. *Journal of Genetics*, **55**, 511–24.
Halligan, D. L., Oliver, F., Eyre-Walker, A., Harr, B., and Keightley, P. D. (2010). Evidence for pervasive adaptive protein evolution in wild mice. *PLoS Genetics*, **6**, e1000825.
Hardouin, E. A., Chapuis, J.-L., Stevens, M. I., *et al.* (2010). House mouse colonization patterns on the sub-Antarctic Kerguelen Archipelago suggest singular primary invasions and resilience against re-invasion. *BMC Evolutionary Biology*, **10**, 325.
Harr, B., Kauer, M., and Schlötterer, C. (2002). Hitchhiking mapping: a population-based fine-mapping strategy for adaptive mutations in *Drosophila melanogaster*. *Proceedings of the National Academy of Sciences of the United States of America*, **99**, 12949–54.
Harr, B., Voolstra, C., Heinen, T. J., *et al.* (2006). A change of expression in the conserved signaling gene MKK7 is associated with a selective sweep in the western house mouse *Mus musculus domesticus*. *Journal of Evolutionary Biology*, **19**, 1486–96.
Heinen, T. J., Staubach, F., Häming, D., and Tautz, D. (2009). Emergence of a new gene from an intergenic region. *Current Biology*, **19**, 1527–31.

Hermisson, J. (2009). Who believes in whole-genome scans for selection? *Heredity*, **103**, 283–4.
Ihle, S., Ravaoarimanana, I., Thomas, M., and Tautz, D. (2006). An analysis of signatures of selective sweeps in natural populations of the house mouse. *Molecular Biology and Evolution*, **23**, 790–7.
Jensen, J. D., Kim, K., Bauer DuMont, V., Aquadro, C. F., and Bustamante, C. D. (2005). Distinguishing between selective sweeps and demography using DNA polymorphism data. *Genetics*, **170**, 1401–10.
Johnsen, J. M., Teschke, M., Pavlidis, P., *et al.* (2009). Selection on *cis*-regulatory variation at *B4galnt2* and its influence on von Willebrand factor in house mice. *Molecular Biology and Evolution*, **26**, 567–78.
Jones, M. G. W. and Ryan, P. G. (2010). Evidence of mouse attacks on albatross chicks on sub-Antarctic Marion Island. *Antarctic Science*, **22**, 39–42.
Karn, R. C., Orth, A., Bonhomme, F., and Boursot, P. (2002). The complex history of a gene proposed to participate in a sexual isolation mechanism in house mice. *Molecular Biology and Evolution* **19**, 462–71.
Kauer, M. O., Dieringer, D., and Schlötterer, C. (2003). A microsatellite variability screen for positive selection associated with the 'Out of Africa' habitat expansion of *Drosophila melanogaster*. *Genetics*, **165**, 1137–48.
Kimura, M. (1983). *The Neutral Theory of Molecular Evolution*. Cambridge: Cambridge University Press.
Le Roux, V., Chapuis, J. -L., Frenot, Y., and Vernon, P. (2002). Diet of the house mouse (*Mus musculus* L.) on Guillou Island, Kerguelen archipelago, Subantarctic. *Polar Biology*, **25**: 49–57.
Lewontin, R. C. (1974). *The Genetic Basis of Evolutionary Change*. New York: Columbia University Press.
Lynch, C. B. (1992). Clinal variation in cold adaptation in *Mus domesticus*: verification of predictions from laboratory populations. *The American Naturalist*, **139**, 1219–36.
Maynard Smith, J. (1968). 'Haldane's dilemma' and the rate of evolution. *Nature*, **219**, 1114–16.
Maynard Smith, J. and Haigh, J. (1974). The hitch-hiking effect of a favourable gene. *Genetical Research*, **23**, 23–35.
Moran, P. A. (1970). 'Haldane's dilemma' and the rate of evolution. *Annals of Human Genetics*, **33**, 245–9.
Mouse Genome Sequencing Consortium (2002). Initial sequencing and comparative analysis of the mouse genome. *Nature*, **420**, 520–62.
Nachman, M. W. (1997). Patterns of DNA variability at X-linked loci in *Mus domesticus*. *Genetics*, **147**, 1303–16.
Nachman, M. W. and Crowell, S. L. (2000). Estimate of the mutation rate per nucleotide in humans. *Genetics*, **156**, 297–304.
Nei, M. (1978). Estimation of average heterozygosity and genetic distance from a small number of individuals. *Genetics*, **89**, 583–90.

Oleksyk, T. K., Smith, M. W., and O'Brien, S. J. (2010). Genome-wide scans for footprints of natural selection. *Philosophical Transactions of the Royal Society B: Biological Sciences*, **12**, 185–205.

Orr, H. A. (2005). The genetic theory of adaptation: a brief history. *Nature Reviews Genetics*, **6**, 119–27.

Payseur, B. A., Krenz, J. G., and Nachman, M. W. (2004). Differential patterns of introgression across the X chromosome in a hybrid zone between two species of house mice. *Evolution*, **58**, 2064–78.

Pennings, P. S. and Hermisson, J. (2006). Soft sweeps III: the signature of positive selection from recurrent mutation. *PLoS Genetics*, **2**, e186.

Pocock, M. J. O., Hauffe, H. C., and Searle, J. B. (2005). Dispersal in house mice. *Biological Journal of the Linnean Society*, **84**, 565–83.

Prescott, C. V. (1996). A preliminary study of the genetics of resistance in house mice. In *Proceedings of the 17th Vertebrate Pest Conference*, ed. R. M. Timm and A. C. Crabb. Davis, CA: University of California, pp. 83–7.

Przeworski, M. (2002). The signature of positive selection at randomly chosen loci. *Genetics*, **160**, 1179–89.

Przeworski, M., Coop, G., and Wall, J. (2005). The signature of positive selection on standing genetic variation. *Evolution*, **59**, 2312–23.

Rost, S., Pelz, H. J., Menzel, S., *et al.* (2009). Novel mutations in the VKORC I gene of wild rats and mice: a response to 50 years of selection pressure by warfarin. *BMC Genetics*, **10**, 4.

Sabeti, P. C., Varilly, P., Fry, B., *et al.* (2007). Genome-wide detection and characterization of positive selection in human populations. *Nature*, **449**, 913–18.

Schlötterer, C. (2002). A microsatellite-based multilocus screen for the identification of local selective sweeps. *Genetics*, **160**, 753–63.

Schlötterer, C. (2003). Hitchhiking mapping: functional genomics from the population genetics perspective. *Trends in Genetics*, **19**, 32–8.

She, J. X., Bonhomme, F., Boursot, P., Thaler, L., and Catzeflis, F. (1990). Molecular phylogenies in the genus *Mus*: comparative analysis of electrophoretic, scnDNA hybridization, and mtDNA RFLP data. *Biological Journal of the Linnean Society*, **41**, 83–103.

Siepel, A., Bejerano, G., Pedersen, J. S., *et al.* (2005). Evolutionarily conserved elements in vertebrate, insect, worm, and yeast genomes. *Genome Research*, **15**, 1034–50.

Slatkin, M. (1995). Hitchhiking and associative overdominance at a microsatellite locus. *Molecular Biology and Evolution*, **12**, 473–80.

Smith, N. G. C. and Eyre-Walker, A. (2002). Adaptive protein evolution in *Drosophila*. *Nature*, **415**, 1022–4.

Southwick, C. H. (1955). Regulatory mechanisms of house mouse populations: social behavior affecting litter survival. *Ecology*, **36**, 627–34.

Staubach, F., Teschke, M., Voolstra, C. R., Wolf, J. B., and Tautz, D. (2010). A test of the neutral model of expression change in natural populations of house mouse subspecies. *Evolution*, **64**, 549–60.

Tautz, D. and Schmid, K. J. (1998). From genes to individuals: developmental genes and the generation of the phenotype. *Philosophical Transactions of the Royal Society B: Biological Sciences*, **353**, 231–40.

Teeter, K. C., Thibodeau, L. M., Gompert, Z., *et al.* (2010). The variable genomic architecture of isolation between hybridizing species of house mice. *Evolution*, **64**, 472–85.

Teschke, M., Mukabayire, O., Wiehe, T., and Tautz, D. (2008). Identification of selective sweeps in closely related populations of the house mouse based on microsatellite scans. *Genetics*, **180**, 1537–45.

Teshima, K. M., Coop, G., and Przeworski, M. (2006). How reliable are empirical genomic scans for selective sweeps? *Genome Research*, **16**, 702–12.

Thomas, M., Möller, F., Wiehe, T., and Tautz, D. (2007). A pooling approach to detect signatures of selective sweeps in genome scans using microsatellites. *Molecular Ecology Notes*, **7**, 400–3.

Thornton, K. R., Jensen, J. D., Becquet, C., and Andolfatto, P. (2007). Progress and prospects in mapping recent selection in the genome. *Heredity*, **98**, 340–8.

Wanless, R. M., Angel, A., Cuthbert, R. J., Hilton, G. M., and Ryan, P. G. (2007). Can predation by invasive mice drive seabird extinctions? *Biology Letters*, **3**, 241–4.

Wiehe, T. (1998). The effect of selective sweeps on the variance of the allele distribution of a linked multiallele locus: hitchhiking of microsatellites. *Theoretical Population Biology*, **53**, 272–83.

Wiehe, T., Nolte, V., Zivkovic, D., and Schlötterer, C. (2007). Identification of selective sweeps using a dynamically adjusted number of linked microsatellites. *Genetics*, **175**, 207–18.

Wolfe, K. H., Sharp, P. M., and Li, W. H. (1989). Mutation rates differ among regions of the mammalian genome. *Nature*, **337**, 283–5.

Wright, S. I. and Andolfatto, P. (2008). The impact of natural selection on the genome: emerging patterns in *Drosophila* and *Arabidopsis*. *Annual Review of Ecology, Evolution, and Systematics*, **39**, 193–213.

Yang, Z. and Nielsen, R. (2008). Mutation-selection models of codon substitution and their use to estimate selective strengths on codon usage. *Molecular Biology and Evolution*, **25**, 568–79.

14

What can the *Mus musculus musculus/M. m. domesticus* hybrid zone tell us about speciation?

STUART J. E. BAIRD AND MILOŠ MACHOLÁN

Introduction

One of the crucial tasks in the study of speciation is to explain the genetic basis of reproductive barriers between diverged taxa. Despite the best endeavours of generations of evolutionary biologists since Darwin and Wallace, our understanding of the genetics of speciation has lagged behind other fields of evolutionary biology. The reason may be simply that the greater the reproductive isolation, the less cross progeny we have to work with, but more probably it is because the genetic factors involved are numerous, with complex interactions. Indeed, it appears that except for a few cases (Coyne and Orr, 2004) reproductive isolation arises in many small steps rather than a few strongly selected mutations, in agreement with theoretical predictions (Walsh, 1982; Barton and Charlesworth, 1984). Obviously, the more genes with small effect that are involved in reproductive isolation, the more difficult will be their identification, making traditional laboratory hybridization experiments and association studies challenging.

It seems only common sense, therefore, to search for alternative approaches. One promising alternative is to study naturally occurring hybrid zones. The seeds for progress were sown by Barton and Hewitt (1985, 1989), who suggested the majority of hybrid zones were 'tension zones' (Box 14.1). While initially received with scepticism in some quarters, the tension zone paradigm now dominates the hybrid zone literature and is arguably the main justification for the view that hybrid zones are 'natural laboratories' (Hewitt, 1988) and 'windows on evolutionary process' (Harrison, 1990). The reason this paradigm is so important is that it unifies hybrid zone process over a diversity of species, geographic scales, and histories of contact into a single analytically tractable framework (Box 14.2) with intuitive, testable results and predictions that have been borne out repeatedly over the subsequent decades. Tension zones are measured relative to the scale of

Evolution of the House Mouse, ed. Miloš Macholán, Stuart J. E. Baird, Pavel Munclinger, and Jaroslav Piálek. Published by Cambridge University Press.

Box 14.1 Hybrid zones and the tension zone paradigm

In this chapter we use a consensus definition of hybrid zones as areas where genetically distinct groups of individuals meet, mate, and leave at least some offspring of mixed ancestry (Barton and Hewitt, 1985, 1989; Harrison, 1990). An advantage of this rather broad notion is its independence of our knowledge of the history or geography of hybrid zones and the evolutionary forces affecting their dynamics. Hybrid zones can be maintained by various kinds of selection – for example, by balancing selection favouring hybrids within a narrow region of intermediate habitat ('bounded hybrid superiority'; Moore, 1977). In this case, a trait at equilibrium may vary gradually from place to place, independent of dispersal. However, as pointed out by Barton and Gale (1993), dispersal has a negligible effect only when character gradients (clines) are much wider than the characteristic scale of selection (i.e. the distance over which selection changes allele frequencies), $l = \frac{\sigma}{\sqrt{s}}$, where σ is the rate of dispersal, defined as the standard deviation of the distance between parent and offspring measured along a linear axis, and s is the magnitude of selection (Slatkin, 1973; Barton and Hewitt, 1989). In fact, clines are usually much narrower than potential environmental gradients and dispersal-independent zones are thus rare. Moreover, most hybrid zones consist of a cluster of coincident clines even for characters with no obvious functional relationship (Barton and Hewitt, 1985, 1989) and with similar width and shape across different portions of the zone, patterns that are unlikely for clines maintained directly by the external environment. Thus it appears that *most hybrid zones are maintained by a balance between dispersal and selection*. This selection can be exogenous (so that alleles are selected against when appearing in foreign habitat) or endogenous (so that alleles are counter-selected when appearing on foreign genetic background). If different alleles are favoured in different habitats, the hybrid zone will be located at a particular environmental gradient, whereas if selection acts against hybrids, the zone is free to move and will stop when trapped by a geographic barrier or in an area of low population density (a 'density trough'; Hewitt, 1975, 1989; Barton, 1979; Barton and Hewitt, 1985). These zones, maintained by a balance between dispersal from each side and central endogenous selection, are called 'tension zones' because, like the surface of a bubble, they minimize their extent while balancing pressure on either side (Key, 1968; Barton, 1979; Barton and Hewitt, 1985) (Fig. 14.1).

dispersal of the organism involved, and on this metric the stronger the selection against admixture the more abrupt are the changes between the taxa (clines) we see as we cross the hybrid zone. This simple contrast (strong selection = narrow cline; weak selection = wide cline) holds even when comparing snail hybrid zones to bird hybrid zones – as long as we measure width relative to their respective dispersal scales – but comparisons across clines at loci *within the same genome* are even more robust. The approach of plotting cline width for a number of loci along a linkage group to see where clines are narrow, and therefore to map genes responsible for reproductive barriers seems obvious given current technology,

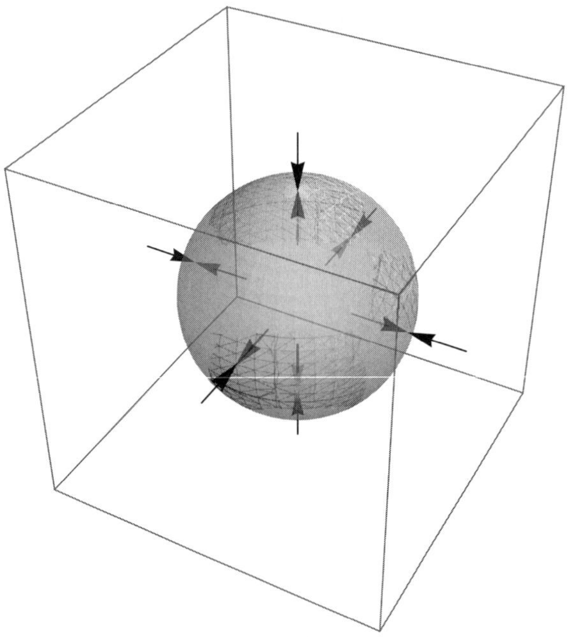

Figure 14.1 When we blow a soap bubble its shape wobbles, then settles into a sphere which balances the internal and external air pressure while minimizing the extent of the soap boundary (the sphere is the shape with minimum surface area for the enclosed volume). Tension zones have similar behaviour: wobbling over the landscape they settle where the pressure of gene flow is equal from either side and the extent of the zone is minimized, smoothing the path of the zone centre towards a straight line (the shortest distance between two points). See the plate section for a colour version of this figure.

but in the 1990s when this was first discussed, multi-locus studies focused on a few unlinked loci, and it took more than a decade to bring the idea to fruition. One of the ground-breaking studies was done on the house mouse hybrid zone (HMHZ) in Europe (Payseur *et al.*, 2004), which suggested a reproductive barrier region on the central X chromosome. In the same year laboratory crosses mapped sterility genes to the same chromosomal region (Oka *et al.*, 2004; Storchová *et al.*, 2004; see Forejt *et al.*, Chapter 19 in this volume, for review). These results demonstrate the efficacy of hybrid zones as tools for speciation studies, complementary to laboratory crosses and positional cloning approaches, as well as the paramount role the house mouse can play in this research.

Despite this positive outlook we suggest caution in hybrid zone speciation research. The intuitive nature of the idea to use narrow clines to locate barriers to gene flow has led to overconfidence in its application. Just because hybrids are sampled in the field, rather than bred in the laboratory, does not mean strict rigour is unnecessary. In fact, *more* rigour is needed if we hope to distinguish signals in the data from stochastic noise; laboratory-based research minimizes this noise by controlling conditions, but when sampling from nature we cannot control conditions and so we must be particularly vigilant to sources of noise (Boxes 14.3, 14.4). Ignoring such sources of uncertainty leads to overconfident inference, and overconfident inference could discredit hybrid zone speciation research before it has shown its full potential. Under these circumstances our chapter title might also have been 'What can the HMHZ *not* tell us about speciation?'. Knowing the limits of what we can infer is central to the scientific endeavour. The better our understanding of (1) the circumstances from which we are sampling in nature; (2) the assumptions underlying our models of those circumstances; and (3) the stochastic effects introduced by sampling and the limitations of our analytical techniques, the better we will be able to modulate and communicate our confidence in the results. Here we review each of these issues in turn with respect to inference about speciation from the HMHZ in an attempt to set a solid foundation for future research. We finish by suggesting promising future directions.

The house mouse hybrid zone in Europe

There is wide agreement that no extant house mice occupied western Eurasia before the end of the last glaciation (Thaler, 1986; Auffray *et al.*, 1990; Auffray and Britton-Davidian, 1992; Cucchi *et al.*, Chapter 3 in this volume; but see Sage *et al.*, 1990 for a different view). The mouse expansion into western Eurasia followed two colonization routes consistent with human historical patterns (Kratochvíl, 1986; Thaler, 1986; Auffray *et al.*, 1990; Cucchi *et al.*, 2005; Bonhomme *et al.*, 2011). *Mus musculus domesticus* arrived from Asia Minor, along the Mediterranean Basin, and now occurs in southern and western Europe; *M. m. musculus* followed a pathway north of the Black Sea and today occupies northern and eastern parts of the continent. Where their ranges abut, a narrow hybrid zone is created, running across the central part of the Jutland Peninsula (approximately from Vejle Fjord westwards) and from the Baltic coast *ca.* 25 km east of Kiel Fjord (East Holstein, northern Germany) through central Europe and the Balkan Peninsula to the Black Sea coast (Sage *et al.*, 1993; Macholán *et al.*, 2003; Fig. 14.2). Recently, a new portion of the zone has been localized in Norway (Jones *et al.*, 2010).

Following the pioneering works of Degerbøl (1949), Zimmermann (1949), and Ursin (1952), this zone has been studied in Denmark (Selander *et al.*, 1969; Hunt and Selander, 1973; Schnell and Selander, 1981; Ferris *et al.*, 1983; Vanlerberghe *et al.*, 1986, 1988b; Nancé *et al.*, 1990; Dod *et al.*, 1993, 2005; Fel-Clair *et al.*, 1996, 1998; Lanneluc *et al.*, 2004; Raufaste *et al.*, 2005); northern Germany (van Zegeren and van Oortmerssen, 1981; Prager *et al.*, 1993); Saxony, eastern

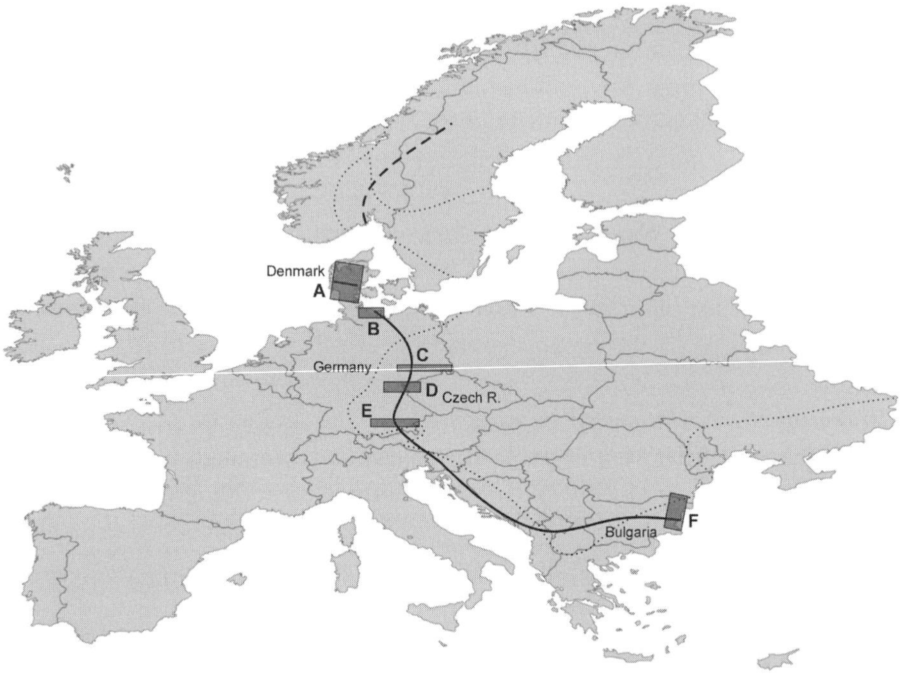

Figure 14.2 The course of the *M. m. musculus/M. m. domesticus* hybrid zone in Europe (bold line). In Norway, its position is only tentative (dashed bold line). Shaded rectangles depict studied transects: A: Denmark; B: East Holstein (Germany); C: Saxony, Saxony-Anhalt, and Thuringia (Germany); D: west Bohemia (Czech Republic) and northeast Bavaria (Germany); E: southeast Bavaria (Germany and northwest Austria); F: east Bulgaria. The dotted line indicates where the eastern humid continental climate zone meets any of four other climate zones: anticlockwise from top these are subarctic (northern Scandinavia); humid oceanic (west); humid and dry subtropical (south); and semi arid (in the east, north of the Black Sea). For clarity the boundary between the western and southern zones is omitted, likewise the highland climate zone of the Alps (simplified from http://printable-maps.blogspot.com/2008/09/map-of-climate-zones-in-europe.html).

Germany (Teeter *et al.*, 2010); southern Bavaria (south-eastern Germany) and northwestern Austria (Sage *et al.*, 1986b; Tucker *et al.*, 1992; Payseur *et al.*, 2004; Payseur and Nachman, 2005; Teeter *et al.*, 2008, 2010; Dufková *et al.*, 2011); northeastern Bavaria and the western part of the Czech Republic (Munclinger *et al.*, 2002; Božíková *et al.*, 2005; Macholán *et al.*, 2007, 2008, 2011); and eastern Bulgaria (Vanlerberghe *et al.*, 1986, 1988a).

Despite the common belief that the *musculus/domesticus* zone has resulted from the secondary contact of the two taxa in the Holocene, the history of initial contact, including exact dating, is unclear. Almost certainly first contact was in

Box 14.2 Tension zone models

Sigmoid clines

The simplest model of a tension zone was proposed by Bazykin (1969). It assumes a homogeneous environment, heterozygote fitness to be $(1 - s)$ relative to both homozygotes, weak selection acting on each locus independently (no cross-locus associations, i.e. linkage disequilibrium), and gene flow approximated by diffusion. The allele frequency is then $p = 1 / (1 + \exp[-(x - c) / l])$, where $(x - c)$ is the distance of a site from the cline centre and l is the scale of selection defined in Box 14.1 (Slatkin, 1973; Barton and Hewitt, 1989) (Fig. 14.3a, solid line). The width of a cline in trait z is usually defined as $w = \Delta z / (\partial z / \partial x)$: this is the inverse of the maximum slope $\partial z/\partial x$ scaled by the total trait change Δz. This definition applies equally to clines in the frequency of diagnostic alleles from 0 to 1 ($\Delta z = 1$), clines in allele frequency at informative loci ($0 < \Delta z < 1$), and clines in quantitative traits where Δz depends only on what is measured. It also allows explicit theoretical predictions impossible with other ad hoc width definitions such as the distance over which z changes from 20% to 80% (Endler, 1977; Barton and Hewitt, 1989). The cline width (w) depends on the selection acting: for example, when selection acts against heterozygotes it is $\sqrt{8\sigma^2 s}$ (Bazykin, 1969; Barton and Hewitt, 1989), whereas when selection favours different alleles on different sides of the hybrid zone the width equals $\sqrt{3\sigma^2 s}$ (Haldane, 1948; if we assume no dominance the cline is slightly wider: $w = 1.782\sqrt{\sigma^2 s}$; Barton and Gale, 1993).

There are several ways of expressing the shape of the single locus clines described by Bazykin (1969), and this has led to some confusion. Bazykin chose to use the hyperbolic tangent function (Tanh), giving a smooth sigmoid curve (or a straight line if plotted on a logit scale). The Tanh cline is *sigmoid* in the sense that it is equivalent to a logistic (sigmoid) curve $[(\tanh(x) + 1) / 2 = \text{logistic}(2x)]$, and this is also why it is a straight line if plotted on a logit scale – the inverse of the logistic function is the logit function. The type of selection has little effect on cline shape, so this model can be used for clines caused by heterozygote disadvantage, extrinsic selection favouring different alleles in different places, or selection acting on quantitative traits (Haldane, 1948; Fisher, 1950; Bazykin, 1969; Slatkin, 1973, 1975; Nagylaki, 1975, 1976; Endler, 1977). If selection is weak the model also approximates clines in structured populations described by the stepping-stone model (Nagylaki, 1975). However, Bazykin's model is a single locus model, while selection acting in a hybrid zone typically affects several or many loci.

Stepped clines

Influx of parental gene combinations into a tension zone causes strong associations between the loci or linkage disequilibrium, D (Li and Nei, 1974; Slatkin, 1975). Since D is proportional to the gradient of allele frequencies, across-locus associations are weak at the edges of the zone, and selection affects each locus separately. However, as we move towards the zone centre, the gradient steepens, causing stronger disequilibria. As associations across loci become stronger, selection no longer affects each locus independently. Selection is strengthened by a 'hitch-hiking' effect: stronger

Box 14.2 (*cont.*)

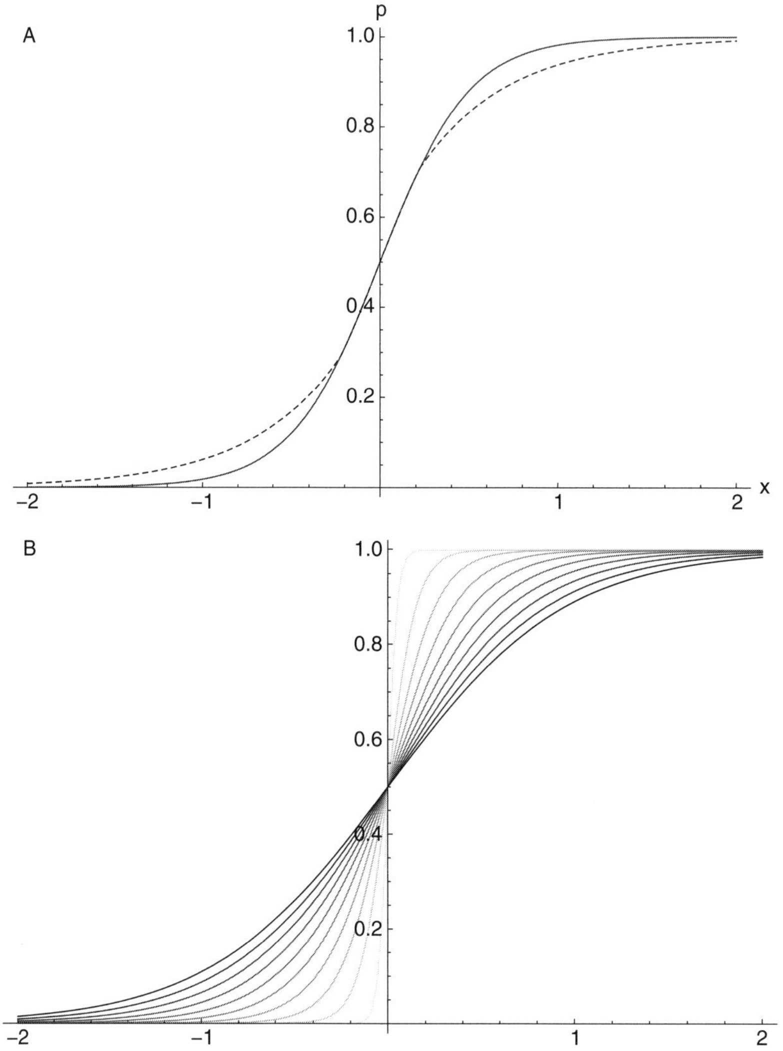

Figure 14.3 (a) Solid: sigmoid cline; dashed: stepped cline with the same centre and width, but more shallow tails of introgression. (b) Progress over time from steep clines at initial contact (light grey) to quasi-equilibrium width (dark grey).

selection generates stronger associations which, in turn, cause yet stronger selection. The result of these associations is a barrier to the (independent) flow of genes across the zone centre, and is manifested as a sharp step at the centre of each cline (Barton, 1983; Barton and Bengtsson, 1986; Barton and Gale, 1993). Barton (1986) models clines with a central barrier to gene flow as tripartite: allele frequencies in the central portion follow the sigmoid model, but on either side (away from the influence of central associations) the

Box 14.2 (*cont.*)

cline decays exponentially as $p \propto \exp(4x\sqrt{\theta}/w)$, where θ is the rate of decay on the right and left side, respectively (Fig. 14.3a, dashed line). Parameter θ can be expressed in terms of the ratio between the selection acting on an individual locus itself (s_e) and the effective selection pressure on the locus at the centre, which is primarily due to association with other loci (s^*): $\theta = s_e / s^*$ (Szymura and Barton, 1986). A stepped cline pattern suggests a zone with a central barrier to gene flow, regardless of whether this is caused by a physical obstacle or by associations with other loci. The strength of the barrier is defined as $B = \Delta p / p'$, where Δp is the height of the central step and $p' = \partial p / \partial x$ is the gradient of allele frequency on either side of this step. B can be estimated separately for the left and right side of the cline (Nagylaki, 1976) so the asymmetrical stepped model comprises six parameters: w, c, θ_L, B_L, θ_L, B_L, where subscripts L and R denote left and right side, respectively. The barrier strength has units of distance, and can be thought of as the distance over which unimpeded gene flow would lead to the same change in allele frequencies. Barrier strength may be different in each direction, leading to asymmetric clines that will have a tendency to move.

Another way of assessing the magnitude of the barrier is to calculate the number of generations that an allele is delayed when crossing the zone. For a neutral allele, this delay can be substantial, from hundreds to tens of thousands of generations ($T \approx (B/\sigma)^2$; Barton and Hewitt, 1985, 1989; Barton and Gale, 1993). In contrast, even slightly advantageous alleles can cross the zone quite rapidly, at $T \approx \log[(B/\sigma)^2 \pi s_a / 2] / 2s_a$ generations if $s_a \gg (B/\sigma)^{-2}$, where s_a is the selective advantage (Barton and Hewitt, 1985, 1989; Barton and Bengtsson, 1986; Barton and Gale, 1993).

Because the central barrier in a tension zone depends so crucially on the association between loci, estimation of D can be used, along with cline shape parameters, for subsequent estimation of the scale of dispersal, effective selection on marker loci, selection on selected loci, the total number of genes under selection, and the mean fitness of hybrids (see Appendix).

Dynamics of secondary contact

The models above are for clines that have reached a quasi-equilibrium balance between dispersal of individuals into the centre of the zone and selection against admixture. However, at the beginning of secondary contact, clines are step-like transitions: it takes some generations for them to settle into their equilibrium shape (Fig. 14.3). At the genomic level progression towards equilibrium depends on the ratio of selection to recombination (or segregation for loci on different chromosomes), and may be very slow leaving intact blocks of the genome of each taxon long after secondary contact, a signature that can be used to date this contact (Baird, 1995, 2006).

southern regions, progressing to central and northern Europe, like a zipper being pulled up through the continent. Therefore, the northern portion of the hybrid zone may be much younger than the southern parts. According to Boursot *et al.* (1993), while mice arrived in the western Mediterranean and

Box 14.3 Spatial sampling noise

When we calculate the likelihood of a set of parameters describing cline shape and make inferences about selection affecting individual loci, several sources of error should be taken into account. These are perhaps especially strong for the house mouse, which has an aggregated spatial distribution and usually lives in small and defended demes in which most offspring are sired by a dominant male. The lifespan of these demes is rather short, adding a further source of stochasticity to the distribution of allele frequencies across localities. Calculating the effective sample size (Phillips *et al.*, 2004; Raufaste *et al.*, 2005; Macholán *et al.*, 2008) takes into account the sampling noise introduced by small samples with deviations from Hardy Weinberg equilibrium within each sampling locality and differences in relatedness across localities, but does not take into account the noise introduced by how the sampled localities fall in relation to the zone centre. This noise is particularly strong when a one-dimensional (1D) transect is used to sample a 2D hybrid zone (Fig. 14.4).

One component of spatial sampling error has recently been more rigorously assessed for the Czech–Bavarian portion of the HMHZ. The underlying orientation of this zone has been estimated with good support (Macholán *et al.*, 2008, 2011). For this known orientation Dufková *et al.* (2011) used a jackknife-based procedure in which one locality at a time was removed from the dataset and cline estimates re-computed for each locus. They showed that exclusion of a single locality can result in increase of cline width estimates across loci up to 84-fold, considerably higher than the 50-fold inter-locus span found in the 39 SNP dataset from 1D sampling in southern Germany (Teeter *et al.*, 2008). When clines are stepped, exclusion of central localities has a more severe impact on cline width estimates than omitting sites far from the centre, whereas for sigmoid clines the impact on the cline width is greater if localities further from the centre are deleted. Dense sampling of central localities is therefore vital for reliable estimates of introgression patterns and inferences of genomic regions important in speciation (Barton and Gale, 1993; Macholán *et al.*, 2007; Dufková *et al.*, 2011). When sampling from the central segment of the zone is sparse relative to cline width, only upper bounds can be placed on the width estimates (Macholán *et al.*, 2007).

central Europe during the Bronze Age (4000–2800 BP; Auffray *et al.*, 1990; Vigne, 1992; Auffray, 1993), northwestern and northern Europe were colonized during the Iron Age (*ca.* 2800 BP; Lepiksaar, 1980; Auffray *et al.*, 1990; O'Connor, 1992; Auffray, 1993). Sage *et al.* (1993) supposed 6000 BP for southern Europe and Gyllensten and Wilson (1987) 5000–6000 BP for central Europe. Estimates of the time of colonization of Scandinavia are even more variable: Prager *et al.* (1993) assumed spread of mice following human demic diffusion 4000–5000 BP (Sokal *et al.*, 1991), whereas Gyllensten and Wilson (1987) concurred with Clark (1975), who dated the advent of agriculture to 3500–4500 BP. On the other hand, a recent careful revision of zooarchaeological data

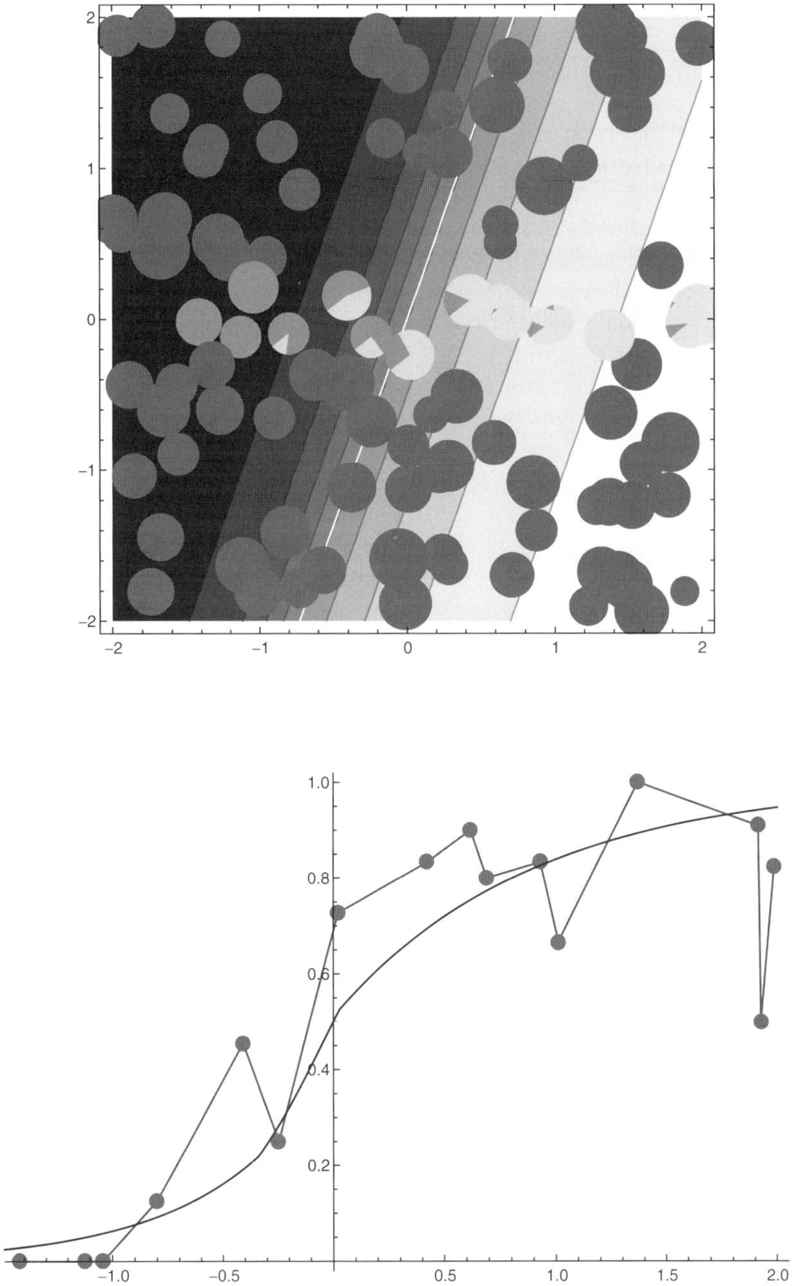

Figure 14.4 Spatial sampling noise. One hundred sampling localities are distributed at random across a hybrid zone of unknown orientation. (a) Assuming effective sample sizes are Poisson distributed with mean 10, pies show binomial allele frequency samples from an underlying cline (shades of grey). Pies close to a horizontal transect line are shaded in orange and cyan (see colour plate of this figure). (b) The shape of the underlying cline is compared to allele frequency estimates from the 14 locality samples close to the horizontal transect line: cline estimation would either displace the centre to the left, or overestimate the steepness of the cline. Without sampling noise there would be no displacement, and an underestimation of the cline steepness (because the horizontal transect is oblique to the path of the zone). See plate section for a colour version of this figure.

Box 14.4 Model choice in hybrid zone analysis

Suppose we choose to analyse hybrid zone samples using a two-parameter sigmoid cline model (Box 14.2), but we later find out that the hybrid zone is a multi-locus tension zone with stepped clines (Box 14.2) due to an asymmetric central barrier to gene flow. What are the consequences of having fitted the wrong model? In particular, what are the consequences of fitting a symmetric sigmoid curve model to an asymmetric cline (Figure 14.6a)? As the likelihood fitting algorithm tries to match the asymmetric data with the symmetric model it shifts the model to one side. The more asymmetric the data, the more the cline centre *estimate* diverges from *the real position of the cline centre* (Fig. 14.6b). This is not a bug in the likelihood approach. Rather, it is exactly what we expect to happen if the inappropriate model is chosen for analysis. The consequences for inference are grave. In the data we fully expect clines at different loci to have different degrees of asymmetry. Analyses of these with symmetric models will shift the centre estimates of each to a different degree, giving the impression that the zone has non-coincident cline centres. We might then mis-infer the very nature of the zone because 'coincident cline centres' is one of the most fundamental expectations of a zone with a central barrier to gene flow. It is for such reasons that we have repeatedly stressed the importance of the model-choice stage of cline analysis (Macholán *et al.*, 2007, 2008; Vošlajerová Bímová *et al.*, 2011). Model choice is an automatic feature of, for example, phylogenetic sequence analyses, and it is a matter of some frustration that the issue is still often ignored in the hybrid zone literature.

has suggested that western and northern Europe had not been colonized by mice before 1000 BC–AD 300 (Cucchi *et al.*, 2005, see also Cucchi *et al.*, Chapter 3). The most extreme of recent estimates are given by Sage *et al.* (1993), assuming an arrival of mice in Denmark 250 years ago, and Hunt and Selander (1973), who pointed out that stable mouse populations in western Jutland could not have been established before the beginning of the programme of reclamation of marshlands in 1850.

Why is the age of the hybrid zone so important? A hybrid zone will get wider over time unless some force acts to maintain the distinction between taxa, in which case the zone will settle into quasi-equilibrium (Box 14.2). Assuming cline widths ~10–20 km and a scale of dispersal ~1 km $gen^{-1/2}$ (see below), a neutral allele will cross the zone after 16–32 generations. Because the zone, at all points, is thought to be much older than this, we can assume it has settled into quasi-equilibrium all along its length. Current estimates suggest the width of the HMHZ is very similar in Jutland and some 700 km to the south in the Czech Republic (Macholán *et al.*, 2007). This similarity in the width of the hybrid zone despite different times of contact, and diverse locations, argues for a similar mechanism maintaining taxon separation whenever and wherever *M. m. musculus* and *M. m. domesticus* meet.

Is the zone maintained by exogenous or endogenous factors?

The European HMHZ is a ~20-km wide complex gradation of late-generation hybrids and backcrosses with F_1 hybrids missing or extremely rare; for most markers the zone structure is unimodal with intermediate genotypes in the centre (Raufaste *et al.*, 2005; Macholán *et al.*, 2007). There has been some debate regarding the origins of the factors that maintain the zone. A seeming coincidence of its position and the boundary between oceanic and continental climate zones led several authors to conclude that the contact of these subspecies is regulated primarily by climatic factors (Zimmermann, 1949; Serafiński, 1965; Hunt and Selander, 1973; Thaler *et al.*, 1981; Boursot *et al.*, 1984; Klein *et al.*, 1987). However, recalling the concept of the characteristic scale of selection (Box 14.2), the scale of this climatic gradient is too wide to maintain such a narrow hybrid zone (Boursot *et al.*, 1993). Further, as human commensals, mice mostly live in artificial habitats buffered from climate influences, and there is an obvious discrepancy between the position of the hybrid zone and the climatic transition, a fact pointed out by Kraft (1985). As shown in Fig. 14.2, the HMHZ does not match the climate transition closely anywhere in Europe, suggesting climate is not an important factor in either its position or width.

Barton and Hewitt's (1985) suggestion that the HMHZ is a tension zone, maintained by a balance between dispersal and endogenous selection against hybrids, has received increasing support from genetic studies over geographically independent transects (e.g. Payseur *et al.*, 2004; Raufaste *et al.*, 2005; Macholán *et al.*, 2007). Stepped clines (Box. 14.2) with stronger linkage disequilibria in the centre have been demonstrated in Denmark (Dod *et al.*, 2005; Raufaste *et al.*, 2005), southern Germany (Payseur *et al.*, 2004), and the Czech–Bavarian portion of the zone (Macholán *et al.*, 2007, 2008). Leaving aside other potential causes of the stepped pattern of the molecular clines, such as geographic barriers or epistasis (see Macholán *et al.*, 2007 for discussion), the stepped shape can be produced either by strong selection acting on a small number of loci or weak selection affecting a moderate to large number of loci (cf. Porter *et al.*, 1997; see Walsh, 1982 and Barton and Charlesworth, 1984 for theoretical arguments in favour of the latter possibility). A stepped cline pattern has been confirmed for allozyme markers in the Danish transect, where the number of loci under selection was estimated at 46–120, with effective selection pressure maintaining the cline $s^* \approx 3$–7% and fitness of hybrids ~45% (Raufaste *et al.*, 2005); and in the Czech–Bavarian transect, with 56–99 selected loci,

$s^* \approx 6$–9%, and fitness of hybrids 25–60% (Macholán *et al.*, 2007). An analysis of five X chromosome markers revealed substantially stronger selection ($s^* \approx 25\%$) and a larger reduction in fitness of hybrids than selection affecting autosomal loci ($\approx$ 23–35%; Macholán *et al.*, 2007, 2008). Estimates of the number of selected X-linked loci were rather incongruent: while the original study of Macholán *et al.* (2007) yielded 380 loci, a later analysis of 17 X chromosome markers resulted in a much lower estimate (eight selected loci; M. Macholán, unpublished results). As the first estimate exceeds all estimates of the total number of selected loci in the genome, we favour the second of these calculations. It should be noted, however, that all these calculations assume that epistatic interactions between loci can be neglected. As suggested by several studies, such interactions may occur between X-linked genes (Oka *et al.*, 2004; Storchová *et al.*, 2004) and between these genes and autosomal loci (Forejt, 1981, 1996; Montagutelli *et al.*, 1996; Britton-Davidian *et al.*, 2005; Payseur and Hoekstra, 2005). In addition, the estimates of selection parameters are highly derived: they are calculated using values which themselves are estimates, and so should be treated with appropriate caution.

In summary, accumulation of data and better understanding of hybrid zones mean exogenous hypotheses for maintenance of the HMHZ are now seen as unlikely. At the same time, this accumulating data consistently matches tension zone predictions suggesting the zone is maintained by selection against hybridization, with causes intrinsic to the mice in contact.

What is the nature of endogenous selection against hybrids?

The genetic analyses of different transects consistently imply that hybrid mice are selectively disadvantaged. But what is the nature of this selection? Strikingly, our knowledge of this issue is still extremely limited. However, in theory we expect sterility selection to be more important than viability selection (Coyne and Orr, 2004), and the results so far are consistent with this idea.

Two potential sources of viability selection against hybrids have been investigated by multiple teams: susceptibility to parasites and developmental instability. Sage *et al.* (1986a), Moulia *et al.* (1991), and Moulia and Joly (2009) suggest increased parasite load is one of the only indications of less fit hybrids, but a much larger recent study (Baird *et al.*, 2012; see also Goüy de Bellocq *et al.*, Chapter 18) not only indicates that this pattern is inconsistent across transects, but also shows strong evidence of *reduced* load in hybrids, consistent with heterosis. Second, decreased fluctuating asymmetry (FA) in natural hybrids captured along the Danish transect, as well as in laboratory-produced hybrids (Alibert *et al.*, 1994, 1997; Debat *et al.*, 2000; Alibert and Auffray, 2003), has been taken as evidence of heterosis. A study of mandible shape in the Czech–Bavarian zone also showed decreased asymmetry in hybrids (Mikula *et al.*, 2010). However, an analysis of FA in the shape of the ventral side of the skull of these mice showed the opposite trend, i.e. higher FA in hybrid populations (Mikula and Macholán, 2008). Thus, the impact of hybridization on FA appears to be trait-specific (Mikula *et al.*, 2010).

Forejt and Iványi (1974) set out to examine sterility, rather than decreased viability, in mice. In their experiment, a wild-derived *M. m. musculus* inbred strain was crossed with the 'classic' laboratory strain C57BL/10 carrying predominantly *M. m. domesticus* alleles, yielding sterile male F_1 hybrids, in agreement with Haldane's rule (Forejt and Iványi, 1974). Moreover, backcross data suggested epistatic interactions with at least two other loci (Forejt, 1981, 1996). This gene, named Hybrid sterility 1 (*Hst1*), was mapped to chromosome 17 (Forejt *et al.*, 1991; Trachtulec *et al.*, 1994; Gregorová *et al.*, 1996), close to the location of the *t*-haplotype (see Herrmann and Bauer, Chapter 12) but not involved with its function. Subsequent analyses led to identification of *Hst1* with the *Prdm9* gene encoding meiotic histone H3 lysine 4-methyltransferase (Mihola *et al.*, 2009). Another one or two sterility genes have been mapped to the *M. m. musculus* X chromosome (*Hstx1*, possibly *Hstx2*; see Forejt *et al.*, Chapter 19 for a review). However, sterility studies of wild or wild-derived inbred mice have revealed a rather complicated picture (Forejt and Iványi, 1974; Britton-Davidian *et al.*, 2005; Vyskočilová *et al.*, 2005, 2009; Good *et al.*, 2008b), suggesting the basis of reproductive isolation in the house mouse is likely to be more complex than can be revealed by the progeny of pairs of laboratory strains.

In summary, given evidence consistent with heterosis (or at least no unequivocal evidence to the contrary) for viability traits, it seems likely that selection against hybridization takes the form of sterility/fertility selection rather than viability selection. Recent studies of sperm morphology and motility and histology of seminiferous tubules across the Czech–Bavarian region of the HMHZ, showing reduced motility of sperm from hybrid males (Albrechtová *et al.*, submitted) and slightly decreased number of round spermatids relative to primary spermatocytes (G. Pražanová and M. Macholán, unpublished data), are consistent with this view.

What is the scale of dispersal and population density of mice?

Since tension zones are maintained by a balance between selection and gene flow, reliable estimates of the scale of dispersal are critical in calculations of absolute (as opposed to relative) zone measures such as effective selection or number of genes implicated (see Appendix). Assuming standardized linkage equilibrium at the zone centre $R_{ij} \approx 0.058$ and a harmonic mean recombination rate $\bar{r} \approx 0.4$ (Macholán *et al.*, 2007), we get a point estimate of the scale of dispersal for the Czech–Bavarian transect $\sigma \approx 1.24\,\text{km gen}^{-1/2}$. This is somewhat higher than the point estimate reported by Raufaste *et al.* (2005) for Denmark ($\sigma \approx 0.7\,\text{km gen}^{-1/2}$), but given the large degree of uncertainty we must associate with such point estimates, they are in fact reassuringly close, and consistent with a dispersal scale of around $1\,\text{km gen}^{-1/2}$. This seems to be at odds with direct observations of mouse movement to the order of several tens of metres (Pocock *et al.*, 2005). However, direct estimates of dispersal will tend to miss long-distance dispersal events that can, along with frequent extinctions and recolonizations, greatly increase effective dispersal (Barton and Hewitt, 1985). Mouse dispersal events exceeding 1000 m have been reported by several authors (see Sage, 1981 for a review) and thus direct estimates of mouse dispersal are likely to be underestimates of the effective dispersal. As with the difference between census population size and effective population size, it is the effective dispersal scale that is the more relevant for understanding the evolutionary process, and it is this we estimate from cline widths. Typically, σ is estimated by contrasting the centre and tails of a consensus stepped multi-locus cline based on genotype data combined from several loci. Displacement of one or more loci relative to the others will result in an overestimate of cline width. Likewise, overestimated width can result from incorrect orientation of transect direction across the zone (Box 14.3). Conversely, if the central step is not just due to associations across loci, but is reinforced by a physical barrier, dispersal will be underestimated.

In terms of evolutionary process, the effective scale of dispersal is inextricably bound up with the effective population density, their product dictating Wright's (1943) *neighbourhood size* $\mathcal{N} = 4\pi\rho\sigma^2$, where ρ is the effective population density. This is the inverse of the probability that two individuals sampled from a locality had the same ancestor in the previous generation, assuming Gaussian dispersal, and is a factor that repeatedly appears in spatial genetic analysis (Barton and Wilson, 1995; Barton *et al.*, 2002; Baird and Santos, 2010). Neighbourhood size should have a central place in our understanding of tension zones because when $\mathcal{N}$ is small drift is strong, and this can affect the width of clines (Polechová and Barton, 2011). When we say tension zones will move towards density troughs or barriers to gene flow, it might be simpler to say they move towards regions of low neighbourhood size (or high drift). It is startling therefore to realize that we have very little information about the effective density of house mouse populations. While Macholán *et al.* (2007) estimated census mouse density at ~$1.69/\text{km}^2$, sampling effort has rarely been recorded in mouse surveys, and so, regarding further direct estimates, all we can say is that there are no data suggesting reduction of mouse population densities at the centre of the HMHZ. The neighbourhood size–drift relationship can, however, be used to give an indirect estimate of house mouse effective density. Applying the Rousset approach for isolation by distance (Rousset, 1997, 2000) to microsatellite data from localities in the centre of the Czech–Bavarian zone yields estimates of ρ between 0.69/km^2 (for F_{ST} data) and 0.89/km^2 (for Rousset's coefficient).

In summary, as human commensals, house mice have a clustered distribution in space, being concentrated predominantly in places providing shelter and food. While tension zone analyses can provide us with estimates of the effective scale of dispersal and effective density sufficiently reliable to justify calculation of point estimates for absolute measures regarding the zone, these point estimates should always be associated with a large degree of uncertainty. Further, the path of the tension zone across Europe and its width from place to place will be influenced by local details of mouse distribution and barriers to movement.

What are the influences of local geography?

Theory predicts that tension zones should follow straight lines through homogeneous environments (Box 14.1). Neither Europe nor the habitats preferred by house mice are homogeneous. The Jutland Peninsula is by its nature a narrow corridor oriented north–south. We might assume any transect across the hybrid zone should follow the same axis. However, as shown by Hunt and Selander (1973), the Danish zone centre line is rotated clockwise and the zone is wider in the western part of the Jutland Peninsula than on the eastern side. Thus, either the peninsula is not a narrow corridor from the point of view of mouse dispersal or there is another process affecting the local zone orientation. Either way, extrapolating the local orientation of the zone from the global geography of the Jutland Peninsula would have introduced an error from the real orientation of between 10° and 20°. It is not difficult to show that this error would result in biased estimates of cline position and shape (Box 14.3). Similar biases will result from guessing the contact zone path anywhere else in Europe, and thus a correct course of a zone centre in the 2D field space should be precisely inferred prior to any cline analyses.

The centre of the HMHZ in Jutland coincides with a river (the Omme; Ursin, 1952; S. J. E. Baird, unpublished results). In southern Germany most of the change in gene frequencies occurs across the floodplain of the Isar River (Sage *et al.*, 1986b). In northern Germany this is true of the Elbe (Zimmermann, 1949), and in Bulgaria the Kamchiya River (Vanlerberghe *et al.*, 1988a). These observations could be coincidental, or a product of the human mind's tendency to see patterns where there are none. We therefore grade possible implications, starting with the most cautious. First, mice are small and warm blooded, and so they may avoid crossing water, a good conductor of heat. Therefore, it seems a reasonable working hypothesis that rivers act as barriers to gene flow for mice, so if one was setting out to map a new stretch of the centre of the HMHZ, one would likely profit by focusing one's efforts around rivers. Second, if rivers are barriers to mouse gene flow, then this will be true not just along the centre of the HMHZ, but wherever there are rivers. A fine-scale study of mouse genetic variation on the Jutland Peninsula suggested that the central Omme River is not the only river on the peninsula across which there are sharp changes in allele frequencies (S. J. E. Baird, unpublished results). The Czech–Bavarian study area is in a corridor between flanking hills. These flanking hills likely form a sufficient barrier to gene flow/mouse density trough to trap the tension zone at this location without the need for any special habitat features in the contact corridor itself. The region can, however, be divided into three strips crossing the contact zone, roughly reflecting different geographic features within it (forests, valleys with steep slopes, and a central water reservoir in the north; the Eger River separates the central and southern part; Macholán *et al.*, 2007). When analysed separately, the width estimated from a compound cline, composed of 17 X-linked markers, was almost two-fold in the southern and central sections (16.4 km and 13.9 km, respectively) when compared with the northern (water reservoir) section (7.3 km), in spite of being separated by less than 10–20 km (M. Macholán, unpublished results).

In summary, the fact that a tension zone is maintained by endogenous selection does not mean the environment has no influence. It is expected that the HMHZ has moved to fall along density troughs (e.g. the hills of the Czech–Germany border with little human population) and barriers to mouse dispersal (e.g. rivers). The estimates of zone characteristics will not be independent of these local circumstances.

What is the influence of prezygotic barriers in the HMHZ?

It has long been suggested that when the postzygotic barrier between hybridizing taxa is incomplete, selection will favour any changes resulting in avoidance of disadvantageous matings, so that eventually reproductive isolation is reinforced by the emergence of prezygotic barriers, this leading to complete speciation (see Coyne and Orr, 2004 and references therein). The idea became well known due to its inclusion in the seminal works of the architects of the Modern Synthesis (Fisher, 1930; Dobzhansky, 1937, 1940), but it can be traced back to Wallace (1889) and hence it is sometimes called the 'Wallace effect' (though as shown by Coyne and Orr (2004) its current notion is slightly different from Wallace's original version).

It is perhaps for this reason that special attention has been paid to the evolution of assortative mating (e.g. Lande, 1981; Butlin, 1987, 1995; Turelli *et al.*, 2001; Servedio and Noor, 2003; Ritchie, 2007). In the context of recognition between *M. m. musculus* and *M. m. domesticus*, the ability to discriminate and choose males or females of the same subspecies has been demonstrated using several types of olfactory stimuli (see Bímová *et al.*, 2009 and references therein). One of the specific mate recognition systems proposed to play an important role in behavioural premating isolation is based on urinary cues (Smadja and Ganem, 2002, 2005; Smadja *et al.*, 2004; Ganem *et al.*, 2008; Bímová *et al.*, 2009; Vošlajerová Bímová *et al.*, 2011). These studies have consistently shown consubspecific preferences in *M. m. musculus* while *M. m. domesticus* displayed no preference, in agreement with experiments using either bedding (Christophe and Baudoin, 1998; Munclinger and Frynta, 2000; Bímová *et al.*, 2009; Smadja and Ganem, 2002) or individuals of the other sex as signal sources (Smadja and Ganem, 2002). Moreover, Smadja and Ganem (2005) found stronger preferences inside the HMHZ than in allopatric populations (see Ganem, Chapter 15 for a review).

Another proposed specific recognition system is based on salivary signals coded for by genes of the androgen binding protein (*Abp*) family (see Laukaitis and Karn, Chapter 7 for a review). One of these genes, *Abpa27*, was found to possess a different allele fixed in each of the three *M. musculus* subspecies ($Abpa27^a$ in *M. m. domesticus*, $Abpa27^b$ in *M. m. musculus*, and $Abpa27^c$ in *M. m. castaneus*; Karn and Dlouhy, 1991; Laukaitis *et al.*, 2008), which can be viewed as a minimum requirement for the gene to be involved in premating isolation. However, the transition of the *Abpa27* gene across the Danish and Czech–Bavarian hybrid zone does not differ from most allozyme loci assumed to be neutral or nearly neutral (Dod *et al.*, 2005; Vošlajerová Bímová *et al.*, 2011), bringing into question its importance for maintaining the HMHZ.

In order to test explicitly for potential reinforcement of postzygotic barriers by behavioural isolation in the mouse hybrid zone, a model of the transition of quantitative traits under reinforcement selection was developed (Vošlajerová Bímová *et al.*, 2011). If the two subspecies prefer cues originating from their own rather than the other taxon, then we can expect reinforcement to modify a simple sigmoid or stepped cline in preference: preference for the same taxon should be amplified close to the centre. This phenomenon is called reproductive character displacement (Butlin, 1995; Lemmon *et al.*, 2004). A reinforced cline then displays 'pulses' of higher consubspecific preferences at the zone edges, and has a shape similar to self-reinforcing solitary waves, or solitons, known from physics (Bullough and Caudrey, 1980; Lakshmanan, 1988; Barton and Hewitt, 1989; Fig. 14.5). For the Czech–Bavarian transect, reinforcement was strongly supported over all scenarios considered (i.e. symmetric *vs* asymmetric preferences; clines coincident *vs* shifted relative to the molecular consensus centre; same *vs* different levels of reinforcement) (Vošlajerová Bímová *et al.*, 2011).

In summary, it appears that there is some evidence for reinforcement selection acting on the HMHZ, mate preferences being modified in a direction consistent

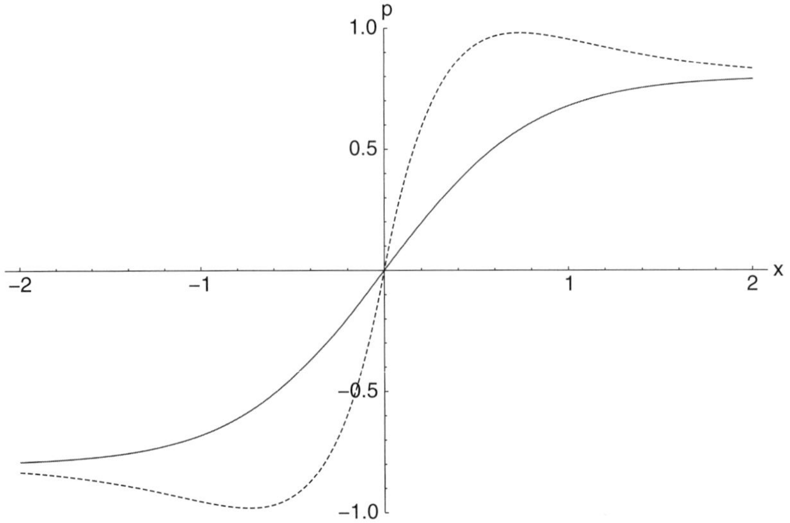

Figure 14.5 An olfactory preference cline modelled as a simple sigmoid curve (solid line). If the two hybridizing taxa prefer cues originating from their own rather than the other taxon, preference is amplified close to the zone centre. A reinforced cline (dashed line) then shows 'pulses' similar to self-reinforcing solitary waves called 'topological solitons' (Barton and Hewitt, 1989).

with avoiding the production of unfit hybrids. Whether such reinforcement selection could ever lead to complete speciation is an entirely different (and unaddressed) question. The signs of reinforcement selection are yet another indirect indication that there is a postzygotic barrier to gene flow.

Use of the tension zone paradigm to infer genomic locations of barriers to gene flow in the HMHZ

The idea that if we score a sufficiently large set of molecular markers with a known position in the genome we can estimate width of their clines and hence detect genomic regions under strong selection (Payseur *et al.*, 2004; Carling and Brumfield, 2009; Payseur, 2010), is intuitively appealing. However, great rigour is necessary if we hope to distinguish signals in the data from stochastic noise. Recent studies (Dod *et al.*, 2005; Raufaste *et al.*, 2005; Macholán *et al.*, 2007, 2008, 2011; Dufková *et al.*, 2011) have shown the HMHZ is *a multi-locus tension zone with stepped clines and asymmetric tails*, but we have also seen that its shape is likely to be perturbed from place to place, with variation in mouse neighbourhood size. Estimates of the number of genes involved in the central barrier to gene flow, in the order of 50–100, are consistent with theoretical expectations that reproductive isolation is more likely to arise in many small steps than a few strongly selected mutations (Walsh, 1982; Barton and Charlesworth, 1984). Many genes of small effect means there will be a low signal-to-noise ratio for any particular gene, making the distinction between signal and stochastic effects a particularly important focus in HMHZ studies.

Sampling the hybrid zone and estimating cline parameters

Barton not only developed much of the model framework for understanding hybrid zones, he also provided the tools for researchers to estimate cline parameters using data collected in the field (ANALYSE: Barton and Baird, 1995). The explicit models of tension zones (Box 14.2) allow the likelihood of a dataset to be calculated for different values of cline parameters. The most likely parameters can be found using computational searches, and these point estimates of the parameters are made meaningful with the addition of a measure of belief in the estimate, the support interval (a point estimate with no measure of belief is no better than a guess). The efficacy of likelihood estimation has led all the geographic cline fitting software that followed Barton's work also to use likelihood inference (CLINEFIT: Porter *et al.*, 1997; CFIT: Gay *et al.*, 2008). Given this consistency of inference framework across hybrid zone studies it might seem unnecessary for us to discuss estimation of cline parameters, but increasingly this uniform inference framework seems also to have led to a thoughtlessness in inference about hybrid zones, which we will illustrate here with reference to the centre and width of a zone.

It is crucial that researchers distinguish between *the centre and width of a hybrid zone* and *estimates of these parameters*. To illustrate the importance of this distinction we use a reference scenario of a multi-locus tension zone with an asymmetric central barrier to genc flow such that in reality clines in many traits have coincident central steps with steep change on one side and more gradual change on the other side. What inference could we hope to make about this reference scenario? It will depend greatly on our spatial sampling of the zone (Box 14.3) and the cline model we use for analysis (Box 14.4). Let us first assume spatial sampling. If we sample localities in a line that is oblique to the path of the zone through the field area, we introduce an upward bias in our estimate of the zone width (Macholán *et al.*, 2008). The stochastic nature of locality samples near the zone centre means such biases in width and centre estimates can be surprisingly large and erratic (Dufková *et al.*, 2011). Both of these issues can be resolved by sampling localities across a 2D area, estimating the course of the hybrid zone through that region, and then using all the locality samples in relation to their orthogonal distance from the zone centre (MacCallum, 1994; Bridle *et al.*, 2001).

Taking the path of the zone centre into account minimizes the upward bias in width estimates, and because the set of localities near the centre of the zone is not limited to one on either side (the case for linear sampling), the erratic stochasticity of centre and width estimates is greatly reduced.

Turning to the second issue, model choice, suppose we choose to analyse field samples from the reference scenario using the two-parameter sigmoid cline model described in Box 14.2. What are the consequences of fitting this model to asymmetric stepped clines? Since the former is by definition symmetric, the resulting fitted cline is shifted to one side to try to match the asymmetry in the data: the more asymmetric the data, the more the cline centre *estimate* diverges from *the real position of the cline centre* (see Fig. 14.6). The consequences for inference are grave. Since the two-parameter analysis will give the impression that the zone has non-coincident cline centres, we might mis-infer the very nature of the zone because 'coincident cline centres' is one of the most fundamental expectations of a zone with a central barrier to gene flow.

Non-geographic alternatives

An alternative approach to spatial cline analysis, first proposed by Szymura and Barton (1986, 1991), and now called Barton's concordance (Macholán *et al.*, 2011), minimizes both of these issues of spatial sampling and model selection to some extent. When the geographic pattern of hybridization within a contact zone is complex it may be rational to disregard information on geography and fit clines against the hybrid index rather than against geographic distance. This approach can be especially useful in cases where conventional geographic cline models are not suitable for describing the spatial pattern, such as in mosaic hybrid zones or when the introgression pattern of some loci strongly deviates from that of other markers (Harrison and Rand, 1989; Arntzen and Wallis, 1991; Bridle *et al.*, 2001; Bierne *et al.*, 2003; Macholán *et al.*, 2008, 2011). A similar approach of fitting clines against the hybrid index rather than against geographic distance was taken by Gompert and Buerkle (2009, 2011), who called the resulting plots 'genomic clines'. The methods are compared in Macholán *et al.* (2011), who note that while these non-spatial analyses may be convenient, they do not entirely escape the issues of sampling and model choice.

Studies on the X chromosome in the HMHZ

When choosing where to start a search for genetic factors causing barriers to gene flow, researchers can take as guidance two strong empirical patterns, also known as the 'two rules of speciation' (Coyne and Orr, 1989).

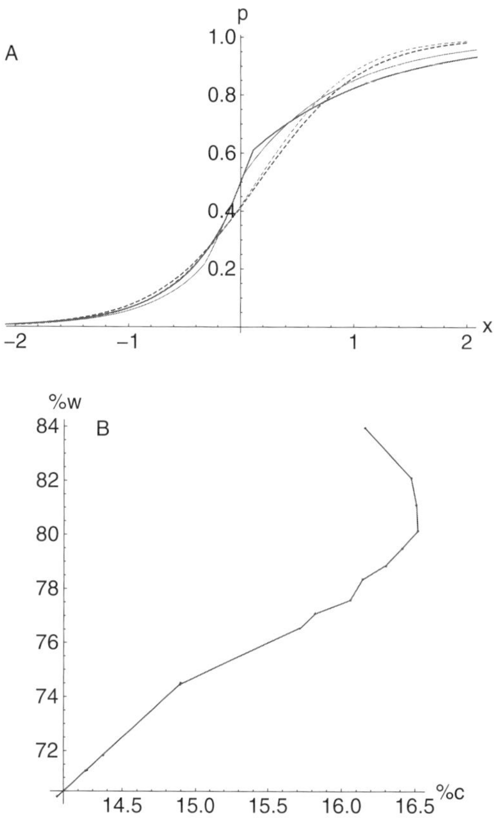

Figure 14.6 Assuming a hybrid zone scenario in which every sampled locus has a stepped asymmetric cline with parameters $\{c, w, B_L, \theta_L, B_R, \theta_R\} = \{0, 1, 0.1, 1, 0.2, \theta_R\}$, such that all clines have identical centre and width, but vary in their asymmetry as θ_1 changes in the interval $0.25 < \theta_R < 0.74$. (a) Solid curves show the cline shapes at the extremes of this interval. The dashed lines are the corresponding MLE cline estimates from data on 100 large locality samples spaced evenly across the zone and analysed using a two-parameter $\{c, w\}$ sigmoid model. (b) The centre and width estimates from this inappropriate model exceed the true centre location and width. The errors are expressed as percentages of the true value.

The first is Haldane's rule, stating that when hybrids of one sex suffer from reduced viability or fertility, it is usually the heterogametic sex (Haldane, 1922; Orr, 1997). The second is that genes contributing to reproductive isolation often map to the X chromosome, the phenomenon known as the large X-effect (Charlesworth *et al.*, 1987; Coyne and Orr, 1989; Coyne, 1992), recently dubbed 'Coyne's rule' (Turelli and Moyle, 2007). Indeed, it has been shown that sex chromosomes harbour more genes causing disruption of fertility and/or viability in hybrids than autosomes and hence will be under stronger selection (see Coyne and Orr, 2004 and references therein; for evidence in the mouse, see Oka *et al.*, 2004; Storchová *et al.*, 2004; Harr, 2006; Good *et al.*, 2008a).

To date, three surveys of X chromosome introgression heterogeneity across the European HMHZ have been carried out. Working from a 1D transect of samples that straddle this zone in Bavaria, to the south of the Czech–Bavarian portion (Fig. 14.2), Payseur *et al.* (2004) fitted geographic stepped clines to data for 13 loci, identifying a central-X region of reduced introgression, and also proposing the X inactivation locus *Xist* as marking an adaptive introgression into *musculus* territory. Teeter *et al.* (2010), working with the same Bavarian dataset, and an additional 1D transect sampling across the zone in Saxony (to the north of the Czech–Bavarian zone, Fig. 14.2), found the resulting likelihood surfaces for the six-parameter geographic cline model impracticably complex, so instead chose to fit sigmoid clines to 41 SNPs (but only three X markers, a subset of Payseur *et al.* (2004): *Emd*, *Pola1*, and *Xist*) drawn from both datasets. However, in recognition of the difficulty introduced by geographic complexity, the authors also used the non-geographic genome clines approach of Gompert and Buerkle (2009). They concluded that 'the differences between transects raise the possibility that there may not be a single genetic architecture of isolation between these species' (Teeter *et al.*, 2010: 10), though two out of the three X markers did not show any difference between transects, but noted that 'it is possible that stochastic variation, differences in sampling between transects, or a combination, could have contributed to these differences' (Teeter *et al.*, 2010: 10).

The third study (Macholán *et al.*, 2011) was of a 24-locus superset of the Payseur *et al.* (2004) study, on a sample of more than 2870 mice from a geographically intensively sampled 145 × 50 km 2D (rectangular) section of the Czech–Bavarian portion of the HMHZ. In order to distinguish geographic stochastic effects from deterministic introgression patterns they compared both non-geographic clines analysis approaches (Barton's concordance and genomic clines) to analysis using a spatially explicit Bayesian model-based clustering algorithm (GENELAND 3.1.4; Guillot *et al.*, 2005). Analysis of the Czech–Bavarian zone (lying between the two focal transects of the Teeter study) found the same central-X-reduced introgression region as Payseur *et al.* (2004) found for one of their transects, and concluded that there is some evidence of common architecture of reproductive isolation over the transect studied, and no reliable evidence to the contrary. We suggest that once it is taken into account that 1D sampling across a 2D zone (Box 14.3) and that fitting a sigmoid model to an asymmetric hybrid zone (Box 14.4) can introduce large sources of stochastic error, the differences between transect estimates in the Teeter *et al.* (2010) study does not justify the suggestion of multiple genetic architectures maintaining the HMHZ.

Two other interesting findings arose from the Macholán *et al.* (2011) X chromosome study. In the GENELAND analysis, several loci showed the geographic pattern known as 'footprints of a moving hybrid zone' (Arntzen and Wallis, 1991), indicating a history of movement of the zone from east to west. More recently, a study of linkage disequilibrium across 1401 loci came to the same conclusion regarding zone movement in Bavaria (Wang *et al.*, 2011; see Tucker *et al.*, Chapter 20). A tendency for *M. m. musculus* to gain territory from *M. m. domesticus* could explain a number of disparate characteristics of the HMHZ. (1) The geographic asymmetry of the zone across multiple transects – the long tails of introgression on the *musculus* side being the result of many small *domesticus* 'footprints' left behind the zone front. (2) The Jutland *musculus* 'colony', which persists despite being essentially a closed system faced by hybridization with a comparatively infinite *domesticus* range. (3) The eight-fold greater incidence of tapeworms in *domesticus* – parasites capable of disrupting reproduction in rodents (see Goüy de Bellocq *et al.*, Chapter 18). The second interesting finding from Macholán *et al.* (2011) is evidence of a proximal-X, male-biased westward introgression, despite the central X barrier to gene flow between the taxa. This proximal X introgression was in fact predicted by Macholán *et al.* (2008) in their study of the Y chromosome invasion across this section of the HMHZ (see below), and an accompanying mtDNA introgression.

Future perspectives

Much as theoretical expectations and empirical evidence for the large X-effect made the X chromosome the natural candidate for the studies summarized above, we should let our current understanding of speciation direct future efforts as we expand the search for genes involved in reproductive barriers across the rest of the genome.

A model independently proposed by Bateson (1909) and later by Dobzhansky (1936) and Muller (1940, 1942) is now widely accepted as a framework for understanding the genetic basis of postzygotic isolation. The Dobzhansky–Muller (DM) model solves Darwin's paradox that postzygotic incompatibilities arising within a population should be counter-selected, making their fixation unlikely. According to the DM model this problem can be circumvented if hybrid sterility is brought about through interaction of loci accumulating different mutations in allopatry that are only revealed to be incompatible at secondary contact. Thus, speciation can occur even if the allopatric populations have not passed through an adaptive valley. (For the sake of completeness it should be noted that the model can also be derived for a single locus.)

Although the DM model has become a cornerstone of the genetics of speciation (Orr, 1996; Coyne and Orr, 2004), it does not address the following questions: (1) What *kind* of genes might produce such epistatic interactions between alleles? (2) Do these genes drift to fixation in allopatry, or is their fixation driven by natural selection? (3) To what degree is their evolution within the diverging subpopulations influenced by intragenomic processes such as genetic conflict? For example, there is growing empirical evidence that genes implicated in postzygotic isolation are also associated with meiotic drive, suggesting their fixation in allopatry could have been through selection at the level of the gene rather than through Darwinian adaptation of individuals to the external environment or by random drift (Tao *et al.*, 2003; Orr and Irving, 2005; Orr *et al.*, 2007; Phadnis and Orr, 2009; Presgraves, 2010). Twenty years ago genetic conflict was suggested as an explanation for Haldane's rule of speciation (Frank, 1991; Hurst and Pomiankowski, 1991), but the idea fell into disfavour, with models invoking faster substitution on the sex chromosomes being preferred (Charlesworth *et al.*, 1993). The recent evidence associating genetic conflict with speciation genes (Presgraves, 2010) therefore adds new interest to an old debate.

Here, in the context of the HMHZ hybrid zone, we are prompted to ask: what is the expected outcome when a genetic conflict system is split into independent isolates, and subsequently comes into secondary contact? In natural hybrid zones we may observe outcomes of many generations of recombination, numbers entirely impractical in the laboratory. Such natural experiments can therefore give high-resolution estimates of the number and locations of genes or genomic regions causing reproductive isolation ('speciation' genes), but also allow us to spot the converse pattern: genes that appear to preferentially cross and break down the taxon boundary ('anti-speciation' genes), *even when the two types of gene are neighbours on the same chromosome* – the central X barrier to introgression and the proximal X invasion, respectively (Macholán *et al.*, 2011).

Introgression of the Y chromosome

In accordance with the idea that introgression of sex chromosomes in a hybrid zone should be severely impeded, virtually no introgression of the mouse Y chromosome has been found in Denmark (Vanlerberghe *et al.*, 1986; Dod *et al.*, 2005), Bulgaria (Vanlerberghe *et al.*, 1988a), or northern Germany (Prager *et al.*, 1997). This picture, seemingly consistent along 2500 km of the zone of secondary contact, began to crumble during a survey of the distribution of several autosomal and sex-linked markers across the Czech and Slovak Republics, which revealed an unexpectedly gradual introgression of the Y chromosome, substantially higher and shifted to the *domesticus* territory compared to the diagnostic autosomal and

X-linked loci analysed (Munclinger *et al.*, 2002). This pattern was later corroborated and quantified using a larger dataset from the Czech–Bavarian transect (Macholán *et al.*, 2008) and is possibly also the case in Saxony (K. C. Teeter and P. K. Tucker, personal communication).

Macholán *et al.* (2008) showed that the most likely direction of change of the Y chromosome is rotated about 45° clockwise relative to other loci. When estimated along this axis, the width of the Y transition was not significantly different from an X chromosome locus (*Btk*), while using the consensus direction based on seven autosomal and one X-linked loci resulted in considerably broader cline and significantly worse fit. Inspection of the spatial introgression pattern showed the *musculus* Y chromosome has spread across the zone up to 22 km into the *domesticus* range, covering – at minimum – a triangular area approximately 330 km^2. This introgression is accompanied by differences in the census sex ratio: while this ratio is significantly female-biased in the *musculus* territory (0.45, $p = 0.0005$) and in *domesticus* localities without the introgressed *musculus* Y (0.44, $p = 0.0065$), in the *domesticus* area with the introgressed *musculus* Y it is not different from parity (0.51, $p = 0.2037$) and significantly different from the former two areas (Macholán *et al.*, 2008). Reconciling the Y chromosome introgression and sex ratio distribution led to a hypothesis that a genetic conflict between the Y and X (and possibly some autosomal) elements is the underlying cause of both the invasion and sex ratio distortion patterns. According to this hypothesis, the *musculus* Y is successful on the naïve *domesticus* background. As a consequence, the former spreads across the zone at the expense of the latter. The recent results of Jones *et al.* (2010), who found the *musculus* Y in Norway far behind the zone in *domesticus* territory, seem consistent with this hypothesis.

What does the mouse hybrid zone tell us about speciation?

The first evidence arising from the HMHZ of genes causing barriers to gene flow came from the X chromosome. But this was the first place we looked. With the accelerating accumulation of large genomic datasets, such as SNP maps now available for the mouse (Lindblad-Toh *et al.*, 2000; Wade *et al.*, 2002; Abe *et al.*, 2004; Pletcher *et al.*, 2004; Shifman *et al.*, 2006; Frazer *et al.*, 2007; Yang *et al.*, 2009), applying similar methods as for the X chromosome we may expect to find more and more such factors. While this work will gradually build up a detailed picture of which genes are involved in the species barrier between two subspecies of house mouse, perhaps the more exciting results concern the speciation process in general. If hybridizing genomes involve complex mixtures of DM incompatibility genes intermingled with universally advantaged and/or invasive conflict elements, as suggested by the contrasting patterns of central and proximal X chromosome variation (Macholán *et al.*, 2011) in association with the Y chromosome invasion (Macholán *et al.*, 2008), then the latter may counter the action of the former. That is: conflict genes may act as anti-speciation genes. While laboratory-based speciation research has focused on DM incompatibilities and the circumstances that favour them, perhaps the most important thing hybrid zone speciation research has shown us is that to understand speciation it is also necessary to consider what might *counter* the action of DM incompatibilities, and what circumstances might *disfavour* their accumulation (Baird *et al.*, 2012; Goüy de Bellocq *et al.*, Chapter 18).

REFERENCES

Abe, K., Noguchi, H., Tagawa, K., *et al.* (2004). Contribution of Asian mouse subspecies *Mus musculus molossinus* to genomic constitution of strain C57BL/6J, as defined by BAC-end sequence-SKIP analysis. *Genome Research*, **14**, 2439–47.

Albrechtová, J., Albrecht, T., Baird, S. J. E., *et al.* (Submitted). Differential sperm performance in hybrids coincides with Y chromosome invasion across the house mouse hybrid zone. *Current Biology*.

Alibert, P. and Auffray, J.-C. (2003). Genomic coadaptation, outbreeding depression, and developmental instability. In *Developmental Instability: Causes and Consequences*, ed. M. Polak. Oxford: Oxford University Press, pp. 116–34.

Alibert, P., Fel-Clair, F., Manolakou, K., Britton-Davidian, J., and Auffray, J.-C. (1997). Developmental stability, fitness, and trait size in laboratory hybrids between European subspecies of the house mouse. *Evolution*, **51**, 1284–95.

Alibert, P., Renaud, S., Dod, B., Bonhomme, F., and Auffray, J.-C. (1994). Fluctuating asymmetry in the *Mus musculus* hybrid zone: a heterotic effect in disrupted co-adapted genomes. *Proceedings of the Royal Society of London B: Biological Sciences*, **258**, 53–9.

Arntzen, J. W. and Wallis, G. P. (1991). Restricted gene flow in a moving hybrid zone of the newts *Triturus cristatus* and *T. Marmoratus* in western France. *Evolution*, **45**, 805–26.

Auffray, J.-C. (1993). Chromosomal evolution in the house mouse in the light of palaeontology: a colonization-related event? *Quaternary International*, **19**, 21–5.

Auffray, J.-C. and Britton-Davidian, J. (1992). When did the house mouse colonize Europe? *Biological Journal of the Linnean Society*, **45**, 187–90.

Auffray, J.-C., Vanlerberghe, F., and Britton-Davidian, J. (1990). The house mouse progression in Eurasia: a palaeontological and archaeozoological approach. *Biological Journal of the Linnean Society*, **41**, 13–25.

Baird, S. J. E. (1995). A simulation study of multilocus clines. *Evolution*, **49**, 1038–45.

Baird, S. J. E. (2006). Phylogenetics: Fisher's markers of admixture. *Heredity*, **97**, 81–3.

Baird, S. J. E., Ribas, A., Macholán, M., *et al.* (2012). Where are the wormy mice? A re-examination of hybrid parasitism in the European house mouse hybrid zone. *Evolution*. doi: 10.1111/j.1558-5646.2012.01633.x.
Baird, S. J. E. and Santos, F. (2010). Monte Carlo integration over stepping stone models for spatial genetic inference using approximate Bayesian computation. *Molecular Ecology Resources*, **10**, 873–85.
Barton, N. H. (1979). Gene flow past a cline. *Heredity*, **43**, 333–9.
Barton, N. H. (1983). Multilocus clines. *Evolution*, **37**, 454–71.
Barton, N. H. (1986). The effects of linkage and density-dependent regulation on gene flow. *Heredity*, **57**, 415–26.
Barton, N. H. and Baird, S. J. E. (1995). *Analyse: An Application for Analyzing Hybrid Zones*. Freeware, Edinburgh, UK. Available at http://helios.bto.ed.ac.uk/evolgen/Mac/Analyse.
Barton, N. H. and Bengtsson, B. O. (1986). The barrier to genetic exchange between hybridising populations. *Heredity*, **56**, 357–76.
Barton, N. H. and Charlesworth, B. (1984). Genetic revolutions, founder effects, and speciation. *Annual Review of Ecology and Systematics*, **15**, 133–64.
Barton, N. H., Depaulis, F., and Etheridge, A. M. (2002). Neutral evolution in spatially continuous populations. *Theoretical Population Biology*, **61**, 31–48.
Barton, N. H. and Gale, K. S. (1993). Genetic analysis of hybrid zones. In *Hybrid Zones and the Evolutionary Process*, ed. R. G. Harrison. Oxford: Oxford University Press, pp. 13–45.
Barton, N. H. and Hewitt, G. M. (1985). Analysis of hybrid zones. *Annual Review of Ecology and Systematics*, **16**, 113–48.
Barton, N. H. and Hewitt, G. M. (1989). Adaptation, speciation and hybrid zones. *Nature*, **341**, 497–503.
Barton, N. H. and Shpak, M. (2000). The effect of epistasis on the structure of hybrid zones. *Genetical Research*, **75**, 179–98.
Barton, N. H. and Wilson, I. (1995). Genealogies and geography. *Philosophical Transactions of the Royal Society of London Series B: Biological Sciences*, **349**, 49–59.
Bateson, W. (1909). Heredity and variation in modern lights. In *Darwin and Modern Science*, ed. A. C. Seward. Cambridge: Cambridge University Press, pp. 85–101.
Bazykin, A. D. (1969). Hypothetical mechanism of speciation. *Evolution*, **23**, 685–7.
Bierne, N., Daguin, C., Bonhomme, F., David, P., and Borsa, P. (2003). Direct selection on allozymes is not required to explain heterogeneity among marker loci across a Mytilus hybrid zone. *Molecular Ecology*, **12**, 2505–10.
Bímová, B., Albrecht, T., Macholán, M., and Piálek, J. (2009). Signalling components of mate recognition system in the house mouse. *Behavioural Processes*, **80**, 20–7.
Bonhomme, F., Orth, A., Cucchi, T., *et al.* (2011). Genetic differentiation of the house mouse around the Mediterranean basin: matrilineal footprints of early and late colonization. *Proceedings of the Royal Society B: Biological Sciences*, **278**, 1034–43.
Boursot, P., Auffray, J. -C., Britton-Davidian, J., and Bonhomme F. (1993). The evolution of house mice. *Annual Review of Ecology and Systematics*, **24**, 119–52.

Boursot, P., Bonhomme, F., Britton-Davidian, J., *et al.* (1984). Differential introgression of the nuclear and mitochondrial genomes between 2 semi-species of European mice. *Comptes Rendus de l'Academie des Sciences III – Sciences de la Vie – Life Sciences*, **299**, 365–70.

Božíková, E., Munclinger, P., Teeter, K. C., *et al.* (2005). Mitochondrial DNA in the hybrid zone between *Mus musculus musculus* and *Mus musculus domesticus*: a comparison of two transects. *Biological Journal of the Linnean Society*, **84**, 363–78.

Bridle, J. R., Baird, S. J. E., and Butlin, R. K. (2001). Spatial structure and habitat variation in a grasshopper hybrid zone. *Evolution*, **55**, 1832–43.

Britton-Davidian, J., Fel-Clair, F., Lopez, J., Alibert, P., and Boursot, P. (2005). Postzygotic isolation between the two European subspecies of the house mouse: estimates from fertility patterns in wild and laboratory-bred hybrids. *Biological Journal of the Linnean Society*, **84**, 379–93.

Bullough, R. K. and Caudrey, P. J. (eds) (1980) *Solitons*. Berlin: Springer-Verlag.

Butlin, R. K. (1987). Speciation by reinforcement. *Trends in Ecology and Evolution*, **2**, 8–13.

Butlin, R. K. (1995). Reinforcement: an idea evolving. *Trends in Ecology and Evolution*, **10**, 432–4.

Carling, M. D. and Brumfield, R. T. (2009). Speciation in *Passerina* buntings: introgression patterns of sex-linked loci identify a candidate gene region for reproductive isolation. *Molecular Ecology*, **18**, 834–47.

Charlesworth, B., Coyne, J. A., and Barton, N. H. (1987). The relative rates of evolution of sex chromosomes and autosomes. *The American Naturalist*, **130**, 113–46.

Charlesworth, B., Coyne, J. A., and Orr, H. A. (1993). Meiotic drive and unisexual hybrid sterility: a comment. *Genetics*, **133**, 421–4.

Christophe, N. and Baudoin, C. (1998). Olfactory preferences in two subspecies of mice *Mus musculus musculus* and *Mus musculus domesticus* and their hybrids. *Animal Behaviour*, **56**, 365–9.

Clark, G. (1975). *The Earlier Stone Age Settlements of Scandinavia*. Cambridge: Cambridge University Press.

Coyne, J. A. (1992). Genetics and speciation. *Nature*, **355**, 511–15.

Coyne, J. A. and Orr, H. A. (1989). Two rules of speciation. In *Speciation and Its Consequences*, ed. D. Otte and J. Endler. Sunderland, MA: Sinauer Associates, pp. 180–207.

Coyne, J. A. and Orr, H. A. (2004). *Speciation*. Sunderland, MA: Sinauer Associates.

Cucchi, T., Vigne, J. -D., and Auffray, J. -C. (2005). First occurrence of the house mouse (*Mus musculus domesticus* Schwarz & Schwarz, 1943) in the western Mediterranean: a zooarchaeological revision of subfossil occurrences. *Biological Journal of the Linnean Society*, **84**, 429–45.

Debat, V., Alibert, P., David, P., Paradis, E., and Auffray, J. -C. (2000). Independence between developmental stability and canalization in the skull of the house mouse. *Proceedings of the Royal Society of London B: Biological Sciences*, **267**, 423–30.

Degerbøl, M. (1949). *Gnavere, vol. 1*. Copenhagen: Vort Lands Dyreliv.

Dobzhansky, T. (1936). Studies on hybrid sterility: II. Localization of sterility factors in *Drosophila pseudoobscura* hybrids. *Genetics*, **21**, 113–35.

Dobzhansky, T. (1937). *Genetics and the Origin of Species*. New York: Columbia University Press.

Dobzhansky, T. (1940). Speciation as a stage in evolutionary divergence. *The American Naturalist*, **74**, 312–21.

Dod, B., Jermiin, L. S., Boursot, P., *et al.* (1993). Counterselection on sex chromosomes in the *Mus musculus* European hybrid zone. *Journal of Evolutionary Biology*, **6**, 529–46.

Dod, B., Smadja, C., Karn, R. C., and Boursot, P. (2005). Testing for selection on the androgen-binding protein in the Danish mouse hybrid zone. *Biological Journal of the Linnean Society*, **84**, 447–59.

Dufková, P., Macholán, M., and Piálek, J. (2011). Inference of selection and stochastic effects in the house mouse hybrid zone. *Evolution*, **65**, 993–1010.

Endler, J. A. (1977). *Geographic Variation, Speciation and Clines*. Princeton, NJ: Princeton University Press.

Fel-Clair, F., Catalan, J., Lenormand, T., and Britton-Davidian, J. (1998). Centromeric incompatibilities in the hybrid zone between house mouse subspecies from Denmark: evidence from patterns of NOR activity. *Evolution*, **52**, 592–603.

Fel-Clair, F., Lenormand, T., Catalan, J., *et al.* (1996). Genomic incompatibilities in the hybrid zone between house mice in Denmark: evidence from steep and non-coincident chromosomal clines for Robertsonian fusions. *Genetical Research*, **67**, 123–34.

Ferris, S. D., Sage, R. D., Huang, C. -M., *et al.* (1983). Flow of mitochondrial DNA across a species boundary. *Proceedings of the National Academy of Sciences of the United States of America*, **80**, 2290–4.

Fisher, R. A. (1930). *The Genetical Theory of Natural Selection*. Oxford: Oxford University Press.

Fisher, R. A. (1950). Gene frequencies in a cline determined by selection and diffusion. *Biometrics*, **6**, 353–61.

Forejt, J. (1981). Hybrid sterility gene located in the T/t-H-2 supergene on chromosome 17. In *Current Trends in Histocompatibility: Immunogenetic and Molecular Profiles*, ed. R. A. Reisfeld and S. Ferrone. London: Plenum Press, pp. 103–31.

Forejt, J. (1996). Hybrid sterility in the mouse. *Trends in Genetics*, **12**, 412–17.

Forejt, J. and Iványi, P. (1974). Genetic studies on male sterility of hybrids between laboratory and wild mice (*Mus musculus* L.). *Genetical Research*, **24**, 189–206.

Forejt, J., Vincek, V., Klein, J., Lehrach, H., and Loudová-Micková, M. (1991). Genetic mapping of the *t*-complex region on mouse chromosome 17 including the Hybrid sterility-1 gene. *Mammalian Genome*, **1**, 84–91.

Frank, S. H. (1991). Divergence of meiotic drive-suppressors as an explanation for sex-biased hybrid sterility and inviability. *Evolution*, **45**, 262–7.

Frazer, K. A., Eskin, E., Kang, H. M., *et al.* (2007). A sequence-based variation map of 8.27 million SNPs in inbred mouse strains. *Nature*, **448**, 1050–3.

Ganem, G., Litel, C., and Lenormand, T. (2008). Variation in mate preference across a house mouse hybrid zone. *Heredity*, **100**, 594–601.

Gay, L., Crochet, P. -A., Bell, D. A., and Lenormand, T. (2008). Comparing clines on molecular and phenotypic traits in hybrid zones: a window on tension zone models. *Evolution*, **62**, 2789–806.

Gompert, Z. and Buerkle, C. A. (2009). A powerful regression-based method for admixture mapping of isolation across the genome of hybrids. *Molecular Ecology*, **18**, 1207–24.
Gompert, Z. and Buerkle, C. A. (2011). Bayesian estimation of genomic clines. *Molecular Ecology*, **20**, 2111–27.
Good, J. M., Dean, M. D., and Nachman, M. W. (2008a). A complex genetic basis to X-linked hybrid male sterility between two species of house mice. *Genetics*, **179**, 2213–28.
Good, J. M., Handel, M. A., and Nachman, M. W. (2008b). Asymmetry and polymorphism of hybrid male sterility during the early stages of speciation in house mice. *Evolution*, **62**, 50–65.
Gregorová, S., Mňuková-Fajdelová, M., Trachtulec, Z., *et al.* (1996). Sub-milli-Morgan map of the proximal part of mouse Chromosome 17 including the hybrid sterility 1 gene. *Mammalian Genome*, **7**, 107–13.
Guillot, G., Mortier, F., and Estoup, A. (2005). Geneland: a program for landscape genetics. *Molecular Ecology Notes*, **5**, 712–15.
Gyllensten, U. and Wilson, A. C. (1987). Interspecific mitochondrial DNA transfer and the colonization of Scandinavia by mice. *Genetical Research*, **49**, 25–9.
Haldane, J. B. S. (1922). Sex ratio and unisexual sterility in animal hybrids. *Journal of Genetics*, **12**, 101–9.
Haldane, J. B. S. (1948). The theory of a cline. *Journal of Genetics*, **48**, 277–84.
Harr, B. (2006). Genomic islands of differentiation between house mouse subspecies. *Genome Research*, **16**, 730–7.
Harrison, R. G. (1990). Hybrid zones: windows on evolutionary process. *Oxford Surveys in Evolutionary Biology*, **7**, 69–128.
Harrison, R. G. and Rand, D. M. (1989). Mosaic hybrid zones and the nature of species boundaries. In *Speciation and Its Consequences*, ed. D. Otte and J. Endler. Sunderland, MA: Sinauer Associations, pp. 111–33.
Hewitt, G. M. (1975). A sex-chromosome hybrid zone in the grasshopper *Podisma pedestris* (Orthoptera: Acrididae). *Heredity*, **35**, 375–87.
Hewitt, G. (1988). Hybrid zones: natural laboratories for evolution studies. *Trends in Ecology and Evolution*, **3**, 158–66.
Hewitt, G. M. (1989). The subdivision of species by hybrid zones. In *Speciation and Its Consequences*, ed. D. Otte and J. A. Endler. Sunderland, MA: Sinauer Associates, pp. 85–110.
Hunt, W. G. and Selander, R. K. (1973). Biochemical genetics of hybridisation in European house mice. *Heredity*, **31**, 11–33.
Hurst, L. D. and Pomiankowski, A. (1991). Causes of sex ratio bias may account for unisexual sterility in hybrids: a new explanation of Haldane's rule and related phenomena. *Genetics*, **128**, 841–58.
Jones, E. P., van der Kooij, J., Solheim, R., and Searle, J. B. (2010). Norwegian house mice (*Mus musculus musculus/domesticus*): distributions, routes of colonization and patterns of hybridization. *Molecular Ecology*, **19**, 5252–64.

Karn, R. C. and Dlouhy, S. R. (1991). Salivary androgen-binding protein variation in *Mus* and other rodents. *Journal of Heredity*, **82**, 453–8.

Key, K. H. (1968). The concept of stasipatric speciation. *Systematic Zoology*, **17**, 14–22.

Klein, J., Tichy, H., and Figueroa, F. (1987). On the origin of mice. *Anales de l'Universidad de Chile*, **5**, 91–120.

Kraft, R. (1985). Merkmale und Verbreitung der Hausmäuse *Mus musculus musculus* L., 1758, und *Mus musculus domesticus* Rutty, 1772 (Rodentia, Muridae) in Bayern. *Säugetierkundliche Mitteilungen*, **31**, 1–13.

Kratochvíl, J. (1986). Die intraspezifische Evolution der Art *Mus domesticus*. *Acta scientiarum naturalium Academiae scientiarum bohemoslovacae – Brno*, **20**, 1–49.

Lakshmanan, M. (ed.) (1988). *Solitons*. New York: Springer-Verlag.

Lande, R. (1981). Models of speciation by sexual selection on polygenic traits. *Proceedings of the National Academy of Sciences of the United States of America*, **78**, 3721–5.

Lanneluc, I., Desmarais, E., Boursot, P., Dod, B., and Bonhomme, F. (2004). Characterization of a centromeric marker on mouse chromosome 11 and its introgression in a *domesticus/musculus* hybrid zone. *Mammalian Genome*, **15**, 924–34.

Laukaitis, C. M., Heger, A., Blakley, T. D., *et al.* (2008). Rapid bursts of androgen-binding protein (Abp) gene duplication occurred independently in diverse mammals. *BMC Evolutionary Biology*, **8**, 46–62.

Lemmon, A. R., Smadja, C., and Kirkpatrick, M. (2004). Reproductive character displacement is not the only possible outcome of reinforcement. *Journal of Evolutionary Biology*, **17**, 177–83.

Lepiksaar, J. (1980). Animal remains at Tornör: a study of a Thanatocoenosis (Late Iron Age to recent times). *Striae*, **10**, 3–41.

Li, W. H. and Nei, M. (1974). Stable linkage disequilibrium without epistasis in subdivided populations. *Theoretical Population Biology*, **6**, 173–83.

Lindblad-Toh, K., Winchester, E., Daly, M. J., *et al.* (2000). Large-scale discovery and genotyping of single-nucleotide polymorphisms in the mouse. *Nature Genetics*, **24**, 381–6.

MacCallum, C. J. (1994). *Adaptation and habitat preference in a hybrid zone between* Bombina bombina *and* Bombina variegata. Unpublished PhD thesis, University of Edinburgh.

Macholán, M., Baird, S. J. E., Dufková, P., *et al.* (2011). Assessing multilocus introgression patterns: a case study on the mouse X chromosome in central Europe. *Evolution*, **65**, 1428–46.

Macholán, M., Baird, S. J. E., Munclinger, *et al.* (2008). Genetic conflict outweighs heterogametic incompatibility in the mouse hybrid zone? *BMC Evolutionary Biology*, **8**, 271–84.

Macholán, M., Kryštufek, B., and Vohralík, V. (2003). The location of the *Mus musculus/M. domesticus* hybrid zone in the Balkans: clues from morphology. *Acta Theriologica*, **48**, 177–88.

Macholán, M., Munclinger, P., Šugerková, M., *et al.* (2007). Genetic analysis of autosomal and X-linked markers across a mouse hybrid zone. *Evolution*, **61**, 746–71.

Mihola, O., Trachtulec, Z., Vlcek, C., Schimenti, J. C., and Forejt, J. (2009). A mouse speciation gene encodes a meiotic Histone H3 Methyltransferase. *Science*, **323**, 373–5.

Mikula, O., Auffray, J.-C., and Macholán, M. (2010). Fluctuating asymmetry in the central European transect across the house mouse hybrid zone. *Biological Journal of the Linnean Society*, **101**, 13–27.

Mikula, O. and Macholán, M. (2008). There is no heterotic effect upon developmental stability in the ventral side of the skull within the house mouse hybrid zone. *Journal of Evolutionary Biology*, **21**, 1055–67.

Montagutelli, X., Turner, R., and Nadeau, J. H. (1996). Epistatic control of non-Mendelian inheritance in mouse specific crosses. *Genetics*, **143**, 1739–52.

Moore, W. S. (1977). An evaluation of narrow hybrid zones in vertebrates. *Quarterly Review of Biology*, **52**, 263–78.

Moulia, C., Aussel, J. P., Bonhomme, F., *et al.* (1991). Wormy mice in a hybrid zone: a genetic control of susceptibility to parasite infection. *Journal of Evolutionary Biology*, **4**, 679–87.

Moulia, C. and Joly, P. (2009). Parasitism and hybrid zones. In *Ecology and Evolution of Parasitism*, ed. F. Thomas, J. F. Guégan, and F. Renaud. New York: Oxford University Press, pp. 69–82.

Muller, H. J. (1940). Bearing of the *Drosophila* work on systematics. In *The New Systematics*, ed. J. S. Huxley. Oxford: Clarendon Press, pp. 185–268.

Muller, H. J. (1942). Isolating mechanisms, evolution, and temperature. *Biological Symposia*, **6**, 71–125.

Munclinger, P., Božíková, E., Šugerková, M., Piálek, J., and Macholán, M. (2002). Genetic variation in house mice (*Mus*, Muridae, Rodentia) from the Czech and Slovak Republics. *Folia Zoologica*, **51**, 81–92.

Munclinger, P. and Frynta, D. (2000). Social interactions within and between two distant populations of house mouse. *Folia Zoologica*, **49**, 1–6.

Nagylaki, T. (1975). Conditions for the existence of clines. *Genetics*, **80**, 595–615.

Nagylaki, T. (1976). Clines with variable migration. *Genetics*, **83**, 867–86.

Nancé, V., Vanlerberghe, F., Nielsen, J. T., Bonhomme, F., and Britton-Davidian, J. (1990). Chromosomal introgression in house mice from the hybrid zone between *M. m. domesticus* and *M. m. musculus*. *Biological Journal of the Linnean Society*, **41**, 215–27.

O'Connor, T. P. (1992). Pets and pests in Roman and medieval Britain. *Mammal Review*, **22** (2), 107–13.

Oka, A., Mita, A., Sakurai-Yamatani, N., *et al.* (2004). Hybrid breakdown caused by substitution of the X chromosome between two mouse subspecies. *Genetics*, **166**, 913–24.

Orr, H. A. (1996). Dobzhansky, Bateson, and the genetics of speciation. *Genetics*, **144**, 1331–5.

Orr, H. A. (1997). Haldane's rule. *Annual Review of Ecology and Systematics*, **28**, 195–218.

Orr, H. A. and Irving, S. (2005). Segregation distortion in hybrids between the Bogota and USA subspecies of *Drosophila pseudoobscura*. *Genetics*, **169**, 671–82.

Orr, H. A., Masly, J. P., and Phadnis, N. (2007). Speciation in *Drosophila*: from phenotypes to molecules. *Journal of Heredity*, **98**, 103–10.

Payseur, B. A. (2010). Using differential introgression in hybrid zones to identify genomic regions involved in speciation. *Molecular Ecology Resources*, **10**, 806–20.

Payseur, B. A. and Hoekstra, H. E. (2005). Signatures of reproductive isolation in patterns of single nucleotide diversity across inbred strains of mice. *Genetics*, **171**, 1905–16.
Payseur, B. A., Krenz, J. G., and Nachman, M. W. (2004). Differential patterns of introgression across the X chromosome in a hybrid zone between two species of house mice. *Evolution*, **58**, 2064–78.
Payseur, B. A. and Nachman, M. W. (2005). The genomics of speciation: investigating the molecular correlates of X chromosome introgression across the hybrid zone between *Mus domesticus* and *Mus musculus*. *Biological Journal of the Linnean Society*, **84**, 523–34.
Phadnis, N. and Orr, H. A. (2009). A single gene causes both male sterility and segregation distortion in *Drosophila* hybrids. *Science*, **323**, 376–9.
Phillips, B. L., Baird, S. J. E., and Moritz, C. (2004). When vicars meet: a narrow contact zone between morphologically cryptic phylogeographic lineages of the rainforest skink, *Carlia rubrigularis*. *Evolution*, **58**, 1536–48.
Pletcher, M. T., McClurg, P., Batalov, S., *et al.* (2004). Use of a dense single nucleotide polymorphism map for in silico mapping in the mouse. *PLoS Biology*, **2**, e393.
Pocock, M. J. O., Hauffe, H. C., and Searle, J. B. (2005). Dispersal in house mice. *Biological Journal of the Linnean Society*, **84**, 565–83.
Polechová, J. and Barton, N. H. (2011). Genetic drift widens the expected cline but narrows the expected cline width. *Genetics*, **189**, 227–35.
Porter, A. H., Wenger, R., Geiger, H., Scholl, A., and Shapiro, A. M. (1997). The *Pontia daplidice-edusa* hybrid zone in northwestern Italy. *Evolution*, **51**, 1561–73.
Prager, E. M., Boursot, P., and Sage, R. D. (1997). New assays for Y chromosome and p53 pseudogene clines among East Holstein house mice. *Mammalian Genome*, **8**, 279–81.
Prager, E. M., Sage, R. D., Gyllensten, U., *et al.* (1993). Mitochrondrial DNA sequence diversity and the colonization of Scandinavia by house mice from East Holstein. *Biological Journal of the Linnean Society*, **50**, 85–122.
Presgraves, D. C. (2010). The molecular evolutionary basis of species formation. *Nature Reviews Genetics*, **11**, 175–80.
Raufaste, N. (2001). *Barrières au flux génique et sélection dans une zone hybride: etude théorique et expérimentale chez la souris domestique*. Unpublished PhD thesis, Université Montpellier II, Montpellier, France.
Raufaste, N., Orth, A., Belkhir, K., *et al.* (2005). Inferences of selection and migration in the Danish house mouse hybrid zone. *Biological Journal of the Linnean Society*, **84**, 593–616.
Ritchie, M. G. (2007). Sexual selection and speciation. *Annual Review of Ecology, Evolution, and Systematics*, **38**, 79–102.
Rousset, F. (1997). Genetic differentiation and estimation of gene flow from F-statistics under isolation by distance. *Genetics*, **145**, 1219–28.
Rousset, F. (2000). Genetic differentiation between individuals. *Journal of Evolutionary Biology*, **58**, 58–62.
Sage, R. D. (1981). Wild mice. In *The Mouse in Biomedical Research*, *vol. 1*, ed. H. L. Foster, J. D. Small, and J. G. Fox. New York: Academic Press, pp. 39–90.
Sage, R. D., Atchley, W. R., and Capanna, E. (1993). House mice as models in systematic biology. *Systematic Biology*, **42**, 523–61.

Sage, R. D., Heyneman, D., Lim, K. -C., and Wilson, A. C. (1986a). Wormy mice in a hybrid zone. *Nature*, **324**, 60–3.

Sage, R. D., Prager, E. M., Tichy, H., and Wilson, A. C. (1990). Mitochondrial DNA variation in house mice, *Mus domesticus* (Rutty). *Biological Journal of the Linnean Society*, **41**, 105–23.

Sage, R. D., Whitney, J. B., III, and Wilson, A. C. (1986b). Genetic analysis of a hybrid zone between *domesticus* and *musculus* mice (*Mus musculus* complex): hemoglobin polymorphisms. *Current Topics in Microbiology and Immunology*, **127**, 75–85.

Schnell, M. H. and Selander, R. K. (1981). Environmental and morphological correlates of genetic variation in mammals. In *Mammalian Population Genetics*, ed. J. Joule and M. H. Smith. Athens, GA: University of Georgia Press, pp. 60–99.

Selander, R. K., Hunt, W. G., and Yang, S. Y. (1969). Protein polymorphism and genic heterozygosity in two European subspecies of the house mouse. *Evolution*, **23**, 379–90.

Serafiński, W. (1965). The subspecific differentiation of the Central European house mouse (*Mus musculus* L.) in the light of their ecology and morphology. *Ekologia Polska: Seria A*, **13**, 305–48.

Servedio, M. R. and Noor, M. A. F. (2003). The role of reinforcement in speciation: theory and data. *Annual Review of Ecology and Systematics*, **34**, 339–64.

Shifman, S., Bell, J. T., Copley, R. R., *et al.* (2006). A high resolution single nucleotide polymorphism genetic map of the mouse genome. *PLoS Biology*, **4**, 2227–37.

Slatkin, M. (1973). Gene flow and selection in a cline. *Genetics*, **75**, 733–56.

Slatkin, M. (1975). Gene flow and selection on a two-locus system. *Genetics*, **81**, 787–802.

Smadja, C., Catalan, J., and Ganem, G. (2004). Strong premating divergence in a unimodal hybrid zone between two subspecies in the house mouse. *Journal of Evolutionary Biology*, **17**, 165–76.

Smadja, C. and Ganem, G. (2002). Subspecies recognition in the house mouse: a study of two populations from the border of a hybrid zone. *Behavioral Ecology*, **13**, 312–20.

Smadja, C. and Ganem, G. (2005). Asymmetrical reproductive character displacement in the house mouse. *Journal of Evolutionary Biology*, **18**, 1485–93.

Sokal, R. R., Oden, N. L., and Wilson, A. C. (1991). Genetic evidence for the spread of agriculture in Europe by demic diffusion. *Nature*, **351**, 143–5.

Storchová, R., Gregorová, S., Buckiová, *et al.* (2004). Genetic analysis of X-linked hybrid sterility in the house mouse. *Mammalian Genome*, **15**, 515–24.

Szymura, J. M. and Barton, N. H. (1986). Genetic analysis of a hybrid zone between the fire-bellied toads, *Bombina bombina* and *B. variegata*, near Cracow in southern Poland. *Evolution*, **40**, 1141–59.

Szymura, J. M. and Barton, N. H. (1991). The genetic structure of the hybrid zone between the firebellied toads *Bombina bombina* and *B. variegata*: comparisons between transects and between loci. *Evolution*, **45**, 237–61.

Tao, Y., Chen, S., Hartl, D. L., and Laurie, C. C. (2003). Genetic dissection of hybrid incompatibilities between *Drosophila simulans* and *D. mauritiana*: I. Differential accumulation of hybrid male sterility effects on the X and autosomes. *Genetics*, **164**, 1383–97.

Teeter, C. K., Payseur, B. A., Harris, L. W., *et al.* (2008). Genome-wide patterns of gene flow across a house mouse hybrid zone. *Genome Research*, **18**, 67–76.

Teeter, C. K., Thibodeau, L. M., Gompert, Z., *et al.* (2010). The variable genomic architecture of isolation between hybridizing species of house mice. *Evolution*, **64**, 472–85.

Thaler, L. (1986). Origin and evolution of mice: an appraisal of fossil evidence and morphological traits. *Current Topics in Microbiology and Immunology*, **127**, 3–11.

Thaler, L., Bonhomme, F., and Britton-Davidian, J. (1981). Processes of speciation and semi-speciation in the house mouse. In *Biology of the House Mouse*, ed. R. J. Berry. London: Academic Press, pp. 27–41.

Trachtulec, Z., Vincek, V., Hamvas, R. M., *et al.* (1994). Physical map of mouse chromosome 17 in the region relevant for positional cloning of the Hybrid sterility 1 gene. *Genomics*, **23**, 132–7.

Tucker, P., Sage, R. D., Warner, J., Wilson, A. C., and Eicher, E. M. (1992). Abrupt cline for sex chromosomes in a hybrid zone between two species of mice. *Evolution*, **46**, 1146–63.

Turelli, M., Barton, N. H., and Coyne, J. A. (2001). Theory and speciation. *Trends in Ecology and Evolution*, **16**, 330–43.

Turelli, M. and Moyle, L. C. (2007). Asymmetric postmating isolation: Darwin's corollary to Haldane's rule. *Genetics*, **176**, 1059–88.

Ursin, E. (1952). Occurrence of voles, mice and rats (Muridae) in Denmark, with a special note on a zone of intergradation between two species of the house mouse (*Mus musculus* L.). *Videnskabelige Meddelelser fra Dansk naturhistorisk Forening i Kobenhavn*, **114**, 217–44.

Vanlerberghe, F., Boursot, P., Catalan, J., *et al.* (1988a). Analyse génétique de la zone d'hybridation entre les deux sous-espèces de souris *Mus musculus domesticus* et *Mus musculus musculus* en Bulgarie. *Genome*, **30**, 427–37.

Vanlerberghe, F., Boursot, P., Nielsen, J. T., and Bonhomme, F. (1988b). A steep cline for mitochondrial DNA in Danish mice. *Genetical Research*, **52**, 185–93.

Vanlerberghe, F., Dod, B., Boursot, P., Bellis, M., and Bonhomme, F. (1986). Absence of Y-chromosome introgression across the hybrid zone between *Mus musculus domesticus* and *Mus musculus musculus*. *Genetical Research*, **48**, 191–7.

van Zegeren, K. and van Oortmerssen, G. A. (1981). Frontier disputes between the West- and East-European house mouse in Schleswig-Holstein, West Germany. *Zeitschrift für Säugetierkunde*, **46**, 363–9.

Vigne, J.-D. (1992). Zooarchaeological and biogeographical history of the mammals of Corsica and Sardinia since the last Ice Age. *Mammal Review*, **22** (2), 87–96.

Vošlajerová Bímová, B., Macholán, M., Baird, S., *et al.* (2011). Reinforcement selection acting on the European house mouse hybrid zone. *Molecular Ecology*, **20**, 2403–24.

Vyskočilová, M., Pražanová, G., and Piálek, J. (2009). Polymorphism in hybrid male sterility in wild-derived *Mus musculus musculus* strains on proximal chromosome 17. *Mammalian Genome*, **20**, 83–91.

Vyskočilová, M., Trachtulec, Z., Forejt, J., and Piálek, J. (2005). Does geography matter in hybrid sterility in house mice? *Biological Journal of the Linnean Society*, **64**, 663–74.

Wade, C. M., Kulbokas, E. J., Kirby, A. W., *et al.* (2002). The mosaic structure of variation in the laboratory mouse genome. *Nature*, **420**, 574–8.

Wallace, A. R. (1889). *Darwinism: An Exposition of the Theory of Natural Selection with some of Its Applications*. London: Macmillan.
Walsh, J. B. (1982). Rate of accumulation of reproductive isolation by chromosome rearrangements. *The American Naturalist*, **120**, 510–32.
Wang, L., Luzynski, K., Pool, J., *et al.* (2011). Measures of linkage disequilibrium among neighboring SNPs indicate asymmetries across the house mouse hybrid zone. *Molecular Ecology*, **20**, 2985–3000.
Wright, S. (1943). Isolation by distance. *Genetics*, **28**, 114–38.
Wright, S. (1978). *Evolution and the Genetics of Populations, vol 4: Variability within and among Natural Populations*. Chicago, IL: University of Chicago Press.
Yang, H., Ding, Y., Hutchins, L. N., *et al.* (2009). A customized and versatile high-density genotyping array for the mouse. *Nature Methods*, **9**, 663–8.
Zimmermann, K. (1949). Zur Kenntnis der mitteleuropäischen Hausmäuse. *Zoologische Jahrbücher, Abteilung für Systematik, Ökologie und Geographie der Tiere*, **78**, 217–322

Appendix

Linkage disequilibrium can be computed as the product of the ratio between the variance of the distance parents to offspring (σ^2) and recombination rate (r) and the gradients at loci i and j ($\partial p_i / \partial x_i$ and $\partial p_j / \partial x_j$, respectively):

$$D_{ij} = \frac{\sigma^2}{r} \frac{\partial p_i}{\partial x_i} \frac{\partial p_j}{\partial x_j}, \tag{14.A1}$$

since at the centre these gradients are the inverse of the cline widths, $D_{ij} = \sigma^2 / (rw_i w_j)$. Linkage disequilibria can also be estimated as the covariance between quantitative traits:

$$\text{cov}(z_i, z_j) = \frac{\sigma^2}{2r} \frac{\Delta z_i}{w_i} \frac{\Delta z_j}{w_j}, \tag{14.A2}$$

where Δz_i and Δz_j are differences between populations on either side of the zone for loci i and j, respectively. These equations are based on the assumption that genetic variance is additive. On the other hand, different loci are not required to coincide in strength and direction and alleles need not be fixed on either side. To overcome a potential problem of varying allele frequencies across loci, which can also cause linkage disequilibria to vary, the standardized measure, $R_{ij} = D_{ij}\sqrt{p_i q_i p_j q_j}$, should be used instead of D_{ij}.

The stepped cline model consists of three functions: the central segment is described by a Tanh ('sigmoid') function, whereas the tails of the cline are modelled with exponential functions. The left tail can be expressed as (Raufaste *et al.*, 2005; Macholán *et al.*, 2007):

$$p(x_i) = \frac{w}{2\sqrt{\theta_L}}\frac{\Delta u}{B_L}\exp\left[\frac{2(x_i - c)}{w}\sqrt{\theta_L}\right], \tag{14.A3}$$

and similarly for the right tail:

$$p(x_i) = 1 - \frac{w}{2\sqrt{\theta_R}}\frac{\Delta u}{B_R}\exp\left[\frac{x - 2(x_i - c)}{w}\sqrt{\theta_R}\right], \tag{14.A4}$$

where

$$\Delta u = \frac{(2/w)B_L\sqrt{\theta_L}B_R\sqrt{\theta_R}}{B_L\sqrt{\theta_L} + B_R\sqrt{\theta_R} + (2/w)B_L\sqrt{\theta_L}B_R\sqrt{\theta_R}}, \tag{14.A5}$$

which simplifies to $\Delta u = B\sqrt{\theta}/\left(w + B\sqrt{\theta}\right)$ when the cline is symmetrical.

If we assume that fitness is not frequency-dependent and linkage disequilibria are weak ($R_{ij} \ll 1$) the rate of dispersal can be estimated as

$$\sigma = w\sqrt{\frac{R_{ij}\bar{r}}{1 + \bar{r}}}, \tag{14.A6}$$

where $\bar{r}$ is the harmonic mean recombination rate among loci (Barton and Gale, 1993; see Szymura and Barton, 1991; Porter *et al.*, 1997; Macholán *et al.*, 2007 for the computation of $\bar{r}$). Assuming selection acting against heterozygotes, the effective selection pressure maintaining the cline is given as $s^* = 8(\sigma/w)^2$ (Szymura and Barton, 1986, 1991). If we further assume that the number of selected loci is large and that linkage disequilibrium is generated predominantly by dispersal, then the barrier B can be approximated as $B = \sigma\sqrt{2ns/\bar{r}}$ (Barton and Shpak, 2000) and $ns/\bar{r} = 2\ln(B/w)$ (Barton and Bengtsson, 1986; Raufaste, 2001), where s is selection at loci responsible for reproductive barrier and n is the number of these loci. From these equations we can derive formulae for s and n (Raufaste *et al.*, 2005; Macholán *et al.*, 2007), with the total selection pressure against hybrids $S = ns$. Finally, assuming fitnesses are not density-dependent, we may use estimates of cline width, the strength of the barrier, and the harmonic mean recombination rate to derive mean fitness of a hybrid population relative to parental populations (Barton and Bengtsson, 1986): $\bar{W}_H/\bar{W}_P = (w/B)^{\bar{r}}$; in the zone centre, where we expect $p = q = 0.5$, $\bar{W}_H/\bar{W}_P$ can be approximated as $\approx \exp(-S/2)$ (Szymura and Barton, 1991; Macholán *et al.*, 2007).

To cope with potential non-independence of samples due to relatedness and deviations from the Hardy–Weinberg equilibrium, we can calculate the effective number of alleles first introduced by Phillips *et al.* (2004) and described in full by Macholán *et al.* (2008). If N is the number of diploid individuals sampled at a site, F_{IS} is the measure of deficit/excess of heterozygosity in an individual relative to a subpopulation (Wright, 1978), and F_{ST} is the measure of relatedness in each subpopulation, then the effective number of alleles is

$$N_e = \frac{2N}{2NF_{ST} + (1 - F_{ST})(1 + |F_{IS}|)}. \qquad (14.A7)$$

If all sampled individuals are completely related, N_e will be 1, whereas if the individuals are completely unrelated N_e will range from N (when $F_{\mathrm{IS}} = 1$) to $2N$ (when $F_{\mathrm{IS}} = 0$). Thus, for high F_{ST} we are avoiding giving undue weight to large samples. For haploid markers $F_{\mathrm{IS}} = 0$ and so Eq. (14.A7) simplifies to:

$$N_e = \frac{N}{2NF_{ST} + 1 - F_{ST}}. \qquad (14.A8)$$

A slightly more complex formula must be used for the X chromosome, which is diploid in females and haploid in males. If we denote the number of sampled males and females as N_m and N_f, respectively, and $N_{\mathrm{IS}f}$ is the measure of deficit/excess of heterozygotes in females, then the effective number of alleles is:

$$N_e = \frac{2N_f + N_m(1 + |F_{ISf}|)}{2N_f F_{ST} + [1 - F_{ST}(1 - N_m)]\left(1 + |F_{IS_f}|\right)}. \qquad (14.A9)$$

As pointed out by Macholán *et al.* (2008), in some cases estimation of F_{IS} may lead to confusing results. For example, if one of the genotype classes equals 0 and the second one is close to 0, then the maximum likelihood estimate of F_{IS} will be 1 (e.g. when $pp = 50$, $pq = 0$, $qq = 1$) or -1 (e.g. when $pp = 0$, $pq = 1$, $qq = 50$) irrespective of the 'true' value. In these cases some arbitrary decisions must be taken – for instance, if one of the counts is 0 and the second lowest count is less than some critical value (say, 3), then we set F_{IS} to 0 (Macholán *et al.*, 2008).

15

Behaviour, ecology, and speciation in the house mouse

GUILA GANEM

Introduction

The house mouse is a model organism that also provides an exciting natural setting to study different mechanisms and steps in the process of speciation. In this chapter I will discuss to what extent the consideration of ecology and behaviour has improved our understanding of the house mouse evolution, and has the potential to contribute to progress in our general understanding of mechanisms of speciation. I will address specifically two evolutionary processes: chromosomal evolution within *Mus musculus domesticus* and incipient speciation between *M. m. domesticus* and *M. m. musculus*.

The house mouse is one of the most widespread mammals. It occurs on all continents and most islands and thrives in diversified habitats. It is a complex of subspecies (see Auffray and Britton-Davidian, Chapter 1 in this volume; Suzuki and Aplin, Chapter 2) and chromosomal races (due to centric fusions and other chromosomal rearrangements; see Hauffe *et al.*, Chapter 16), and its association with human activities and dwellings strongly impacts its population dynamics and structure, resulting in patchily distributed small to large populations with connections that vary in intensity both in space and time.

Its geographic and ecological characteristics are expected to favour diversification and make the house mouse an ideal model to test hypotheses related to behavioural and ecological speciation in a mammal. Moreover, chromosomal mutations and secondary contacts after evolution in allopatry provide natural evolutionary settings to test the role of selection in speciation.

Research into the ecology and behaviour of wild house mice started in the early 1940s in Europe, first driven by its pest status (Laurie, 1946; Southern and Laurie, 1946), and ~30 years later in Australia (Newsome, 1971; Singleton and Redhead, 1990; Singleton *et al.*, 2007), a status that is still relevant today for conservation as the mouse has become a predator of and a threat to endangered species (Simberloff, 2009).

Evolution of the House Mouse, ed. Miloš Macholán, Stuart J. E. Baird, Pavel Munclinger, and Jaroslav Piálek. Published by Cambridge University Press.

In parallel, the pest has slowly become a model in systematic biology (Boursot *et al.*, 1993; Sage *et al.*, 1993) following the discovery of its chromosomal variation (Gropp *et al.*, 1969), the description of a hybrid zone between the two European subspecies (Selander *et al.*, 1969), and several other clarifications of the systematic of the house mouse: a complex of distinct races and subspecies occurring in wild and commensal habitats (Thaler *et al.*, 1981).

To date, major progress has been made in describing and understanding the genetic mechanisms involved in house mice diversification (e.g. Dumas and Britton-Davidian, 2002; Lundrigan *et al.*, 2002; Raufaste *et al.*, 2005; Payseur and Place, 2007; Geraldes *et al.*, 2008; Macholán *et al.*, 2008; Teeter *et al.*, 2008, 2010; Mihola *et al.*, 2009; White *et al.*, 2009) and the general ecology and behaviour of the house mouse (Brain, 1989; Berry and Bronson, 1992; Novotny, 2002; Beynon and Hurst, 2003; Hurst and Beynon, 2004; Pocock *et al.*, 2004; Singleton *et al.*, 2005, 2007; Cheetham *et al.*, 2007), while much less research has been devoted to understanding the ecological and behavioural drivers of incipient radiation in the house mouse (e.g. Capanna *et al.*, 1985; Ganem and Searle, 1996; Christophe and Baudoin, 1998; Chatti *et al.*, 1999; Castiglia and Caporioni, 2005; Nunes *et al.*, 2005, 2009; Smadja and Ganem, 2005; Bímová *et al.*, 2009). Specifically, factors such as quality, continuity, and stability of habitats, social structure, and reproduction system of populations as well as sexual recognition signals and mate choice are known to be drivers of diversification in many model systems (Coyne and Orr, 2004; Rundle and Nosil, 2005; Ritchie, 2007; Smadja and Butlin, 2009; Elmer *et al.*, 2010; Hoskin and Higgie, 2010). They have been regularly evoked by mouse evolutionary biologists (Boursot *et al.*, 1993; Sage *et al.*, 1993; Capanna and Castiglia, 2004) and behavioural ecologists (Ganem, 1996), but, until recently, did not lead to strong interactions between the two. Nevertheless, things are changing with, on the one hand, accumulation of evidence for the role of ecology and behaviour in mouse evolutionary processes (Chatti *et al.*, 1999; Nunes *et al.*, 2005, 2009; Smadja and Ganem, 2005; Bímová *et al.*, 2009) and, on the other, the development of functional genomics with the prerequisites to make the link between complex phenotypes, such as most behavioural traits, and 'genes of speciation' (Robinson, 2002; Carroll and Potts, 2006).

What have investigations of house mouse behaviour and ecology taught us about the house mouse chromosomal radiation?

On the influence of adaptation on chromosomal radiation

M. m. domesticus is the only subspecies of house mouse exhibiting chromosomal radiation, a phenomenon mainly observed in Europe and

Tunisia (reviewed in Piálek *et al.*, 2005). One type of chromosomal mutation, centric or Robertsonian fusion (hereafter Rb mutation), is predominant in this subspecies: a fusion of two chromosomes by their centromeres – although other mechanisms such as hybridization and WARTs (whole-arm reciprocal translocations) are known to generate novel Rb variants (Piálek *et al.*, 2005).

The fixation of Rb mutations in a large number of natural populations despite at least slight underdominance (Searle, 1993; Nachman and Searle, 1995; Hauffe and Searle, 1998; Castiglia and Capanna, 2000) has motivated research into mechanisms that could compensate for this disadvantage, one of them being that the mutation may produce an adaptive phenotype conferring a selective advantage to the carrier.

A variety of studies have addressed this question, involving laboratory and natural populations of mice (see Table 15.1). The traits studied were mainly behavioural, but also include ecophysiology and morphology. Among the studies compiled in Table 15.1, only five concluded that the Rb mutation under study could have conferred an adaptive advantage to the carrier. Among these studies, I believe that only one presented convincing results (synthesized in Dyban and Baranov, 1987), being the only one to have replicated the investigation in several groups of mice and to have found consistent results. Dyban and collaborators worked on laboratory mice carrying an Rb mutation between chromosomes 8 and 17 (hereafter Rb(8.17)). This type of Rb can be found in several natural populations (14 out of the 97 compiled by Piálek *et al.*, 2005) from various geographical origins (Greece, Switzerland, Croatia, Germany, Austria, and Italy). Dyban and collaborators compared laboratory mice carrying the above mutation with mice carrying either none or one of two other Rb mutations (Rb(6.15) or Rb(5.19)). The results indicated, consistently in different genetic backgrounds, that mice carrying Rb(8.17) had higher cognitive abilities (i.e. extrapolation of movement direction of food stimuli), leading the authors to propose that genes on chromosome 17 may be involved and may produce the advantageous behavioural phenotype *only* when chromosome 17 is fused with chromosome 8 (see Dyban and Baranov, 1987 for a review). These results are intriguing, and could be further tested in natural populations carrying this fusion. Nevertheless, although there is theoretical support for the possible influence of such mutations on recombination (Dumas and Britton-Davidian, 2002), and/or variation of gene expression due to differential chromosomal organization within the interphase nucleus (Spilianakis *et al.*, 2005), it still does not explain the diversity of fusions and combinations of fusions that exist in natural conditions. Indeed, if some Rb mutations have detectable behavioural phenotypic effects, it is most probable that the effects differ among Rb types and their combinations occurring in nature (e.g. Gazave *et al.*, 2003). Further, even if some of the Rb mutations provide an

Table 15.1 A compilation of studies comparing traits that may impact fitness of *M. m. domesticus* mice involved in chromosomal radiation.

Trait	No. of populations compared (no. mice)	Origin	Results and characteristics of Rb involved	Authors interpretation of results	References
Male aggression and dominance	2Rb (20, 20)	Laboratory progeny of natural populations, Migiondo, Italy	*CHPO(2n = 24) > IUVA (2n = 26) The two races differ by CHPO having Rb (8.12) and IUVA Rb(7.18), (2.8), (10.12)	By-product of isolation. No role for Rb per se	Capanna *et al.* (1984)
Male aggression and dominance	1AC 1Rb (28, 18)	Laboratory mice. X-rays → Rb	Rb > AC Rb(10.13)	Adaptive impact of Rb mutation	Schröder and Otten (1985)
Male aggression and dominance	1AC, 1Rb (10, 10)	Natural populations, central Italy	Rb > AC ICDE(2n = 22)	By-product of competition and selection against hybridization	Corti *et al.* (1989)
Male aggression and dominance	2Rb (13, 8)	Natural populations, central Italy	IACR (2n = 24) ICDE (2n = 22) All their Rb are different	By-product of selection against hybridization	Carpineti and Castiglia (2004)
Male aggression, dominance, and reproductive success	2AC, 2 Rb (29, 20 and 30)	Laboratory progeny of natural populations, central Tunisia	TUMO (2n = 22) > AC	By-product of chromosomal divergence	Ould Brahim (2005)

Extrapolation (direction of movement of food stimulus)	2AC, 5Rb (155, 60 and 27, 23, 193, 26, 27)	Laboratory mice	Rb(8.17) > AC, Rb(6.15), Rb(5.19). Laboratory and natural origin of Rb(8.17) in laboratory genetic background Results confirmed in strains where Rb(8.17) and AC mice had similar genetic background	Presence of Rb(8.17) → developmental impact on central nervous system (cognition) Possible role for genes on chromosome 17	Krushinsky *et al.* (1986); Dyban and Baranov (1987)
Swimming and navigation Spatial memory Reactivity to external stimuli in the open field	2AC, 2Rb (8, 7 and 5, 9)	Laboratory mice	Rb < AC Rb < AC Rb > AC Laboratory origin of Rb (8.17)	Rb may result in slight alterations in limbic system	Leitinger *et al.* (1994)
Basal levels of corticosterone Stress levels of corticosterone	2AC (commensal) 1AC (feral) and 1Rb (22, 24 and 28, 28)	Natural populations, Montpellier, France; Torino and Binasco, Italy	Feral < commensal Feral > commensal IBIN (2n = 24) intermediate levels	By-product of adaptation to the commensal habitat No Rb replicate	Ganem (1991)
Body mass Testes mass Social interactions Corticosterone	2AC, 3Rb (10–30, 10–32 and 8–26, 10–52, 4–13)	Natural populations, Orkney, Scotland	Variable not different between Rb and AC 2 UKED (2n = 34) and 1 UKWE Differ by 3 Rb	No role for Rb	Ganem (1998)

Table 15.1 (*cont.*)

Trait	No. of populations compared (no. mice)	Origin	Results and characteristics of Rb involved	Authors interpretation of results	References
Parasite load	4AC, 6Rb (~160)	Natural populations, Caithness, Sutherland and Orkney, Scotland	UKLY,UKWE (2n = 36) < AC, UKJO (2n = 32), and UKED (34) The two Rb(2n = 36) differ by 1 Rb	Hybrid origin of insular and mainland Rb 2n = 36.	Ressouche *et al.*, (1998)
Skull and mandible shape in relation to aggression	2Rb (~20, ~30)	Natural populations, Migiondo, Italy	CHPO (2n = 26) ≠ IUVA (2n = 24) Shape changes correlate with intensity of aggressive behaviour The two differ by 3 Rb	By-product of divergence in aggressive behaviour No role for Rb per se	Corti and Rohlf (2001)
Mean daily motor activity	1AC, 1Rb (9, 9)	Natural population, Barcelona, Spain	AC > EBAR (2n = 29–32)	Rb affect gene regulation related to metabolic pathways	Sans-Fuentes *et al.* (2005)
Ultradian rhythms			EBAR (2n = 29–32) > AC		
Resting oxygen consumption	2AC, 6Rb (>41)	Natural populations, Madeira, Lisbon, Portugal	Rb < AC Lisbon Rb ♯ AC Madeira archipelago (PLDB, PSAN, PEDC, PADC, PSVI).	Adaptation to insular conditions	Mathias *et al.* (2006)

Trappability in enclosure and in the wild	1Rb (~300)	Laboratory-born progeny and natural populations, Belgium	Enclosure 1: Rb heterozygote < Rb homozygote Enclosure 2 and in the wild: Rb heterozygote ♯ Rb homozygote BBEL(2n = 36)	Adaptive advantage of Rb heterozygote which could favour Rb fixation	Gérard *et al.* (1994)
Social interactions within a population	2AC, 2Rb (20, 18 and 16, 20)	Natural populations, Caithness and Sutherland, Scotland	Variable no differences between Rb and AC. UKLY (2n = 36), UKJO (2n = 32) The two differ by 2 Rb	No role for Rb	Ganem and Searle (1996)
Social amicable behaviour during between-population encounters	1AC, 2Rb (8, 12 and 16, 20)	Natural populations, Caithness and Sutherland, Scotland	AC > UKLY(2n = 36) or UKJO(2n = 32)	By-product of geographical isolation between populations	Ganem and Searle (1996)
Race recognition signal	AC/Rb and Rb/Rb comparisons with or without replicates	Natural populations or laboratory progeny of natural populations, Tunisia, Scotland, Italy, Madeira	Discrimination exists between all tested races (AC/Rb and Rb/Rb) *TUMO, ICDE, IACR, UKJO, UKLY, IUVA, CHPO, PADC, PEDC	By-product of isolation or local adaptation	Capanna *et al.* (1985); Corti *et al.* (1989); Ganem and Searle (1996); Benzekri (2000); Benzekri *et al.* (2002); Carpineti and Castiglia (2004); Ould Brahim (2005); Nunes *et al.* (2009)

Note: **Notes:* AC: all acrocentric standard mice 2n = 40. Rb: mice carrying chromosomal centric fusions.
‡* Rb populations coding refers to Piálek *et al.* (2005).

adaptive advantage when carrying specific genes, it seems reasonable not to expect that all of them or all the different combinations they are involved in do so. Consequently, if a general explanation for the paradox of fixation of a slightly underdominant chromosomal mutation is to be found, we may need to look for other causes than a behavioural advantage (e.g. Pardo-Manuel de Villena and Sapienza, 2001).

The cascade of influence of the commensal habitat on mice behaviour and chromosomal evolution

Male aggressiveness and dominance is a quality trait in mice (Mossman and Drickamer, 1996; Gosling *et al.*, 2000; Roberts and Gosling, 2003) and has been investigated in many Rb populations (Table 15.1). When compared to standard mice, Rb mice were often found to be more aggressive and to dominate males with a standard karyotype (e.g. Corti *et al.*, 1989). In a study performed in natural conditions in which commensal Rb mice (from one of the Orkney Islands) were introduced to the Isle of May (United Kingdom) and were allowed to interact with a local outdoor population of mice carrying a standard karyotype, dominance of Rb mice was also reported (Berry *et al.*, 1991; Jones *et al.*, 1995). Moreover, a rapid genetic take-over by males of the Rb population was observed (introgression of markers on the Y chromosome) and interpreted as implying a competitive advantage of the Rb mice compared to the standard mice (Sage *et al.*, 1993). However, this was not the interpretation of the authors of the study, who proposed that relatively high levels of inbreeding on the Isle of May – before the introduction of the Rb mice – could have disadvantaged the local mice (Berry *et al.*, 1991; Jones *et al.*, 1995; Gray and Hurst, 1998). The introduction was not aimed at testing the competitive ability of Rb mice and as such the study contained several confounding effects precluding any conclusion about dominance of Rb mice. One of the factors that was not controlled and that may be important is the characteristics of the habitat in which the introduced population had evolved: unlike the May population, the Rb population originated from a commensal habitat, and its dominance could have been a by-product of adaptation to a type of habitat where between-individual interactions and competition can be high (Christian, 1970; Bronson, 1979) and selective for dominant aggressive mice (Franks and Lenington, 1986; Benus *et al.*, 1991).

Most Rb populations occupy indoor commensal habitats (Ganem, 1993; Castiglia and Caporioni, 2005). When Rb populations establish permanent populations outdoors, it is either in the absence of competitors on islands such as Faray in Scotland (Ganem, 1998) or Madeira (Nunes *et al.*, 2005), or in human-modified habitats (e.g. walls, agricultural fields, or very close to human dwellings). The more frequent occurrence of standard mice relative to Rb mice in outdoor habitats has led some authors to propose that Rb mice may be better adapted to the indoor habitat than standard mice (Castiglia and Caporioni, 2005). However, when standard and Rb mice occur in the same type of habitats they display similar characteristics (Ganem *et al.*, 1989; Ganem, 1991; Mathias *et al.*, 2004, 2006), suggesting that they are not commensally adapted relative to the standard-type population. In fact, the very rare occurrence of Rb populations outdoors can be simply explained by the facts that, first, Rb populations are relatively rare, second, most available outdoor habitats contain competitors with which the house mouse does not seem to cope well (Berry *et al.*, 1982; Fox and Pople, 1984; Boitani *et al.*, 1985), and third, outdoor habitats are not very widespread where Rb populations occur. In line with this, habitat selection was hypothesized to explain occupation by Rb populations of the most stable commensal habitats in the central part of Tunisia (Saïd and Britton-Davidian, 1991), and was rejected on the grounds of a detailed micro-geographical and historical investigation of this Rb system. This study revealed that a plausible and more parsimonious explanation for habitat partitioning may be stochastic processes related to the history of urbanization in central Tunisia and classical mechanisms giving advantage to the first colonizer of high-quality habitats, which were probably Rb mice in central Tunisia (Chatti *et al.*, 1999; see also Hardouin *et al.*, 2010 for a discussion on constraints on displacement of settled populations).

It has long been hypothesized that human commensal habitats could have created favourable conditions, in terms of population structure and dynamics, for the fixation and maintenance of Rb populations (Bush *et al.*, 1977; Capanna and Redi, 1994; Capanna and Castiglia, 2004; but see Nachman and Searle, 1995). Human commensal habitats are aggregated patches of more or less suitable habitat where mouse reproduction can take place all year round, with frequent extinction/recolonization events (Pocock *et al.*, 2004). Mouse social structure is also patchy, with extended family reproductive units more or less isolated from each other over various periods of time (Selander, 1970; Baker, 1981; Cox, 1984; Baker and Petras, 1986). However, even if the available data indicate that many commensal mouse populations are actually not strictly structured in isolated reproductive demes (Pocock *et al.*, 2005), some of the data do suggest that variations occur and that such conditions (isolated demes) could exist in some commensal habitats (e.g. Cox, 1984). Further, more isolated and stable commensal habitats could have prevailed in historical times, when chromosomal radiation took place. Finally, commensal mouse social structure may still be more favourable for Rb fixation than the structure of wild/feral populations of mice in which densities are generally low, or fluctuate dramatically, the reproductive unit is a territorial pair or a triad, and the rate of dispersal is high (Berry and Jakobson, 1974; Cassaing and Croset, 1985; Cassaing, 1988; Singleton and Krebs, 2007).

Commensal habitats include buildings, agricultural fields, and gardens. The three types of environments are different and each of them can show tremendous variation. They are all heterogeneous in terms of stability of food and shelter and hence their quality for mice (see discussion in Nunes *et al.*, 2005). High-quality habitats are also those where density tends to be the highest, because they attract more mice and reproduction occurs all year round. However, there may also be increased competition, favouring hierarchical social structure where both males and females have delayed reproduction, and dominant animals have the majority of access to reproduction (van Zegeren, 1980).

Evolution in high-quality habitats may explain higher male aggressiveness of Rb mice in Tunisia (Ould Brahim, 2005) and other places (see Table 15.1), or female preference for odours of Rb males occurring in high-quality habitats if such habitat selects for males of better quality (see discussion in Nunes *et al.*, 2009).

Variation in dominance has also influenced interactions between Rb populations. For example, dominance and assortative mating is suspected to have resulted in competitive exclusion of one population by another in an Italian village (Capanna and Corti, 1982; Corti *et al.*, 1989; Hauffe and Searle, 1992). In this example, differential aggressiveness has probably evolved in allopatry, and competitive exclusion took place in a geographical context similar to insularity (Fraguedakis-Tsolis *et al.*, 1997).

The evolution of mate recognition systems in the context of chromosomal radiation

Mate recognition systems (MRS) are communication devices allowing sexual partners to identify and choose their mate (Butlin and Ritchie, 1994; White *et al.*, 1995). Such devices allow recognition from the population to the species level, where they are expected to facilitate species cohesion. How these systems evolve during the speciation process has been the subject of much debate (Dobzhansky, 1940; Paterson, 1980; Eldredge, 1995; White *et al.*, 1995) and fruitful research (reviews in Coyne and Orr, 2004; Ritchie, 2007; Smadja and Butlin, 2009).

Research on the house mouse strongly suggests that its main MRS communication channel is olfactory (see Stopka *et al.*, Chapter 8; Penn and Musolf, Chapter 9), although other sensory systems may be involved (see Musolf and Penn, Chapter 10). Studies involving chromosomal races have assessed divergence in mate preference or signals by testing for positive or negative discriminatory behaviour in within-population, as compared to between-population, trials (e.g. Capanna *et al.*, 1985; Ganem and Searle, 1996). Divergence of mate recognition signals and preference was found to occur in several instances (see Table 15.1). As with differences in aggressive behaviour, population processes could account for divergence in MRS between some chromosomal races (Ganem and Searle, 1996). In other instances, 'assortative mating' loci located close to the centromeres of fused chromosomes might be fixed in a population despite gene flow, as hypothesized for the only case of putative local (i.e. at the scale of a village) speciation by reinforcement described for two chromosomal races in Italy (Hauffe and Searle, 1992; Hauffe *et al.*, 2004; Panithanarak *et al.*, 2004), however, to date the existence of such loci has not been documented.

Investigations of mate signalling and preference in the context of chromosomal evolution have revealed differences between chromosomal races (e.g. Fig. 15.1; Table 15.1). However, these could only rarely or weakly be related to chromosomal divergence *per se*, and other mechanisms can be evoked such as

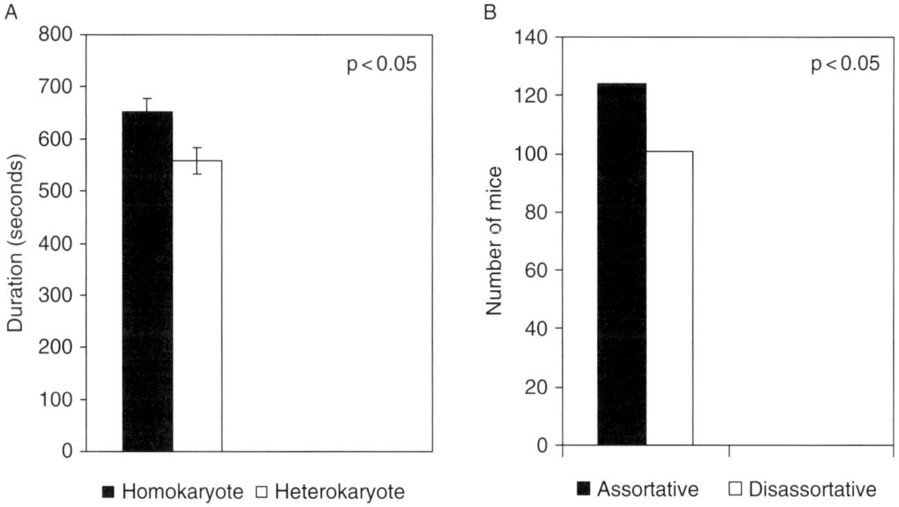

Figure 15.1 Mate preference assessed in wild mice from the Tunisian Rb system. (a) Time spent sniffing a mouse of the opposite sex with the same karyotype (homokaryote) *vs* with a different karyotype (heterokaryote). (b) Number of mice sniffing a mouse of the same chromosomal type more (assortative) *vs* number of mice sniffing a different mouse more (disassortative). A total of 225 adult male and female mice trapped in five 2n = 22 and five 2n = 40 geographical locations in central Tunisia was tested. Testing took place in a two-way choice maze involving a chooser and two mice of the opposite sex. Each test lasted 30 minutes and all females were tested while sexually receptive (modified from Benzekri, 2000).

sexual selection (e.g. Nunes *et al.*, 2009). When several populations carrying the same combination of fusions could be compared, as is the case for the race occurring in Tunisia, it appeared that most divergence could also be explained by population processes in allopatry, or in parapatry in the presence of ecological barriers (Benzekri, 2000).

To conclude the first part of this chapter, as far as behavioural ecology of the house mouse chromosomal radiation is concerned, the distribution and the relative success of some Rb populations relative to other Rb or standard populations may not require the involvement of adaptive advantages other than those selected via classical stochastic population processes related to the colonization of commensal habitats of good quality. Further, behavioural isolation between some of the Rb populations or between Rb and standard populations could be explained by divergence in allopatry, or after a secondary contact with maladaptive hybridization, or sexual selection favouring divergence of the MRS in different environmental conditions. Structural mechanisms inherent to the Rb mutation may participate in, but are not required for, the behavioural diversification between chromosomal races.

Evolution in the context of maladaptive hybridization between two subspecies of the house mouse

The two European subspecies of the house mouse hybridize wherever they meet (i.e. in Europe and Asia), a process partially counter-selected and investigated both in laboratory and natural conditions (e.g. Fel-Clair *et al.*, 1998; Raufaste *et al.*, 2005; Payseur and Place, 2007; Macholán *et al.*, 2008; Teeter *et al.*, 2010). Secondary contacts provide a unique opportunity to test the role of selection in speciation, and assessment of the extent of divergence preceding contact may help to disentangle the mechanisms involved.

The two subspecies have distinct distributions (see Auffray and Britton-Davidian, Chapter 1) and occur, for a large part of their distribution, in areas with different environmental characteristics (e.g. climate, human culture). Even the routes that ancestral populations of the two subspecies followed during their 'diffusion' into Europe several thousand years ago (Cucchi *et al.*, 2005; see also Cucchi *et al.*, Chapter 3) – the Mediterranean route for *domesticus* and the Danubian route for *musculus* – were different, and hence could have affected their evolution from the very beginning of their geographic separation. Nevertheless, studies comparing ecological features of their distribution are lacking; one reason may be their association with human dwellings, which is considered to protect them from factors such as climate variations (e.g. Bronson, 1979; Pocock *et al.*, 2004); however, the relative protection may not suffice to erase all potential influence of climate differences. Moreover, biological factors (e.g. competitors and predators) and even human culture may have had different effects on the environment in which the two subspecies have evolved. Such comparative studies of the ecological characteristics of habitats of the two subspecies do not exist and may be particularly relevant for our understanding of the drivers involved in the house mouse diversification.

Ecological characteristics of the two subspecies

It has been suggested that *M. m. domesticus* may be more strictly commensal than *M. m. musculus* (Sage, 1981; Orsini *et al.*, 1983; but see Hunt and Selander, 1973). However, this does not seem to have ever been properly documented. Indeed, the rare studies on the subject indicate that both subspecies have populations displaying seasonal movements between indoor and outdoor habitats (e.g. Pelikán, 1981; Rowe *et al.*, 1987; Walkowa *et al.*, 1989; Carlsen, 1993; Gomez *et al.*, 2008), and both *M. m. domesticus* (Berry, 1968; Berry and Tricker, 1969; Orsini *et al.*, 1982; Boitani *et al.*, 1985; Ganem *et al.*, 1989) and *M. m. musculus* (Pelikán, 1981; Kotenkova *et al.*, 1994) also present permanent outdoor populations.

The main limit for the occurrence of *M. m. domesticus* outdoors seems to be the presence of competitors, mainly other rodent species (Lidicker, 1966; Berry *et al.*, 1982; Orsini *et al.*, 1982; Fox and Pople, 1984; Boitani *et al.*, 1985; Fairley and Smal, 1987; Auffray *et al.*, 1990; Busch *et al.*, 2005). As far as *M. m. musculus* is concerned there are indications that seasonal movements of mice from farm buildings to nearby fields coincide with a decrease in density of the rodent community present outdoors, and that their return to buildings coincides with an increase in density of the same community (Carlsen, 1993), further suggesting that in *M. m. musculus*, too, presence outdoors may be mainly limited by competition (but see Pelikán, 1981).

In the absence of competitors, the house mouse seems to choose outdoor habitats rather than indoor habitats – for example in Madeira where *M. m. domesticus* occurs outside buildings in gardens and close by vineyards all year round (Nunes *et al.*, 2005), or in southeastern Denmark where *M. m. musculus* occurs outdoors when competition is low (Carlsen, 1993), which suggests that the two subspecies may prefer natural habitats. In such habitats within-subspecies competition (and the resulting social stress) may be lower and more easily coped with (Ganem *et al.*, 1989; Ganem, 1991), fecundity might be higher (Pelikán, 1981), and mortality lower (Pocock *et al.*, 2004). Whether the two subspecies differ in the way they thrive in similar habitats remains an unexplored question. Only anecdotal information is available that their life history strategies could be different. For example, male *musculus* are slightly lighter than male *domesticus* from the same geographical areas (van Zegeren and van Oortmerssen, 1981; Piálek *et al.*, 2008; G. Ganem, unpublished data), and female *musculus* tend to produce a smaller litter size (Britton-Davidian *et al.*, 2005; Piálek *et al.*, 2008; G. Ganem, unpublished data).

Behavioural characteristics of the two subspecies

Behavioural flexibility and, more specifically, social flexibility is a remarkable feature of the house mouse and has been suggested as one explanation of its success (Mackintosh, 1981; Krebs *et al.*, 1995; Singleton and Krebs, 2007). The social structure of the house mouse varies with habitat, resource partitioning, and density, and many studies have investigated this in *M. m. domesticus* (e.g. Stueck and Barrett, 1978; Butler, 1980; Noyes *et al.*, 1982; Pocock *et al.*, 2004; Singleton and Krebs, 2007).

Comparative studies of behaviour of the two subspecies have mainly involved mice from commensal populations and primarily concern aggressive behaviour and mate choice (Table 15.2). Results suggest that the two subspecies differ in their aggressive behaviour, and that their MRS allows mice of the two subspecies to discriminate against each other without complete assortative mating. Candidate odorant molecules that may be involved in the recognition signal and aggression divergence have been proposed (Stopková *et al.*, 2007; Mucignat-Caretta *et al.*, 2010).

M. m. domesticus males are more aggressive and dominate *M. m. musculus* males (Thuesen, 1977; van Zegeren and van Oortmerssen, 1981; Munclinger and Frynta, 2000; Frynta *et al.*, 2005; Davidian, 2008; Piálek *et al.*, 2008; Ďureje *et al.*, 2010). This is consistent with higher levels of brevicomin (2,3-dehydro-exo-brevicomin) and thiazolin (2-sec-butyl-4,5-dihydrothiazole) found in urine of *domesticus* males as compared to *musculus* (Mucignat-Caretta *et al.*, 2010): two male pheromones involved in signalling aggressiveness and dominance (Jemiolo *et al.*, 1985; Novotny *et al.*, 1985; Harvey *et al.*, 1989). Aggressive behaviour is a heritable trait, with a complex genetic, epigenetic, and environmental determinism (van Oortmerssen and Bakker, 1981; van Oortmerssen *et al.*, 1992; Sluyter *et al.*, 1996; Maxson and Canastar, 2003). Moreover, it was recently shown that divergence in aggressive behaviour between the two subspecies had a strong genetic component (Ďureje *et al.*, 2010). Aggression is likely subject to natural and sexual selection, hence it may be particularly relevant to understand what promoted such divergence between the two subspecies.

Variation in aggressiveness between species can be the result of evolution in different environmental conditions (e.g. Nevo *et al.*, 1975; Huntingford, 1976).

Table 15.2 A compilation of studies comparing behavioural-related traits in *M. m. musculus*, *M. m. domesticus*, and their hybrids.

Studied trait	Type and origin of mice	Results	Reference
Subspecies mate preference	Outbred strains of the two subspecies. Denmark, border hybrid zone	Assortative preference in *musculus*, not detected in *domesticus*. Soiled bedding and urine contain signals involved in subspecies recognition	Christophe and Baudoin (1998)
Subspecies mate recognition signals		Subspecies recognition may require contact	Ganem *et al.* (2005)
Subspecies mate preference	Outbred strains of the two subspecies. Nine different allopatric locations	Weak assortative preference: *musculus* tend to be more choosy than *domesticus*, patterns of discrimination suggest signal divergence	Ginane (1998)
Subspecies mate preference Subspecies mate recognition signals	Wild mice of the two subspecies. Denmark, border hybrid zone	Assortative preference in *musculus* Not detected or weak preference in *domesticus* Urine contain signals for subspecies recognition	Smadja and Ganem (2002, 2005); Ganem *et al.* (2008)
Subspecies mate preference	Wild mice of the two subspecies. Denmark and Czech–Bavarian border hybrid zone and different allopatric locations	Preference divergence between subspecies: contact zone > allopatry; *musculus* > *domesticus*	Smadja and Ganem (2005)
Subspecies mate recognition signals		Urinary signals divergence between subspecies: contact zone > allopatry	Smadja and Ganem (2008)
Subspecies mate preference Subspecies mate recognition signals	Outbred strains of the two subspecies. Czech–Bavarian border hybrid zone	Assortative preference in *musculus*, less marked in *domesticus*. Involvement of urine and faeces in signalling	Bímová *et al.* (2009)

Table 15.2 (*cont.*)

Studied trait	Type and origin of mice	Results	Reference
Mate preference for saliva containing androgen-binding proteins (ABP)	Inbred and outbred strains of the two subspecies and congenic mice	Divergence involves molecules present in saliva (ABP)	Laukaitis *et al.* (1997); Talley *et al.* (2001)
Subspecies mate preference for saliva containing androgen-binding proteins (ABP)	Wild mice of the two subspecies and their hybrids. Czech–Bavarian hybrid zone and allopatry Saliva obtained from congenic mice	Mice discriminate between the two saliva types Time spent investigating the two stimuli: contact zone populations > allopatric populations.	Bímová *et al.* (2005)
Perception of potential mate odours of the two subspecies by their hybrids	Wild mice of the two subspecies and their hybrids. Denmark, hybrid zone. Urine: outbred strains; allopatry	Natural hybrids discriminate between urine of the two subspecies. *M. m. domesticus*-like preference dominates in the hybrid zone	Ganem *et al.* (2008)
Subspecies mate preference Subspecies mate recognition signals Male aggressiveness and dominance in neutral arena and in intruder tests	Outbred strains of the two subspecies. Czech–Bavarian, border hybrid zone.	Assortative preference in *musculus*, not detected or weak in *domesticus* Urine contains signals for subspecies recognition Male *domesticus* more aggressive and dominant despite territorial defence by *musculus* males	Piálek *et al.* (2008)
Aggression: the role of genetic versus postnatal maternal environment	Outbred strains of the two subspecies. Czech–Bavarian, border hybrid zone.	Aggression has a significant genetic component	Ďureje *et al.* (2010)
Aggression: paternal effect	Outbred strains of the two subspecies and their hybrids Allopatry	No paternal effects detected, hybrid F_1 are more aggressive than F_2 and on average more aggressive than parental forms	Volfová *et al.* (2002)

Aggression and dominance	Wild mice of the two subspecies. Denmark, border hybrid zone	Male *domesticus* are more aggressive than male *musculus*. Male *domesticus* win two-thirds of fights	Thuesen (1977)
Aggression, dominance, mate preference	Wild-derived mice of the two subspecies Allopatry	Male *domesticus* more aggressive than *musculus* No assortative mate preference	Munclinger and Frynta (1997, 2000)
Aggression	Wild-derived mice of the two subspecies Allopatry	Variation within subspecies Male *musculus* less aggressive than *domesticus*. Females do not differ	Volfová *et al.* (2002); Frynta *et al.* (2005)
Tolerance to cross-fostering	Outbred strains of the two subspecies and their hybrids. Denmark, border hybrid zone	Female *domesticus* more tolerant to progeny of the other subspecies than female *musculus*	Christophe-Ligonnière (1999)
Exploratory behaviour Aggression and dominance	Outbred strains of the two subspecies. Germany, border hybrid zone and allopatry	No difference in exploratory behaviour Male *domesticus* are more aggressive Male *domesticus* win 90% of fights	van Zegeren and van Oortmerssen (1981)
Exploratory behaviour	Wild-derived mice of the two subspecies. Denmark, border hybrid zone.	*domesticus* explore more than *musculus* *musculus* hoard more than *domesticus*	Brubaker (1970) cited by Hunt and Selander (1973)
Exploratory behaviour	Wild-derived mice of the two subspecies Allopatry	Slight difference may exist between the two subspecies (geographic effect not controlled)	Kotenkova *et al.* (1994)
Foraging behaviour in the presence of foreign odour	Outbred strains of the two subspecies. Denmark, border hybrid zone.	Foraging rate is higher in *musculus*. Influence of other subspecies odours is not conclusive	Jacquot and Baudoin (2002)
Major urinary proteins: candidate signals for subspecies recognition	Wild mice of the two subspecies. Czech–Bavarian border hybrid zone.	Male *musculus* express larger quantities of MUP than females or mice of the other subspecies	Stopková *et al.* (2007)
Male pheromones related to aggression and dominance: brevicomin and thiazolin	Outbred strains of the two subspecies, and their hybrids. Denmark, border hybrid zone	Higher quantities in the urine of male *domesticus* than in *musculus* Hybrids: intermediate pattern	Mucignat-Caretta *et al.* (2010)

However, as mentioned before, environmental variation was not properly compared between the two subspecies and across their ranges. Aggression in *M. m. domesticus*, the most studied subspecies, appears to be higher in wild populations than in commensal ones (e.g. Frynta *et al.*, 2005); however, in this study habitat is partly confounded with latitude and the results might also be interpreted as higher aggressiveness in mice from southern than northern latitudes. It was also hypothesized that aggressive males (e.g. with a short attack latency) could be favoured by selection in stable habitats while less aggressive phenotypes would thrive better in unstable ones (Benus *et al.*, 1991). Given the habitat and geographical heterogeneity of the distribution of the two subspecies, one may also expect high variability in patterns of aggression within each subspecies, which may mask differences between them. However, this is not what the results of studies compiled in Table 15.2 indicate: selection does not seem to have resulted in convergent evolution of aggressiveness in *musculus* and *domesticus*, these differences seem to be genetically fixed (Ďureje *et al.*, 2010) and the causes for such divergence, and their implication for speciation, remain to be investigated.

The two male pheromones, brevicomin and thiazolin, that seem central to aggression and dominance in the house mouse have very low (to nil) concentrations in *M. m. musculus* populations from Denmark (Mucignat-Caretta *et al.*, 2010) and in two allopatric locations (G. Ganem, unpublished data). If, as shown in laboratory strains so far, levels of thiazolin and brevicomin reflect male dominance, and female mice prefer males presenting higher levels of these pheromones (Jemiolo *et al.*, 1985), sexual selection could have played a major role in quantitative divergence of aggression between the two subspecies.

Another paper (Soini *et al.*, 2009) pointed out the absence of thiazolin in *M. spicilegus*, which is a non-commensal mound-building species described as monogamous (Patris and Baudoin, 2000; Patris *et al.*, 2002), that nevertheless presents levels of aggression similar to *M. m. domesticus* (Ivantcheva and Cassaing, 1999; Frynta *et al.*, 2005). Hence, low levels of thiazolin may not explain lower aggressiveness of *M. m. musculus* as compared to *M. m. domesticus*, but low levels of brevicomin might be part of the explanation. Further, *M. m. musculus* and *M. spicilegus* share a large part of their southeastern distribution and may have experienced similar selective pressures resulting partly in convergent changes in their pheromonal patterns, although the similarities might also have evolved by chance.

Quantitative differences in chemical urinary molecules and in urinary proteins (MUPs; see Stopka *et al.*, Chapter 8) between the two subspecies (Stopková *et al.*, 2007; Mucignat-Caretta *et al.*, 2010) may be involved in mate recognition and signal divergence. Remarkably, the MUPs, one function of which is to carry pheromones such as brevicomin and thiazolin, are present in larger quantities in urine of *musculus* than in that of *domesticus*, while the latter displays larger quantities of pheromones, further suggesting that the quantitative divergence of MUPs may not relate to their carrier function (Stopková *et al.*, 2007; Janotová and Stopka, 2009). Finally, *M. m. musculus*, unlike *M. m. domesticus*, shows a sexual dimorphism in quantitative variation of MUPs, which could suggest differences in patterns of sexual selection in the two subspecies, and a role for these proteins in signalling male quality within *M. m. musculus* (Stopková *et al.*, 2007; Janotová and Stopka, 2009).

Whenever tested, *musculus* mice (and unexpectedly the males) display higher mate choosiness than *domesticus* (see references in Table 15.2). Low choosiness in *M. m. domesticus* is consistent with their polygynous/promiscuous reproduction system (Levine and Lascher, 1965; Levine *et al.*, 1966; Levine, 1967; Rolland *et al.*, 2003; Dean *et al.*, 2006). Would high choosiness in *M. m. musculus* indicate that their mating system is different? Based on results involving a study assessing attraction of females of one *musculus* population to familiar versus unfamiliar males, it was concluded that *M. m. musculus* may not be as monogamous as *M. spicilegus* (Frynta *et al.*, 2010). However, intermediate systems exist between these two extreme types of mating system – monogamy versus promiscuity – and an explanation for the contrasting pattern of choosiness in the two subspecies may be partly explained by slight differences in their mating systems.

Divergence under the prediction of adaptive speciation

Secondary contacts provide a unique opportunity to test the role of selection in speciation, and investigations in the European contact zone of the house mouse have provided the first evidence that adaptive speciation could occur in a mammal (Bímová *et al.*, 2005; Smadja and Ganem, 2005, 2008; Ganem *et al.*, 2008).

Both the contact zones in Denmark and at the Czech–Bavarian border are subject to behavioural investigations and have provided data showing differences between the two house mouse subspecies. As mentioned above, *M. m. musculus* is choosier than *M. m. domesticus* (Smadja *et al.*, 2004; Ganem *et al.*, 2008; Bímová *et al.*, 2009). *M. m. musculus* populations from the border of the hybrid zone consistently show assortative sexual preference, while the pattern is more variable in *domesticus*, with some populations showing assortative preference (Smadja, 2003; Ganem *et al.*, 2008) and others not (Christophe and Baudoin, 1998; Smadja *et al.*, 2004; Piálek *et al.*, 2008). Further, the signals involved in sexual recognition may be of different origin in the two subspecies – indeed, recently the role of faeces in recognition by *domesticus* has been hypothesized (Bímová *et al.*, 2009). Nevertheless, patterns of discrimination and preference in *M. m. musculus* suggest that *M. m. domesticus* urine contains enough information to allow discrimination against that odour (Smadja and Ganem, 2002, 2005, 2008). Another potential origin for signals allowing potential mate discrimination between the two subspecies is saliva – more specifically the androgen-binding proteins (ABPs) shown in laboratory strains to be involved in mate preference (Laukaitis *et al.*, 1997) and also shown in outbred strains from different geographical origins, including the Danish contact zone (Talley *et al.*, 2001). However, these results have not been confirmed in naturally occurring populations yet (Bímová *et al.*, 2005, 2009, G. Ganem, personal observations), the only detected pattern being that mice closer to the Czech–Bavarian hybrid zone (Bímová *et al.*, 2005) investigated the two stimuli more than mice distant from the contact zone (for a different interpretation see Laukaitis and Karn, Chapter 7). Nevertheless, given that the different alleles of the gene coding for the protein (*Abpa27*; Laukaitis *et al.*, 2008) are fixed over the entire range of the two subspecies (Hwang *et al.*, 1997), its role in the contact zone may be questionable unless epigenetic differences, e.g. in the expression of these alleles in the contact zone versus in allopatry, may have evolved (see Laukaitis and Karn, Chapter 7).

Variation and a lower level of preference in *M. m. domesticus* despite apparently similar signal divergence between the two subspecies (Smadja and Ganem, 2008) suggest that selection on preference may not be symmetric. Such discordance in the evolution of the signal and the receptor component of the MRS is difficult to understand in the framework of theories based on concerted evolution of the two components of a recognition system (but see Endler and Basolo, 1998). Nonetheless, divergence of the signals involved in subspecies recognition could be a by-product of a different process affecting the same sensory system (Boughman, 2002). In the previous section I mentioned that candidate signals for subspecies recognition include molecules such as pheromones involved in advertising male quality. Subspecies recognition signals could have partly evolved in response to competitive interactions and evolution of dominance within and between the two subspecies and/or between the house mouse and other competitors (e.g. Grether *et al.*, 2009) at different stages of their colonization of Europe.

The greater choosiness of *musculus* mice is not specific to populations from the contact zone (see Table 15.2), so it may not be relevant to evoke asymmetric selective pressures occurring in the hybrid zone. Nevertheless, it was hypothesized that dominance of *domesticus* over *musculus* could have contributed to stronger selective pressures on *musculus* in the contact zone to avoid incompatible mating (Smadja *et al.*, 2004). When the two subspecies first met in Europe it is probable that *musculus* was choosier and the males less aggressive than *domesticus*. In that context, there may have been conflict between quality (male dominance) and compatibility information during mate choice in the first phases of the contact (Stopková *et al.*, 2007; Davidian, 2008; Janotová and Stopka, 2009), and how this was coped with deserves a careful investigation as it could have contributed to MRS divergence (both preference and signals) between the two subspecies.

As far as aggression and dominance are concerned, the patterns do not seem to differ significantly between the contact zone and allopatry (see references in Table 15.2). Moreover, recombined hybrids from the centre of the Czech–Bavarian contact zone display levels of aggression intermediate between the two parental forms (Frynta *et al.*, 2005), which contrasts with first-generation hybrids (obtained in the laboratory), which show high levels of aggression similar or larger than those observed for *M. m. domesticus* (Volfová *et al.*, 2002). Consistent with the above, patterns of preference differ between natural hybrids from Denmark and F_1 hybrids obtained in the laboratory from parental mice of the same geographical origin. Indeed, the first category shows a *domesticus*-like preference beyond the *domesticus* side and the centre of the zone (Ganem *et al.*, 2008), while the two reciprocal F_1 crosses do not show a consistent preference, and when a preference is detected it is *musculus*-like (Christophe and Baudoin, 1998).

Discrepancies between naturally occurring hybrids and laboratory crosses indicate that patterns of behavioural interactions may have been very different at the beginning of the formation of the hybrid zone, when the majority of hybrids were F_1 and backcrosses, as compared to today's highly recombined genotypes. Comparative studies of behaviour of laboratory-bred hybrids and natural hybrids should improve insight into the genetic determinism of behavioural traits involved in divergence and isolation between the two subspecies (Piálek *et al.*, 2008), and a better understanding of the process that took place when the secondary contact happened.

Conclusion

The diversified evolutionary settings characterizing the house mouse complex provide an exceptional playground for evolutionary biologists with a pluridisciplinary mind. We are now getting closer to an *integrative* understanding of house mouse evolution thanks to blossoming and growing interactions between researchers in different disciplines with the same aim: understanding what drives speciation.

The study of speciation and radiation in the house mouse from the behavioural ecology point of view has helped to clarify several key questions, but obviously many others remain partially or totally unsolved. The response to some of the questions requires a better assessment of the environmental correlates of behavioural divergence, and a better understanding of the role of potential earlier secondary contacts between the two subspecies on primary divergence (Duvaux *et al.*, 2011). Clearly, we need a better understanding of the behavioural ecology of *M. m. musculus* and an explanation for some puzzling results pointing out the success of the *musculus* Y chromosome in several instances of hybridization with *domesticus* (Macholán *et al.*, 2008; Jones *et al.*, 2010), despite low aggressiveness and sub-dominance of male *M. m. musculus*.

This review points out the important role played by the commensal habitat in the house mouse evolution: a heterogeneous and spatially structured environment providing replicates of heterogeneous settings where diversifying selection and local adaptation occur. The role of sexual selection as a potential mechanism of divergence and isolation between mice populations has also been suggested, and investigations in this direction should be encouraged at several integration levels: from testing the plausibility of a sexual genomic conflict in the house mouse hybrid zone (Macholán *et al.*, 2008; see also Baird and Macholán, Chapter 14), to understanding how mice cope with a conflict of information between mate quality and compatibility, to the investigation of the role of sperm competition and female control in isolation between the two house mouse subspecies (Dean and Nachman, 2009).

Finally, the study of the evolution of the house mouse MRS in the context of speciation has opened the road to better understanding of an olfactory-based species recognition system, although the role of other sensory systems would need to be tested (auditory, visual). Today knowledge of the house mouse MRS points to the involvement of signals of different origin: urinary, faecal, and salivary. The molecules involved in signalling are under investigation and several candidates have been suggested that can be tested with bioassays, and investigations of the receptor component of the recognition system have started (e.g. Karn *et al.*, 2010).

With populations presenting different degrees of postzygotic differentiation, geographically separated or in contact, a broad range of evolutionary questions related to speciation and radiation can be tested in the house mouse both in natural and controlled conditions. The house mouse being a model organism, we can now envision using the most modern genomic tools to dissect the proximal and distal mechanisms underlying behavioural variation and speciation (Stapley *et al.*, 2010).

REFERENCES

Auffray, J. -C., Tchernov, E., Bonhomme, F., *et al.* (1990). Presence and ecological distribution of *Mus 'spretoides'* and *Mus musculus domesticus* in Israel: circum-Mediterranean variance in the genus *Mus*. *Zeitschrift für Säugetierkunde*, **55**, 1–10.

Baker, A. E. M. (1981). Gene flow in house mice: introduction of a new allele into free-living populations. *Evolution*, **35**, 243–58.

Baker, A. E. M. and Petras, M. L. (1986). The fate of *Mus domesticus* demes after destruction of their habitats. *Biological Journal of the Linnean Society*, **29**, 81–8.

Benus, R. F., Bohus, B., Koolhaas, J. M., and Van Oortmerssen, G. A. (1991). Heritable variation for aggression as a reflection of individual coping strategies. *Experientia*, **47**, 1008–19.

Benzekri, K. (2000). *Aspects comportementaux de la biologie de la reproduction et analyse de la fertilité en laboratoire chez deux races chromosomiques de la souris domestique (*Mus musculus domesticus*) en Tunisie*. Unpublished PhD thesis, Université Tunis II, Faculté des Sciences.

Benzekri, K., Britton-Davidian, J., and Ganem, G. (2002). Sexual preference between two chromosomal races of the house mouse (Rodentia, Muridae): a preliminary study comparing wild-caught parents and laboratory-born offspring. *Folia Zoologica*, **51**, 15–22.

Berry, R. J. (1968). The ecology of an island population of the house mouse. *Journal of Animal Ecology*, **37**, 445–70.

Berry, R. J. and Bronson, F. H. (1992). Life history and bioeconomy of the house mouse. *Biological Review*, **67**, 519–50.

Berry, R. J., Cuthbert, A., and Peters, J. (1982). Colonization by house mice: an experiment. *Journal of Zoology* **198**, 329–36.

Berry, R. J. and Jakobson, M. E. (1974). Vagility in an island population of the house mouse. *Journal of Zoology, London*, **173**, 341–54.

Berry, R. J. and Tricker, B. J. K. (1969). Competition and extinction: the mice of Foula, with notes on those of Fair Isle and St. Kilda. *Journal of Zoology*, **158**, 247–65.

Berry, R. J., Triggs, G. S., King, P., Nash, H. R., and Noble, L. R. (1991). Hybridization and gene flow in house mice introduced into an existing population on an island. *Journal of Zoology*, **225**, 615–32.

Beynon, R. J. and Hurst, J. L. (2003). Multiple roles of major urinary proteins in the house mouse, *Mus domesticus*. *Biochemical Society Transactions*, **31**, 142–6.

Bímová, B., Albrecht, T., Macholán, M., and Piálek, J. (2009). Signalling components of the house mouse mate recognition system. *Behavioural Processes*, **80**, 20–7.

Bímová, B., Karn, R. C., and Piálek, J. (2005). The role of salivary androgen-binding protein in reproductive isolation between two subspecies of house mouse: *Mus musculus musculus* and *Mus musculus domesticus*. *Biological Journal of the Linnean Society*, **84**, 349–61.

Boitani, L., Loy, A., and Molinari, P. (1985). Temporal and spatial displacement of two sympatric rodents (*Apodemus sylvaticus* and *Mus musculus*) in a Mediterranean coastal habitat. *Oikos*, **45**, 246–52.

Boughman, J. W. (2002). How sensory drive can promote speciation. *Trends in Ecology & Evolution*, **17**, 571–7.

Boursot, P., Auffray, J. -C., Britton-Davidian, J., and Bonhomme, F. (1993). The evolution of house mice. *Annual Review of Ecology and Systematics*, **24**, 119–52.

Brain, P. F. (1989). The adaptiveness of house mouse aggression. In *House Mouse Aggression: A Model for Understanding the Evolution of Social Behaviour*, ed. P. F. Brain, D. Mainardi, and S. Parmigiani. Chur: Harwood Academic Publishers, pp. 1–22.

Britton-Davidian, J., Fel-Clair, F., Lopez, J., Alibert, P., and Boursot, P. (2005). Postzygotic isolation between the two European subspecies of the house mouse: estimates from fertility patterns in wild and laboratory-bred hybrids. *Biological Journal of the Linnean Society*, **84**, 379–93.

Bronson, F. H. (1979). The reproductive ecology of the house mouse. *The Quarterly Review of Biology*, **54**, 265–99.

Brubaker, L. L. (1970). *A behaviour-genetic study of race and deme differences in* Mus musculus. Unpublished PhD thesis, University of Texas at Austin.

Busch, M., Bilenca, D. N., Cittadino, E. A., and Cueto, G. R. (2005). Effect of removing a dominant competitor, *Akodon azarae* (Rodentia, Sigmodontinae) on community and population parameters of small rodent species in Central Argentina. *Austral Ecology*, **30**, 168–78.

Bush, G. L., Case, S. M., Wilson, A. C., and Patton, J. L. (1977). Rapid speciation and chromosomal evolution in mammals. *Proceedings of the National Academy of Sciences of the United States of America*, **74**, 3942–6.

Butler, R. G. (1980). Population size, social behaviour, and dispersal in house mice: a quantitative investigation. *Animal Behaviour*, **28**, 78–85.

Butlin, R. K. and Ritchie, M. G. (1994). Behaviour and speciation. In *Behaviour and Evolution*, ed. P. J. B. Slater and T. R. Halliday. Cambridge: Cambridge University Press, pp. 43–79.

Capanna, E. and Castiglia, R. (2004). Chromosomes and speciation in *Mus musculus domesticus*. *Cytogenetic and Genome Research*, **105**, 375–84.

Capanna, E. and Corti, M. (1982). Reproductive isolation between two chromosomal races of *Mus musculus* in the Rhaetian Alps (Northern Italy). *Mammalia*, **46**, 107–9.

Capanna, E., Corti, M., Mainardi, D., Stefano, P., and Brain, P. F. (1984). Karyotype and intermale aggression in wild house mice: ecology and speciation. *Behavior Genetics*, **14**, 195–208.

Capanna, E., Corti, M., Nascetti, G., and Bullini, L. (1985). Pre- and post-mating isolating mechanisms in the speciation of the European long-tailed house mouse *Mus musculus domesticus*. *Acta Zoologica Fennica*, **170**, 115–20.

Capanna, E. and Redi, C. A. (1994). Chromosomes and microevolutionary processes. *Bolletino di Zoologia*, **61**, 285–94.

Carlsen, M. (1993). Migrations of *Mus musculus musculus* in Danish farmland. *Zeitschrift für Säugetierkunde*, **58**, 172–80.

Carpineti, M. and Castiglia, R. (2004). Analysis of behavioural discrimination mechanisms in a contact zone between two metacentric races of the house mouse, *Mus musculus domesticus*, in central Italy. *Rendiconti Lincei: Scienze Fisiche e Naturali*, **15**, 31–41.

Carroll, L. S. and Potts, W. K. (2006). Functional genomics requires ecology. In *Advances in the Study of Behavior, vol. 36*. San Diego, CA: Elsevier Academic Press, pp. 173–215.
Cassaing, J. (1988). Structure spatiale de diverses populations de souris ouest-méditerranéennes. *Sciences et Techniques de l'Animal de Laboratoire*, **13**, 111–14.
Cassaing, J. and Croset, H. (1985). Organisation spatiale, compétition et dynamique des populations sauvages de souris (*Mus spretus* Lataste et *Mus musculus domesticus* Rutty) du Midi de la France. *Zeitschrift für Säugetierkunde*, **50**, 271–84.
Castiglia, R. and Capanna, E. (2000). Contact zone between chromosomal races of *Mus musculus domesticus*: II. Fertility and segregation in laboratory-reared and wild mice heterozygous for multiple Robertsonian rearrangements. *Heredity*, **85**, 147–56.
Castiglia, R. and Caporioni, M. (2005). Altitudinal distribution and outdoor occurrence in chromosomal races of the house mouse (*Mus musculus domesticus*) in central Italy. *Folia Zoologica*, **54**, 225–39.
Chatti, N., Ganem, G., Benzekri, K., *et al.* (1999). Microgeographical distribution of two chromosomal races of house mice in Tunisia: pattern and origin of habitat partitioning. *Proceedings of the Royal Society B: Biological Sciences*, **266**, 1561–9.
Cheetham, S. A., Thom, M. D., Jury, F., *et al.* (2007). The genetic basis of individual-recognition signals in the mouse. *Current Biology*, **17**, 1771–7.
Christian, J. J. (1970). Social subordination, population density, and mammalian evolution. *Science*, **168**, 84–90.
Christophe, N. and Baudoin, C. (1998). Olfactory preferences in two sub-species of mice *Mus musculus musculus* and *Mus musculus domesticus* and their hybrids. *Animal Behaviour*, **56**, 365–9.
Christophe-Ligonnière, N. (1999). Echanges possibles entre les deux sous-espèces de souris *Mus musculus musculus et Mus musculus domesticus* et leurs hybrides: étude du rôle des comportements sociaux et préférences sexuelles. Unpublished PhD thesis, Université Paris XIII.
Corti, M., Parmigiani, S., Mainardi, D., Capanna, E., and Brain, P. F. (1989). The role of intermale aggression in speciation processes in chromosomal races of house mice. In *House Mouse Aggression: A Model for Understanding the Evolution of Social Behaviour*, ed. P. F. Brain, D. Mainardi, and S. Parmigiani. Chur: Harwood Academic Publishers, pp. 49–69.
Corti, M. and Rohlf, F. J. (2001). Chromosomal speciation and phenotypic evolution in the house mouse. *Biological Journal of the Linnean Society*, **73**, 1–14.
Cox, T. P. (1984). Ethological isolation between local populations of house mice (*Mus musculus*) based on olfaction. *Animal Behaviour*, **32**, 1068–77.
Coyne, J. A. and Orr, H. A. (2004). *Speciation*. Sunderland, MA: Sinauer Associates.
Cucchi, T., Vigne, J. -D., and Auffray, J. -C. (2005). First occurrence of the house mouse (*Mus musculus domesticus* Schwarz & Schwarz, 1943) in the western mediterranean: a zooarcheological revision of subfossil occurrences. *Biological Journal of the Linnean Society*, **84**, 429–45.
Davidian, C. -E. (2008). Signaux de qualité versus de compatibilité: interférences et impact sur la préférence des femelles. Unpublished MSc thesis, University Montpellier II.

Dean, M. D., Ardlie, K. G., and Nachman, M. W. (2006). The frequency of multiple paternity suggests that sperm competition is common in house mice (*Mus domesticus*). *Molecular Ecology*, **15**, 4141–51.

Dean, M. D. and Nachman, M. W. (2009). Faster fertilization rate in conspecific versus heterospecific matings in house mice. *Evolution*, **63**, 20–8.

Dobzhansky, T. (1940). Speciation as a stage in evolutionary divergence. *The American Naturalist*, **74**, 312–21.

Dumas, D. and Britton-Davidian, J. (2002). Chromosomal rearrangements and evolution of recombination: comparison of chiasma distribution patterns in standard and Robertsonian populations of the house mouse. *Genetics*, **162**, 1355–66.

Ďureje, L., Bímová, B. V., and Piálek, J. (2010). No postnatal maternal effect on male aggressiveness in wild-derived strains of house mice. *Aggressive Behavior*, **35**, 1–8.

Duvaux, L., Belkhir, K., Boulesteix, M., and Boursot, P. (2011). Isolation and gene flow: inferring the speciation history of European house mice. *Molecular Ecology*, **20**, 5248–64.

Dyban, A. P. and Baranov, V. S. (1987). *Cytogenetics of Mammalian Embryonic Development*. Oxford: Oxford University Press.

Eldredge, N. (1995). Species, selection, and Paterson's concept of the specific-mate recognition system. In *Speciation and the Recognition Concept: Theory and Application*, ed. D. M. Lambert and H. G. Spencer. Baltimore, MD and London: Johns Hopkins University Press, pp. 157–74.

Elmer, K., Lehtonen, T., Kautt, A., Harrod, C., and Meyer, A. (2010). Rapid sympatric ecological differentiation of crater lake cichlid fishes within historic times. *BMC Biology*, **8**, 60.

Endler, J. A. and Basolo, A. L. (1998). Sensory ecology, receiver biases and sexual selection. *Trends in Ecology & Evolution*, **13**, 415–20.

Fairley, J. S. and Smal, C. M. (1987). Feral house mice in Ireland. *Irish Naturalists' Journal*, **22**, 284–90.

Fel-Clair, F., Catalan, J., Lenormand, T., and Britton-Davidian, J. (1998). Centromeric incompatibilities in the hybrid zone between house mouse subspecies from Denmark: evidence from patterns of NOR activity. *Evolution*, **52**, 592–603.

Fox, B. J. and Pople, A. R. (1984). Experimental confirmation of interspecific competition between native and introduced mice. *Australian Journal of Ecology*, **9**, 323–34.

Fraguedakis-Tsolis, S., Hauffe, H. C., and Searle, J. B. (1997). Genetic distinctiveness of a village population of house mice: relevance to speciation and chromosomal evolution. *Proceedings of the Royal Society B: Biological Sciences*, **264**, 355–60.

Franks, P. and Lenington, S. (1986). Dominance and reproductive behavior of wild house mice in a seminatural environment correlated with T-locus genotype. *Behavioural Ecology and Sociobiology*, **18**, 395–404.

Frynta, D., Slábová, M., Váchová, H., Volfová, R., and Munclinger, P. (2005). Aggression and commensalism in house mouse: a comparative study across Europe and the Near East. *Aggressive Behavior*, **31**, 283–93.

Frynta, D., Volfová, R., Fraňková-Nováková, M., and Stejskal, V. (2010). Oestrous females investigate the unfamiliar male more than the familiar male in both commensal and non-commensal populations of house mice. *Behavioural Processes*, **83**, 54–60.

Ganem, G. (1991). A comparative study of different populations of *Mus musculus domesticus*: emotivity as an index of adaptation to commensalism. *Comparative Biochemistry and Physiology: Part A – Molecular & Integrative Physiology*, **99**, 531–6.

Ganem, G. (1993). Ecological characteristics of the house mouse Robertsonian populations: is their habitat relevant to their evolution? *Mammalia*, **57**, 349–57.

Ganem, G. (1996). Chromosomal variation and behavioural ecology: a challenge for mouse geneticists? *Trends in Ecology & Evolution*, **11**, 130–1.

Ganem, G. (1998). A behavioural and ecophysiological study of house mice populations from the Orkney archipelago (Scotland) in relation to chromosomal divergence. *Acta Theriologica*, **43**, 23–38.

Ganem, G., Boissin, J., and Croset, H. (1989). Plasma corticosterone response to a stress with reference to the chromosomal evolution of house mice. *Comptes Rendus de l'Académie des Sciences – Series III – Sciences de la Vie*, **309**, 735–40.

Ganem, G., Ginane, C., Ostrowski, M.-F., and Orth, A. (2005). Assessment of mate preference in the house mouse with reference to investigations on assortative mating. *Biological Journal of the Linnean Society* **84**, 461–71.

Ganem, G., Litel, C., and Lenormand, T. (2008). Variation in mate preference across a house mouse hybrid zone. *Heredity*, **100**, 594–601.

Ganem, G. and Searle, J. B. (1996). Behavioural discrimination among chromosomal races of the house mouse (*Mus musculus domesticus*). *Journal of Evolutionary Biology*, **9**, 817–31.

Gazave, E., Catalan, J., Ramalhinho, M., *et al.* (2003). The non-random occurrence of robertsonian fusion in the house mouse. *Genetical Research*, **81**, 33–42.

Geraldes, A., Basset, P., Gibson, B., *et al.* (2008). Inferring the history of speciation in house mice from autosomal, X-linked, Y-linked and mitochondrial genes. *Molecular Ecology*, **17**, 5349–63.

Gérard, D., Bauchau, V., and Smets, S. (1994). Reduced trappability in wild mice, *Mus musculus domesticus*, heterozygous for Robertsonian translocations. *Animal Behaviour*, **47**, 877–83.

Ginane, G. (1998). *Estimation de la variabilité intrasubspécifique et de la divergence intersubspécifique des systèmes de reconnaissance chez deux sous-espèces de la souris domestique* Mus musculus musculus *et* Mus musculus domesticus. Masters thesis, Montpellier II Sciences et Techniques du Languedoc.

Gomez, M. D., Priotto, J., Provensal, M. C., *et al.* (2008). A population study of house mice (*Mus musculus*) inhabiting different habitats in an Argentine urban area. *International Biodeterioration & Biodegradation*, **62**, 270–3.

Gosling, L. M., Roberts, S. C., Thornton, E. A., and Andrew, M. J. (2000). Life history costs of olfactory status signalling in mice. *Behavioral Ecology and Sociobiology*, **48**, 328–32.

Gray, S. J. and Hurst, J. L. (1998). Competitive behaviour in an island population of house mice, *Mus domesticus*. *Animal Behaviour*, **56**, 1291–9.

Grether, G. F., Losin, N., Anderson, C. N., and Okamoto, K. (2009). The role of interspecific interference competition in character displacement and the evolution of competitor recognition. *Biological Reviews*, **84**, 617–35.

Gropp, A., Tettenborn, U., and von Lehmann, E. (1969). Chromosomenuntersuchungen bei der tabakmaus (*M. poschiavinus*) und bei den Hybriden mit der Laboratorumsmaus. *Experientia*, **25**, 875–6.

Hardouin, E. A., Chapuis, J. L., Stevens, M. I., *et al.* (2010). House mouse colonization patterns on the sub-Antarctic Kerguelen Archipelago suggest singular primary invasions and resilience against re-invasion. *BMC Evolutionary Biology*, **10**, 325.

Harvey, S., Jemiolo, B., and Novotny, M. (1989). Pattern of volatile compounds in dominant and subordinate male mouse urine. *Journal of Chemical Ecology*, **15**, 2061–72.

Hauffe, H. C., Panithanarak, T., Dallas, J. F., *et al.* (2004). The tobacco mouse and its relatives: a 'tail' of coat colors, chromosomes, hybridization and speciation. *Cytogenetic and Genome Research*, **105**, 395–405.

Hauffe, H. C. and Searle, J. B. (1992). A disappearing speciation event? *Nature*, **357**, 26.

Hauffe, H. C. and Searle, J. B. (1998). Chromosomal heterozygosity and fertility in house mice (*Mus musculus domesticus*) from northern Italy. *Genetics*, **150**, 1143–54.

Hoskin, C. J. and Higgie, M. (2010). Speciation via species interactions: the divergence of mating traits within species. *Ecology Letters*, **13**, 409–20.

Hunt, W. G. and Selander, R. K. (1973). Biochemical genetics of hybridisation in European house mice. *Heredity*, **31**, 11–33.

Huntingford, F. A. (1976). The relationship between inter- and intra-specific aggression. *Animal Behaviour*, **24**, 485–97.

Hurst, J. L. and Beynon, R. J. (2004). Scent wars: the chemobiology of competitive signalling in mice. *Bioessays*, **26**, 1288–98.

Hwang, J. M., Hofstetter, J. R., Bonhomme, F., and Karn, R. C. (1997). The microevolution of mouse salivary Androgen-Binding Protein (ABP) paralleled subspeciation of *Mus musculus*. *Journal of Heredity*, **88**, 93–7.

Ivantcheva, A. and Cassaing, J. (1999). Male–female interactions and socio-sexual isolation of *Mus spicilegus* toward two other species of east european mice, *M. macedonicus* and *M. m. musculus*. *Comptes Rendus de l'Académie des Sciences – Series III – Sciences de la Vie*, **322**, 597–605.

Jacquot, C. and Baudoin, C. (2002). Foraging behavioural changes induced by conspecific and heterosubspecific odours in two strains of wild mice. *Behavioural Processes*, **58**, 115–23.

Janotová, K. and Stopka, P. (2009). Mechanisms of chemical communication: the role of Major Urinary Proteins. *Folia Zoologica*, **58** (Suppl. 1), 41–55.

Jemiolo, B., Alberts, J., Sochinski-Wiggins, S., Harvey, S., and Novotny, H. M. (1985). Behavioral and endocrine responses of female mice to synthetic analogues of volatile compounds in male urine. *Animal Behaviour*, **33**, 1114–18.

Jones, C. S., Noble, L. R., Jones, J. S., *et al.* (1995). Differential male genetic success determines gene flow in an experimentally manipulated mouse population. *Proceedings of the Royal Society B: Biological Sciences*, **260**, 251–6.

Jones, E. P., van der Kooij, J., Solheim, R., and Searle, J. B. (2010). Norwegian house mice (*Mus musculus musculus/domesticus*): distributions, routes of colonization and patterns of hybridization. *Molecular Ecology*, **19**, 5252–64.

Karn, R. C., Young, J. M., and Laukaitis, C. M. (2010). A candidate subspecies discrimination system involving a vomeronasal receptor gene with different alleles fixed in *M. m. domesticus* and *M. m. musculus*. *PLoS ONE*, **5**, e12638.

Kotenkova, E., Meshkova, N., and Zagoruiko, N. (1994). Exploratory behaviour in synanthropic and outdoor mice of superspecies complex *Mus musculus*. *Polish Ecological Studies*, **20**, 375–81.

Krebs, C. J., Kenney, A. J., and Singleton, G. R. (1995). Movements of feral house mice in agricultural landscapes. *Australian Journal of Zoology*, **43**, 293–302.

Krushinsky, L. V., Dyban, A. P., Baranov, V. S., *et al.* (1986). Behavior of mice with Robertsonian translocations of chromosomes. *Genetika*, **22**, 434–41.

Laukaitis, C. M., Crister, E. S., and Karn, R. C. (1997). Salivary androgen-binding protein (ABP) mediates sexual isolation in *Mus musculus*. *Evolution*, **51**, 2000–5.

Laukaitis, C., Heger, A., Blakley, T., *et al.* (2008). Rapid bursts of androgen-binding protein (Abp) gene duplication occurred independently in diverse mammals. *BMC Evolutionary Biology*, **8**, 46.

Laurie, E. M. O. (1946). The reproduction of the house mouse (*Mus musculus*) living in different environments. *Proceedings of the Royal Society B: Biological Sciences*, **133**, 248–81.

Leitinger, B., Poletaeva, I. I., Wolfer, D. P., and Lipp, H.-P. (1994). Swimming navigation, open-field activity, and extrapolation behavior of two inbred mouse strains with Robertsonian translocation of chromosomes 8 and 17. *Behavior Genetics*, **24**, 273–84.

Levine, L. (1967). Sexual selection in mice: IV. Experimental demonstration of selective fertilization. *American Naturalist*, **101**, 289–94.

Levine, L., Barsel, G. E., and Diakow, C. A. (1966). Mating behaviour of two inbred strains of mice. *Animal Behaviour*, **14**, 1–6.

Levine, L. and Lascher, B. (1965). Studies on sexual selection in mice: II. Reproductive competition between black and brown males. *American Naturalist*, **99**, 67–72.

Lidicker, W. Z. (1966). Ecological observations on a feral house mouse population declining to extinction. *Ecological Monographs*, **36**, 27–50.

Lundrigan, B. L., Jansa, S. A., and Tucker, P. K. (2002). Phylogenetic relationships in the genus *Mus*, based on paternally, maternally, and biparentally inherited characters. *Systematic Biology*, **51**, 410–31.

Macholán, M., Baird, S., Munclinger, P., *et al.* (2008). Genetic conflict outweighs heterogametic incompatibility in the mouse hybrid zone? *BMC Evolutionary Biology*, **8**, 271.

Mackintosh, J. H. (1981). Behaviour of the house mouse. *Symposium of the Zoological Society of London* **47**: 337–65.

Mathias, M. L., Nunes, A. C., Marques, C. C., *et al.* (2004). Adaptive energetics in house mice, *Mus musculus domesticus*, from the island of Porto Santo (Madeira archipelago, North Atlantic). *Comparative Biochemistry and Physiology: Part A – Molecular & Integrative Physiology of Comparative Biochemistry and Physiology*, **137**, 703–9.

Mathias, M. L., Nunes, A. C., Marques, C., *et al.* (2006). Effects of climate on oxygen consumption and energy intake of chromosomal divergent populations of the house mouse (*Mus musculus domesticus*) from the island of Madeira (North Atlantic, Portugal). *Functional Ecology*, **20**, 330–9.

Maxson, S. C. and Canastar, A. (2003). Conceptual and methodological issues in the genetics of mouse agonistic behavior. *Hormones and Behavior*, **44**, 258–62.

Mihola, O., Trachtulec, Z., Vlcek, C., Schimenti, J. C., and Forejt, J. (2009). A mouse speciation gene encodes a meiotic Histone H3 methyltransferase. *Science*, **323**, 373–5.

Mossman, C. A. and Drickamer, L. C. (1996). Odor preferences of female house mice (*Mus domesticus*) in seminatural enclosures. *Journal of Comparative Psychology*, **110**, 131–8.

Mucignat-Caretta, C., Redaelli, M., Orsetti, A., *et al.* (2010). Urinary volatile molecules vary in males of the 2 European subspecies of the house mouse and their hybrids. *Chemical Senses*, **35**, 647–54.

Munclinger, P. and Frynta, D. (1997). Relations between distant populations of *Mus musculus* sensu lato: is there any odour-based discrimination? *Folia Zoologica*, **46**, 193–9.

Munclinger, P. and Frynta, D. (2000). Social interactions within and between two distant populations of house mouse. *Folia Zoologica*, **49**, 1–6.

Nachman, M. W. and Searle, J. B. (1995). Why is the house mouse karyotype so variable? *Trends in Ecology & Evolution*, **10**, 397–402.

Nevo, E., Naftali, G., and Guttman, R. (1975). Aggression patterns and speciation. *Proceedings of the National Academy of Sciences of the United States of America*, **72**, 3250–4.

Newsome, A. E. (1971). The ecology of house-mice in cereal haystacks. *Journal of Animal Ecology*, **40**, 1–15.

Novotny, M. V. (2002). Biochemical aspects of pheromone communication in the house mouse. *Biology of Reproduction*, **66**, 92.

Novotny, M., Harvey, S., Jemiolo, B., and Alberts, J. (1985). Synthetic pheromones that promote inter-male aggression in mice. *Proceedings of the National Academy of Sciences of the United States of America*, **82**, 2059–61.

Noyes, R. F., Barrett, G. W., and Taylor, D. H. (1982). Social structure of feral house mouse (*Mus musculus* L.) populations: effects of resource partitioning. *Behavioral Ecology and Sociobiology*, **10**, 157–63.

Nunes, A. C., Britton-Davidian, J., Catalan, J., *et al.* (2005). Influence of physical environmental characteristics and anthropogenic factors on the position and structure of a contact zone between two chromosomal races of the house mouse on the island of Madeira (North Atlantic, Portugal). *Journal of Biogeography*, **32**, 2123–34.

Nunes, A. C., Mathias, M. da Luz and Ganem, G. (2009). Odor preference in house mice: influences of habitat heterogeneity and chromosomal incompatibility. *Behavioral Ecology*, **20**, 1252–61.

Orsini, P., Bonhomme, F., Britton-Davidian, J., *et al.* (1983). Le complèxe d'espèces du genre Mus en Europe Centrale et Orientale. *Zeitschrift für Säugertierkunde*, **48**, 86–95.

Orsini, P., Cassaing, J., Duplantier, J. M., and Croset, H. (1982). Premières données sur l'écologie des populations naturelles de souris, *Mus spretus* lataste et *Mus musculus domesticus* Rutty dans le midi de la France. *Revue d'Ecologie (la Terre et la Vie)*, **36**, 321–36.

Ould Brahim, I. (2005). *Origine du phénomène robertsonien chez la souris domestique (*Mus musculus domesticus*) en Tunisie et étude de l'isolement pré-zygotique entre les deux races chromosomiques*. Unpublished PhD thesis, Université Tunis el Mannar.

Panithanarak, T., Hauffe, H. C., Dallas, J. F., *et al.* (2004). Linkage-dependent gene flow in a house mouse chromosomal hybrid zone. *Evolution*, **58**, 184–92.

Pardo-Manuel de Villena, F. and Sapienza, C. (2001). Female meiosis drives karyotypic evolution in mammals. *Genetics*, **159**, 1179–89.

Paterson, H. E. (1980). A comment on 'mate recognition systems'. *Evolution*, **34**, 330–1.

Patris, B. and Baudoin, C. (2000). A comparative study of parental care between two rodent species: implications for the mating system of the mound-building mouse *Mus spicilegus*. *Behavioral Processes*, **51**, 35–43.

Patris, B., Gouat, P., Jacquot, C., Christophe, N., and Baudoin, C. (2002). Agonistic and sociable behaviors in the mound-building mice, *Mus spicilegus*: a comparative study with *Mus musculus domesticus*. *Aggressive Behavior*, **28**, 75–84.

Payseur, B. A. and Place, M. (2007). Searching the genomes of inbred mouse strains for incompatibilities that reproductively isolate their wild relatives. *Journal of Heredity*, **98**, 115–22.

Pelikán, J. (1981). Patterns of reproduction in the house mouse. In *Biology of the House Mouse*, ed. R. J. Berry. London: Academic Press, pp. 205–24.

Piálek, J., Hauffe, H. C., and Searle, J. B. (2005). Chromosomal variation in the house mouse: a review. *Biological Journal of the Linnean Society*, **84**, 535–63.

Piálek, J., Vyskočilová, M., Bímová, B., *et al.* (2008). Development of unique house mouse resources suitable for evolutionary studies of speciation. *Journal of Heredity*, **99**, 34–44.

Pocock, M. J. O., Hauffe, H. C., and Searle, J. B. (2005). Dispersal in house mice. *Biological Journal of the Linnean Society*, **84**, 565–83.

Pocock, M. J. O., Searle, J. B., and White, P. C. L. (2004). Adaptations of animals to commensal habitats: population dynamics of house mice *Mus musculus domesticus* on farms. *Journal of Animal Ecology*, **73**, 878–88.

Raufaste, N., Orth, A., Belkhir, K., *et al.* (2005). Inference of selection and migration in the Danish house mouse hybrid zone. *Biological Journal of the Linnean Society*, **84**, 593–616.

Ressouche, L., Ganem, G., Deroth, J. M., *et al.* (1998). Host chromosomal evolution and parasites of the house mouse *Mus musculus domesticus* in Scotland. *Zeitschrift für Säugetierkunde*, **63**, 52–7.

Ritchie, M. G. (2007). Sexual selection and speciation. *Annual Review of Ecology and Systematics*, **38**, 79–102.

Roberts, S. C. and Gosling, L. M. (2003). Genetic similarity and quality interact in mate choice decisions by female mice. *Nature Genetics*, **35**, 103–6.

Robinson, G. E. (2002). Sociogenomics takes flight. *Science*, **297**, 204–5.

Rolland, C., MacDonald, D. W., De Fraipont, M., and Berdoy, M. (2003). Free female choice in house mice: leaving best for last. *Behaviour*, **140**, 1371–88.

Rowe, F. P., Quy, R. J., and Swinney, T. (1987). Recolonization of the buildings on a farm by house mice. *Acta Theriologica*, **32**, 3–19.

Rundle, H. D. and Nosil, P. (2005). Ecological speciation. *Ecology Letters*, **8**: 336–52.

Sage, R. D. (1981). Wild mice. In *The Mouse in Biomedical Research, vol. 1*, ed. H. L. Foster, J. D. Small, and J. G. Fox. New York: Academic Press, pp. 39–90.

Sage, R. D., Atchley, W. R., and Capanna, E. (1993). House mice as models in systematic biology. *Systematic Biology*, **42**, 523–61.

Saïd, K. and Britton-Davidian, J. (1991). Genetic differentiation and habitat partition of Robertsonian house mouse populations (*Mus musculus domesticus*) of Tunisia. *Journal of Evolutionary Biology*, **4**, 409–27.

Sans-Fuentes, M. A., Lopez-Fuster, M. J., Ventura, J., Diez-Noguera, A., and Cambras, T. (2005). Effect of Robertsonian translocations on the motor activity rhythm in the house mouse. *Behavior Genetics*, **35**, 603–13.

Schröder, J. H. and Otten, I. S. (1985). Increase in aggressiveness of male mice carrying a reciprocal translocation, T(10,13), in the heterozygous state. *Behavior Genetics*, **15**, 43–51.

Searle, J. B. (1993). Chromosomal hybrid zones in eutherian mammals. In *Hybrid Zones and the Evolutionary Process*, ed. R. G. Harrison. New York: Oxford University Press, pp. 309–53.

Selander, R. K. (1970). Behavior and genetic variation in natural populations. *American Zoologist*, **10**, 53–66.

Selander, R. K., Hunt, W. G., and Yang, S. Y. (1969). Protein polymorphism and genic heterozygosity in two European subspecies of the house mouse. *Evolution*, **23**, 379–90.

Simberloff, D. (2009). Rats are not the only introduced rodents producing ecosystem impacts on islands. *Biological Invasions*, **11**, 1735–42.

Singleton, G. R., Brown, P. R., Pech, R. P., *et al.* (2005). One hundred years of eruptions of house mice in Australia: a natural biological curio. *Biological Journal of the Linnean Society*, **84**, 617–27.

Singleton, G. and Krebs, C. J. (2007). The secret world of wild mice. In *The Mouse in Biomedical Research, vol. 1.* History, Wild Mice, and Genetics, ed. J. G. Fox, S. W. Barthold, M. T. Davisson *et al.*,. Oxford: Elsevier, pp. 25–52.

Singleton, G. R. and Redhead, T. D. (1990). Structure and biology of mouse populations that plague irregularly: an evolutionary perspective. *Biological Journal of the Linnean Society*, **41**, 285–300.

Singleton, G. R., Tann, C. R., and Krebs, C. J. (2007). Landscape ecology of house mouse outbreaks in south-eastern Australia. *Journal of Applied Ecology*, **44**, 644–52.

Sluyter, F., van Oortmersen, G. A., Ruiter, A. J. H., and Koolhaas, J. M. (1996). Aggression in wild house mice: current state of affairs. *Behavior Genetics*, **26**, 489–96.

Smadja, C. (2003). *Système de reconnaissance et speciation: considerations théoriques et approche experimentale chez la souris domestique.* Unpublished PhD thesis, Montpellier II.

Smadja, C. and Butlin, R. K. (2009). On the scent of speciation: the chemosensory system and its role in premating isolation. *Heredity*, **102**, 77–97.

Smadja, C., Catalan, J., and Ganem, G. (2004). Strong premating divergence in a unimodal hybrid zone between two subspecies of the house mouse. *Journal of Evolutionary Biology*, **17**, 165–76.

Smadja, C. and Ganem, G. (2002). Subspecies recognition in the house mouse: a study of two populations from the border of a hybrid zone. *Behavioral Ecology*, **13**, 312–20.

Smadja, C. and Ganem, G. (2005). Asymmetrical reproductive character displacement in the house mouse. *Journal of Evolutionary Biology*, **18**, 1485–93.

Smadja, C. and Ganem, G. (2008). Divergence of odorant signals within and between the two European subspecies of the house mouse. *Behavioral Ecology*, **19**, 223–30.
Soini, H., Wiesler, D., Koyama, S., *et al.* (2009). Comparison of urinary scents of two related mouse species, *Mus spicilegus* and *Mus domesticus*. *Journal of Chemical Ecology*, **35**, 580–9.
Southern, H. N. and Laurie, E. M. O. (1946). The house mouse (*Mus musculus*) in corn ricks. *Journal of Animal Ecology*, **15**, 135–49.
Spilianakis, C. G., Lalioti, M. D., Town, T., Lee, G. R., and Flavell, R. A. (2005). Interchromosomal associations between alternatively expressed loci. *Nature*, **435**, 637–45.
Stapley, J., Reger, J., Feulner, P. G. D., *et al.* (2010). Adaptation genomics: the next generation. *Trends in Ecology & Evolution*, **25**, 705–12.
Stopková, R., Stopka, P., Janotová, K., and Jedelský, P. L. (2007). Species-specific expression of major urinary proteins in the house mice (*Mus musculus musculus* and *Mus musculus domesticus*). *Journal of Chemical Ecology*, **33**, 861–9.
Stueck, K. L. and Barrett, G. W. (1978). Effects of resource partitioning on the population dynamics and energy utilization strategies of feral house mice (*Mus musculus*) populations under experimental field conditions. *Ecology*, **59**, 539–51.
Talley, H. M., Laukaitis, C. M., and Karn, R. C. (2001). Female preference for male saliva: implications for sexual isolation of *Mus musculus* species. *Evolution*, **55**, 631–4.
Teeter, K. C., Payseur, B. A., Harris, L. W., *et al.* (2008). Genome-wide patterns of gene flow across a house mouse hybrid zone. *Genome Research*, **18**, 67–76.
Teeter, K. C., Thibodeau, L. M., Gompert, Z., *et al.* (2010). The variable genomic architecture of isolation between hybridizing species of house mice. *Evolution*, **64**, 472–85.
Thaler, L., Bonhomme, F., and Britton-Davidian, J. (1981). Processes of speciation and semi-speciation in the house mouse. In *Biology of the House Mouse*, ed. R. J. Berry. London: Academic Press, pp. 27–41.
Thuesen, P. (1977). A comparison of the agonistic behaviour of the *Mus musculus* L. and *Mus musculus domesticus* Rutty (Mammalia and Rodentia). *Videnskabelige Meddelelser fra Dansk naturhistorisk Forening i Kobenhavn*, **140**, 117–28.
van Oortmerssen, G. A. and Bakker, T. C. M. (1981). Artificial selection for high and low attack latencies in wild *Mus musculus domesticus*. *Behaviour Genetics*, **11**, 115–26.
van Oortmerssen, G. A., Benus, R. F., and Sluyter, F. (1992). Studies on wild house mice: IV. On the heredity of testosterone and readiness to attack. *Aggressive Behavior*, **18**, 143–8.
van Zegeren, K. (1980). Variation in aggressiveness and the regulation of numbers in house mouse populations. *Netherlands Journal of Zoology*, **30**, 635–770.
van Zegeren, K. and van Oortmerssen, G. A. (1981). Frontier disputes between the West- and East-European house mouse in Schleswig-Holstein, West Germany. *Zeitschrift für Säugetierkunde*, **46**, 363–9.
Volfová, R., Munclinger, P., and Frynta, D. (2002). Aggression in reciprocal crosses of two subspecies of wild house mouse. *Folia Zoologica*, **51**, 17–22.
Walkowa, W., Adamczyk, K., and Chelkowska, H. (1989). Characteristics of migrants in a free-living population of the house mouse. *Acta Theriologica*, **34**, 305–13.

White, C. S., Lambert, D. M., and Foster, S. P. (1995). Chemical signals and the recognition concept. In *Speciation and the Recognition Concept*, ed. D. M. Lambert and H. G. Spencer. Baltimore, MD and London: John Hopkins University Press, pp. 301–26.
White, M. A., Cécile, A., Dewey, C. N., Larget, B. R., and Payseur, B. A. (2009). Fine-scale phylogenetic discordance across the house mouse genome. *PLoS Genetics*, 5, e1000729.

16

Chromosomal hybrid zones in the house mouse

HEIDI C. HAUFFE, MABEL D. GIMÉNEZ, AND JEREMY B. SEARLE

Introduction

Genetically distinct forms within species may generically be called 'races'. The areas where such races meet and interbreed to produce hybrid offspring are known as hybrid zones (Barton and Hewitt, 1985; Harrison, 1990). In the western subspecies of the house mouse (*Mus musculus domesticus*) there are many geographically local races that differ by chromosomal rearrangements from the widespread ancestral chromosomal form (Piálek *et al.*, 2005). Therefore, *chromosomal* hybrid zones in this species are the geographical locations where two or more of these chromosomal races make contact and interbreed, and/or where such races interact with the ancestral chromosomal form.

Hybrid zones between races differing by chromosomal rearrangements are expected to be 'tension zones', i.e. areas maintained by hybrid unfitness, with well-defined properties including a tendency to minimize their length (Barton and Hewitt, 1985). Pairing and segregation abnormalities associated with heterozygous meiotic configurations in hybrid offspring often lead to reduced fertility in chromosomal hybrid zones, including those of the house mouse (Searle, 1993). However, as we will show in this chapter, there is far more to the chromosomal hybrid zones of the house mouse than their simple description as 'typical' tension zones. We discuss past and future work to develop house mouse chromosomal hybrid zones as models to study the role of chromosomal rearrangements in genetic subdivision, race-formation, and speciation, and to understand chromosomal hybrid zones in the context of chromosomal evolution.

Before considering what has been discovered in chromosomal hybrid zones, we will present relevant background on the nature of the chromosomal variation in the house mouse and on the chromosome heterozygosity that defines F_1 and other hybrids formed at the meeting of chromosomal races.

Evolution of the House Mouse, ed. Miloš Macholán, Stuart J. E. Baird, Pavel Munclinger, and Jaroslav Piálek. Published by Cambridge University Press.

Background: chromosomal variation in the house mouse

The ancestral chromosomal form of the house mouse has a karyotype that consists of 20 pairs of telocentrics (single-armed chromosomes): autosomes numbered from 1 to 19 plus a pair of sex chromosomes (Piálek *et al.*, 2005). The chromosomal races are defined by the presence of one or more pairs of metacentrics (bi-armed chromosomes) in a fixed state. The ancestral chromosomal form is often referred to as the standard race (abbreviated 40ST), and the chromosomal races as 'metacentric races' (reflecting their defining chromosomal characteristic). The metacentrics are formed by the joining together of two non-homologous telocentrics at their centromeres (a so-called Robertsonian fusion). However, the arm combination of a particular metacentric may also reflect subsequent rearrangement by another type of mutation: a whole-arm reciprocal translocation (WART), whereby one of the arms of one metacentric swaps with one of the arms of another metacentric or with a free telocentric (with the emblematic example being that described in Garagna *et al.*, 1997). Clearly, the metacentric condition is also associated with reduced diploid numbers, and metacentric races with diploid numbers between 22 (homozygous for nine metacentrics) and 38 (homozygous for one metacentric) have been described (Piálek *et al.*, 2005).

Nachman and Searle (1995) tallied just over 40 metacentric races (89 metacentric combinations) in the house mouse, where a metacentric race may be precisely defined as a population of individuals over a defined geographical area with a particular metacentric-containing karyotype that is predominantly present in a homozygous state. However, it is probable that there is a much greater number of races than this, given the large geographical areas that have still not been investigated. In fact, just ten years later, Piálek *et al.* (2005) listed more than double that number, a total of 97 metacentric 'populations', the vast majority of which are 'races' as defined above, although some are not, at present, well characterized (i.e. described in less than five individuals). For simplicity we will refer hereafter to all metacentric races and populations as 'races'. Four more metacentric races have been described since the Piálek *et al.* review (see the note for Fig. 16.1). Piálek *et al.* (2005) also introduced a widely accepted four-letter nomenclature for each metacentric race (starting with a country code and followed by letters relating to the geographical location of the races; e.g. CHPO for the *Po*schiavo race first described in Switzerland [CH]). In wild house mice, only autosomes are involved in the chromosomes that define metacentric races; the full set is shown in Fig. 16.1, including their frequency of occurrence. While metacentrics may evolve into other metacentrics by WARTs (and some of the metacentrics in Fig. 16.1 will represent instances of this), it is believed to be unlikely that new telocentrics will evolve from metacentrics by Robertsonian

	Telocentric arm 2																		
Telocentric arm 1	**2**	**3**	**4**	**5**	**6**	**7**	**8**	**9**	**10**	**11**	**12**	**13**	**14**	**15**	**16**	**17**	**18**	**19**	**Total[a]**
1	3	17	1	1	1	1			1	3							4		32
2			8	10			5				1		7	4	1	1	1	1	42
3			6		10		13	3	1		1	1	5						57
4				1	6				4	1	38	2	5	1	2	1			76
5						11			1		3	7	8	19	1	1	3		66
6						13		3	3	1	4	3	1	6	2	1			54
7							4			1				4			4		38
8								2	1	4	9	1	2	2	2	14			59
9									3	3	9	4	20		9		1		57
10										14	14	3	14	4	6	2			71
11											3	15	1	1	1	3	2	1	54
12												3	1						86
13													4	5	10	6			64
14																2			70
15																5	3		54
16																16			50
17																			52
18																			18
19																			2

Figure 16.1 Matrix showing all metacentric combinations of the 19 autosomes described in the karyotypes of metacentric races to date. The number in each cell represents the number of metacentric races with that particular combination (data from Piálek *et al.*, 2005, and more recent literature): [a] Total number of metacentric races in which the telocentric arm is found as a metacentric (written as 'telocentric arm 1.telocentric arm 2' as in 1.3): includes data from Piálek *et al.*, (2005), as well as reports of new metacentric races; three from the Lipari Islands, Italy: IVUL: 1.2, 3.9, 4.13, 5.14, 8.12, 10.16, 15.17 (Solano *et al.*, 2007); IALC: 5.14, 8.12, 10.15 and IPAN: 3.4, 5.13 (Solano *et al.*, 2009) and one from northern Italy, ILVC: 1.18, 2.4, 3.8, 5.15, 6.7, 9.14, 10.12, 11.13, 16.17 (Burt *et al.*, 2009), bringing the total to 101 metacentric combinations in 101 metacentric races. The metacentrics 3.17 in two mice in the FBIS × FSAN hybrid zone (see Table 16.1; Britton-Davidian *et al.*, 2002), 1.15 and 9.15 (Gazave *et al.*, 2003), 7.17 (in the EGAR × 40ST hybrid zone (see Table 16.1; Sans-Fuentes *et al.*, 2007) and 2.13, 3.15, 5.6, and 12.17 (in Sicily, Castiglia *et al.*, 2011) have also been noted in wild populations of house mice, but not been confirmed as representative of metacentric races.

fission because of the loss of telomeres and centromeric DNA during fusion (Garagna *et al.*, 1995). Out of a total of 171 possible autosomal telocentric combinations, 101 have already been described (59.1%; Fig. 16.1). Although some telocentrics are found combined in a large number of metacentrics (numbers 4, 10, 12, and 14 are involved in metacentrics in 70 or more races), and certain metacentrics are found in more than 15 different races (i.e. 1.3, 4.12, 5.15, 9.14, 16.17), this may reflect the intensity of study of certain geographical areas of Europe, rather than a structural or genetic propensity for occurrence of particular arm combinations. Only one study thus far has addressed the multiple origins of a particularly common metacentric: 5.15 (Riginos and Nachman, 1999). However, the relative rarity of involvement of telocentrics 18 and 19 in metacentrics suggests that either these telocentrics undergo chromosomal rearrangements far less frequently than other telocentrics or that fixation of metacentrics involving these very small chromosomes is less likely (see also Gazave *et al.*, 2003).

Some of the metacentric races of the house mouse are found in geographical isolation (e.g. the Monastir race (TUMO), Tunisia), but the majority are located in geographical groupings known as metacentric systems (Piálek *et al.*, 2005). The largest of these are found in the vicinity of the central Alps, with the northern Europe – northern Switzerland (NENSS), southern Switzerland (SSS), and northern Italy (NIS) – systems defined by Piálek *et al.* (2005) consisting of 25, 14, and 15 (with the addition of the Lower Valchiavenna race; Burt *et al.*, 2009) metacentric races, respectively. Within metacentric systems there is the potential for hybrid zones between different metacentric races. Given the small geographical area occupied by any particular race, there is also ample opportunity within one area to have more than one race in contact and hybridizing.

Background: chromosomal heterozygosity in the house mouse

When two chromosomal forms of the house mouse come into contact, the F_1 hybrids will be chromosomal heterozygotes; that is, there will be one or more metacentrics found in at least one of the races that are not found in the other race, and those metacentrics will be present in a heterozygous state in the F_1 hybrid. The types of configurations produced at meiosis I in relation to this heterozygosity affect the fitness of the F_1 hybrid, which in turn influences the properties of the hybrid zone. The configurations produced at meiosis I reflect the synapsis of homologous chromosomal regions. Therefore, if there is a metacentric and homologous telocentric in the F_1 hybrid karyotype (e.g. telocentrics 1 and 2 and metacentric 1.2), a chain-of-three (CIII) configuration is formed at meiosis I (1–1.2–2 for the example above). Individuals with such a karyotype and meiotic expectation are known as 'simple heterozygotes' (Searle, 1993). If the F_1 hybrid karyotype includes metacentrics that share a common chromosome arm (e.g. metacentrics 1.2 and 2.3, with homologous telocentrics 1 and 3 also in the karyotype), then chains or rings longer than three elements are formed at meiosis

I (1–1.2–2.3–3 for the example above). These individuals are known as 'complex heterozygotes' (Searle, 1993).

Heterozygotes may suffer reduced fertility due to errors of synapsis leading to inappropriate gene expression and germ cell death, and errors of segregation leading to gametic and zygotic aneuploidy and zygotic death (Searle, 1993). Simple heterozygotes for one or a few metacentrics generally show little or no reduction in fertility (Britton-Davidian *et al.*, 1990; Wallace *et al.*, 1992), which may help explain the rapidity with which metacentrics accumulate (one at a time) in populations (Nachman and Searle, 1995). However, for both simple and complex heterozygotes, the greater the degree of heterozygosity, the greater the infertility, with complex heterozygotes showing the more extreme unfitness (Searle, 1993). The most dramatic fertility problems are shown by males with long meiotic chain configurations. For some interracial crosses, even a CV meiotic configuration may be associated with complete sterility due to germ cell death in all males tested (e.g. ICRE × IBIN laboratory cross: Piálek *et al.*, 2001), and for crosses that result in hybrids that produce even longer chains, male sterility is likely to be the norm. Germ cell death is also promoted by such configurations in females, and will most likely influence female reproductive lifespan (Garagna *et al.*, 1990). In both sexes, heterozygotes forming long meiotic chains and rings or large numbers of CIII configurations are likely to suffer a substantial fertility reduction due to segregation errors. Also, genetic background is likely to have an impact on fertility (Wallace *et al.*, 2002) – for example, the reduced fertility in TUMO × 40ST hybrids (nine CIIIs at meiosis I) is believed to partially reflect genetic incompatibility (Saïd *et al.*, 1993; Chatti *et al.*, 2005).

Hybrid fertility affects fitness and, consequently, gene flow across a hybrid zone. Gene flow for particular chromosomal regions is also influenced by the location of meiotic recombination events in the F_1 hybrid. Recombination can occur in regions of homologous synapsis and is visualized at late prophase/ metaphase I as chiasmata. In the house mouse, the presence of metacentric chromosomes, either in a homozygous or a heterozygous state, affects the position of recombination events along the chromosomes. Non-homologous pairing revealed by synaptonemal complexes suggests that there may be no recombination in the vicinity of the centromere of heterozygous configurations in simple heterozygotes (Wallace *et al.*, 2002). Genetic analysis of laboratory stocks also suggests 'recombination suppression' near the centromere of heterozygous configurations in simple heterozygotes (Davisson and Akeson, 1993). Interestingly, analysis at diakinesis suggests that metacentric homozygosity has as strong or a stronger impact on the distal localization of chiasmata in meiotic configurations (Bidau *et al.*, 2001; Castiglia and Capanna, 2002; Dumas and Britton-Davidian, 2002).

Evidence: data from chromosomal hybrid zones of the house mouse and associated inferences

The same researchers who have described metacentric races in the house mouse have also discovered hybrid zones involving those races, either fortuitously or through systematic survey. To date, 19 hybrid zones have been described in the house mouse, as well as ten additional areas of polymorphism indicative of an unconfirmed hybrid zone (Fig. 16.2). Particularly detailed work has been carried out in Italy, but also in Belgium, Denmark, France, Greece, Madeira, Scotland, Spain, and Tunisia. Of the 19 hybrid zones, 12 involve contacts between a metacentric race and 40ST (Table 16.1), while the remaining seven hybrid zones are contacts between metacentric races. Metacentric systems may have both types of hybrid zone. Obviously there is the possibility of contact between metacentric races, and often the systems are surrounded by 40ST and may have 40ST populations embedded within them. For example, within the NIS system five distinct hybrid zones have been analysed (Fig. 16.2), of which three involve 40ST; there are two additional unconfirmed hybrid zones within that system.

One of the first questions to ask about a particular hybrid zone is how it was formed, e.g. whether it is 'primary' (formed within a continuous distribution of the species) or 'secondary' (formed when two geographically isolated [allopatric] populations came into contact). Deciding whether a chromosomal hybrid zone in the house mouse is primary or secondary is not straightforward from chromosomal characteristics alone (Gündüz *et al.*, 2010). For example, for the contacts between a metacentric race and 40ST, even when the metacentric race has a low diploid number (as true for many of the contacts: Table 16.1), it is possible that this large accumulation may have occurred one Robertsonian fusion at a time while the two races were parapatric. Also, for contacts between metacentric races, a difference in karyotype could accumulate in parapatry through fixation of WARTs.

If the hybrid zone between a metacentric race and the 40ST race is secondary, we would expect all the chromosomal clines that make up the zone to be located at the same position (Barton and Hewitt, 1985), and there may be associated genic differences between the two races. From Table 16.1 it can be seen that both coincidence of chromosomal clines and genic differences have been recorded in the IUVA– and ILVA–40ST and ICDE–40ST contacts in Italy and the TUMO–40ST contact in Tunisia, and chromosomal coincidence has also been well described in the Italian ICRE–40ST contact with genic studies in progress (Giménez *et al.*, in prep.). Genic differences have more commonly been described in hybrid zones involving the 40ST race than in metacentric–metacentric hybrid zones (Table 16.1). This is not surprising given that metacentric–metacentric hybrid zones will be in metacentric systems where all races

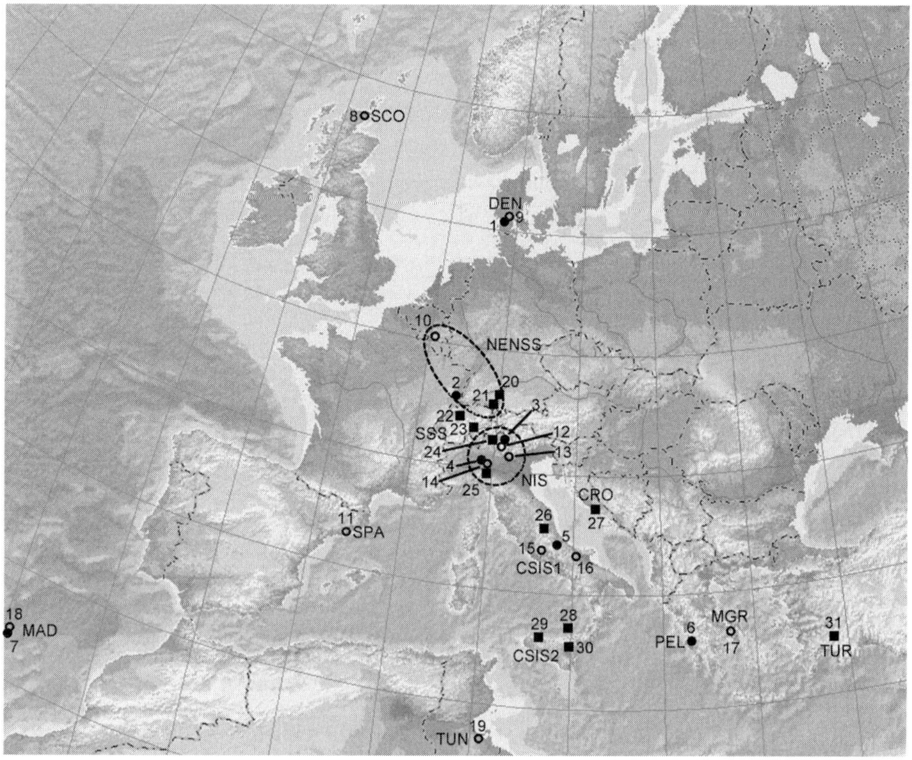

Figure 16.2 Map showing the locality of each chromosomal hybrid zone of the house mouse described to date. The system of metacentric race (if mentioned in Piálek *et al.*, 2005) or country to which each belongs is indicated (CRO: Croatia system; CSIS: central southern Italy system [CSIS1 for mainland Italy, CSIS2 for Sicily]; DEN: Denmark system; MAD: Madeira; MGR: mainland Greece system; NENSS: northern Europe northern Switzerland system; NIS: northern Italy system; PEL: Peloponnesus system; SCO: Scotland system; SPA: Spain system; SSS: southern Switzerland system; TUN: Tunisia; TUR: Turkey). Circles represent well-characterized hybrid zones (characteristics and references in Table 16.1). Closed circles (1–7) are hybrid zones between two metacentric races; open circles (8–19) are hybrid zones between one metacentric race and the standard race (40ST). Black squares represent areas where polymorphic karyotypes have been described and/or where hybrid zones are assumed to exist given the distribution of karyotypes, but are not yet well characterized (geographic area: metacentric race(s) involved): 20 (southern Germany: DDUD, DNEU, DWEN, DROT, DFEL, DSTJ, DHOL, DHOF, DIND; Adolph and Klein, 1983); 21 (southern Germany: DBAR, DMOC, DEYB, DBAI, DAIL, DLAN; Adolph and Klein, 1983); 22 (Switzerland: CHRM; Hübner, 1992); 23 (Switzerland: CHE2; Hübner, 1992); 24 (northern Italy: ICHI, CHBO; Gropp *et al.*, 1971, 1982; ILVC, ICHI: Burt *et al.*, 2009; ILVA: Giménez, 2009); 25 (northern Italy: ICAN; Capanna and Riscassi, 1978; Capanna, 1980; Gropp *et al.*, 1982); 26 (central Italy: ICOL, IFOR; Sadoyan *et al.*, 2003); 27 (Croatia: HRSP, HRZA; Tichy and Vucak, 1987); 28 (Lipari Islands, southern Italy: ILIP, IPAN; Amori *et al.*, 1983; Solano *et al.*, 2007, 2009); 29 (Sicily: IPAL; von Lehmann and Radbruch, 1977); 30 (Sicily: unidentified; Castiglia *et al.*, 2011); 31 (Turkey: TRDE; Gündüz *et al.*, 2000).

Table 16.1 List of metacentric races making contact and forming a hybrid zone confirmed to date (only those indicated by circles in Fig. 16.2).

Geographic location (number on Fig. 16.2)	Metacentric races (2*n*)[a]	Expected F_1 meiotic configuration[[b]]	Intermediate homozygous form (2*n*) [process]	Additional mutations/ polymorphisms	Cline widths (km)[c]	Structure	Geographical barrier?	Genic differences?	References[f]
Two metacentric races:									
Denmark (1)	DKEA (34) x DKWE (34)	1 CIV [no]	No	10.11	NK	Unknown	No	Unknown	1
France (2)	FBIS (34) x FSAN (32)	1 CVI, 1 CIII [no: up to 3 CIII]	BBEL (36)[ZR][d]	Telocentrics	*ca.* 4	Telocentric peak	No	No (allozymes)	2, 3
Italy (3)	ILVA (22) x CHPO (26)	1 CIII, 1 CV [yes: het = hom if < 3 CIII; otherwise het < hom]	IUVA (24), IMVA (24) [ZR]	Telocentrics	0	Patchy; races in separate villages; recombinant peak	Area of low density of suitable habitat	Yes (STRs[e]); No (D-loop; allozymes); villages rather than races distinct; pre-mating isolation in one village	4–10
Italy (4)	IBIN (24) x IGAL (24)	1 RVI [no]	ISEV (24) [WART]	No	NK	Unknown	No	Unknown	11, 12
Italy (5)	IACR (24) x ICDE (22)	1 CV, 1 RXIV [no]	No	WART	0	Parapatric contact area in one village	No	Yes (mtDNA; STRs; behaviour; ecology)	13–17
Greece (6)	GRP1 (30) x GRKA (28) or GROL (24)	1 CIII or 3 CIII [1 CIII yes, 3 CIII no]	GROL (24) [WART]	No	NK	Simple transition (both)	No	Unknown	18–21
Madeira (7)	PEDC (24) x PADC (24)	1 CIV [yes]	New telocentric population? [ZR]	Telocentrics	*ca.* 6	Telocentric peak	Found in different ecological conditions	No (mtDNA)	22–27

Metacentric race x standard race (40ST):									
Scotland (8)	UKJO (32) x 40ST (1 transect west; 1 south)	4 CIII [yes: 1 CIII het = hom]	UKCA (34), UKLY (36), UKWE (36) [ZR?]	No	18–38; 19	Staggered	No	Yes (mtDNA; allozymes)	28–32
Denmark (9)	DKEA (34) or DKWE (34) x 40ST	3 CIII [no: up to 2 CIII]	DKSL (38); DKOD (36) [accumulation of Rbs or ZR?]	No	5–12	Staggered	No: primary contact?	Unknown	1, 33
Belgium (10)	BBEL(36) x 40ST (1 transect northwest; 1 southeast)	2 CIII [yes]	BNIV (38) [ZR?]	WART	15; 56	Staggered	No	Unknown	32, 34–37
Spain (11)	EGAR (28) x 40ST	6 CIII [no]	EAVI (30), EBAR (32) [accumulation of Rbs?]	7.17	13–49	Staggered	No: primary contact?	Yes (morphology); No (mtDNA; allozymes)	38–43
Italy (12)	IUVA (24) or ILVA (22) x 40ST	8 CIII [yes]; 9 CIII [no: up to 3 CIII: het = hom]	Unknown	XXY, XO	NK	Simple transition, unknown	No: secondary contact?; perhaps areas of low density	Yes (mtDNA)	6, 7, 44, 45
Italy (13)	ICRE (22) x 40ST	9 CIII [no: up to 5 CIII]	No	XXY	5–8	Simple transition	No: secondary contact?; areas of low density	Unknown	11, 32, 45
Italy (14)	IBIN (24) x 40ST	7 CIII [no]	ICAN (36), ICCH (38) [unknown]	No	NK	Simple transition, unknown	No	Unknown	11, 46, 47
Italy (15)	ICDE (22) x 40ST	9 CIII [no: up to 5 CIII: het = hom to het < hom with increasing number of CIII]	No	No	3–11	Simple transition	No	Yes (STRs; behaviour)	13, 32, 48–55

Table 16.1 (*cont.*)

Geographic location (number on Fig. 16.2)	Metacentric races (2*n*)[a]	Expected F_1 meiotic configuration[[b]]	Intermediate homozygous form (2*n*) [process]	Additional mutations/ polymorphisms	Cline widths (km)[c]	Structure	Geographical barrier?	Genic differences?	References[f]
Italy (16)	ICBO (22) x 40ST	9 CIII [no]	No	WART	NK	Unknown	No	Unknown	56–60
Greece (17)	GRPY (38) or GRAM (34) or GRT2 (28) or GRT1 (26) x 40ST	1–7 CIII [some]	Unknown	No	NK	Unknown	No	Unknown	20, 39, 61
Madeira (18)	PSAN (22) x 40ST	9 CIII [no]	No	No	0.4–1.2	Simple transition	No	No (mtDNA)	25, 62
Tunisia (19)	TUMO (22) x 40ST	9 CIII [yes: het < hom]	No	XO	0.4	Patchy	Urban vs rural habitat	Yes (allozymes; morphology; behaviour; ecology)	63–68

Notes: [a] See Piálek *et al.* (2005) for full race names.
[b] No/yes means that F_1 hybrids were not found/found in natural populations; het=/<hom means that the fertility of wild heterozygous hybrids found in the hybrid zone was equivalent to/less than that of homozygous individuals.
[c] Standardized width (e.g. Hauffe *et al.*, 2011). NK = not known.
[d] Zonal raciation.
[e] Simple tandem repeats (same as microsatellites).
[f] 1: Fel-Clair *et al.*, 1996; 2: Belkhir *et al.*, 1991; 3: Britton-Davidian *et al.*, 2002; 4: Capanna and Corti, 1982; 5: Hauffe and Searle, 1992; 6: Hauffe and Searle, 1993; 7: Hauffe and Searle, 1998; 8: Hauffe *et al.*, 2002; 9: Hauffe *et al.*, 2004; 10: Panithanarak *et al.*, 2004; 11: Gropp *et al.*, 1982; 12: Garagna *et al.*, 1997; 13: Castiglia and Capanna, 1999a; 14: Castiglia and Capanna, 1999b; 15: Castiglia *et al.*, 2002; 16: Capanna and Castiglia, 2004; 17: Franchini *et al.*, 2008; 18: Giagia *et al.*, 1987; 19: Bauchau, 1990; 20: Mitsainas and Giagia-Athanasopoulou, 2005; 21: Mitsainas and Giagia-Athanasopoulou, 2009; 22: Britton-Davidian *et al.*, 2000; 23: Nunes *et al.*, 2005; 24: Nunes *et al.*, 2011; 25: Förster *et al.*, 2009; 26: Gündüz *et al.*, 2001a; 27: Gündüz *et al.*, 2010; 28: Searle, 1991; 29: Searle *et al.*, 1993; 30: Searle, 1993; 31: Wallace *et al.*, 1992; 32: Hauffe *et al.*, 2011; 33: Nancé *et al.*, 1990; 34: Hübner, 1988; 35: Hübner and Koulischer, 1990; 36: Bauchau *et al.*, 1990; 37: Auffray *et al.*, 2001; 38: Adolph and Klein, 1981; 39: Nachman *et al.*, 1994; 40: Gündüz *et al.*, 2001b; 41: Moñoz-Moñoz *et al.*, 2003; 42: San-Fuentes *et al.*, 2007; 43: San-Fuentes *et al.*, 2009; 44: Giménez, 2009; 45: Hauffe *et al.*, 2010; 46: Capanna and Riscassi, 1978; 47: Capanna, 1980; 48: Capanna, 1976; 49: Capanna, 1984; 50: Capanna *et al.*, 1973; 51: Capanna *et al.*, 1977a; 52: Spirito *et al.*, 1980; 53: Corti *et al.*, 1990; 54: Castiglia and Capanna, 2000; 55: Franchini *et al.*, 2010; 56: Capanna *et al.*, 1977b; 57: Capanna *et al.*, 1994; 58: Capanna and Redi, 1994; 59: Capanna and Redi, 1995; 60: Civitelli *et al.*, 1991; 61: Tichy and Vucak, 1987; 62: Gazave *et al.*, 2003; 63: Saïd *et al.*, 1986; 64: Saïd *et al.*, 1993; 65: Saïd and Britton-Davidian, 1991; 66: Chatti *et al.*, 1999; 67: Chatti *et al.*, 2005; 68: Benzekri *et al.*, 2002.

have a recent common ancestor (Piálek *et al.*, 2005). However, the metacentric races in two of the well-studied contacts in Italy (ILVA–CHPO and IACR–ICDE) do show genic differences and the contacts are probably secondary (Table 16.1). For the IACR–ICDE contact, there are also coincident chromosomal clines. For the ILVA–CHPO contact, the chromosomal clines do not coincide (there are intermediate homozygous forms within the zone: Table 16.1); however, this clinal separation can be explained by processes that occurred after contact (see below). There are phylogenetic grounds for thinking that the ILVA–CHPO contact is secondary (Piálek *et al.*, 2005).

The most compelling case for a primary contact is the area of chromosomal polymorphism in the vicinity of Barcelona (Gündüz *et al.*, 2010). Here, there are staggered chromosomal clines and no genic (allozymes or mtDNA) differences between the metacentric and 40ST populations (although differences in morphology have been noted: Table 16.1).

The hybrid zones between chromosomal races that have accumulated genic differences are interesting for another reason: chromosomal mutations (Shaw *et al.*, 1983). Chromosomal mutations are one expression of hybrid dysgenesis, which is a response to genic hybridity (Kidwell, 2001). Chromosomal mutations have been well demonstrated in chromosomal hybrid zones in common shrews (*Sorex araneus*) which, like mice, display chromosomal variation based on Robertsonian rearrangements and WARTs (Bulatova *et al.*, 2011). As in shrews, there are several chromosomal hybrid zones in the house mouse where novel WARTs have been demonstrated (Table 16.1). It has been argued that in shrews telocentric (or more strictly acrocentric) chromosomes may also arise in hybrid zones through mutation (Robertsonian fission; Bulatova *et al.*, 2011). However, in the house mouse, this would appear less likely (see above), and instances of telocentrics being observed in hybrid zones, such as the FBIS–FSAN and PEDC–PADC contacts, may reflect long-distant migrant mice coming into the hybrid zone with the relevant telocentrics in their karyotypes (see also Hauffe and Searle, 1993). However, whether the telocentrics within these hybrid zones are intrinsic or extrinsic needs to be tested.

As well as novel autosomal elements, it is notable that sex chromosome aneuploids have been found in several hybrid zones (Table 16.1). If there are genic incompatibilities between the races, then meiocytes may be more prone to abnormalities such as non-disjunction, as well as a greater tendency for chromosomal mutation (Bulatova *et al.*, 2011).

On the basis of their ecology (as human commensals), it might be expected that house mouse hybrid zones show a patchy structure (following the patchy distribution of human settlements). Actually researchers more typically report a basic linear structure (Table 16.1), similar to that recorded in the hybrid zone between *M. m. musculus* and *M. m. domesticus* (e.g. Macholán *et al.*, 2007). However, the simplicity of the recorded contacts may reflect a lack of detailed sampling. Scale is important: hybrid zones may appear patchy at fine detail and linear at larger scales (Harrison and Rand, 1989). Among chromosomal hybrid zones in the house mouse, the Valtellina hybrid zone (Geographic location (3) in Table 16.1) is the most obviously patchy, with different races (ILVA, CHPO, IUVA, IMVA) in different settlements (villages in an Alpine river valley) in a fairly haphazard way, with some gene and chromosome flow between populations. The TUMO–40ST hybrid zone is patchy in a different way, with the metacentric race occupying the dense housing of ancient city centres of coastal Tunisia, and the 40ST race occurring in more peripheral areas. This is the hybrid zone where there are most clearly ecological differences between the races.

Considering the chromosomal hybrid zones in the house mouse as tension zones, there is an expectation that the greater the chromosomal difference between the races (and hence the lower the fertility of F_1 hybrids), the narrower the hybrid zone and the chromosomal clines within it. This is what is found for the hybrid zones between the metacentric races and 40ST (Hauffe *et al.*, 2011), with hybrid zones involving a 22-chromosome race consisting of narrower clines (on average) than those involving a race with 28 or more chromosomes (Table 16.1). There is considerable variation in cline width among hybrid zones involving metacentric races with the same chromosome number and even among transects of the same hybrid zone. This probably reflects differences in the distribution of suitable habitat, which in turn affects dispersal (Hauffe *et al.*, 2011).

The hybrid zone with the most extreme difference in karyotype between the races (IACR–ICDE) has no recorded chromosomal hybrids (hence an estimated hybrid zone width of 0 km) and, at most, only a small amount of gene exchange (Castiglia *et al.*, 2002; Franchini *et al.*, 2008). In this case, the F_1 hybrids should produce a CV and an RXIV configuration at meiosis and it would be expected that males will be sterile, and with probably only extremely low fecundity in females.

Two of the other metacentric–metacentric hybrid zones (FBIS–FSAN, PEDC–PADC) have chromosomal clines that are wider than in some of the hybrid zones involving a metacentric race and 40ST contacts (Table 16.1). This may appear surprising, given that the F_1 hybrids would be complex heterozygotes and therefore likely to suffer low fertility (Searle, 1993). However, in both cases there is a high frequency of telocentrics within the hybrid zone, such that the metacentric clines are, in essence, making contact with telocentric clines, not other metacentric clines, and complex heterozygotes are at low frequency.

The width and clinal structure of the chromosomal hybrid zones in the house mouse relate in interesting ways to the formation and reproductive isolation of metacentric races. As we have seen, the IACR–ICDE contact involves two metacentric races that have extreme differences in karyotype, are expected to produce an F_1 hybrid with very low fertility, and which show very little gene exchange. The two races make contact in one village, but different farms are characterized by different races with the complete absence of chromosomal hybrids (Franchini *et al.*, 2008). It is not clear whether parapatric processes (i.e. behavioural isolation) have contributed to this absence of hybrids. Essentially, these two races are reproductively isolated, presumably largely due to the accumulation of different metacentrics in allopatry and the impact this has on fertility, but there may also have been reinforcement in parapatry.

Within the Valtellina hybrid zone (Geographic location (3) in Table 16.1) there was also an absence of hybrids in samples collected at the contact of two of the races (CHPO and one of the intermediate homozygous forms, IUVA) in a single village (Migiondo) in the late 1970s/early 1980s (Capanna and Corti, 1982). In this case the two races occurred together in the same buildings and there appears to have been behavioural isolation, apparently the result of reinforcement arising from selection against complex heterozygous F_1 hybrids (Hauffe and Searle, 1992). Again, chromosomal difference appears to have led to local reproductive isolation. Unfortunately, this situation no longer exists because one of the races (IUVA) became extinct in the village (Hauffe and Searle, 1992).

In these two cases, the width of the hybrid zone was estimated as 0 km; or, more accurately, the zone is non-existent. Another very narrow hybrid zone is that between TUMO and 40ST in Tunisia (cline widths: 0.4 km; Table 16.1). In this example, F_1 and higher generation hybrids are detected, but the predominance of the F_1 hybrid karyotype at the contact of the races indicates rather infrequent hybridization of the races in nature. This differs from other contacts between 22-chromosome races and 40ST (listed in Table 16.1), where hybridization occurs readily and backcross and higher generation hybrids are found rather than F_1s. It appears that the TUMO and 40ST races have become adapted to different ecological niches in Tunisia (old city centres *vs* suburban habitat) and appear to differ genetically in other ways (the reduced fertility of hybrids reflects genic as well as chromosomal differences: see above). Although these two races are not reproductively isolated at present, some combination of the chromosomal and genic differences may lead to that in the future. Chatti *et al.* (1999) suggest that there may also be a degree of behavioural isolation.

For the IACR–ICDE, CHPO–IUVA, and TUMO–40ST contacts there are strong indications of reproductive isolation and that chromosomal differences have contributed to that. In these cases it is also likely that the hybrid zones themselves were important in the development of that reproductive isolation. This is significant because chromosomal hybrid zones in the house mouse have long been a model for speciation (White, 1978; Capanna, 1982; King, 1993; Coyne and Orr, 2004). However, there are other examples where mouse metacentric races that are currently not in contact have been shown to be sterile or have very low fertility on hybridization in the laboratory (Searle, 1993), suggesting that reproductive isolation between such races could develop entirely in allopatry. For this reason it is worthwhile to examine further the build-up of reproductive isolation in hybrid zones where chromosomal races are undoubtedly hybridizing and exchanging genes.

We have been characterizing the gene flow in Valtellina between the CHPO and IMVA races *vs* ILVA and IUVA races. These are the forms that on hybridization would produce an F_1 with a CV meiotic configuration. Proximal regions of the chromosomes differing between the races show more differentiation than distal ones due to some combination of hybrid unfitness and recombination suppression (Panithanarak *et al.*, 2004; Giménez *et al.*, in prep.), and these may be regions of the genome where incompatibilities between the races could accumulate, leading to reproductive isolation in parapatry (Piálek *et al.*, 2001). Franchini *et al.* (2010) demonstrated the same phenomenon in a hybrid zone between the ICDE and 40ST races. Thus, among chromosomal hybrid zones in the house mouse there are examples of races that are reproductively isolated (or close to being so) and others where an interruption to gene flow may lead to reproductive isolation in the future.

We turn now to the formation of new metacentric races within chromosomal hybrid zones, a process described as 'zonal raciation' (Searle, 1993; Piálek *et al.*, 2005). Among the 19 hybrid zones that we surveyed (Table 16.1), there are several examples in which intermediate homozygous forms are found in the hybrid zone, and the presence of these intermediate forms widens the hybrid zone. The zones have a 'staggered' structure, with the chromosomal clines within them separated from each other (non-coincident), generating a wider zone than one formed from coincident clines. With future range changes, these intermediate forms could become widely distributed races. If the intermediate form is produced after contact of the parental forms, then this would be considered zonal raciation. This is the most reasonable explanation for the presence of intermediate forms in contacts between two metacentric races and may also apply to some contacts between a metacentric race and 40ST. However, intermediate forms between a metacentric race and 40ST could also represent accumulation of metacentrics in a primary contact (see above; Gündüz *et al.*, 2010).

The best studied example of intermediate forms within a hybrid zone is the Valtellina zone (Geographic location (3) in Table 16.1), where the IMVA and

IUVA metacentric races have homozygous karyotypes that are, on the basis of molecular studies (Giménez *et al.*, in prep.), hybrid products of the original hybridizing forms (CHPO and ILVA). This is what is known as a 'recombinant peak', where recombinant products of a hybridization event dominate within a hybrid zone (Searle, 1993). The recombinant races IMVA and IUVA could have been produced by historical processes (e.g. population bottlenecking) and genetic drift. However, simulation studies have shown how the alternative process of natural selection may have led to the local fixation of these races (Piálek *et al.*, 2001). Thus, the F_1 hybrids of the races that originally came into contact (ILVA and CHPO) would have formed a CV and CIII configuration at meiosis. However, IMVA generates CIII-forming hybrids on hybridization with CHPO and CV-forming hybrids on hybridization with ILVA. Therefore, IMVA does not produce hybrids with such deleterious meiotic configurations as in the original parental cross. If the sum of fitness reductions of the separate CV- and CIII-forming hybrids is less than the fitness reduction of the hybrid that forms both a CV and a CIII configuration, then the IMVA race would be favoured on contact with the CHPO and ILVA and could spread and become fixed within the hybrid zone (Piálek *et al.*, 2001). Similar arguments would apply to fixation of the IUVA race.

Comparable historical and selective explanations can be made to explain the intermediate forms at the contact of metacentric races and the 40ST race, as found in Belgium (BBEL–40ST) and Scotland (UKJO–40ST). Genomic studies have great potential to distinguish between these alternatives (Gündüz *et al.*, 2010).

Finally, another type of structure involving staggered chromosomal clines is found at the contact of the FBIS and FSAN races in France, and the PEDC and PADC races in Madeira. Here, the F_1 hybrids are expected to produce CVI and CIV meiotic configurations, respectively (Table 16.1). However, such complex heterozygous hybrids are uncommon because in the centre of these hybrid zones there is a population of individuals characterized by telocentric chromosomes instead of the metacentrics that distinguish the parental races. Based on detailed studies of chromosomal hybrid zones with a similar structure in the common shrew (Hatfield *et al.*, 1992), it is likely that selection favours the individuals characterized by telocentrics, because they will, whoever they interbreed with, produce homozygotes or simple heterozygous (CIII-forming) progeny. CIII-forming individuals will have higher fertility than CVI- or CIV-forming individuals, and simulations show that the homozygous telocentric individuals will increase in frequency in the hybrid zone as a result (Hatfield *et al.*, 1992). Hybrid zones of this type have been termed as 'acrocentric peaks' in common shrews, and so appropriately called 'telocentric peaks' in the house mouse. Again, this may lead to zonal raciation, because a novel homozygous form has originated in the hybrid zone and could widen its distribution. Because telocentrics are unlikely to arise by Robertsonian fission in the house mouse, it can be suggested that migration events are the agent bringing telocentrics into the hybrid zones (see above).

The occurrence of novel homozygous forms in chromosomal hybrid zones of the house mouse is of considerable interest. Given that such forms may become races (zonal raciation), then the occurrence of this process may go some way to explaining the extraordinary diversity of metacentric races in the house mouse. In phylogenetic analysis of the common shrew, inclusion of zonal raciation substantially reduces the number of evolutionary steps required to generate the entire set of known races (White *et al.*, 2010).

The process of increasing the number of metacentric races in the house mouse is not trivial. Zonal raciation, by generating new races within the species, is increasing the number of entities that may ultimately become reproductively isolated from all other entities in the species, i.e. that may generate new species. In this way, zonal raciation may be considered to contribute to the speciation process (Gündüz *et al.*, 2010).

Outlook: what more can chromosomal hybrid zones in the house mouse tell us?

As pointed out at the beginning of this chapter, house mouse chromosomal hybrid zones are not merely typical examples of tension zones. They show an extraordinary diversity, and through studying that diversity, we have the opportunity to develop general principles about hybrid zones. Studies of house mouse chromosomal hybrid zones so far have very substantially been based on cytogenetic analysis. However, the molecular resources for the mouse are extraordinary, and those studies conducted thus far that have made use of mapped molecular markers have already had a major impact on our understanding of the hybrid zones (Panithanarak *et al.*, 2004; Franchini *et al.*, 2010). Studies that will make use of the mouse genome arrays (e.g. Yang *et al.*, 2009) and other genomic resources are likely to vastly increase our knowledge of aspects such as gene flow across the zones.

First and foremost, house mouse chromosomal hybrid zones are model systems for the study of speciation in mammals, to complement the detailed studies being conducted on the hybrid zone between *M. m. musculus* and *M. m. domesticus* (e.g. Raufaste *et al.*, 2005; Macholán *et al.*, 2007; Teeter *et al.*, 2010). The interesting observations on hybrid fitness and behavioural and ecological isolation in house mouse chromosomal hybrid zones are well worth pursuing with more sophisticated approaches than hitherto, including genomic studies.

However, speciation is not the only thing of interest in relation to chromosomal hybrid zones. There is apparently a tendency towards zonal raciation, and this needs to be investigated more fully. How many of the house mice chromosomal races have originated in this way? Phylogenetic analysis (White *et al.*, 2010) will help to answer this question, but it should be coupled with molecular analysis to test the chromosomal phylogenies generated.

It is striking that one subspecies of house mouse (*M. m. domesticus*) is subdivided into more than 100 metacentric races, but we know very little about other ways in which the subspecies may be genetically subdivided. Perversely, in order to fully understand the significance of chromosomal variation in creating genetic subdivisions in the house mouse, it is important to search thoroughly for other factors that may promote such subdivisions, with detailed studies comparing different populations within the 40ST race. Molecular studies are important, but a range of analyses are needed, including breeding studies. The house mouse is a fantastic evolutionary model and although the chromosomal variation is extraordinary, it will be better appreciated and understood if we know more about other factors that can create barriers to gene exchange and promote speciation.

REFERENCES

Adolph, S. and Klein, J. (1981). Robertsonian variation in *Mus musculus* from central Europe, Spain, and Scotland. *Journal of Heredity*, **72**, 219–21.

Adolph, S. and Klein, J. (1983). Genetic variation of wild mouse populations in southern Germany: I. Cytogenetic study. *Genetical Research*, **41**, 117–34.

Amori, G., Cristaldi, M., and Federici, R. (1983). Faunal and cytotaxonomic observations on Aeolian island rodents. *Bollettino di Zoologia*, **50**, 9–14.

Auffray, J.-C., Fontanillas, P., Catalan, J., and Britton-Davidian, J. (2001). Developmental stability in house mice heterozygous for single Robertsonian fusion. *Journal of Heredity*, **92**, 23–9.

Barton, N. H. and Hewitt, G. M. (1985). Analysis of hybrid zones. *Annual Review of Ecology and Systematics*, **16**, 113–48.

Bauchau, V. (1990). Phylogenetic analysis of the distribution of chromosomal races of *Mus musculus domesticus* Rutty in Europe. *Biological Journal of the Linnean Society*, **41**, 171–92.

Bauchau, V., Smets, S., Viroux, M.-C., Nootens, D., and De Caritat, A.-K. (1990). Robertsonian translocations in free-living populations of the house mouse in Belgium. *Biological Journal of the Linnean Society*, **41**, 193–201.

Belkhir, K., Britton-Davidian, J., and König, B. (1991). De nouvelles populations robertsoniennes de souris (*Mus musculus domesticus*) au nord des Alpes. *Genome*, **34**, 658–60.

Benzekri, K., Britton-Davidian, J., and Ganem, G. (2002). Sexual preference between two chromosomal races of the house mouse (Rodentia, Muridae): a preliminary study comparing wild-caught parents and laboratory-born offspring. *Folia Zoologica*, **51**, 15–22.

Bidau, C. J., Giménez, M. D., Palmer, C. L., and Searle, J. B. (2001). The effects of Robertsonian fusions on chiasma frequency and distribution in the house mouse (*Mus musculus domesticus*) from a hybrid zone in northern Scotland. *Heredity*, **87**, 305–13.

Britton-Davidian, J., Catalan, J., and Belkhir, K. (2002). Chromosomal and allozyme analysis of a hybrid zone between parapatric Robertsonian races of the house mouse: a case of monobrachial homology. *Cytogenetic and Genome Research*, **96**, 75–84.

Britton-Davidian, J., Catalan, J., Ramalhinho, M. G., *et al.* (2000). Rapid chromosomal evolution in island mice. *Nature*, **403**, 158.

Britton-Davidian, J., Sonjaya, H., Catalan, J., and Cattaneo-Berrebi, G. (1990). Robertsonian heterozygosity in wild mice: fertility and transmission rates in Rb (16.17) translocation heterozygotes. *Genetica*, **80**, 171–4.

Bulatova, N., Jones, R. M., White, T. A., *et al.* (2011). Natural hybridization between extremely divergent chromosomal races of the common shrew (*Sorex araneus*, Soricidae, Soricomorpha): hybrid zone in European Russia. *Journal of Evolutionary Biology*, **24**, 573–86.

Burt, G., Hauffe, H. C., and Searle, J. B. (2009). New metacentric population of the house mouse (*Mus musculus domesticus*) found in Valchiavenna, northern Italy. *Cytogenetic and Genome Research*, **125**, 260–5.

Capanna, E. (1976). Gametic aneuploidy in mouse hybrids. *Chromosomes Today*, **5**, 83–9.

Capanna, E. (1980). Chromosomal rearrangement and speciation in progress in *Mus musculus*. *Folia Zoologica*, **29**, 43–57.

Capanna, E. (1982). Robertsonian numerical variation in animal speciation: *Mus musculus*, an emblematic model. In *Mechanisms of Speciation*, ed. C. Barigozzi. New York: Alan R. Liss, pp. 155–77.

Capanna, E. (1984). Karyotype variability and chromosome transilience in rodents: the case of the genus *Mus*. In *Evolutionary Relationships among Rodents: A Multi-disciplinary Analysis*, ed. W. P. Luckett and J. L. Hartenberger. New York: Plenum Press, pp. 643–70.

Capanna, E. and Castiglia, R. (2004). Chromosomes and speciation in *Mus musculus domesticus*. *Cytogenetic and Genome Research*, **105**, 375–84.

Capanna, E., Civitelli, M. V., and Cristaldi, M. (1973). Una popolazione appennimca di *Mus musculus* caratterizzata da un cariotipo a 22 cromosomi. *Atti Accademia Nazionale dei Lincei, Rendiconti della Classe di Scienze Fisiche, Matematiche e Naturali, Serie VIII*, **54**, 981–4.

Capanna, E., Civitelli, M. V., and Cristaldi, M. (1977a). Chromosomal rearrangement, reproductive isolation and speciation in mammals: the case of *Mus musculus*. *Bollettino di Zoologia*, **44**, 213–46.

Capanna, E., Civitelli, M. V., Cristaldi, M., and Noack, G. (1977b). New Robertsonian metacentrics in another 22-chromosome mouse population in central Apennines. *Experientia*, **33**, 173–5.

Capanna, E. and Corti, M. (1982). Reproductive isolation between two chromosomal races of *Mus musculus* in the Rhaetian Alps (northern Italy). *Mammalia*, **46**, 107–9.

Capanna, E. and Redi, C. A. (1994). Chromosomes and microevolutionary processes. *Bollettino di Zoologia*, **61**, 285–94.

Capanna, E. and Redi, C. A. (1995). Whole-arm reciprocal translocation (WART) between Robertsonian chromosomes: finding of a Robertsonian heterozygous mouse with karyotype derived through WARTs. *Chromosome Research*, **3**, 135–7.

Capanna, E., Redi, C., Valeri, S., and Gentili, G. (1994). Robertsonian chromosomes with new arm combination in a natural hybrid zone between two chromosomal races of *Mus domesticus*. *Atti Accademia Nazionale dei Lincei, Rendiconti della Classe di Scienze Fisiche, Matematiche e Naturali, Serie IX*, **5**, 269–76.

Capanna, E. and Riscassi, E. (1978). Robertsonian karyotype variability in natural *Mus musculus* populations in the Lombardy area of the Po valley. *Bollettino di Zoologia*, **45**, 63–71.

Castiglia, R., Annesi, F., and Capanna, E. (2002). Contact zones between chromosomal races of *Mus musculus domesticus*: III. Molecular and chromosomal evidence of restricted gene flow between the CD race (2n=22) and the ACR race (2n=24). *Heredity*, **89**, 219–24.

Castiglia, R. and Capanna, E. (1999a). Contact zones between chromosomal races of *Mus musculus domesticus*: I. Temporal analysis of a hybrid zone between the CD chromosomal race (2n=22) and populations with standard karyotype. *Heredity*, **83**, 319–26.

Castiglia, R. and Capanna, E. (1999b). Whole-arm reciprocal translocation (WART) in a feral population of mice. *Chromosome Research*, **7**, 493–5.

Castiglia, R. and Capanna, E. (2000). Contact zone between chromosomal races of *Mus musculus domesticus*: II. Fertility and segregation in laboratory-reared and wild mice heterozygous for multiple Robertsonian rearrangements. *Heredity*, **85**, 147–56.

Castiglia, R. and Capanna, E. (2002). Chiasma repatterning in a hybrid zone between chromosomal races of *Mus musculus domesticus*. *Genetica*, **114**, 35–40.

Castiglia, R., Gornung, E., Cividin, M., and Cristaldi, M. (2011). High diversity of centric fusions with monobrachial homology in an area of chromosomal polymorphism of *Mus musculus domesticus*. *Biological Journal of the Linnean Society*, **103**, 722–31.

Chatti, N., Britton-Davidian, J., Catalan, J., Auffray, J. C., and Saïd, K. (2005). Reproductive trait divergence and hybrid fertility patterns between chromosomal races of the house mouse in Tunisia: analysis of wild and laboratory-bred males and females. *Biological Journal of the Linnean Society*, **84**, 407–16.

Chatti, N., Ganem, G., Benzekri, K., *et al.* (1999). Microgeographical distribution of two chromosomal races of house mice in Tunisia: pattern and origin of habitat partitioning. *Proceedings of the Royal Society of London B: Biological Sciences*, **266**, 1561–9.

Civitelli, M. -V., Claar Montalto Di Tocco, R., and Caldarini, G. (1991). Variability of nucleolus organizer regions active in two *Mus musculus domesticus* populations (Rodentia, Muridae). *Atti della Accademia Nazionale dei Lincei, Rendiconti della Classe di Scienze Fisiche, Matematiche e Naturali, Serie IX*, **2**, 69–78.

Corti, M., Ciabatti, C. M., and Capanna, E. (1990). Parapatric hybridisation in the chromosomal speciation of the house mouse. *Biological Journal of the Linnean Society*, **41**, 203–14.

Coyne, J. A. and Orr, H. A. (2004). *Speciation*. Sunderland, MA: Sinauer Associates.

Davisson, M. T. and Akeson, E. C. (1993). Recombination suppression by heterozygous Robertsonian chromosomes in the mouse. *Genetics*, **133**, 649–67.

Dumas, D. and Britton-Davidian, J. (2002). Chromosomal rearrangements and evolution of recombination: comparison of chiasma distribution patterns in standard and Robertsonian populations of the house mouse. *Genetics*, **162**, 1355–66.

Fel-Clair, F., Lenormand, T., Catalan, J., *et al.* (1996). Genomic incompatibilities in the hybrid zone between house mice in Denmark: evidence from steep and non-coincident chromosomal clines for Robertsonian fusions. *Genetical Research*, **67**, 123–34.

Förster, D. W., Gündüz, I., Nunes, A. C., *et al.* (2009). Molecular insights into the colonization and chromosomal diversification of Madeiran house mice. *Molecular Ecology*, **18**, 4477–94.

Franchini, P., Castiglia, R., and Capanna, E. (2008). Reproductive isolation between chromosomal races of the house mouse *Mus musculus domesticus* in a parapatric contact area revealed by an analysis of multiple unlinked loci. *Journal of Evolutionary Biology*, **21**, 502–13.

Franchini, P., Colangelo, P., Solano, E., *et al.* (2010). Reduced gene flow at pericentromeric loci in a hybrid zone involving chromosomal races of the house mouse *Mus musculus domesticus*. *Evolution*, **64**, 2020–32.

Garagna, S., Broccoli, D., Redi, C. A., *et al.* (1995). Robertsonian metacentrics of the house mouse lose telomeric sequences but retain some minor satellite DNA in the pericentromeric area. *Chromosoma*, **103**, 685–92.

Garagna, S., Redi, C. A., Zuccotti, M., Britton-Davidian, J., and Winking, H. (1990). Kinetics of oogenesis in mice heterozygous for Robertsonian translocation. *Differentiation*, **42**, 167–71.

Garagna, S., Zuccotti, M., Redi, C. A., and Capanna, E. (1997). Trapping speciation. *Nature*, **390**, 241–2.

Gazave, E., Catalan, J., Ramalhinho, M. G., *et al.* (2003). The non-random occurrence of Robertsonian fusion in the house mouse. *Genetical Research*, **81**, 33–42.

Giagia, E. B., Fraguedakis-Tsolis, S. E., and Chondropoulos, B. P. (1987). Contribution to the study of the taxonomy and zoogeography of wild house mouse, genus *Mus* L. (Mammalia, Rodentia, Muridae) in Greece: II. Karyological study of two populations from southern Greece. *Mammalia*, **51**, 111–16.

Giménez, M. D. (2009). A molecular and chromosomal study of raciation and speciation in house mice. Unpublished PhD thesis, University of York.

Gropp, A., Olert, J., and Maurizio, R. (1971). Robertsonian chromosomal polymorphism in the mouse (*M. musculus domesticus*). *Experientia*, **27**, 1226–7.

Gropp, A., Winking, H., Redi, C., *et al.* (1982). Robertsonian karyotype variation in wild house mice from Rhaeto-Lombardia. *Cytogenetics and Cell Genetics*, **34**, 67–77.

Gündüz, I., Auffray, J.-C., Britton-Davidian, J., *et al.* (2001a). Molecular studies on the colonization of the Madeiran archipelago by house mice. *Molecular Ecology*, **10**, 2023–9.

Gündüz, I., López-Fuster, M. J., Ventura, J., and Searle, J. B. (2001b). Clinal analysis of a chromosomal hybrid zone in the house mouse. *Genetical Research*, **77**, 41–51.

Gündüz, I., Pollock, C. L., Giménez, M. D., *et al.* (2010). Staggered chromosomal hybrid zones in the house mouse: relevance to reticulate evolution and speciation. *Genes*, **1**, 193–209.

Gündüz, I., Tez, C., and Searle, J. B. (2000). House mice with metacentric chromosomes in the Middle East. *Hereditas*, **133**, 175–7.

Harrison, R. G. (1990). Hybrid zones: windows on the evolutionary process. In *Oxford Surveys in Evolutionary Biology, vol. 7*, ed. D. Futuyma and J. Antonovics. Oxford: Oxford University Press, pp. 69–128.

Harrison, R. G. and Rand, D. M. (1989). Mosaic hybrid zones and the nature of species boundaries. In *Speciation and Its Consequences*, ed. D. Otte and J. A. Endler. Sunderland, MA: Sinauer Associates, pp. 111–33.

Hatfield, T., Barton, N., and Searle, J. B. (1992). A model of a hybrid zone between two chromosomal races of the common shrew (*Sorex araneus*). *Evolution*, **46**, 1129–45.

Hauffe, H. C., Fraguedakis-Tsolis, S., Mirol, P. M., and Searle, J. B. (2002). Studies of mitochondrial DNA, allozyme and morphometric variation in a house mouse hybrid zone. *Genetical Research*, **80**, 117–29.

Hauffe, H. C., Giménez, M. D., Garagna, S., and Searle, J. B. (2010). First wild XXY house mice. *Chromosome Research*, **18**, 599–604.

Hauffe, H. C., Giménez, M. D., Vega, R., White, T. A., and Searle, J. B. (2011). Properties of a hybrid zone between highly distinct chromosomal races of the house mouse (*Mus musculus domesticus*) in northern Italy, and comparisons with other hybrid zones. *Cytogenetic and Genome Research*, **134**, 191–9.

Hauffe, H. C., Panithanarak, T., Dallas, J. F., *et al.* (2004). The tobacco mouse and its relatives: a 'tail' of coat colours, chromosomes, hybridisation and speciation. *Cytogenetic and Genome Research*, **105**, 395–405.

Hauffe, H. C. and Searle, J. B. (1992). A disappearing speciation event? *Nature*, **357**, 26.

Hauffe, H. C. and Searle, J. B. (1993). Extreme karyotypic variation in a *Mus musculus domesticus* hybrid zone: the tobacco mouse story revisited. *Evolution*, **47**, 1374–95.

Hauffe, H. C. and Searle, J. B. (1998). Chromosomal heterozygosity and fertility in house mice (*Mus musculus domesticus*) from northern Italy. *Genetics*, **150**, 1143–54.

Hübner, R. (1988). Populations Robertsoniennes chez la souris 'sauvage' (*Mus domesticus* Rutty, 1772) en Belgique. *Annales de la Société Royale de Zoologie de Belgique*, **118**, 69–75.

Hübner, R. (1992). Chromosomal and biochemical variation in wild mice from Switzerland: relevance for models of chromosomal evolution in European house mice. Unpublished PhD thesis, University of Oxford.

Hübner, R. and Koulischer, L. (1990). Cytogenetic studies on wild house mice from Belgium. *Genetica*, **80**, 93–100.

Kidwell, M. G. (2001). Hybrid dysgenesis. In *Encyclopedia of Genetics*, ed. S. Brenner and J. H. Miller. San Diego, CA: Academic Press, pp. 984–5.

King, M. (1993). *Species Evolution: The Role of Chromosome Change*. Cambridge: Cambridge University Press.

Macholán, M., Munclinger, P., Šugerková, M., *et al.* (2007). Genetic analysis of autosomal and X-linked markers across a mouse hybrid zone. *Evolution*, **61**, 746–71.

Mitsainas, G. P. and Giagia-Athanasopoulou, E. (2005). Studies on the Robertsonian chromosomal variation of *Mus musculus domesticus* (Rodentia, Muridae) in Greece. *Biological Journal of the Linnean Society*, **84**, 503–13.

Mitsainas, G. P. and Giagia-Athanasopoulou, E. (2009). Possible involvement of whole-arm reciprocal translocations (WARTs) in the evolution of a *Mus musculus domesticus* Robertsonian system from Greece. *Rendiconti Lincei: Scienze Fisiche e Naturali*, **20**, 153–62.

Muñoz-Muñoz, F., Sans-Fuentes, M. A., López-Fuster, M. J., and Ventura, J. (2003). Non-metric morphological divergence in the western house mouse, *Mus musculus domesticus*, from the Barcelona chromosomal hybrid zone. *Biological Journal of the Linnean Society*, **80**, 313–22.

Nachman, M. W., Boyer, S. N., Searle, J. B., and Aquadro, C. F. (1994). Mitochondrial DNA variation and the evolution of Robertsonian chromosomal races of house mice, *Mus domesticus*. *Genetics*, **136**, 1105–20.

Nachman, M. W. and Searle, J. B. (1995). Why is the house mouse karyotype so variable? *Trends in Ecology and Evolution*, **10**, 397–402.

Nancé, V., Vanlerberghe, F., Nielsen, J. T., Bonhomme, F., and Britton-Davidian, J. (1990). Chromosomal introgression in house mice from the hybrid zone between *M. m. domesticus* and *M. m. musculus* in Denmark. *Biological Journal of the Linnean Society*, **41**, 215–27.

Nunes, A. C., Britton-Davidian, J., Ramalhinho, M. G., *et al.* (2005). Influence of physical environmental characteristics and anthropogenic factors on the position and structure of a contact zone between two chromosomal races of the house mouse on the island of Madeira (North Atlantic, Portugal). *Journal of Biogeography*, **32**, 2123–34.

Nunes, A. C., Catalan, J., Lopez, J., *et al.* (2011). Fertility assessment in hybrids between monobrachially homologous Rb races of the house mouse from the island of Madeira: implications for modes of chromosomal evolution. *Heredity*, **106**, 348–56.

Panithanarak, T., Hauffe, H. C., Dallas, J. F., *et al.* (2004). Linkage-dependent gene flow in a house mouse chromosomal hybrid zone. *Evolution*, **58**, 184–92.

Piálek, J., Hauffe, H. C., Rodríguez-Clark, K. M., and Searle, J. B. (2001). Raciation and speciation in house mice from the Alps: the role of chromosomes. *Molecular Ecology*, **10**, 613–25.

Piálek, J., Hauffe, H. C., and Searle, J. B. (2005). Chromosomal variation in the house mouse. *Biological Journal of the Linnean Society*, **84**, 535–63.

Raufaste, N., Orth, A., Belkhir, K., *et al.* (2005). Inference of selection and migration in the Danish house mouse hybrid zone. *Biological Journal of the Linnean Society*, **84**, 593–616.

Riginos, C. and Nachman, M. W. (1999). The origin of a Robertsonian chromosomal translocation in house mice inferred from linked microsatellite markers. *Molecular Biology and Evolution*, **16**, 1763–73.

Sadoyan, T., Castiglia, R., Capanna, E., and Serva, L. (2003). Robertsonian polymorphism in house mouse *Mus musculus domesticus* from an area of intense seismic activity. *Acta Theriologica*, **48**, 189–95.

Saïd, K. and Britton-Davidian, J. (1991). Genetic differentiation and habitat partition of Robertsonian house mouse populations (*Mus musculus domesticus*) of Tunisia. *Journal of Evolutionary Biology*, **4**, 409–27.

Saïd, K., Jacquart, T., Montgelard, C., *et al.* (1986). Robertsonian house mouse populations in Tunisia: a karyological and biochemical study. *Genetica*, **68**, 151–6.

Saïd, K., Saad, A., Auffray, J. -C., and Britton-Davidian, J. (1993). Fertility estimates in the Tunisian all-acrocentric and Robertsonian populations of the house mouse and their chromosomal hybrids. *Heredity*, **71**, 532–8.

Sans-Fuentes, M. A., Muñoz-Muñoz, F., Ventura, J., and López-Fuster, M. J. (2007). Rb (7.17), a rare Robertsonian fusion in wild populations of the house mouse. *Genetical Research*, **89**, 207–13.

Sans-Fuentes, M. A., Ventura, J., López-Fuster, M. J., and Corti, M. (2009). Morphological variation in house mice from the Robertsonian polymorphism area of Barcelona. *Biological Journal of the Linnean Society*, **97**, 555–70.

Searle, J. B. (1991). A hybrid zone comprising staggered chromosomal clines in the house mouse (*Mus musculus domesticus*). *Proceedings of the Royal Society of London B: Biological Sciences*, **246**, 47–52.

Searle, J. B. (1993). Chromosomal hybrids in eutherian mammals. In *Hybrid Zones and the Evolutionary Process*, ed. R. G. Harrison. New York: Oxford University Press, pp. 309–53.

Searle, J. B., Navarro, Y. N., and Ganem, G. (1993). Further studies of a staggered hybrid zone in *Mus musculus domesticus* (the house mouse). *Heredity*, **71**, 523–31.

Shaw, D. D., Wilkinson, P., and Coates, D. J. (1983). Increased chromosomal mutation rate after hybridization between two subspecies of grasshoppers. *Science*, **220**, 1165–7.

Solano, E., Castiglia, R., and Capanna, E. (2009). Chromosomal evolution of the house mouse in the Aeolian archipelago (Sicily–Italy). *Biological Journal of the Linnean Society*, **96**, 194–202.

Solano, E., Castiglia, R., and Corti, M. (2007). A new chromosomal race of the house mouse, *Mus musculus domesticus*, in the Vulcano Island–Aeolian Archipelago, Italy. *Hereditas*, **144**, 75–7.

Spirito, F., Modesti, A., Perticone, P., *et al.* (1980). Mechanisms of fixation and accumulation of centric fusions in natural populations of *Mus musculus* L: I. Karyological analysis of a hybrid zone between two populations in the Central Apennines. *Evolution*, **34**, 453–66.

Teeter, K. C., Thibodeau, L. M., Gompert, Z., *et al.* (2010). The variable genomic architecture of isolation between hybridizing species of house mice. *Evolution*, **64**, 472–85.

Tichy, H. and Vucak, I. (1987). Chromosomal polymorphism in the house mouse (*Mus domesticus*) of Greece and Yugoslavia. *Chromosoma*, **95**, 31–6.

von Lehmann, E. and Radbruch, A. (1977). Robertsonian translocations in *Mus musculus* from Sicily. *Experientia*, **33**, 1025–6.

Wallace, B. M. N., Searle, J. B., and Everett, C. A. (1992). Male meiosis and gametogenesis in wild house mice (*Mus musculus domesticus*) from a chromosomal hybrid zone; a comparison between 'simple' Robertsonian heterozygotes and homozygotes. *Cytogenetics and Cell Genetics*, **61**, 211–20.

Wallace, B. M. N., Searle, J. B., and Everett, C. A. (2002). The effect of multiple simple Robertsonian heterozygosity on chromosome pairing and fertility of wild-stock house mice (*Mus musculus domesticus*). *Cytogenetic and Genome Research*, **96**, 276–86.

White, M. J. D. (1978). *Modes of Speciation*. San Francisco, CA: Freeman.

White, T. A., Bordewich, M., and Searle, J. B. (2010). A network approach to study karyotypic evolution: the chromosomal races of the common shrew (*Sorex araneus*) and house mouse (*Mus musculus*) as model systems. *Systematic Biology*, **59**, 262–76.

Yang, H., Ding, Y. M., Hutchins, L. N., *et al.* (2009). A customized and versatile high-density genotyping array for the mouse. *Nature Methods*, **6**, 663–6.

17

The role of the X chromosome in house mouse speciation

AYAKO OKA AND TOSHIHIKO SHIROISHI

Introduction

When animals from genetically divergent populations breed, sterility or inviability is often observed in the F_1 hybrids or in their descendants. This phenomenon, known as 'postzygotic reproductive isolation', accelerates irreversible genetic divergence by preventing free gene flow across the diverging populations, and thereby plays a pivotal role in speciation. Understanding the genetic basis underlying postzygotic reproductive isolation remains a challenge in evolutionary biology. In recent decades, ecological and evolutionary studies of wild populations in nature, and experimental genetic studies with inbred and/or outbred strains established from wild populations have led to marked progress in advancing understanding of how postzygotic reproductive isolation evolved.

Postzygotic reproductive isolation is primarily caused by disharmonious interactions between nuclear genes derived from their respective parental species or subspecies. This causative deleterious epistasis is called Dobzhansky–Muller (DM) incompatibility (Bateson, 1909; Dobzhansky, 1936; Muller, 1942). The DM model is widely accepted in animal and plant biology, and is also applicable to the sterility of hybrid animals in the F_2 or other later-generation hybrids generated from fertile F_1 hybrids. This form of hybrid sterility is also known as 'hybrid breakdown' to distinguish it from the hybrid sterility in F_1 and is found in cases in which one or more of the genes causing postzygotic reproductive isolation are recessive (Orr, 2005). Many genetic studies in animal models have sought to identify the chromosomal loci involved in DM incompatibilities in an effort to unveil the genetic architecture of postzygotic reproductive isolation. Accumulating data from experimental genetic studies of animals, such as the house mouse (*Mus musculus*) and fruit fly (*Drosophila*), suggest that the X chromosome contributes disproportionately to postzygotic reproductive isolation. The genetic loci that cause hybrid disruptions appear to be located preferentially on the X chromosome. This phenomenon is referred to as the

Evolution of the House Mouse, ed. Miloš Macholán, Stuart J. E. Baird, Pavel Munclinger, and Jaroslav Piálek. Published by Cambridge University Press.

'large X-effect' (Coyne and Orr, 1989). Since the mechanism underlying hybrid sterility likely differs from that of hybrid inviability (Orr, 1993; Wu and Davis, 1993), we focus here on hybrid sterility.

In this chapter we first review genetic studies of hybrid sterility conducted in mice, focusing on the role of the X chromosome, and then discuss why the X chromosome is frequently involved in hybrid sterility. In particular, we shed light on the hemizygosity of the X chromosome in males and the accumulation of genes expressed in early spermatogenesis on the X chromosome. We also discuss the unique cytological behaviour of the sex chromosomes in relation to X-linked incompatibility, including the process of transcriptional inactivation of sex chromosomes at the meiotic and postmeiotic stages. Finally, we examine the link between the large X-effect and 'Haldane's rule', which predicts that the sterility or inviability of a hybrid occurs first in the heterogametic sex, characteristic of postzygotic reproductive isolation in almost all animals (Haldane, 1922).

Experimental genetic studies of postzygotic reproductive isolation

Obtaining a sufficiently large number of progeny may be difficult in experimental genetic studies of hybrid sterility, because the experiment is examining a refractory trait, sterility (Presgraves, 2003). In spite of this difficulty, many genetic studies using well-characterized inbred strains of mice and fruit flies have detected chromosomal loci responsible for postzygotic reproductive isolation. Recently, some of the responsible genes have been successfully identified (Ting *et al.*, 1998; Masly *et al.*, 2006; Bayes and Malik, 2009; Mihola *et al.*, 2009). Studies of postzygotic reproductive isolation in mice are summarized in Tables 17.1 and 17.2.

In many cases of F_1 hybrid sterility in recently diverged species or subspecies, backcross or intercross progeny show a wide range of continuous variation in fertility. Other reproductive phenotypes, such as reduced weight of testes and seminiferous tubules and reduced sperm number, are often observed in the progeny as well. Quantitative trait loci (QTL) analysis is often carried out to map the chromosomal loci affecting reproductive phenotypes. Results of the QTL analysis provide an overview of the genetic components that affect reproductive phenotypes (Wu and Davis, 1993). However, QTL analyses are not always successful because certain genotypes cannot be obtained or are rarely obtained in the backcross and intercross progeny due to the limited size of experimental populations. This is particularly problematic when recessive DM incompatibility is involved in the sterility. For example, if homozygous alleles of both species are necessary to give rise to male sterility, working with an

Table 17.1 Summary of F_1 hybrid male sterility in interspecific and intersubspecific crosses.

Maternal strains	Paternal strains	Symmetry of phenotype[a]	Stages disrupted	Chromosomes associated (names of responsible loci)	Reference
Interspecific F_1					
M. m. domesticus (C57BL/6)	*M. spretus*	NT	Meiosis (pachynema and MI)	Chr X (*Hst3*, *Ihtw1*)	Elliott *et al.* (2001); Guénet *et al.* (1990); Hale *et al.* (1993); Matsuda *et al.* (1991, 1992); Oka *et al.* (2010)
M. m. domesticus (ICR, C57BL/6[b])	*M. hortulanus*	NT	Metaphase I	ND	Matsuda and Chapman (1994)
M. m. domesticus (C57BL/6[b])	*M. abbotti*	NT	Meiosis (no MI spermatocytes)	ND	Matsuda and Chapman (1994)
M. m. domesticus (C57BL/6[b])	*M. macedonicus*	NT	Meiosis (no MI spermatocytes), spermiogenesis (from the analysis of backcross progeny)	Chr 17 and X for reduced testis weight, chr X for lack of XY pairing	Elliott *et al.* (2004); Matsuda and Chapman (1994)
Intersubspecific F_1					
M. m. musculus (PWD, PWK)	*M. m. domesticus* (C57BL/10, BALB/c)[b]	NT	Meiosis (pachynema and MI)	Chr 17 (*Hst1*) and undefined chromosomes	Chubb and Nolan (1987); Forejt (1981, 1996); Forejt and Iványi (1974); Iványi *et al.* (1969); Mihola *et al.* (2009)

Table 17.1 *(cont.)*

Maternal strains	Paternal strains	Symmetry of phenotype[a]	Stages disrupted	Chromosomes associated (names of responsible loci)	Reference
M. m. domesticus (C57BL/10[b])	*M. m. musculus* (STUS)	Symmetric	Meiosis, spermiogenesis	Chr 17 (for sperm count), chr X (for sperm count and testis weight)	Vyskočilová *et al.* (2009)
M. m. domesticus (BALB/c[b])	*M. m. musculus* (NJL)	NT	Meiosis (zygonema)	Chr 17	Yoshiki *et al.* (1993)
M. m. domesticus (C57BL/6[b])	*M. m. musculus* (NJL)	NT	Early meiosis (leptonema, zygonema, and pachynema)	ND	Kaku *et al.* (1995); Oka *et al.* (2010)
M. m. domesticus (LEWES, WSB)	*M. m. musculus* (CZECHII)	Symmetric	Spermiogenesis	ND	Good *et al.* (2008a)
M. m. musculus (PWK)	*M. m. domesticus* (LEWES, WSB)	Asymmetric	Spermiogenesis	Chr X and undefined autosomes	Good *et al.* (2008a)
M. m. musculus (PWK)	*M. m. domesticus* (WLA)	Asymmetric	Meiosis	Chr X and undefined autosomes	Britton-Davidian *et al.* (2005)

Note: [a] Symmetry means appearance of F_1 hybrid male sterility in both directional crosses. Symmetric and asymmetric mean that F_1 hybrid males in both directional or only in one directional cross are sterile, respectively. MI: metaphase of meiosis I; NT: not tested; ND: not defined.

[b] These strains are classical laboratory strains predominantly derived from *M. m. domesticus*.

experimental population large enough to obtain sterile homozygous male progeny is often impractical.

The above-mentioned difficulty with QTL analyses may be overcome by genetic analyses using 'consomic strains', also known as 'chromosome substitution strains'. Consomic strains are an ordered series of inbred strains that share the same inbred genetic background and differ only by one entire chromosome introgressed from another inbred strain by repeated backcrossing (Nadeau *et al.*, 2000). Any phenotypic variation between a consomic strain and the background strain can be readily attributed to one or more genes on the substituted chromosome. Interspecific or intersubspecific consomic strains allow researchers to evaluate the influence of each chromosome on fertility. Genetic analyses using consomic strains are a powerful method for finding recessive DM incompatibility. Moreover, congenic lines from consomic strains with a phenotype of interest can be generated quickly to clone the responsible genes. One limitation of this approach is that finding the complex DM incompatibilities in cases where two or more substituted chromosomes are necessary for full sterility may be difficult.

F_1 hybrid male sterility in interspecific and intersubspecific crosses

Postzygotic reproductive isolation in mice is observed at two levels of hybrid animals, interspecific and intersubspecific hybrid males. In most studies of interspecific hybrid male sterility, F_1 hybrid males are generated between classical laboratory strains, which are predominantly derived from *Mus musculus domesticus* (Yonekawa *et al.*, 1980; Ferris *et al.*, 1982; Bonhomme *et al.*, 1984; Moriwaki, 1994; Frazer *et al.*, 2007; Yang *et al.*, 2011), and strains from other congeneric species, such as *M. spretus* (Guénet *et al.*, 1990; Matsuda *et al.*, 1991, 1992; Hale *et al.*, 1993; Elliott *et al.*, 2001), *M. macedonicus* (Elliott *et al.*, 2004; Matsuda and Chapman, 1994), and *M. spicilegus* (Matsuda and Chapman, 1994). *M. spretus*, *M. macedonicus*, and *M. spicilegus* were separated from *M. musculus* about one million years ago, and their reproductive isolation is completed in nature by premating isolating mechanisms (Bonhomme *et al.*, 1978; Bonhomme and Guénet, 1996), although some exceptions to this rule exist (Orth *et al.*, 2002).

In the laboratory, it is possible to generate viable F_1 hybrids between classical laboratory strains such as C57BL/6 (B6) and strains from the above-mentioned congeneric species. Following Haldane's rule, F_1 hybrid males are often sterile, while F_1 hybrid females are often fertile (Guénet *et al.*, 1990; Matsuda and Chapman, 1994; Bonhomme and Guénet, 1996). Spermatogenic breakdown during metaphase of the first meiosis (metaphase I) has been reported in spermatocytes of (B6 × *M. spretus*)F_1 (Matsuda *et al.*, 1991). During metaphase I, X and Y chromosomes establish non-homologous pairing by exploiting the pseudoautosomal regions (PARs) that share common DNA sequences on the distal regions of these chromosomes. Cytological studies have demonstrated that there is an abnormally high frequency of X–Y chromosome dissociation and association of asynapsed autosomal segments with the X chromosome in the spermatocytes of (B6 × *M. spretus*)F_1 males during the pachytene stage to metaphase I (Matsuda *et al.*, 1991; Hale *et al.*, 1993; Matsuda and Chapman, 1994). A locus responsible for the abnormal X–Y dissociation, X-linked Hybrid sterility 3 (*Hst3*), was mapped to PARs by a linkage analysis with backcross to *M. spretus* (Guénet *et al.*, 1990; Matsuda *et al.*, 1991; Hale *et al.*, 1993). Another locus, Interspecific hybrid testis weight 1 (*Ihtw1*), which is responsible for reduced testis weight, was mapped to the central region of the X chromosome (Elliott *et al.*, 2001). Spermatogenic defects in (B6 × *M. macedonicus*)F_1 males are more severe, lacking meiotic spermatocytes in metaphase I (Matsuda and Chapman, 1994; Elliott *et al.*, 2004). Linkage analysis revealed that the proximal chromosome 17 and X chromosome are responsible for reduced testis weight, which may result from the lack of meiotic spermatocytes. These studies also revealed that the distal X chromosome is responsible for the lack of X–Y pairing (Elliott *et al.*, 2004).

In these interspecific crosses, the X chromosome appears to be predominantly involved in hybrid male sterility, and is associated with abnormal X–Y paring. Though evidence of polymorphic chromosomal structure in PAR between species is still limited (Kipling *et al.*, 1996), the location of the responsible genes in PAR and the high frequency of the X–Y dissociation phenotype suggest that structural incompatibility of PAR is the major cause of hybrid sterility in (B6 × *M. spretus*)F_1 and (B6 × *M. macedonicus*)F_1 males.

The second level of postzygotic reproductive isolation is observed in intersubspecific F_1 hybrid males. In contrast to interspecific hybrid males, reproductive isolation between *M. musculus* subspecies is incomplete in nature. This level of postzygotic reproductive isolation is suited for the study of early stages of speciation (Forejt, 1996). Two mouse subspecies, *M. m. domesticus* and *M. m. musculus*, diverged from their common ancestor about 0.5 million years ago (Geraldes *et al.*, 2008). *M. m. domesticus* ranges across western Europe and the Middle East, whereas *M. m. musculus* ranges throughout eastern Europe and northern Asia (Guénet and Bonhomme, 2003; see also Auffray and Britton-Davidian, Chapter 1 in this volume; Suzuki and Aplin, Chapter 2). The two subspecies meet in a narrow hybrid zone extending more than 2500 km across Europe, which is most likely maintained by a balance between reduced fitness and gene flow (Hunt and Selander, 1973; Bonhomme and Guénet, 1996; Payseur *et al.*, 2004; see also Baird and Macholán, Chapter 14). When a laboratory strain such as C57BL/10 (B10) is crossed with PWD or PWK strains derived from eastern European *M. m. musculus*, the resulting F_1 hybrid males are sterile, whereas F_1 hybrid females are fertile (Forejt, 1981; see also Forejt *et al.*, Chapter 19). The male sterility is characterized by a complete arrest of spermatogenesis in the pachytene stage of meiotic prophase (Forejt and Iványi, 1974; Yoshiki *et al.*, 1993). One of the loci responsible for this hybrid male sterility, Hybrid sterility 1 (*Hst1*), was mapped to the proximal region of chromosome 17 (Forejt and Iványi, 1974; Forejt, 1981; Forejt *et al.*, 1991), and the PR-domain 9 (*Prdm9*) gene, which encodes a meiotic histone H3 lysine 4-trimethyltransferase, was recently identified as *Hst1* (Mihola *et al.*, 2009; for a review, see Forejt *et al.*, Chapter 19).

While the F_1 males from crosses between PWK females and males of other wild *M. m. domesticus*-derived strains (WLA, LEWES, and WSB) are sterile, it is notable that F_1 males from the reciprocal crosses are fertile. This asymmetric hybrid sterility also suggests that the *M. m. musculus* X chromosome is involved in causing sterility (Britton-Davidian *et al.*, 2005; Good *et al.*, 2008b). In the European hybrid zone, the X chromosome shows lower levels of gene flow relative to the autosomes, suggesting that there is selection against X-linked incompatibilities (Tucker *et al.*, 1992; Dod *et al.*, 1993; Payseur *et al.*, 2004; Macholán *et al.*, 2007, but see Macholán *et al.*, 2011). Asymmetric introgression has been noted across several hybrid zone transects, with a tendency for more introgression of *M. m. domesticus* alleles into the *M. m. musculus* population, though it is unknown whether this is due to a shift in the boundary over time or asymmetric genetic incompatibilities (Hunt and Selander, 1973; Vanlerberghe *et al.*, 1986; Tucker *et al.*, 1992; Dod *et al.*, 1993; for a review, see Baird and Macholán, Chapter 14).

X-linked incompatibility in the consomic strains

The incompleteness of reproductive isolation between *M. musculus* subspecies allowed us to construct intersubspecific consomic strains. At present, two series of full-set intersubspecific consomic strains are available. *M. m. musculus*-derived PWD and *M. m. molossinus*-derived MSM strains were used as chromosome donors, and B6 served as the chromosome recipient genetic background strain (Gregorová *et al.*, 2008; Takada *et al.*, 2008). Notably, some strains in both consomic series exhibit varying degrees of reproductive problems. The consomic strains with PWD Chr 10, 11, and the X chromosome have severe reproductive impairments. Other strains with PWD Chr 2, 3, 4, 7, 8, 9, 13, and 17 have lower reproductive performance compared to the B6 background strain. Likewise, the consomic strains with MSM Chr 10, 13, and the X chromosome have severe reproductive problems, and those with MSM Chr 4, 5, 7, 8, 9, and 11 have lower reproductive performance. These facts suggest that the extant *M. musculus* subspecies have many genetic incompatibilities that have accumulated since they diverged genetically. Although it is difficult to infer the exact nature of the 'first incompatibility' that occurred during the early stage of subspecies divergence, the X chromosome undoubtedly played a significant role in the reproductive isolation between *M. m. domesticus* and *M. m. musculus/molossinus*. Studies of intersubspecific X chromosomal consomic strains are summarized in Table 17.2.

The C57BL/6J-ChrX^{PWD} strain, which carries the PWD-derived X chromosome, shows complete male sterility with significantly reduced testes size, low sperm count, and abnormal sperm-head morphology (Storchová *et al.*, 2004). Heterozygous females of the C57BL/6J-ChrX^{PWD} strain are fertile. In contrast to (PWD × B6)F_1 hybrids, whose major defect is apoptosis during meiosis, the sterile C57BL/6J-ChrX^{PWD} males show defective spermatogenesis during the postmeiotic stage. The locus primarily responsible for this phenotype, Hybrid

Table 17.2 Summary of male sterility in intersubspecific X chromosomal consomic strains.

Recipient strains	Donor strains	Stages disrupted	Chromosomes associated (names of responsible loci)	Reference
M. m. domesticus (C57BL/6[a])	*M. m. musculus* (PWD)	Spermiogenesis	Chr X (*Hstx1*) and undefined chromosomes	Storchová *et al.* (2004)
M. m. domesticus (LEWES)	*M. m. musculus* (PWK)	Spermiogenesis	Chr X and undefined chromosomes	Good *et al.* (2008a)
M. m. domesticus (C57BL/6[a])	*M. m. molossinus* (MSM)	Premeiotic stage, meiosis (pachytene), spermiogenesis	Chr X (*sha1*, *sha2*, *sha3*), Chr 1 (*Ilx1*), Chr 11 (*Ilx2*, *Ilx3*) and undefined autosomes	Oka *et al.* (2004, 2007, 2010)
M. m. domesticus (PGN)	*M. m. molossinus* (MSM)	Meiosis (MI), spermiogenesis	Chr X and undefined chromosomes	Oka *et al.* (2010)

Note: [a] These strains are classical laboratory strains predominantly derived from *M. m. domesticus*.

sterility-X1 (*Hstx1*), was mapped to the central region of the X chromosome. In addition to *Hstx1*, the PWD alleles of one proximal gene and/or one distal gene are needed for complete male sterility in the C57BL/6J-ChrX^{PWD} strain (Storchová *et al.*, 2004).

Similar complete male sterility is observed in the subconsomic strain carrying a large segment (<146.4 Mb) of the PWD X chromosome in the genetic background of the wild *M. m. musculus*-derived LEWES strain (Good *et al.*, 2008a). This observation indicates that reproductive isolation is not limited to the cross between *M. m. musculus* and the B6 laboratory strain, but that it is a more general phenomenon due to X-linked incompatibility between the two subspecies, *M. m. musculus* and *M. m. domesticus*. Further genetic studies showed that at least four loci responsible for male sterility are distributed along the wide range of the X chromosome (Good *et al.*, 2008a).

Reproductive isolation between *M. m. domesticus* and Japanese *M. m. molossinus* wild mice was shown in the C57BL/6J-ChrX^{MSM} strain that has the X chromosome of the *M. m. molossinus*-derived MSM strain (Oka *et al.*, 2004). *M. m. molossinus* is derived from the hybridization of *M. m. castaneus* and *M. m. musculus*, and its nuclear genome is predominantly derived from *M. m. musculus* (Yonekawa *et al.*, 1980, 1988; Moriwaki, 1994; Sakai *et al.*, 2005). The C57BL/6J-ChrX^{MSM} strain shows male-specific sterility, and is characterized by cell-cycle arrest at the premeiotic stage and apoptosis at the pachytene stage, which result in reduced testis weight and abnormal morphology and low motility in mature sperm (Oka *et al.*, 2004, 2010). Genetic studies indicate that the abnormal morphology of the sperm head is caused by DM incompatibility between the MSM alleles of at least three X-linked loci and B6 alleles of loci on autosomes, including Chr 1 and 11 (Oka *et al.*, 2007) (Fig. 17.1a, b). Males carrying the MSM X chromosome and Chr 11 and a large segment of MSM Chr 1 in the B6 background have partially restored fertility, suggesting that genetic components of the DM incompatibility between *M. m. domesticus* and *M. m. molossinus* are located on these chromosomes, and that MSM alleles of additional component(s) on other chromosome(s) are necessary for complete recovery of fertility. This is the first report showing the locations of interacting DM incompatibility components in mice. The reduced testis weight phenotype is caused by DM incompatibility between the MSM alleles of a single or a few genes on the distal region of the X chromosome and B6 alleles on undefined autosome(s) (Oka *et al.*, 2004). Since PARs in the C57BL/6J-ChrX^{MSM} strain are derived from the B6 strain, structural incompatibility in PARs is not the cause of the reduced testis weight characteristic of this strain (Oka *et al.*, 2010).

PGN-ChrX^{MSM}, a consomic strain with the same MSM X chromosome in the background of a wild *M. m. domesticus*-derived PGN strain, shows similar

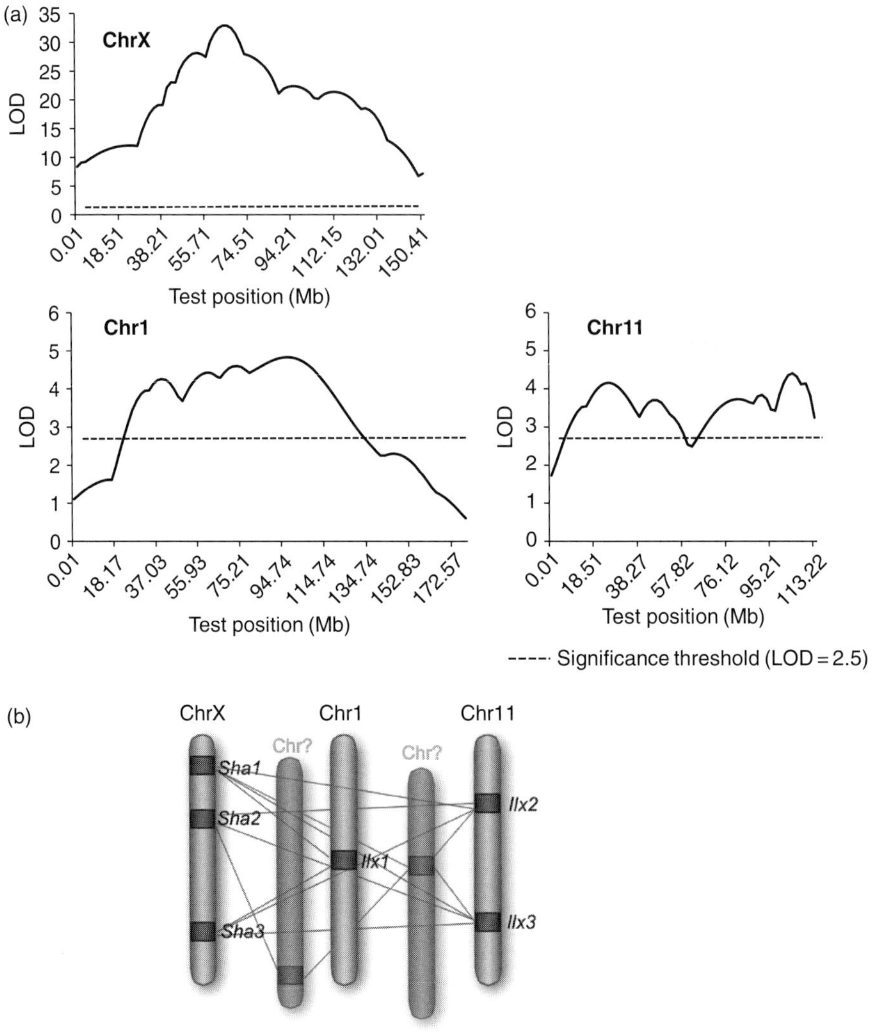

Figure 17.1 Chromosomal regions involved in DM incompatibility in the C57BL/6J-ChrX^{MSM} strain. (a) QTLs affecting sperm-head morphology in C57BL/6J-ChrX^{MSM} males are located on the X chromosome, Chr 1, and Chr 11. LOD scores indicate the probability of the existence of responsible genes. (b) Epistatic interaction between the QTLs. Dark regions depict chromosomal segments where the QTLs reside. Males that carry MSM-derived chromosomal segments in the C57BL/6J background have partially restored fertility. *Sha1* to *Sha3* and *Ilx1* to *Ilx3* are the names of responsible loci.

phenotypes, including reduced testis weight and abnormal sperm morphology (Oka *et al.*, 2010). Thus, X-linked incompatibility is not specific to the B6 background, but is observed more generally in the background of *M. m. domesticus*. In contrast to X–Y structural incompatibility, multiple components on the X chromosome could be the cause of the reproductive isolation between the two subspecies.

Evolutionary timing of reproductive isolation mutations

In general, a pair of species or subspecies will have multiple DM incompatibilities, because reproductive isolation evolves gradually as a by-product of genetic divergence, and problems associated with hybrid fitness are the result of an accumulation of deleterious epistatic interactions of genes across species or subspecies (Mayr, 1942; Coyne, 1992). As shown in Fig. 17.2, spermatogenic disruptions, such as cell-cycle arrest and apoptosis, may occur at different stages in cases of reproductive isolation. Extensive apoptosis has been reported to occur during metaphase I in interspecific (B6 × *M. spretus*)F_1 and (B6 × *M. hortulanus*)F_1 males (Matsuda *et al.*, 1991; Matsuda and Chapman, 1994). The major phenotype of intersubspecific (B6 × NJL)F_1 and (PWD × B6)F_1 hybrid males is apoptosis of meiotic spermatocytes, which occurs during the zygotene–pachytene stage (Kaku *et al.*, 1995; Forejt, 1996; Oka *et al.*, 2010). Other *M. m. domesticus* and *M. m. musculus* intersubspecific F_1 hybrid males, such as (PWK × LEWES)F_1, (CZECH × LEWES)F_1, (LEWES × CZECH)F_1, (CZECH × WSB)F_1 and (WSB × CZECH)F_1, suffer spermatogenic disruptions during spermiogenesis (Good *et al.*, 2008b), as does the C57BL/6J-ChrX^{PWD} consomic strain (Storchová *et al.*, 2004). Disruptions occur in the C57BL/6J-ChrX^{MSM} strain at three different stages of spermatogenesis: the premeiotic stage, the pachytene stage of meiotic prophase, and spermiogenesis (Oka *et al.*, 2004, 2010). In another consomic strain, PGN-ChrX^{MSM}, spermatogenic disruption occurs during metaphase of meiosis I (Oka *et al.*, 2010).

Over a substantial time period of evolutionary divergence, it is conceivable that mutations involved in DM incompatibilities have accumulated in both diverging lineages, which may have stochastically increased the chance of mutations leading to reproductive defects during early spermatogenesis. Spermatogenic disruptions occurring during the early stages of spermatogenesis sometimes mask disruptions occurring in the later stages. However, there is no explicit correlation between the genetic distance between two parental strains and the ordering of the stages of spermatogenic disruptions. This implies that *M. musculus* is in the early stages of speciation, and that the divergence time has not been long enough to accumulate DM incompatibility mutations (Oka *et al.*, 2010). In spite of the relatively small genetic distance between species or subspecies, it is notable that many of the hybrid males involve X-linked incompatibilities (Table 17.1 and Table 17.2).

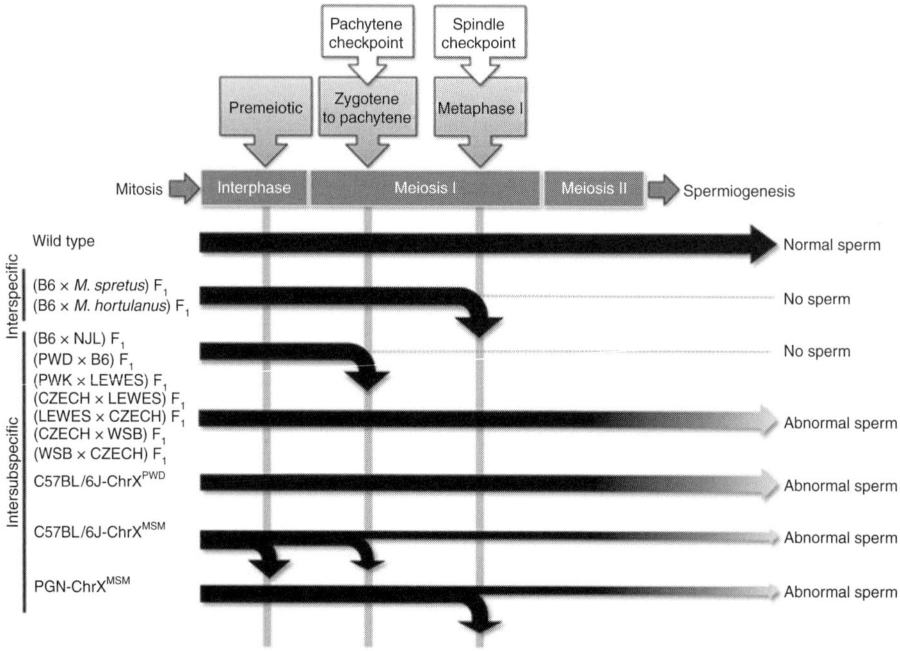

Figure 17.2 The stages at which spermatogenic disruptions occur in different cases of hybrid male sterility. Downward arrows indicate spermatogenic disruptions and grey arrows indicate abnormal spermiogenesis. Only predominant spermatogenic disruptions are indicated here. Note that meiotic disruptions occur preferentially during the two meiotic checkpoint stages.

Postzygotic reproductive isolation and Haldane's rule

Haldane's rule is partly explained by the hemizygosity of the X chromosome in males. One explanation along this line is referred to as the 'dominance theory' (Orr, 1993; Turelli and Orr, 1995), which posits that the alleles causing hybrid sterility or inviability are X-linked, and typically behave as recessives. Thus, hybrid males preferentially suffer X-linked incompatibility defects due to hemizygosity (Muller, 1942; Turelli and Orr, 1995, 2000). The dominance theory predicts that hybrid females will be sterile, as is the case in hybrid males, if they carry homozygous incompatible X-linked alleles. Although females homozygous for the incompatible X chromosome can never be obtained by natural mating due to complete male sterility caused by the incompatible X chromosome, homozygous females of C57BL/6J-ChrX^{MSM} can be produced by a reproductive manipulation, intracytoplasmic sperm injection (ICSI). The resulting homozygous females are fertile, but they have low fertility and a shortened reproductive lifespan (A. Oka and T. Shiroishi, unpublished data), which is consistent with the dominance effect of X-linked responsible genes. Additional support for the dominance theory comes from (laboratory strain × *M. spretus*) F_1 females with the monosomic *M. spretus* X chromosome, which is generated by crossing fertile XO females with the heterogeneous genetic background of B6 and C3H strains and *M. spretus* males. Like the F_1 hybrid males, the hybrid XO females with the *M. spretus* X chromosome are sterile (Biddle *et al.*, 1994).

Accumulation of male reproductive genes on the X chromosome

The disproportionately large effect of the X chromosome on reproductive isolation can be explained in part by the hemizygosity of the X chromosome in males, which may allow for accumulation of recessive mutations on the X chromosome that are beneficial to males' X chromosomes (Rice, 1984; Vicoso and Charlesworth, 2006; Gurbich and Bachtrog, 2008). Several studies reported that spermatogonial genes and/or male reproductive genes expressed in the premeiotic stage are generally enriched on the mouse X chromosome (Wang *et al.*, 2001; Khil *et al.*, 2004). However, transcriptional inactivation during spermatogenesis influences the genetic content of the X chromosome. During male meiosis, sex chromosomes are incorporated into the heterochromatinized region called the 'XY body' (also known as the 'sex body') and become transcriptionally inactivated. This process is referred to as meiotic sex chromosome inactivation (MSCI) (Turner, 2007). The transcriptional repression of sex chromosomes continues postmeiotically (Namekawa *et al.*, 2006), and thus the expression of X-linked genes is generally restricted to early spermatogenesis. Therefore, genes that function in a male-specific manner in the late stage of spermatogenesis are expected to be selectively favoured on autosomes, where they can avoid inactivation (Wu and Xu, 2003). It was reported that male-specific genes from the X chromosome have undergone retrotransposition onto autosomes in the human and mouse genomes, enabling male-specific genes to escape the influence of X-inactivation (Emerson *et al.*, 2004).

Preferential accumulation of early spermatogenic genes on the X chromosome may predict the high frequency of premeiotic disruptions in hybrid males. However, only one study has reported cell-cycle arrest at the premeiotic stage in sterile hybrid males of the mouse subspecies (Oka *et al.*, 2010). In many cases of reproductive isolation between mouse species or subspecies, disruptions at the meiotic and/or postmeiotic stages are observed more frequently, suggesting that genes involved in reproductive isolation are acting at these stages. A recent study indicated that multi-copy genes on the mouse X chromosome are expressed during postmeiotic stages. Increased copy number of an X-linked gene likely compensates for a repressive chromatin environment (Mueller *et al.*, 2008). Thus, X-linked genes might function longer during spermatogenesis than previously thought, making the X chromosome highly susceptible to DM incompatibility. Other plausible genetic bases underlying disruptions at the meiotic and post-meiotic stages are described below.

Meiotic checkpoints and X-linked incompatibility

Meiosis is specialized cell division to produce gametes, and includes the induction of DNA double-strand breaks (DSBs), homologous chromosome pairing, homologous recombination, repair of DSBs, and segregation of homologous chromosomes to each daughter cell (Handel and Schimenti, 2010). Since these intricate chromosomal processes are highly error-prone, most organisms have evolved surveillance systems known as 'checkpoints' to ensure that the processes have been accurately completed (Tarsounas and Moens, 2001; Wolgemuth *et al.*, 2002). The meiotic checkpoint responses induce cell-cycle arrest and/or apoptosis of defective meiocytes to prevent the generation of defective gametes. Studies of mutant mice with meiotic impairments show the presence of two meiotic checkpoints during meiosis: the pachytene checkpoint and the spindle checkpoint. Notably, meiotic disruptions observed in sterile hybrid males from multiple crosses occur at specific stages (Fig. 17.2). These stages might be subjected to the two meiotic checkpoints (Oka *et al.*, 2010).

The pachytene checkpoint is known to monitor unrepaired DSBs and/or chromosome asynapsis in yeast (Roeder and Bailis, 2000). Though the molecules involved in monitoring in the pachytene checkpoint have not been identified in mice, many mutant mice with severe recombination defects show meiotic arrest and apoptosis at the pachytene stage, implying that a similar checkpoint system exists in mammals (Roeder and Bailis, 2000). Cytological studies indicate the presence of extensive unrepaired DSBs in pachytene spermatocytes of (B6 × *M. m. musculus*-NJL)F_1 males. These spermatocytes exhibit extensive apoptosis during the zygotene–pachytene stage, suggesting the pachytene checkpoint responds (Oka *et al.*, 2010). Since asynapsis normally triggers the pachytene checkpoint, mammalian males have evolved a heterochromatinized XY body on the sex chromosomes to suppress checkpoint surveillance for asynapsis on these chromosomes (Solari, 1974; Turner, 2007). In this context, a correlation between lack of the XY bodies and male sterility has been implicated (Lifschytz and Lindsley, 1972; Forejt, 1996). A lack of XY bodies and apoptosis of pachytene spermatocytes were observed in (B6 × NJL)F_1 and (PWK × B6)F_1 sterile males (Mihola *et al.*, 2009; Oka *et al.*, 2010). Thus, the connection between the XY body and apoptosis is likely due to the responses of the pachytene checkpoint.

Alternatively, failure of MSCI could result in improper expression of sex-linked genes, potentially triggering apoptosis of defective spermatocytes by a mechanism other than the pachytene checkpoint response (Lifschytz and Lindsley, 1972; Homolka *et al.*, 2007; Burgoyne *et al.*, 2009).

During the metaphase stage of meiosis, the spindle checkpoint monitors segregation failures to ensure high fidelity of meiotic chromosome segregation to each daughter cell, as occurs during mitosis (Malmanche *et al.*, 2006; Yin *et al.*, 2008). Extensive apoptosis of spermatocytes at metaphase I may be induced by the spindle checkpoint responses due to the abnormal X–Y dissociation, as observed in (B6 × *M. spretus*)F_1 males.

Interestingly, many meiotic mouse mutants have checkpoint responses that are associated with sexual dimorphism (Morelli and Cohen, 2005; Cohen *et al.*, 2006). In most cases, the severity of defects is considerably greater for mutant males than females. Since postzygotic reproductive isolation in mice is often caused by meiotic disruptions, it is an intriguing possibility that more stringent checkpoint responses in males than females may explain Haldane's rule (Li *et al.*, 2009).

Postmeiotic gene expression related to X-linked incompatibility

Spermiogenesis is a highly differentiated process in males. Round spermatids undergo dynamic morphologic changes, including nuclear condensation and clongation, formation of flagella and the acrosome, reorganization of organelles, and elimination of the cytoplasm upon spermiation. Approximately one-half of the testis-specific genes are exclusively expressed during spermiogenesis. Genes required for late spermiogenesis have to be transcribed in round spermatids, because transcription ceases when nuclear elongation and condensation starts. The transcripts are stabilized and stored until they are translated in later steps, when their encoded proteins are needed (Yan, 2009). Thus, defective postmeiotic heterochromatinization of the sex chromosomes and resulting improper expression of the sex-linked genes could influence the cellular processes occurring during late spermiogenesis. Sterile F_1 hybrid males from an interspecific cross and X-chromosomal substitution between two different subspecies often produce morphologically or functionally abnormal sperm, despite having relatively normal testicular histology (Oka *et al.*, 2004; Storchová *et al.*, 2004; Good *et al.*, 2008a), suggesting that excess or improper expression of sex-linked genes at an earlier stage could disturb sperm morphogenesis. This model is consistent with our observation that the C57BL/6J-ChrX^{MSM} males showed the overexpression of X-linked genes in the round spermatids (A. Oka and

T. Shiroishi, unpublished data). Genes expressed in later stages of spermatogenesis generally evolve faster and are more likely to be under positive selection than genes expressed earlier in spermatogenesis, possibly due to stronger sexual selection acting on the genes that influence sperm structure and function (Good and Nachman, 2005). Thus, the genes responsible for spermiogenesis disruption in hybrid males might be affected by positive selection.

Possible molecular basis underlying DM incompatibilities

Whether speciation genes are ordinary genes that have normal cellular functions within a species, or are unusual genetic factors such as repetitive DNA sequences or transposable elements, has been the subject of considerable debate (Coyne and Orr, 2004). This question has not been fully answered because, as yet, only a small number of the genes responsible for reproductive isolation have been identified in model animals. However, some recent studies provide clues regarding the underlying molecular basis of genetic incompatibilities. These studies suggest that heterochromatinization of sex chromosomes is mediated by repeat sequences and/or proteins binding to these sequences, many of which are sex-linked, and that failure of heterochromatinization in hybrid animals eventually leads to the misexpression of other genes or defective chromosome segregation.

In crosses between *D. simulans* females and *D. melanogaster* males, the F_1 hybrid females die at the embryonic stage. The Zygotic hybrid rescue (*Zhr*) element, which is a block of 359-bp satellite repeats located in the centric region of the X chromosome, is responsible for this inviability (Sawamura and Yamamoto, 1997). In *D. melanogaster*, undefined maternal factors (e.g. proteins or RNAs) bind to *Zhr* to induce the heterochromatic state of the X chromosome, which is necessary for proper chromosome segregation during mitosis. The *D. simulans* genome has no *Zhr*, however, and F_1 hybrid females suffer a mitotic breakdown due to the failure of heterochromatinization of the X chromosome, which results in lagging chromatids and missegregation during cell division. Thus, failure of heterochromatinization in F_1 hybrid females is likely caused by the lack of appropriate maternal factors from *D. simulans*, which are necessary to regulate the *D. melanogaster*-specific *Zhr* element (Ferree and Barbash, 2009). These observations suggest that evolutionary change in the quantity and/or composition of species-specific satellite DNA can cause genetic incompatibilities.

Another example is male sterility caused by the X-linked Odysseus-site homeobox (*OdsH*) gene in F_1 hybrids of *D. simulans* and *D. mauritiana* crosses (Ting *et al.*, 1998). In *D. simulans*, the ODSH protein associates with repeat-rich regions on the pericentric regions of the X and the fourth chromosomes in primary spermatocytes, but such expression has not been observed in *D. mauritiana*. In sterile males of the introgression line that carries the *D. mauritiana*-derived *OdsH*-containing region in the *D. simulans* background, increased expression of the ODSH protein and its ectopic localization on the Y chromosome, which eventually results in decondensation of the Y chromosome, have been observed (Bayes and Malik, 2009). The authors suggest that the genetic conflict arising between host organisms and rapidly evolving DNA repeat elements might be an important evolutionary force driving hybrid incompatibility.

A recent study identified the first mammalian speciation gene, *Prdm9*, which encodes a histone H3 lysine 4-trimethyltransferase (Mihola *et al.*, 2009; see Forejt *et al.*, Chapter 19). The involvement of *Prdm9* in speciation implies that modification of histones during meiosis is involved in F_1 hybrid male sterility. The PRDM9 protein also regulates meiotic recombination by binding to a specific DNA motif and inducing the epigenetic mark H3K4me3 to specify recombination hotspot locations (Baudat *et al.*, 2010). (PWD × B6)F_1 hybrid sterile males lack XY bodies, as has been observed in *Prdm9*-deficient mice, indicating failure of MSCI (Hayashi *et al.*, 2005; Mihola *et al.*, 2009). The PRDM9 protein and its DNA binding motif are rapidly evolving by keeping compatible interaction with each other. Thus, incompatibilities between *Prdm9*-binding specificity and its binding motifs may be the molecular basis of *Prdm9*-mediated hybrid male sterility (Oliver *et al.*, 2009).

In our QTL analysis of the male sterility phenotype of the C57BL/6J-ChrX^{MSM} strain, we detected a significantly high LOD score along the long stretch of the X chromosome (Fig. 17.1a; Oka *et al.*, 2004). We found it difficult to narrow down the QTL to a single genetic region affecting the phenotype. Upon decreasing the length of the MSM-derived X chromosome by generating subconsomic strains, the phenotype became less pronounced (A. Oka and T. Shiroishi, unpublished data). Notably, genome-wide comparative analysis of single nucleotide polymorphisms (SNPs) demonstrated that the X chromosome is highly differentiated among subspecies along the entire X chromosome length (Harr, 2006; Yang *et al.*, 2011).

These findings suggest that multiple loci/regions along the long stretch of the X chromosome synergistically or additively contribute to postzygotic reproductive isolation. Alternatively, some genetic components on the X chromosome, such as repetitive sequences or non-coding sequences, may mediate heterochromatinization of the sex chromosome, the components from different subspecies leading to development of an abnormal sex chromosome chromatin structure, in turn resulting in improper expression of X-linked genes by *cis*-regulation and/or misexpression of autosomal genes by *trans*-regulation. Song *et al.* (2009) reported that many X-linked miRNAs escape sex chromosome inactivation during meiotic and early postmeiotic stages, suggesting that X-linked miRNAs may contribute to the process of MSCI and/or postmeiotic sex chromosome inactivation. Another study showed that the density of miRNAs is two-fold higher on X chromosomes than it is on autosomes in mammalian species, and that X-linked miRNAs evolve faster than those on autosomes (Guo *et al.*, 2009). Notably, the QTLs that we detected are located in the highly repetitive regions (Mueller *et al.*, 2008). This model is consistent with our observation of overexpression of X-linked genes and misexpression of autosomal genes in the pachytene spermatocytes and round spermatids of the C57BL/6J-ChrX^{MSM} strain (A. Oka and T. Shiroishi, unpublished data; see also Good *et al.*, 2010).

Conclusion

In this chapter we have reviewed the disproportionately large contribution of the X chromosome to postzygotic reproductive isolation, and in light of the unique evolution and cytological behaviour of the X chromosome, discussed possible reasons why X-linked incompatibility frequently evokes hybrid male sterility. Finally, we reiterated the importance of X: autosome imbalance in postzygotic reproductive isolation. The X: autosome imbalance hypothesis was originally proposed by Muller (1942) to explain Haldane's rule, and states that the hemizygous sex has the X chromosome from only one species, potentially resulting in expression imbalance (Muller, 1942). Recent studies have indicated that defective heterochromatinization of the sex chromosomes leads to misexpression of genes not only on the sex chromosome, but also on autosomes, which may eventually result in disruption of spermatogenesis. Accumulating data suggest that the large X-effect in mice is most likely a composite phenomenon constituting different components.

REFERENCES

Bateson, W. (1909). Heredity and variation in modern lights. In *Darwin and Modern Science*, ed. A. Seward. Cambridge: Cambridge University Press, pp. 85–101.

Baudat, F., Buard, J., Grey, C., *et al.* (2010). PRDM9 is a major determinant of meiotic recombination hotspots in humans and mice. *Science*, **327**, 836–40.

Bayes, J. and Malik, H. (2009). Altered heterochromatin binding by a hybrid sterility protein in *Drosophila* sibling species. *Science*, **326**, 1538–41.

Biddle, F., Eales, B., and Dean, W. (1994). Haldane's rule and heterogametic female and male sterility in the mouse. *Genome*, **37**, 198–202.

Bonhomme, F., Catalan, J., Britton-Davidian, J., *et al.* (1984). Biochemical diversity and evolution in the genus *Mus*. *Biochemical Genetics*, **22**, 275–303.
Bonhomme, F. and Guénet, J.-L. (1996). The laboratory mouse and its wild relatives. In *Genetic Variants and Strains of the Laboratory Mouse*, ed. M. F. Lyon, S. Rastan, and S. D. M. Brown. Oxford: Oxford University Press, pp. 1577–96.
Bonhomme, F., Martin, S., and Thaler, L. (1978). Hybridization between *Mus musculus* L. and *Mus spretus* Lataste under laboratory conditions. *Experientia*, **34**, 1140–1.
Britton-Davidian, J., Fel-Clair, F., Lopes, J., Alibert, P., and Boursot, P. (2005). Postzygotic isolation between the two European subspecies of the house mouse: estimates from fertility patterns in wild and laboratory-bred hybrids. *Biological Journal of the Linnean Society*, **84**, 379–93.
Burgoyne, P. S., Mahadevaiah, S. K., and Turner, J. M. (2009). The consequences of asynapsis for mammalian meiosis. *Nature Reviews Genetics*, **10**, 207–16.
Chubb, C. and Nolan, C. (1987). Mouse hybrid sterility and testicular function. *Biology of Reproduction*, **36**, 1343–8.
Cohen, P. E., Pollack, S. E., and Pollard, J. W. (2006). Genetic analysis of chromosome pairing, recombination, and cell cycle control during first meiotic prophase in mammals. *Endocrine Reviews*, **27**, 398–426.
Coyne, J. (1992). Genetics and speciation. *Nature*, **355**, 511–15.
Coyne, J. A. and Orr, H. A. (1989). Two rules of speciation. In *Speciation and Its Consequences*, ed. D. Otte and J. Endler. Sunderland, MA: Sinauer Associates, pp. 180–207.
Coyne, J. A. and Orr, H. A. (2004). *Speciation*. Sunderland, MA: Sinauer Associates.
Dobzhansky, T. (1936). Studies on hybrid sterility: II. Localization of sterility factors in *Drosophila pseudoobscura* hybrids. *Genetics*, **21**, 113–35.
Dod, B., Jermiin, L. S., Boursot, P., *et al.* (1993). Counterselection on sex chromosomes in the *Mus musculus* European hybrid zone. *Journal of Evolutionary Biology*, **6**, 529–46.
Elliott, R. W., Miller, D. R., Pearsall, R. S., *et al.* (2001). Genetic analysis of testis weight and fertility in an interspecies hybrid congenic strain for Chromosome X. *Mammalian Genome*, **12**, 45–51.
Elliott, R. W., Poslinski, D., Tabaczynski, D., Hohman, C., and Pazik, J. (2004). Loci affecting male fertility in hybrids between *Mus macedonicus* and C57BL/6. *Mammalian Genome*, **15**, 704–10.
Emerson, J., Kaessmann, H., Betrán, E., and Long, M. (2004). Extensive gene traffic on the mammalian X-chromosome. *Science*, **303**, 537–40.
Ferree, P. and Barbash, D. (2009). Species-specific heterochromatin prevents mitotic chromosome segregation to cause hybrid lethality in *Drosophila*. *PLoS Biology*, **7**, e1000234.
Ferris, S. D., Sage, R. D., and Wilson, A. C. (1982). Evidence from mtDNA sequences that common laboratory strains of inbred mice are descended from a single female. *Nature*, **295**, 163–5.
Forejt, J. (1981). Hybrid sterility gene located in the T/t-H-2 supergene on chromosome 17. In *Current Trends in Histocompatibility: Immunogenetic and Molecular Profiles*, ed. R. Reisfeld, and S. Ferrone. London: Plenum Press, pp. 103–31.
Forejt, J. (1996). Hybrid sterility in the mouse. *Trends in Genetics*, **12**, 412–17.

Forejt, J. and Iványi, P. (1974). Genetic studies on male sterility of hybrids between laboratory and wild mice (*Mus musculus* L.). *Genetical Research*, **24**, 189–206.

Forejt, J., Vincek, V., Klein, J., Lehrach, H., and Loudová-Micková, M. (1991). Genetic mapping of the t-complex region on mouse chromosome 17 including the Hybrid sterility-1 gene. *Mammalian Genome*, **1**, 84–91.

Frazer, K. A., Eskin, E., Kang, H. M., *et al.* (2007). A sequence-based variation map of 8.27 million SNPs in inbred mouse strains. *Nature*, **448**, 1050–3.

Geraldes, A., Basset, P., Gibson, B., *et al.* (2008). Inferring the history of speciation in house mice from autosomal, X-linked, Y-linked and mitochondrial genes. *Molecular Ecology*, **17**, 5349–63.

Good, J. M., Dean, M. D., and Nachman, M. W. (2008a). A complex genetic basis to X-linked hybrid male sterility between two species of house mice. *Genetics*, **179**, 2213–28.

Good, J., Giger, T., Dean, M., and Nachman, M. (2010). Widespread over-expression of the X chromosome in sterile F_1 hybrid mice. *PLoS Genetics*, **6**, e1001148.

Good, J. M., Handel, M. A., and Nachman, M. W. (2008b). Asymmetry and polymorphism of hybrid male sterility during the early stages of speciation in house mice. *Evolution*, **62**, 50–65.

Good, J. M. and Nachman, M. W. (2005). Rates of protein evolution are positively correlated with developmental timing of expression during mouse spermatogenesis. *Molecular Biology and Evolution*, **22**, 1044–52.

Gregorová, S., Divina, P., Storchová, R., *et al.* (2008). Mouse consomic strains: exploiting genetic divergence between *Mus m. musculus* and *Mus m. domesticus* subspecies. *Genome Research*, **18**, 509–15.

Guénet, J. -L. and Bonhomme, F. (2003). Wild mice: an ever-increasing contribution to a popular mammalian model. *Trends in Genetics*, **19**, 24–31.

Guénet, J. -L., Nagamine, C., Simon-Chazottes, D., Montagutelli, X., and Bonhomme, F. (1990). Hst-3: an X-linked hybrid sterility gene. *Genetical Research*, **56**, 163–5.

Guo, X., Su, B., Zhou, Z., and Sha, J. (2009). Rapid evolution of mammalian X-linked testis microRNAs. *BMC Genomics*, **10**, 97.

Gurbich, T. and Bachtrog, D. (2008). Gene content evolution on the X-chromosome. *Current Opinion in Genetics & Development*, **18**, 493–8.

Haldane, J. B. S. (1922). Sex ratio and unisexual sterility in hybrid animals. *Journal of Genetics*, **12**, 101–9.

Hale, D. W., Washburn, L. L., and Eicher, E. M. (1993). Meiotic abnormalities in hybrid mice of the C57BL/6J × *Mus spretus* cross suggest a cytogenetic basis for Haldane's rule of hybrid sterility. *Cytogenetics and Cell Genetics*, **63**, 221–34.

Handel, M. and Schimenti, J. (2010). Genetics of mammalian meiosis: regulation, dynamics and impact on fertility. *Nature Reviews Genetics*, **11**, 124–36.

Harr, B. (2006). Genomic islands of differentiation between house mouse subspecies. *Genome Research*, **16**, 730–7.

Hayashi, K., Yoshida, K., and Matsui, Y. (2005). A histone H3 methyltransferase controls epigenetic events required for meiotic prophase. *Nature*, **438**, 374–8.

Homolka, D., Ivanek, R., Capkova, J., Jansa, P., and Forejt, J. (2007). Chromosomal rearrangement interferes with meiotic X-chromosome inactivation. *Genome Research*, **17**, 1431–7.

Hunt, W. G. and Selander, R. K. (1973). Biochemical genetics of hybridisation in European house mice. *Heredity*, **31**, 11–33.

Iványi, P., Vojtísková, M., Démant, P., and Micková, M. (1969). Genetic factors in the ninth linkage group influencing reproductive performance in male mice. *Folia Biologica (Praha)*, **15**, 401–21.

Kaku, Y., Kon, Y., Takagi, N., *et al.* (1995). Histological analysis of male hybrid sterility induced by the *Hst-1* gene in mice. *Journal of Veterinary Medical Science*, **57**, 973–5.

Khil, P., Smirnova, N., Romanienko, P., and Camerini-Otero, R. (2004). The mouse X-chromosome is enriched for sex-biased genes not subject to selection by meiotic sex chromosome inactivation. *Nature Genetics*, **36**, 642–6.

Kipling, D., Wilson, H., Thomson, E., *et al.* (1996). Structural variation of the pseudoautosomal region between and within inbred mouse strains. *Proceedings of the National Academy of Sciences of the United States of America*, **93**, 171–5.

Li, X. C., Barringer, B. C., and Barbash, D. A. (2009). The pachytene checkpoint and its relationship to evolutionary patterns of polyploidization and hybrid sterility. *Heredity*, **102**, 24–30.

Lifschytz, E. and Lindsley, D. (1972). The role of X-chromosome inactivation during spermatogenesis. *Proceedings of the National Academy of Sciences of the United States of America*, **69**, 182–6.

Macholán, M., Baird, S. J. E., Dufková, P., *et al.* (2011). Assessing multilocus introgression patterns: a case study on the mouse X chromosome in central Europe. *Evolution*, **65**, 1428–46.

Macholán, M., Munclinger, P., Šugerková, M., *et al.* (2007). Genetic analysis of autosomal and X-linked markers across a mouse hybrid zone. *Evolution*, **61**, 746–71.

Malmanche, N., Maia, A., and Sunkel C. E. (2006). The spindle assembly checkpoint: preventing chromosome mis-segregation during mitosis and meiosis. *FEBS Letters*, **580**, 2888–95.

Masly, J., Jones, C., Noor, M., Locke, J., and Orr, H. (2006). Gene transposition as a cause of hybrid sterility in *Drosophila*. *Science*, **313**, 1448–50.

Matsuda, Y. and Chapman, V. M. (1994). X–Y chromosome pairing and hybrid sterility of mice. In *Genetics in Wild Mice: Its Application to Biomedical Research*, ed. K. Moriwaki, T. Shiroishi, and H. Yonekawa. Tokyo: Japan Scientific Press and Karger, pp. 257–75.

Matsuda, Y., Hirobe, T., and Chapman, V. M. (1991). Genetic basis of X–Y chromosome dissociation and male sterility in interspecific hybrids. *Proceedings of the National Academy of Sciences of the United States of America*, **88**, 4850–4.

Matsuda, Y., Moens, P. B., and Chapman, V. M. (1992). Deficiency of X and Y chromosomal pairing at meiotic prophase in spermatocytes of sterile interspecific hybrids between laboratory mice (*Mus domesticus*) and *Mus spretus*. *Chromosoma*, **101**, 483–92.

Mayr, E. (1942). *Systematics and the Origin of Species*. New York: Columbia University Press.

Mihola, O., Trachtulec, Z., Vlcek, C., Schimenti, J. C., and Forejt, J. (2009). A mouse speciation gene encodes a meiotic histone h3 methyltransferase. *Science*, **323**, 373–5.

Morelli, M. A. and Cohen, P. E. (2005). Not all germ cells are created equal: aspects of sexual dimorphism in mammalian meiosis. *Reproduction*, **130**, 761–81.

Moriwaki, K. (1994). Wild mice from the geneticist's viewpoint. In *Genetics in Wild Mice: Its Application to Biomedical Research*, ed. K. Moriwaki, T. Shiroishi, and H. Yonekawa. Tokyo: Japan Scientific Press and Karger, pp. viii–xxv.

Mueller, J. L., Mahadevaiah, S. K., Park, P. J., *et al.* (2008). The mouse X-chromosome is enriched for multicopy testis genes showing postmeiotic expression. *Nature Genetics*, **40**, 794–9.

Muller, H. J. (1942). Isolating mechanisms, evolution and temperature. *Biological Symposia*, **6**, 71–125.

Nadeau, J., Singer, J., Matin, A., and Lander, E. (2000). Analysing complex genetic traits with chromosome substitution strains. *Nature Genetics*, **24**, 221–5.

Namekawa, S. H., Park, P. J., Zhang, L. F., *et al.* (2006). Postmeiotic sex chromatin in the male germline of mice. *Current Biology*, **16**, 660–7.

Oka, A., Aoto, T., Totsuka, Y., *et al.* (2007). Disruption of genetic interaction between two autosomal regions and the X-chromosome causes reproductive isolation between mouse strains derived from different subspecies. *Genetics*, **175**, 185–97.

Oka, A., Mita, A., Sakurai-Yamatani, N., *et al.* (2004). Hybrid breakdown caused by substitution of the X-chromosome between two mouse subspecies. *Genetics*, **166**, 913–24.

Oka, A., Mita, A., Takada, Y., Koseki, H., and Shiroishi, T. (2010). Reproductive isolation in hybrid mice due to spermatogenesis defects at three meiotic stages. *Genetics*, **186**, 339–51.

Oliver, P., Goodstadt, L., Bayes, J., *et al.* (2009). Accelerated evolution of the *Prdm9* speciation gene across diverse metazoan taxa. *PLoS Genetics*, **5**, e1000753.

Orr, H. A. (1993). Haldane's rule has multiple genetic causes. *Nature*, **361**, 532–3.

Orr, H. A. (2005). The genetic basis of reproductive isolation: insights from *Drosophila*. *Proceedings of the National Academy of Sciences of the United States of America*, **102**, 6522–6.

Orth, A., Belkhir, K., Britton-Davidian, J., *et al.* (2002). Natural hybridisation between two sympatric species of mice *Mus musculus domesticus* L. and *Mus spretus* Lataste. *Comptes Rendus Biologies*, **325**, 89–97.

Payseur, B. A., Krenz, J. G., and Nachman, M. W. (2004). Differential patterns of introgression across the X-chromosome in a hybrid zone between two species of house mice. *Evolution*, **58**, 2064–78.

Presgraves, D. (2003). A fine-scale genetic analysis of hybrid incompatibilities in *Drosophila*. *Genetics*, **163**, 955–72.

Rice, W. (1984). Sex chromosomes and the evolution of sexual dimorphism. *Evolution*, **38**, 735–42.

Roeder, G. S. and Bailis, J. M. (2000). The pachytene checkpoint. *Trends in Genetics*, **16**, 395–403.

Sakai, T., Kikkawa, Y., Miura, I., *et al.* (2005). Origins of mouse inbred strains deduced from whole-genome scanning by polymorphic microsatellite loci. *Mammalian Genome*, **16**, 11–19.

Sawamura, K. and Yamamoto, M. (1997). Characterization of a reproductive isolation gene, zygotic hybrid rescue, of *Drosophila melanogaster* by using minichromosomes. *Heredity*, **79**, 97–103.
Solari, A. J. (1974). The behavior of the XY pair in mammals. *International Review of Cytology*, **38**, 273–317.
Song, R., Ro, S., Michaels, J., *et al.* (2009). Many X-linked microRNAs escape meiotic sex chromosome inactivation. *Nature Genetics*, **41**, 488–93.
Storchová, R., Gregorová, S., Buckiová, D., *et al.* (2004). Genetic analysis of X-linked hybrid sterility in the house mouse. *Mammalian Genome*, **15**, 515–24.
Takada, T., Mita, A., Maeno, A., *et al.* (2008). Mouse intersubspecific consomic strains for genetic dissection of quantitative complex traits. *Genome Research*, **18**, 500–8.
Tarsounas, M. and Moens, P. B. (2001). Checkpoint and DNA-repair proteins are associated with the cores of mammalian meiotic chromosomes. *Current Topics in Developmental Biology*, **51**, 109–34.
Ting, C., Tsaur, S., Wu, M., and Wu, C. (1998). A rapidly evolving homeobox at the site of a hybrid sterility gene. *Science*, **282**, 1501–4.
Tucker, P. K., Sage, R. D., Warner, J., Wilson, A. C., and Eicher, E. M. (1992). Abrupt cline for sex chromosomes in a hybrid zone between two species of mice. *Evolution*, **46**, 1146–63.
Turelli, M. and Orr, H. A. (1995). The dominance theory of Haldane's rule. *Genetics*, **143**, 609–11.
Turelli, M. and Orr, H. A. (2000). Dominance, epistasis and the genetics of postzygotic isolation. *Genetics*, **154**, 1663–79.
Turner, J. M. A. (2007). Meiotic sex chromosome inactivation. *Development*, **134**, 1823–31.
Vanlerberghe, F., Boursot, P., Bellis, M., Thaler, L., and Bonhomme, F. (1986). Genetical aspect of the semi-specific interaction between *Mus musculus domesticus* and *Mus musculus musculus* in eastern Europe. *Genetical Research*, **47**, 222–3.
Vicoso, B. and Charlesworth, B. (2006). Evolution on the X-chromosome: unusual patterns and processes. *Nature Reviews Genetics*, **7**, 645–53.
Vyskočilová, M., Pražanová, G., and Piálek, J. (2009). Polymorphism in hybrid male sterility in wild-derived *Mus musculus musculus* strains on proximal chromosome 17. *Mammmalian Genome*, **20**, 83–91.
Wang, P. J., McCarrey, J. R., Yang, F., and Page, D. C. (2001). An abundance of X-linked genes expressed in spermatogonia. *Nature Genetics*, **27**, 422–6.
Wolgemuth, D. J., Laurion, E., and Lele, K. M. (2002). Regulation of the mitotic and meiotic cell cycles in the male germ line. *Recent Progress in Hormone Research*, **57**, 75–101.
Wu, C. and Davis, A. (1993). Evolution of postmating reproductive isolation: the composite nature of Haldane's rule and its genetic bases. *The American Naturalist*, **142**, 187–212.
Wu, C. and Xu, E. (2003). Sexual antagonism and X inactivation: the SAXI hypothesis. *Trends in Genetics*, **19**, 243–7.
Yan, W. (2009). Male infertility caused by spermiogenic defects: lessons from gene knock-outs. *Molecular and Cellular Endocrinology*, **306**, 24–32.

Yang, H., Wang, J. R., Didion, J. P., *et al.* (2011). Subspecific origin and haplotype diversity in the laboratory mouse. *Nature Genetics*, **43**, 648–55.

Yin, S., Sun, X. F., Schatten, H., and Sum, Q. Y. (2008). Molecular insights into mechanisms regulating faithful chromosome separation in female meiosis. *Cell Cycle*, **7**, 2997–3005.

Yonekawa, H., Moriwaki, K., Gotoh, O., *et al.* (1980). Relationship between laboratory mice and the subspecies *Mus musculus domesticus* based on restriction endonuclease cleavage sites. *Japanese Journal of Genetics*, **55**, 289–96.

Yonekawa, H., Moriwaki, K., Gotoh, O., *et al.* (1988). Hybrid origin of Japanese mice '*Mus musculus molossinus*': evidence from restriction analysis of mitochondrial DNA. *Molecular Biology and Evolution*, **5**, 63–78.

Yoshiki, A., Moriwaki, K., Sakakura, T., and Kusakabe, M. (1993). Histological studies on male sterility of hybrids between laboratory and wild mouse strains. *Development, Growth and Differentiation*, **35**, 271–81.

18

New insights into parasitism in the house mouse hybrid zone

JOËLLE GOÜY DE BELLOCQ, ALEXIS RIBAS, AND STUART J. E. BAIRD

Introduction

Parasitism is a selective force that can drive several aspects of host evolution, including genetic diversity, life history traits and behaviour. Studies of parasitism in animal hybrid zones are still scarce. The European house mouse hybrid zone (HMHZ) is likely the most studied hybrid zone system to date in regard to parasitism. In this hybrid zone, parasites have been cited as one of the rare phenotypic traits indicating lower fitness of hybrids (e.g. Thomas *et al.*, 1996; Derothe *et al.*, 2004; Teeter *et al.*, 2009) and have been proposed as a potential factor involved in the limitation of gene flow between the two taxa (Sage *et al.*, 1986a; Moulia *et al.*, 1991, 1993). These ideas date from two pioneering studies in the German and Danish parts of the secondary contact of the house mouse that showed that hybrids harboured a higher helminth parasites load than their parental counterparts (Sage *et al.*, 1986a; Moulia *et al.*, 1991). Recently, a study based on a much larger sample size in the Czech–Bavarian part of the zone challenged these previous findings, showing that hybrids harbour lower parasite diversity and load than pure individuals (Baird *et al.*, 2012). In this chapter we will re-examine the potential role parasites can play in the house mouse hybrid zone. We will focus mainly on the helminths – the only type of parasites investigated in detail so far in this context (though see Sak *et al.*, 2011). As the context of the HMHZ in Europe has been discussed at length in Chapter 14, we focus first on the helminth parasites of the house mouse before describing the different patterns of helminth infection in natural and experimental mouse studies. We then discuss explanations for these different patterns and introduce some theoretical models and expectations for host–parasite interaction on host secondary contact. We finish with some perspectives for future work.

Evolution of the House Mouse, ed. Miloš Macholán, Stuart J. E. Baird, Pavel Munclinger, and Jaroslav Piálek. Published by Cambridge University Press.

What are helminth parasites?

The designation 'helminth parasites' groups together parasitic worms from two phyla: Nematoda (roundworms) and Platyhelminthes (flatworms, i.e. flukes, tapeworms, and thorny-headed worms). These phyla diverged at least 600 million years ago (Knoll and Carroll, 1999). Helminth parasite species have evolved to occupy a diverse range of niches within their hosts, using a wide range of infection strategies. Helminth infections are quite commonly asymptomatic, but can cause high rates of morbidity by establishing chronic infections and can occur repeatedly throughout the lifetime of an individual (Behnke *et al.*, 1992). They parasitize all known vertebrates (from fish to mammals), invertebrates (molluscs, arthropods), and plants. In mammals most helminths do not replicate within the host, and so morbidity and pathogenicity only occur when the host suffers a repeated or large burden of worms. This has been described, for example, with the pinworms of the house mouse, *Aspiculuris tetraptera* and *Syphacia obvelata* (Taffs, 1976). However, the host immune and inflammatory responses can also cause some pathology and it is the capacity of helminths to modulate the host immune system that underpins their longevity in the mammalian host (Behnke *et al.*, 1992; Maizels *et al.*, 2004).

Helminth infection triggers the T-helper 2 (Th2) response in mammals, involving the production of cytokines interleukin-4 (Il-4), Il-5, Il-10, and Il-13, immunoglobulin E, and the expansion and mobilization of specific effector cells (mast cells, eosinophils, basophils). The fact that mammal hosts respond to parasites from two distantly related phyla in a consistent and even stereotypical manner is quite remarkable (Maizels *et al.*, 2004). The induction of this strongly Th2 polarized response can also influence immunological responses to other pathogens such as viruses, bacteria, or protozoa (Cox, 2001).

Helminth parasites have been shown to reduce the fitness of their hosts, affecting reproduction and/or survival (Gulland *et al.*, 1993; Vandegrift *et al.*, 2008) and thus influencing evolutionary and ecological processes such as host population dynamics (Gulland *et al.*, 1993; Hudson *et al.*, 1998; Pedersen and Greives, 2008), host life history (Sorci *et al.*, 1997; Kristan, 2004), and host genetic polymorphism (Gulland *et al.*, 1993; Paterson *et al.*, 1998). Although a negative impact of helminths on host fitness has been shown in natural host populations treated with antihelminthics (e.g. Vandegrift *et al.*, 2008), it seems such impacts are difficult to demonstrate under experimental laboratory conditions. For example, wild-derived mice infected with the intestinal nematode *Heligmosomoides polygyrus* show a *higher* reproductive investment in the laboratory (2.8 more pups per litter at weaning) compared to non-parasitized mice

(Kristan, 2004), while Porcherie (2005) found no clear effect of the whipworm *Trichuris muris* on the reproduction of different strains of house mice.

Helminth occurrence and abundance show spatial and seasonal heterogeneity (Kisielewska, 1970; Haukisalmi *et al.*, 1988; Montgomery and Montgomery, 1989; Behnke *et al.*, 1999), which is further structured within hosts: some host individuals harbour many worms but most individuals none or few (Shaw and Dobson, 1995; Haukisalmi and Henttonen, 1999). This aggregation is generated by the stochastic nature of infection and by differences between individuals in their susceptibility once the parasite is encountered. Host factors (e.g. sex, age, genetic background, body condition, behaviour) and parasite factors (e.g. genetics) have been shown to influence the degree of aggregation (Haukisalmi and Henttonen, 1999; reviewed in Wilson *et al.*, 2002).

Host resistance/susceptibility to helminths has an important heritable genetic component (Wakelin, 1988; Behnke *et al.*, 2003; Quinnell, 2003). This has been clearly demonstrated in mice with respect to helminth models, including the whipworm *Trichuris muris* (Wakelin, 1975a, b; Else and Wakelin, 1988), the pinworm *Aspiculuris tetraptera* (Moulia *et al.*, 1993), the trichostrongylid worm *Heligmosomoides polygyrus* (Behnke *et al.*, 2006), and the tapeworm *Taenia crassiceps* (Fragoso *et al.*, 1996).

The diversity of helminth parasites in natural house mouse populations has already been reviewed by Tattersall *et al.* (1994) and Singleton and Krebs (2007), but without distinction between the different subspecies of the house mouse. In Table 18.1, we present a review of the genera of helminth parasites recorded in natural populations of house mice in the geographical range neighbouring the HMHZ. This table shows that several nematode (*Trichuris*, *Aspiculuris*, and *Syphacia*) and cestode (*Hymenolepis*, *Catenotaenia*, and *Taenia*) genera are very common in both subspecies. These helminths are thus good candidates for investigation of distribution and load in the hybrid zone. In contrast, trematodes and acanthocephala are rare, often specific to a given location of a given mouse subspecies, and thus of limited interest in our framework of parasitism in a hybrid zone.

The helminths in the mouse hybrid zone: 25 years of studies

The German and Danish field studies

The first study investigating the distribution of helminth parasites in the HMHZ dates from 1986 and focused on samples from 22 localities along a 200-km transect in southern Germany and adjacent Austria (Sage *et al.*, 1986a) (Fig. 18.1). Ninety-three individuals were genotyped at four enzyme loci and

Table 18.1 Helminth parasite genera of the house mouse in Europe.

	Mus musculus domesticus range										*Mus musculus musculus* range									
Localizations	France	Mallorca/ Menorca	Hyeres Island	Formentera/ Eivissa	Great Britain/ Ireland	Italy	Germany	Sicily	Spain	Corsica	Lithuania	Poland	Hungary	Romania	Serbia	Czech Republic	Bulgaria	Ukraine	Moldavia	Former Soviet Union
References[b]	3	8	10	11, 12	3, 6	6	3	8, 13	9	8	5	1, 3	3	3	2	3, 14	3	3, 4	3	3, 7
N hosts (if available)	n.a.	29; 16	20	69; 7	102, na	51	106	44; 73	376	180	6	34; 154	590	379	429	n.a., 149	627	105	n.a	n.a.
Nematodes																				
Aonchotheca										X										
Eucoleus										X										
Gallegostrongylus				X																
Physaloptera				X																
Nippostrongylus								X	X	X										
Heligmosomum																			X	
Trichosimoides																		X		
Protospirura						X														
Calodium[a]								X	X	X										X
Trichinella									X									X	X	X
Heterakis[a]							X	X	X						X		X	X		X
Gongylonema		X		X				X	X					X	X	X		X		X
Heligmosomoides					X			X	X	X			X	X	X		X		X	X
Mastophorus[a]		X		X		X	X	X	X	X	X	X	X		X	X	X	X	X	X
Trichuris[a]				X	X		X	X	X	X		X	X	X	X	X	X	X	X	X
Aspiculuris[a]	X			X	X	X	X	X	X	X	X	X	X	X	X	X	X	X	X	X
Syphacia[a]	X	X	X	X	X	X	X	X	X	X	X	X	X		X	X	X	X	X	X
Cestodes																				
Paranoplocephala										X										
Skrjabinotaenia																X				
Echinococcus	X																		X	X
Mesocestoides[a]				X					X					X					X	X
Mathevotaenia													X					X	X	X
Rodentolepis[a]							X	X	X	X			X		X		X	X	X	X
Hymenolepis[a]				X	X	X	X	X	X	X				X		X	X	X		X
Catenotaenia		X		X	X	X		X	X	X			X	X	X	X	X	X	X	X
Taenia[a]		X		X	X	X	X	X	X	X		X	X	X	X	X	X	X	X	X

Localizations	*Mus musculus domesticus* range										*Mus musculus musculus* range									
	France	Mallorca/ Menorca	Hyeres Island	Formentera/ Eivissa	Great Britain/ Ireland	Italy	Germany	Sicily	Spain	Corsica	Lithuania	Poland	Hungary	Romania	Serbia	Czech Republic	Bulgaria	Ukraine	Moldavia	Former Soviet Union
Trematodes																				
Echinostoma									X											
Fasciola										X										
Lecithodendriidae									X											
Zonorchis										X										
Brachylaima		X	X	X				X	X	X							X			X
Corrigia			X							X							X			
Psilotrema																	X			
Skrjabinoplagiorchis																X				
Plagiorchis														X		X		X		X
Acanthocephala																				
Moniliformis																			X	X
Acanthocephalan										X										

Note: [a] Helminth genera recorded in the Czech–Bavarian study (Baird *et al.*, 2012); n.a.: not available.

[b] 1: Izdebska and Rolbiecki (2006); 2: Kataranovski *et al.* (2008); 3: Kriska (1993); 4: Gumenyuk (1953); 5: Mažeika (1992); 6: Tattersall *et al.* (1994); 7: Petrochenko and Romanovskii (1966); 8: Milazzo *et al.* (2003); 9: Feliu *et al.* (1997); 10: Miquel *et al.* (1996); 11: Mas-Coma *et al.* (1998); 12: Mas-Coma *et al.* (2000); 13: Milazzo *et al.* (2010); 14: Prokopič and Genov (1974).

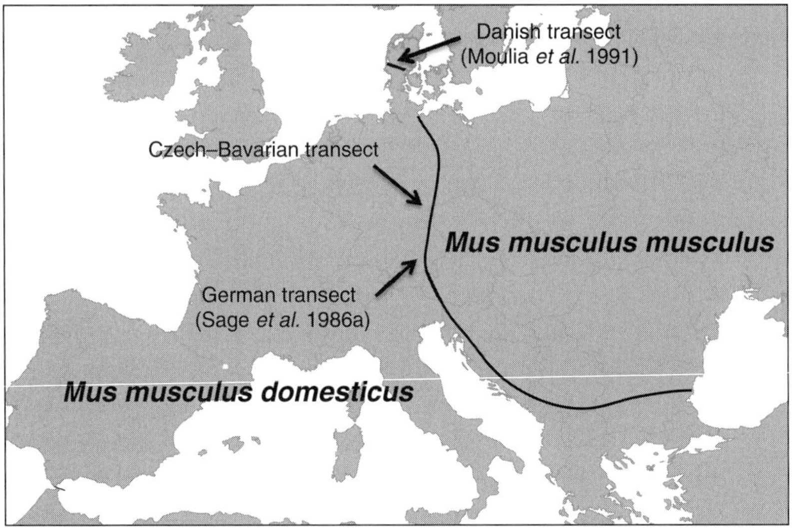

Figure 18.1 The European house mouse hybrid zone with the locations of the three transects investigated for helminth parasites. The black line depicts the approximate course of the zone.

dissected for nematode and cestode parasites. In the 46 individuals classified as hybrids, the mean nematode load per gut was greater than 197 (with 14 of the 46 hybrid mice harbouring over 500 pinworms), far exceeding the means of 59 and 15 nematodes per gut in the two parental taxa. A similar trend of higher hybrid parasite load was also reported for tapeworms. At that time, heritable susceptibility to the bacterium *Salmonella typhimurium* had been mapped to different chromosomes in the two mouse taxa and resistance to nematodes was thought to be mainly due to dominant genes. Given this context, the authors suggested two simple genetic models to explain the observed helminth infection pattern: either the two subspecies had different genes for resistance, and hyper-infected hybrid individuals possessed none after the shuffling of segregation and recombination, or recombination had broken up advantageous combinations of alleles at two or more loci conferring resistance to parasites and differing in the two species (Sage *et al.*, 1986a).

Subsequently, Moulia *et al.* (1991) described a similar pattern of higher hybrid parasite load in 120 mice sampled from 12 localities along a 150-km transect of the Danish house mouse hybrid zone of the Jutland Peninsula (Fig. 18.1). To exclude ecological effects arising from heteroxenous life cycles (when several hosts are required for the completion of the parasite life cycle) they examined only the direct life cycle pinworms (*Aspiculuris tetraptera* and *Syphacia obvelata*). In the 35 individuals classified as hybrids, based on ten enzyme loci, the mean pinworm load per gut was 135, exceeding the mean of 58 and 55 nematodes per gut in the parental taxa. The authors also performed a preliminary infection experiment with *A. tetraptera* on the offspring (seven and eight individuals, respectively) of two pairs of mice sampled contemporaneously from a locality in the hybrid zone. The two groups showed very different pinworm load (mean 223 *vs* 2 worms), and from this the authors concluded that pinworm load was probably under genetic determination. The authors concurred with the second Sage *et al.* (1986a) hypothesis: that 'coadapted genetic systems' controlling the helminth load in the parental taxa were disrupted in hybrids due to meiotic recombination, stating: 'the greater parasitic loads observed in the recombinant genotypes are so far the only phenotypic trait to suggest that selection is acting against the hybrids' (Moulia *et al.*, 1991: 685).

Laboratory studies motivated by parasitism in the HMHZ

Following these observations based on samples taken from nature, three research directions were pursued: to evaluate the extent of hybrid susceptibly to parasites; to estimate the effect of helminths on host fitness; and to demonstrate that hybrid susceptibility is due to disruption of coadapted genetic systems due to recombination. Experimental infections of strains developed from wild mice showed hybrid susceptibility (i.e. higher load in hybrids relative to parentals) to a coccidian protozoan parasite, *Sarcocystis muris* (Derothe *et al.*, 2001), but not to the blood protozoan *Trypanosoma musculi* (Derothe *et al.*, 1999) nor to the mouse whipworm *Trichuris muris* (Porcherie, 2005). Moreover, the mouse whipworm was found to have no significant effect on host fitness over different strains of mice, as estimated by reproductive success (percentage of reproductive couples, numbers of litters per reproductive couple, size of litters and weight of pups at weaning) (Porcherie, 2005). Finally, susceptibility to the pinworm *A. tetraptera* was compared between parental mouse strains and experimental hybrids (from F_1 to F_4) to test if recombination was involved in hybrid susceptibility (Derothe *et al.*, 2004). As recombination does not break-up parental genes in F_1s, these were expected to show the same or lower pinworm load than the parental mice, susceptibility only appearing in F_2s and increasing in F_3s and F_4s. In fact, no F generation showed increased load relative to parentals: they all showed either reduced relative load (F_1, F_3, F_4) or no significant effect (F_2). The hypothesis that disruption of coadapted allelic associations due to recombination causes susceptibility was therefore discarded and the authors proposed an alternative hypothesis: a pleiotropic factor inhibiting some resistance component with an antagonist effect. Early-generation hybrids would have a lower selective value because of some unidentified (non-parasite-related) dysgeny. Subsequently, rare alleles will arise that compensate for the negative effect of this dysgeny at the expense of other functions, such as resistance to parasites. The authors based this pleiotropic hypothesis on the idea that hybrid zones present peculiar selective conditions that could favour the emergence of rare alleles that restore impaired function in hybrids (Derothe *et al.*, 2004; for more details see Moulia and Joly, 2009). It was further proposed that the hypothetical pleiotropic factor might affect Th2 immunity (Porcherie, 2005), regarded as the component of the immune response most strongly regulating helminth infections. However, Porcherie (2005) showed that hybrids have no difficulty mounting a Th2 immune response.

Lessons to be learned from the German and Danish field studies

As mentioned above, helminths are aggregated in their host populations. This aggregation is a characteristic of the two mouse pinworm species (Scott and Gibbs, 1986). An important implication of this pattern is that an accurate determination of parasite abundance in the host population requires sampling a large number of hosts. In the context of the house mouse hybrid zone, Klein (1988) pointed out that the findings of Sage *et al.* (1986a) should be regarded as awaiting confirmation with 'a truly representative sample'. Given helminth aggregation and spatiotemporal heterogeneity, the sampling of both previous studies appears limited in terms of both numbers of mice and the distribution of localities investigated: the German study is based on 93 mice from 22 localities spanning over 200 km (Sage *et al.*, 1986a). The number of localities was even lower in the second study (Moulia *et al.*, 1991), with 120 mice captured in 12 localities spanning over 150 km. Such geographically patchy sampling makes it difficult to ignore the possibility that a spuriously hyper-infected region overlaps with localities near the centre of the zone. This would cause a spurious correlation between hybrids and parasite load.

There are also lessons to be learned about temporal sampling design. The Sage *et al.* (1986a) study consisted of an initial sampling of 42 mice in October 1984, and an additional sampling of 51 mice in June 1985. Pinworms have been shown to display significant seasonal variations in prevalence and abundance in rodents (Haukisalmi *et al.*, 1988; Abu-Madi *et al.*, 2000). Although Sage *et al.* stated that 'the heavy infections are widely distributed within the hybrid zone and present during both collecting periods' (1986a: 62), no additional data about the sampling are available to allow these statements to be quantified. In the Danish study the temporal sampling weakness does not lie in a potential seasonal effect since all mice were captured during the same month, but in the fact that mice were kept for two months in the laboratory before being inspected for parasites (Moulia *et al.*, 1991). Pinworms are highly contagious in laboratory conditions because eggs, both in bedding and airborne, result in continual re-exposure of mice to pinworms (Taffs, 1976; Clifford and Watson, 2008). The life cycles of pinworms require as little as 12 and 24 days for completion for *S. obvelata* and *A. tetraptera*, respectively (Taffs, 1976). Moreover, mice develop an age-related resistance to pinworm infections (reviewed in Taffs, 1976), with immunocompetent mice becoming partially immune and generally harbouring fewer pinworms as they age (Scott and Gibbs, 1986). Under these circumstances, it is difficult to know how much the pattern described for the mice taken from the Danish transect reflects the conditions of mice in nature.

Regarding analyses, it should be noted that the definition of a hybrid is not controlled across studies. A hybrid index (*HI*) can be used to place any mouse on a linear scale from *M. m. musculus* to *M. m. domesticus*, depending on the count of *domesticus* alleles at assayed loci. Sage *et al.* (1986a) and Moulia *et al.* (1991) respectively assayed four and ten enzyme loci for this purpose. However, the point at which 'hybrids' fall along this linear scale is defined differently from study to study. Expressing *HI* as a percentage *domesticus*, 'hybrids' have $12.5\% < HI < 87.5\%$ in Sage *et al.* (1986a), $20\% < HI < 60\%$ in Moulia *et al.* (1991), and $2\% < HI < 97\%$ in Moulia *et al.* (1993). Moreover, because the *HI* was based on a limited number of genetic markers, Moulia *et al.* (1991) chose to characterize individuals by the mean genetic composition over individuals sampled at the same locality, further obscuring the relationship between individuals and their parasite load.

Many of these limitations of the two previous field studies are easily understandable. Improvements to field sampling design almost always appear with hindsight, and it seems likely the primary purpose of the Danish sampling was not a parasitological study at all (given the long delay before inspection for parasites). The studies were performed 25 years ago: at that time genotyping a high number of mice and/or loci was more laborious. But the fact that these kinds of issues are to be expected in studies of this sort does not mean the patterns they report are beyond question. Twenty-five years' more knowledge about the HMHZ, about the expected outcome of hybridization, and of technical advances motivated a recent large-scale study of helminth parasites in the HMHZ in a region stretching from northeastern Bavaria (Germany) to western Bohemia (Czech Republic), with a sampling design drawing on all the lessons to be learned from these pioneering studies (Baird *et al.*, 2012).

The study in Bohemia–Bavaria

This new study (Baird *et al.*, 2012) in a region of 145 × 50 km of northeastern Bavaria and western Bohemia (Fig. 18.1) was designed specifically in order to avoid the limitations of the two previous studies: 689 mice were captured at 107 localities within a given season (ten days at the beginning of autumn 2006 to 2009) and dissected for helminth parasites the day after capture. Hybrid indices (Fig. 18.2a) were calculated using 1401 SNPs (Wang *et al.*, 2011). Rather than dividing individuals into arbitrarily constructed pure/hybrid categories, each individual was considered in relation to its *HI*. To test whether hybrid individuals have greater or lesser diversity and load than additive expectations, an individual-based model was developed allowing for hybridization effects on parasite load and diversity (hybrid resistance *vs* susceptibility) as a function of expected heterozygosity. The expected heterozygosity increases with the degree to which new combinations of genes have been brought together in the hybrid genotype of an individual. The model is parameterized to take into account the degree of aggregation of helminths and the potential density trough that often characterizes the location of tension zones such as the HMHZ. Finally, the model allows for subspecies- and sex-specific effects on parasite diversity and load.

The field sampling found ten different helminth species corresponding, save one, to common helminths of the house mouse (see notes, Table 18.1). Parasite loads of these various helminth species were consistent with negative binomial distributions. The most prevalent helminths were pinworms (*Aspiculuris tetraptera* and *Syphacia obvelata*), whipworms (*Trichuris muris*), tapeworms (*Taenia taeniformis*), and spirurid nematodes (*Mastophorus muris*). The two pinworms species, *A. tetraptera* and *S. obvelata*, were considered together, as in the Moulia *et al.* (1991) study. The model was tested separately for load of the four most prevalent and widespread helminth parasites (Fig. 18.2), which infect different parts of the host (caecum, stomach, and liver), to gauge the generality of the load patterns. Allowing non-additive effects significantly increased the likelihood of the observations for the two most prevalent types of helminths, the pinworms and the whipworms, and also for the overall diversity of helminth parasites in individual mice (Fig. 18.3b, c, a, respectively). In each case these effects were consistent with hybrid resistance, and contradictory to hybrid susceptibility. The results were robust to potential confounding factors (sex- and taxon-specific loads). While the lower-prevalence parasites (tapeworms and *M. muris*) further support this pattern, each individually holds insufficient information to exclude zero hybridization effect, as might be expected given less information about the distribution of load (Fig. 18.3d, e). In all cases this study found significant taxon differences.

Since one goal of the study was to compare the results to the two previous studies on helminth parasites in the HMHZ, the model was also applied to the original dataset of Sage *et al.* (1986a), using the data collected from figure 3

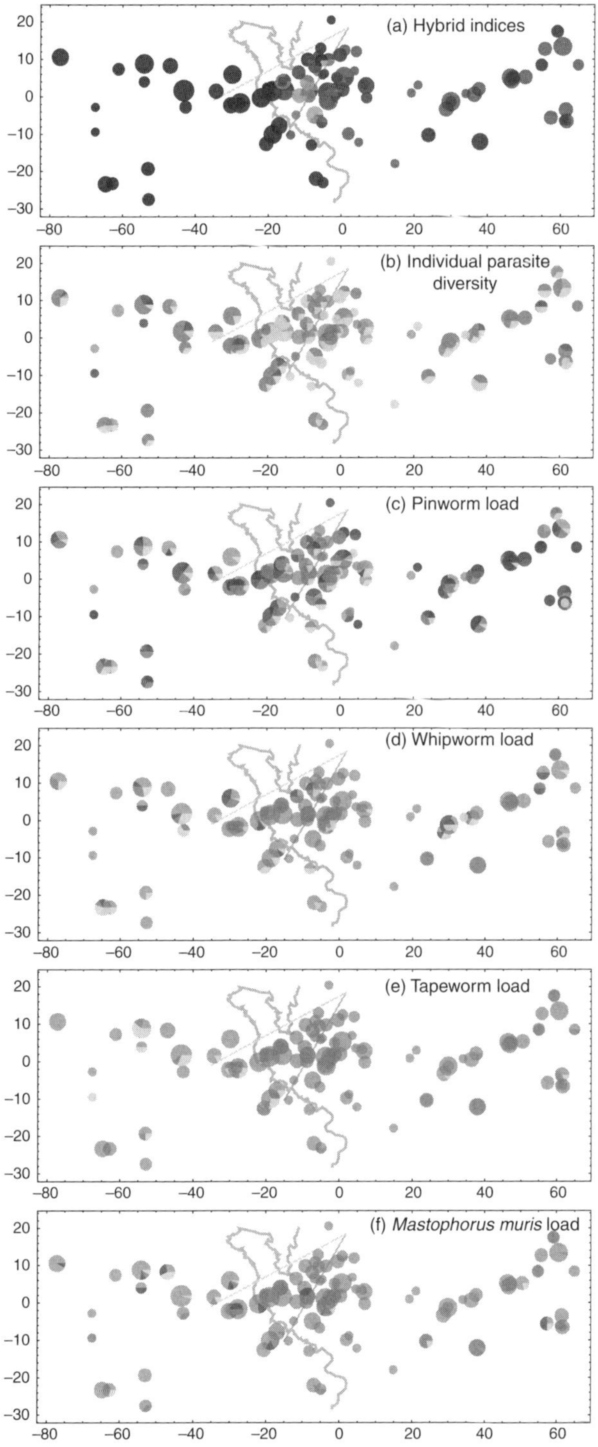

Figure 18.2 Geography of hybrids and parasites across the Czech–Bavarian region of the HMHZ. Thick grey line = Czech–German border (northeastern Bavaria to the west, western Bohemia to the east). Data for 107 localities are shown (axes in kilometres, after gnomonic projection onto the plane around their centroid). The consensus centre of the hybrid zone is shown in green, and the extent of the Y introgression is indicated by the dashed yellow line, following Macholán *et al.* (2008) (see plate section for a colour version of this figure). Pie areas are proportional to the $\log_{10}$ (locality mouse sample sizes + 1).

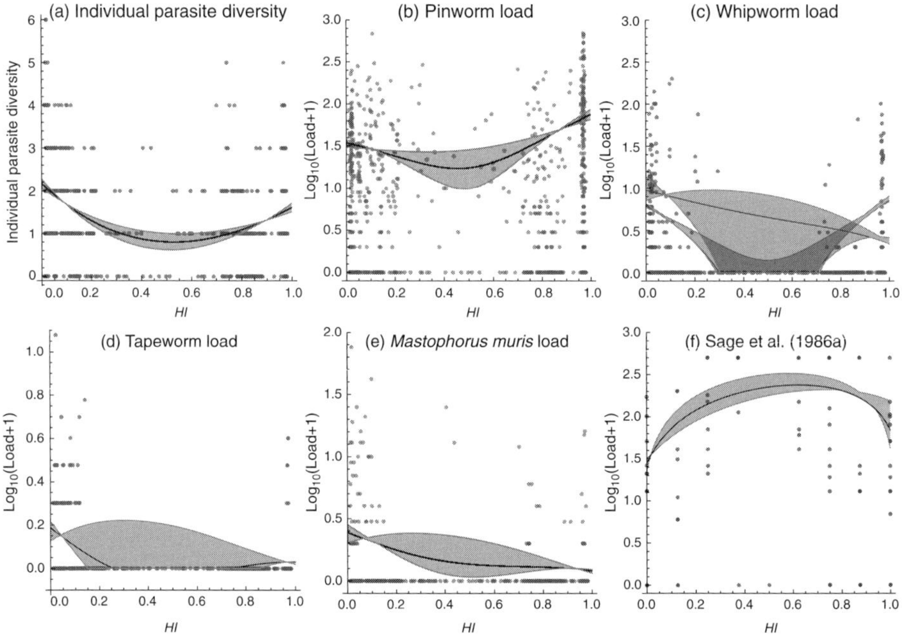

Figure 18.3 Distributions of individual parasite diversity and mean load with two-unit support envelopes across individual HIs compared to the ML parameterization of the best model. (a) Individual parasite diversity; (b) pinworms; (c) whipworms; (d) tapeworms; (e) *Mastophorus muris;* (f) Sage dataset. HI ranges from HI = 0 (*domesticus*) to HI = 1 (*musculus*). Blue: males; red: females; purple: no sex distinction. The scale of the y-axis is individual parasite diversity for (a) and $\log_{10}$ (load + 1) for the remaining figures. The two-unit support envelopes show that there is increased uncertainty of mean load for the most hybrid individuals, corresponding to the simple fact that F1s and equal-source-contribution hybrids are very rarely observed in the HMHZ. For whipworm load (c), mean loads for males and females are shown separately because of a significant taxon × sex interaction (modified from Baird *et al.*, 2012). See plate section for a colour version of this figure.

Caption for figure 18.2 (*cont.*)

(a) Distribution of individuals' hybrid indices (HIs) within localities. Colours are taken from a continuous blend of blue (*domesticus*) through yellow (HI = 0.5) to red (*musculus*). (b) Distribution of parasite diversity over individuals within localities. Grey: individuals with no parasites; yellow: 1; orange: 2–3; red: 4–5; purple: 6–8 parasite species. (c–f) Distribution of individuals' parasite loads for localities; load bins [a–b] are designed to divide non-zero expectations equiprobably, relative to the (H0) MLE parameterization for the negative binomial distribution. Grey: individuals with none of the focal parasite. Yellow, orange, red, purple: load bins 1–4, respectively. (c) Distribution of pinworm load over individuals within localities; load bins: [1–4], [5–23], [24–83], [>83]. (d) Distribution of whipworm load over individuals within localities; load bins [1–5], [6–20], [>20]. (e) Distribution of tapeworm load over individuals within localities; load bins [1], [>1]. (f) Distribution of *Mastophorus muris* load over individuals within localities; load bins [1], [2–6], [>6] (modified from Baird *et al.*, 2012).

of Sage *et al.* (1986a), which represents nematode load (with 90% of nematodes being pinworms) in relation to individual *HI*s based on four enzymatic loci (hereafter referred to as the Sage dataset). The pattern in this dataset was the opposite of the above: parasite load across individuals was not distributed following a negative binomial distribution; deviation from additivity for hybrids was very significantly *negative*, consistent with hybrid susceptibility; there was no evidence for a taxon difference in parasite load (Fig. 18.3f).

In summary, the pattern of hybrid resistance in the Czech–Bavarian study clearly contradicts the pattern of hybrid susceptibility found in the two previous field studies (Sage *et al.*, 1986a; Moulia *et al.*, 1991). However, the Czech–Bavarian pattern of hybrid resistance is consistent with the result of the experimental infection of hybrid and parental strains with *Aspiculuris tetraptera*, which shows hybrid resistance over several generations of hybrids (Derothe *et al.*, 2004). A question then arises: why does the pattern of helminth infection differ between studies?

Explaining the reversal in outcomes of studies

Moving across the mouse *HI* from one subspecies through hybrids to the other subspecies, the German and Danish studies found a pattern of low–high–low parasite load. The larger, more recent Czech–Bavaria study found a high–low–high pattern. Assuming both patterns are well supported, explanations for this heterogeneity between studies will revolve around parasite heterogeneity, host heterogeneity, or some combination of the two.

Hybridization gives rise to susceptible individuals in one part of the hybrid zone, but resistant individuals in another part?

The two initial studies conclude that the low–high–low pattern is due to the genetics of the host: specifically, hybrids are hyper-infected because they are hybrids. The heterogeneity between studies might then be explained by proposal A: *hybridization gives rise to susceptible individuals in one part of the hybrid zone, but resistant individuals in another part (or at another time)*. While proposal A, if accepted, at least partially rescues the conclusions of the initial studies, it severely undermines the suggestion that parasites play an important role in the maintenance of the mouse species barrier (Sage *et al.*, 1986a; Moulia *et al.*, 1991, 1993), because for the helminth role to be important, their effects should be consistent. If helminths were maintaining the species barrier in some regions and not in others we would expect to find substantial differences in the width of the zone from place to place (we do not). It could be argued that similar zone widths would arise due to the selective spread of host resistance genes from region to region – this is true, but it would result in a uniform and minimal role of parasites in the zone, not an important and heterogeneous role. If helminths were maintaining the species barrier at some points in time (seasons, years) and not at others, their influence integrated over time would again be minimal, akin to closing the stable door after the horse has bolted.

We find proposal A unsatisfactory not because it weakens the argument for parasites as mediators of species boundaries, but because it requires that either the nature of hybrids with respect to their helminth parasites, or the converse (parasites with respect to hybrids), is radically different in different parts of the zone (or at different times), and we feel both of these alternatives are unlikely. Analyses of the HMHZ over several regional transects (e.g. Raufaste *et al.*, 2005; Macholán *et al.*, 2007) suggest that this is a tension zone (Key, 1968; Bazykin, 1969), where selection against hybrids is dominated by genetic rather than environmental factors, and that the contact process is broadly similar over a very long front (>2500 km), despite the extremes having quite different times of initial contact (Macholán *et al.*, 2007). While Teeter *et al.* (2009) speculate that the zone might have different mechanisms maintaining it in different places, Macholán *et al.* (2011) point out that the differences in observations motivating this speculation are very likely due to different sampling strategies, concluding that 'there is some evidence of common architecture of reproductive isolation, and no reliable evidence to the contrary' (Macholán *et al.*, 2011: 1443). Occam's razor would therefore suggest the nature of hybrids with respect to their helminth parasites is similar along the contact front.

Likewise, from the parasite point of view, host-specific direct life cycle parasites such as the pinworms are likely to have a history closely following that of their hosts (Hugot, 1988). For example, the pinworm, *Syphacia obvelata*, has followed the synanthropic (human-commensal) spread of their mouse hosts all the way to the subantarctic Kerguelen Archipelago (Pisanu *et al.*, 2001). It thus seems unlikely that the pinworms in one part of the HMHZ are of a type that preferentially infect hybrids while at the same time elsewhere the pinworms preferentially fail to infect hybrids.

Inconsistent infection patterns due to frequency-dependent selection?

It has been pointed out that hybrid study systems repeatedly screened for infection often show variable infection scenarios (reviewed in Wolinska *et al.*, 2008). Wolinska and colleagues proposed that (hereafter proposal B) *inconsistent patterns of infection observed over space and time in hybridizing communities may be due to frequency-dependent selection*, i.e. Red Queen dynamics, but extrapolated from the level of alternative genotypes to the level of alternative taxa (Wolinska *et al.*, 2006, 2008). The core of proposal B is 'the dynamic scenario predicts continuous, bidirectional changes in the relative infection of hybrid and parental taxa. The change is expected to result from frequency-dependent selection against common genotypes contained within each of the taxa' (Wolinska *et al.*, 2008: 124). The relevance of this model to systems other than *Daphnia* (Wolinska *et al.*, 2006), its inspiration, is unclear however. Although common host genotypes might be expected to occur in parental species, the notion of hybrids as a taxon with common genotypes is restricted to systems where only F_1s (and perhaps early backcrosses) are produced, and where, for example, clonal reproduction can maintain such hybrid 'taxa'. In contrast, the defining feature of post-F_1 hybridization (the case for the mouse hybrid zone) is an explosion of new possible genotypes (two raised to the power of the number of loci considered). As the number of loci considered increases, every single hybrid genotype quickly becomes rare, making the taxon 'hybrids' an extremely unlikely target for parasites in search of the most common host genotype. We do not, therefore, think it likely that this kind of Red Queen model could drive heterogeneity of the hybrid–parasite interaction along the span of the HMHZ.

Environmental heterogeneity associated with the hybrid zone centre?

Thus far, having assumed first that both reported hybrid parasitism patterns are well supported, and second that hybrids were hyper-infected in previous studies because they were hybrids, the heterogeneity over studies leads us to unsatisfactory and seemingly overcomplicated proposals – A and B – regarding the hybrid–parasite interaction. This leads us to question the initial assumptions: were hybrids hyper-infected in previous studies because they were hybrids? Are both hybrid parasitism patterns equally supported? We will discuss each of these points in turn.

If both hybrid parasitism patterns do indeed occur in nature, is there a reasonable alternative to the idea that in the initial studies hybrids were hyper-infected because they were hybrids? We suggest there is: hybrids may be hyper-infected not because they are hybrids, but because they live at the centre of a tension zone. Tension zones tend to move towards and be trapped by regions of poor habitat suitability and barriers to gene flow (Key, 1968; Bazykin, 1969). This movement therefore tends to maximize contrast between environmental factors at the centre of the zone and in the flanks. As hybrids are only found at the centre of the zone, contrasts between hybrids and parentals may therefore reveal environmental rather than genetic effects. If we found highly parasitized hybrids in such circumstances it could be because of *where* they are, not *what* they are. Are there any reasons to suspect that the geographical locations of the three studies may have had such an influence? The centre of the HMHZ on the east coast of the Jutland Peninsula is a marsh traversed by a river at the head of the Vejle Fjord (Moulia *et al.*, 1991). Sage's sampling region in southern Bavaria is also centred on a floodplain traversed by a river (Sage *et al.*, 1986b). In contrast, the Czech–Bavarian mouse contact that is the subject of our study occurs in a well-drained corridor between hills (Macholán *et al.*, 2007). These flanking hills likely form a sufficient barrier to gene flow to trap the tension zone at this location without the need for any special habitat features in the contact corridor itself. If this is the case, the situation of the Czech contact corridor controls at least to some extent for correlation of environmental effects with the location of hybrids, making it more likely that hybrid/parental contrasts will reveal the genetic nature of the hybrid–parasite interaction.

Models and expectations for host–parasite interaction on host secondary contact

In the 25 years since scientists started examining hybrid parasitism in the mouse, there has been much progress in developing both models of the outcome of secondary contact, and models of the host–parasite interaction. Here we put some of these developments into the context of hybrid parasitism and find our expectations quite different from those of 25 years ago. In particular, a pattern of reduced load in hybrids would appear a natural expectation.

A Red Queen recombination model

While host taxa are isolated they diverge, but their parasites may diverge faster due to shorter generation time. If parasites reduce host fitness significantly, then one part of the host genome must somehow keep up with this faster parasite evolution: the part controlling host immune response. Recombination is one way in which a set of host genes can quickly produce new combinations of alleles as yet unseen by contemporary parasites. This simple Red Queen recombination (RQR) model has a number of notable implications:

(1) If recombination is the important source of new immune gene combinations, then coadaptation across those loci is unlikely – coadaptation requires

combinations to be held together long enough for natural selection to act. This is unlikely if recombination is continually re-shuffling them.

(2) On secondary contact, recombination between host immune genes from each source should produce rare combinations, thus allowing hybrids to avoid parasites targeting common genotypes.

(3) Vicariant sister taxa need not show high divergence for immune-related genes – the Red Queen can be run by producing new (re)combinations of existing alleles, rather than fixing new mutations.

These implications are notable because the first argues against the existence of 'coadapted gene complexes', the break-up of which would explain hybrid susceptibility with respect to parasite load (Sage *et al.*, 1986a; Moulia *et al.*, 1991, 1993). The second is consistent with the results of the Czech–Bavarian study, and argues against the reduced hybrid fitness due to parasite interactions proposed by Sage *et al.* (1986a) and Moulia *et al.* (1991). The third is consistent with evidence from genes implicated in the immune response. These often show trans-species polymorphism – for example, the immunoglobulin heavy-chain variable region (IgVH) (Su and Nei, 1999; Esteves *et al.*, 2005); the proteasome subunit β-type 8 (PSMB8) (Nonaka *et al.*, 2000; Miura *et al.*, 2010); the major histocompatibility complex (MHC) in general (Klein, 1987) and across the two hybridizing mouse subspecies in particular (Čížková *et al.*, 2011). Trans-species polymorphism is quite the opposite of high divergence in sister taxa (point 3 above).

Returning to the first of the three implications of the RQR model, we have suggested coadaptation among immune genes is unlikely if they are using recombination to run a Red Queen race against parasites, arguing against the existence of the 'coadapted gene complexes'. Even if recombination is not breaking up coadapted gene complexes (Derothe *et al.*, 2004), there is still, however, reason to expect hybridization to produce suboptimal genotypes. This is because hybridization involves the admixture of alleles that have never coexisted in the same population. In a single panmictic population a new allele that is incompatible with a common allele is unlikely to increase in frequency, because it will mostly occur in individuals with the incompatible common allele. When a population is split in two, this filtering out of new incompatible alleles continues within each subunit but, as Dobzhansky (1936) and Muller (1942) pointed out, pairs of new alleles arising across the two isolated populations are, by definition, never tested against each other by natural selection. Secondary contact between the populations is the first time such pairs come together. During isolation these alleles may have reached high frequency, and in this way unfit (unfiltered) Dobzhansky–Muller incompatibilities (DMIs) can be revealed at secondary contact, without either isolated subunit having had reduced fitness. While the alleles concerned are usually thought of as being at different loci, the logic works similarly for alleles at the same locus, and the outcome is the same: admixture within/across loci is expected to reveal previously unseen (untested) incompatibilities. Here, then, is a clear potential cause of hybrid susceptibility that does not rely on coadaptation across sets of genes with high recombination. A further question then arises: why is all the evidence in the large Czech–Bavarian study consistent with an improvement in hybrid immune response? In other words, why do we see no evidence of DMIs affecting host immune genes in this study?

Expectations: is the immune system immune to DMIs? Transitivity of allele compatibility

Equality (=) is a transitive binary relation: if A = B and A = C, then B = C. Suppose allele compatibility (♥) was also a transitive binary relation: if A ♥ B and A ♥ C, then B ♥ C. The Dobzhansky–Muller model breaks this rule: if B and C are a pair of new alleles that arise across isolated subpopulations, while A denotes their original partner(s), maintained in each vicariant population during subdivision, then transitive allele compatibility implies that when B and C are brought together for the first time on secondary contact, they would be compatible (because both are guaranteed to be compatible with common resident allele A). This is one reason to favour the existence of DMIs between new alleles at different loci, rather than at the same locus: the constraints imposed by the common function of alleles at the same locus (encoding the same gene) make it more difficult to imagine how a new allele (B) could be incompatible with some alternatives (C), while compatible with others (A). Is there a reason to suspect that allele compatibility should be strongly transitive for immune genes? If this was the case, it would make immune gene DMIs unlikely. One circumstance favouring strongly transitive compatibility is if many alleles are maintained at a locus. This is certainly true of the MHC (Klein, 1986; Piertney and Oliver, 2006). Then, to increase in frequency, a new allele must be compatible with very many alternatives. For N maintained alleles the transitive rule then becomes:

If $\{A_1 ♥ B, A_2 ♥ B, \ldots, A_{N-1} ♥ B, A_N ♥ B\}$,
and $\{A_1 ♥ C, A_2 ♥ C, \ldots, A_{N-1} ♥ C, A_N ♥ C\}$, then B ♥ C.

The greater compatibility constraints on B and C with respect to overlapping sets of partners have two implications: in each vicariant population new alleles are less likely to reach high frequency, and, if they do, they are more likely to be co-compatible on secondary contact. That is: *systems where many alleles are already maintained are resistant to the development of DMIs during vicariance*. This vicariant balancing selection scenario forms a natural counterpoint to Orr's argument regarding the preferential *accumulation* of DMIs in vicariant epistatic systems (Orr, 1995; Kondrashov, 2003).

Conclusions and perspectives

Effect of parasites on host fitness

Parasites have the potential to affect the outcome of hybridization by differentially impacting the fitness of host lineages and their hybrid descendants (Coustau *et al.*, 1991; Fritz, 1999). While the initial studies in the HMHZ implicitly assume parasite load as a proxy for fitness (Sage *et al.*, 1986a; Moulia *et al.*, 1991), this appears to be a dangerous assumption when considering hybrids because (1) the fitness of a sterile hybrid is zero, no matter what its parasite load; (2) fit hosts can be either resistant or tolerant (Raberg *et al.*, 2007) – a tolerant individual maintains high fitness regardless of its parasite load; (3) chronic helminth infection has been shown to protect against allergy and autoimmune diseases (Wilson and Maizels, 2004; van Riet *et al.*, 2007). Taking parasite load as a proxy for fitness assumes a strong correlation between the two. The three points above illustrate why this is questionable in general and particularly so in a hybrid zone. The segue between a pattern of high parasite load in hybrids and parasites defining species boundaries becomes less clear if the parasites do not have a significant adverse effect on hybrid fitness.

Is there still a role for parasites in hybrid zones and species boundaries?

The findings of the Czech–Bavarian study indicate that it is extremely unlikely that helminth parasites, through high load, contribute to the maintenance of the species boundary for the house mouse, contrary to what has been proposed over the past 25 years (Sage *et al.*, 1986a; Moulia *et al.*, 1991; Moulia and Joly, 2009). This should not be taken, however, to mean helminths have no influence on hybrid zones.

If there *is* a parasite influence on the outcome of mouse hybridization, the Czech–Bavarian data would suggest it will be directional selection: with the exception of pinworms and male whipworms in male mice, *domesticus* mice are more heavily parasitized than *musculus* mice – most strikingly, tapeworm load is eight-fold lower on the *musculus* genetic background (Fig. 18.2e and 18.3d): introgression of any genes responsible into the *domesticus* genome may therefore be favoured. This may be related to a history of zone movement, *musculus* replacing *domesticus*, indicated by spatial X chromosome marker variation

(Macholán *et al.*, 2011). Tapeworm susceptibility differences have been shown across experimental infections of mouse laboratory strains (Olivier, 1962; Orihara, 1962). Larval tapeworm can affect the reproductive function of male and female rats, with a strong decrease in fertility for both sexes (Lin *et al.*, 1990). Excretory-secretory products of tapeworm larvae have a direct inhibitory effect on Leydig cell testosterone production of rats (Rikihisa *et al.*, 1985), and a recent study shows that tapeworm (*T. crassiceps*) can affect the seminiferous epithelium of male mice and lead to infertility (Zepeda *et al.*, 2011). In the Czech–Bavarian transect, the Y chromosome of *musculus* has introgressed into the territory of *domesticus*, covering more than 330 km^2 of Bavaria (Macholán *et al.*, 2008). Resistance to a parasite capable of impacting male fitness by affecting reproductive function, such as tapeworms, could facilitate this Y chromosome invasion. It appears there are dominant loci of major effect on tapeworm (*T. crassiceps*) resistance (Fragoso *et al.*, 1996). Dominance would further aid the rapid spread of resistance alleles, suggesting the role of tapeworms in the movement of the Czech–Bavarian hybrid zone is worth further investigation.

The hybrid–parasite interaction and the host's genetics

All evidence in the Czech–Bavarian study points to reduced parasitism of hybrids, and the model developed above leads us to pose the question: are immune genes immune to DMIs? This proposal is immediately testable for the house mouse system: a survey of natural variation at two MHC loci has already been carried out in the two taxa (Čížková *et al.*, 2011), and this is the necessary prerequisite for work in progress to examine MHC gene flow across the same region of the European HMHZ for which, here, we have shown significantly reduced parasite load in hybrids. If these immune genes are comparatively free of DMIs and introgressing alleles are favoured, while at the same time the sharp distinction between the mouse taxa is maintained by DMIs at other loci, clear patterns should be observable. In the simplest case: wide clines at the MHC compared to other loci; in the more complex moving hybrid zone case suggested by Macholán *et al.* (2011): large geographic inclusions of introgressed MHC alleles behind the moving wave front.

The hybrid–parasite interaction and the parasite's genetics

One factor that has been markedly missing from hybrid parasitism studies in general, and of the HMHZ in particular, is the genetics of the parasites (with the exception of Jackson and Tinsley, 2003). Many empirical studies have shown that the host–parasite interaction (host resistance or parasite infectivity) is often based on genotype-by-genotype (G × G) interactions: some hosts are 'compatible' with a subset of parasite genotypes while other hosts are 'compatible' with another subset (review in Lambrechts *et al.*, 2006). In such circumstances, to predict the influence of parasitism on a host secondary contact it would be necessary to know both the genetic structure of the parasite across the host hybrid zone and the host genotype by parasite genotype interactions. In the experimental infections of parental and hybrid mouse strains with nematode or protozoa parasites (Derothe *et al.*, 1999, 2001, 2004; Porcherie, 2005) a single parasite strain was used, often maintained for multiple generations in laboratory mice. While the heterogeneity in parasite load across the different mouse strains demonstrates clearly that the host genotype is important in the host–parasite interaction, the effect of the parasite genotype and the existence of a G × G effect is unknown. As a concrete example, if G × G interactions are important, it is unsatisfactory to assume that the pinworms in one subspecies of mouse are host–parasite-interaction equivalent to the pinworms in another subspecies merely because taxonomists assign the pinworms to the same species. Since parasites often evolve faster then their hosts due to shorter generation times, we may expect that diverged host taxa will bring their own specific parasite strains to a secondary contact. The match between the host and parasite genetic structure after contact will partially depend on this G × G interaction of hosts and parasites.

In this chapter we have examined some new insights into parasitism in the European HMHZ. In thinking carefully about this single system, we in turn gain insight into the evolutionary role of host–parasite interactions in general. For this reason, although we do not believe helminths are maintaining the mouse species barrier, we believe the parasites of the HMHZ will continue to be a model system for understanding the limits of parasite specificity, the genetic basis of resistance/susceptibility, and the behaviour of immune genes at species boundaries.

REFERENCES

Abu-Madi, M. A., Behnke, J. M., Lewis, J. W., and Gilbert, F. S. (2000). Seasonal and site specific variation in the component community structure of intestinal helminths in *Apodemus sylvaticus* from three contrasting habitats in south-east England. *Journal of Helminthology*, **74**, 7–15.

Baird, S. J. E., Ribas, A., Macholán, M., *et al.* (2012). Where are the wormy mice? A re-examination of hybrid parasitism in the European house mouse hybrid zone. *Evolution*. doi: 10.1111/j.1558-5646.2012.01633.x.

Bazykin, A. D. (1969). Hypothetical mechanism of speciation. *Evolution*, **23**, 685–7.

Behnke, J. M., Barnard, C. J., and Wakelin, D. (1992). Understanding chronic nematode infections: evolutionary considerations, current hypotheses and the way forward. *International Journal for Parasitology*, **22**, 861–907.

Behnke, J. M., Iraqi, F., Menge, D., *et al.* (2003). Chasing the genes that control resistance to gastrointestinal nematodes. *Journal of Helminthology*, **77**, 99–110.

Behnke, J. M., Iraqi, F., Mugambi, J., *et al.* (2006). High resolution mapping of chromosomal regions controlling resistance to gastrointestinal nematode infections in an advanced intercross line of mice. *Mammalian Genome*, **17**, 584–97.

Behnke, J. M., Lewis, J. W., Zain, S. N., and Gilbert, F. S. (1999). Helminth infections in *Apodemus sylvaticus* in southern England: interactive effects of host age, sex and year on the prevalence and abundance of infections. *Journal of Helminthology*, **73**, 31–44.

Čížková, D., Goüy de Bellocq, J., Baird, S. J. E., Piálek, J., and Bryja, J. (2011). Genetic structure and contrasting selection pattern at two major histocompatibility complex genes in wild house mouse populations. *Heredity*, **106**, 727–40.

Clifford, C. B., and Watson, J. (2008). Old enemies, still with us after all these years. *ILAR Journal*, **49**, 291–302.

Coustau, C., Renaud, F., Maillard, C., Pasteur, N., and Delay, B. (1991). Differential susceptibility to a trematode parasite among genotypes of the *Mytilus edulis/galloprovincialis* complex. *Genetics Research*, **57**, 207–12.

Cox, F. E. (2001). Concomitant infections, parasites and immune responses. *Parasitology*, **122**, S23–38.

Derothe, J.-M., Le Brun, N., Loubes, C., Perriat-Sanguinet, M., and Moulia, C. (2001). Susceptibility of natural hybrids between house mouse subspecies to *Sarcocystis muris*. *International Journal for Parasitology*, **31**, 15–19.

Derothe, J.-M., Loubes, C., Perriat-Sanguinet, M., Orth, A., and Moulia, C. (1999). Experimental trypanosomiasis of natural hybrids between house mouse subspecies. *International Journal for Parasitology*, **29**, 1011–16.

Derothe, J.-M., Porcherie, A., Perriat-Sanguinet, M., Loubès, C., and Moulia, C. (2004). Recombination does not generate pinworm susceptibility during experimental crosses between two mouse subspecies. *Parasitology Research*, **93**, 356–63.

Dobzhansky, T. (1936). Studies on hybrid sterility: II. Localization of sterility factors in *Drosophila pseudoobscura* hybrids. *Genetics*, **21**, 113–35.

Else, K. and Wakelin, D. (1988). The effects of H-2 and non-H-2 genes on the expulsion of the nematode *Trichuris muris* from inbred and congenic mice. *Parasitology*, **96**, 543–50.

Esteves, P., Lanning, D., Ferrand, N., *et al.* (2005). The evolution of the immunoglobulin heavy chain variable region *(IgVH)* in Leporids: an unusual case of transspecies polymorphism. *Immunogenetics*, **57**, 874–82.

Feliu, C., Renaud, F., Catzeflis, F., *et al.* (1997). A comparative analysis of parasite species richness of Iberian rodents. *Parasitology*, **115**, 453–66.

Fragoso, G., Lamoyi, E., Mellor, A., *et al.* (1996). Genetic control of susceptibility to *Taenia crassiceps* cysticercosis. *Parasitology*, **112**, 119–24.

Fritz, R. S. (1999). Resistance of hybrid plants to herbivores: genes, environment, or both? *Ecology*, **80**, 382–91.

Gulland, F. M. D., Albon, S. D., Pemberton, J. M., Moorcroft, P. R., and Clutton-Brock, T. H. (1993). Parasite-associated polymorphism in a cyclic ungulate population. *Proceedings of the Royal Society B: Biological Sciences*, **254**, 7–13.

Gumenyuk, T. U. (1953). Investigation of the helminth fauna of cats, dogs, rats and mice in Chernovtsy. In *Papers on Helminthology Presented to Academician K. I. Skryabin on His 75th Birthday*. Moscow: Izdatel'stvo Akademii Nauk SSSR, pp. 184–7. (In Russian.)

Haukisalmi, V. and Henttonen, H. (1999). Determinants of helminth aggregation in natural host populations: individual differences or spatial heterogeneity? *Ecography*, **22**, 629–36.

Haukisalmi, V., Henttonen, H., and Tenora, F. (1988). Population dynamics of common and rare helminths in cyclic vole populations. *Journal of Animal Ecology*, **57**, 807–25.

Hudson, P. J., Dobson, A. P., and Newborn, D. (1998). Prevention of population cycles by parasite removal. *Science*, **282**, 2256–8.

Hugot, J. P. (1988). Les nématodes Syphaciinae parasites de rongeurs et de lagomorphes: taxonomie, zoogéographie, évolution. *Mémoires du Muséum National d'Histoire Naturelle, Paris, Série A Zoologie*, **141**, 1–153.

Izdebska, J. N. and Rolbiecki, L. (2006). Correlation between the occurrence of mites (Demodex spp.) and nematodes in house mice (*Mus musculus* Linnaeus, 1758) in the Gdańsk urban agglomeration. *Biological Letters*, **43**, 175–8.

Jackson, J. A. and Tinsley, R. C. (2003). Parasite infectivity to hybridising host species: a link between hybrid resistance and allopolyploid speciation? *International Journal for Parasitology*, **33**, 137–44.

Kataranovski, D. S., Vukićević-Radić, O. D., Kataranovski, M. V., Radović, D. L., and Mirkov, I. I. (2008). Helminth fauna of *Mus musculus* Linnaeus, 1758 from the suburban area of Belgrade, Serbia. *Archives of Biological Sciences*, **60**, 609–17.

Key, K. H. L. (1968). The concept of stasipatric speciation. *Systematic Zoology*, **17**, 14–22.

Kisielewska, K. (1970). Ecological organization of intestinal helminth groupings in *Clethrionomys glareolus* (Schreb.) (Rodentia): I. Structure and seasonal dynamics of helminth groupings in a host population in the Bialowieza National Park. *Acta Parasitologica Polonica*, **18**, 121–47.

Klein, J. (1986). *Natural History of the Major Histocompatibility Complex*. New York: Wiley.

Klein, J. (1987). Origin of major histocompatibility complex polymorphism: the trans-species hypothesis. *Human Immunology*, **19**, 155–62.

Klein, J. (1988). Debate about three pages. *Immunogenetics*, **28**, 67–70.

Knoll, A. H. and Carroll, S. B. (1999). Early animal evolution: emerging views from comparative biology and geology. *Science*, **284**, 2129–37.

Kondrashov, A. S. (2003). Accumulation of Dobzhansky–Muller incompatibilities within a spatially structured population. *Evolution*, **57**, 151–3.

Kriska, T. (1993). Parasitic helminths of house mouse (*Mus musculus* Linnaeus, 1758) in Hungary. *Miscellanea Zoologica Hungarica*, **8**, 13–23.

Kristan, D. M. (2004). Intestinal nematode infection affects host life history and offspring susceptibility to parasitism. *Journal of Animal Ecology*, **73**, 227–38.

Lambrechts, L., Fellous, S., and Koella, J. C. (2006). Coevolutionary interactions between host and parasite genotypes. *Trends in Parasitology*, **22**, 12–16.

Lin, Y. C., Rikihisa, Y., Kono, H., and Gu, Y. (1990). Effects of larval tapeworm (*Taenia taeniaeformis*) infection on reproductive functions in male and female host rats. *Experimental Parasitology*, **70**, 344–52.

Macholán, M., Baird, S. J. E., Dufková, P., *et al.* (2011). Assessing multilocus introgression patterns: a case study on the house mouse X chromosome in Central Europe. *Evolution*, **65**, 1428–46.

Macholán, M., Baird, S. J. E., Munclinger, P., *et al.* (2008). Genetic conflict outweighs heterogametic incompatibility in the mouse hybrid zone? *BMC Evolutionary Biology*, **8**, 271.

Macholán, M., Munclinger, P., Šugerková, M., *et al.* (2007). Genetic analysis of autosomal and X-linked markers across a mouse hybrid zone. *Evolution*, **61**, 746–71.

Maizels, R. M., Balic, A., Gomez-Escobar, N., *et al.* (2004). Helminth parasites: masters of regulation. *Immunological Reviews*, **201**, 89–116.

Mas-Coma, S., Esteban, J. G., Fuentes, M. V., *et al.* (1998). Helminth parasites of small mammals (insectivores and rodents) on the Pityusic island of Formentera (Baleartic archipelago). *Research and Reviews in Parasitology*, **58**, 125–33.

Mas-Coma, S., Esteban, J. G., Fuentes, M. V., *et al.* (2000). Helminth parasites of small mammals (insectivores and rodents) on the Pityusic island of Eivissa (Balearic Archipelago). *Research and Reviews in Parasitology*, **60**, 41–9.

Mažeika, V. (1992). Helminth fauna of mouse rodents in Lithuania. *Ekologija*, **4**, 53–68.

Milazzo, C., Cagnin, M., Di Bella, C., Geraci, F., and Ribas, A. (2010). Helminth fauna of commensal rodents, *Mus musculus* (Linnaeus, 1758) and *Rattus rattus* (Linnaeus, 1758) (Rodentia, Muridae) in Sicily (Italy). *Revista Ibero-latinoamericana de Parasitología*, **69**, 194–8.

Milazzo, C., Goüy de Bellocq J., Cagnin, M., *et al.* (2003). Helminths and ectoparasites of *Rattus rattus* and *Mus musculus* from Sicily, Italy. *Comparative Parasitology*, **70**, 199–204.

Miquel, J., Casanova, J. C., Fons, R., *et al.* (1996). Helminthes parasites des Rongeurs muridés des îles d'Hyères (Var, France): aspects écologiques. *Vie et Milieu*, **46**, 219–23.

Miura, F., Tsukamoto, K., Mehta, R. B., *et al.* (2010). Transspecies dimorphic allelic lineages of the proteasome subunit β-type 8 gene (*PSMB8*) in the teleost genus *Oryzias*. *Proceedings of the National Academy of Sciences of the United States of America*, **107**, 21599–604.

Montgomery, S. S. and Montgomery, W. I. (1989). Spatial and temporal variation in the infracommunity structure of helminths of *Apodemus sylvaticus* (Rodentia: Muridae). *Parasitology*, **98**, 145–50.

Moulia, C., Aussel, J. P., Bonhomme, F., *et al.* (1991). Wormy mice in a hybrid zone: a genetic control of susceptibility to parasite infection. *Journal of Evolutionary Biology*, **4**, 679–87.

Moulia, C. and Joly, P. (2009). Parasitism and hybrid zones. In *Ecology and Evolution of Parasitism*, ed. F. Thomas, J. F. Guégan, and F. Renaud. New York: Oxford University Press, pp. 69–82.

Moulia, C., Le Brun, N., Dallas, J., Orth, A., and Renaud, F. (1993). Experimental evidence of genetic determinism in high susceptibility to intestinal pinworm infection in mice: a hybrid zone model. *Parasitology*, **106**, 387–93.

Muller, H. J. (1942). Isolating mechanisms, evolution, and temperature. *Biological Symposia*, **6**, 71–125.

Nonaka, M., Yamada-Namikawa, C., Flajnik, M. F., and Du Pasquier, L. (2000). Trans-species polymorphism of the major histocompatibility complex-encoded proteasome subunit LMP7 in an amphibian genus, *Xenopus*. *Immunogenetics*, **51**, 186–92.

Olivier, L. (1962). Natural resistance of *Taenia taeniaeformis*: I. Strain differences in susceptibility of rodents. *Journal of Parasitology*, **48**, 373–8.

Orihara, M. (1962). Studies on *Cysticercus fasciolaris*, especially on differences of susceptibility among uniform strains of the mouse. *Japanese Journal of Veterinary Research*, **10**, 37–56.

Orr, H. A. (1995). The population genetics of speciation: the evolution of hybrid incompatibilities. *Genetics*, **139**, 1805–13.

Paterson, S., Wilson, K., and Pemberton, J. M. (1998). Major histocompatibility complex variation associated with juvenile survival and parasite resistance in a large unmanaged ungulate population (*Ovis aries* L.). *Proceedings of the National Academy of Sciences of the United States of America*, **95**, 3714–19.

Pedersen, A. B. and Greives, T. J. (2008). The interaction of parasites and resources cause crashes in a wild mouse population. *Journal of Animal Ecology*, **77**, 370–7.

Petrochenko, V. I. and Romanovskii, A. B. (1966). On a role of rodents in distribution of ovine coenurosis and rodent helminths on the steppes of Kyrgyzstan. *Materialy Nauchnoi Konferentsii Vsesoyuznogo Obshchestva Gel'mintologov*, **3**, 222–8. (In Russian.)

Piertney, S. B. and Oliver, M. K. (2006). The evolutionary ecology of the major histocompatibility complex. *Heredity*, **96**, 7–21.

Pisanu, B., Chapuis, J. L., and Durette-Desset, M. C. (2001). Helminths from introduced small mammals on Kerguelen, Crozet, and Amsterdam Islands (southern Indian Ocean). *Journal of Parasitology*, **87**, 1205–8.

Porcherie, A. (2005). Susceptibilité aux parasites des hybrides entre *Mus musculus musculus* et *Mus musculus domesticus*: *origine du phénomène et rôle dans la contre-sélection des hybrides*. Unpublished PhD thesis, University Montpellier II.

Prokopič, J. and Genov, T. (1974). *Distribution of Helminths in Micromammals (Insectivora and Rodentia) under Different Ecological and Geographical Conditions*, Prague: Studie Československé akademie.

Quinnell, R. J. (2003). Genetics of susceptibility to human helminth infection. *International Journal for Parasitology*, **33**, 1219–31.

Raberg, L., Sim, D., and Read, A. F. (2007). Disentangling genetic variation for resistance and tolerance to infectious diseases in animals. *Science*, **318**, 812–14.

Raufaste, N., Orth, A., Belkhir, K., *et al.* (2005). Inferences of selection and migration in the Danish house mouse hybrid zone. *Biological Journal of the Linnean Society*, **84**, 593–616.

Rikihisa, Y., Lin, Y. C., and Fukaya, T. (1985). *Taenia taeniaeformis*: inhibition of rat testosterone production by excretory-secretory product of the cultured metacestode. *Experimental Parasitology*, **59**, 390–7.

Sage, R. D., Heyneman, D., Lim, K. C., and Wilson, A. C. (1986a). Wormy mice in a hybrid zone. *Nature*, **324**, 60–3.

Sage, R. D., Whitney, J. B., III., and Wilson, A. C. (1986b). Genetic analysis of a hybrid zone between *domesticus* and *musculus* mice (*Mus musculus* complex): hemoglobin polymorphisms. *Current Topics in Microbiology and Immunology*, **127**, 75–85.

Sak, B., Kváč, M., Květoňová, D., Albrecht, T., and Piálek, J. (2011). The first report on natural *Enterocytozoon bieneusi* and *Encephalitozoon* spp. infections in wild East-European House Mice (*Mus musculus musculus*) and West-European House Mice (*M. m. domesticus*) in a hybrid zone across the Czech Republic–Germany border. *Veterinary Parasitology*, **178**, 246–50.

Scott, M. E. and Gibbs, H. C. (1986). Long-term population dynamics of pinworms (*Syphacia obvelata* and *Aspiculuris tetraptera*) in mice. *Journal of Parasitology*, **72**, 652–62.

Shaw, D. J. and Dobson, A. P. (1995). Patterns of macroparasite abundance and aggregation in wildlife populations: a quantitative review. *Parasitology*, **111**, S111–27.

Singleton, G. R. and Krebs, C. J. (2007). The secret world of wild mice. In *The Mouse in Biomedical Research, 2nd edn*, ed. J. G. Fox *et al.*, London: Academic Press, pp. 25–51.

Sorci, G., Morand, S., and Hugot, J.-P. (1997). Host–parasite coevolution: comparative evidence for covariation of life history traits in primates and oxyurid parasites. *Proceedings of the Royal Society B: Biological Sciences*, **264**, 285–9.

Su, C. and Nei, M. (1999). Fifty-million-year-old polymorphism at an immunoglobulin variable region gene locus in the rabbit evolutionary lineage. *Proceedings of the National Academy of Sciences of the United States of America*, **96**, 9710–15.

Taffs, L. F. (1976). Pinworm infections in laboratory rodents: a review. *Laboratory Animals*, **10**, 1–13.

Tattersall, F. H., Nowell, F., and Smith, R. H. (1994). A review of the endoparasites of wild house mice *Mus domesticus*. *Mammal Review*, **24**, 61–71.

Teeter, K. C., Thibodeau, L. M., Gompert, Z., *et al.* (2009). The variable genomic architecture of isolation between hybridizing species of house mice. *Evolution*, **64**, 472–85.

Thomas, F., Verneau, O., De Meeus, T., and Renaud, F. (1996). Parasites as host evolutionary prints: insights into host evolution from parasitological data. *International Journal for Parasitology*, **26**, 677–86.

van Riet, E., Hartgers, F. C., and Yazdanbakhsh, M. (2007). Chronic helminth infections induce immunomodulation: consequences and mechanisms. *Immunobiology*, **212**, 475–90.

Vandegrift, K. J., Raffel, T. R., and Hudson, P. J. (2008). Parasites prevent summer breeding in white-footed mice, *Peromyscus leucopus*. *Ecology*, **89**, 2251–8.

Wakelin, D. (1975a). Genetic control of immune responses to parasites: selection for responsiveness and non-responsiveness to *Trichuris muris* in random-bred mice. *Parasitology*, **71**, 377–84.

Wakelin, D. (1975b). Genetic control of immune responses to parasites: immunity to *Trichuris muris* in inbred and random-bred strains of mice. *Parasitology*, **71**, 51–60.

Wakelin, D. (1988). Genetic control of immunity to helminth infections. In *Parasitology in Focus*, ed. H. Mehlhorn. Berlin: Springer, pp. 651–70.

Wang, L., Luzynski, K., Dufková, P., *et al.* (2011). Measures of linkage disequilibrium among neighboring SNPs indicate dispersal asymmetry across the house mouse hybrid zone. *Molecular Ecology*, **20**, 2985–3000.

Wilson, K., Bjornstad, A. P., Dobson, A. P., *et al.* (2002). Heterogeneities in macroparasite infections: patterns and processes. In *The Ecology of Wildlife Diseases*, ed. P. J. Hudson, A. Rizzoli, B. T. Grenfell, H. Heesterbeek, and A. P. Dobson. New York: Oxford University Press, pp. 6–44.
Wilson, M. S. and Maizels, R. M. (2004). Regulation of allergy and autoimmunity in helminth infection. *Clinical Reviews in Allergy and Immunology*, **26**, 35–50.
Wolinska, J., Bittner, K., Ebert, D., and Spaak, P. (2006). The coexistence of hybrid and parental *Daphnia*: the role of parasites. *Proceedings of the Royal Society B: Biological Sciences*, **273**, 1977–83.
Wolinska, J., Lively, C. M., and Spaak, P. (2008). Parasites in hybridizing communities: the red queen again? *Trends in Parasitology*, **24**, 121–6.
Zepeda, N., Copitin, N., Solano, S., *et al.* (2011). *Taenia crassiceps*: infections of male mice lead to severe disruption of seminiferous tubule cells and increased apoptosis. *Experimental Parasitology*, **127**, 153–9.

19

Hybrid male sterility genes in the mouse subspecific crosses

JIŘÍ FOREJT, JAROSLAV PIÁLEK, AND ZDENĚK TRACHTULEC

Pure species have of course their organs of reproduction in a perfect condition, yet when intercrossed they produce either few or no offspring. Hybrids, on the other hand, have their reproductive organs functionally impotent, as may be clearly seen in the state of the male element in both plants and animals.

Charles Darwin, *On the Origin of Species by Means of Natural Selection*, 1859

Introduction

Hybrid sterility has been described in various eukaryotic species of animal and plant kingdoms. Studies in *Drosophila* and other species revealed that hybrid sterility belongs to reproductive isolation mechanisms, which contribute to speciation by restricting gene flow between diverging taxa. Theodosius Dobzhansky defined hybrid sterility as a situation in which two parental forms, each of which is fertile *inter se*, produce a hybrid, which is sterile (Dobzhansky, 1951; Maheshwari *et al.*, 2008).

In the house mouse *Mus musculus*, as in *Drosophila* species, hybrid sterility predominantly affects males, following Haldane's rule: when in the F_1 offspring of two different animal races one sex is absent, rare, or sterile, that sex is the heterozygous [heterogametic] sex (Haldane, 1922). In spite of 90 years of genetic studies, neither the genetic architecture nor the molecular mechanism of hybrid sterility or Haldane's rule has been elucidated in any species. In fact, the first hybrid sterility genes in *Drosophila* (*OdsH*, *JYAlpha*, and *Ovd*) and in the house mouse (*Hst1/Prdm9*) were identified only recently (Ting *et al.*, 1998; Masly *et al.*, 2006; Mihola *et al.*, 2009; Phadnis and Orr, 2009). Our understanding of the molecular mechanisms of hybrid sterility had been stalled for decades by resistance of the controlling oligogenic network to high-resolution genetic mapping. Furthermore, validation of candidate genes, once they are identified, is more

Evolution of the House Mouse, ed. Miloš Macholán, Stuart J. E. Baird, Pavel Munclinger, and Jaroslav Piálek. Published by Cambridge University Press.

difficult compared to classical null mutations because the function of each hybrid sterility gene within the context of its own species is not affected.

Dobzhansky and Muller (Muller, 1942; Dobzhansky, 1951) elaborated a model of gene – gene incompatibilities, known as the Dobzhansky–Muller (DM) model, explaining reproductive isolation between populations. In its simplest form it envisages the consequences of independent evolution of a pair of interacting genes. After the new alleles, untested together by natural selection, meet in a hybrid, their epistatic interaction is suboptimal and results in a deleterious effect on fitness. So far, none of the cloned hybrid sterility genes in *Drosophila* or in the mouse have a known interacting partner that would allow the DM incompatibility hypothesis to be tested at the molecular level. However, *Hmr* and *Lhr*, the hybrid lethality genes in *Drosophila melanogaster* × *D. simulans* crosses, fulfil the criterion by a still unknown molecular mechanism (Brideau *et al.*, 2006; but see Maheshwari *et al.*, 2008). Not all models of hybrid sterility that have been studied in *Drosophila* or in mice actually represent the speciation process. Some genetic incompatibilities might have evolved long after reproductive isolation had been established, thus representing a consequence of, not the cause of, speciation. We assume many hybrid sterility genes described in the laboratory crosses of *Mus musculus domesticus* × *M. spretus* (Bonhomme *et al.*, 1978; Guénet *et al.*, 1990; Pilder *et al.*, 1991, 1993, 1997; Fossella *et al.*, 2000; Elliott *et al.*, 2001) and C57BL/6J × *M. macedonicus* (Elliott *et al.*, 2004) to be the by-products of speciation, not its cause.

Hybrid male sterility in *M. m. musculus* × *M. m. domesticus* crosses

M. m. musculus and *M. m. domesticus* (abbreviated hereafter *Mmm* and *Mmd*) are the two most widespread subspecies of the house mouse in the world. They diverged from a common ancestor 0.3 – 0.5 million years (Myr) ago (She *et al.*, 1990; Boursot *et al.*, 1996; Geraldes *et al.*, 2008) and have developed efficient yet incomplete reproductive isolation evidenced by a narrow hybrid zone in the regions of their secondary contact (see Baird and Macholán, Chapter 14 in this volume). This review is focused on the genetic analysis of hybrid sterility in *Mmm* × *Mmd* crosses, in particular on mapping and identification of the first hybrid sterility gene in the mouse (and, in fact, in vertebrates), Hybrid sterility 1 (*Hst1*), and on the interacting X chromosomal genetic factors. Because reproductive isolation between *Mmm* and *Mmd* is still evolving, there is a chance that these factors are true 'speciation genes' and not speciation by-products.

Based on the experimental setup, hybrid sterility can be studied in mice from wild populations or in wild-derived inbred strains. In the former approach, the wild mice trapped within or outside the hybrid zone can be further bred in the laboratory and their progeny can be studied in more detail. Such a strategy can capture a significant part of the genetic variation of the studied populations, but does not allow high-resolution genetic mapping and functional genomics studies. The latter approach, discussed in this chapter, is based on the wild-derived inbred strains and their derivatives. Here the resolution power of genetic studies is much higher, but so is the risk that an inbred strain, a single 'snapshot' of a diverging genome, will not capture all relevant variants.

How many genes control hybrid sterility in *Mmm* × *Mmd* crosses?

According to the DM incompatibility model (Muller, 1942; Dobzhansky, 1951), two interacting genes are minimally necessary to create reproductive isolation barriers. Determination of the number of hybrid sterility genes and their genomic locations is the basic requirement in studying the genetic architecture of hybrid sterility. It can be achieved by quantitative trait locus (QTL) analysis, or by designing crosses or introgressions in which only one hybrid sterility gene segregates. In the QTL approach, backcross populations can be prepared via fertile F_1 hybrid females and analysed for surrogate fertility phenotypes, such as the number of sperm cells in the *ductus epididymis* and testes weight. QTL mapping can provide an estimate of the number of hybrid sterility genes, but is not sensitive enough to permit their positional cloning. The first described *Mmm* × *Mmd* backcross experiment used C57BL/10 or C57BL/6 (hereafter B10 and B6) as a laboratory substitute of *Mmd*. The first backcross – (wild *Mmm* × B10) × B10 – yielded 12% of sterile male progeny (Forejt and Iványi, 1974), which corresponds to dominant epistatic incompatibilities of at least three hybrid sterility genes. In a recent, more detailed study (Dzúr-Gejdošová *et al.*, in prep.) the 6% incidence of sterile males indicated even more than three segregating hybrid sterility genes in the (PWD × B6) × B6 BC_1 cross. However, genome-wide mapping has revealed only two strong QTLs, on chromosome 17 (Chr 17) and on Chr X. Genome-reconstitution experiments using chromosome substitution strains (Gregorová *et al.*, 2008) demonstrated genes on Chr 17 and Chr X are necessary but not sufficient to reproduce the full phenotype of F_1 hybrid sterility – pachytene arrest and lack of sperm. There must be a few minor loci involved that are more difficult to identify by QTL mapping (Dzúr-Gejdošová *et al.*, in prep.). Thus, in the *Mmm* × *Mmd* genetic crosses, as in *Drosophila* studies of closely related species, hybrid sterility behaves as an oligogenic DM incompatibility. This conclusion is in accord with genetic analysis of autosomal and X-linked markers of wild mice in the *Mmd*/*Mmm* hybrid zone (Macholán *et al.*, 2007).

Piálek and co-workers (Vyskočilová *et al.*, 2005, 2009) reported intraspecific variation of the genes controlling *Mmm* × *Mmd* hybrid sterility. They confirmed previous classification of *Mmm* wild mice into three types – as mice producing only sterile, only fertile, or both sterile and fertile males after crosses with the B10 laboratory (predominantly *Mmd*) strain (Forejt and Iványi, 1974). Moreover, Piálek and co-workers (Piálek *et al.*, 2008; Vyskočilová *et al.*, 2009) prepared new *Mmm* inbred strains – STUF and STUS – producing only fertile and only sterile males with B10 mice, respectively. In a search for polymorphic hybrid sterility genes using the (STUS × STUF) × B10 cross, they found that 90% of genetic variance between STUF and STUS can be explained by a QTL in the proximal part of Chr 17 and 10% by another QTL in the central part of Chr X. It is tempting to speculate that these intra-subspecific polymorphic loci are identical with the Hst^{ws}/*Hst1* and *Hstx2* (Bhattacharyya *et al.*, in prep.) loci identified in intersubspecific crosses. Similar intraspecific polymorphism of hybrid male sterility genes, marking the early stage of hybrid sterility, was reported in *Drosophila* crosses between *D. mojavensis* females and *D. arizonae* males (Reed and Markow, 2004; Reed *et al.*, 2008).

Spermatogenesis-related phenotypes of hybrid sterility

In accord with the studies of hybrid sterility in *Drosophila*, the only apparent pathology in *Mmm* × *Mmd* mouse intersubspecific hybrids is spermatogenic breakdown. Fertility of hybrid mice was originally evaluated by OMU (offspring per mating unit), in which one mating unit (MU) is one male caged with one female for one month). OMU correlated well with the wet mass of paired testes: (*Mmm* × B10)F_1 or (PWD × B6)F_1 hybrid males with testes smaller than 90 mg were invariably sterile. Histological sections revealed arrest of spermatogenesis in primary spermatocytes and cytogenetic examination of rare primary spermatocytes at diakinesis/metaphase I indicated incomplete autosomal synapsis (Iványi *et al.*, 1969; Forejt and Iványi, 1974; Forejt *et al.*, 1981; Forejt, 1985). Recently, more detailed studies of spermatogenesis were conducted in inter(sub)specific F_1 hybrids and in consomics with the X chromosome introgressed from other species (Mihola *et al.*, 2009; Oka *et al.*, 2010; Bhattacharyya *et al.*, in prep.). It was found that the beginning and the extent of spermatogenic arrest differed considerably among different crosses and genotypes. After introgression of *M. m. molossinus* (MSM) Chr X into the B6 strain, the B6.MSM-ChrX males produced some sperm and revealed two incomplete blocks of spermatogenesis; the first occurred unexpectedly before onset of the first meiotic division and the second at the pachytene stage (see also Oka and Shiroishi, Chapter 17). In PGN.MSM-ChrX intersubspecific consomics (PGN is a wild-derived *Mmd* inbred strain) meiosis was arrested between pachynema and metaphase I (Oka *et al.*, 2010). Incomplete synapsis of a few autosomes was reported at pachynema of B6.MSM-ChrX pachytene spermatocytes (Oka *et al.*, 2010). Asynapsis was even more apparent in (PWD × B6) males

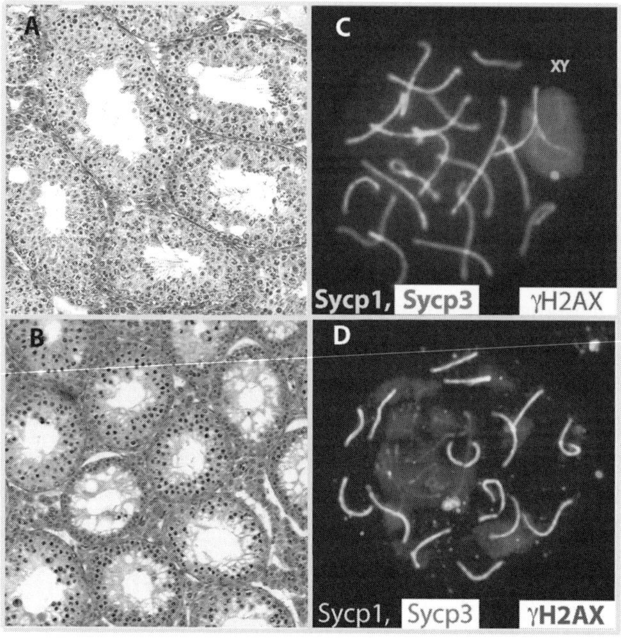

Figure 19.1 Male germ cell differentiation in sterile and fertile intersubspecific hybrids. (a, b) Histological sections of (B6 × PWD)F1 and (PWD × B6)F1 fertile and sterile hybrids, respectively. Note the lack of spermatids and sperm and the presence of vacuoles in sterile testis. (c, d) Pachytene spermatocyte spreads from fertile and sterile hybrids. In sterile males unsynapsed autosomal pairs (synaptonemal complex proteins SYCP3+, SYCP1−) display phosphorylated histone γH2AX while lacking the XY-containing sex body characteristic for normal pachynema. The asymmetric phenotypes of reciprocal F1 hybrids document at molecular level the Darwin corollary to Haldane's rule (see text) (T. Bhattacharyya, unpublished). See the plate section for a colour version of this figure.

(Mihola *et al.*, 2009; Bhattacharyya *et al.*, in prep.). As evidenced by the lack of specific immunostaining with antibodies against γH2AX, ATR, and Rad51, most of their pachytene spermatocytes failed to form the sex body, a nuclear compartment containing heterochromatic X and Y chromosomes (Fig. 19.1). The recombination nodules visualized as MLH1 foci were missing in asynapsed autosomes (Bhattacharyya *et al.*, in prep.). Similar abnormalities were observed in (B6 × NJL/Ms) intersubspecific *Mmd* × *Mmm* hybrids by Oka and co-workers (Oka *et al.*, 2010). Unexpectedly, spermatogenic arrest in (B6 × *M. spretus*)F_1 hybrids occurred much later, at metaphase I of primary spermatocytes. The combined pieces of cytological and gene expression evidence point towards male sex chromosome inactivation as a surveillance mechanism. Thus the predicted role of the X chromosome inactivation of pachytene spermatocytes in chromosomal and genic hybrid sterility (Forejt, 1982, 1985, 1996) can finally be verified.

Hybrid sterility 1 (*Hst1*) gene

The first sterile male hybrids of the mouse were reported in studies seeking new *H2* histocompatibility locus polymorphisms (Iványi *et al.*, 1969). Iványi and co-workers crossed wild mice trapped at different places near Prague, Czech Republic with laboratory strains and observed sterile males with B10 and A/Ph strains but not with the C3H/Di strain. The authors hypothesized that an *H2*-linked genetic factor associated with male sterility could be a *t*-haplotype sterility factor without *T* (Brachyury) modifying effect. Later it was realized that the male-limited sterility shows all attributes of hybrid sterility, and the locus, designated by then as Hybrid sterility 1, *Hst1*, was localized between the *T* and *H2* markers on Chr 17 (Forejt and Iványi, 1974). The first experimental setup that enabled fine genetic mapping of a single hybrid sterility gene was based on the cross (B10 × C3H) × wild *Mmm* mice. A single hybrid sterility gene (*Hst1*) segregated in the progeny of the B10 × C3H laboratory strain F_1 hybrids. This and later mapping studies used wild *Mmm* mice or wild-derived inbred strains for high-resolution genetic mapping of the *Hst1* locus using wild × (B10 × C3H) or PWD × ((B10 × C3H) × B10) types of test cross (Forejt, 1981, 1985; Forejt *et al.*, 1981; Gregorová *et al.*, 1996). This type of genetic mapping does not provide information on the number and location of epistatically interacting hybrid sterility genes because it detects only *Hst1* as the sole difference between B10 and C3H strains.

Hst1 maps within the radius of meiotic drive of *t*-haplotypes

Mouse *t*-haplotypes are the natural variants of the proximal part of Chr 17 that are transmitted from males up to more than 90% of the offspring (Lyon, 2003; see Herrmann and Bauer, Chapter 12, for a review). Three genes controlling transmission ratio distortion (TRD) of *t*-haplotypes have been positionally cloned: the *t*-complex responder *Smok* (Herrmann *et al.*, 1999) and two *t*-complex distorters, *Tagap1* and *Fgd2* (Bauer *et al.*, 2007; see also Herrmann and Bauer, Chapter 12). The *Hst1* gene is situated within the *t*-haplotype region (Forejt and Iványi, 1974; Gregorová *et al.*, 1996); therefore, its spread in natural populations might be affected by meiotic drive exerted by *t*-haplotypes. The question remains whether *Hst1* by itself could be a part of the *t*-haplotype TRD genetic mechanism. Since the *t*-haplotypes examined so far carry the $Hst1^f$ allele (see below), and $Hstw^f$ (the supposed allelic form of *Hst1* in *Mmm* species) does not interfere with *t*-haplotype TRD, direct involvement of *Hst1* in *t*-haplotype TRD seems unlikely.

High-resolution genetic and haplotype mapping of *Hst1*

The introduction of methods of molecular genetics into the studies of hybrid sterility allowed new designs for positional cloning of the genes involved. Gene cloning based on genomic location starts with genetic mapping of QTL and proceeds with physical mapping and gene identification and characterization. The final step is validation of the candidate gene, e.g. by transgenesis or gene targeting. Genetic mapping of *Hst1* at sub-milliMorgan resolution was accomplished with RFLP and SSLP DNA markers on a panel of 2020 (B10-*T* × C3H)*T* × B10 backcross animals (Forejt *et al.*, 1991; Gregorová *et al.*, 1996). To construct the physical map, yeast artificial chromosomes (YACs) carrying mouse genomic DNA were isolated using probes specific for loci previously mapped into the region (Trachtulec *et al.*, 1994, 1997). When genomic libraries in bacterial artificial chromosomes (BACs) became available, clones from the *Hst1* region were isolated from the library of the mouse strain 129S1/SvImJ (carrying $Hst1^s$; Trachtulec *et al.*, 2005) and from the C3H/HeJ library ($Hst1^f$; Mihola *et al.*, 2009). Large genomic rearrangements associated with $Hst1^s$ and $Hst1^f$ alleles were excluded by physical mapping of YACs, BACs, and genomic DNAs. Subsequent genetic mapping of markers isolated from these clones resulted in restriction of the *Hst1* critical region to 252 kb harbouring seven protein-coding genes (Table 19.1).

To further reduce the interval and to identify the *Hst1* gene, haplotype-association mapping (Trachtulec *et al.*, 2008) was undertaken. The

Table 19.1 Characteristics of seven *Hst1* candidate genes identified by genetic mapping.

Gene	Function	Expression	B × C
Gm10510	Unknown	Early embryo	UTR
Dll1	Notch signalling	Various tissues, including embryo	Syn
Fam120b	PPARγ signalling	Adipocytes and other cells	Syn
Psmb1	ATP-dependent proteolysis	Various tissues, including embryo	Syn, 3′ UTR
Tbp	RNA polymerase I-, II-, and III-dependent transcription	All tissues	5′ UTR, Syn, Alt
Pdcd2	Apoptosis	Many tissues	Alt
Prdm9	Meiotic recombination and gene transactivation	Germ cells	5′ UTR, Alt, CTZFs

Note: B × C: polymorphisms between B6 and C3H; Syn: synonymous coding; Alt: alternative mRNA; UTR: untranslated region; CTZF: C-terminal zinc fingers.

haplotype-association mapping takes advantage of the historical recombination events that accumulated during construction of mouse inbred strains. Classical laboratory strains were phenotyped for the *Hst1* allele and genotyped by re-sequencing for the co-segregating polymorphisms distinguishing the B6, C3H, and 129S2/SvPas strains. Three haplotypes within the 252-kb region were identified in 15 classical laboratory (Castle's) strains or strains derived from them. Ten strains carrying two of the three haplotypes produced sterile males with PWD, while five strains with the third haplotype produced fertile hybrids (Trachtulec *et al.*, 2008, see also Table 19.2). The haplotype analysis of laboratory strains thus confirmed the genetic mapping. The two haplotypes bearing the $Hst1^s$ allele were identical in the region extending from *Pgcc1* to *Prdm9*, which was therefore more likely to contain *Hst1*. Re-sequencing included a series of wild and wild-derived inbred *Mmd* and *Mmm* mice, as well as other mouse species. It was shown that all three *Hst1* haplotypes of classical laboratory strains originated in *Mmd*, while the PWD haplotype came from *Mmm*. The next step was the analysis of candidate genes, namely identification of their isoforms and allelic differences, as well as their expression levels and tissue specificity. The candidate genes *Tbp*, *Pdcd2*, and *Prdm9* revealed alternative isoforms, but none of them could be associated with hybrid sterility (Trachtulec *et al.*, 2005; Mihola *et al.*, 2007, 2009). Moreover, allelic sequencing of six of the seven candidate genes (Table 19.1) did not show any difference in the predicted protein sequence of B6 ($Hst1^s$) and C3H ($Hst1^f$) alleles. The coding region of the seventh gene, *Prdm9*, differed in the number of

Table 19.2 Phenotypes of mice with disrupted *Prdm9* and of sterile hybrids. The phenotype of $Prdm9^{-/-}$ males (Hayashi *et al.*, 2005) was compared to the wild-type and the phenotype of sterile hybrids (Mihola *et al.*, 2009) to fertile hybrids. *Morc2b* is a germ-cell-specific gene transactivated by *Prdm9*.

Phenotype/genotype	$Prdm9^{-/-}$	(PWD × B6)F_1
Testis weight	Reduced	Reduced
Number of pachytene SCs	Reduced	Reduced
Spermatids	None	None or a few
Sperm	None	None
Chromosomal pairing	Partial	Frequent univalents
Sex body formation	Impaired	Impaired (66%)
Morc2b gene induction	Failed	Reduced
H3K4m3 in *Morc2b* promoter	Decreased	Decreased
Females sterile	Yes	No

Note: H3K4m3: trimethylation of lysine 4 of histone 3; SCs: spermatocytes.

C_2H_2 zinc fingers. Moreover, *Prdm9* was the only gene of the seven candidates with germ-cell-specific expression (Hayashi *et al.*, 2005) (Table 19.1).

Hst1/Prdm9 encodes meiotic histone H3K4 methyltransferase

In parallel with restriction of the *Hst1* region by haplotype association mapping, a functional hybrid sterility rescue experiment was designed by C3H BAC transgenesis (Mihola *et al.*, 2009). The idea was to add to the *Hst*^{ws}/*Hst1*^{s} sterility genotype 'fertility' *Hst1*^{f} alleles in one or more copies, and to check for sterility rescue. Four overlapping C3H/HeJ (*Hst1*^{f}) BACs from the CHORI-34 library, prepared by Pieter J. de Jong's laboratory (http://bacpac.chori.org/library.php?id=235), were selected to cover the critical *Hst1* region. The clone 45 F17 (henceforth BAC5) encompassed the *Pgcc1*, *Psmb1*, *Tbp*, *Pdcd2*, and *Prdm9* genes; the clone 255E14 (BAC19) the *Dll1* and *Pgcc1* genes; the clone 289M8 (BAC21) carried the complete genes *Psmb1*, *Tbp*, and *Pdcd2*, but it contained neither the first exon of *Pgcc1* nor the first nine coding exons of *Prdm9*. The clone 331G23 (BAC24) carried intact *Psmb1*, *Tbp*, *Pdcd2*, and *Prdm9* genes, but no *Pgcc1* (Fig. 19.2). Embryonic stem cell (ES) line v6.5 F1 (129/S4SvJae × C57BL/6) (Howell *et al.*, 2005) harbouring *Hst1*^{s} was transfected with the BACs and microinjected in C57BL/6 blastocysts. BAC19 chimeras made from two ES cell lines suffered from tail and leg defects and did not transmit the transgene. BAC5 (two copies), BAC21 (two copies), and BAC24 (≈ six copies) transgenics and their offspring were fully fertile when crossed to

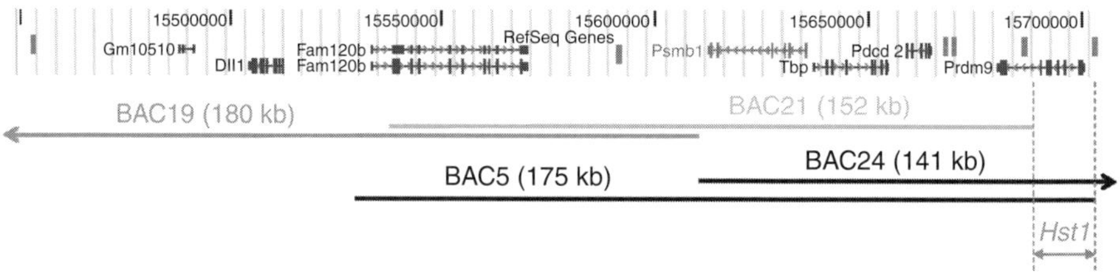

Figure 19.2 Transgenic rescue of *Hst1* sterility. The *Hst1* co-segregating region, 252 kb in size, is displayed as a UCSC Genome Browser track. The numbers at the top refer to the mm9 mouse genome assembly positions on Chr 17. Exons are shown as boxes connected by arrows pointing from 5′ to 3′; the grey boxes symbolize pseudogenes. The BAC clones used for transgenesis are shown as horizontal lines (their size in parentheses). The chimeras carrying BAC19 (grey line) did not transmit the transgene. The black and green lines show the extent of BAC transgenes rescuing and not rescuing hybrid sterility, respectively. The region necessary for the rescue is denoted as *Hst1*. See the plate section for a colour version of this figure.

B6. After outcrossing to PWD females (*Mmm*), the F_1 males without the transgene were sterile with complete meiotic arrest, but their transgenic littermates carrying BAC5 and BAC24 were fertile. The same results were obtained after transferring BAC5 and BAC24 to the B6 background by repeated backcrosses, to eliminate possible unlinked mutations coming from the ES cells. Intersubspecific hybrids carrying BAC21 were sterile. The only interval common to BAC5 and BAC24 but missing in BAC21 was a 16-kb region containing only one known gene, *Prdm9*, and a likely pseudogene *Mrps21-rs*. The pseudogene is not polymorphic between B10 and C3H. Based on the positional cloning and transgenic rescue, the PR domain-containing 9 (*Prdm9*) gene, also called *Meisetz* (*Me*iosis-*i*nduced factor containing a *SET*/PR domain and *z*inc fingers), remained as the only *Hst1* candidate.

Meiotic anomalies in sterile F_1 hybrids and *Prdm9* null mice

Positional cloning identified the *Prdm9* gene as *Hst1* but did not provide unequivocal evidence for polymorphic nucleotide(s) that would change the sterile allele into fertile (*Hst1*^{s}–*Hst1*^{f}) and *vice versa*. The number of zinc fingers was the best guess as it differs between *Hst1/Prdm9* alleles; however, the relation may not be straightforward. Thus, comparison of meiotic phenotypes of mice with a *Prdm9* null mutation (Hayashi *et al.*, 2005) with meiotic abnormalities seen in sterile F_1 hybrids was informative. In accordance with *Prdm9* germ-cell-specific expression, the null mutation did not cause any obvious abnormality in somatic tissues but resulted in meiotic arrest in male and female gametogenesis. In males, incomplete pairing of homologous chromosomes and failure of sex body formation was observed in pachytene spermatocytes (Hayashi *et al.*, 2005). As described above and summarized in Table 19.3, the meiotic phenotypes of (PWD × B6)F_1 sterile male hybrids closely followed what was observed in *Prdm9*$^{-/-}$ males.

The product of *Hst1/Prdm9/Meisetz* exerts histone H3 lysine 4 trimethylation (H3K4me3)-activity-dependent transactivation (Hayashi *et al.*, 2005). One of the genes directly transactivated, *Morc2b/4932411A10Rik*, is specifically down-regulated not only in *Prdm9*$^{-/-}$ testis (Hayashi *et al.*, 2005), but also in sterile (PWD × B6)F_1 males. Moreover, the same gene is upregulated in a *Prdm9*-copy-dependent manner in transgenic fertile BAC-rescued hybrids (Mihola *et al.*, 2009). The analysis of the promoter of *Morc2b* by chromatin immunoprecipitation (ChIP) indicated a lower level of H3K4me3 in both *Prdm9* knock-out and sterile (both adult and prepubertal) hybrids when compared to fertile testis. Moreover, a ChIP-qPCR experiment using an antibody against PRDM9 supports the view that PRDM9 directly binds the promoter of *Morc2b*

Table 19.3 Hybrid sterility phenotypes, zinc finger alleles, and haplotypes in mouse strains.

Strain	Fertility of hybrids (allele)	No. CTZFs	Haplotype
C57BL/6J*, C57BL/10SnPh*, FVB/NCrl*	Sterile (*Hst1^s^*)	12	B
129S2/SvPas*, 129/S4SvJae*	Sterile (*Hst1^s^*)	12	A
BALB/c*, BALB/cCrSlc*, BALB/cAJcl*	Sterile (*Hst1^s^*)	12	A
AKR/J*, DBA1/J*, DBA/2J*	Sterile	12	A
A/Ph*, A/J*	Sterile (*Hst1^s^*)	12	A
C3H/DiSnPh*, C3H/HeN*, C3H/HeJ*	Fertile (*Hst1^f^*)	13	C
P/J*, B10.P**	Fertile (*Hst1^f^*)	13	C
CBA/CaHN*, CBA/J*, BTBR-*Ttf/+tf*	Fertile	13	C
F/St, B10.F(13R)**	Fertile (*Hst1^f^*)	11	F
t^{p4}, t^{w12}, t^{0}	Fertile (*Hst1^f^*)	11	t
I/St	Fertile	11	I
B10.KPA132**, B10.STC77**	Fertile (*Hst1^f^*)	14	W1, W2
B10.CHR51**	Fertile (*Hst1^f^*)	14	O
B10.CAA2**	Sterile	12	W3

Note: *classical inbred strain.
**MHC-congenic strain; CTZFs: C-terminal C_2H_2 zinc fingers.

(Z. Trachtulec, unpublished data). Normal fertility of hybrid females (five offspring per female per month) in contrast to sterile *Prdm9*$^{-/-}$ females was the only difference in phenotypes of *Prdm9*$^{-/-}$ and (PWD × B6)F_1 mice. It has to be stressed that contrary to the sterility caused by the *Prdm9* null mutation, hybrid sterility is not a single-gene fertility defect under the sole control of *Prdm9*. Hybrid male sterility of *Mmm* × *Mmd* crosses results from genetic incompatibility of a particular combination of alleles of several genes on autosomes and the X chromosome, including *Hst1/Prdm9*.

Does the number of *Prdm9* zinc fingers determine *Hst1* alleles?

In the mouse reference B6 genome, *Prdm9* encodes a KRAB domain, an SSXRD motif, a PR/SET domain, and 1 internal and 12 C-terminal C_2H_2 zinc finger repeats (Fig. 19.3). The terminal zinc fingers are encoded by a single exon and are highly similar to each other even at the DNA level, forming a minisatellite-like array with an 84-bp repeat unit. The C3H allele encompasses one additional zinc finger (Mihola *et al.*, 2009) of the same sequence as both of its neighbours. This variation in copy number of zinc fingers could affect binding of

```
MNTNKLEENSPEEDTGKFEWKPKVKDEFKDISIYFSKEEWAEMGEWEKI
RYRNVKRNYKMLISIGLRAPRPAFMCYQRQAMKPQINDSEDSDEEWTPK
QQVSPPWVPFRVKHSKQQKESSRMPFSGESNVKEGSGIENLLNTSGSEH
VQKPVSSLEEGNTSGQHSGKKLKLRKKNVEVKMYRLRERKGLAYEEVSE
PQDDDYLYCEKCQNFFIDSCPNHGPPLFVKDSMVDRGHPNHSVLSLPPG
LRISPSGIPEAGLGVWNEASDLPVGLHFGPYEGQITEDEEAANSGYSWL
ITKGRNCYEYVDGQDESQAHWMRYVNCARDDEEQNLVAFQYHRKIFYRT
CRVIRPGCELLVWYGDEYGQELGIKWGSKMKKGFTAGRELRTEIHP
CLLCSLAFSSQKFLTQHMEWNHRTEIFPGTSARINPKPGDPCSDQL
QEQHVDSQNKNDKASNEVKRKSKPRQRISTTFPSTLKEQMRSEESK
RTVEELRTGQTTNTEDTVKSFIASEISSI
ERQCGQYFSDKSNVNEHQKTHTGEKPYV
CRECGRGFTQNSHLIQHQRTHTGEKPYV
CRECGRGFTQKSDLIKHQRTHTGEKPYV
CRECGRGFTQKSDLIKHQRTHTGEKPYV
CRECGRGFTQKSVLIKHQRTHTGEKPYV
CRECGRGFTQKSVLIKHQRTHTGEKPYV
CRECGRGFTAKSVLIQHQRTHTGEKPYV
CRECGRGFTAKSNLIQHQRTHTGEKPYV
CRECGRGFTAKSVLIQHQRTHTGEKPYV
CRECGRGFTAKSVLIQHQRTHTGEKPYV ←
CRECGRGFTAKSVLIQHQRTHTGEKPYV
CRECGRGFTQKSNLIKHQRTHTGEKPYV
CRECGWGFTQKSDLIQHQRTHTREK
          ▲  ▲  ▲
```

Figure 19.3 The protein sequence of the C3H allele of PRDM9 (constitutive form). The domains are marked by background colours: KRAB, blue; SSXRD, grey; PR/SET, green; C_2H_2 zinc finger, yellow. The arrow indicates one C_2H_2 domain that is inserted in this allele (*Hst1*^{f}) in comparison to the *Hst1*^{s} allele. The red arrowheads point to the coloured residues of zinc fingers that affect DNA binding (see text). See the plate section for a colour version of this figure.

PRDM9 to DNA, RNA, or protein (Brayer and Segal, 2008), but its relevance to hybrid sterility remains to be tested. The number of the zinc fingers corresponds to the *Hst1* allele in classical (Castle's) laboratory strains tested (Table 19.2) and agrees with the haplotype map of this region in Castle's strains (Trachtulec *et al.*, 2008). The data suggest that the copy number of zinc fingers has not changed since the establishment of these strains 70 years ago. The pre-classical strains I/St, F/St, and the *t*-haplotypes t^{P4}, t^{w12}, and t^{0} produce fertile hybrids with wild *Mmm* or PWD strains, but their *Prdm9* carries only 11 zinc fingers, and thus differs from both B6 (sterile) and C3H (fertile) *Prdm9* alleles. Wild *Mmd*-derived

MHC-congenic strains B10.STC77, B10.CHR51, and B10.KPA132 also produce fertile hybrids with PWD, but bear 15 zinc fingers. Therefore, the number of C_2H_2 zinc fingers of PRDM9 can differ in different mouse populations and its relation to *Hst1* alleles may not be straightforward. Because of the multigenic control of hybrid sterility, a strain producing fertile hybrids with PWD could carry a true sterility allele of *Hst1/Prdm9*, provided that one or more fertility alleles of interacting hybrid sterility loci would disrupt the DM hybrid sterility gene network. The opposite situation would be more difficult to explain – namely an *Mmd* strain producing sterile hybrids with PWD but carrying a *Prdm9* allele indiscernible from the fertility allele of C3H origin. Clearly, an allelic swap of the zinc fingers is needed to see whether it would be sufficient to change hybrid sterility to fertility or vice versa.

Hst1/Prdm9 controls recombination hotspots in mice and humans

Meiotic crossovers and conversions do not occur along chromosomes randomly, but cluster at recombination hotspots. It has been shown that the chromatin of mouse recombination hotspots is marked by trimethylation on histone H3K4 at the leptotene stage – preceding meiotic recombination repair (Buard *et al.*, 2009) – and genetic mapping of a gene acting in *trans* on recombination hotspots has revealed a locus in the proximal part of mouse Chr 17 (Grey *et al.*, 2009; Parvanov *et al.*, 2009, 2010; Baudat *et al.*, 2010). The candidate region contained *Prdm9*, a known gene encoding H3K4 trimethyltransferase that was polymorphic between two mouse strains differing in the *trans*-activation of recombination hotspots. Moreover, the human PRDM9 protein binds to a consensus recombination hotspot sequence *in silico* and *in vitro* (Baudat *et al.*, 2010; Myers *et al.*, 2010). The role of *Prdm9* in the regulation of meiotic recombination hotspots was recently reviewed (Cheung *et al.*, 2010; Hochwagen and Marais, 2010; Neale, 2010; Paigen and Petkov, 2010; Sandovici and Sapienza, 2010); hence here we discuss only the aspects related to hybrid sterility. The level of H3K4me3 is reduced but not absent in *Prdm9*$^{-/-}$ spermatocytes in comparison to wild type (Hayashi *et al.*, 2005), suggesting that other H3K4 methyltransferases may play a role in the meiotic prophase. Gametogenesis of *Prdm9*$^{-/-}$ animals is arrested at the pachytene stage of primary spermatocytes, which display defects in chromosomal pairing and perhaps also in recombination-related repair (Bhattacharyya *et al.*, in prep.). A possible reason for these defects is the altered expression of a target gene(s) of PRDM9 (Hayashi and Matsui, 2006). The wild-type and *Prdm9*$^{-/-}$ testis appear to be similar in the initiation stage of recombination as measured by γH2AX, Rad51, and ATR staining, but the differences could be below the detection limit of immunofluorescence assay and the residual H3K4me3 levels may be sufficient to initiate recombination. Clearly, a more sensitive assay of meiotic recombination in *Prdm9*$^{-/-}$ animals and intersubspecific F_1 hybrids is needed. It would be helpful, though technically challenging, to define *in vivo* binding sites of PRDM9.

Fast evolution of PRDM9 zinc fingers: is *Prdm9* a universal speciation gene?

The C-terminal zinc finger domain of PRDM9 is the fastest evolving part of the molecule in a variety of species, with signs of positive selection restricted to the same three amino acid positions, −1, 3, and 6, numbered relative to the start of the zinc finger alpha helix. These are the predicted sites that provide the DNA-binding specificity of zinc fingers (Oliver *et al.*, 2009; Thomas *et al.*, 2009; Berg *et al.*, 2010). A formal parallel to rapid evolution of the PRDM9 DNA-binding domain can be found in *Drosophila* hybrid sterility gene *OdsH*. In this case the positive selection and rapid evolution of the *OdsH* homeodomain that interacts with the satellite DNA in centromeric heterochromatin of *Drosophila* chromosomes could reflect the known rapid evolution of satellite repeats. In support of this idea, decondensed centromeric heterochromatin and altered binding of OdsH were observed in sterile hybrids (Bayes and Malik, 2009; Oliver *et al.*, 2009). Thomas and co-workers (Thomas *et al.*, 2009) suggested an alternative scenario for *Prdm9* function in hybrid sterility that would even better mimic the *OdsH* function in *Drosophila* hybrids. They noticed that transition from di- to tri-methyl histone H3K4 not only marks transcriptional activation, but also distinguishes H3 nucleosomes of centromeric and pericentromeric heterochromatin. The authors propose that PRDM9 could directly bind rapidly evolving tandem repeats of satellite DNA and intersubspecific incompatibility of satellite DNA and PRDM9 could interfere with meiosis in male hybrids, as is the case of OdsH in *Drosophila*. However, experimental evidence does not support the idea, because anti-PRDM9 antibody does not label centromeric heterochromatin during meiosis (O. Mihola *et al.*, unpublished data) and the *Hst1*/*Prdm9* interacting hybrid sterility loci do not map to the centromeric regions (M. Dzúr-Gejdošová *et al.*, unpublished data). Moreover, an additional mechanism would be necessary to explain the fertility of female hybrids. Thus, most compatible with the *Prdm9* role in hybrid sterility could be its function in meiotic recombination. The fast evolution of C-terminal zinc fingers of PRDM9 may represent an adaptive response to the meiotic hotspot paradox (Coop and Myers, 2007, and references therein), which refers to disappearance of hotspots as a result of the repair of double-strand breaks by gene conversion-like events. Hybrid sterility could be a consequence of such fast evolution resulting in inter(sub)specific meiotic incompatibility.

Large X-effect in the mouse

Studies in *Drosophila* and many other species point to a major role of the X chromosome in genetic control of hybrid male sterility, the large X-effect (Dobzhansky, 1951; Coyne and Orr, 2004; Presgraves, 2010). The rule holds true also for *Mmm* and *Mmd* species as follows from the inspection of the fertility parameters in B6.PWD-Chr♯ chromosome substitution strains (figure 2 in Gregorová *et al.*, 2008). While none of the introgressed PWD autosomes caused hybrid male sterility in the B6 genetic background, the PWD Chr X caused abnormal spermiogenesis and male infertility (Storchová *et al.*, 2004; Bhattacharyya *et al.*, in prep.). Homosomic introgressions of individual autosomes (PWD/PWD) decreased the average litter size but did not result in infertile crosses, with the exception of Chr 10 and Chr 11. The incross of B6.PWD-Chr10 homosomic males and females rarely produced offspring. However, the B6.PWD-Chr10 homosomic males were fully fertile when outcrossed to females of unrelated inbred strains. Also, B6.PWD-Chr11 homosomic incrosses were most often infertile. Following the classification of DM incompatibilities as suggested by Turelli and Orr (2000), H_0 (dominant/dominant) and H_1 (dominant/recessive) incompatibility loci were absent in PWD autosomes, and two recessive (H_2) interactions affected viability but not spermatogenesis.

QTL mapping of genetic factors responsible for spermatogenic impairment after introgression of *Mmm* Chr X into the *Mmd* background (Storchová *et al.*, 2004; Good *et al.*, 2008b; J. Piálek *et al.*, unpublished data) indicated involvement of large portions of Chr X and suggested multiple hybrid sterility loci. Forejt and co-workers (Storchová *et al.*, 2004; R. Storchová *et al.*, unpublished data) mapped one of these genes as Hybrid sterility-X1 (*Hstx1*) locus to a 6-Mb interval between *DXMit119* and *DXMit60* markers (70–76 Mb) in B6.PWD-ChrX consomic males. An overlapping interval harbours the *Hstx2* locus that is part of the epistatic network responsible for the early meiotic arrest in (B6.PWD-ChrX × PWD)F_1 sterile hybrid males (Bhattacharyya *et al.*, in prep.). The *Hstx1* and *Hstx2* labels may refer to the same gene although they describe different phenotypes.

Infertility as observed in B6.PWD-ChrX was described in consomic males B6.MSM-ChrX carrying Chr X of the Japanese wild-derived strain MSM/Ms. The Japanese wild mouse, *M. m. molossinus*, arose by hybridization of *Mmm* and *M. m. castaneus*, yet its nuclear genome is predominantly of *Mmm* origin (see Yonekawa *et al.*, Chapter 4). The F_1 hybrids between MSM/Ms and C57BL/6 are fertile in both sexes, but consomic males B6.MSM-ChrX are sterile with incomplete intrameiotic arrest at pachynema and abnormal, dysfunctional sperm (Oka *et al.*, 2004, 2007, 2010; see also Oka and Shiroishi, Chapter 17). Introduction of MSM/Ms Chr 11 and Chr 1 into the Chr X^{MSM} strain partially rescued infertility, indicating the presence of epistatic suppressors of hybrid sterility. Good *et al.* (2008a) repeated and extended the Chr X introgression studies by using PWK and LEWES wild-derived strains as models for *Mmm* and *Mmd* subspecies, respectively, and constructed reciprocal consomics: Chr X^{PWK} on LEWES background and Chr X^{LEWES} on PWK background. While *Mmd* X chromosome did not interfere with male fertility on *Mmm* background, the PWK Chr X on LEWES background caused arrest of spermatogenesis. Four broad QTL for sperm head morphology may overlap with some of those described previously (Oka *et al.*, 2004; Storchová *et al.*, 2004), but their resolution was too low for a more detailed comparison.

Mus macedonicus is a relative of the *M. musculus* complex, having evolved separately for more than 1.5 Myr (Guénet, 2005). Laboratory hybrids between B6 females and *M. macedonicus* are viable and display male-limited hybrid sterility. In a backcross of fertile F_1 hybrid females to B6 males, about three-quarters of male progeny were infertile. QTL analysis of testes weight revealed one strong hybrid sterility locus on Chr 17, near the *D17Mit198* marker and another one near the *Hprt* gene on Chr X (Elliott *et al.*, 2004). In both cases the hybrid sterility loci seemed to map to different sites than in the case of *Hst1* and *Hstx1*/*Hstx2* genes identified in *Mmm* × *Mmd* hybrids (see above). The importance of X-linked factors was also confirmed in crosses between laboratory *Mmd* strains and inbred strains derived from *M. spretus* (Guénet *et al.*, 1990; Matsuda *et al.*, 1991; Elliott *et al.*, 2001, 2004).

The enigma of reciprocal crosses: Darwin's corollary to Haldane's rule

Hybrid sterility in both *Drosophila* and mouse can be asymmetrical so that male fertility differs in reciprocal crosses. Good *et al.* (2008b) compared reproductive parameters of reciprocal crosses between two *Mmm*-derived inbreds (PWK/PhJ and CZECHII/EiJ) and two *Mmd*-derived strains (LEWES/EiJ and WSB/EiJ). PWK produced sterile male hybrids when crossed to *Mmd*-derived strains only as a female parent. Such asymmetry in fertility effects was not found in reciprocal crosses of CZECHII with *Mmd*-derived strains. The asymmetry of PWD or PWK reciprocal crosses with classical *Mmd* laboratory strains was documented in our laboratory in 1987, soon after the construction of these inbred strains (Gregorová and Forejt, 2000, and unpublished). The finding was unexpected because construction of PWD and PWK inbred strains was initiated using *Mmm* wild males for progeny testing with B10 females; only those *Mmm* males that produced 100% of sterile male offspring were used as founders for inbreeding. Because it was very difficult to test wild females in the progeny test, the female partners for the preparation of PWK and PWD inbred strains were randomly selected from the same locality where the progeny-tested male was trapped. In the initial inbreeding crosses, the males were tested for sterility of their male progeny with B10 females. Thus the change to asymmetry of the reciprocal crosses occurred during the inbreeding process and was fixed already before 1987.

The asymmetry of reciprocal crosses could be due to sex chromosome–autosome incompatibilities, X–Y incompatibility, genomic imprinting, or to incompatibilities in mitochondrial–nuclear interactions. Using the chromosome substitution strains (Gregorová *et al.*, 2008) we could exclude genomic imprinting, X–Y incompatibility, and mitochondrial genome (S. Gregorová *et al.*, unpublished data) as a cause of asymmetry of (PWD × B6) and (B6 × PWD) crosses, leaving sex chromosome–autosome incompatibilities as the preferred candidate.

Conclusions and future trends

Hybrid sterility is the best-studied postzygotic reproductive isolation mechanism between closely related taxa. The analysis of hybrid sterility in mouse *Mmm* × *Mmd* crosses added some new information, previously not known from *Drosophila* studies; nevertheless, three main questions concerning the genetics of hybrid sterility still remain unanswered:

(1) How many genes cause hybrid sterility and where are they located? Mouse studies confirm the preferential involvement of the X chromosome in hybrid sterility, the large X-effect. Analysis of subspecific backcrosses indicates three or more genes involved. Thus it seems that polygenic control of hybrid sterility, known from some *Drosophila* studies, is limited to distantly related species, where most of the sterility genes evolved after reproductive isolation had been established. However, a final conclusion will require combining the knowledge from genetic analyses of laboratory crosses with field studies in hybrid zones (Macholán *et al.*, 2007, 2011; Teeter *et al.*, 2008, 2010).

(2) Which genes are involved in hybrid sterility? It is still too early to make conclusions since so few hybrid sterility genes have been identified at the sequence level. Nevertheless, fast evolution by positive Darwinian selection and DNA–protein interaction are phenomena that may be useful for spotting new hybrid sterility gene candidates.

(3) Which molecular mechanisms lead to hybrid sterility? It is generally accepted that reproductive isolation between species is a consequence of independent evolution of their genomes. This could mean that each evolving pair of species

could randomly acquire a unique set of genetic incompatibilities. However, this intuitive conclusion is in conflict with many known features of hybrid sterility, such as preferential involvement of the X chromosome, Haldane's rule, or the asymmetry of reciprocal crosses. Recent studies (Good *et al.*, 2010; Lu *et al.*, 2010) support our previous hypothesis (Forejt, 1982, 1984, 1996) proposing meiotic sex chromosome inactivation (MSCI) (Turner, 2007) as a possible control mechanism linking different forms of chromosomal and genic hybrid sterility. The MSCI can be part of the pachytene checkpoint that ensures proper meiotic double-strand break repair and chromosome pairing in primary spermatocytes (Handel and Schimenti, 2010).

REFERENCES

Baudat, F., Buard, J., Grey, C., *et al.* (2010). PRDM9 is a major determinant of meiotic recombination hotspots in humans and mice. *Science*, **327**, 836–40.

Bauer, H., Véron, N., Willert, J., and Herrmann, B. G. (2007). The *t*-complex-encoded guanine nucleotide exchange factor *Fgd2* reveals that two opposing signaling pathways promote transmission ratio distortion in the mouse. *Genes & Development*, **21**, 143–7.

Bayes, J. J. and Malik, H. S. (2009). Altered heterochromatin binding by a hybrid sterility protein in *Drosophila* sibling species. *Science*, **326**, 1538–41.

Berg, I. L., Neumann, R., Lam, K. W., *et al.* (2010). PRDM9 variation strongly influences recombination hot-spot activity and meiotic instability in humans. *Nature Genetics*, **42**, 859–63.

Bonhomme, F., Martin, S., and Thaler, L. (1978). Hybridization between *Mus musculus* L. and *Mus spretus* Lataste under laboratory conditions. *Experientia*, **34**, 1140–1.

Boursot, P., Din, W., Anand, R., *et al.* (1996). Origin and radiation of the house mouse: mitochondrial DNA phylogeny. *Journal of Evolutionary Biology*, **9**, 391–415.

Brayer, K. J. and Segal, D. J. (2008). Keep your fingers off my DNA: protein–protein interactions mediated by C_2H_2 zinc finger domains. *Cell Biochemistry and Biophysics*, **50**, 111–31.

Brideau, N. J., Flores, H. A., Wang, J., *et al.* (2006). Two Dobzhansky–Muller genes interact to cause hybrid lethality in *Drosophila*. *Science*, **314**, 1292–5.

Buard, J., Barthes, P., Grey, C., and de Massy, B. (2009). Distinct histone modifications define initiation and repair of meiotic recombination in the mouse. *EMBO Journal*, **28**, 2616–24.

Cheung, V. G., Sherman, S. L., and Feingold, E. (2010). Genetic control of hotspots. *Science*, **327**, 791–2.

Coop, G. and Myers, S. R. (2007). Live hot, die young: transmission distortion in recombination hotspots. *PLoS Genetics*, **3**, e35.

Coyne, J. A. and Orr, H. A. (2004). *Speciation*. Sunderland, MA: Sinauer Associates.

Darwin, C. (1859). *On the Origin of Species by Means of Natural Selection*. London: John Murray.

Dobzhansky, T. (1951). *Genetics and the Origin of Species*. New York: Columbia University Press.

Elliott, R. W., Miller, D. R., Pearsall, R. S., *et al.* (2001). Genetic analysis of testis weight and fertility in an interspecies hybrid congenic strain for Chromosome X. *Mammalian Genome*, **12**, 45–51.

Elliott, R. W., Poslinski, D., Tabaczynski, D., Hohman, C., and Pazik, J. (2004). Loci affecting male fertility in hybrids between *Mus macedonicus* and C57BL/6. *Mammalian Genome*, **15**, 704–10.

Forejt, J. (1981). Hybrid sterility gene located in the *T/t – H-2* supergene on chromosome 17. In *Current Trends in Histocompatibility*, ed. R. A. Reisfeld and S. Ferrone. New York: Plenum Press, pp. 103–31.

Forejt, J. (1982). X–Y involvement in male sterility caused by autosome translocations: a hypothesis. In *Genetic Control of Gamete Production and Function*, ed. M. Fraccaro and B. Rubin. New York: Academic Press, pp. 135–51.

Forejt, J. (1984). X-inactivation and its role in male sterility. In *Chromosomes Today*, ed. M. Bennett, A. Gropp, and U. Wolf. London: George Allen and Unwin, pp. 117–27.

Forejt, J. (1985). Chromosomal and genic sterility of hybrid type in mice and men. *Experimental and Clinical Immunogenetics*, **2**, 106–19.

Forejt, J. (1996). Hybrid sterility in the mouse. *Trends in Genetics*, **12**, 412–17.

Forejt, J., Gregorová, S., and Goetz, P. (1981). XY pair associates with the synaptonemal complex of autosomal male-sterile translocations in pachytene spermatocytes of the mouse (*Mus musculus*). *Chromosoma*, **82**, 41–53.

Forejt, J. and Iványi, P. (1974). Genetic studies on male sterility of hybrids between laboratory and wild mice (*Mus musculus* L.). *Genetical Research*, **24**, 189–206.

Forejt, J., Vincek, V., Klein, J., Lehrach, H., and Loudová-Micková, M. (1991). Genetic mapping of the *t*-complex region on mouse chromosome 17 including the *Hybrid sterility-1* gene. *Mammalian Genome*, **1**, 84–91.

Fossella, J., Samant, S., Silver, L., *et al.* (2000). An axonemal dynein at the *Hybrid Sterility 6* locus: implications for *t* haplotype-specific male sterility and the evolution of species barriers. *Mammalian Genome*, **11**, 8–15.

Geraldes, A., Basset, P., Gibson, B., *et al.* (2008). Inferring the history of speciation in house mice from autosomal, X-linked, Y-linked and mitochondrial genes. *Molecular Ecology*, **17**, 5349–63.

Good, J. M., Dean, M. D., and Nachman, M. W. (2008a). A complex genetic basis to X-linked hybrid male sterility between two species of house mice. *Genetics*, **179**, 2213–28.

Good, J. M., Giger, T., Dean, M. D., and Nachman, M. W. (2010). Widespread over-expression of the X chromosome in sterile F_1 hybrid mice. *PLoS Genetics*, **6**, e1001148.

Good, J. M., Handel, M. A., and Nachman, M. W. (2008b). Asymmetry and polymorphism of hybrid male sterility during the early stages of speciation in house mice. *Evolution*, **62**, 50–65.

Gregorová, S., Divina, P., Storchová, R., *et al.* (2008). Mouse consomic strains: exploiting genetic divergence between *Mus m. musculus* and *Mus m. domesticus* subspecies. *Genome Research*, **18**, 509–15.

Gregorová, S. and Forejt, J. (2000). PWD/Ph and PWK/Ph inbred mouse strains of *Mus m. musculus* subspecies: a valuable resource of phenotypic variations and genomic polymorphisms. *Folia Biologica (Praha)*, **46**, 31–41.

Gregorová, S., Mnuková-Fajdelová, M., Trachtulec, Z., *et al.* (1996). Sub-milliMorgan map of the proximal part of mouse Chromosome 17 including the hybrid sterility 1 gene. *Mammalian Genome*, **7**, 107–13.

Grey, C., Baudat, F., and de Massy, B. (2009). Genome-wide control of the distribution of meiotic recombination. *PLoS Biology*, **7**, e1000035.

Guénet, J. L. (2005). The mouse genome. *Genome Research*, **15**, 1729–40.

Guénet, J. L, Nagamine, C., Simon-Chazottes, D., Montagutelli, X., and Bonhomme, F. (1990). *Hst-3*: an X-linked hybrid sterility gene. *Genetical Research*, **56**: 163–5.

Haldane, J. (1922). Sex ratio and unisexual sterility in animal hybrids. *Journal of Genetics*, **12**, 101–9.

Handel, M. A. and Schimenti, J. C. (2010). Genetics of mammalian meiosis: regulation, dynamics and impact on fertility. *Nature Reviews Genetics*, **11**, 124–36.

Hayashi, K. and Matsui, Y. (2006). Meiosis-specific histone methylation is essential for meiotic progression. *Tanpakushitsu Kakusan Koso*, **51**, 441–5.

Hayashi, K., Yoshida, K., and Matsui, Y. (2005). A histone H3 methyltransferase controls epigenetic events required for meiotic prophase. *Nature*, **438**, 374–8.

Herrmann, B. G., Koschorz, B., Wertz, K., McLaughlin, K. J., and Kispert, A. (1999). A protein kinase encoded by the *t* complex responder gene causes non-Mendelian inheritance. *Nature*, **402**, 141–6.

Hochwagen, A. and Marais, G. A. B. (2010). Meiosis: a PRDM9 guide to the hotspots of recombination. *Current Biology*, **20**, R271–4.

Howell, G. R., Munroe, R. J., and Schimenti, J. C. (2005). Transgenic rescue of the mouse *t* complex haplolethal locus Thl1. *Mammalian Genome*, **16**, 838–46.

Iványi, P., Vojtísková, M., Démant, P., and Micková, M. (1969). Genetic factors in the ninth linkage group influencing reproductive performance in male mice. *Folia Biologica (Praha)*, **15**, 401–21.

Lu, X., Shapiro, J. A., Ting, C.-T., *et al.* (2010). Genome-wide misexpression of X-linked versus autosomal genes associated with hybrid male sterility. *Genome Research*, **20**, 1097–102.

Lyon, M. F. (2003). Transmission ratio distortion in mice. *Annual Review of Genetics*, **37**, 393–408.

Macholán, M., Baird, S. J. E., Dufková, P., *et al.* (2011). Assessing multilocus introgression patterns: a case study on the mouse X chromosome in Central Europe. *Evolution*, **65**, 1428–46.

Macholán, M., Munclinger, P., Šugerková. M., *et al.* (2007). Genetic analysis of autosomal and X-linked markers across a mouse hybrid zone. *Evolution*, **61**, 746–71.

Maheshwari, S., Wang, J., and Barbash, D. A. (2008). Recurrent positive selection of the *Drosophila* hybrid incompatibility gene *Hmr*. *Molecular Biology and Evolution*, **25**, 2421–30.

Masly, J. P., Jones, C. D., Noor, M. A., Locke, J., and Orr, H. A. (2006). Gene transposition as a cause of hybrid sterility in *Drosophila*. *Science*, **313**, 1448–50.

Matsuda, Y., Hirobe, T., and Chapman, V. M. (1991). Genetic basis of X–Y chromosome dissociation and male sterility in interspecific hybrids. *Proceedings of the National Academy of Sciences of the United States of America*, **88**, 4850–4.

Mihola, O., Forejt, J., and Trachtulec, Z. (2007). Conserved alternative and antisense transcripts at the programmed cell death 2 locus. *BMC Genomics*, **8**, 20.

Mihola, O., Trachtulec, Z., Vlcek, C., Schimenti, J. C., and Forejt, J. (2009). A mouse speciation gene encodes a meiotic histone H3 methyltransferase. *Science*, **323**, 373–5.

Muller, H. J. (1942). Isolating mechanisms, evolution, and temperature. *Biological Symposia*, **6**, 71–125.

Myers, S., Bowden, R., Tumian, A., *et al.* (2010). Drive against hotspot motifs in primates implicates the *PRDM9* gene in meiotic recombination. *Science*, **327**, 876–9.

Neale, M. J. (2010). PRDM9 points the zinc finger at meiotic recombination hotspots. *Genome Biology*, **11**, 104.

Oka, A., Aoto, T., Totsuka, Y., *et al.* (2007). Disruption of genetic interaction between two autosomal regions and the X chromosome causes reproductive isolation between mouse strains derived from different subspecies. *Genetics*, **175**, 185–97.

Oka, A., Mita, A., Sakurai-Yamatani, N., *et al.* (2004). Hybrid breakdown caused by substitution of the X chromosome between two mouse subspecies. *Genetics*, **166**, 913–24.

Oka, A., Mita, A., Takada, Y., Koseki, H., and Shiroishi, T. (2010). Reproductive isolation in hybrid mice due to spermatogenesis defects at three meiotic stages. *Genetics*, **186**, 339–51.

Oliver, P. L., Goodstadt, L., Bayes, J. J., *et al.* (2009). Accelerated evolution of the *Prdm9* speciation gene across diverse metazoan taxa. *PLoS Genetics*, **5**, e1000753.

Paigen, K. and Petkov, P. (2010). Mammalian recombination hot spots: properties, control and evolution. *Nature Reviews Genetics*, **11**, 221–33.

Parvanov, E. D., Ng, S. H. S., Petkov, P. M., and Paigen, K. (2009). *Trans*-regulation of mouse meiotic recombination hotspots by *Rcr1*. *PLoS Biology*, **7**, e1000036.

Parvanov, E. D., Petkov, P. M., and Paigen, K. (2010). *Prdm9* controls activation of mammalian recombination hotspots. *Science*, **327**, 835.

Phadnis, N. and Orr, H. A. (2009). A single gene causes both male sterility and segregation distortion in *Drosophila* hybrids. *Science*, **323**, 376–9.

Piálek, J., Vyskočilová, M., Bímová, B., *et al.* (2008). Development of unique house mouse resources suitable for evolutionary studies of speciation. *Journal of Heredity*, **99**, 34–44.

Pilder, S. H., Hammer, M. F., and Silver, L. M. (1991). A novel mouse chromosome 17 hybrid sterility locus: implications for the origin of t haplotypes. *Genetics*, **129**, 237–46.

Pilder, S. H., Olds-Clarke, P., Orth, J. M., Jester, W. F., and Dugan, L. (1997). *Hst7*: a male sterility mutation perturbing sperm motility, flagellar assembly, and mitochondrial sheath differentiation. *Journal of Andrology*, **18**, 663–71.

Pilder, S. H., Olds-Clarke, P., Phillips, D. M., and Silver, L. M. (1993). *Hybrid sterility-6*: a mouse *t* complex locus controlling sperm flagellar assembly and movement. *Developmental Biology*, **159**, 631–42.

Presgraves, D. C. (2010). Darwin and the origin of interspecific genetic incompatibilities. *American Naturalist*, **176**(S1), S45–60.

Reed, L. K., LaFlamme, B. A., and Markow, T. A. (2008). Genetic architecture of hybrid male sterility in *Drosophila*: analysis of intraspecies variation for interspecies isolation. *PLoS ONE*, **3**, e3076.

Reed, L. K. and Markow, T. A. (2004). Early events in speciation: polymorphism for hybrid male sterility in *Drosophila*. *Proceedings of the National Academy of Sciences of the United States of America*, **101**, 9009–12.

Sandovici, I. and Sapienza, C. (2010). PRDM9 sticks its zinc fingers into recombination hotspots and between species. *F1000 Biology Reports*, **2**, 37.

She, J. X., Bonhomme, F., Boursot, P., Thaler, L., and Catzeflis, F. (1990). Molecular phylogenies in the genus *Mus*: comparative analysis of electrophoretic, scnDNA hybridization, and mtDNA RFLP data. *Biological Journal of the Linnean Society*, **41**: 83–103.

Storchová, R., Gregorová, S., Buckiová, D., *et al.* (2004). Genetic analysis of X-linked hybrid sterility in the house mouse. *Mammalian Genome*, **15**, 515–24.

Teeter, K. C., Payseur, B. A., Harris, L. W., *et al.* (2008). Genome-wide patterns of gene flow across a house mouse hybrid zone. *Genome Research*, **18**, 67–76.

Teeter, K. C., Thibodeau, L. M., Gompert, Z., *et al.* (2010). The variable genomic architecture of isolation between hybridizing species of house mice. *Evolution*, **64**, 472–85.

Thomas, J. H., Emerson, R. O., and Shendure, J. (2009). Extraordinary molecular evolution in the PRDM9 fertility gene. *PLoS ONE*, **4**, e8505.

Ting, C.-T., Tsaur, S.-C., Wu, M.-L., and Wu, C.-I. (1998). A rapidly evolving homeobox at the site of a hybrid sterility gene. *Science*, **282**, 1501–4.

Trachtulec, Z., Mihola, O., Vlcek, C., *et al.* (2005). Positional cloning of the hybrid sterility 1 gene: fine genetic mapping and evaluation of two candidate genes. *Biological Journal of the Linnean Society*, **84**, 637–41.

Trachtulec, Z., Mnuková-Fajdelová, M., Hamvas, R., *et al.* (1997). Isolation of candidate hybrid sterility 1 genes by cDNA selection in a 1.1 megabase pair region on mouse chromosome 17. *Mammalian Genome*, **8**, 312–16.

Trachtulec, Z., Vincek, V., Hamvas, R., *et al.* (1994). Physical map of mouse chromosome 17 in the region relevant for positional cloning of the Hybrid sterility 1 gene. *Genomics*, **23**, 132–7.

Trachtulec, Z., Vlcek, C., Mihola, O., *et al.* (2008). Fine haplotype structure of a chromosome 17 region in the laboratory and wild mouse. *Genetics*, **178**, 1777–84.

Turelli, M. and Orr, H. A. (2000). Dominance, epistasis and the genetics of postzygotic isolation. *Genetics*, **154**, 1663–79.

Turner, J. M. A. (2007). Meiotic sex chromosome inactivation. *Development*, **134**, 1823–31.

Vyskočilová, M., Pražanová, G., and Piálek, J. (2009). Polymorphism in hybrid male sterility in wild-derived *Mus musculus musculus* strains on proximal chromosome 17. *Mammalian Genome*, **20**, 83–91.

Vyskočilová, M., Trachtulec, Z., Forejt, J., and Piálek, J. (2005). Does geography matter in hybrid sterility in house mice? *Biological Journal of the Linnean Society*, **64**, 663–74.

Linkage disequilibrium approaches for detecting hybrid zone movement: a study of the house mouse hybrid zone in southern Bavaria

PRISCILLA K. TUCKER, LIUYANG WANG, KEN LUZYNSKI, AND KATHERINE TEETER

Introduction

In a series of papers spanning over a decade (e.g. Hewitt, 1975; Barton, 1979; Barton and Hewitt, 1985, 1989), Barton and Hewitt explored the nature of hybrid zones that are formed when genetically distinct populations overlap and reproduce (Barton and Hewitt, 1985, 1989). They viewed most hybrid zones as 'tension zones' (Key, 1968), i.e. clines maintained by a balance between dispersal and selection against hybrids, and they proposed that tension zones can be stationary or they can move across geographic space due to demographic differences between populations or as a result of asymmetric selection (Hewitt, 1975; Barton, 1979; Barton and Hewitt, 1985). Determining whether a hybrid zone is stationary or moving has important implications for understanding the evolutionary processes affecting interactions in hybrid populations. However, in a recent review of the hybrid zone literature, Buggs (2007) discussed the difficulties in detecting hybrid zone movement, especially from patterns of gene flow. For example, although a moving tension zone will leave a signature tail of neutral clines in its wake (Arntzen and Wallis, 1991), this pattern could also be interpreted as the adaptive introgression of alleles away from a stationary front. Buggs (2007) concluded that evidence for moving hybrid zones was not well established in the literature.

The well-studied hybrid zone between genetically distinct populations of house mice, *Mus musculus musculus* and *M. m. domesticus* (also referred to in the literature as *Mus musculus* and *Mus domesticus*) (Hunt and Selander, 1973; Sage *et al.*, 1986b; Vanlerberghe *et al.*, 1986, 1988a, b; Tucker *et al.*, 1992; Fel-Clair *et al.*, 1996; Orth *et al.*, 1996; Boissinot and Boursot, 1997; Prager *et al.*, 1997; Munclinger *et al.*, 2002;

Evolution of the House Mouse, ed. Miloš Macholán, Stuart J. E. Baird, Pavel Munclinger, and Jaroslav Piálek. Published by Cambridge University Press.

Payseur *et al.*, 2004; Božíková *et al.*, 2005; Raufaste *et al.*, 2005; Macholán *et al.*, 2007, 2008, 2011; Teeter *et al.*, 2008, 2010), offers an opportunity to explore the possibility of hybrid zone movement. The hybrid zone is likely a tension zone with selection against hybrids due to genetic incompatibilities, including hybrid male sterility resulting from spermatogenic failure (Forejt, 1996; Storchová *et al.*, 2004; Britton-Davidian *et al.*, 2005; Good *et al.*, 2008a, b, 2010; Vyskočilová *et al.*, 2009), hybrid female sterility (Britton-Davidian *et al.*, 2005), hybrid inviability (Sage *et al.*, 1986a; Moulia *et al.*, 1993), and reinforcement selection affecting mate choice preferences (Smadja and Ganem, 2005; Vošlajerová Bímová *et al.*, 2011). The history of initial contact is unknown, but the hybrid zone is not older than the establishment of *M. m. domesticus* in western Europe around the first millennium BC (Cucchi *et al.*, 2005).

Multiple genetic analyses across the hybrid zone have identified asymmetric introgression from *M. m. domesticus* to *M. m. musculus* for unlinked autosomal markers and/or for mtDNA (Hunt and Selander, 1973; Vanlerberghe *et al.*, 1986, 1988a, b; Tucker *et al.*, 1992; Prager *et al.*, 1998; Raufaste *et al.*, 2005; Božíková *et al.*, 2005; Teeter *et al.*, 2008, 2010). The pattern has been variably explained as resulting from the advance of advantageous alleles from *M. m. domesticus* to *M. m. musculus* or to movement of the hybrid zone from east to west due to dispersal and/or mating asymmetries (Raufaste *et al.*, 2005; Teeter *et al.*, 2008). A recent study (Macholán *et al.*, 2011) found evidence for asymmetric introgression of X-linked markers in both directions, but with different patterns between east- and west-introgressing markers. In particular, the west-introgressing markers displayed significant linkage disequilibria, a pattern that was interpreted as being a result of westward zone movement (Macholán *et al.*, 2011).

The availability of a genome-wide high-density SNP array for the house mouse (Yang *et al.*, 2009, 2011) has provided an opportunity to more fully consider linkage disequilibrium approaches for detecting hybrid zone movement. In a recent study, Wang *et al.* (2011) developed over 1400 SNP markers spaced approximately 1.8 Mb apart to investigate genome-wide patterns of linkage disequilibrium (LD) in naturally occurring hybrid populations across two transects, one spanning a portion of the state of Bavaria in Germany and Austria (Tucker *et al.*, 1992; Payseur *et al.*, 2004; Božíková *et al.*, 2005; Teeter *et al.*, 2008, 2010) and a second, located 200 km to the north, spanning a portion of Bavaria and the Czech Republic (Munclinger *et al.*, 2002; Božíková *et al.*, 2005; Macholán *et al.*, 2007, 2008, 2011; Dufková *et al.*, 2011). LD is created when two genetically distinct populations first hybridize, and then decays over time as a result of segregation and recombination (Reich *et al.*, 2001), so a comparison of genome-wide measures of LD between adjacent diallelic SNPs across hybrid populations can, in principle, provide evidence of hybrid zone movement, assuming minimal effects on LD from other population genetic forces, such as mutation or drift. If genome-wide LD increases asymmetrically across a hybrid zone, one might infer hybrid zone movement in the direction of the higher estimates. Indeed, from simulation analyses Wang *et al.* (2011) showed that some models of hybrid zone movement result in asymmetrical patterns of LD with higher estimates in the direction of the movement.

In this chapter we review the analyses of Wang *et al.* (2011), who assessed genome-wide measures of LD, alongside the more traditional methods of cline analysis, to detect hybrid zone movement in the southern transect and we discuss the advantages as well as limitations of these analyses.

Genetic resources and methods of analysis

The transect across southern Bavaria was originally sampled by Richard Sage, who collected a total of 437 mice in a roughly linear east–west manner from 39 populations in 1984, 1985, and 1992 (Fig. 20.1). The genetic markers used in Wang *et al.* (2011) were selected from a beta version of the Mouse Diversity Array (Yang *et al.*, 2011) and are diagnostic for *M. m. musculus* and *M. m. domesticus* based on sampling European populations of the two subspecies. While the authors aimed for a density of approximately 1 SNP every 2 Mb across the genome, there are some gaps in coverage. Genotype data from 1316 autosomal SNPs and 85 X-linked SNPs were recovered (Fig. 20.2). The mean and median gap size between two SNPs for these markers is about 1.8 Mb. Of the 437 mice genotyped, SNP data were recovered for 432 samples.

These same mice were previously typed for a mitochondrial DNA marker located in the *Nd1* gene (Božíková *et al.*, 2005) and a subset of mice collected in 1984 and 1985 were previously typed for the non-recombining portion of the Y chromosome (Tucker *et al.*, 1992). Along with these data, we include here the genotypes for the non-recombining portion of the Y chromosome for mice collected in 1992, using a marker based on an 18 bp deletion in the Y-linked *Zinc finger protein 2* gene (*Zfy2*). The deletion, located in the last exon of the *Zfy2* gene, is present and fixed in *M. m. musculus*, but absent in *M. m. domesticus* (Boissinot and Boursot, 1997; Prager *et al.*, 1997, 1998). The *Zfy1* and *Zfy2* genes were co-amplified using the primers listed in these references. In *M. m. domesticus*, a single band of 202 bp is observed, and in *M. m. musculus*, two bands are observed, one at 202 bp (*Zfy1*) and a second, smaller band at 184 bp (*Zfy2*), which has the 18 bp deletion.

Wang *et al.* (2011) employed a series of population genetic and cline analyses to interrogate the data. These included: (1) calculating hybrid indices as measures of ancestry for all samples using the R package INTROGRESS (Buerkle, 2005;

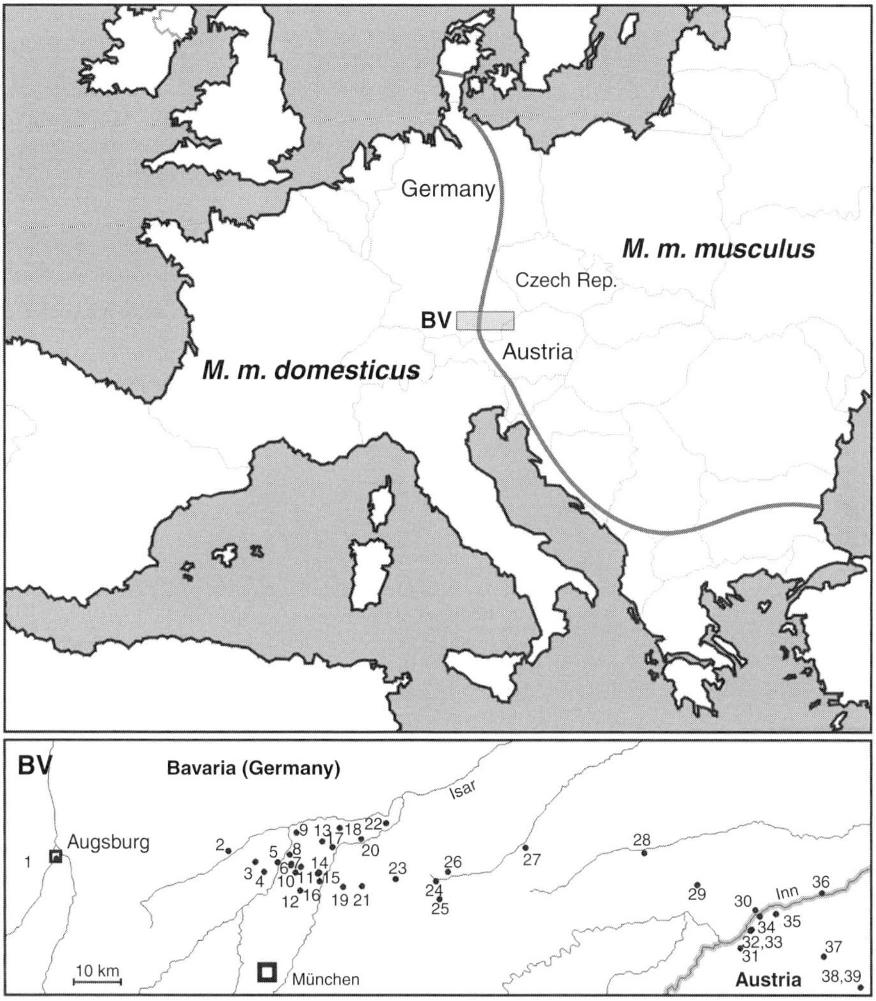

Figure 20.1 The location of the house mouse hybrid zone in Europe and collecting localities for the Bavaria (BV) transect located in southern Bavaria, Germany (adapted from Wang *et al.*, 2011). The black line demarcates the hybrid zone and the shaded rectangle indicates the location of the transect (top map). Collecting localities for the BV transect are shown in the bottom map. Numbers refer to localities listed in Table 20.2.

Gompert and Buerkle, 2010); (2) estimating geographic clines (Barton and Hewitt, 1985; Barton and Bengtsson, 1986; Barton and Gale, 1993) using the software ClineFit (Porter *et al.*, 1997); (3) determining whether populations in close geographic proximity were genetically differentiated (Hudson *et al.*, 1992), using the program DnaSP v.5.10 (Librado and Rozas, 2009); and (4) estimating r^2, the squared correlation of allelic indicators at two loci in a haploid gamete

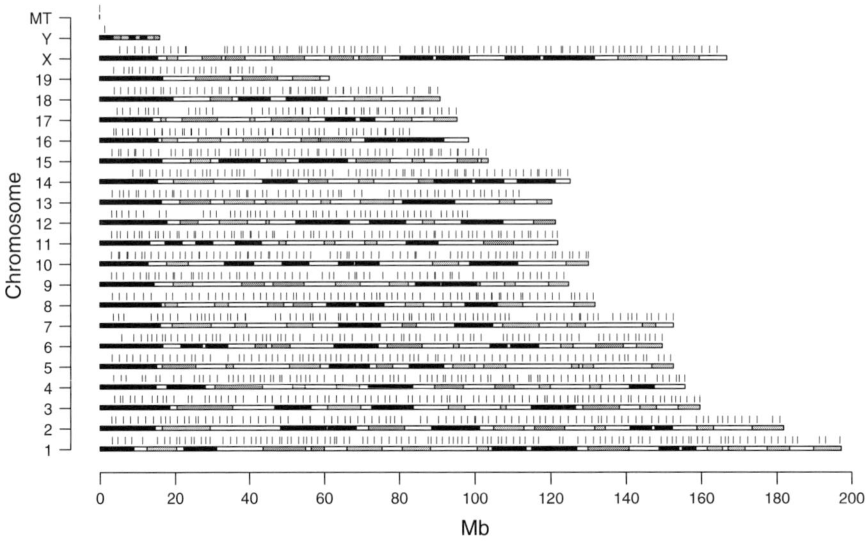

Figure 20.2 Genomic location of genetic markers used in this study (adapted from Wang *et al.*, 2011). Markers diagnostic for *M. m. musculus* and *M. m. domesticus* were identified from the Mouse Diversity Array (Yang *et al.*, 2009, 2011) and 1536 SNPs from autosomes, and the X chromosome were selected. Of these, 135 were not recovered from the genotyping assay, resulting in an SNP dataset that includes 1316 autosomal and 85 X-linked SNP markers (indicated by bars). One Y-linked marker and one mitochondrial marker were also included in this study. The sequence location of autosomal and X-linked SNP markers is given in Wang *et al.* (2011).

(Weir *et al.*, 2004), from phased data. Because LD can be affected by sample size and allele frequency, r^2 was calculated from each locality using the same number of randomly chosen individuals collected in the same year and by removing alleles from each locality with a minor allele frequency less than 0.05. For the r^2 analysis, the resulting dataset included eight localities. For two localities, Gesselthausen-Warta (GessW) and Neufahrn (Neuf), a sufficient number of individuals ($N \geq 10$) were collected in multiple years, making it possible to generate r^2 estimates for different years from the same locality. For one locality (Neufahrn), sample size was large enough to also calculate r^2 from a randomly selected sample of 15 and 25 individuals. The estimates of r^2 were also used to compare haplotype block size across populations and to estimate age, in generations, of each population. The former analysis involved comparing the physical distance (Mb) at which r^2 drops below 0.5, or 'half length of LD', following Reich *et al.* (2001). The latter analysis used a genome-wide average recombination rate of 0.63 cM/Mb for mice (Shifman *et al.*, 2006) over time.

Results and discussion

Calculating hybrid indices for all samples and plotting them against degree of heterozygosity captures both the distribution of hybrid genotypes and degree to which they represent multigenerational backcrosses or intercrosses. In the southern Bavaria transect, most hybrid mice are the result of multiple generations of intercrossing and/or backcrossing (Fig. 20.3a; adapted from Wang *et al.*, 2011). One interesting exception is a mouse collected from the eastern edge of the transect. It has a hybrid index score near 0.5 and is almost completely heterozygous. This mouse is likely only a few generations removed from a dispersal event involving an *M. m. domesticus* mouse to the eastern edge of the hybrid zone. Plotting the hybrid index of each mouse against the locality in which it was collected (Fig. 20.3b) reveals the location of an abrupt change in the genetic composition of mice over a very short geographical distance. This phenomenon is also captured from geographical cline analyses (Fig. 20.4).

Although cline analyses show varying estimates of cline centre and width across loci, there are some general patterns in the data (Table 20.1, Fig. 20.4). Mean cline centre and width estimates for X-linked markers are 57.26 km and 10.23 km, respectively, and the cline centre and width for the non-recombining portion of the Y chromosome is 57.11 km and 7.73 km, respectively. The simple two-parameter model returns larger estimates of cline width than the six-parameter model (Macholán *et al.*, 2007; Teeter *et al.*, 2010). However, the

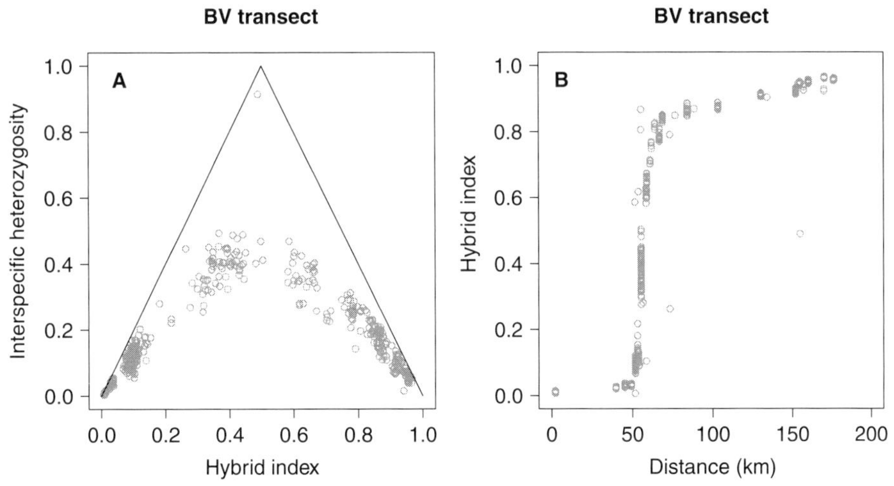

Figure 20.3 (a) Hybrid index plotted against heterozygosity of individual mice. (b) Hybrid index plotted against locality (distance from the western edge of the hybrid zone) where individual mice were collected (adapted from Wang *et al*, 2011).

Table 20.1 Two-parameter estimates of cline centre and width for autosomal, X- and Y- linked, and mtDNA genetic markers.

	No. markers	Cline centre mean (median) range	Cline width mean (median) range
Autosomal SNPs	1281	66.21 (61.05) 0.00–176.40	45.99 (24.03) 3.60–175.33
X-linked SNPs	84	57.26 (56.23) 54.16–108.70	10.23 (5.87) 3.35–168.09
Zfy (Y-linked)	1	57.11 –	7.73 –
mtDNA	1	95.35 –	103.11 –

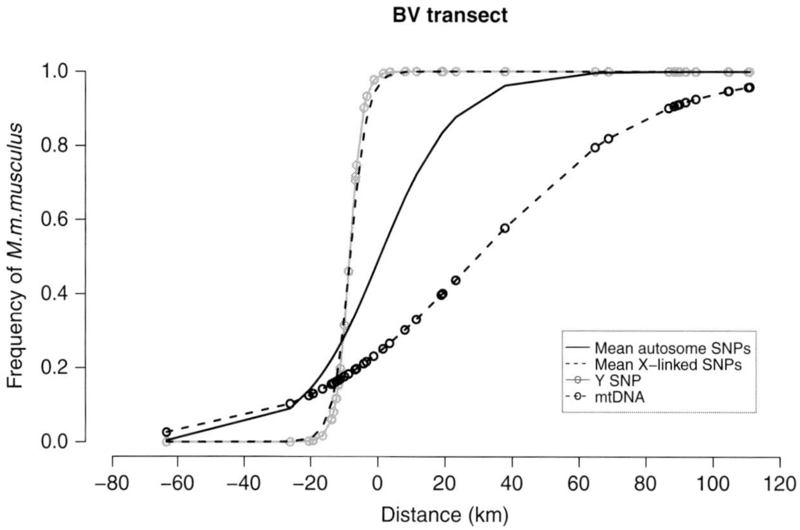

Figure 20.4 Geographical clines estimated using a two-parameter model for mean autosomal and X-linked SNPs and for one Y-linked and one mtDNA marker. Circles indicate location of sampled populations.

mean cline centre for the X-linked SNPs is very similar to the mean cline centre of 55.10 km calculated for 13 X-linked markers for the same transect analysed with the six-parameter model (Payseur *et al.*, 2004). In contrast to the sex chromosomes, mean cline centre for autosomal SNPs is shifted to the east (66.21 km) and mean cline width is considerably larger (45.99 km). The shift in cline centre

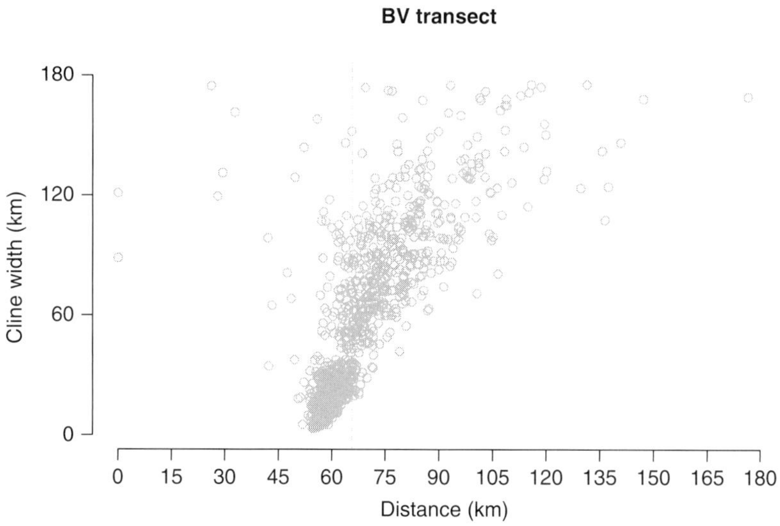

Figure 20.5 Estimates of cline width plotted against estimates of cline centre for all markers. Dashed line indicates mean cline centre based on the autosomal loci (from Wang *et al.*, 2011).

estimates results from asymmetry in cline shape, most easily visualized when cline width is plotted against cline centre for each SNP marker (Fig. 20.5). As cline width increases, cline centre estimates shift to the east, a pattern resulting from the introgression of *M. m. domesticus* alleles onto an *M. m. musculus* background. This pattern is consistent with previous studies using many fewer markers (Teeter *et al.*, 2008). A similar but even more pronounced pattern is found for mtDNA, in which centre and width are 95.35 km and 103.11 km, respectively (Table 20.1; Fig. 20.4). As mentioned earlier, these patterns of introgression for autosomal and mitochondrial DNA markers could be interpreted in two distinct ways. If the hybrid zone is moving, the introgressed markers are likely to be neutral and, if the hybrid zone is stationary, the introgressed markers are likely to confer fitness advantages.

To distinguish between these two hypotheses, Wang *et al.* (2011) measured LD by calculating r^2 from adjacent SNPs in populations from across the hybrid zone transect. As discussed earlier, if genome-wide LD increases asymmetrically across a hybrid zone, one can infer hybrid zone movement in the direction of the higher estimates. To perform these analyses, it was first necessary to discover whether, in hybrid populations, LD decayed with physical distance. A single population (Neufahrn) was used for this purpose and the results are summarized in Fig. 20.6.

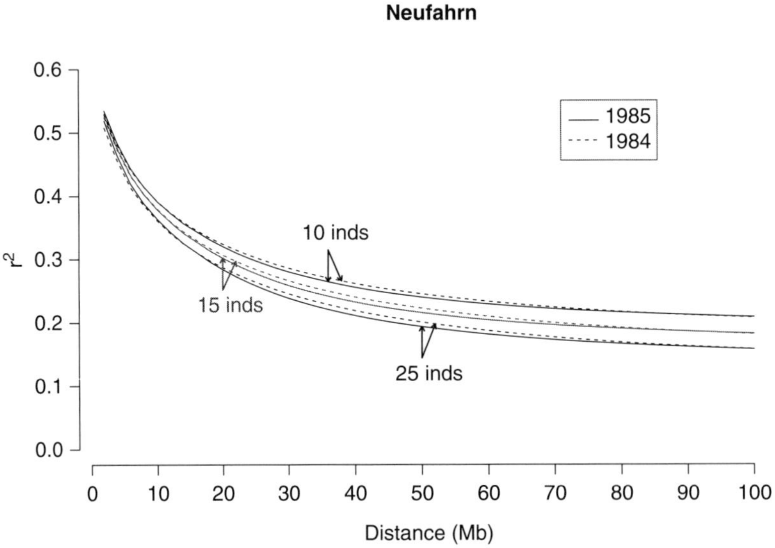

Figure 20.6 Decay of LD for samples sizes of 10, 15, and 25 individuals from a single population (Neufahrn) sampled in two separate years. Dashed lines are mice from 1984, solid lines are mice from 1985.

This population was also used to determine that LD, measured separately for samples collected in 1984 and 1985, showed similar patterns of decay for similar sample sizes ($N = 10$, 15, or 25) between years and was slightly lower when sample size was increased by five or ten individuals. Wang *et al.* (2011) chose to make comparisons between populations using r^2 values of all pair-wise SNPs with a genomic distance ≤2 Mb as this measure included most of the adjacent SNPs. The results are shown in Table 20.2. Mean r^2 ranged from 0.2679 to 0.7093, with values generally increasing from east to west (Table 20.2), and when plotted against distance from median cline centre they are greatest in populations to the west of cline centre (Fig. 20.7). Other measures derived from r^2 estimates, the physical distance (Mb) at which r^2 drops below 0.5 ('half length of LD' or HLD), and population age (in generations) (Table 20.2; Fig. 20.7) are consistent with these trends, e.g. HLD increases and population age (the relative stability of populations) decreases from east to west. For some populations – Gesselthausen-Warta and Neufahrn – age estimates for mice collected in 1984 and 1985 are older than estimates for mice collected in 1992, suggesting these populations experienced episode(s) of extinction and/or colonization, a finding also consistent with other studies that have found house mouse populations to comprise small demes subject to extinction and (re)colonization events (e.g. Dallas *et al.*, 1998; Hauffe *et al.*, 2000). The older age estimates for populations on the eastern side of the BV

Table 20.2 Values of r^2, half-length of LD (HLD) or block size in Mb, estimated age in generations, mean hybrid index, and mean heterozygosity (*He*) based on samples (N = 10) collected at specific localities in specific years.

Population	Collection date	Distance (km)[a]	Number	SNP pairs[b]	r^2	SE of r^2	HLD (Mb)	Age[c]	Hybrid index (mean)	*He* (mean)
Appe	1992	49.20	5	3697	0.5979	0.0067	7.2704	41	0.0312	0.3199
GessZ	1992	51.80	7	20227	0.6401	0.0028	9.2058	35	0.0990	0.1310
GessW	1984	51.80	6	19762	0.6754	0.0027	13.7778	31	0.1001	0.1276
GessW	1985	51.80	6	20018	0.6136	0.0029	7.4431	39	0.0862	0.1174
GessW	1992	51.80	6	27064	0.7093	0.0022	24.7567	27	0.1057	0.1048
Neuf	1984	55.60	12	95382	0.5296	0.0011	3.4506	50	0.3881	0.3958
Neuf	1985	55.60	12	96334	0.5345	0.0011	3.6275	49	0.4050	0.3716
Rudl	1984	67.20	20	44161	0.2680	0.0014	0.5021	104	0.7833	0.2591
Sonn	1984	84.60	24	24320	0.2679	0.0019	0.6093	104	0.8655	0.1775
Brun	1992	103.40	27	18627	0.2908	0.0023	0.7532	97	0.8777	0.1502
Simb	1985	152.40	30	10511	0.3384	0.0036	1.0802	85	0.9225	0.0966

Notes: [a]Distance is calculated from the western-most locality.

[b]Total number of SNP comparisons at ≤2 Mb distance is based on 100 permutations of samples at each locality. SNPs with minor allele frequency <0.05 and missing values greater than 50% were excluded.

[c]The age, in generations, of each population was calculated based on the decline of r^2 using the equation $\frac{r_t^2}{r_0^2} = (1 - c)^t$ (Saunders *et al.*, 2002) as described in the methods.

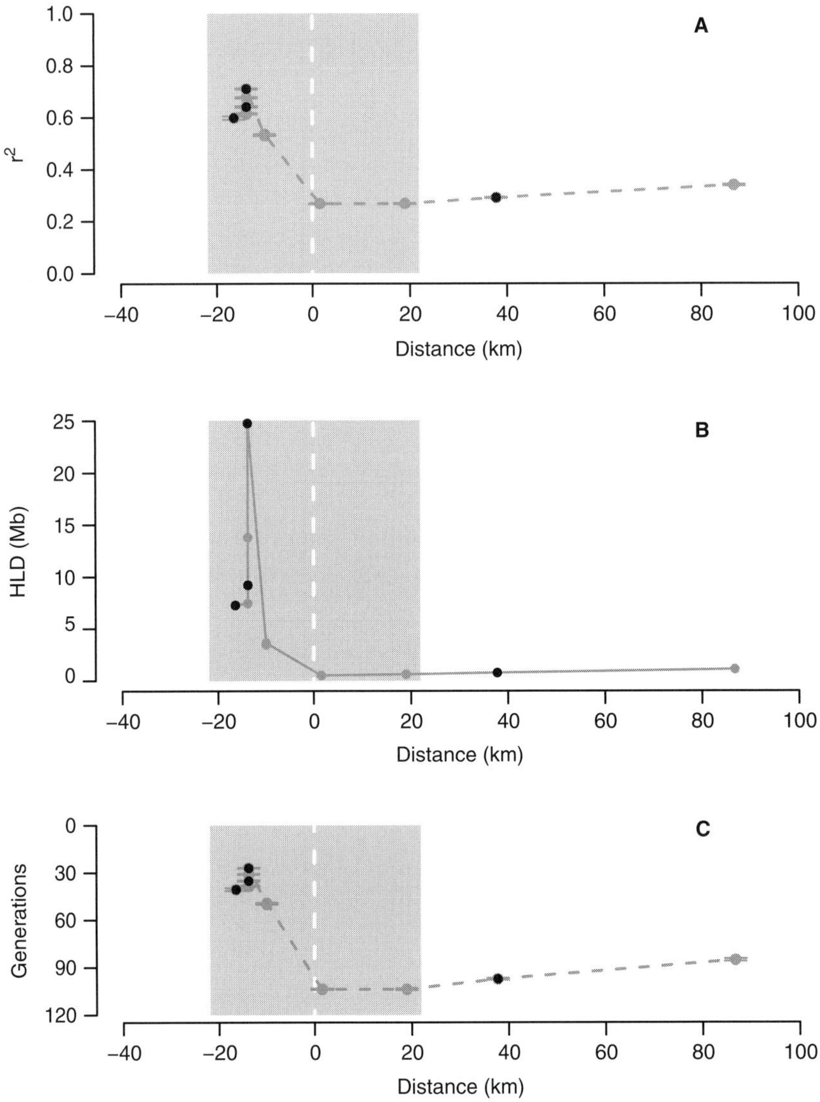

Figure 20.7 Mean gametic r^2 values of all pairwise SNPs with a genomic distance 2 Mb and half-length of LD (HLD) in Mb is plotted for eight populations sampled in the BV transect (a). HLD (b) and population age (c) in generations is plotted for the same populations (adapted from Wang *et al.*, 2011). 95% confidence intervals (bars) are also shown for each r^2 estimate. The shaded regions indicate mean cline widths of each transect based on geographical cline analyses and the dashed line indicates mean cline centre. Width and centre estimates are based on autosomal SNPs. Black circles = populations sampled in 1992 in BV. Grey circles = populations sampled in 1984 and/or 1985. Two populations were sampled over multiple years. See Table 20.2 for details.

transect suggest that these populations are more stable than populations on the western side of the BV transect. As indicated by Wang *et al.* (2011), these measures are a snapshot in time, documenting the history of the sampled populations. They provide evidence for past movement of the hybrid zone. With this information the asymmetrical patterns of introgression documented from the cline analyses can be interpreted as resulting from hybrid zone movement.

What is driving the movement of the house mouse hybrid zone in southern Bavaria is unknown. Asymmetric behavioural selection has been proposed as a possible explanation for movement (Raufaste *et al.*, 2005; Teeter *et al.*, 2008, 2010; Vošlajerová Bímová *et al.*, 2011). However, movement in the northern transect, though less well established, appears to be in the opposite direction (Wang *et al.*, 2011) and, thus, it is difficult to imagine how this phenomenon could result in different trends across different transects. Asymmetric sterility due to genetic incompatibilities between the X chromosome and autosomes (Storchová *et al.*, 2004; Britton-Davidian *et al.*, 2005; Good *et al.*, 2008a, b; Vyskočilová *et al.*, 2009) is also a possible cause of asymmetry. In this case, differences in direction of hybrid movement between transects may result from known instances of polymorphism of hybrid sterility alleles (Forejt and Iványi, 1974; Britton-Davidian *et al.*, 2005; Good *et al.*, 2008b; Vyskočilová *et al.*, 2009). Given that house mice live in close association with human populations, it is also quite possible that hybrid zone movement is geography-dependent, resulting from human-mediated transportation of mice and/or changes in land use over time. This possibility was suggested as an explanation for clinal patterns in the Danish hybrid zone transect (Raufaste *et al.*, 2005) and was simulated in Wang *et al.* (2011). Finally, it is important to keep in mind that whether the hybrid zone has moved since the populations were sampled in the 1980s and early 1990s is unknown. It could, however, be determined with additional sampling from the same localities. Updated information on movement would help to inform our understanding of the underlying mechanisms for the observed patterns examined here. For example, one might predict continued movement over time if movement is driven by selection. If, however, hybrid zone dynamics are geography-dependent, one might expect that documented changes in the landscape or in human activities over time may alter or even halt movement.

REFERENCES

Arntzen, J. W. and Wallis, G. P. (1991). Restricted gene flow in a moving hybrid zone of the newts *Triturus cristatus* and *T. marmoratus* in western France. *Evolution*, **45**, 805–26.

Barton, N. H. (1979). The dynamics of hybrid zones. *Heredity*, **43**, 341–59.

Barton, N. and Bengtsson, B. O. (1986). The barrier to genetic exchange between hybridizing populations. *Heredity*, **57**, 357–76.

Barton, N. H. and Gale, K. S. (1993). Genetic analysis of hybrid zones. In *Hybrid Zones and the Evolutionary Process*, ed. R. G. Harrison. New York: Oxford University Press, pp. 13–45.

Barton, N. H. and Hewitt, G. M. (1985). Analysis of hybrid zones. *Annual Review of Ecology and Systematics*, **16**, 113–48.

Barton, N. H. and Hewitt, G. M. (1989). Adaptation, speciation and hybrid zones. *Nature*, **341**, 497–503.

Boissinot, S. and Boursot, P. (1997). Discordant phylogeographic patterns between the Y chromosome and mitochondrial DNA in the house mouse: selection on the Y chromosome? *Genetics*, **146**, 1019–34.

Božíková, E., Munclinger, P., Teeter, K. C., *et al.* (2005). Mitochondrial DNA in the hybrid zone between *Mus musculus musculus* and *Mus musculus domesticus*: a comparison of two transects. *Biological Journal of the Linnean Society*, **84**, 363–78.

Britton-Davidian, J., Fel-Clair, F., Lopez, J., Alibert, P., and Boursot, P. (2005). Postzygotic isolation between the two European subspecies of the house mouse: estimates from fertility patterns in wild and laboratory-bred hybrids. *Biological Journal of the Linnean Society*, **84**, 379–93.

Buerkle, C. A. (2005). Maximum-likelihood estimation of a hybrid index based on molecular markers. *Molecular Ecology Notes*, **5**, 684–7.

Buggs, R. J. A. (2007). Empirical study of hybrid zone movement. *Heredity*, **99**, 301–12.

Cucchi, T., Vigne, J. D., and Auffray, J. C. (2005). First occurrence of the house mouse (*Mus musculus domesticus* Schwarz & Schwarz, 1943) in the Western Mediterranean: a zooarchaeological revision of subfossil occurrences. *Biological Journal of the Linnean Society*, **84**, 429–45.

Dallas, J. F., Bonhomme, F., Boursot, P., Britton-Davidian, J., and Bauchau, V. (1998). Population genetic structure in a Robertsonian race of house mice: evidence from microsatellite polymorphism. *Heredity*, **80**, 70–7.

Dufková, P., Macholán, M., and Piálek, J. (2011). Inference of selection and stochastic effects in the house mouse hybrid zone. *Evolution*, **65**, 993–1010.

Fel-Clair, F., Lenormand, T., Catalan, J., *et al.* (1996). Genomic incompatibilities in the hybrid zone between house mice in Denmark: evidence from steep and non-coincident chromosomal clines for Robertsonian fusions. *Genetical Research*, **67**, 123–34.

Forejt, J. (1996). Hybrid sterility in the mouse. *Trends in Genetics*, **12**, 412–17.

Forejt, J. and Iványi, P. (1974). Genetic studies on male sterility of hybrids between laboratory and wild mice (*Mus musculus* L.). *Genetical Research*, **24**, 189–206.

Gompert, Z. and Buerkle, C. A. (2010). INTROGRESS: a software package for mapping components of isolation in hybrids. *Molecular Ecology Resources*, **10**, 378–84.

Good, J. M., Dean, M. D., and Nachman, M. W. (2008a). A complex genetic basis to X-linked hybrid male sterility between two species of house mice. *Genetics*, **179**, 2213–28.

Good, J. M., Giger, T., Dean, M. D., and Nachman, M. W. (2010). Widespread over-expression of the X chromosome in sterile F_1 hybrid mice. *PLoS Genetics*, **6**, e1001148.

Good, J. M., Handel, M. A., Nachman, M. W., and Feder, J. (2008b). Asymmetry and polymorphism of hybrid male sterility during the early stages of speciation in house mice. *Evolution*, **62**, 50–65.

Hauffe, H. C., Piálek, J., and Searle, J. B. (2000). The house mouse chromosomal hybrid zone in Valtellina (SO): a summary of past and present research. *Hystirx*, **11**, 17–25.

Hewitt, G. M. (1975). A sex-chromosome hybrid zone in the grasshopper *Podisma pedestris* (Orthoptera: Acrididae). *Heredity*, **35**, 375–87.

Hudson, R. R. Boos, D. D., and Kaplan, N. L. (1992). A statistical test for detecting geographic subdivision. *Molecular Biology Evolution*, **9**, 138–51.

Hunt, W. G. and Selander, R. K. (1973). Biochemical genetics of hybridisation in European house mice. *Heredity*, **31**, 11–33.

Key, K. H. L. (1968). The concept of stasipatric speciation. *Systematic Zoology*, **17**, 14–22.

Librado, P. and Rozas, J. (2009). DnaSP v5: a software for comprehensive analysis of DNA polymorphism data. *Bioinformatics*, **25**, 1451–2.

Macholán, M., Baird, S. J. E., Dufková, P., *et al.* (2011). Assessing multilocus introgression patterns: a case study on the mouse X chromosome in central Europe. *Evolution*, **65**, 1428–46.

Macholán, M., Baird, S. J. E., Munclinger, P., *et al.* (2008). Genetic conflict outweighs heterogametic incompatibility in the mouse hybrid zone? *BMC Evolutionary Biology*, **8**, 271.

Macholán, M., Munclinger, P., Šugerková, M., *et al.* (2007). Genetic analysis of autosomal and X-linked markers across a mouse hybrid zone. *Evolution*, **61**, 746–71.

Moulia, C., LeBrun, N., Dallas, J., Orth, A., and Renaud, F. (1993). Experimental evidence of genetic determinism in high susceptibility to intestinal pinworm infection in mice: a hybrid zone model. *Parasitology*, **106**, 387–93.

Munclinger, P., Božíková, E., Šugerková, M., Piálek, J., and Macholán, M. (2002). Genetic variation in house mice (*Mus*, Muridae, Rodentia) from the Czech and Slovak Republics. *Folia Zoologica*, **51**, 81–92.

Orth, A., Lyapunova, E., Kandaurov, A., *et al.* (1996). Polytypic species *Mus musculus* in Transcaucasia. *Comptes Rendus de l'Académie des Sciences – Series III – Sciences de la Vie*, **319**, 435–41.

Payseur, B. A., Krenz, J. G., and Nachman, M. W. (2004). Differential patterns of introgression across the X chromosome in a hybrid zone between two species of house mice. *Evolution*, **58**, 2064–78.

Porter, A. H., Wenger, R., Geiger, H., Scholl, A., and Shapiro, A. M. (1997). The *Pontia daplidice–edusa* hybrid zone in northwestern Italy. *Evolution*, **51**, 1561–73.

Prager, E. M., Boursot, P., and Sage, R. D. (1997). New assays for Y chromosome and p53 pseudogene clines among East Holstein house mice. *Mammalian Genome*, **8**, 279–81.

Prager, E. M., Orrego, C., and Sage, R. D. (1998). Genetic variation and phylogeography of central Asian and other house mice, including a major new mitochondrial lineage in Yemen. *Genetics*, **150**, 835–61.

Raufaste, N., Orth, A., Belkhir, K., *et al.* (2005). Inferences of selection and migration in the Danish house mouse hybrid zone. *Biological Journal of the Linnean Society*, **84**, 593–616.

Reich, D. E., Cargill, M., Bolk, S., *et al.* (2001). Linkage disequilibrium in the human genome. *Nature*, **411**, 199–204.

Sage, R. D., Heyneman, D., Lim K. C., and Wilson, A. C. (1986a). Wormy mice in a hybrid zone. *Nature*, **324**, 60–3.

Sage, R. D., Whitney, J. B., III., and Wilson A. C. (1986b). Genetic analysis of a hybrid zone between *domesticus* and *musculus* mice (*Mus musculus* complex): hemoglobin polymorphisms. *Current Topics in Microbiology and Immunology*, **127**, 75–85.

Saunders, M. A., Hammer, M. F., and Nachman, M. W. (2002). Nucleotide variability at G6pd and the signature of malarial selection in humans. *Genetics*, **162**, 1849–61.

Shifman, S., Bell, J. T., Copley, R. R., *et al.* (2006). High-resolution single nucleotide polymorphism genetic map of the mouse genome. *PLoS Biology*, **4**, 2227–37.

Smadja, C. and Ganem, G. (2005). Asymmetrical reproductive character displacement in the house mouse. *Journal of Evolutionary Biology*, **18**, 1485–93.

Storchová, R., Gregorová, S., Buckiová, D., *et al.* (2004). Genetic analysis of X-linked hybrid sterility in the house mouse. *Mammalian Genome*, **15**, 515–24.

Teeter, K. C., Payseur, B. A., Harris, L. W., *et al.* (2008). Genome-wide patterns of gene flow across a house mouse hybrid zone. *Genome Research*, **18**, 67–76.

Teeter, K. C., Thibodeau, L. M., Gompert, Z., *et al.* (2010). The variable genomic architecture of isolation between hybridizing species of house mice. *Evolution*, **64**, 472–85.

Tucker, P. K., Sage, R. D., Warner, J. H., Wilson, A. C., and Eicher, E. M. (1992). Abrupt cline for sex chromosomes in a hybrid zone between two species of mice. *Evolution*, **46**, 1146–63.

Vanlerberghe, F., Boursot, P., and Catalan, J. (1988a). Analyse genetique de la zone d'hybridation entre les deux sous-especes de souris *Mus musculus domesticus* et *Mus musculus musculus* en Bulgaire. *Genome*, **30**, 427–37.

Vanlerberghe, F., Boursot, P., Nielsen, J. T., and Bonhomme, F. (1988b). A steep cline for mitochondrial DNA in Danish mice. *Genetical Research*, **52**, 185–93.

Vanlerberghe, F., Dod, B., Boursot, P., Bellis, M., and Bonhomme, F. (1986). Absence of Y chromosome introgression across the hybrid zone between *Mus musculus domesticus* and *Mus musculus musculus*. *Genetical Research*, **48**, 191–7.

Vošlajerová Bímová, B., Macholán, M., Baird, S. J. E., *et al.* (2011). Reinforcement selection acting on the European house mouse hybrid zone. *Molecular Ecology*, **20**, 2403–24.

Vyskočilová, M., Pražanová, G., and Piálek, J. (2009). Polymorphism in hybrid male sterility in wild-derived *Mus musculus musculus* strains on proximal chromosome 17. *Mammalian Genome*, **20**, 83–91.

Wang, L., Luzynski, K., Pool, J., *et al.* (2011). Measures of linkage disequilibrium among neighboring SNPs indicate asymmetries across the house mouse hybrid zone. *Molecular Ecology*, accepted.

Weir, B. S., Hill, W. G., and Cardon, L. R. (2004). Allelic association patterns for a dense SNP map. *Genetic Epidemiology*, **27**, 442–50.

Yang, H., Ding, Y. M., Hutchins, L. N., *et al.* (2009). A customized and versatile high-density genotyping array for the mouse. *Nature Methods*, **6**, 663–6.

Yang, H., Wang, J., Didion, J. P., *et al.* (2011). Subspecific origin and haplotype diversity in the laboratory mouse. *Nature Genetics*, **43**, 648–55.

Index